Objective CHEMISTRY

Chapter-wise MCQs

for

NTA JEE Main/ BITSAT/ NEET/ AIIMS

• **Corporate Office :** 45, 2nd Floor, Maharishi Dayanand Marg, Corner Market,
Malviya Nagar, New Delhi-110017
Tel. : 011-49842349 / 49842350

By
Dheeraj Pandey

Typeset by Disha DTP Team

DISHA PUBLICATION

For further information about the books from DISHA,

Log on to **www.dishapublication.com** or email to **info@dishapublication.com**

CONTENTS

1 Some Basic Concepts of Chemistry

1. If the true value for an experimental result is 6.23 and the results reported by three students X, Y and Z are :
   X: 6.18 and 6.28
   Y: 6.20 and 6.023
   Z: 6.22 and 6.24
   Which of the following option is correct?
   (a) X precise, Y accurate, Z precise and accurate.
   (b) X precise and accurate, Y not precise, Z precise
   (c) Both X & Z precise & accurate, Y not precise.
   (d) Both X & Y neither precise nor accurate, Z both precise and accurate.

2. Irrespective of the source, pure sample, of water always yields 88.89% mass of oxygen and 11.11% mass of hydrogen. This is explained by the law of

   Tricky

   (a) conservation of mass
   (b) multiple proportions
   (c) constant composition
   (d) constant volume

3. The percentage of Se in peroxidase anhydrous enzyme is 0.5% by weight (atomic weight = 78.4). Then minimum molecular weight of peroxidase anhydrous enzyme is
   (a) 1.568×10^3 (b) 1.568×10^4
   (c) 15.68 (d) 3.136×10^4

4. The amount of zinc required to produce 224 mL of H_2 at STP on treatment with dil. H_2SO_4 will be
   (a) 6.5 g (b) 0.65 g
   (c) 65 g (d) 0.065 g

5. Assuming fully decomposed, the volume of CO_2 released at STP on heating 9.85 g of $BaCO_3$ (Atomic mass, Ba = 137) will be
   (a) 1.12 L (b) 2.24 L (c) 4.06 L (d) 0.84 L

6. The number of water molecules present in a drop of water (volume 0.0018 mL) density = 18 mL^{-1} at room temperature is
   (a) 1.084×10^{18} (b) 6.023×10^{19}
   (c) 4.84×10^{17} (d) 6.023×10^{23}

7. 1.44 g of titanium (At. mass = 48) reacted with excess of O_2 and produce x g of non stoichiometric compound $Ti_{1.44}O$. The value of x is:

   (a) 2 (b) 1.77
   (c) 1.44 (d) none of these

8. Silver oxide (Ag_2O) decomposes at temperature 300 °C yielding metallic silver and oxygen gas. A 1.60 g sample of impure silver oxide yields 0.104 g of oxygen gas. What is the per cent by mass of the silver oxide in the sample?
   (a) 5.9 (b) 47.125 (c) 94.25 (d) 88.2

9. A sample of AlF_3 contains 3.0×10^{24} F^- ions. The number of formula unit of this sample are
   (a) 9×10^{24} (b) 3×10^{24}
   (c) 0.75×10^{24} (d) 1.0×10^{24}

10. Dissolving 120 g of a compound of mol. wt. 60 in 1000 g of water gave a solution of density 1.12 g/mL. The molarity of the solution is:
    (a) 1.00 M (b) 2.00 M
    (c) 2.50 M (d) 4.00 M

11. A mixture of O_2 and gas "Y" mol. mass 80 in the mole ratio a : b has a mean molecular mass 40. What would be mean molecular mass, if the gases are mixed in the ratio b : a under, identical conditions? (Assume that gases are non-reacting):

    Toughnut

    (a) 40 (b) 48 (c) 62 (d) 72

12. If 3.01×10^{20} molecules are removed from 98 mg of H_2SO_4, then the number of moles of H_2SO_4 left are
    (a) 0.1×10^{-3} (b) 0.5×10^{-3}
    (c) 1.66×10^{-3} (d) 9.95×10^{-2}

13. The number of molecules in 8.96 litre of a gas at 0 °C and 1 atm. pressure is approximately
    (a) 6.023×10^{23} (b) 12.04×10^{23}
    (c) 18.06×10^{23} (d) 24.08×10^{22}

14. The density of 3M solution of sodium chloride is 1.252 g mL^{-1}. The molality of the solution will be: (molar mass, NaCl = 585 g mol^{-1}) Tricky

(a) 2.60m (b) 2.18m (c) 2.79m (d) 3.00m

15. Indium (atomic mass = 114.82) has two naturally occurring isotopes, the predominant one form has isotopic mass 114.9041 and abundance of 95.72%. Which of the following isotopic mass is the most likely for the other isotope?

(a) 112.94 (b) 115.90
(c) 113.90 (d) 114.90

16. A compound contains 54.55 % carbon, 9.09% hydrogen, 36.36% oxygen. The empirical formula of this compound is:

(a) C_3H_5O (b) $C_4H_8O_2$
(c) $C_2H_4O_2$ (d) C_2H_4O

17. An organic compound whose empirical and molecular formula are same, contains 20% carbon, 6.7% hydrogen, 46.7% nitrogen and the rest oxygen. On heating it yields ammonia, leaving a solid residue. The solid residue gives a violet colour with dilute solution of alkaline copper sulphate. The organic compound is

Toughnut

(a) NH_2COONH_4 (b) $HCOONH_4$
(c) NH_2NHCHO (d) NH_2CONH_2

18. 10 mL of 2 M NaOH solution is added to 200 mL of 0.5 M of NaOH solution. What is the final concentration ?

(a) 0.57 M (b) 5.7 M
(c) 11.4 M (d) 1.14 M

19. Arrange the following in the order of increasing mass (atomic mass: O = 16, Cu = 63, N = 14)

I. one atom of oxygen
II. one atom of nitrogen
III. 1×10^{-10} mole of oxygen
IV. 1×10^{-10} mole of copper

(a) II < I < III < IV (b) I < II < III < IV
(c) III < II < IV < I (d) IV < II < III < I

20. The number of moles of oxygen in one litre of air containing 21% oxygen by volume, under standard conditions are

(a) 0.0093 mole (b) 0.21 mole
(c) 2.10 mole (d) 0.186 mole

21. The concentrated sulphuric acid that is peddled commercial is 95% H_2SO_4 by weight. If the density of this commercial acid is 1.834 g cm^{-3}, the molarity of this solution is

(a) 17.8 M (b) 12.0 M
(c) 10.5 M (d) 15.7 M

22. 9 moles of "D" and 14 moles of E are allowed to react in a closed vessel according to given reactions. Calculate number of moles of B formed in the end of reaction, if 4 moles of G are present in reaction vessel. (Percentage yield of reaction is mentioned in the reaction)

Step-1 $3D + 4E \xrightarrow{80\%} 5C + A$

Step-2 $3C + 5G \xrightarrow{50\%} 6B + F$

(a) 2.4 (b) 30 (c) 4.8 (d) 1

23. A mixture of NH_4NO_3 and $(NH_4)_2HPO_4$ contain 30.40% mass per cent of nitrogen. What is the mass ratio of the two components in the mixture?

(a) 2 : 1 (b) 1 : 2 (c) 3 : 4 (d) 4 : 1

24. The mass of N_2F_4 produced by the reaction of 2.0 g of NH_3 and 8.0 g of F_2 is 3.56 g. What is the per cent yield?

$$2NH_3 + 5F_2 \longrightarrow N_2F_4 + 6HF$$

(a) 79.0 (b) 71.2
(c) 84.6 (d) None of these

25. The ppm level of F^- in a 500 g sample of a tooth paste containing 0.2 g F^- is

(a) 400 (b) 1000 (c) 250 (d) 200

26. Two glucose solutions are mixed. One has a volume of 480 mL and a concentration of 1.50 M and the second has a volume of 520 mL and concentration 1.20 M. The molarity of final solution is

(a) 1.20 M (b) 1.50 M
(c) 1.34 M (d) 2.70 M

27. What is the volume of CO_2 liberated (in litres) at 1 atmosphere and 0 °C when 10 g of 100% pure calcium carbonate is treated with excess dilute sulphuric acid?
(Atomic mass Ca : 40, C : 12, O : 16)

(a) 0.224 (b) 2.24 (c) 22.4 (d) 224

28. Complete combustion of 0.858 g of compound X gives 2.64 g of CO_2 and 1.26 g of H_2O. The lowest molecular mass X can have:

(a) 43 g (b) 86 g (c) 129 g (d) 172 g

29. The impure 6 g of NaCl is dissolved in water and then treated with excess of silver nitrate solution. The mass of precipitate of silver chloride is found to be 14 g. The % purity of NaCl solution would be: Toughnut

(a) 95% (b) 85% (c) 75% (d) 65%

30. How many of 0.1N HCl are required to react completely with 1 g mixture of Na_2CO_3 and $NaHCO_3$ containing equimolar amounts of two ?

(a) 157.7mL (b) 15.77mL
(c) 147.7mL (d) 14.77mL

31. 5 g of benzene on nitration gave 6.6 g of nitrobenzene. The theoretical yield of the nitrobenzene will be

(a) 4.5 g (b) 5.6 g (c) 8.09 g (d) 6.6 g

32. An aqueous solution of oxalic acid dihydrate contains its 6.3 g in 250 mL. The volume of 0.1 N NaOH required to completely neutralize 10 mL of this solution

(a) 4mL (b) 20mL (c) 2mL (d) 40mL

33. Choose the incorrect formula out of the four compounds for an element X below :

(a) X_2O_3 (b) X_2Cl_3
(c) $X_2(SO_4)_3$ (d) XPO_4

34. A gaseous compound of nitrogen and hydrogen contains 12.5% (by mass) of hydrogen. The density of the compound relative to hydrogen is 16. The molecular formula of the compound is:

(a) NH_2 (b) N_3H (c) NH_3 (d) N_2H_4

35. In an organic compound of molar mass 108 g mol^{-1} C, H and N atoms are present in 9 : 1 : 3.5 by mass. Molecular formula can be:

(a) $C_6H_8N_2$ (b) $C_7H_{10}N$
(c) $C_5H_6N_3$ (d) $C_4H_{18}N_3$

36. What is the empirical formula of vanadium oxide, if 2.74 g of the metal oxide contains 1.53 g of metal?

(a) V_2O_3 (b) VO (c) V_2O_5 (d) V_2O_7

37. 100 cm^3 of 0.1 N HCl is mixed with 100 cm^3 of 0.2 N NaOH solution. The resulting solution is Tricky

(a) 0.1 N and the solution is basic
(b) 0.1 N and the solution is acidic
(c) 0.05 N and the solution is basic
(d) 0.05 N and the solution is acidic

38. On subjecting 10 mL mixture of N_2 and CO to repeated electro spark, 7 mL of O_2 was required for combustion. What was the mole percent of CO in the mixture? (All volumes were measured under identical conditions)

(a) 60 (b) 40 (c) 6 (d) 4

39. When burnt in air, 14.0 g mixture of carbon and sulphur gives a mixture of CO_2 and SO_2 in the volume ratio of 2:1, volume being measured at the same conditions of temperature and pressure moles of carbon in the mixture is

(a) 0.75 (b) 0.5 (c) 0.40 (d) 0.25

40. What is the molarity of SO_4^{2-} ion in aqueous solution that contain 34.2 ppm of $Al_2(SO_4)_3$? (Assume complete dissociation and density of solution 1g/mL) Critical Thinking

(a) 3×10^{-4} M (b) 2×10^{-4} M
(c) 10^{-4} M (d) None of these

41. A transition metal *M* forms a volatile chloride which has a vapour density of 94.8. If it contains 74.75% of chlorine the formula of the metal chloride will be

(a) MCl_3 (b) MCl_2 (c) MCl_4 (d) MCl_5

42. 50 mL 10 N H_2SO_4, 25 mL 12 N HCl and 40 mL 5 N HNO_3 were mixed together and the volume of the mixture was made 1000 mL by adding water. The normality of the resultant solution will be

(a) 2 N (b) 1 N (c) 3 N (d) 4 N

43. 250 mL of a sodium carbonate solution contains 2.65 g of Na_2CO_3. If 10 mL of this solution is diluted to one litre, what is the concentration of the resultant solution?
(mol. wt. of Na_2CO_3 = 106)

(a) 0.1 M (b) 0.01 M
(c) 0.001 M (d) 10^{-4} M

44. 3.92 g of ferrous ammonium sulphate crystals are dissolved in 100 mL of water. 20 mL of this solution requires 18 mL of potassium permaganate during titration for complete oxidation. The weight of $KMnO_4$ present in one litre of the solution of

(a) 3.47 g (b) 12.38 g
(c) 1.23 g (d) 34.76 g

45. 1.12 mL of a gas is produced at S.T.P. by the action of 4.12 mg of alcohol ROH with methyl magnesium Iodide. The molecular mass of alcohol is

(a) 16.0 (b) 41.2 (c) 82.4 (d) 156.0

46. How many moles of P_4 can be produced by reaction of 0.10 moles $Ca_5(PO_4)_3F$, 0.36 moles SiO_2 and 0.90 moles C according to the following reaction?

$$4\,Ca_5(PO_4)_3F + 18\,SiO_2 + 30\,C \longrightarrow 3P_4 + 2CaF_2 + 18CaSiO_3 + 30\,CO$$

(a) 0.060 (b) 0.030 (c) 0.045 (d) 0.075

47. 1 mole of mixture of CO and CO_2 requires exactly 28 g KOH in solution for complete conversion of all the CO_2 into K_2CO_3. How much amount more of KOH will be required for conversion into K_2CO_3. If one mole of mixture is completely oxidized to CO_2

Critical Thinking

(a) 112 g (b) 84 g (c) 56 g (d) 28 g

48. 2.0 g of a sample contains mixture of SiO_2 and Fe_2O_3. On very strong heating, it leaves a residue weighing 1.96 g. The reaction responsible for loss of mass is given below.

$Fe_2O_3\,(s) \rightarrow Fe_3O_4\,(s) + O_2\,(g)$, (unbalance equation)

What is the percentage by mass of SiO_2 in original sample?

(a) 100% (b) 20% (c) 40% (d) 60%

49. 20 g of $CaCO_3$ on heating gave 8.8 g of CO_2 and 11.2 g of CaO. This is in accordance with

(a) The law of conservation of mass.
(b) The law of constant composition.
(c) The law of reciprocal proportion.
(d) None of these

50. If 224 mL of a triatomic gas has a mass of 1 g at 273K and 1 atmospheric pressure then the mass of one atom is

(a) 8.30×10^{-23} g (b) 2.08×10^{-23} g
(c) 5.53×10^{-23} g (d) 6.24×10^{-23} g

51. Specific volume of cylindrical virus particle is 6.02×10^{-2} cc/g. whose radius and length are 7 Å & 10 Å respectively.

If $N_A = 6.02 \times 10^{23}\,mol^{-1}$, find molecular weight of virus

Toughnut

(a) 3.08×10^3 kg/mol (b) 3.08×10^4 kg/mol
(c) 1.54×10^4 kg/mol (d) 15.4 kg/mol

52. 1.575 g of oxalic acid $(COOH)_2.xH_2O$ are dissolved in water and the volume made upto 250 mL. On titration 16.68 mL of this solution requires 25 mL of N/15 NaOH solution for complete neutralization, calculate x.

(a) 3 (b) 2 (c) 4 (d) 5

53. 2 g of a mixture of CO and CO_2 on reaction with excess I_2O_5 produced 2.54 g of I_2. What will be the mass % of CO_2 in the original mixture?

(a) 35 (b) 70 (c) 30 (d) 60

54. Which statement is false for the balanced equation given below?

$$CS_2 + 3O_2 \longrightarrow 2SO_2 + CO_2$$

(a) One mole of CS_2 will produce one mole of CO_2
(b) The reaction of 16 g of oxygen produces 7.33 g of CO_2
(c) The reaction of one mole of O_2 will produce 2/3 mole of SO_2
(d) Six molecules of oxygen requires three molecules of CS_2

55. Number of moles of $KMnO_4$ required to oxidize one mole of $Fe(C_2O_4)$ in acidic medium is

Toughnut

(a) 0.167 (b) 0.6 (c) 0.2 (d) 0.4

56. Haemoglobin contains 0.33% of iron by weight. The molecular weight of haemoglobin is approximately 67200. The number of iron atoms (at. wt. of Fe = 56) present in one molecule of haemoglobin is

(a) 6 (b) 1 (c) 2 (d) 4

57. A 25.0 mm × 40.0 mm piece of gold foil is 0.25 mm thick. The density of gold is 19.32 g/cm³. How many gold atoms are in the sheet? (Atomic weight : Au = 197.0)

Tricky

(a) 7.7×10^{23} (b) 1.5×10^{23}
(c) 4.3×10^{21} (d) 1.47×10^{22}

58. What volume of oxygen gas (O_2) measured at 0°C and 1 atm, is needed to burn completely 1L of propane gas (C_3H_8) measured under the same conditions ?

(a) 7L (b) 6L (c) 5L (d) 10L

59. 25.4 g of I_2 and 14.2 g of Cl_2 are made to react completely to yield a mixture of ICl and ICl_3. Calculate moles of ICl and ICl_3 formed

(a) 0.1, 0.1 (b) 0.2, 0.2
(c) 0.1, 0.2 (d) 0.2, 0.1

60. 10 moles SO_2 and 15 moles O_2 were allowed to react over a suitable catalyst. 8 moles of SO_3 were formed. The remaining moles of SO_2 and O_2 respectively are -
(a) 2 moles, 11 moles (b) 2 moles, 8 moles
(c) 4 moles, 5 moles (d) 8 moles, 2 moles

61. In the reaction

$$4NH_3(g) + 5O_2(g) \longrightarrow 4NO(g) + 6H_2O(l),$$

when 1 mole of ammonia and 1 mole of O_2 are made to react to completion
(a) 1.0 mole of H_2O is produced
(b) 1.0 mole of NO will be produced
(c) all the ammonia will be consumed
(d) all the oxygen will be consumed

62. An ideal gaseous mixture of ethane (C_2H_6) and ethene (C_2H_4) occupies 28 litre at 1 atm and 273 K. The mixture reacts completely with 128 g O_2 to produce CO_2 and H_2O. Mole fraction at C_2H_6 in the mixture is: Toughnut
(a) 0.6 (b) 0.4 (c) 0.5 (d) 0.8

63. Wood's metal contains 50.0% bismuth, 25.0% lead, 12.5% tin and 12.5% cadmium by mass. What is the mole fraction of tin?
(Atomic mass : Bi=209, Pb=207, Sn=199, Cd=112)
(a) 0.202 (b) 0.158 (c) 0.176 (d) 0.221

64. The mass of $BaCO_3$ produced when excess CO_2 is bubbled through a solution of 0.205 mol $Ba(OH)_2$ is :
(a) 81 g (b) 40.5 g
(c) 20.25 g (d) 162 g

65. Sulfuryl chloride (SO_2Cl_2) reacts with water to give a mixture of H_2SO_4 and HCl. How many moles of baryta would be required to neutralize the solution formed by adding 4 mole of SO_2Cl_2 to excess of water ?
(a) 1 (b) 2 (c) 3 (d) 4

66. 1 g mixture of equal number of mole of Li_2CO_3 and other metal carbonate (M_2CO_3) required 21.6 mL of 0.5 N HCl for complete neutralisation reaction. What is the approximate atomic mass of the other metal? Critical Thinking
(a) 25 (b) 23 (c) 51 (d) 118

67. When 2.5 g of a sample of Mohr's salt reacts completely with 50 mL of $\frac{N}{10}$ $KMnO_4$ solution. The % purity of the sample of Mohr's salt is:
(a) 78.4 (b) 70 (c) 37 (d) 40

68. If potassium chlorate is 80% pure, then 48 g of oxygen would be produced from (atomic mass of K=39)
(a) 153.12 g of $KClO_3$ (b) 122.5 g of $KClO_3$
(c) 245 g of $KClO_3$ (d) 98 g of $KClO_3$

69. 12 g of Mg (atomic mass 24) will react completely with hydrochloric acid to give
(a) One mol of H_2
(b) 1/2 mol of H_2
(c) 2/3 mol of O_2
(d) both 1/2 mol of H_2 and 1/2 mol of O_2

70. 2.76 g of silver carbonate (at mass of Ag 108) on being heated strongly yield a residue weighing

(a) 2.16 g (b) 2.48 g
(c) 2.32 g (d) 2.64 g

Answer KEY

1	(d)	8	(c)	15	(a)	22	(a)	29	(a)	36	(c)	43	(c)	50	(c)	57	(d)	64	(b)
2	(c)	9	(d)	16	(d)	23	(a)	30	(a)	37	(c)	44	(a)	51	(d)	58	(c)	65	(b)
3	(b)	10	(b)	17	(d)	24	(d)	31	(c)	38	(a)	45	(c)	52	(b)	59	(a)	66	(d)
4	(b)	11	(d)	18	(a)	25	(a)	32	(d)	39	(b)	46	(a)	53	(c)	60	(a)	67	(a)
5	(a)	12	(b)	19	(a)	26	(c)	33	(b)	40	(a)	47	(b)	54	(d)	61	(d)	68	(a)
6	(b)	13	(d)	20	(a)	27	(b)	34	(d)	41	(c)	48	(c)	55	(b)	62	(b)	69	(b)
7	(b)	14	(c)	21	(a)	28	(a)	35	(a)	42	(b)	49	(a)	56	(d)	63	(c)	70	(a)

Hints & Solutions

1. (d) Both Y and X are neither precise nor accurate as the two values in each of them are not close. With respect to X & Y, the values of Z are close & agree with the true value. Hence, both precise & accurate.

2. (c) The H : O ratio in water is fixed, irrespective of its source. Hence it is law of constant composition.

3. (b) 0.5% by weight means if Mol. wt. is 100 then mass of Se is 0.5. If at least one atom of Se is present in the molecule then

$$\text{M. Wt} = \frac{100}{0.5} \times 78.4 = 1.568 \times 10^4$$

4. (b) $Zn + H_2SO_4 \longrightarrow ZnSO_4 + H_2$

65 g Zn gives 1 mole of H_2 = 22400 mL of H_2
224 mL of H_2 will be obtained from 0.65 g Zn.

5. (a) $BaCO_3 \longrightarrow BaO + CO_2$

192 g of $BaCO_3$ gives 1 mol of CO_2 = 22.4 L
9.85 g of $BaCO_3$ will give 0.05 mol of CO_2 which is equal to 1.12 litre.

6. (b) 0.0018 mL = 0.0018 g = 0.0001 mole of water

$= 10^{-4}$ mole

∴ Number of water molecules

$= 6.023 \times 10^{23} \times 10^{-4} = 6.023 \times 10^{19}$

7. (b) $Ti + O_2 \longrightarrow Ti_{1.44}O$

$$\frac{1.44}{48}\text{ mole} \qquad \frac{x}{48(1.44)+16}\text{ mole}$$

$$\therefore \frac{1.44}{48} = \frac{1.44x}{48(1.44)+16}$$

x = 1.77 g

8. (c) Let mass of Ag_2O = x g

$$Ag_2O \longrightarrow 2Ag + \frac{1}{2}O_2$$

xg

232 g $\qquad \frac{1}{2} \times 32 = 16 \text{ g } O_2$

$$\frac{16}{232} \times x = 0.104 \Rightarrow x = 1.508 \text{ g}$$

$$\% \text{ of } Ag_2O = \frac{1.508}{1.6} \times 100 = 94.25\%$$

9. (d) An, AlF_3 the number of F is 3 for one AlF_3 molecule $3F^- \equiv$ 1 formula unit of AlF_3

$$3.0 \times 10^{24}\, F^- \equiv \frac{1}{3} \times 3.0 \times 10^{24}\ AlF_3 \text{ units}$$

10. (b) Given
mass of solute (w) = 120 g
mass of solvent (w) = 1000 g
Mol. mass of solute = 60 g
density of solution = 1.12 g/ mL
From the given data,
Mass of solution = 1000 + 120 = 1120 g

$$\because\ d = \frac{\text{Mol. mass}}{V} \text{ or } V = \frac{\text{Mol. mass}}{d}$$

$$\text{Volume of solution } V = \frac{1120}{1.12} = 1000 \text{ mL}$$

or = 1 litre

$$\text{Now molarity (M)} = \frac{W}{\text{Mol. mass} \times V(\text{lit})}$$

$$= \frac{120}{60 \times 1} = 2\text{M}$$

11. (d) Let mole fraction of O_2 is x
$40 = 32 \times x + 80(1-x)$
or $x = 5/6$

$$a : b = x : (1-x) = \frac{5}{6} : \frac{1}{6}$$

When ratio is changed

$$M_{mixture} = 32 \times \frac{1}{60} + 80 \times \frac{5}{6} = 72$$

12. (b) Moles of H_2SO_4 in 98 mg of H_2SO_4

$$= \frac{1}{98} \times 0.098 = 0.001$$

Moles of H_2SO_4 removes

$= \frac{3.01\times10^{20}}{6.02\times10^{23}} = 0.5\times10^{-3} = 0.0005$

Moles of H_2SO_4 left
$= 0.001 - 0.0005 = 0.5 \times 10^{-3}$

13. (d) At S.T.P. 22.4 litre of gas contains 6.023×10^{23} molecules
∴ molecules in 8.96 litre of gas

$= \frac{6.023\times10^{23}\times8.96}{22.4} = 24.08\times10^{22}$

14. (c) The relation between molarity (M) and molality (m) is

$d = M\left(\frac{1}{m}+\frac{M_2}{1000}\right)$, M_2 = Mol. mass of solute

On putting value

$1.252 = 3\left(\frac{1}{m}+\frac{58.5}{1000}\right)$

on solving $m = 2.79$

15. (a) Let atomic weight of other isotope is M

$114.82 = \frac{114.9041\times95.72 + M\times4.28}{100}$

$M = 112.94$

16. (d)

C	54.55	54.55/12 = 4.5	4.5/2.27 = 2
H	9.09	9.09/1 = 9.09	9.09/2.27 = 4
O	36.36	36.36/16 = 2.27	2.27/2.27 = 1

Hence empirical formula of the compound = C_2H_4O

17. (d)

Element	Percentage	Atomic mass	Relative no. of atoms	Simplest ratio
C	20%	12	1.66	1
H	6.7%	1	6.7	4
N	46.7%	14	3.33	2
O	26.6%	16	1.66	1

Empirical formula = Molecular formula

$= CH_4N_2O$ or NH_2CONH_2

$$H_2NCONH_2 + H_2NCONH_2 \xrightarrow{\Delta} \underset{\text{biuret,}}{H_2NCONHCONH_2} + NH_3$$

When an aqueous solution of biuret is treated with dilute sodium hydroxide and a drop of copper sulphate, a violet colour is produced. This test is known as biuret test, and is characteristic of compounds having the group –CONH–

18. (a) From molarity equation

$M_1V_1 + M_2V_2 = MV_{(total)}$

$2\times\frac{10}{1000} + 0.5\times\frac{200}{1000} = M\times\frac{210}{1000}$

$120 = M \times 210$

$M = \frac{120}{210} = 0.57$ M

19. (a) Mass of 6.023×10^{23} atoms of oxygen = 16 g
Mass of one atom of oxygen

$= \frac{16}{6.023\times10^{23}} = 2.66\times10^{-23}$ g

Mass of 6.023×10^{23} atoms of nitrogen = 14 g
Mass of one atom of nitrogen

$= \frac{14}{6.023\times10^{23}} = 2.32\times10^{-23}$ g

Mass of 1 mole of oxygen = 16 g
Mass of 1×10^{-10} mole of oxygen = 16×10^{-10}
Mass of 1 mole of copper = 63 g
Mass of 1×10^{-10} mole of copper = $63 \times 1 \times 10^{-10}$
$= 63 \times 10^{-10}$

So, the order of increasing mass is II < I < III < IV.

20. (a) 21% of 1 litre is 0.21 litre.
22.4 litres = 1 mole at STP

∴ 0.21 litre $= \frac{0.21}{22.4} = 0.0093$ mol

21. (a) 95% H_2SO_4 by weight means 100g H_2SO_4 solution contains 95g H_2SO_4 by mass.
Molar mass of H_2SO_4 = 98g mol^{-1}

Moles in 95g $= \frac{95}{98} = 0.969$ mole

Volume of 100g H_2SO_4

$= \frac{\text{mass}}{\text{density}} = \frac{100g}{1.834g\ cm^{-3}}$

$= 54.52\ cm^3 = 54.52 \times 10^{-3}$ L

$$\text{Molarity} = \frac{\text{Moles of solute}}{\text{Volme of solute in L}}$$

$$= \frac{0.969}{54.52 \times 10^{-3}} = 17.8 \text{ M}$$

22. (a) $3D + 4E \xrightarrow{80\%} 5C + A$

9 mole 14 mole $\frac{5}{3} \times 9 \times 0.8 = 12$ mole

$3C + 5G \xrightarrow{50\%} 6B + F$

12 mole 4 mole

Limiting Reagent is G

$\therefore$ Moles of B formed $= \frac{6}{5} \times 4 \times 0.5 = 2.4$

23. (a) Let wt. of NH_4NO_3 and $(NH_4)_2HPO_4$ are x and y gram respectively

$$\frac{\frac{x}{80} \times 2 \times 14 + \frac{y}{132} \times 2 \times 14}{x + y} \times 100 = 30.4$$

$\Rightarrow x : y = 2 : 1$

24. (d) $2NH_3 + 5F_2 \longrightarrow N_2F_4 + 6HF$

2 × 17 g NH_3 gives 66 g N_2F_4

2 g will give $- \frac{66}{34} \times 2 = 3.88$ g N_2F_4

$$\% \text{ yield} = \frac{3.56}{3.88} \times 100 = 91.75\%$$

25. (a) $\text{ppm} = \frac{\text{wt. of solute}}{\text{wt. of solvent}} \times 10^6$

$$= \frac{0.2}{500} \times 10^6 = 400$$

26. (c) $M_1V_1 + M_2V_2 = M_3V_3 \quad (V_3 = V_1 + V_2)$

$1.5 \times 480 + 1.2 \times 520 = M_3 \times 1000$

$M_3 = 1.34$ M

27. (b) $CaCO_3 + H_2SO_4 \rightarrow CaSO_4 + H_2O + CO_2$

1 mol	100 g	22.4 litre
1 mol	10 g	2.24 litre

28. (a) Moles of C $= n_{CO_2} = \frac{2.64}{44} = 0.06$

$\Rightarrow$ mass of C = 0.72

Mole of H = 2 × Moles of H_2O =

$2 \times \frac{1.26}{18} = 0.14 \Rightarrow$ mass of H = 0.14

Compound does not contains oxygen.

So EF $\rightarrow C_{0.06}H_{0.14} \Rightarrow C_3H_{17}$

$\Rightarrow$ Lowest M.M. = 43

29. (a) The reaction that takes place is

$NaCl + AgNO_3 \longrightarrow AgCl \downarrow + NaNO_3$

$\therefore$ 143.5 g of AgCl is produced from 58.5 g NaCl

$\therefore$ 14 g of AgCl will be produced from

$\frac{58.5 \times 14}{143.5} = 5.70$ g NaCl

This is the amount of NaCl in common salt;

$$\% \text{ purity} = \frac{5.70}{6} \times 100 = 95\%$$

30. (a)

Na_2CO_3	$NaHCO_3$
x	(1 – x)

$\frac{x}{106} = \frac{1-x}{84}$ given (moles are equal)

x = 0.557

$$\frac{0.557}{53} + \frac{0.443}{84} = \frac{V \times 0.1}{1000}$$

V = 157.7 mL

31. (c) $C_6H_6 + HNO_3 \rightarrow C_6H_5NO_2 + H_2O$

78g 123g

Now since 78g of benzene on nitration give = 123g nitrobenzene

hence 5g of benzene on nitration give

$$= \frac{123}{78} \times 5 = 7.88\text{g}$$

nearest answer is (c) i.e. theoritical yield = 7.88 g

32. (d) Normality of oxalic acid solution

$$= \frac{6.3 \times 1000}{63 \times 250} = 0.4 \text{ N}$$

Now from

$N_1V_1 = N_2V_2$

$0.4 \times 10 = 0.1 \times V_2$

$V_2 = 40$ mL

33. (b) $X_2O_3 \Rightarrow X^{3+}O^{2-}$

$X_2Cl_3 \Rightarrow X^{3+}Cl^{2-}$

$X_2(SO_4)_3 \Rightarrow X^{3+}SO_4^{\;2-}$

$XPO_4 \Rightarrow X^{3+}PO_4^{\;3-}$

Because Cl^{2-} does not exist. So, X_2Cl_3 is incorrect. The correct formula should be XCl_3.

34. (d) In an unknown compounds containing N and H

given % of H = 12.5%

$\therefore$ % of N = 100 − 12.5 = 87.5%

Element	Percentage	Atomic ratio	Simple ratio
H	12.5%	$\frac{12.5}{1}=12.5$	$\frac{12.5}{6.25}=2$
N	87.5	$\frac{87.5}{14}=6.25$	$\frac{6.25}{6.25}=1$

2 × vapour density = Mol. wt = mol wt.
= 16 × 2 = 32.

Molecular formula = n × empirical formula

mass $n=\frac{32}{16}=2$

$\therefore$ Molecular formula of the compound will be $=(NH_2)_2 = N_2H_4$

35. (a)

$$\begin{array}{lcccc} & C & H & N & \\ \text{Moles} & \frac{9}{12} & \frac{1}{1} & \frac{3.5}{14} & n=\frac{108}{54}=2 \\ & \frac{3}{4} : & 1 : & \frac{1}{4} & \\ & 3 : & 4 : & 1 & \end{array}$$

$\Rightarrow C_3H_4N \Rightarrow M.F. = (C_3H_4N)_n$

$MF = (C_3H_4N)_2 = C_6H_8N_2$

36. (c) Mass of oxide = Mass of metal + Mass of oxygen

$2.74 = 1.53 + W_{Oxygen} \Rightarrow W_{Oxygen} = 1.21$ g

Moles of V $=\frac{1.53}{51}=0.03$

Moles of O $=\frac{1.21}{16}=0.075$

$V_{0.03}O_{0.075}$

$VO_{2.5} \Rightarrow V_2O_5$

37. (c) Normality

$$=\frac{N_1V_1 - N_2V_2}{V_1+V_2} = \frac{0.2\times100 - 0.1\times100}{100+100}$$

$$=\frac{10}{200} = 0.05\text{N NaOH}$$

38. (a)

$$\begin{array}{ccc} N_2 + & O_2 \rightarrow & 2NO \\ a & a & 2a \end{array}$$

$$\begin{array}{ccc} CO & +\ \frac{1}{2}O_2 & \rightarrow CO_2 \\ (10-a) & \frac{(10-a)}{2} & (10-a) \end{array}$$

$$a+\frac{(10-a)}{2}=7$$

$\therefore$ a = 4

volume of CO = 6 mL

Mole % of CO $=\frac{6\times100}{10}=60$

39. (b) Let weight of C be x g, then S will be (14 − x)g

$$\frac{x/12}{(14-x)/32}=\frac{2}{1}$$

$\therefore$ x = 6 g; Moles of C $=\frac{6}{12}=0.5$

40. (a) $\frac{\text{mass of } Al_2(SO_4)_3}{\text{mass of water}}\times10^6 = 34.2$

1 litre solution contains 1000 g of water

$\Rightarrow$ In 1 litre solution, mass of

$Al_2(SO_4)_3 = \frac{34.2\times1000}{10^6} = 34.2$ mg

molarity of

$Al_2(SO_4)_3 = \frac{34.2\times10^{-3}}{342}\text{M} = 10^{-4}\text{M}$

$$\begin{array}{lcc} Al_2(SO_4)_3(aq) \longrightarrow & 2Al^{3+}(aq) + & 3SO_4^{2-}(aq) \\ 10^{-4}\text{M} & 2\times10^{-4}\text{M} & 3\times10^{-4}\text{M} \end{array}$$

$\left[SO_4^{2-}\right] = 3\times10^{-4}$M

41. (c) 74.75% of chlorine means 74.75g chlorine is present in 100g of metal chloride.

Weight of metal $= 100g - 74.75g$
$= 25.25g$

Equivalent weight

$$= \frac{\text{weight of metal}}{\text{weight of chlorine}} \times 35.5$$

$$= \frac{25.25}{74.75} \times 35.5 = 12$$

Valency of metal

$$= \frac{2 \times V.D.}{\text{Equivalent wt. of metal} + 35.5}$$

$$= \frac{2 \times 94.8}{12 + 35.5} = 4$$

$\therefore$ Formula of compound = MCl_4

42. (b) $N = \dfrac{N_1V_1 + N_2V_2 + N_3V_3}{\text{Total volume}}$

$$= \frac{50 \times 10 + 25 \times 12 + 40 \times 5}{1000} = 1N$$

43. (c) Molarity

$$= \frac{Wt \times 1000}{MW \times V} = \frac{2.65 \times 1000}{106 \times 250} = 0.1M$$

$M_1V_1 = M_2V_2$

$\therefore 10 \times 0.1 = 1000 \times M_2 = 0.001\,M$

44. (a) Normality of ferrous amm. sulphate

$$= \frac{3.92 \times 1000}{392 \times 100} = 0.1 \text{ (Eq. wt of FAS is 392)}$$

$N_1V_1 = N_2V_2$

$20 \times 0.1 = 18 \times N_2$ $\quad N_2 = 0.111$

1 g ev. of $KMnO_4 = 31.6\,g$

0.111 g ev. of $KMnO_4 = 31.6 \times 0.111 = 3.5$ g.

45. (c) Let the alcohol be ROH and x its molecular weight

$$\underset{x\,g}{ROH} + CH_3MgI \rightarrow \underset{16\,g}{CH_4} + ROMgI$$

$\dfrac{4.12}{1000}$ g of alcohol will produce

$\dfrac{16}{x} \times \dfrac{4.12}{1000}$ g of methane

Methane actually obtained is $= \dfrac{16 \times 1.12}{22400}$ g

$$\therefore \frac{16 \times 4.12}{x \times 1000} = \frac{16 \times 1.12}{22400} = x = 82.4$$

46. (a) $4Ca_5\underset{0.1}{(PO_4)_3}F + 18\underset{0.36}{SiO_2} + 30\underset{0.9}{C} \xrightarrow{L.R.} 3P_4 + 2CaF_2 + 18CaSiO_3 + 30CO$

18 moles of SiO_2 gives 3 moles of P_4

0.36 moles of SiO_2 will give

$$= \frac{3}{18} \times 0.36 = 0.06 \text{ mole}$$

47. (b) $CO + \frac{1}{2}O_2 \rightarrow CO_2$;

$CO_2 + 2KOH \rightarrow K_2CO_3 + H_2O$

Moles of KOH $= \dfrac{28}{56} = 0.50$

It corresponds to 0.25 mol of CO_2

Hence mol of CO $= 1 - 0.25 = 0.75 \equiv$ mole of CO_2 formed

Mol of KOH requred $= 2 \times 0.75 = 1.5$
$= 1.5 \times 56 = 84$ g

48. (c) $3Fe_2O_3(s) \rightarrow 2Fe_3O_4 + \frac{1}{2}O_2$

480 g Fe_2O_3 provide 16 g O_2. For loss of

$$0.04 \text{ g } O_2 \rightarrow 0.04 \times \frac{480}{16} = 1.2 \text{ g } Fe_2O_3$$

$$\% \text{ by mass of } SiO_2 = \frac{0.8}{2.0} \times 100 = 40\%$$

49. (a) $CaCO_3 \rightarrow CaO + CO_2$

20 g $\quad$ 8.8 g $\quad$ 11.2 g

mass of reactant = mass of products = 20g. Hence the law of conservation of mass is obeyed.

50. (c) The conditions given are standard conditions

224 mL has mass = 1g;

22400 mL will have mass = 100g. This is mol. wt of gas

6.023×10^{23} molecules have $3 \times 6.023 \times 10^{23}$ atoms since gas is triatomic

$\therefore$ weight of one atom

$= \frac{100}{3 \times 6.023 \times 10^{23}} = 5.5 \times 10^{-23}$ g

51. (d) Specific volume (volume of 1 g) of cylindrical virus particle $= 6.02 \times 10^{-2}$ cc/g

Radius of virus (r) = 7 Å = 7×10^{-8} cm

Length of virus = 10×10^{-8} cm

Volume of virus =

$\pi r^2 l = \frac{22}{7} \times (7 \times 10^{-8})^2 \times 10 \times 10^{-8}$

$= 154 \times 10^{-23}$ cc

Wt. of one virus particle $= \frac{\text{volume}}{\text{specific volume}}$

∴ Mol. wt. of virus = Wt. of N_A particle

$= \frac{154 \times 10^{-23}}{6.02 \times 10^{-2}} \times 6.02 \times 10^{23} = 15400$ g/mol

= 15.4 kg/mol

52. (b) Meq of oxalic acid in 16.68 ml = Meq of NaOH

$= 25 \times \frac{1}{15}$

Meq of oxalic acid in 250 ml

$= 25 \times \frac{1}{15} \times \frac{250}{16.68} = 24.98$

$\frac{1.575}{(90+18x)/2} \times 1000 = 24.98$

∴ x=2

53. (c) $5CO + I_2O_5 \rightarrow 5CO_2 + I_2$

Moles of $I_2O_5 = \frac{25.4}{254}$

= 0.01 ≡ 0.05 moles of CO

Weight of CO = 0.05 × 28 = 1.4 g;

Weight of $CO_2 = 2 - 1.4 = 0.6$ g

Hence % of $CO_2 = \frac{0.6}{2} \times 100 = 30\%$

54. (d) 3 molecules of O_2 = 1 molecules of CS_2
6 molecules of O_2 = 2 molecules of CS_2

55. (b) The required equation is

$2KMnO_4 + 3H_2SO_4 \longrightarrow$

$K_2SO_4 + 2MnSO_4 + 3H_2O + \underset{\text{nascent oxygen}}{5[O]}$

$2Fe(C_2O_4) + 3H_2SO_4 + 3[O] \longrightarrow$

$Fe_2(SO_4)_3 + 2CO_2 + 3H_2O$

O required for 1 mol. of $Fe(C_2O_4)$ is 1.5, 5 O are obtained from 2 moles of $KMnO_4$

∴ 1.5 [O] will be obtained from

$= \frac{2}{5} \times 1.5 = 0.6$ moles of $KMnO_4$.

56. (d) Weight of Iron in 67200

$= \frac{0.33}{100} \times 67200 = 221.76$

Number of atoms of Iron

$= \frac{221.76}{56} = 3.96 \equiv 4$

57. (d) Volume of gold foil = $25 \times 40 \times 0.25$ mm^3
$= 250 \times 10^{-3}$ cm^3

Mass of gold foil = $19.32 \times 250 \times 10^{-3}$ g
= 4.83 g

No. of gold atoms $= \frac{4.83}{197} \times N_A$

$= 1.47 \times 10^{22}$

58. (c) Writing the equation of combustion of propane (C_3H_8), we get

$\underset{\substack{1\text{vol} \\ 1\text{L}}}{C_3H_8} + \underset{\substack{5\text{vol} \\ 5\text{L}}}{5O_2} \rightarrow 3CO_2 + 4H_2O$

From the above equation we find that we need 5 L of oxygen at NTP to completely burn 1 L of propane at N.T.P.

If we change the conditions for both the gases from N.T.P. to same conditions of temperature and pressure. The same results are obtained. i.e. 5 L is the correct answer.

59. (a)

	I_2 +	$2Cl_2$ ⟶	ICl +	ICl_3
No. of moles	$\frac{25.4}{254}$	$\frac{14.2}{71}$	0	0
initially	0.1	0.2	0	0
No. of moles after reaction	0	0	0.1	0.1

60. (a) $2SO_2 + O_2 \longrightarrow 2SO_3$

	$2SO_2$	O_2	$2SO_3$
	10	15	0
	$10-2x$	$15-x$	$2x$

$\because 2x=8 \quad x=4$

Hence, remaining, $SO_2 = 10-8 = 2$ moles,

$O_2 = 15-4 = 11$ moles

61. (d)

$$4NH_3(g)+5O_2(g) \longrightarrow 4NO(g)+6H_2O(l)$$

	$4NH_3$	$5O_2$	$4NO$	$6H_2O$
	4 moles	5 moles	4 moles	6 moles
Given	1 Mole	1 Mole		
Reacting	0.8	1	0.8	1.2

All O_2 consumed being limiting reagent.

62. (b) $C_2H_6 + 3.5O_2 \longrightarrow 2CO_2 + 3H_2O$;

$C_2H_4 + 3O_2 \longrightarrow 2CO_2 + 2H_2O$

Let volume of ethane is x litre,

$22.4 \times 4 = 3.5x + 3(28-x)$

$\Rightarrow x = 11.2$ litre

at constant T and P, $V \propto n$;

$\Rightarrow$ Mole fraction of C_2H_6 in mixture

$$= \frac{11.2}{28} = 0.4$$

63. (c) $X_{Sn} = \dfrac{12.5/119}{\dfrac{50}{209}+\dfrac{25}{207}+\dfrac{12.5}{119}+\dfrac{12.5}{112}} = 0.176$

64. (b) $\underset{1\,mol}{Ba(OH)_2} + CO_2 \longrightarrow \underset{1\,mol}{BaCO_3} + H_2O$

1 mol $Ba(OH)_2$ = 1 mol $BaCO_3$

$\therefore$ 0.205 mol $Ba(OH)_2 \equiv$ 0.205 mol $BaCO_3$

Wt. of substance = No. of moles × Molecular mass = 0.205 × 197.3 = 40.5 g

65. (b) $SO_2Cl_2 + 2H_2O \rightarrow H_2SO_4 + 2HCl$

$H_2SO_4 + Ba(OH)_2 \rightarrow BaSO_4 + 2H_2O$

$2HCl + Ba(OH)_2 \rightarrow BaCl_2 + 2H_2O$

Total moles of $Ba(OH)_2$ required = 2

66. (d) Let x g of Li_2CO_3 and (1 – x) g of M_2CO_3 present in given mixture

2 × Total moles of carbonates = moles of HCl and

$$2\times\left(\frac{x}{74}+\frac{1-x}{2M+60}\right) = 21.6\times0.5\times10^{-3}$$

$x = 0.20$

$$\because \frac{x}{74} = \frac{1-x}{2M+60}$$

$M = 118$

67. (a) m-eq. $FeSO_4(NH_4)_2SO_4.6H_2O$ = m-eq. of $KMnO_4$ (n = 1)

$$\frac{W}{392}\times1\times1000 = 0.1\times50$$

Hence, % purity of Mohr's salt $\qquad W = 1.96$ g

$$= \frac{1.96}{2.5}\times100 = 78.4\%$$

68. (a) $\underset{\substack{2\times122.5g \\ 245}}{2KClO_3} \xrightarrow{heat} 2KCl + \underset{\substack{3\times32g \\ 96}}{3O_2}$

48 g of oxygen will be produced from 122.5 g of $KClO_3$

$\therefore$ Amount of 80% $KClO_3$ needed

$$= \frac{100}{80}\times122.5 = 153.12g$$

69. (b) $\underset{\substack{1\text{ mole} \\ \frac{1}{2}\text{ mole}}}{Mg} + 2HCl \rightarrow MgCl_2 + \underset{\substack{1\text{ mole} \\ \frac{1}{2}\text{ mole}}}{H_2} \uparrow$ (12g of Mg $= \frac{1}{2}$ mol)

70. (a) Decomposition of Ag_2CO_3 gives silver

$$\underset{\substack{2\times276g \\ 2.76g}}{2Ag_2CO_3} \rightarrow \underset{\substack{4\times108g \\ 2.16g}}{4Ag} + 2CO_2 + O_2$$

2 Structure of Atom

1. What is the angular velocity (ω) of an electron occupying second orbit of Li^{2+} ion?

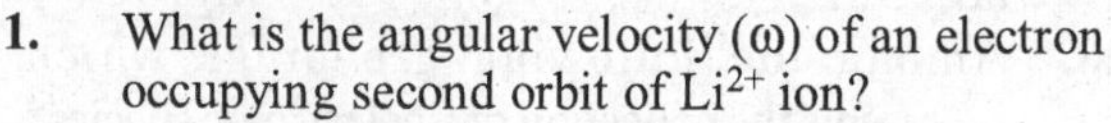

(a) $\frac{8\pi^3 me^4}{h^3}K^2$ (b) $\frac{8\pi^3 me^4}{9h^3}K^2$

(c) $\frac{64}{9}\times\frac{\pi^3 me^4}{h^3}K^2$ (d) $\frac{9\pi^3 me^4}{h^3}K^2$

2. In a collection of H-atoms, all the electrons jump from n = 5 to ground level finally (directly or indirectly), without emitting any line in Balmer series. The number of possible different radiations is : **Tricky**

(a) 10 (b) 8 (c) 7 (d) 6

3. The wavelength of H_α line of Balmer series is X Å. What is the X of H_β line of Balmer series.

(a) $X\frac{108}{80}$Å (b) $X\frac{80}{108}$Å

(c) $\frac{1}{X}\frac{80}{108}$Å (d) $\frac{1}{X}\frac{108}{80}$Å

4. The number of photons of light having wave number 'x' in 10 J of energy source is :

(a) 10hcx (b) $\frac{hc}{10x}$

(c) $\frac{10}{hcx}$ (d) None of these

5. The values of Planck's constant is 6.63×10^{-34} Js. The velocity of light is 3.0×10^8 m s^{-1}. Which value is closest to the wavelength in nanometres of a quantum of light with frequency of 8×10^{15} s^{-1}?

(a) 5×10^{-18} (b) 4×10^1

(c) 3×10^7 (d) 2×10^{-25}

6. According to law of photochemical equivalence the energy absorbed (in ergs/mole) is given as (h = 6.62×10^{-27} ergs, c = 3×10^{10} cm s^{-1}, $N_A = 6.02 \times 10^{23}$ mol^{-1})

(a) $\frac{1.956\times10^{16}}{\lambda}$ (b) $\frac{1.19\times10^8}{\lambda}$

(c) $\frac{2.859\times10^5}{\lambda}$ (d) $\frac{2.859\times10^{16}}{\lambda}$

7. The energy of an electron in first Bohr orbit of H-atom is – 13.6 eV. The energy value of electron in the excited state of Li^{2+} is:

(a) –27.2 eV (b) 30.6 eV

(c) –30.6 eV (d) 27.2 eV

8. Based on the equation:

$$\Delta E = -2.0\times10^{-18}\,J\left(\frac{1}{n_2^2}-\frac{1}{n_1^2}\right)$$

the wavelength of the light that must be absorbed to excite hydrogen electron from level n = 1 to level n = 2 will be:

(h = 6.625×10^{-34} Js, C = 3×10^8 ms^{-1})

(a) 1.325×10^{-7} m (b) 1.325×10^{-10} m

(c) 2.650×10^{-7} m (d) 5.300×10^{-10} m

9. In an atom, an electron is moving with a speed of 600 m/s with an accuracy of 0.005%. Certainity with which the position of the electron can be located is (h = 6.6×10^{-34} kg m^{2}s^{-1}, mass of electron, $e_m = 9.1 \times 10^{-31}$ kg) **Toughnut**

(a) 5.10×10^{-3} m (b) 1.92×10^{-3} m

(c) 3.84×10^{-3} m (d) 1.52×10^{-4} m

10. Which of the following is **not** permissible arrangement of electrons in an atom?

(a) $n = 5, l = 3, m = 0, s = +1/2$

(b) $n = 3, l = 2, m = -3, s = -1/2$

(c) $n = 3, l = 2, m = -2, s = -1/2$

(d) $n = 4, l = 0, m = 0, s = -1/2$

11. A dye absorbs a photon of wavelength λ and re-emits the same energy into two photons of wavelength λ_1 and λ_2 respectively. The wavelength λ is related with λ_1 and λ_2 as:

(a) $\lambda = \frac{\lambda_1 + \lambda_2}{\lambda_1 \lambda_2}$ (b) $\lambda = \frac{\lambda_1 \lambda_2}{\lambda_1 + \lambda_2}$

(c) $\lambda = \frac{\lambda_1^2 \lambda_2^2}{\lambda_1 + \lambda_2}$ (d) $\lambda = \frac{\lambda_1 \lambda_2}{(\lambda_1 + \lambda_2)^2}$

12. The Bohr's energy equation for H atom reveals that the energy level of a shell is given by $E = -13.58/n^2 eV$. The smallest amount that an H atom will absorb if in ground state is

(a) 1.0 eV (b) 3.39 eV
(c) 6.79 eV (d) 10.19 eV

13. If *m* and *e* are the mass and charge of the revolving electron in the orbit of radius *r* for hydrogen atom, the total energy of the revolving electron will be:

(a) $\frac{1}{2}\frac{e^2}{r}$ (b) $-\frac{e^2}{r}$

(c) $\frac{me^2}{r}$ (d) $-\frac{1}{2}\frac{e^2}{r}$

14. Excited hydrogen atom emits light in the ultraviolet region at 2.47×10^{15} Hz. With this frequency, the energy of a single photon is: ($h = 6.63 \times 10^{-34}$ Js)

(a) 8.041×10^{-40} J (b) 2.680×10^{-19} J
(c) 1.640×10^{-18} J (d) 6.111×10^{-17} J

15. The orbital diagram in which the Aufbau principle is violated is:

	2s	2p		
(a)	↑↓	↑↓	↑	
(b)	↑	↑↓	↑	↑
(c)	↑↓	↑	↑	↑
(d)	↑↓	↑↓	↑	↑

16. Calculate the velocity of ejected electron from the metal surface when light of frequency 2×10^{15} Hz fall on the metal surface and the threshold frequency is 7×10^{14} Hz for metal?

(a) 1.37×10^6 (b) 1.26×10^6
(c) 1.45×10^7 (d) 1.37×10^7

17. An electron, e_1 is moving in the fifth stationary state, and another electron e_2 is moving in the fourth stationary state. The radius of orbit of electron, e_1 is five times the radius of orbit of electron, e_2 calculate the ratio of velocity of electron e_1 (v_1) to the velocity of electron e_2 (v_2).

Toughnut

(a) 5 : 1 (b) 4 : 1 (c) 1 : 5 (d) 1 : 4

18. Among the following groupings which represents the collection of isoelectronic species?

(a) $NO^+, C_2^{2-}, O_2^-, CO$
(b) N_2, C_2^{2-}, CO, NO
(c) CO, NO^+, CN^-, C_2^{2-}
(d) NO, CN^-, N_2, O_2^-

19. The wave number of the first emission line in the Balmer series of H-Spectrum is: (R = Rydberg constant):

(a) $\frac{5}{36}R$ (b) $\frac{9}{400}R$ (c) $\frac{7}{6}R$ (d) $\frac{3}{4}R$

20. What is the maximum wavelength line in the Lyman series of He^+ ion?

(a) 3R (b) $\frac{1}{3R}$
(c) $\frac{4}{4R}$ (d) None of these

21. The position of both, an electron and a helium atom is known within 1.0 nm. Further the momentum of the electron is known within 5.0×10^{-26} kg ms^{-1}. The minimum uncertainty in the measurement of the momentum of the helium atom is

(a) 50 kg ms^{-1} (b) 80 kg ms^{-1}
(c) 8.0×10^{-26} kg ms^{-1} (d) 5.0×10^{-26} kg ms^{-1}

22. What will be the difference between electromagnetic radiation shown in A and B respectively?

Tricky

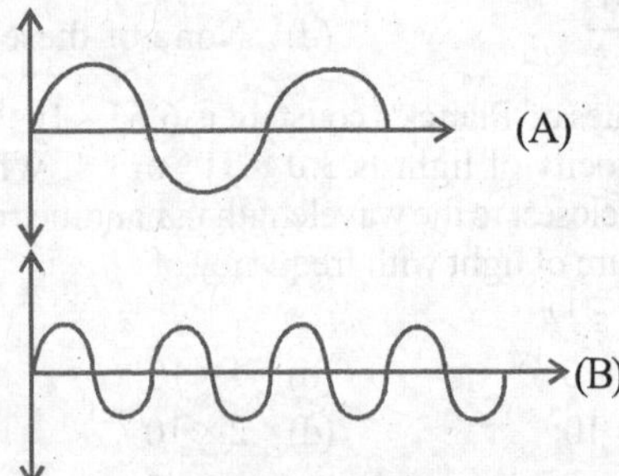

(i) Velocity (ii) Wavelength
(iii) Frequency (iv) Energy
(a) (ii) only (b) (ii) and (iv)
(c) (ii), (iii) and (iv) (d) (iv) only

23. Arrange the electromagnetic radiations a, b, c, d and e in increasing order of energy. Frequencies of a, b and c are 10^{15}, 10^{14} and 10^{17} respectively whereas wavelength of (d) and (e) are 350 nm and 100 nm respectively ?
(a) a, b, c, d, e (b) a, b, d, e, c
(c) a, d, b, e, c (d) b, a, d, e, c

24. When electronic transition occurs from higher energy state to lower energy state with energy difference equal to ΔE electron volts, the wavelength of the line emitted is approximately equal to
(a) $\frac{12395}{\Delta E}\times 10^{-10}$ m (b) $\frac{12395}{\Delta E}\times 10^{10}$ m
(c) $\frac{12395}{\Delta E}\times 10^{-10}$ cm (d) $\frac{12395}{\Delta E}\times 10^{10}$ cm

25. Which of the following should be the number of electrons present in X^{2+} on the basis of electronic configuration, if the ion X^{3-} has 14 protons ?
(a) 12 (b) 14 (c) 16 (d) 18

26. In a hydrogen atom, if energy of an electron in ground state is 13.6. ev, then that in the 2nd excited state is
(a) 1.51 eV (b) 3.4 eV
(c) 6.04 eV (d) 13.6 eV.

27. At temperature T, the average kinetic energy of any particle is $\frac{3}{2}$KT. The de Broglie wavelength follows the order : Toughnut
(a) Visible photon > Thermal neutron > Thermal electron
(b) Thermal proton > Thermal electron > Visible photon
(c) Thermal proton > Visible photon > Thermal electron
(d) Visible photon > Thermal electron > Thermal neutron

28. Which one of the following set of quantum numbers is not possible for 4*p* electron?
(a) $n=4, l=1, m=-1, m_s=+\frac{1}{2}$
(b) $n=4, l=1, m=0, m_s=+\frac{1}{2}$
(c) $n=4, l=1, m=2, m_s=+\frac{1}{2}$
(d) $n=4, l=1, m=-1, m_s=-\frac{1}{2}$

29. What is the maximum number of electrons that can be accomodated in an atom in which the highest principal quantum number value is 4?
(a) 10 (b) 18 (c) 36 (d) 54

30. The de Broglie wavelength of a car of mass 1000 kg and velocity 36 km/hr is :
(a) 6.626×10^{-34} m (b) 6.626×10^{-38} m
(c) 6.626×10^{-31} m (d) 6.626×10^{-30} m

31. If the radius of first orbit of H atom is a_0, the de-Broglie wavelength of an electron in the third orbit is Critical Thinking
(a) $4\pi a_0$ (b) $8\pi a_0$ (c) $6\pi a_0$ (d) $2\pi a_0$

32. Which of the following radial distribution graphs correspond to $l = 2$ for the H atom ?

(a) 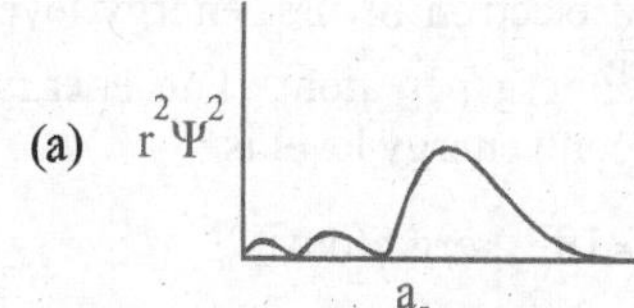

(b) 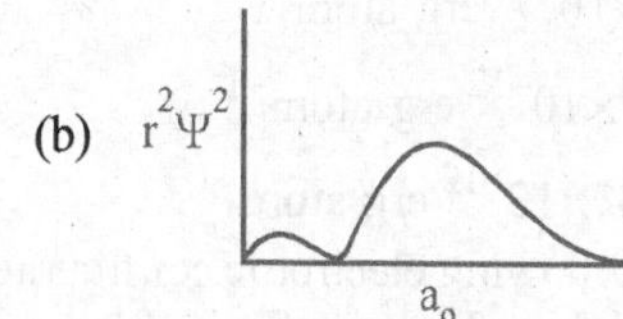

(c) 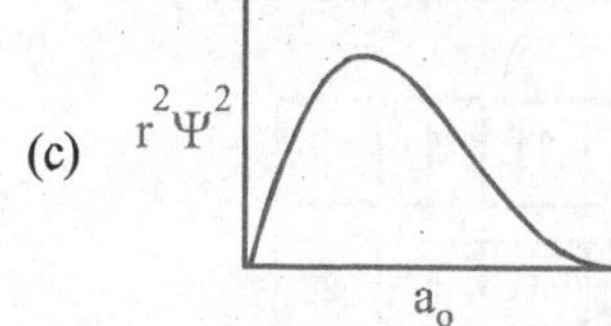

(d) 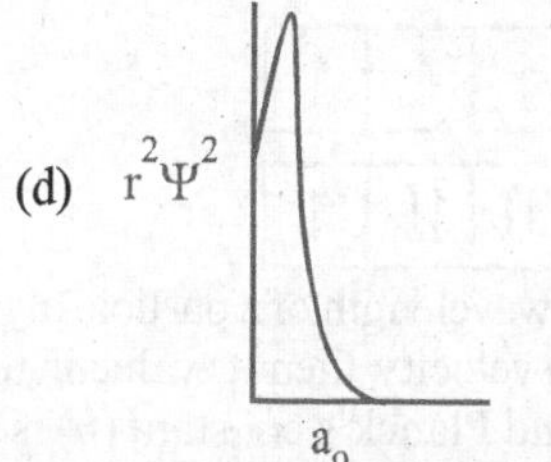

33. In ground state of Cu^+. The no. of shells occupied, subshells, filled orbitals and unpaired electrons respectively are :
(a) 4, 8, 15, 0 (b) 3, 6, 15, 1
(c) 3, 6, 14, 0 (d) 4, 7, 14, 2

34. The Li^{2+} ion is moving in the third stationary state, and its linear momentum is 7.3×10^{-34} kgms^{-1}. Angular momentum is.

(a) 1.158×10^{-45} kg m^{2}s^{-1}
(b) 11.58×10^{-48} kg m^{2}s^{-1}
(c) 11.58×10^{-47} kg m^{2}s^{-1}
(d) 12×10^{-45} kg m^{2}s^{-1}

35. The de Broglie wavelength of a tennis ball of mass 60 g moving with a velocity of 10 metres per second is approximately

(a) 10^{-31} (b) 10^{-16} (c) 10^{-25} (d) 10^{-33}

36. In an atom how many orbital(s) will have the quantum numbers; $n = 3$, $l = 2$ and $m_l = +2$?

(a) 5 (b) 3 (c) 1 (d) 7

37. According to Bohr's theory, the angular momentum of an electron in 5th orbit is

(a) 10 h/π (b) 2.5 h/π

(c) 25 h/π (d) 1.0 h/π

38. The energy of electron in first energy level is -21.79×10^{-12} erg per atom. The energy of electron in second energy level is :

(a) -54.47×10^{-12} erg atom^{-1}

(b) -5.447×10^{-12} erg atom^{-1}

(c) -0.5447×10^{-12} erg atom^{-1}

(d) -0.05447×10^{-12} erg atom^{-1}

39. Which of the following electronic configuration of *d*-orbital will have highest affinity for gaining an electron?

(a) 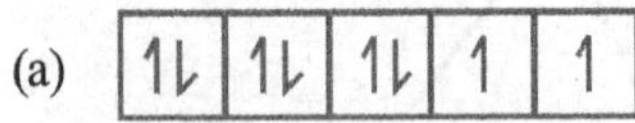

(b) | ↑↓ | ↑↓ | ↑ | ↑ | ↑ |

(c) | ↑↓ | ↑ | ↑ | ↑ | ↑ |

(d) | ↑↓ | ↑↓ | ↑↓ | ↑↓ | ↑ |

40. If the de-Broglie wavelength of a particle of mass *m* is 100 times its velocity, then its value in terms of its mass (*m*) and Planck's constant (*h*) is

(a) $\frac{1}{10}\sqrt{\frac{m}{h}}$ (b) $10\sqrt{\frac{h}{m}}$

(c) $\frac{1}{10}\sqrt{\frac{h}{m}}$ (d) $10\sqrt{\frac{m}{h}}$

41. An electron in an atom jumps in such a way that its kinetic energy changes from x to $\frac{x}{4}$. The change in potential energy will be:

(a) $+\frac{3}{2}x$ (b) $-\frac{3}{8}x$

(c) $+\frac{3}{4}x$ (d) $-\frac{3}{4}x$

42. The kinetic and potential energy (in eV) of an electron present in third Bohr's orbit of hydrogen atom are respectively:

(a) –1.51, –3.02 (b) 1.51, –3.02

(c) –3.02, 1.51 (d) 1.51, –1.51

43. What atomic number of an element "X" would have to become so that the 4th orbit around X would fit inside the I Bohr orbit of H atom?

(a) 3 (b) 4 (c) 16 (d) 25

44. Which combinations of quantum numbers, n, *l*, m and s for the electron in an atom does not provide a permissible solution of the wave equation ?

(a) $3, 2, 1, +\frac{1}{2}$ (b) $3, 1, 1, -\frac{1}{2}$

(c) $3, 3, 1, -\frac{1}{2}$ (d) $3, 2, -2, +\frac{1}{2}$

45. If the nitrogen atom had electronic configuration $1s^7$ it would have energy lower than that of the normal ground state configuration $1s^2\ 2s^2\ 2p^3$ because the electrons would be closer to the nucleus. Yet $1s^7$ is not observed. It violates

(a) Heisenberg's uncertainty principle

(b) Hund's rule

(c) Pauli exclusion principle

(d) Bohr postulate of stationary orbits

46. The position of both, an electron and a helium atom is known within 1.0 nm. Further the momentum of the electron is known within 5.0×10^{-26} kg ms^{-1}. The minimum uncertainty in the measurement of the momentum of the helium atom is

(a) 50 kg ms^{-1}

(b) 80 kg ms^{-1}

(c) 8.0×10^{-26} kg ms^{-1}

(d) 5.0×10^{-26} kg ms^{-1}

47. Two fast moving particles X and Y are associated with de Broglie wavelengths 1 nm and 4 nm respectively. If mass of X is nine times the mass of Y, the ratio of kinetic energies of X and Y would be

(a) 3 : 1 (b) 9 : 1 (c) 5 : 12 (d) 16 : 9

48. In a multi-electron atom, which of the following orbitals described by the three quantum members will have the same energy in the absence of magnetic and electric fields?
(A) $n=1, l=0, m=0$ (B) $n=2, l=0, m=0$
(C) $n=2, l=1, m=1$ (D) $n=3, l=2, m=1$
(E) $n=3, l=2, m=0$
(a) (D) and (E) (b) (C) and (D)
(c) (B) and (C) (d) (A) and (B)

49. A system irradiated for 10 min. is found to absorb 3×10^{18} quanta per sec. If the amount of substance decomposed is 3×10^{-3} mol ($N_A = 6 \times 10^{23}$). The quantum efficiency of the reaction is
(a) 2.5 (b) 2 (c) 1.5 (d) 1

50. The radii of maximum probability for $3s$, $3p$ and $3d$ electrons are in the order
(a) $(r_{max})_{3d} > (r_{max})_{3p} > (r_{max})_{3s}$
(b) $(r_{max})_{3d} > (r_{max})_{3s} > (r_{max})_{3p}$
(c) $(r_{max})_{3s} > (r_{max})_{3p} > (r_{max})_{3d}$
(d) none of these

51. If in Bohr's model, for unielectronic atom, time period of revolution is represented as $T_{n,z}$ where n represents shell no. and Z represents atomic number then the value of $T_{1,2} : T_{2,1}$ will be :
(a) 8 : 1 (b) 1 : 8 (c) 1 : 1 (d) 1 : 32

52. Find the value of wave number ($\bar{v}$) in terms of Rydberg's constant, when transition of electron takes place between two levels of He^+ ion whose sum is 4 and difference is 2. Tricky
(a) $\frac{8R}{9}$ (b) $\frac{32R}{9}$
(c) $\frac{3R}{9}$ (d) None of these

53. The uncertainty in the position of an electron (mass $= 9.1 \times 10^{-28}$ g) moving with a velocity of 3.0×10^4 cm s^{-1} accurate upto 0.011% will be
(a) 1.92 cm (b) 7.68 cm
(c) 0.175 cm (d) 3.84 cm

54. The energy of e^- in first orbit of He^+ is -871.6×10^{-20} J. The energy of e^- in first orbit of H is:
(a) -871.6×10^{-20} (b) -435.8×10^{-20} J
(c) -217.9×10^{-20} J (d) -108.9×10^{-20}

55. If a proton and α-particle are accelerated through the same potential difference, the ratio of de-Broglie wavelengths λ_p and λ_α is
(a) 3 (b) $2\sqrt{2}$ (c) 1 (d) 2

56. The velocity of particle A is 0.1 ms^{-1} and that of particle B is 0.05 ms^{-1}. If the mass of particle B is five times that of particle A, then the ratio of de-Broglie wavelengths associated with the particles A and B is Toughnut
(a) 2 : 5 (b) 3 : 4 (c) 6 : 4 (d) 5 : 2

57. Let m_p be the mass of a proton, m_n that of a neutron, M_1 that of a $^{20}_{10}Ne$ nucleus and M_2 that of a $^{40}_{20}Ca$ nucleus. Then
(a) $M_2 = 2M_1$ (b) $M_1 < 10(m_p + m_n)$
(c) $M_2 > 2M_1$ (d) $M_1 = M_2$

58. If radius of second stationary orbit (in Bohr's atom) is R then radius of third orbit will be :
(a) R/3 (b) 9R (c) R/9 (d) 2.25R

59. The dissociation energy of H_2 is 430.53 kJmol^{-1}. If hydrogen is dissociated by illumination with radiation of wavelength 253.7 nm the fraction of the radiant energy which will be converted into kinetic energy is given by
(a) 100% (b) 8.82% (c) 2.22% (d) 1.22%

60. The threshold frequency of a metal is 1×10^{15} s^{-1}. The ratio of the maximum kinetic energies of the photoelectrons when the metal is irradiated with radiations of frequencies 1.5×10^{15}s^{-1} and 2.0×10^{15}s^{-1} respectively would be
(a) 4 : 3 (b) 1 : 2 (c) 2 : 1 (d) 3 : 4

61. The wave number of electromagnetic radiation emitted during the transition in between two energy levels of Li^{2+} ion whose principal quantum number sum is 4 and difference is 2, is Toughnut
(a) $3.5 R_H$ (b) $4 R_H$ (c) $8 R_H$ (d) $\frac{8}{9} R_H$

62. An electron travels with a velocity of x ms^{-1}. For a proton to have the same de-Broglie wavelength, the velocity will be approximately :
(a) $\frac{1840}{x}$ (b) $\frac{x}{1840}$ (c) 1840x (d) x

63. The total spin and magnetic moment for the atom with atomic number 24 are:
(a) $\pm 3, \sqrt{48}$ BM (b) $\pm 3, \sqrt{35}$ BM
(c) $\pm \frac{3}{2}, \sqrt{48}$ BM (d) $\pm \frac{3}{2}, \sqrt{35}$ BM

64. Select the incorrect graph for velocity of e^- in an orbit vs. Z, $\frac{1}{n}$ and n :

(a) 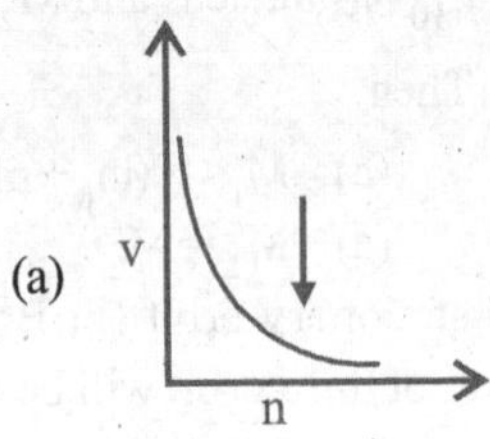

(b) 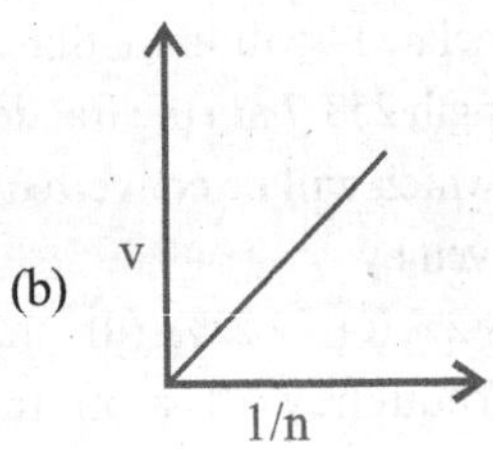

(c) 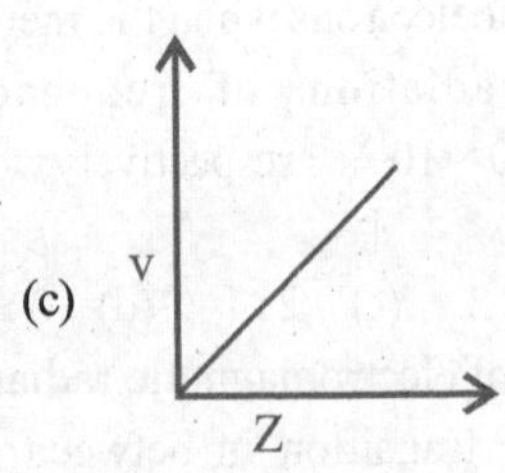

(d) 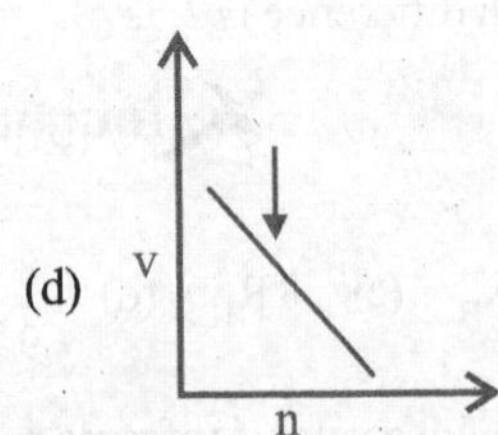

65. An electron has been excited from the first to the fourth energy state in an atom. Which of the following transitions are possible when the electron comes back to the ground state ?
(a) $4 \rightarrow 1$
(b) $4 \rightarrow 2, 2 \rightarrow 1$
(c) $4 \rightarrow 3, 3 \rightarrow 2, 2 \rightarrow 1$
(d) All of the above

66. What should be the kinetic energy and total energy of the electron present in hydrogen atom, if its potential energy is –5.02 eV?
(a) –2.51 eV, +2.51eV
(b) +0.251 eV, –0.251eV
(c) –2.51 eV, –2.51 eV
(d) +2.51 eV, +2.51eV

67. Ionization potential of hydrogen atom is 13.6eV. Hydrogen atom in ground state are excited by monochromatic light of energy 12.1 eV. The spectral lines emitted by hydrogen according to Bohr's theory will be

(a) one (b) two (c) three (d) four

68. If the subsidiary quantum number of a subenergy level is 4, the maximum and minimum values of the spin multiplicities are :
(a) 9, 1 (b) 10, 1 (c) 10, 2 (d) 4, –4

69. If the wavelength of the electromagnetic radiation is increased to thrice the digital value, then what will be the percent change in the value of frequency of the electromagnetic radiation.

(a) Increases by 33% (b) Decreases by 33%
(c) Increases by 66% (d) Decreases by 66%

70. According to Bohr's theory the energy required for the transition of H-atom from n = 6 to n = 8 state is
(a) equal to the energy required for the transition from n = 5 to n = 6 state
(b) larger than in (a)
(c) less than in (a)
(d) equal to the energy required for the transition from n = 7 to n = 9 state.

Answer KEY

1	(d)	8	(a)	15	(b)	22	(c)	29	(c)	36	(c)	43	(c)	50	(c)	57	(a)	64	(d)
2	(d)	9	(b)	16	(a)	23	(d)	30	(b)	37	(b)	44	(c)	51	(d)	58	(d)	65	(d)
3	(b)	10	(b)	17	(d)	24	(a)	31	(c)	38	(b)	45	(c)	52	(b)	59	(b)	66	(a)
4	(c)	11	(b)	18	(c)	25	(a)	32	(c)	39	(d)	46	(d)	53	(c)	60	(b)	67	(c)
5	(b)	12	(d)	19	(a)	26	(a)	33	(c)	40	(b)	47	(d)	54	(c)	61	(c)	68	(c)
6	(b)	13	(d)	20	(b)	27	(d)	34	(b)	41	(a)	48	(a)	55	(b)	62	(b)	69	(d)
7	(c)	14	(c)	21	(d)	28	(c)	35	(d)	42	(b)	49	(d)	56	(d)	63	(a)	70	(c)

Hints & Solutions

1. **(d)** $v_n = r_n \omega$ where $r_n = \frac{n^2h^2}{4\pi^2me^2Z\cdot K}$

and $v_n = \frac{2\pi\cdot Z\cdot e^2\cdot K}{n\cdot h}$

$$\therefore \quad \frac{2\pi Ze^2\cdot K}{n\cdot h} = \frac{n^2h^2}{4\pi^2me^2Z\cdot K}\times\omega;$$

$$\omega = \frac{8\pi^3me^4\cdot Z^2\cdot K^2}{n^3\cdot h^3}$$

$$= \frac{9\pi^3me^4\cdot K^2}{h^3}(\because n = 2 \text{ and } Z = 3)$$

2. **(d)**

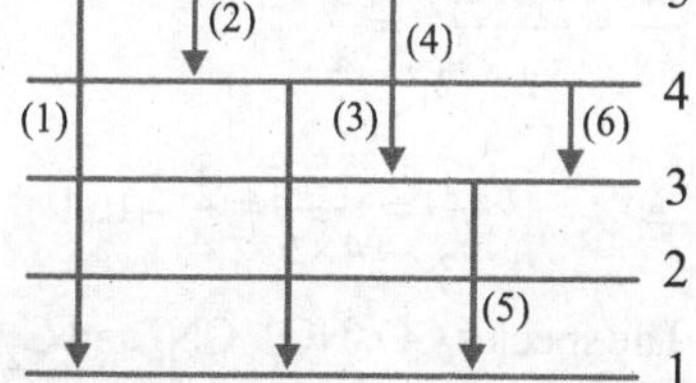

Total radiations are = 6

3. **(b)** H_α line of Balmer series means first line of Balmer series.

$n_1 = 2, n_2 = 3$

$$\bar{\nu} = \frac{1}{\lambda_\alpha} = R\left(\frac{1}{2^2} - \frac{1}{3^2}\right) = \frac{5R}{36}$$

$$\therefore \lambda_\alpha = \frac{36}{5R} = X$$

H_β line of Balmer series means, second line of Balmer series, $n_1 = 2, n_2 = 4$.

$$\bar{\nu} = \frac{1}{\lambda_\beta} = R\left(\frac{1}{2^2} - \frac{1}{4^2}\right) = \frac{3R}{16}$$

$$\therefore \lambda_\beta = \frac{16}{3R} = X$$

when $\frac{36}{3R} = X$

Then $\frac{16}{3R} = \frac{X\times 5R\times 16}{36\times 3R} = \frac{80X}{108}$ Å

4. **(c)** $E = \frac{nhc}{\lambda} = nhc\bar{v}\left(\because \bar{v} = \frac{1}{\lambda}\right)$

$\therefore 10 = nhc\bar{v}$ or $n = \frac{10}{hc\bar{v}} = \frac{10}{hcx}$

5. **(b)** $E = h\nu = \frac{ch}{\lambda}$; and $\nu = \frac{c}{\lambda}$

$$8\times 10^{15} = \frac{3.0\times 10^8}{\lambda}$$

$$\therefore \lambda = \frac{3.0\times 10^8}{8\times 10^{15}} = 0.37\times 10^{-7}$$

$$= 37.5\times 10^{-9}\ \text{m} = 4\times 10^1\ \text{nm}$$

6. **(b)** $E = \frac{hc}{\lambda}\times N_A$

$$= \frac{6.62\times 10^{-27}\times 3\times 10^{10}\times 6.02\times 10^{23}}{\lambda}$$

$$= \frac{1.19\times 10^8}{\lambda}$$

7. (c) For Li^{2+} ion

$$E = -13.6 \times \frac{Z^2}{n^2} = -13.6 \times \frac{(3)^2}{(2)^2}$$

$$= \frac{-13.6 \times 9}{4} = -30.6 \text{ eV}$$

8. (a) $\Delta E = -2.0 \times 10^{-18} \times \left(\frac{1}{2^2} - \frac{1}{1^2}\right)$

$$= -2.0 \times 10^{-18} \times \frac{-3}{4}$$

$$= 1.5 \times 10^{-18}$$

$$\Delta E = \frac{hc}{\lambda}$$

$$\lambda = \frac{hc}{\Delta E} = \frac{6.6 \times 10^{-34} \times 3 \times 10^8}{1.5 \times 10^{-18}}$$

$= 1.325 \times 10^{-7}$ m

9. (b) According to Heisenberg uncertainty principle.

$$\Delta x . m \Delta v = \frac{h}{4\pi}$$

$$\Delta x = \frac{h}{4\pi m \Delta v}$$

Here $\Delta v = \frac{600 \times 0.005}{100} = 0.03$

So, $\Delta x = \frac{6.6 \times 10^{-34}}{4 \times 3.14 \times 9.1 \times 10^{-31} \times 0.03}$

$= 1.92 \times 10^{-3}$

10. (b) $m = -l$ to $+l$, through zero thus for $l = 2$, values of m will be $-2, -1, 0, +1, +2$. Therefore for $l = 2$, m cannot have the value -3.

11. (b) $E = E_1 + E_2$; $\frac{hc}{\lambda} = \frac{hc}{\lambda_1} + \frac{hc}{\lambda_2}$

$$\Rightarrow \frac{hc}{\lambda} = hc\left(\frac{\lambda_2 + \lambda_1}{\lambda_1 \lambda_2}\right); \quad \lambda = \frac{\lambda_1 \lambda_2}{\lambda_1 + \lambda_2}$$

12. (d) The smallest value of energy of an electron in H atom in ground state can absorb is $= E_2 - E_1$.

$$= \frac{-13.58}{4} - \left(\frac{-13.58}{1}\right) = 10.19$$

13. (d) Total energy of a revolving electron is the sum of its kinetic and potential energy.

Total energy = K.E. + P. E.

$$= \frac{e^2}{2r} + \left(-\frac{e^2}{r}\right) = -\frac{e^2}{2r}$$

14. (c) $E = h\nu$

$= 6.63 \times 10^{-34} \times 2.47 \times 10^{15}$

$= 1.640 \times 10^{-18}$ J

15. (b) According to Aufbau principle, the orbital of lower energy (2*s*) should be fully filled before the filling of orbital of higher energy starts.

16. (a) $\frac{1}{2} m v^2 = h\nu - h\nu_0$

$$\Rightarrow \frac{1}{2} m v^2 = h(\nu - \nu_0)$$

$$\Rightarrow v = \sqrt{\frac{2h}{m}(\nu - \nu_0)}$$

$$= \sqrt{\frac{2 \times 6.6 \times 10^{-34}}{9.1 \times 10^{-31}}\left(2 \times 10^{15} - 7 \times 10^{14}\right)}$$

$= 1.37 \times 10^6$ ms^{-1}

17. (d) From the expression of Bohr's theory, we know that

$$m_e v_1 r_1 = n_1 \frac{h}{2\pi}$$

$$\& \; m_e v_2 r_2 = n_2 \frac{h}{2\pi}$$

$$\frac{m_e v_1 r_1}{m_e v_2 r_2} = \frac{n_1}{n_2} \frac{h}{2\pi} \times \frac{2\pi}{h}$$

Given, $r_1 = 5 r_2, n_1 = 5, n_2 = 4$

$$\frac{m_e \times v_1 \times 5 r_2}{m_e \times v_2 \times r_2} = \frac{5}{4}$$

$$\Rightarrow \frac{v_1}{v_2} = \frac{5}{4 \times 5} = \frac{1}{4} = 1:4$$

18. (c) The species CO, NO^+, CN^- and C_2^{2-} contain 14 electrons each.

19. (a) $\bar{\nu} = RZ^2\left(\frac{1}{2^2} - \frac{1}{3^2}\right)$

$$= R\left(\frac{1}{4} - \frac{1}{9}\right) = \frac{5R}{36}$$

20. (b) $\frac{1}{\lambda} = RZ^2\left[\frac{1}{n_1^2} - \frac{1}{n_2^2}\right] = R \times 2^2\left[\frac{1}{1^2} - \frac{1}{2^2}\right]$

$$\Rightarrow 3R; \; \lambda = \frac{1}{3R}$$

21. **(d)** According to Heisenberg uncertainty principle,

$$\Delta x \times \Delta p = \frac{h}{4\pi} \text{ (which is constant).}$$

As Δx for electron and helium atom is same, thus momentum of electron and helium will also be same, therefore the momentum of helium atom is equal to 5×10^{-26} kg ms^{-1}.

22. **(c)** e/m waves shown in figure A has higher wavelength in comparison to e/m waves shown in figure B.
Thus these waves also differ in frequency and energy. $\nu = \frac{c}{\lambda}$

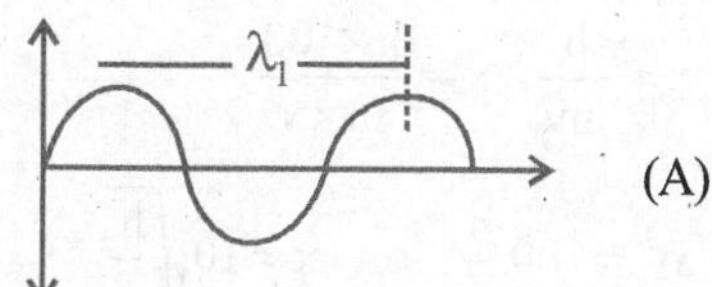

$$\Rightarrow E_1 = \frac{hc}{\lambda_1}$$

λ_2 (B)

$$\Rightarrow E_2 = \frac{hc}{\lambda_2} \quad \lambda_1 > \lambda_2 \Rightarrow E_1 < E_2$$

23. **(d)** $E = h\nu$

and $\nu = \left(\frac{c}{\lambda}\right)$

$\nu_a = 10^{15}$, $\nu_b = 10^{14}$,
$\nu_c = 10^{17}$, $\nu_d = 0.85 \times 10^{15}$

and $\nu_e = 3 \times 10^{15}$,

24. **(a)** $\lambda = \frac{hc}{\Delta E}$

$$= \frac{6.62 \times 10^{-34} \text{ J s} \times 3 \times 10^8 \text{ ms}^{-1}}{E \times 1.602 \times 10^{-19} \text{ J}}$$

$$= \frac{12395}{E} \times 10^{-10} \text{ m}$$

25. **(a)** X^{3-} has 14 protons, i.e., X also has 14 protons and therefore 14 electrons.

$X = 14 = 1s^2, 2s^2 2p^6, 3s^2 3p^2$

$X^{2+} = 1s^2, 2s^2 2p^6, 3s^2 = 12$ electrons

26. **(a)** 2nd excited state will be the 3rd energy level.

$$E_n = \frac{13.6}{n^2} \text{eV or } E = \frac{13.6}{9} \text{ eV} = 1.51 \text{ eV.}$$

27. **(d)** Kinetic energy of any particle $= \frac{3}{2}$KT

Also K.E. $= \frac{1}{2}mv^2$

$$\frac{1}{2}mv^2 = \frac{3}{2}KT \Rightarrow v^2 = \frac{3KT}{m}$$

$$v = \sqrt{\frac{3KT}{m}}$$

De-broglie wavelength

$$= \lambda = \frac{h}{mv} = \frac{h}{m\sqrt{\frac{3KT}{m}}}$$

$$\lambda = \frac{h}{\sqrt{3KTm}} \quad \lambda \propto \frac{1}{\sqrt{m}}$$

Mass of electron < mass of neutron

λ (electron) > λ (neutron)

28. **(c)** For 4*p* electron n = 4, l = 1, m = –1, 0 + 1 and s = +½ or –½

29. **(c)** Energy of subshells follow the order $1s^2$, $2s^2$, $2p^6$, $3s^2$, $3p^6$, $4s^2$, $3d^{10}$, $4p^6$, $5s$, $4d$, $5p$, $6s$..... we can fill them by electrons upto 4p hence 36 electrons.

30. **(b)** $\lambda = \frac{h}{mv}$

$h = 6.6 \times 10^{-34}$ J-s

$m = 1000$ kg

$$v = 36 \text{ km/hr} = \frac{36 \times 10^3}{60 \times 60} \text{ m/sec} = 10 \text{ m/sec}$$

$$\therefore \quad \lambda = \frac{6.6 \times 10^{-34}}{10^3 \times 10} = 6.6 \times 10^{-38} \text{ m}$$

31. **(c)** $r_n = a_0 n^2$

$r = a_0 \times (3)^2 = 9a_0$

$$mvr = \frac{nh}{2\pi};$$

$$mv = \frac{nh}{2\pi r} = \frac{3h}{2\pi \times 9a_0} = \frac{h}{6\pi a_0}$$

$$\lambda = \frac{h}{mv} = \frac{h}{h} \times 6\pi a_0 = 6\pi a_0$$

32. (c) $l = 2$ represent d orbital for which

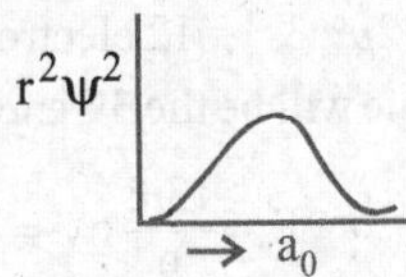

33. (c) The ground state electronic configuration, of Cu^+ ion is
$1s^2, 2s^2, 2p^6, 3s^2, 3p^6, 3d^{10}$
n = 3 (number of shells) ; Number of subshells occupied = 6, Number of filled orbitals = 14 ; There is no unpaired electron.

34. (b) Z = 3 for Li^{2+} ions

So $r_n = \dfrac{52.9 \times n^2}{Z}$ pm

$n = 3, Z = 3$

$r_n = \dfrac{52.9 \times (3)^2}{3}$ pm

= 158.7 pm

Also, linear momentum (mv) = 7.3×10^{-34} kg ms^{-1}

Then angular momentum will be

$\omega = (mv) \times r$

$= (7.3 \times 10^{-34}\ kg\ ms^{-1})(158.7\ pm)$

$= 7.3 \times 10^{-34}\ kg\ ms^{-1} \times (158.7 \times 10^{-12}\ m)$

$= 11.58 \times 10^{-48}\ kg\ m^2\ s^{-1}$

35. (d) $\lambda = \dfrac{h}{mv} = \dfrac{6.6 \times 10^{-34}}{60 \times 10^{-3} \times 10} = 10^{-33}$ m

36. (c) $n = 3, l = 2$ means $3d$ orbital

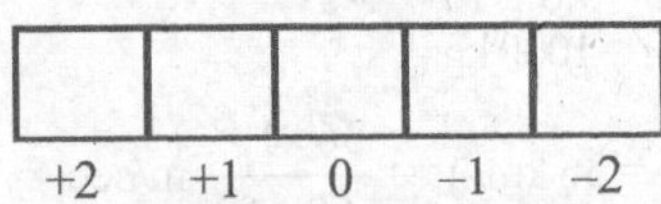

i.e. in an atom only one orbital can have the value $m_l = +2$

37. (b) Angular momentum of an electron in n^{th} orbit is given by

$mvr = \dfrac{nh}{2\pi}$

For n = 5, we have

Angular momentum of electron

$= \dfrac{5h}{2\pi} = \dfrac{2.5h}{\pi}$

38. (b) If we assume the atom to be hydrogen like, energy of nth energy level

$E_n = -\dfrac{E_1}{n^2}$

where E_1 is energy of first energy level

$E_2 = -\dfrac{E_1}{2^2} = -\dfrac{E_1}{4} = \dfrac{-21.79 \times 10^{-12}}{4}$

$= -5.447 \times 10^{-12}$ erg per atom.

39. (d) The d-orbital represented by option (d) will become completely filled after gaining an electron. Therefore option (d) is correct.

40. (b) Let wavelength of particle be x

So, velocity (v) $= \dfrac{x}{100}$

$\lambda = \dfrac{h}{mv};\ x = \dfrac{h \times 100}{m \times x}$

$x^2 = 100\dfrac{h}{m}$ or $x = 10\sqrt{\dfrac{h}{m}}$

41. (a) Change in P.E. $= -\dfrac{2x}{4} + (2x) = \dfrac{3}{2}x$

42. (b) Total energy of third shell $= \dfrac{-13.6}{3^2}$

$= -1.51$ eV

K.E. = − Total energy ⇒ 1.51 eV

P.E. = 2 × T.E. = −3.02 eV

43. (c) $r_1 = 0.529$ Å; $R_{4(x)} = R_1 \times \dfrac{n^2}{Z}$;

$R_{4(x)} \Rightarrow \dfrac{0.529 \times (4)^2}{Z}$; Z = 16

44. (c) Possible values of l and m depend upon the value of n

$l = 0$ to $(n - 1)$

m = $-l$ to $+l$ through zero

$s = +\dfrac{1}{2}$ and $-\dfrac{1}{2}$

Thus for n = 3,

l may be 0, 1 or 2; but not 3

m may be −2, −1, 0, +1 or +2

s may be $+\dfrac{1}{2}$ or $-\dfrac{1}{2}$

45. (c) Not more than two electrons can be present in same atomic orbital. This is Pauli's exclusion principle.

46. (d) By Heisenberg uncertainty Principle

$$\Delta x \times \Delta p = \frac{h}{4\pi} \text{ (which is constant)}$$

As Δx for electron and helium atom is same thus momentum of electron and helium will also be same therefore the momentum of helium atom is equal to 5×10^{-26} kg. m.s^{-1}.

47. (d) de Broglie wavelength $\lambda = \frac{h}{mv}$

$$\frac{\lambda_1}{\lambda_2} = \frac{m_2 v_2}{m_1 v_1}; \ \frac{1}{4} = \frac{1}{9} \times \frac{v_2}{v_1}$$

$$\frac{v_2}{v_1} = \frac{9}{4}; \ \frac{v_1}{v_2} = \frac{4}{9}$$

$$KE = \frac{1}{2} mv^2$$

$$\frac{KE_1}{KE_2} = \frac{m_1}{m_2} \times \frac{v_1^2}{v_2^2} = \frac{9}{1} \times \left(\frac{4}{9}\right)^2 = \frac{16}{9}$$

48. (a) The energy of an orbital is given by $(n + l)$. In (D) and (E), $(n + l)$ value is $(3 + 2) = 5$, hence they will have same energy, since here n values are also same.

49. (d) Quantum efficiency

$$= \frac{\text{Number of moles of substance reacted}}{\text{Number of Quanta absorbed}}$$

$$= \frac{3 \times 10^{-3} \times 6 \times 10^{23}}{3 \times 10^{18} \times 60 \times 10} = 1$$

50. (c) Radius of maximum probability decrease with increase in azimuthal quantum no.

51. (d) $T \propto \frac{n^3}{Z^2}; \ \frac{T_{1,2}}{T_{2,1}} = \frac{1}{4} \times \frac{1}{8} = \frac{1}{32}$

52. (b) $n_1 + n_2 = 4; n_2 - n_1 = 2; \therefore n_1 = 1, n_2 = 3$

$$\bar{v} = R(2)^2 \left[\frac{1}{1^2} - \frac{1}{3^2}\right] = \frac{32R}{9}$$

53. (c) $\Delta x \cdot \Delta p = \frac{h}{4\pi}$ or $\Delta x \cdot m\Delta v = \frac{h}{4\pi}$;

$$\Delta v = \frac{0.011}{100} \times 3 \times 10^4 = 3.3 \text{ cms}^{-1}$$

$$\Delta x = \frac{6.6 \times 10^{-27}}{4 \times 3.14 \times 9.1 \times 10^{-28} \times 3.3} = 0.175 \text{ cm}$$

54. (c) $E_{1_{He^+}} = E_{1_H} \times Z^2$

$$\therefore -871.6 \times 10^{-20} = E_{1_H} \times 4$$

$$\therefore E_{1_H} = -217.9 \times 10^{-20} \text{ J}$$

55. (b) $\lambda_p = \frac{h}{\sqrt{2eVm_p}}$;

$$\lambda_{He^{2+}} = \frac{h}{\sqrt{2 \times 2eVm_{He^{2+}}}} = \frac{h}{\sqrt{2 \times 2eV \times 4m_p}}$$

$$\therefore \frac{\lambda_p}{\lambda_{He^{2+}}} = 2\sqrt{2}$$

56. (d) Given, $v_A = 0.1$ ms^{-1} and $v_B = 0.05$ ms^{-1} also, $m_B = 5m_A$

de-Broglie wavelength, $\lambda = \frac{h}{mv}$

$$\therefore \frac{\lambda_A}{\lambda_B} = \frac{h / m_A v_A}{h / m_B v_B} = \frac{m_B v_B}{m_A v_A}$$

$$= \frac{5m_A \times 0.05}{m_A \times 0.1} = 5 \times 0.5 = 2.5 = 5/2$$

$$\therefore \lambda_A : \lambda_B = 5 : 2$$

57. (a) $^{20}_{10}Ne$ contains 10 protons and 10 neutrons

$$\therefore M_1 = 10m_p + 10m_n$$

$^{40}_{20}Ca$ contains 20 protons and 20 neutrons

$$\therefore M_2 = 20\,m_p + 20\,m_n$$

$$\therefore M_2 = 2M_1$$

58. (d) $\frac{r_1}{r_2} = \left(\frac{n_1}{n_2}\right)^2 = \frac{4}{9}$

59. (b) Energy of 1 mole of photons,

$$E = N_0 \times h\upsilon$$

$$= \frac{N_0 \times h \times c}{\lambda}$$

$$= \frac{6.023 \times 10^{23} \times 6.63 \times 10^{-34} \times 3 \times 10^8}{253.7 \times 10^{-9}}$$

Energy converted into KE = (472.2 – 430.53) kJ
% of energy converted into

$$KE = \frac{(472.2 - 430.53)}{472.2} \times 100 = 8.82\%$$

60. (b) $\frac{(KE)_1}{(KE)_2} = \frac{\nu_1 - \nu_0}{\nu_2 - \nu_0}$

$$= \frac{(1.5\times10^{15} - 1\times10^{15})}{(2.0\times10^{15} - 1\times10^{15})} = \frac{1}{2}$$

61. (c) $n_1 + n_2 = 4$ ...... (i)

$n_2 - n_1 = 2$ ..... (ii)

$n_1 = 1,\ n_2 = 3$

$$\bar{\upsilon} = R_H Z^2 \left(\frac{1}{n_1^2} - \frac{1}{n_2^2}\right)$$

$$= R_H \times 3^2 \times \left[\frac{1}{1^1} - \frac{1}{3^2}\right] = 8\ R_H$$

62. (b) $\lambda = \frac{h}{m_e x} = \frac{h}{m_p V} = \frac{h}{1840\ m_e V}$

$$[\therefore m_p = 1840\ m_e]$$

Hence, $V = \frac{x}{1840}$

63. (a) Atomic number = 24
Electronic configuration is $3d^5, 4s^1$
No. of valence e^{-1}(s) = 6

Total spin = $\pm\frac{1}{2} \times 6 = \pm 3$

Magnetic moment

$$= \sqrt{n(n+2)} = \sqrt{6(6+2)}$$

$$= \sqrt{48}\ BM$$

64. (d) $v = 2.188 \times 10^6\ Z/n$

65. (d) Electron can undergo transition from higher state to all lower states by loss of energy.

66. (a) Total energy = $\frac{\text{Potential energy}}{2} = \frac{-5.02}{2}$

Total energy = – 2.51 eV
Kinetic energy = – Total energy
= –(–2.51 eV) = + (2.51 eV)

67. (c) After absorbing 12.1 eV the electron in H atom is excited to 3 shell.

$$E_n - E_1 = \frac{-13.6}{n^2} - (-13.6) = 12.1$$

$$\frac{-13.6}{n^2} + 13.6 = 12.1$$

$\therefore n = 3$

The possible transitions are = 3

68. (c) $l = 4$;

number of degenerate orbitals = $2l + 1 = 9$;

maximum total spins = $9 \times \frac{1}{2}$

maximum multiplicity = 2s + 1

$$= 2 \times \frac{9}{2} + 1 = 10$$

minimum total spins = $\frac{1}{2}$

minimum multiplicity = $2 \times \frac{1}{2} + 1 = 2$

69. (d) $\nu_1 = \frac{c}{\lambda_1}$

$$\nu_2 = \frac{c}{\lambda_2} = \frac{c}{3\lambda_1}$$

% change in frequency = $\frac{\nu_2 - \nu_1}{\nu_1} \times 100$

$$= \frac{\frac{c}{3\lambda_1} - \frac{c}{\lambda_1}}{\frac{c}{\lambda_1}} \times 100 = \frac{-\frac{2c}{3\lambda_1} \times 100}{\frac{c}{\lambda_1}}$$

= – 66%

70. (c) The difference between the energy of adjacent energy levels decreases on moving away from nucleus.

3 Classification of Elements and Periodicity in Properties

1. Na^+, Mg^{2+}, Al^{3+} and Si^{4+} are isoelectronic. The order of their ionic size is
(a) $Na^+ > Mg^{2+} < Al^{3+} < Si^{4+}$
(b) $Na^+ < Mg^{2+} > Al^{3+} > Si^{4+}$
(c) $Na^+ > Mg^{2+} > Al^{3+} > Si^{4+}$
(d) $Na^+ < Mg^{2+} > Al^{3+} < Si^{4+}$

2. Consider the following changes

$$A \rightarrow A^+ + e^- : E_1 \text{ and } A^+ \rightarrow A^{2+} + e^- : E_2$$

The energy required to pull out the two electrons are E_1 and E_2 respectively. The correct relationship between two energies would be
(a) $E_1 < E_2$ (b) $E_1 = E_2$
(c) $E_1 > E_2$ (d) $E_1 \geq E_2$

3. The elements with zero electron affinity are
(a) Boron and Carbon
(b) Beryllium and Helium
(c) Lithium and Sodium
(d) Fluorine and Chlorine

4. Ionization energies of atoms A and B are 400 and 300 k cal mol^{-1} respectively. The electron affinities of these atoms are 80.0 and 85.0 k cal mol^{-1} respectively. Then which is the correct statement regarding electronegativity χ ?

(a) $\chi_A < \chi_B$ (b) $\chi_A > \chi_B$
(c) $\chi_A = \chi_B$ (d) none of these

5. The set of three elements having successive atomic numbers and having the ionization energies of 2372, 520 and 890 kJ per mol is
(a) H, He, Li (b) He, Li, Be
(c) Li, Be, B (d) B, C, N

6. Similarity in chemical properties of the atoms of elements in a group of the Periodic table is most closely related to:
(a) atomic numbers
(b) atomic masses
(c) number of principal energy levels
(d) number of valence electrons

7. In the long form of the periodic table, the valence shell electronic configuration of $5s^25p^4$ corresponds to the element present in :
(a) Group 16 and period 6
(b) Group 17 and period 6
(c) Group 16 and period 5
(d) Group 17 and period 5

8. Consider the following ionization enthalpies of two elements 'A' and 'B'.

Element	Ionization	enthalpy (kJ/mol)
	1st	2nd	3rd
A	899	1757	14847
B	737	1450	7731

Which of the following statements is correct ?

Toughnut

(a) Both 'A' and 'B' belong to group–1 where 'B' comes below 'A'.
(b) Both 'A' and 'B' belong to group–1 where 'A' comes below 'B'.
(c) Both 'A' and 'B' belong to group–2 where 'B' comes below 'A'.
(d) Both 'A' and 'B' belong to group–2 where 'A' comes below 'B'.

9. Sum of first three ionization energies of Al is 53.0 eV $atom^{-1}$ and the sum of first two ionization energies of Na is 52.2 eV $atom^{-1}$. Out of Al(III) and Na(II)

Critical Thinking

(a) Na (II) is more stable than Al (III)
(b) Al (III) is more stable than Na (II)
(c) Both are equally stable
(d) Both are equally unstable

10. Which pair of elements belongs to same group ?
(a) Elements with atomic no. 17 and 38
(b) Elements with atomic no. 20 and 40
(c) Elements with atomic no. 17 and 53
(d) Elements with atomic no. 11 and 33

11. In a given energy level, the order of penetration effect of different orbitals is
(a) $f<d<p<s$ (b) $s=p=d=f$
(c) $s<p<d<f$ (d) $p>s>d>f$

12. Consider the following changes : Tricky
(1) $M(s) \longrightarrow M(g)$
(2) $M(s) \longrightarrow M^{2+}(g) + 2e^-$
(3) $M(g) \longrightarrow M^+(g) + e^-$
(4) $M^+(g) \longrightarrow M^{2+}(g) + e^-$
(5) $M(g) \longrightarrow M^{2+}(g) + 2e^-$
The second ionization energy of M could be calculated from the energy values associated with:
(a) $1+3+4$ (b) $2-1+3$
(c) $1+5$ (d) $5-3$

13. The correct order of decreasing electronegativity values among the elements I-beryllium, II-oxygen, III-nitrogen and IV-magnesium is
(a) II > III > I > IV (b) III > IV > II > I
(c) I > II > III > IV (d) I > II > IV > III

14. What is the order of ionisation energies of the coinage metal
(a) Cu > Ag < Au (b) Cu > Ag > Au
(c) Cu < Ag < Au (d) Au > Ag < Cu

15. First three ionisation energies (in kJ/mol) of three representative elements are given below :

Element	IE_1	IE_2	IE_3
P	495.8	4562	6910
Q	737.7	1451	7733
R	577.5	1817	2745

Then incorrect option is : Critical Thinking
(a) Q : Alkaline earth metal
(b) P : Alkali metals
(c) R : *s*-block element
(d) They belong to same period

16. What is the correct order of electronegativity?
(a) $M^- < M^{2-} < M^{3-} < M^{4-}$
(b) $M^- > M^{2-} > M^{3-} > M^{4-}$
(c) $M^- < M^{2-} > M^{3-} < M^{4-}$
(d) $M^{4-} < M^{2-} < M^{3-} < M^-$

17. The chemistry of lithium is very similar to that of magnesium even though they are placed in different groups. Its reason is:
(a) Both are found together in nature
(b) Both have nearly the same size
(c) Both have similar electronic configuration
(d) The ratio of their charge and size (i.e. charge density) is nearly the same

18. Which of the following sequence correctly represents the decreasing acidic nature of oxides ?
(a) $Li_2O > BeO > B_2O_3 > CO_2 > N_2O_3$
(b) $N_2O_3 > CO_2 > B_2O_3 > BeO > Li_2O$
(c) $CO_2 > N_2O_3 > B_2O_3 > BeO > Li_2O$
(d) $B_2O_3 > CO_2 > N_2O_3 > Li_2O > BeO$

19. Correct order of first IP among following elements Be, B, C, N, O is
(a) B < Be < C < O < N
(b) B < Be < C < N < O
(c) Be < B < C < N < O
(d) Be < B < C < O < N

20. In which of the following process highest energy is absorbed?
(a) $Cu \rightarrow Cu^+$ (b) $Br \rightarrow Br^-$
(c) $I \rightarrow I^-$ (d) $Li \rightarrow Li^+$

21. The correct order of acidic strength :

(a) $Cl_2O_7 > SO_2 > P_4O_{10}$
(b) $K_2O > CaO > MgO$
(c) $CO_2 > N_2O_5 > SO_3$
(d) $Na_2O > MgO > Al_2O_3$

22. An element of atomic weight 40 has 2, 8, 8, 2 as the electronic configuration. Which one of the following statements regarding this element is not correct?
(a) it belongs to II group of the periodic table
(b) it has 20 neutrons
(c) the formula of its oxide is MO_2
(d) it belongs to 4th period of the periodic table

23. Which of the following series correctly represents relations between the elements from X to Y?
$X \rightarrow Y$
(a) $_3Li \rightarrow {}_{19}K$ Ionization enthalpy increases
(b) $_9F \rightarrow {}_{35}Br$ Electron gain enthalpy (negative sign) increases
(c) $_6C \rightarrow {}_{32}Ge$ Atomic radii increases
(d) $_{18}Ar \rightarrow {}_{54}Xe$ Noble character increases

24. Which of the following arrangements represents the increasing order (smallest to largest) of ionic radii of the given species O^{2-}, S^{2-}, N^{3-}, P^{3-}?
(a) $O^{2-} < N^{3-} < S^{2-} < P^{3-}$
(b) $O^{2-} < P^{3-} < N^{3-} < S^{2-}$
(c) $N^{3} < O^{2-} < P^{3-} < S^{2-}$
(d) $N^{3-} < S^{2-} < O^{2-} < P^{3-}$

25. Which one of the following has largest ionic radius?

(a) Li^+ (b) O_2^{2-} (c) B^{3+} (d) F^-

26. Elements X, Y and Z have atomic numbers 19, 37 and 55 respectively. Which of the following statements is true about them ? Toughnut

(a) Their ionization potential would increase with increasing atomic number
(b) 'Y' would have an ionisation potential between those of 'X' and 'Z'
(c) 'Z' would have the highest ionization potential
(d) 'Y' would have the highest ionization potential.

27. An element X occurs in short period having configuration $ns^2\ np^1$. The formula and nature of its oxide is

(a) XO_3, basic (b) XO_3 acidic
(c) X_2O_3, amphoteric (d) X_2O_3 basic

28. The decreasing order of the ionization potential of the following elements is

(a) Ne > Cl > P > S > Al > Mg
(b) Ne > Cl > P > S > Mg > Al
(c) Ne > Cl > S > P > Mg > Al
(d) Ne > Cl > S > P > Al > Mg

29. The incorrect statement among the following is

(a) the first ionization potential of Al is less than the first ionization potential of Mg
(b) the second ionization potential of Mg is greater than the second ionization potential of Na
(c) the first ionization potential of Na is less than the first ionization potential of Mg
(d) the third ionization potential of Mg is greater than the third ionization potential of Al.

30. The screening effect of inner electrons of the nucleus causes

(a) decrease in the ionization energy
(b) increase in the ionization energy
(c) no effect on the ionization energy
(d) increases the attraction of the nucleus for the electrons

31. The order of increasing sizes of atomic radii among the elements O, S, Se and As is :

(a) As < S < O < Se (b) Se < S < As < O
(c) O < S < As < Se (d) O < S < Se < As

32. Which is the correct order of second ionization potential of C, N, O and F in the following ?

(a) O > N > F > C (b) O > F > N > C
(c) F > O > N > C (d) C > N > O > F

33. The first (Δ_iH_1) and second (Δ_iH_2) ionization enthalpies (in kJ mol^{-1}) and the electron gain enthalpy ($\Delta_{eg}H$) (in kJ mol^{-1}) of the elements I, II, III, IV and V are given below. Critical Thinking

Element	Δ_iH_1	Δ_iH_2	$\Delta_{eg}H$
I	520	7300	– 60
II	419	3051	– 48
III	1681	3374	– 328
IV	1008	1846	– 295
V	2372	5251	+ 48

The most reactive metal and the least reactive non-metal of these are respectively

(a) I and V (b) V and II
(c) II and V (d) IV and V

34. Aqueous solutions of two compounds $M_1 - O - H$ and $M_2 - O - H$ are prepared in two different beakers. If, the electronegativity of M_1 = 3.4, M_2 = 1.2, O = 3.5 and H = 2.1 then the nature of two solutions will be respectively :

(a) acidic, basic (b) acidic, acidic
(c) basic, acidic (d) basic, basic

35. An element having electronic configuration $1s^2 2s^2 2p^6 3s^2 3p^6 4s^1$ forms

(a) Acidic oxide (b) Basic oxide
(c) Amphoteric oxide (d) Neutral oxide

36. The van der Waal and covalent radii of fluorine atom respectively from the following figure are.

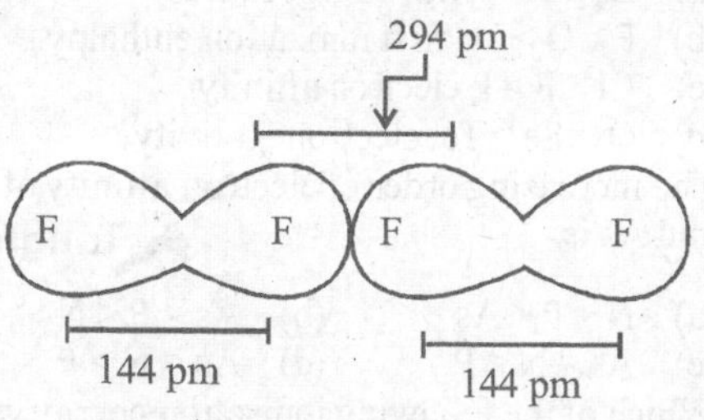

(a) 219 pm, 72 pm (b) 75 pm, 72 pm
(c) 147 pm, 72 pm (d) 147 pm, 144 pm

37. Consider the following four elements, which are represented according to long form of periodic table.

Here W, Y and Z are left, up and right elements with respect to the element 'X' and 'X' belongs to 16th group and 3rd period. Then according to given information the incorrect statement regarding given elements is: Toughnut

(a) Maximum electronegativity : Y
(b) Maximum catenation property : X
(c) Maximum electron affinity : Z
(d) Y exhibits variable covalency

38. The electron affinity of chlorine is 3.7 eV. 1 g of chlorine is completely converted to Cl^- ion in a gaseous state. (1 eV = 23.06 kcal mol^{-1}).
Energy released in the process is

(a) 4.8 kcal (b) 7.2 kcal
(c) 8.2 kcal (d) 2.4 kcal

39. In which of the following arrangements, the sequence is not strictly according to the property written against it?
(a) $CO_2 < SiO_2 < SnO_2 < PbO_2$: increasing oxidising power
(b) $NH_3 < PH_3 < AsH_3 < SbH_3$: increasing basic strength
(c) HF < HCl < HBr < HI: increasing acid strength
(d) B < C < O < N: increasing first ionisation enthalpy.

40. Which among the following elements has the highest first ionization enthalpy?
(a) Nitrogen (b) Boron
(c) Carbon (d) Oxygen

41. Which of the following is not the correct order for the stated property ?
(a) Ba > Sr > Mg; atomic radius
(b) F > O > N : first ionization enthalpy
(c) Cl > F > I; electron affinity
(d) O > Se > Te; electronegativity

42. The increasing order of electron affinity of N, P and As is Toughnut

(a) N < P < As (b) As < P < N
(c) As < N < P (d) As < N > P

43. Which of the following ionisation energy values for calcium show a sudden increase?
(a) Third (b) Second
(c) First (d) Fourth

44. The electronegativity follows the order
(a) F > O > Cl > Br (b) F > Cl > Br > O
(c) O > F > Cl > Br (d) Cl > F > O > Br

45. The first ionisation potential of aluminium is smaller than that of magnesium because
(a) Atomic size of Al > Atomic size of Mg.
(b) Atomic size of Al < Atomic size of Mg.
(c) Al has one electron in p - orbital
(d) None of these

46. Which one of the following statements is incorrect in relation to ionisation enthalpy?
(a) Ionisation enthalpy increases for each successive electron.
(b) The greatest increase in ionisation enthalpy is experienced on removal of electron from core noble gas configuration.
(c) End of valence electrons is marked by a big jump in ionisation enthalpy.
(d) Removal of electron from orbitals bearing lower n value is easier than from orbital having higher n value.

47. Highest electron affinity is shown by
(a) O^- (b) F^- (c) Cl_2 (d) F_2

48. Which of the following order is wrong?
(a) $NH_3 < PH_3 < AsH_3$ — Acidic
(b) Li < Be < B < C — IE_1
(c) $Al_2O_3 < MgO < Na_2O < K_2O$ — Basic
(d) $Li^+ < Na^+ < K^+ < Cs^+$ — Ionic radius

49. The electronic configuration of an element is . The atomic number and the group number of the element 'X' which is just below the above element in the periodic table are respectively
(a) 23 and 5 (b) 23 and 15
(c) 33 and 15 (d) 33 and 5

50. Consider the following information about element P and Q: Tricky

	Period number	Group number
P	2	15
Q	3	2

(a) QP (b) Q_2P_3 (c) Q_3P_2 (d) Q_2P

51. Which one of the following statements is incorrect ?
(a) Greater the nuclear charge, greater is the electron affinity
(b) Nitrogen has zero electron affinity
(c) Electron affinity decreases from fluorine to iodine in 17th group
(d) Chlorine has highest electron affinity

52. Which of the following statements is wrong ?
(a) van der Waal's radius of iodine is more than its covalent radius
(b) All isoelectronic ions belong to same period of the periodic table
(c) I.E.$_1$ of N is higher than that of O while I.E.$_2$ of O is higher than that of N
(d) The electron gain enthalpy of N is almost zero while that of P is 74.3 kJ mol^{-1}

53. Electron affinity is positive for
(a) $O(g) + e^- \rightarrow O^-(g)$
(b) $S(g) + e^- \rightarrow S^-(g)$
(c) $O^+(g) + e^- \rightarrow O(g)$
(d) $O^-(g) + e^- \rightarrow O^{2-}(g)$

54. Which one of the following ionic species has the greatest proton affinity to form stable compound?
(a) Cl^- (b) F^- (c) I^- (d) Br^-

55. Which property decreases from left to right across the periodic table and increases from top to bottom?
(i) Atomic radius (ii) Electronegativity
(iii) Ionisation energy (iv) Metallic character
(a) (i) only (b) (i), (ii) and (iii)
(c) (i), (iii) and (iv) (d) (i) and (iv)

56. The basic character of MgO, BaO, Na_2O and FeO follow the order
(a) $MgO < FeO < BaO < Na_2O$
(b) $FeO < MgO < Na_2O < BaO$
(c) $FeO < MgO < BaO < Na_2O$
(d) $Na_2O < MgO < FeO < BaO$

57. If the ionization enthalpy and electron gain enthalpy of an element are 275 and 86 kcal mol^{-1} respectively, then the electronegativity of the element on the Pauling scale is: **Tricky**
(a) 2.8 (b) 0.0 (c) 4.0 (d) 2.6

58. Which transition involves maximum amount of energy ?
(a) $M^-(g) \longrightarrow M(g) + e^-$
(b) $M^-(g) \longrightarrow M^+(g) + 2e^-$
(c) $M^+(g) \longrightarrow M^{2+}(g) + e^-$
(d) $M^{2+}(g) \longrightarrow M^{3+}(g) + e^-$

59. Which ionisation potential *(IP)* in the following equations involves the greatest amount of energy ?
(a) $Na \rightarrow Na^+ + e^-$ (b) $K^+ \rightarrow K^{2+} + e^-$
(c) $C^{2+} \rightarrow C^{3+} + e^-$ (d) $Ca^+ \rightarrow Ca^{2+} + e^-$

60. Consider the following ionisation reactions :

I.E. (kJ mol^{-1}) I.E. (kJ mol^{-1})

$A(g) \rightarrow A^+(g) + e^-, A_1$ $A(g) \rightarrow A^+(g) + e^-, A_1$

$B^+(g) \rightarrow B^{2+}(g) + e^-, B_2$ $C(g) \rightarrow C^+(g) + e^-, C_1$

$C^+(g) \rightarrow C^{2+}(g) + e^-, C_1$ $C^{2+}(g) \rightarrow C^{3+}(g) + e^-, C_3$

If monovalent positive ion of A, divalent positive ion of B and trivalent positive ion of C have zero electron. Then incorrect order of corresponding I.E. is : **Toughnut**
(a) $C_3 > B_2 > A_1$ (b) $B_1 > A_1 > C_1$
(c) $C_3 > C_2 > B_2$ (d) $B_2 > C_3 > A_1$

Answer KEY

1	(c)	7	(c)	13	(a)	19	(a)	25	(a)	31	(d)	37	(d)	43	(a)	49	(c)	55	(d)
2	(a)	8	(c)	14	(a)	20	(a)	26	(b)	32	(b)	38	(d)	44	(a)	50	(c)	56	(c)
3	(b)	9	(b)	15	(c)	21	(a)	27	(c)	33	(c)	39	(b)	45	(c)	51	(c)	57	(a)
4	(b)	10	(c)	16	(b)	22	(c)	28	(b)	34	(a)	40	(a)	46	(d)	52	(b)	58	(d)
5	(b)	11	(a)	17	(d)	23	(c)	29	(b)	35	(b)	41	(b)	47	(a)	53	(d)	59	(b)
6	(a)	12	(d)	18	(b)	24	(a)	30	(a)	36	(c)	42	(c)	48	(b)	54	(b)	60	(d)

Hints & Solutions

1. (c) Among the isoelectronic species, the cation having more positive charge would be the smaller in size.
2. (a) IE_1 is always less than IE_2.
3. (b) Fully filled electronic configuration.
4. (b) Electronegativity of an atom $= \frac{IE + EA}{2 \times 62.5}$ where IE and EA are in k cal mol^{-1}.
5. (b) He (Z = 2) is a noble gas and has highest I.E. The I.E. of Be (Z=4) is more than that of Li (Z=3)
6. (a) If elements are arranged in order of their increasing atomic numbers, element coming at intervals of 2, 8, 8, 18, 18, 32 and 32 will have similar physical and chemical properties and thus grouped in one particular group.
7. (c) Tellurium (Te) has $5s^2 5p^4$ valence shell configuration. It belongs to group 16 and present in period 5 of the periodic table.
8. (c) Generally, the ionization enthalpies or energy increases from left to right in a period and decreases from top to bottom in a group. Several factor such as atomic radius, nuclear charge, shielding effect are responsible for change of ionization enthalpies.

   Here, Ist ionization enthalpy of *A* and *B* is greater than group I (Li = 520 kJmol^{-1} to Cs = 374 kJmol^{-1}), which means element *A* and *B* belong to group –2 and all three given ionization enthalpy values are less for element *B* means *B* will come below *A*.
9. (b) Ionization energy is not the only the criteria for the stability of an oxidation state.
10. (c) Atomic no. 17 (Cl) and 53 (I) are present in the same group
11. (a) Penetration effect order is $s > p > d > f$.
12. (d) Second ionization energy is amount of energy required to take out an electron from the monopositive cation.
13. (a) Electronegativity values of given elements are as follows:

    Be – 1.5 (I) Mg – 1.2 (IV)
    O – 3.5 (II) N – 3.0 (III)
    i.e. II > III > I > IV
14. (a) Cu $= 3d^{10} 4s^1$, Ag $= 4d^{10} 5s^1$, Au $= 4f^{14} 5d^{10} 6s^1$
    In all the above given cases, unpaired *s*-electron has to be removed. In the case of Cu, a 4*s* electron is to be removed which is closer to the nucleus than the 5*s* electron of Ag.
    So, IE_1 of Cu > IE_1 of Ag.
    However, in case of Au, due to imperfect screening effect of 14 e$^-$*s* of 4*f* orbitals, the nuclear charge increases and therefore 5*s* e$^-$ of Au is more tightly held.
    Thus, the order of IE_1 is Cu > Ag < Au.
15. (c) R is *p*-block element, because difference between IE_2 and IE_3 is not very high as compared to between IE_1 and IE_2; hence stable oxidation state of R will be higher than +2.
16. (b) Higher the positive charge the greater is the EN and higher the negative charge the lesser is the EN. So, EN order is
    $M^- > M^{2-} > M^{3-} > M^{4-}$
17. (d) Due to diagonal relationship.
18. (b) On passing from left to right in a period, acidic character of the normal oxides of the elements increases with increase in electronegativity.
19. (a) Be – $1s^2 2s^2$; B – $1s^2 2s^2 2p^1$; C – $1s^2 2s^2 2p^2$; N – $1s^2 2s^2 2p^3$; O – $1s^2 2s^2 2p^4$. IP increases along the period. But IP of Be > B. Further IP of O < N because atoms with fully or partly filled orbitals are most stable and hence have high ionisation energy.
20. (a) Cu has completely filled *d*-orbital, so highest energy is absorbed when it converts to Cu$^+$ ion.
21. (a) Acidic character of oxide Non-metallic nature of element.

Non-metallic character increases along the period. Hence order of acidic character is

$Cl_2O_7 > SO_2 > P_4O_{10}$.

22. (c) Its valency is 2. So it will form MO type compound.

23. (c) On moving down a group atomic radii increases.

24. (a) For isoelectronic species ionic radii decreases as the charge on ion decreases. Further on moving down in a group ionic radii increases. Hence the correct order is $O^{2-} < N^{3-} < S^{2-} < P^{3-}$

25. (a) On moving along a period ionic radii decreases due to increase in effective nuclear charge.

26. (b) Elements X, Y, Z with atomic numbers 19, 37, 55 lie in group 1 (alkali metals). On moving down a group from the size of atoms increases, the outermost electrons become less strongly held. So the ionization energy decreases. Therefore, IE of Y could be between those of X and Z.

	X	Y	Z
	K(19)	Rb(37)	Cs(55)
I.E.	4.3	4.2	3.9

27. (c) $ns^2\,np^1$ is the electronic configuration of III A period. Al_2O_3 is amphoteric oxide

28. (b) Closed shell (Ne), half filled (P) and completely filled configuration (Mg) are the cause of higher value of I.E.

29. (b) IE_2 of Mg is lower than that of Na because in case of Mg^+, 3*s*-electron has to be removed where as in case of Na^+, an electron is removed from the stable inert gas configuration which is difficult.

30. (a) Higher the screening effect, lower is the I.E.

31. (d) On moving down in a group atomic radii increases due to successive addition of extra shell hence

$O < S < Se$

Further As is in group 15 having one less electron in its *p* orbital hence have higher atomic radii than group 16 elements.

i.e., $O < S < Se < As$

32. (b) The second ionization potential means removal of electron from cation

$C^+ = 1s^2 2s^2 2p^1$, $N^+ = 1s^2 2s^2 2p^2$

$O^+ = 1s^2 2s^2 2p^3$, $F^+ = 1s^2 2s^2 2p^4$

Therefore $O > F > N > C$

33. (c) I represents Li, II represents K
III represents Br, IV represents I
V represents He
So, amongst these, II represents most reactive metal and V represents least reactive non-metal.

34. (a) The electronegativity difference between M_1 and O is 0.1, which indicates M_1– O bond will be covalent, since O–H bond having more ionic character thus bond will break and H^+ ions gets release and acidic solution is formed and whereas difference between electronegativity of M_2O is 2.3, thus, M_2–OH bond will break. Hence, solution will be basic in nature.

35. (b) It is electronic configuration of alkali metal. Hence it will form basic oxide.

36. (c) Covalent radius is radius of an atom in its bound state i.e., in fluorine it is half of distance between two covalently bonded fluorine atoms; van der Waal radii is one-half of the distance between the nuclei of two identical non-bonded isolated atoms. These atoms are attracted toward each other through weak van der Waal's force hence van der Waal radii are very large.

37. (d)

W : Phosphorus	Y : Oxygen	X : Sulphur	Z : Chlorine
Electronegativity	O > Cl > S > P	Catenation :	S > P > O > Cl
Electron Affinity :	Cl > O > S > P	Oxygen exhibits covalency of two only	

38. (d) Number of moles $= \frac{1}{35.5}$

Given, $1\text{ eV} = 23.06\text{ kcal mol}^{-1}$

$3.7\text{ eV} = 3.7 \times 23.06\text{ kcal mol}^{-1}$

i.e. 1 mole realease energy

$= 3.7 \times 23.06\text{ kcal}$

$\therefore$ Energy released

$= \frac{1}{35.5} \times 3.7 \times 23.06\text{ kcal} = 2.4\text{ kcal}$

39. **(b)** Correct order of increasing basic strength is $NH_3 > PH_3 > AsH_3 > SbH_3 > BiH_3$

40. **(a)** Due to stable $2s^2\ 2p^3$ configuration (half filled *p*-orbital). Nitrogen atom has highest energy.

41. **(b)** On moving along the period, ionization enthalpy increases.
In second period, the order of ionization enthalpy should be as follows : F > O > N
But N has half-filled structure, therefore, it is more stable than O. That's why its ionization enthalpy is higher than O. Thus, the correct order of IE is
F > N > O.

42. **(c)** Phosphorus has vacant '*d*'-orbitals due to which it has higher electron affinity than nitrogen. As also has vacant *d*-orbitals but its atomic size is very big thus its electron affinity will be lesser than N.

43. **(a)** $Ca = 3s^2\ 3p^6\ 4s^2$
$Ca^{2+} = 3s^2\ 3p^6\ 4s^0$
(Stable noble gas configuration)
Hence, there would be sudden increase in IE_3 value.

44. **(a)** F and O belong to 2nd period whereas Cl and Br belong to 3rd and 4th periods respectively. Hence the sequence of the E.N. is F > O > Cl > Br

45. **(c)** Al ($3s^2\ p^1$) and Mg ($3s^2$). Lower energy is required to remove $3p^1$ electron than $3s^1$ electron (penetrating effect is $s > p > d > f$). Secondly Mg has stable electronic configuration than Al

46. **(d)** As the value of *n* increases, the nuclear attraction over the outermost shell decreases, therefore removal of an electron will be easier.

47. **(a)** Due to inert gas configuration molecules F_2, Cl_2 and anion F^- have almost zero electron affinity.

48. **(b)** The right sequence of $I.E_1$ of Li < B < Be < C.

49. **(c)** Atomic number of the given element = 15 ; atomic number of element 'X' will be
18 + 15 = 33;
group no. of 'X' = 10 + 5 (valence electrons) = 15 , period = 4th

50. **(c)** P is a trivalent non-metal and Q is a divalent metal.
Hence formula of compound is Q_3P_2.

51. **(c)** Electron affinity of ^9F is less than that of ^{17}Cl

52. **(b)** In the isoelectronic species, all isoelectronic anions belong to the same period and cations to the next period.

53. **(d)** O^- ion repel the incoming electron thus energy is required to add incoming electron.

54. **(b)** Proton affinity decreases in moving across the period from left to right due to increase in charge, within a group the proton affinities decreases from top to bottom.

55. **(d)** Atomic radius and metallic character decreases from left to right across the period and increases from top to bottom down the group.

56. **(c)** The basic character of metal oxides decreases from left to right in a period and increases down the group.

57. **(a)** I.E. + E.A. = 275 + 86 = 361 kcal mol^{-1}
= 361 × 4.184 = 1510.42 kJ mol^{-1}

$$\therefore \text{ Electronegativity} = \frac{1510.42}{540}$$

= 2.797 = 2.8

58. **(d)** The energy involved is ionisation energy (I.E.). Further the 3rd ionisation energy will be greater than the 2nd and 1st.

59. **(b)** $K^+ \rightarrow K^2 + e^-$. Since e^- is to be removed from stable configuration.

60. **(d)** $A \Rightarrow H(1s^1)$
$B \Rightarrow He(1s^2)$
$C \Rightarrow Li(1s^2 2s^1)$

$A_1 = IE_1(A)$	$B_2 = IE_2(B)$
$B_1 = IE_1(B)$	$C_2 = IE_2(C)$
$C_1 = IE_1(C)$	$C_3 = IE_3(C)$
$B_1 > A_1 > C_1$	$C_3 > B_2 > A_1$
He > H > Li	Li^{2+} He^+H
$1s^2$ $1s^1$ $2s^1$	$1s^1$ $1s^1$ $1s^1$
$C_3 > C_2 > B_2$	Li^{2+} Li^+ He^+
$1s^2$ $1s^2$ $1s^1$	

4 Chemical Bonding and Molecular Structure

1. Which of the following contains both covalent and ionic bond?
   (a) NH_4Cl (b) H_2O
   (c) CCl_4 (d) $CaCl_2$
2. Which of the following is not a correct statement?
   (a) Ionic compounds are electrically neutral.
   (b) Boiling point of an ionic compound is more than a covalent compound.
   (c) Melting point of a covalent compound is more than an ionic compound.
   (d) Ionic compounds are soluble in polar solvent.
3. Main axis of a diatomic molecule is Z. AO's p_x and p_y overlap to form which of the following orbitals?
   (a) π-MO (b) σ-MO
   (c) δ-MO (d) No bond will form

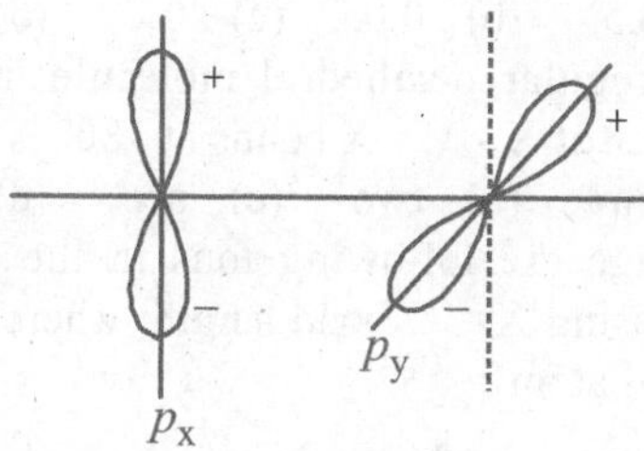

4. The group having triangular planar structures is :
   (a) BF_3, NF_3, CO_3^{2-}
   (b) CO_3^{2-}, NO_3^-, SO_3
   (c) NH_3, SO_3, CO_3^{2-}
   (d) NCl_3, BCl_3, SO_3
5. In piperidine N–H, the hybrid state assumed by N is
   (a) sp (b) sp^2 (c) sp^3 (d) dsp^2
6. The compound MX_4 is tetrahedral. The number of ∠XMX formed in the compound are

   (a) three (b) four (c) five (d) six
7. The correct order of hybridization of the central atom in the following species NH_3, $[PtCl_4]^{2-}$, PCl_5 and BCl_3 is
   (a) dsp^2, dsp^3, sp^2 and sp^3
   (b) sp^3, dsp^2, sp^3d, sp^2
   (c) dsp^2, sp^2, sp^3, dsp^3
   (d) dsp^2, sp^3, sp^2, dsp^3
8. The bond dissociation energy of B – F in BF_3 is 646 kJ mol^{-1} whereas that of C – F in CF_4 is 515 kJ mol^{-1}. The correct reason for higher B – F bond dissociation energy as compared to that of C – F is

   Toughnut

   (a) stronger σ bond between B and F in BF_3 as compared to that between C and F in CF_4.
   (b) significant $p\pi - p\pi$ interaction between B and F in BF_3 whereas there is no possibility of such interaction between C and F in CF_4.
   (c) lower degree of $p\pi - p\pi$ interaction between B and F in BF_3 than that between C and F in CF_4.
   (d) Smaller size of B-atom as compared to that of C-atom.
9. Amongst LiCl, RbCl, $BeCl_2$ and $MgCl_2$ the compounds with the greatest and the least ionic character, respectively are:

   Tricky

   (a) LiCl and RbCl (b) RbCl and $BeCl_2$
   (c) $MgCl_2$ and $BeCl_2$ (d) RbCl and $MgCl_2$
10. Which of these statements is not true?
   (a) NO^+ is not isoelectronic with O_2
   (b) B is always covalent in its compounds
   (c) In aqueous solution, the Tl^+ ion is much more stable than Tl (III)
   (d) $LiAlH_4$ is a versatile reducing agent in organic synthesis.

11. Which of the following does not apply to metallic bond ?
(a) Overlapping valence orbitals
(b) Mobile valency electrons
(c) Delocalized electrons
(d) Highly directed bonds.

12. In PO_4^{3-} , the formal charge on each oxygen atom and the P - O bond order respectively are

(a) –0.75, 0.6 (b) –0.75, 1.0
(c) –0.75, 1.25 (d) –3, 1.25

13. Which one of the following formulae does not correctly represent the bonding capacities of the two atoms involved?

(a) $\left[\begin{array}{c} H \\ | \\ H-P-H \\ | \\ H \end{array} \right]^+$

(b) F–O–F (bent: F, F, O)

(c) O←N(=O)–O–H

(d) H–C(H)=C(=O)–O–H

14. The dipole moments of diatomic molecules AB and CD are 10.41D and 10.27 D, respectively while their bond distances are 2.82 and 2.67 Å, respectively. This indicates that

Toughnut

(a) bonding is 100% ionic in both the molecules
(b) AB has more ionic bond character than CD
(c) AB has lesser ionic bond character than CD
(d) bonding is nearly covalent in both the molecules

15. Among the following, the species having the smallest bond is
(a) NO^- (b) NO^+ (c) O_2 (d) NO

16. Which compound exhibits maximum dipole moment among the following ?

(a) NO_2, NH_2 (b) NO_2

(c) NO_2, NH_2 (d) NO_2, NH_2

17. Molecular AB has a bond length of 1.61Å and a dipole moment of 0.38 D. The fractional charge on each atom (absolute magnitude) is: (e_0 = 4.802 × 10^{-10} esu)

Critical Thinking

(a) 0.5 (b) 0.05 (c) 0 (d) 1.0

18. In a regular octahedral molecule, MX_6 the number of X - M - X bonds at 180° is
(a) three (b) two (c) six (d) four

19. Arrange the following ions in the order of decreasing X – O bond length, where X is the central atom
(a) $ClO_4^-, SO_4^{2-}, PO_4^{3-}, SiO_4^-$
(b) $SiO_4^{4-}, PO_4^{3-}, SO_4^{2-}, ClO_4^-$
(c) $SiO_4^{4-}, PO_4^{3-}, ClO_4^-, SO_4^{2-}$
(d) $SiO_4^{4-}, SO_4^{2-}, PO_4^{3-}, ClO_4^-$

20. Which of the following statements is correct in the context of the allene molecule, C_3H_4?
(a) The central carbon is *sp* hybridized
(b) The terminal carbon atoms are sp^2 hybridized
(c) The planes containing the CH_2 groups are mutually perpendicular to permit the formations two separate π-bonds
(d) All are correct

21. Which of the following pairs have identical bond order ?

(a) N_2, O_2^{2+} (b) N_2, O_2^-

(c) N_2^-, O_2 (d) O^{2+}, N_2

22. Which of the following molecular orbitals has two nodal planes ?

(a) $\sigma 2s$ (b) $\pi 2p_y$ (c) $\pi^* 2p_y$ (d) $\sigma^* 2p_x$

23. Which one of the following molecules is polar ?

(a) XeF_4 (b) IF_5 (c) SbF_5 (d) CF_4

24. Bond distance in HF is 9.17×10^{-11} m. Dipole moment of HF is 6.104×10^{-30} Cm. The percentage ionic character in HF will be : (electron charge = 1.60×10^{-19} C)

(a) 61.0% (b) 38.0%

(c) 35.5% (d) 41.5%

25. Dipole moment of H_2O is 1.85 D. If the bond angle is 105° and O — H bond length is 0.94 Å, what is the magnitude of charge on the oxygen atom in water molecule?

Toughnut

(a) 2×10^{-10} esu (b) 4.28×10^{-10} esu

(c) 3.22×10^{-10} esu (d) 1.602×10^{-19} C

26. Which of the correct increasing order of lone pair of electrons on the central atom?

(a) $IF_7 < IF_5 < ClF_3 < XeF_2$

(b) $IF_7 < XeF_2 < ClF_2 < IF_5$

(c) $IF_7 < ClF_3 < XeF_2 < IF_5$

(d) $IF_7 < XeF_2 < IF_5 < ClF_3$

27. The molecules BF_3 and NF_3 are both covalent compounds, but BF_3 is non polar whereas NF_3 is polar. The reason for this is

(a) atomic size of boron is larger than nitrogen

(b) Boron is metal while nitrogen is gas

(c) B – F bonds are non-polar while N – F bonds are polar

(d) BF_3 is planar but NF_3 is pyramidal

28. Among the following chloro-compound having the lowest dipole moment is

(a) CH_3Cl

(b) $(Cl)(H_3C)C=C(H)(Cl)$

(c) CH_2Cl_2

(d) $(Cl)(H_3C)C=C(Cl)(H)$

29. Although CN^- ion and N_2 molecule are isoelectronic, yet N_2 molecule is chemically inert because of

(a) presence of more number of electrons in bonding orbitals

(b) lone bond energy

(c) absence of bond polarity

(d) uneven electron distribution.

30. The number and type of bonds in ion in CaC_2 are:

(a) One σ bond and one π-bond

(b) One σ bond and two π-bond

(c) Two σ bond and two π-bond

(d) Two σ bond and one π-bond

31. Among the following compounds the one that is polar and has the central atom with sp^2 hybridisation is

(a) H_2CO_3 (b) SiF_4

(c) BF_3 (d) $HClO_2$

32. Among the following ions the $p\pi$–$d\pi$ overlap could be present in

(a) NO_2^- (b) NO_3^-

(c) PO_4^{3-} (d) CO_3^{2-}

33. In the reaction $2PCl_5 \rightleftharpoons PCl_4^+ + PCl_6^-$, the change in hybridisation is from

(a) sp^3d to sp^3 and sp^3d^2

(b) sp^3d to sp^2 and sp^3

(c) sp^3d to sp^3d^2 and sp^3d^3

(d) sp^3d^2 to sp^3 and sp^3d

34. In an octahedral structure, the pair of *d* orbitals involved in d^2sp^3 hybridization is

(a) $d_{x^2-y^2}, d_{z^2}$ (b) $d_{xz}, d_{x^2-y^2}$

(c) d_{z^2}, d_{xz} (d) d_{xy}, d_{yz}

35. N_2 and O_2 are converted into monocations, N_2^+ and O_2^+ respectively. Which of the following statements is wrong?

(a) In N_2^+, N - N bond weakens

(b) In O_2^+, the O - O bond order increases

(c) In O_2^+, paramagnetism decreases

(d) N_2^+ becomes diamagnetic

36. The correct order of bond dissociation energy among N_2, O_2, O_2^- is shown in which of the following arrangements? Tricky

(a) $N_2 > O_2^- > O_2$ (b) $O_2^- > O_2 > N_2$

(c) $N_2 > O_2 > O_2^-$ (d) $O_2 > O_2^- > N_2$

37. In XeF_2, XeF_4 and XeF_6, the number of lone pairs on Xe are respectively

(a) 2, 3, 1 (b) 1, 2, 3 (c) 4, 1, 2 (d) 3, 2, 1

38. Which of the following represents the given mode of hybridisation $sp^2 - sp^2 - sp - sp$ from left to right ?

(a) $H_2C=CH-C\equiv N$

(b) $HC\equiv C-C\equiv CH$

(c) $H_2C=C=C=CH_2$

(d) 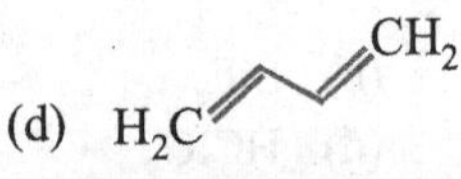

39. In BrF_3 molecule, the lone pairs occupy equatorial positions to minimize

(a) lone pair - bond pair repulsion only

(b) bond pair - bond pair repulsion only

(c) lone pair - lone pair repulsion and lone pair - bond pair repulsion

(d) lone pair - lone pair repulsion only

40. The group of molecules having identical shape is :

(a) PCl_5, IF_5, XeO_2F_2

(b) BF_3, PCl_3, XeO_3

(c) SF_4, XeF_4, CCl_4

(d) ClF_3, $XeOF_2$, XeF_3^+

41. SF_2, SF_4 and SF_6 have the hybridisation at sulphur atom respectively as :

(a) sp^2, sp^3, sp^2d^2 (b) sp^3, sp^3, sp^3d^2

(c) sp^3, sp^3d, sp^3d^2 (d) sp^3, spd^2, d^2sp^3

42. Which of the following molecules has two sigma (σ) and two pi (π) bonds?

(a) C_2H_4 (b) N_2F_2

(c) $C_2H_2Cl_2$ (d) HCN

43. The shape of IF_6^- is :

(a) Trigonally distorted octahedron

(b) Pyramidal

(c) Octahedral

(d) Square antiprism

44. Which of the following statement is correct?

(a) $FeCl_2$ is more covalent than $FeCl_3$.

(b) $FeCl_3$ is more covalent than $FeCl_2$.

(c) Both $FeCl_2$ and $FeCl_3$ are equally covalent.

(d) $FeCl_2$ and $FeCl_3$ do not have any covalent character.

45. The maximum number of 90° angles between bond pair-bond pair of electrons is observed in

(a) dsp^2 hybridization

(b) sp^3d hybridization

(c) dsp^3 hybridization

(d) sp^3d^2 hybridization

46. The true statements from the following are

1. PH_5 and $BiCl_5$ do not exist
2. $p\pi - d\pi$ bond is present in SO_2
3. Electrons travel with the speed of light
4. SeF_4 and CH_4 have same shape
5. I_3^+ has bent geometry Toughnut

(a) 1, 3 (b) 1, 2, 5

(c) 1, 3, 5 (d) 1, 2, 4

47. KF combines with HF to form KHF_2. The compound contains the species

(a) K^+, F^- and H^+ (b) K^+, F^- and HF

(c) K^+ and $[HF_2]^-$ (b) $[KHF]^+$ and F_2

48. Which of the following statements is incorrect?

(a) NH_3 is more basic than PH_3.

(b) NH_3 has a higher boiling point than that of HF.

(c) N_2 is less reactive than P_4.

(d) The dipole moment of NH_3 is less than that of SO_2.

49. In which of the following sets, all the given species are isostructural ?

(a) CO_2, NO_2, ClO_2, SiO_2

(b) PCl_3, $Al Cl_3$, BCl_3, $SbCl_3$

(c) BF_3, NF_3, PF_3, $Al F_3$

(d) BF_4^-, CCl_4, NH_4^+, PCl_4^+

50. The formation of molecular complex BF_3-NH_3 results in a change in hybridization of boron

(a) from sp^2 to dsp^2 (b) from sp^2 to sp^3

(c) from sp^3 to sp^2 (d) from sp^3 to sp^3d

51. Which of the following substances has the least ionic character ?
(a) $FeCl_2$ (b) $ZnCl_2$
(c) $CdCl_2$ (d) $MgCl_2$

52. Sulphur reacts with chlorine in 1 : 2 ratio and forms X. Hydrolysis of X gives a sulphur compound Y. What is the hybridisation state of central atom in the compound. Toughnut
(a) sp^3 (b) sp^2 (c) sp (d) dsp^2

53. CaO and NaCl have the same crystal structure and approximately the same ionic radii. If U is the lattice energy of NaCl, the approximate lattice energy of CaO is Tricky
(a) U/2 (b) U (c) 4 C (d) 2 U

54. Dipole moment is shown by
(a) 1, 4-dichlorobenzene
(b) *cis* -1, 2-dichlorobenzene
(c) *trans* -1, 3-dichlorobenzene
(d) *trans* -2, 3-dichloro -2- butene

55. Which of the following has the square planar structure?
(a) XeF_4 (b) NH_4^+ (c) BF_4^- (d) CCl_4

56. Which one of the following molecules is paramagnetic?
(a) N_2 (b) NO (c) CO (d) O_3

57. Which one of the following pairs of molecules will have permanent dipole moments for both members ?
(a) NO_2 and CO_2 (b) NO_2 and O_3
(c) SiF_4 and CO_2 (d) SiF_4 and NO_2

58. In which of the following ionization processes the bond energy has increased and also the magnetic behaviour has changed from paramagnetic to diamagnetic ? Tricky
(a) $NO \rightarrow NO^+$ (b) $N_2 \rightarrow N_2^+$
(c) $C_2 \rightarrow C_2^+$ (d) $O_2 \rightarrow O_2^+$

59. Bond order normally gives idea of stability of a molecular species. All the molecules viz. H_2, Li_2 and B_2 have the same bond order yet they are not equally stable. Their stability order is
(a) $H_2 > Li_2 > B_2$ (b) $Li_2 > H_2 > B_2$
(c) $Li_2 > B_2 > H_2$ (d) $B_2 > H_2 > Li_2$

60. The internuclear distances in O – O bonds for O_2^+, O_2, O_2^- and O_2^{2-} respectively are :
(a) 1.30 Å, 1.49 Å, 1.12 Å, 1.21 Å
(b) 1.49 Å, 1.21 Å, 1.12 Å, 1.30 Å
(c) 1.21 Å, 1.12 Å, 1.49 Å, 1.30 Å
(d) 1.12 Å, 1.21 Å, 1.30 Å, 1.49 Å

Answer KEY

1	(a)	7	(b)	13	(d)	19	(b)	25	(c)	31	(a)	37	(d)	43	(a)	49	(d)	55	(a)
2	(c)	8	(b)	14	(c)	20	(d)	26	(a)	32	(c)	38	(a)	44	(b)	50	(b)	56	(b)
3	(d)	9	(b)	15	(b)	21	(a)	27	(d)	33	(a)	39	(c)	45	(d)	51	(b)	57	(b)
4	(b)	10	(a)	16	(c)	22	(c)	28	(c)	34	(a)	40	(d)	46	(b)	52	(a)	58	(a)
5	(c)	11	(d)	17	(b)	23	(b)	29	(c)	35	(d)	41	(c)	47	(c)	53	(c)	59	(a)
6	(d)	12	(c)	18	(a)	24	(d)	30	(b)	36	(c)	42	(d)	48	(b)	54	(b)	60	(d)

Hints & Solutions

1. (a) NH_4Cl contains both covalent and ionic bonds $\left[\begin{array}{c} H \\ | \\ H-N-H \\ | \\ H \end{array}\right]^+ Cl^-$

2. (c) Due to electrostatic force of attraction, the melting points and boiling points of ionic compounds are high.

3. (d) p_x and p_y orbitals do not have proper orientation to overlap and hence no bond is formed..

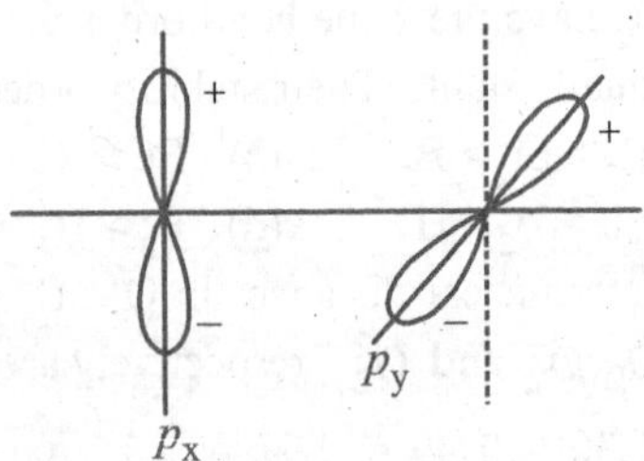

4. (b)

	Group	Hybridization	Shape
(1)	BF_3	sp^2	Triangular Planar (TP)
	NF_3	sp^3	Tetrahedral (T)
	CO_3^{2-}	sp^2	T.P.
(2)	CO_3^{2-}	sp^2	T.P.
	NO_3^-	sp^2	T.P. (All have same hybridization)
	SO_3	sp^2	T.P.
(3)	NH_3	sp^3	T
	SO_3	sp^2	T.P.
	CO_3^{2-}	sp^2	T.P.
(4)	NCl_3	sp^3	T
	BCl_3	sp^2	T.P.
	SO_3	sp^2	T.P.

5. (c) Hybridisation of N = ½ [5 + 3 + 0 – 0] = 4 hence sp^3

6. (d) X–M(–X)(–X)(–X) three angle below M and three above M hence = 6

7. (b) Hybridisation in $NH_3 = sp^3$, $[PtCl_4]^{2-} = dsp^2$ (inner complex); $PCl_5 = sp^3d$ and BCl_3 is sp^2.

8. (b) The delocalised $p\pi - p\pi$ bonding between filled p-orbital of F and vacant p-orbital of B leads to shortening of B–F bond length which results in higher bond dissociation energy of the B–F bond.

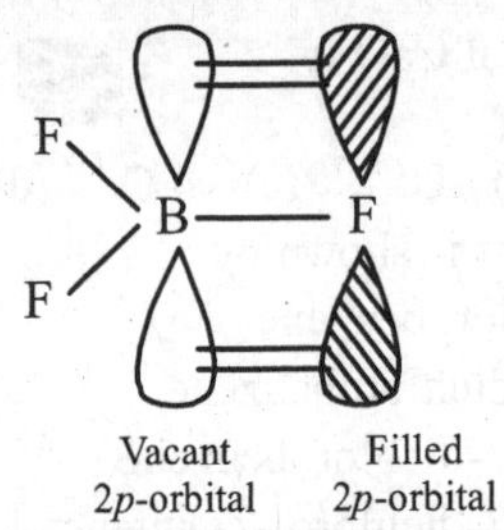

$F_2B=\overset{+}{F} \longleftrightarrow F(\overset{+}{F}=)B-F \longleftrightarrow \overset{+}{F}=B(F)-F \equiv$ B with three F (each $+1/3$)

9. (b) According to Fajan's rule smaller, highly charged cation has greatest covalent character while large cation with smaller charge has greatest ionic character.

10. (a) $NO^+ = 7 + 8 - 1 = 14\ e^-$.
$O_2 = 16\ e^-$
i.e not iso-electronic

(b) Boron forms only covalent compounds. This is due to its extremely high ionisation energy.

(c) Compounds of Tl^+ are much more stable than those of Tl^{3+}.

(d) $LiAlH_4$ is a versatile reducing agent in organic synthesis

11. (d) In metallic bonds, each ion is surrounded by equal number of oppositely charged electrons, hence have equal electrostatic attraction from all sides and hence do not have directional characteristics.

12. (c) $\left[\begin{array}{c} O \\ \| \\ O-P-O \\ | \\ O \end{array}\right]^{3-} \leftrightarrow \left[\begin{array}{c} O \\ | \\ O=P-O \\ | \\ O \end{array}\right]^{3-}$

Bond order

$$=\frac{\text{Number of bonds}}{\text{Number of Resonating structures}}$$

$$=\frac{5}{4}=1.25$$

Three unit negative charge is being shared by four O atoms. Formal charge $=-3/4$ $=-0.75$

13. (d) $\begin{array}{c} H \quad\; O \\ | \quad\;\; \| \\ H-C=C^{*}-O-H \end{array}$

The star marked carbon has a valency of 5 and hence this formula is not correct.

14. (c) As dipole moment = electric charge × bond length

D. M. of AB molecule

$$=4.8\times10^{-10}\times2.82\times10^{-8}=13.53D$$

D.M. of CD molecule

$$=4.8\times10^{-10}\times2.67\times10^{-8}=12.81D$$

Now % ionic character

$$=\frac{\text{Actual dipole moment of the bond}}{\text{Dipole moment of pure ionic compound}}\times100$$

then % ionic character in AB

$$=\frac{10.41}{13.53}\times100=76.94\%$$

% ionic character in CD

$$=\frac{10.27}{12.81}\times100=80.17\%$$

15. (b) $NO^-(16)$ – B.O. – 2

$O_2(16)$ – B.O. – 2

$NO^+(14)$ – B.O. – 3

$NO(15)$ – B.O. – 2.5

Higher the bond order lower is the bond length. Hence NO^+ will have smallest bond.

16. (c) NO_2 (benzene ring) NH_2

Dipole moment = (Distance between opposite charges) × (charge, q)

$$\mu=q\times d$$

So, greater the distance between the opposite charges higher the dipole. Due to the resonance the greater charge separation occurs between charges due to linearity.

17. (b) $1\ e=1.602\times10^{-19}C$

$1 esu=3.33\times10^{-10}C$

$$\frac{1e}{1esu}=\frac{1.602\times10^{-19}C}{3.33\times10^{-10}C}$$

$1\ e=4.802\times10^{-10}$ esu

Dipole moment = q × distance

$\Rightarrow 1\ D\approx10^{-18}$ esu cm

0.38×10^{-18} esu cm $=q\times(1.61\times10^{-8}cm)$

$q=2.36\times10^{-11}$esu

$$q=\frac{2.36\times10^{-11}esu}{4.802\times10^{-10}esu}$$

$q=0.049$

$q\approx0.05$ fractional charge

18. (a)

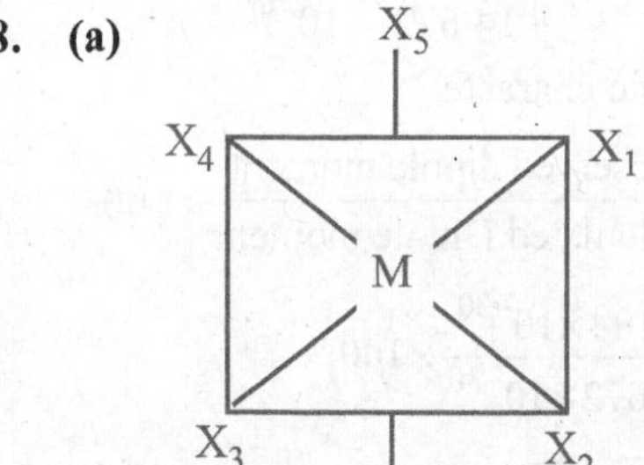

Thus here bond angles between

$X_4-M-X_2=180°$

$X_1-M-X_3=180°$

$X_5-M-X_6=180°$

19. (b) More will be the electronegativity of X, lesser will be the bond length of X-O bond.

20. **(d)**

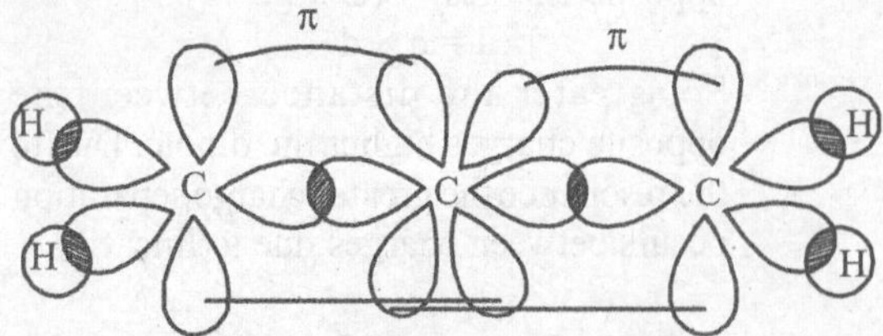

21. **(a)** Bond order in N_2 and O_2^{2+} is 3 (calculate by energy level diagram)

22. **(c)** 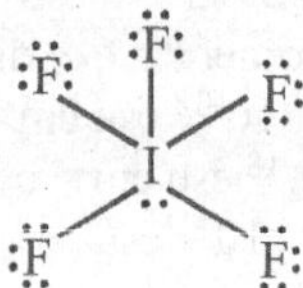 It has two nodal planes. It is π^*2p_y

23. **(b)** The geometry of IF_5 is square Pyramidal with an unsymmetric charge distribution therefore this molecule is polar.

$:\ddot{F}:$ structure of IF_5 (I with five F atoms and one lone pair)

24. **(d)** Given $e = 1.60 \times 10^{-19}$ C

$d = 9.17 \times 10^{-11}$ m

From $\mu = e \times d$

$\mu = 1.60 \times 10^{-19} \times 9.17 \times 10^{-11}$

$= 14.672 \times 10^{-30}$

% ionic character

$$= \frac{\text{Observed dipole moment}}{\text{Calculated Dipole moment}} \times 100$$

$$= \frac{6.104 \times 10^{-30}}{14.672 \times 10^{-30}} \times 100$$

$= 41.5\%$

25. **(c)** $\mu = 1.85$ D $= 1.85 \times 10^{-18}$ esu cm $= q \times d$

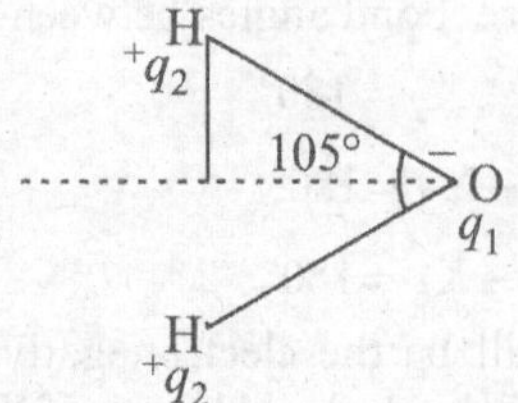

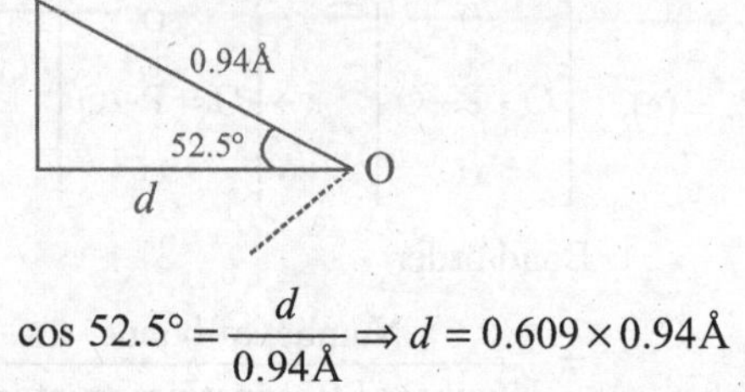

$$\cos 52.5° = \frac{d}{0.94\text{Å}} \Rightarrow d = 0.609 \times 0.94\text{Å}$$

$= 0.572$Å

$\therefore \mu = q \times d$

$$q_1 = \frac{\mu}{d} = \frac{1.85\text{D}}{0.572\text{Å}}$$

$$= \frac{1.85 \times 10^{-18} \text{ esu cm}}{0.572 \times 10^{-8} \text{ cm}}$$

$= 3.2 \times 10^{-10}$ esu

26. **(a)** The number of lone pairs of electrons on central atom in various given species are

Species	**Number of lone pairs on central atom**
IF_7	nil
IF_5	1
ClF_3	2
XeF_2	3

Thus the correct increasing order is

$IF_7 < IF_5 < ClF_3 < XeF_2$

0 1 2 3

27. **(d)** The shape of BF_3 is trigonal planar

$F^{\delta-}$, $B^{\delta+}$, $F^{\delta-}$, $F^{\delta-}$ (trigonal planar BF_3) and $\mu = 0$ hence it is non polar.

The shape of NF_3 is pyramidal

$\ddot{N}^{\delta+}$ with three $F^{\delta-}$ (pyramidal NF_3) and $\mu \neq 0$ hence it is polar.

28. **(c)** Cl(H)C = C(H)Cl (trans)

Dipole moment (μ) = 0

29. **(c)** In nitrogen molecule, both the nitrogen atoms have same electronegativity. So it has zero polarity and hence less tendency to break away and form ions.

30. **(b)** The structure of CaC_2 is $Ca^{2+}[:C \equiv C:]^{2-}$ i.e, one σ and two π bonds

31. (a) H_2CO_3 and BF_3 both have triangular planar structure

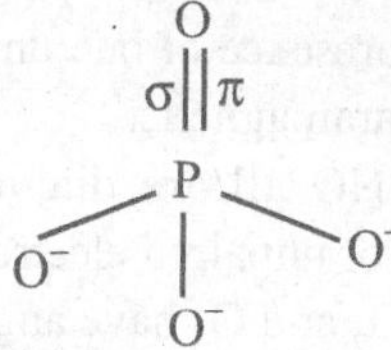

BF_3 is symmetrical

$\therefore \mu = 0$ hence non polar, H_2CO_3 is not symmetrical hence polar in nature.

32. (c) In P–O bond, π bond is formed by the sidewise overlapping of *d*-orbital of P and *p*-orbital of oxygen. Hence it is formed by pπ and dπ overlapping.

33. (a) $2PCl_5 \rightleftharpoons PCl_4^+ + PCl_6^-$

$\uparrow sp^3d \quad \uparrow sp^3 \quad \uparrow sp^3d^2$

34. (a) Only those *d* orbitals whose lobes are directed along X, Y and Z directions hybridise with *s* and *p* orbitals. In other three, *d* orbitals namely d_{xy}, d_{yz} and d_{xz}, the lobes are at an angle of 45° from both axis, hence the extent of their overlap with *s* and *p* orbitals is much lesser than $d_{x^2-y^2}$ and d_{z^2} orbitals.

35. (d) $\sigma_b^2 \sigma^{*2}_a \sigma_b^2 \sigma_a^{*2} (\pi_b^2 = \pi_b^2) \sigma_b^1 (N_2^+ = 13\,\text{electrons})$ it contains one unpaired electron hence paramagnetic

36. (c) The bond order of N_2, O_2, and O_2^- are respectively 3, 2 and 1.5
Since higher bond order implies higher bond dissociation energy hence the correct order will be $N_2 > O_2 > O_2^-$

37. (d) In XeF_2 Total number of valence electrons of Xe = 8, two electrons shared with 2F atoms, 6 electrons left hence 3 lone pairs, in XeF_4 4 shared with 4 F atoms 4 left hence 2 lone pairs; in XeF_6 6 shared with 6 F atoms 2 left hence 1 lone pair.

38. (a) $\underset{1}{CH_2} = \underset{2}{CH} - \underset{3}{C} \equiv \underset{4}{N}$

3 σ bonds (sp^2 hybridisation);
2 σ bonds (*sp*-hybridisation)
C_1 = 3 σ bonds, C_2 = 3 σ bonds,
C_3 = 2 σ bonds

39. (c)

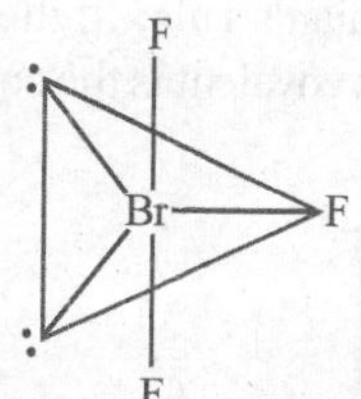

In BrF_3, both bond pairs as well as lone pairs of electrons are present. Due to the presence of lone pairs of electrons (*lp*) in the valence shell, the bond angle is contracted and the molecule takes the T-shape. This is due to greater repulsion between two lone pairs or between a lone pair and a bond pair than between the two bond pairs.

40. (d) $ClF_3 \longrightarrow$ Hybridisation

$= 3 + \frac{1}{2}[7-3] = 5(sp^3d)$

$XeOF_2 \longrightarrow$ Hybridisation

$= 3 + \frac{1}{2}[8-4] = 5(sp^3d)$

$XeF_3^+ \longrightarrow$ Hybridisation

$= 3 + \frac{1}{2}[8-3-1] = 5(sp^3d)$

All molecules have sp^3d hybridization and 2 lone pairs. Hence all have identical (T-shape).

41. (c) Hybridisation :

1. $\ddot{:SF_2} \Rightarrow \frac{1}{2}(6+2) = 4 = sp^3$

2. $:SF_4 \Rightarrow \frac{1}{2}(6+4) = 5 = sp^3d$

3. $SF_6 \Rightarrow \frac{1}{2}(6+6) = 6 = sp^3d^2$

42. (d) $H \underset{\sigma}{-} C \overset{\pi,\,\sigma}{\underset{\pi}{\equiv}} N$

i.e 2 π and 2 σ bonds.

43. (a) The structure of IF_6^- is distorted octahedral This is due to presence of a lone pair.

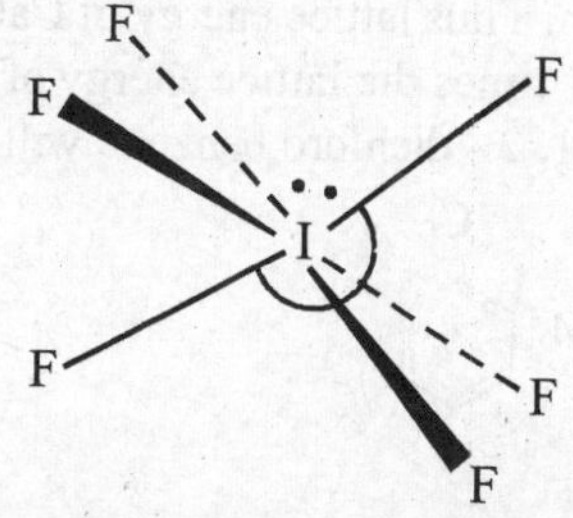

44. (b) According to Fajan's rule, higher charge on the ions, more covalent is the compound.

45. (d)

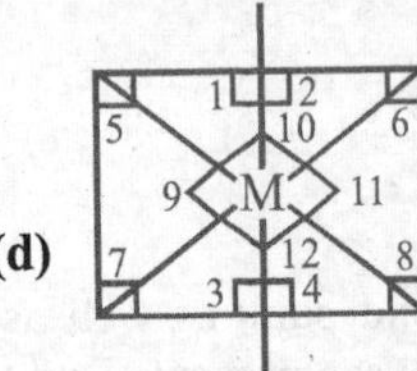

sp^3d^2 hybridisation

Number of 90° angle between bonds = 12

46. (b) SeF_4 has distorted tetrahedral geometry while CH_4 has tetrahedral geometry.

47. (c) Since F form H-bond $[HF_2]^-$ exists. Therefore KHF_2 gives $K^+ + HF_2^-$

48. (b) Due to high EN of F than N, H-bonding is stronger in HF than NH_3 thus b.pt. of HF > NH_3.

49. (d) All have tetrahedral structure.

50. (b) In BF_3, B is sp^2 hybridized with one empty p_z orbital. The empty p_z orbital of BF_3 can be filled by lone pair of molecules such as NH_3. When this occurs a tetrahedral molecule or ion is formed which is sp^3 hybridized.

51. (b) Polarizing power of Zn^{2+} is more than others, hence $ZnCl_2$ is least ionic in nature.

52. (a) $S + 2Cl_2 \rightarrow SCl_4 \xrightarrow{H_2O}$ $S(OH)_4$ (HO, HO, OH, OH on S; loss of H_2O) $\rightarrow$ HO—S(OH)=O (H_2SO_3)

Hybridisation of

$$H_2SO_3 = \frac{1}{2}(6+2+0) = 4 = sp^3$$

53. (c) Lattice energy $= \dfrac{\text{Product of charges}}{\text{interionic distance}}$

In NaCl the product of charges = 1 × 1; In CaO product of charges = 2 × 2 = 4 while the inter ionic distance is almost same in both. Thus lattice energy of CaO is almost four times the lattice energy of NaCl.

54. (b) *Cis*1, 2 - dichloro benzene will have some D.M. (benzene ring with two adjacent Cl)

55. (a) XeF_4 has square planar structure, while NH_4^+, BF_4^- and CCl_4 have tetrahedral structure.

56. (b) The molecular orbital configuration of the molecules given is

Total no. of electrons in NO = 7(N) + 8(O) = 15

Hence E.C. of NO

$= KK\sigma(2s)^2\sigma^*(2s)^2\sigma(2p_z)^2$

$\pi(2p_x)^2 = \pi(2p_y)^2\pi^*(2p_x)^1 = \pi^*(2p_y)^0$

Due to presence of one unpaired electron NO is paramagnetic.

Except NO all are diamagnetic due to absence of unpaired electrons.

57. (b) Both NO_2 and O_3 have angular shape and hence will have net dipole moment.

58. (a) For NO

Total no. of electrons = 15

B.O = 2.5

Mag. Behaviour = Paramagnetic

For NO^+ .

Total no. of electrons = 14

B.O = 3

Mag. Behaviour = Diamagnetic

59. (a) The molecular orbital configuration of the given molecules is

$H_2 = \sigma(1s)^2$ (no electron anti-bonding)

$Li_2 = \sigma(1s)^2\,\sigma^*(1s)^2\,\sigma(2s)^2$

(two anti-bonding electrons)

$B_2 = \sigma(1s)^2\,\sigma^*(1s)^2\,\sigma(2s)^2\,\sigma^*(2s)^2$

$\pi(2p_x)^1 = \pi(2p_y)^1$

(4 anti-bonding electrons)

Though the bond order of all the species are same (B.O = 1) but stability is different. This is due to difference in the presence of no. of anti-bonding electron.

Higher the no. of anti-bonding electron lower is the stability hence the correct order is $H_2 > Li_2 > B_2$

60. (d) The bond length follows the order

$O_2^+ < O_2 < O_2^- < O_2^{2-}$

According to this the possible values are 1.12Å, 1.21Å, 1.30Å, 1.49Å

5 States of Matter

1. Densities of two gases are in the ratio 1:2 and their temperatures are in the ratio 2:1 then the ratio of their respective pressures is
(a) 1:1 (b) 1:2 (c) 2:1 (d) 4:1

2. Which of the following volume (V) - temperature (T) plots represents the behaviour of one mole of an ideal gas at one atmospheric pressure ? Tricky

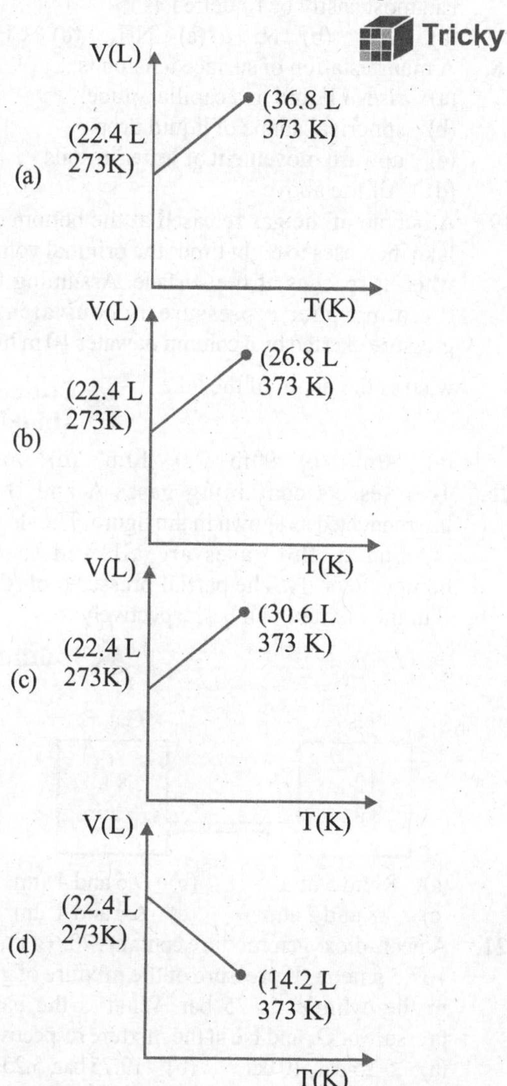

3. Which gas shows real behaviour?
(a) 16 g O_2 at 1 atm and 273 K occupies 11.2 L
(b) 1 g H_2 in 0.5 L flask exerts pressure of 24.63 atm at 300 K
(c) 1 mole NH_3 at 300 K and 1 atm occupies volume 22.4 L
(d) 5.6 L of CO_2 at 1 atm and 273 K is equal to 11 g

4. For a real gas (mol. mass = 60) if density at critical point is 0.80 g/cm³ and its $T_c = \frac{4\times10^5}{821}$ K, then van der waals' constant a (in atm L^2 mol^{-2}) is
(a) 0.3375 (b) 3.375 (c) 1.68 (d) 0.025

5. 600 c.c. of a gas at a pressure of 750 mm of Hg is compressed to 500 c.c. Taking the temperature to remain constant, the increase in pressure, is
(a) 150 mm of Hg (b) 250 mm of Hg
(c) 350 mm of Hg (d) 450 mm of Hg

6. The volume of 0.0168 mol of O_2 obtained by decomposition of $KClO_3$ and collected by displacement of water is 428 mL at a pressure of 754 mm Hg at 25 °C. The pressure of water vapour at 25 °C is
(a) 18 mm Hg (b) 20 mm Hg
(c) 22 mm Hg (d) 24 mm Hg.

7. Three different gases X, Y and Z of molecular masses 2, 16 and 64 were enclosed in a vessel at constant temperature till equilibrium is reached. Which of the following statement is correct?
(a) Gas Z will be at the top of the vessel
(b) Gas Y will be at the top of the vessel
(c) Gas Z will be at the bottom and X will be at the top
(d) Gases will form homogenous mixture

8. When 2 g of a gas A is introduced into an evaluated flask kept at 25 °C, the pressure is found to be one atmosphere. If 3 gm of another gas B is then added to the same flask, the total pressure becomes 1.5 atm. Assuming ideal gas behaviour, calculate the ratio of the molecular weights $M_A : M_B$. Toughnut
(a) 1 : 3 (b) 1 : 1 (c) 2 : 1 (d) 3 : 1

9. When does a gas deviate the most from its ideal behaviour?
(a) At low pressure and low temperature
(b) At low pressure and high temperature
(c) At high pressure and low temperature
(d) At high pressure and high temperature

10. The initial volume of a gas cylinder is 750.0 mL. If the pressure of gas inside the cylinder changes from 840.0 mm Hg to 360.0 mm Hg, the final volume the gas will be:
(a) 1.750 L (b) 3.60 L
(c) 4.032 L (d) 7.50 L

11. Which of the following assumption of kinetic molecular theory states that gases do not have fixed shape ?
(a) Particles of a gas move in all possible directions in straight line.
(b) Particles of a gas are always in constant and random motion.
(c) Total energy of molecules before and after the collision remains same.
(d) None of these

12. Which one of the following is the wrong assumption of kinetic theory of gases ?
(a) Momentum and energy always remain conserved.
(b) Pressure is the result of elastic collision of molecules with the container's wall.
(c) Molecules are separated by great distances compared to their sizes.
(d) All the molecules move in straight line between collision and with same velocity.

13. Equal masses of methane and oxygen are mixed in an empty container at 25°C. The fraction of the total pressure exerted by oxygen is

(a) 1/2 (b) 2/3
(c) $\frac{1}{3}\times\frac{273}{298}$ (d) 1/3

14. A He atom at 300 K is released from the surface of the earth to travel upwards. Assuming that it undergoes no collision with other molecules, how high will it be before coming to the rest?
(a) 9.53 m (b) 95.3 m
(c) 953 m (d) 9.53×10^4 m

15. Sulphur dioxide and oxygen were allowed to diffuse through a porous partition. 20 dm^3 of SO_2 diffuses through the porous partition in 60 seconds. The volume of O_2 in dm^3 which diffuses under the similar condition in 30 seconds will be (atomic mass of sulphur = 32 u):
(a) 7.09 (b) 14.1 (c) 10.0 (d) 28.2

16. For 1 mol of an ideal gas at a constant temperature T, the plot of log P against log V is a (P: Pressure, V: Volume)
(a) Straight line parallel to x-axis.
(b) Straight line with a negative slope.
(c) Curve starting at origin.
(d) Straight line passing through origin.

17. The value of van der waals constant 'a' for gases O_2, N_2, NH_3 and CH_4 are 1.360, 1.390, 4.170 and 2.253 litre^2 atm mol^{-2} respectively. The gas which can most easily be liquefied is :
(a) O_2 (b) N_2 (c) NH_3 (d) CH_4

18. A manifestation of surface tension is :
(a) rise of liquid in a capillary tube
(b) spherical shape of liquid drops
(c) upward movement of water in soils
(d) All the above

19. A bubble of the gas released at the bottom of a lake increases to eight times the original volume when it reaches at the surface. Assuming that the atmospheric pressure is equivalent to pressure exerted by a column of water 10 m high, what is the depth of the lake

(a) 80m (b) 90m (c) 10m (d) 70m

20. Two vessels containing gases A and B are interconected as shown in the figure. The stopper is opened, the gases are allowed to mix homogeneously. The partial pressures of A and B in the mixture will be, respectively

Toughnut

Gas A Gas B

12 L 8 atm	8 L 5 atm

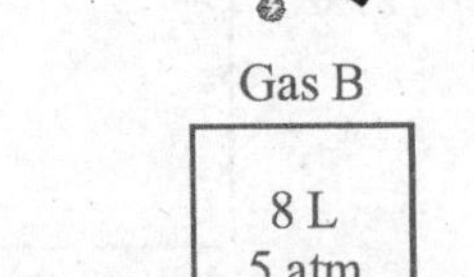

(a) 8 and 5 atm (b) 9.6 and 4 atm
(c) 4.8 and 2 atm (d) 6.4 and 4 atm

21. A neon-dioxygen mixture contains 70.6 g O_2 and 167.5 g neon. If pressure of the mixture of gases in the cylinder is 25 bar. What is the partial pressure of O_2 and Ne in the mixture respectively?
(a) 5.25 bar, 10 bar (b) 19.75 bar, 5.25 bar
(c) 19.75 bar, 10 bar (d) 5.25 bar, 19.75 bar

22. When $CO_2(g)$ is passed over red hot coke it partially gets reduced to $CO(g)$. Upon passing 0.5 L of $CO_2(g)$ over red hot coke, the total volume of the gases increased to 700 mL.
The composition of the gaseous mixture at STP is **Toughnut**

(a) $CO_2 = 300$ mL; $CO = 400$ mL
(b) $CO_2 = 0.0$ mL; $CO = 700$ mL
(c) $CO_2 = 200$ mL; $CO = 500$ mL
(d) $CO_2 = 350$ mL; $CO = 350$ mL

23. Which of the following is not an assumption of the kinetic theory of gases?
(a) Gas particles have negligible volume.
(b) A gas consists of many identical particles which are in continual motion.
(c) At high pressure, gas particles are difficult to compress.
(d) Collisions of gas particles are perfectly elastic.

24. Two vessels of volumes 16.4 L and 5 L contain two ideal gases of molecular existence at the respective temperature of 27 °C and 227 °C and exert 1.5 and 4.1 atmospheres respectively. The ratio of the number of molecules of the former to that of the later is **Toughnut**

(a) 2 (b) 1 (c) $\frac{1}{2}$ (d) $\frac{1}{3}$

25. If 10^{-4} dm^3 of water is introduced into a 1.0 dm^3 flask at 300 K, how many moles of water are in the vapour phase when equilibrium is established ? (Given : Vapour pressure of H_2O at 300 K is 3170 Pa; $R = 8.314$ J K^{-1} mol^{-1})
(a) 5.56×10^{-3} mol (b) 1.53×10^{-2} mol
(c) 4.46×10^{-2} mol (d) 1.27×10^{-3} mol

26. At STP, a container has 1 mole of He, 2 mole Ne, 3 mole O_2 and 4 mole N_2. without changing total pressure if 2 mole of O_2 is removed. The partial pressure of O_2 will be decreased by: **Tricky**

(a) 26% (b) 40%
(c) 58.33% (d) 66.66%

27. A mixture of Ne and Ar kept in a closed vessel at 250 K has a total K.E. = 3kJ. The total mass of Ne and Ar is 30 g. Find mass % of Ne in gaseous mixture at 250 K.
(a) 61.63 (b) 38.37 (c) 50 (d) 28.3

28. The temperature at which oxygen molecules have the same root mean square speed as helium atoms have at 300 K is:
Atomic masses: He = 4 u, O = 16 u)
(a) 300 K (b) 600 K
(c) 1200 K (d) 2400 K

29. By how many folds the temperature of a gas would increase when the root mean square velocity of the gas molecules in a container of fixed volume is increased from 5×10^4 cm/s to 10×10^4 cm/s ?
(a) Two (b) Three (c) Six (d) Four

30. A high-altitude balloon is filled with 1.41×10^4 L of hydrogen at a temperature of 21°C and a pressure of 745 torr. What is the volume of the balloon at a height of 20 km, where the temperature is – 48 °C and the pressure is 63.1 torr?
(a) 1.247×10^5 L (b) 1.66×10^5 L
(c) 1.66×10^4 L (d) None of these

31. A spherical balloon of 21 cm diameter is to be filled with H_2 at NTP from a cylinder containing the gas at 20 atm at 27°C. The cylinder can hold 2.82 L of water at NTP. The number of balloons that can be filled up is
(a) 15 (b) 10 (c) 20 (d) 25

32. Equal volumes of the gases which do not react together are confined in separate vessels. The pressure is 200 mm and 400 mm of Hg respectively. If the two gases are mixed together what will be the pressure of the resulting mixture (temperature remaining constant)
(a) 400 mm (b) $\sqrt{400}$ mm
(c) 300 mm (d) 200 mm

33. In case of CO and CH_4 curve goes to minima then increases with increase in pressure but in case of H_2 and He the curve is linear because:

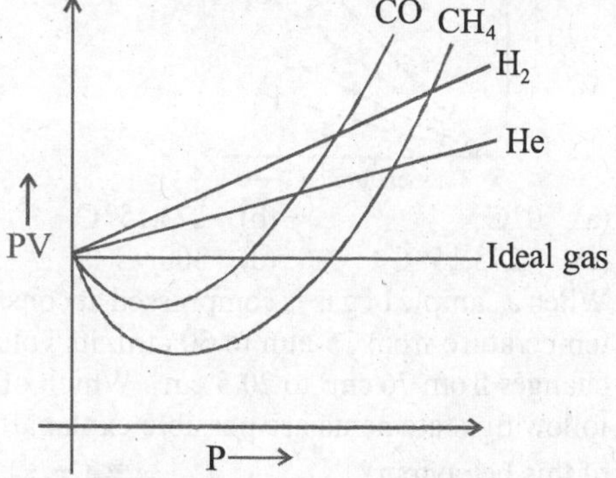

(a) Intermolecular interactions for H_2 and He are very low.
(b) Molecular size or atomic size for H_2 and He is small.
(c) Both (a) and (b)
(d) Neither (a) nor (b)

34. Following table represents critical temperature of some gases. Arrange these gases in their increasing order of liquifaction.

Gas	H_2	He	N_2	O_2
T_c / K	33.2	5.3	126	154.3

(a) $He < N_2 < H_2 < O_2$ (b) $H_2 < He < N_2 < O_2$
(c) $He < H_2 < N_2 < O_2$ (d) $O_2 < N_2 < H_2 < He$

35. α, *v* and *u* represent most probable velocity, average velocity and root mean square velocity respectively of a gas at a particular temperature. The correct order among the following is
(a) $u > v > \alpha$ (b) $v > u > \alpha$
(a) $\alpha > u > v$ (d) $u > \alpha > v$

36. van der Waal's equation for a gas is stated as,

$$P = \frac{nRT}{V - nb} - a\left(\frac{n}{V}\right)^2.$$

This equation reduces to the perfect gas equation, $P = \frac{nRT}{V}$ when,

Critical Thinking

(a) temperature is sufficient high and pressure is low.
(b) temperature is sufficient low and pressure is high.
(c) both temperature and pressure are very high.
(d) both temperature and pressure are very low.

37. What is the value of X in °C for given volume vs temperature curve ?

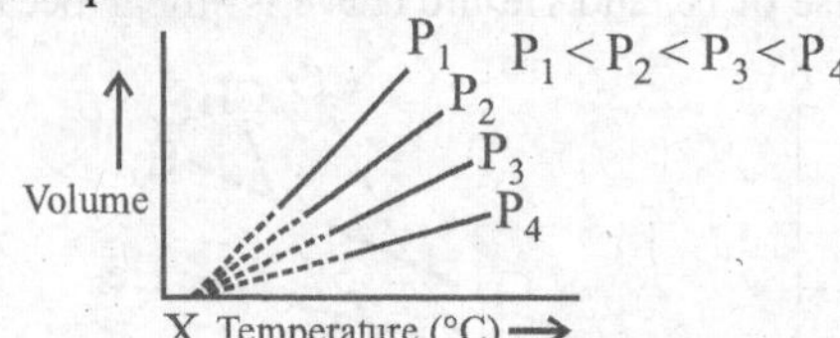

(a) 0°C (b) 273.15°C
(c) −273.15°C (d) 300°C

38. When a sample of gas is compressed at constant temperature from 15 atm to 60 atm, its volume changes from 76 cm^3 to 20.5 cm^3. Which of the following statements are possible explanations of this behaviour?

Tricky

(1) The gas behaves non-ideally
(2) The gas dimerises
(3) The gas is adsorbed into the vessel walls
(a) 1, 2 and 3 (b) 1 and 2 only
(c) 2 and 3 only (d) 1 only

39. What is the ratio of pressure of the 2 g of hydrogen to that of 4 g of helium at temperature of 298 K, 20 mL volume? (consider the ideal behaviour)
(a) 1 : 2 (b) 2 : 1 (c) 1 : 1 (d) 2 : 2

40. A drop of oil is placed on the surface of water. Which of the following statement is correct ?
(a) It will remain on it as a sphere
(b) It will spread as a thin layer
(c) It will be partly as spherical droplets and partly as thin film
(d) It will float as a distorted drop on the water surface

41. The correct order of viscosity of the following liquids will be
(a) Water < methyl alcohol < dimethyl ether < glycerol
(b) methyl alcohol < glycerol < water < dimethyl ether
(c) dimethyl ether < methyl alcohol < water < glycerol
(d) glycerol < dimethyl ether < water < methyl alcohol

42. At 0°C and one atm pressure, a gas occupies 100 cc. If the pressure is increased to one and a half-time and temperature is increased by one-third of absolute temperature, then final volume of the gas will be:
(a) 80 cc (b) 88.9 cc
(c) 66.7 cc (d) 100 cc

43. At 27 °C a sample of ammonia gas exerts a pressure of 5.3 atm. What is the pressure when the volume of the gas is reduced to one-tenth of the original value at the same temperature?
(a) 0.53 atm (b) 5.3 atm
(c) 53 atm (d) None of these

44. The most probable speed of 8 g of H_2 is 200 ms^{-1}. Average kinetic energy (neglect rotational and vibrational energy) of H_2 gas is :
(a) 480 J (b) 240 J
(c) 120 J (d) none of these

45. Oxidation of succinate ion produces ethylene and carbon dioxide gases. On passing 0.2 Faraday electricity through an aqueous solution of potassium succinate, the total volume of gases (at both cathode and anode) at STP (1 atm and 273 K) is :
(a) 8.96 L (b) 4.48 L
(c) 6.72 L (d) 2.24 L

46. A compressed cylinder of gas contains 1.50×10^3 g of N_2 gas at a pressure of 2.0×10^7 Pa and a temperature of 17.1 °C. What volume of gas has been released into the atmosphere if the final pressure in the cylinder is 1.80×10^5 Pa? Assume ideal behaviour and that the gas temperature is unchanged.
(a) 1264 L (b) 126 L
(c) 12600 L (d) 45 L

47. Dimethyl ether decomposes as

$$CH_3OCH_3(g) \longrightarrow CH_4(g) + CO(g) + H_2(g)$$

when CH_3OCH_3 decomposes to 20% extent under certain conditions, what is the ratio of diffusion of pure CH_3OCH_3 with methane?
(a) 0.59 : 1 (b) 1.18 : 1
(c) 2.36 : 1 (d) 1.77 : 1

48. Air entering the lungs ends up in tiny sacs called alveoli. From the alveoli, the oxygen diffuses into the blood. The average radius of the alveoli is 0.0050 cm and the air inside contains 14 per cent oxygen. Assuming that the pressure in the alveoli is 1.0 atm and the temperature is 37°C, calculate the number of oxygen molecules in one of the alveoli. Toughnut
(a) 6×10^{13} (b) 10^{24}
(c) 1.7×10^{22} (d) 1.7×10^{12}

49. A graph is plotted between PV_m (along Y-axis) and P (along X–axis) at 27° (V_m being the molar volume of an ideal gas). The intercept of the graph at Y–axis is (R = 0.082 atm LK^{-1} mol^{-1})
(a) 22.4 atm L mol^{-1}
(b) 24.6 atm L mol^{-1}
(c) 24.6 atm LK^{-1} mol^{-1}
(d) None of these

50. A flask containing air (open to the atmosphere) is heated from 300 K to 500 K. The percentage of the air escaped into the atmosphere is
(a) 16.6 (b) 40 (c) 60 (d) 20

51. Reducing the pressure from 1.0 atm to 0.5 atm would change the number of molecules in one mole of ammonia to
(a) 25% of its initial value
(b) 50% of its initial value
(c) 75% of its initial value
(d) None of the above

52. For real gases van der Waals equation is written as

$$\left(p + \frac{an^2}{V^2}\right)(V - nb) = nRT$$

where '*a*' and '*b*' are van der Waals constants.
Two sets of gases are :
(I) O_2, CO_2, H_2 and He
(II) CH_4, O_2 and H_2
The gases given in set-I in increasing order of '*b*' and gases given in set-II in decreasing order of '*a*', are arranged below. Select the correct order from the following :
(a) (I) $He < H_2 < CO_2 < O_2$ (II) $CH_4 > H_2 > O_2$
(b) (I) $O_2 < He < H_2 < CO_2$ (II) $H_2 > O_2 > CH_4$
(c) (I) $He < H_2 < O_2 < CO_2$ (II) $CH_4 > O_2 > H_2$
(d) (I) $H_2 < O_2 < He < CO_2$ (II) $O_2 > CH_4 > H_2$

53. Which of the following statements are correct?
(i) Hydrogen bonding is a special case of dipole - dipole interaction.
(ii) Energy of hydrogen bond varies between 10 to 100 kJ mol^{-1}.
(iii) Hydrogen bonds are powerful force in determining the structure and properties of compounds like proteins, nucleic acids etc.
(iv) Strength of the hydrogen bond is determined by the coulombic interaction between the lone-pair electrons of the electronegative atom of one molecule and the hydrogen atom of other molecule.
(a) (i) and (ii) (b) (i), (ii) and (iii)
(c) (ii), (iii) and (iv) (d) All of these

54. On a ship sailing in pacific ocean where temperature is 23.4°C, a balloon is filled with 2 L air. What will be the volume of the balloon when the ship reaches Indian ocean, where temperature is 26.1°C ? Tricky
(a) 2.018 L (b) 2.8 L
(c) 3.5 L (d) 1.5 L

55. A bubble of air is underwater at temperature 15°C and the pressure 1.5 bar. If the bubble rises to the surface where the temperature is 25°C and the pressure is 1.0 bar, what will happen to the volume of the bubble ?
(a) Volume will become greater by a factor of 1.6.
(b) Volume will become greater by a factor of 1.1.
(c) Volume will become smaller by a factor of 0.70.
(d) Volume will become greater by a factor of 2.5.

56. Cyclopropane and oxygen at partial pressures 170 torr and 570 torr respectively are mixed in a gas cylinder. What is the ratio of the number of moles of cyclopropane to the number of moles of oxygen (nC_3H_6/nO_2)? **Tricky**

(a) $\frac{170\times42}{570\times32}=0.39$

(b) $\frac{170}{42}\Big/\left(\frac{170}{42}+\frac{570}{32}\right)\approx0.19$

(c) $\frac{170}{740}=0.23$

(d) $\frac{170}{570}=0.30$

57. There are three closed containers in which equal amount of the gas are filled. **Tricky**

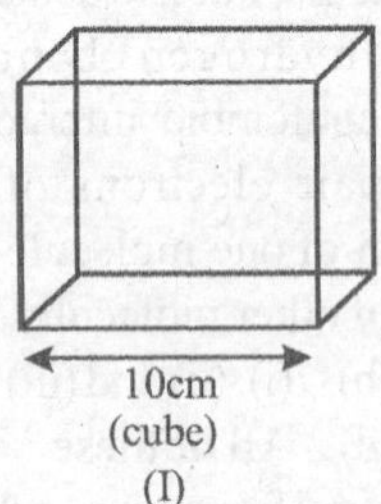
10cm
(cube)
(I)

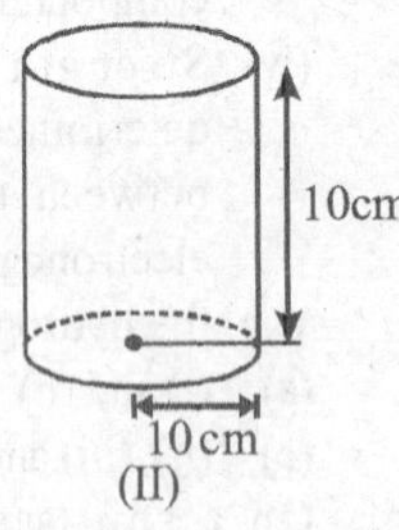

10 cm
(II)

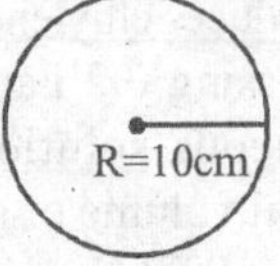

Sphere
(hollow)
(III)

If all the containers are placed at the same temperatures, then find the incorrect options –
(a) Pressure of the gas is minimum in (III) container
(b) Pressure of the gas is equal in I and II container
(c) Pressure of the gas is maximum in (I)
(d) The ratio of pressure in II and III container is 4 : 3

58. Helium atom is two times heavier than a hydrogen molecule at. 298K. The average KE of helium is
(a) 2 times of H_2 molecule
(b) same as that of H_2 molecule
(c) 4 times that of hydrogen molecule
(d) $\frac{1}{2}$ that of H_2 molecule

59. A 3 : 2 molar mixture of N_2 and CO is present in a vessel at 500 bar pressure. Due to hole in the vessel, the gas mixture leaks out. The composition of mixture effusing out initially is

(a) $n_{N_2} : n_{CO} :: 1 : 2$ (b) $n_{N_2} : n_{CO} :: 6 : 1$
(c) $n_{CO} : n_{N_2} :: 1 : 2$ (d) $n_{CO} : n_{N_2} :: 2 : 3$

60. The molecular velocities of two gases at the same temperature are u_1 and u_2 and their masses are m_1 and m_2 respectively. Which of the following expressions are correct?

(a) $\frac{m_1}{u_1^2}=\frac{m_2}{u_2^2}$ (b) $m_1u_1=m_2u_2$

(c) $\frac{m_1}{u_1}=\frac{m_2}{u_2}$ (d) $m_1u_1^2=m_2u_2^2$

Answer KEY

1	(a)	7	(d)	13	(d)	19	(d)	25	(d)	31	(b)	37	(c)	43	(c)	49	(b)	55	(a)
2	(c)	8	(a)	14	(d)	20	(c)	26	(c)	32	(c)	38	(d)	44	(b)	50	(b)	56	(d)
3	(c)	9	(c)	15	(b)	21	(d)	27	(d)	33	(c)	39	(c)	45	(a)	51	(d)	57	(b)
4	(b)	10	(a)	16	(b)	22	(a)	28	(d)	34	(c)	40	(b)	46	(a)	52	(c)	58	(b)
5	(a)	11	(b)	17	(c)	23	(c)	29	(d)	35	(a)	41	(c)	47	(c)	53	(d)	59	(d)
6	(d)	12	(d)	18	(d)	24	(a)	30	(a)	36	(a)	42	(b)	48	(d)	54	(a)	60	(d)

Hints & Solutions

1. **(a)** $P \propto d$ and T,

$$\frac{P_1}{P_2} = \frac{d_1 T_1}{d_2 T_2} = \frac{1}{2} \times \frac{2}{1} \Rightarrow P_1 : P_2 = 1:1$$

2. **(c)** $\frac{V_1}{T_1} = \frac{V_2}{T_2}$ at const. pressure

$$\Rightarrow \frac{22.4}{273} = \frac{V_2}{373}, V_2 = 30.6 \text{ litre}$$

3. **(c)** For ideal gas PV = nRT
Volume of 1 mole gas at 1 atm pressure and 273 K is 22.4 L.

4. **(b)** $V_c = \frac{60}{0.80} = 75 cm^3 mol^{-1}$;

$$b = \frac{V_c}{3} = 25 cm^3 mol^{-1}; = 0.25 L mol^{-1}$$

$$\therefore T_c = \frac{8a}{27Rb}$$

$$\frac{4 \times 10^5}{821} = \frac{2.45}{0.425} = 5.76 gL^{-1}$$

$$\Rightarrow a = 3.375$$

5. **(a)** Given initial volume $(V_1) = 600$ c.c.; Initial pressure (P_1) = 750 mm of Hg and final volume $(V_2) = 500$ c.c. according to Boyle's law,
$P_1V_1 = P_2V_2$

or $750 \times 600 = P_2 \times 500$

or $P_2 = \frac{750 \times 600}{500} = 900$ mm of Hg
Therefore increase of pressure
$= (900 - 750) = 150$ mm of Hg

6. **(d)** Volume of 0.0168 moles at STP
$= 0.0168 \times 22400 = 376.3$mL.

$$\frac{P_1V_1}{T_1} = \frac{P_2V_2}{T_2}$$

$$\text{or } \frac{760 \times 376.3}{273} = \frac{P_2 \times 428}{298}$$

or $P_2 = 730$mm
Pressure of water = 754 – 730 = 24 mm Hg

7. **(d)** All the gases occupy the available volume and will form homogeneous mixture.

8. **(a)** From ideal gas equation, PV = nRT

$$PV = \left(\frac{m}{M}\right)RT \text{ or } M = m\frac{RT}{PV}$$

Let the molecular wt. of A and B be M_A and M_B respectively.

$$\text{Then } M_A = 2 \times \frac{RT}{1 \times V}; M_B = \frac{3 \times RT}{0.5 \times V}$$

$$\therefore \frac{M_A}{M_B} = \frac{2RT}{V} \times \frac{0.5V}{3RT} = \frac{2 \times 0.5}{3} = \frac{1}{3}$$

Therefore, the ratio $M_A : M_B = 1 : 3$

9. **(c)** At high pressure and low temperature, gaseous atoms or molecules get closer to each other and van der Waal forces operates.

So molecules or atoms start attracting each other. Hence a gas deviate the most from its ideal behaviour. While in ideal behaviour we consider that gaseous molecules do not attract each other, i.e., there is no intermolecular forces of attraction.

10. **(a)** According to Boyle's law

$$\frac{V_1}{V_2} = \frac{P_2}{P_1};\ \frac{750}{V_2} = \frac{360}{840}$$

$V_2 = 1750\,\text{mL} = 1.750\,\text{L}$

11. **(b)** Particles of a gas are always in constant and random motion. If the particles were at rest and occupy fixed positions, then a gas would have a fixed shape which is not observed.

12. **(d)** Molecules move very fast in all directions in a straight line by colliding with each other but with different velocity.

13. **(d)** Let the mass of methane and oxygen = m gm.

Mole fraction of O_2

$$= \frac{\text{Moles of } O_2}{\text{Moles of } O_2 + \text{Moles of } CH_4}$$

$$= \frac{m/32}{m/32 + m/16} = \frac{m/32}{3m/32} = \frac{1}{3}$$

Partial pressure of O_2 =
Total pressure × mole fraction of O_2,

$$P_{O_2} = P \times \frac{1}{3} = \frac{1}{3}P$$

14. **(d)** $\frac{3}{2}k_BT = mgh$

$$h = \frac{3}{2} \times \frac{1.38\times10^{-23}\times300}{9.81\times4\times1.66\times10^{-27}} = 9.53\times10^4\,\text{m}$$

15. **(b)** According to Graham's Law Diffusion:

$$\frac{r_1}{r_2} = \sqrt{\frac{d_2}{d_1}}$$

Since rate of diffusion

$$= \frac{\text{Vol. of gas diffused } (V)}{\text{Time taken for diffusion}(t)}$$

$$\therefore \frac{r_1}{r_2} = \frac{V_1/t_1}{V_2/t_2}$$

or $\frac{r_1}{r_2} = \frac{V_1/t_1}{V_2/t_2} = \sqrt{\frac{d_2}{d_1}}$

$$= \frac{20/60}{V_2/30} = \sqrt{\frac{16/2}{32/2}} = \sqrt{\frac{1}{2}}$$

$\because$ Mol. wt = 2 × V.D

$\therefore$ V.D $= \frac{\text{Mol.wt}}{2}$

On calculating,

$V_2 = 14.1$

16. **(b)** According to Boyle's law, PV = constant

$\therefore$ $\log P + \log V$ = constant

$\log P = -\log V$ + constant

Hence, the plot of $\log P$ vs $\log V$ is straight line with negative slope.

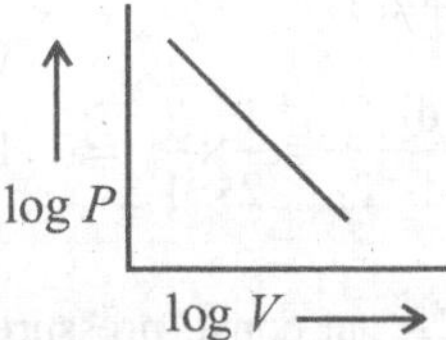

17. **(c)** The higher the value of 'a', more the value of T_c, easy is the liquefaction

18. **(d)** All the given phenomenon occurs due to surface tension

19. **(d)** Let V be the original volume of bubble. The final volume will be 8 V. Let p be the atmospheric pressure and p_1 the pressure at the bottom

$\therefore$ $p \times 8V = p_1 \times V$;

$8p = p_1$

p_1 = atmospheric pressure + pressure due to water lake

$= p + 7p$

The p = 10 m high, the 7p will be = 70 m high

So the depth = 70 m

20. **(c)** Moles of A, $(n_A) = \frac{P_AV_A}{RT} = \frac{8\times12}{RT} = \frac{96}{RT}$

Moles of B, $(n_B) = \frac{P_BV_B}{RT} = \frac{8\times5}{RT} = \frac{40}{RT}$

Total pressure × total volume = $(n_A + n_B) \times$ RT

$$p\times(12+8)=\frac{1}{RT}(96+40)RT$$

$$P=6.8$$

Partial pressure of A = P × mole fraction of A

$$=6.8\left(\frac{96}{RT}\Big/\frac{96+40}{RT}\right)$$

$$=4.8 \text{ atm}$$

Partial pressure of B = 6.8 – 4.8 = 2 atm.

21. **(d)** Number of moles of O_2

$$=\frac{70.6\,g}{32\,g\,mol^{-1}}=2.21\text{ mole}$$

Number of moles of Ne

$$=\frac{167.5\,g}{20\,g\,mol^{-1}}=8.375\text{ mole}$$

Mole fraction of O_2

$$=\frac{2.21}{2.21+8.375}=0.21\text{ mole}$$

Mole fraction of Ne = 1 – 0.21 = 0.79

Partial pressure of a gas

= Mole fraction × total pressure

Partial pressure of O_2 = 0.21 × 25 = 5.25 bar

Partial pressure of Ne = 0.79 × 25

= 19.75 bar

22. **(a)** $CO_2 + C \longrightarrow 2CO$

Stoichoimetry ratio is 1 : 2

AT STP, $P = 1$ atm, $T = 273$ K, $R = 0.0821$

Initial moles of CO_2; $n(CO_2 \text{initial}) = \frac{PV}{RT}$

$$=\frac{1\times0.5}{0.0821\times273}=0.022\text{ mole}$$

In final mixture no. of moles; $n(CO_2/CO$ mixture)

$$=\frac{1\times0.7}{0.0821\times273}=0.031$$

Increase in volume is by = 0.031 – 0.022

= 0.009 mole of gas

Final no. of moles of CO i.e. $n_{(CO\ final)}$

$n_{(CO\ final)} = 2n_{(CO_2\ initial)} - n_{(CO_2\ final)}$

$= 2(0.022 - n_{(CO_2\ final)}$ ...(i)

$n_{(CO\ final)} = 0.044 - 2n_{(CO_2\ final)}$ ...(ii)

∴ Now, $n_{(CO\ final)} + n_{(CO\ final)} = 0.031$

$n_{(CO_2\ final)} = 0.031 - n_{(CO\ final)}$ ...(iii)

Substituting (ii) in eq. (i)

$n_{(CO\ final)} = 0.044 - 2[0.031 - n_{(CO\ final)}]$

$n_{(CO\ final)} = 0.044 - 0.062 + 2n_{(CO\ final)}$

$n_{(CO\ final)} = 0.018$ mol.

Volume of

$$CO = V = \frac{nRT}{P} = \frac{0.018\times0.0821\times273}{1}$$

$$= 0.40 \text{ Litre}$$

and volume of CO_2 = 0.7 litre – 0.4 litre = 0.3 litre

∴ CO_2 = 300 mL, CO = 400 mL

23. **(c)** At high pressure real gas particles are easily compressed.

24. **(a)** Given conditions

$V_1 = 16.4$ L, $V_2 = 5$ L

$P_1 = 1.5$ atm, $P_2 = 4.1$ atm

$T_1 = 273 + 27 = 300$ K,

$T_2 = 273 + 227 = 500$ K

Applying gas equation, $\frac{P_1V_1}{P_2V_2}=\frac{n_1T_1}{n_2T_2}$

$$\frac{n_1}{n_2}=\frac{P_1V_1T_1}{P_2V_2T_2}$$

$$\therefore\ \frac{1.5\times16.4\times500}{4.1\times5\times300}=\frac{2}{1}$$

25. **(d)** From the ideal gas equation :

PV = nRT

$$\text{or } n=\frac{PV}{RT}=\frac{3170\times10^{-3}}{8.314\times300}=1.27\times10^{-3}$$

26. **(c)** $P_{O_2}=\frac{3}{10}\times P_T$;

After removing 2 mole of O_2,

$$P'_{O_2}=\frac{1}{10}\times P_T$$

Decrease in partial pressure of O_2

$$=\frac{\frac{3P_T}{10}-\frac{P_T}{8}}{\frac{3P_T}{10}}\times100=58.33\,\%$$

27. (d) $K = \frac{3}{2} n_T RT \Rightarrow n_T = \frac{3000 \times 2}{3 \times 250 \times 8.314}$

$3 \times 10^3 = \frac{3}{2} n_T \times 8.314 \times 250$

$\frac{x}{20} + \frac{30 - x}{40} = \frac{2}{8.314}$

% Ne = 28.3

28. (d) $V_{rms} = \sqrt{\frac{3RT}{M}}$

$V_{rms(O_2)} = V_{rms(He)}$

$\sqrt{\frac{3RT_{O_2}}{M_{O_2}}} = \sqrt{\frac{3RT_{He}}{M_{He}}}$

or $\frac{T_{O_2}}{M_{O_2}} = \frac{T_{He}}{M_{He}}$

$\therefore \quad T_{O_2} = \frac{300 \times 32}{4} = 2400K$

29. (d) r.m.s. velocity $V_{rms} = \sqrt{\frac{3RT}{M}}$

i.e., $\frac{V_1}{V_2} = \sqrt{\frac{T_1}{T_2}}$

$\frac{5 \times 10^4}{10 \times 10^4} = \frac{1}{2} = \sqrt{\frac{T_1}{T_2}}$

$\therefore \quad T_2 = 4T_1$

30. (a) $\frac{P_1V_1}{T_1} = \frac{P_2V_2}{T_2}$

$V_2 = \frac{745 \times 1.41 \times 10^4 \times 225}{294 \times 63.1} = 1.274 \times 10^5 L$

31. (b) The volume of the balloon $= \frac{4}{3}\pi r^3$

$= \frac{4}{3} \times \frac{22}{7} \times \left(\frac{21}{2}\right)^3 = 4851 \text{ mL}$

Volume of the cylinder = 2820 mL

Volume of H_2 at NTP

$= \frac{20 \times 2820 \times 273}{300 \times 1} \text{ mL} = 51324 \text{ mL}$

After filling the cylinder will have H_2 equal to its volume = 2820 mL

$\therefore$ Volume of H_2 for filling balloons

= 51324 – 2820 = 48504 mL

Hence no. of balloon to be filled

$= \frac{48504}{4851} \approx 10$

32. (c) When vessels are joined the volume is doubled and pressure is reduced to half \ Pressure of mixture

$= \frac{200}{2} + \frac{400}{2} = 300 \text{ mm}$

33. (c) Due to small size of these species (H_2 and He) intermolecular interactions (van der Waal forces) are very low, therefore it is difficult to compress these.

34. (c) More will be critical temperature easier is the liquifaction of the gas. Hence correct order will be

$He < H_2 < N_2 < O_2$

35. (a) $u : v : \alpha = \sqrt{\frac{3RT}{M}} : \sqrt{\frac{8RT}{\pi M}} : \sqrt{\frac{2RT}{M}}$

$= \sqrt{3} : \sqrt{\frac{8}{\pi}} : \sqrt{2}$

36. (a) Given $P = \frac{nRT}{V - nb} - a\left(\frac{n}{V}\right)^2$

Which can also be written as

$\left[P + \frac{n^2 a}{V^2}\right](V - nb) = nRT$

At low pressure and high temperature the effect of $\frac{a}{V^2}$ and b is negligible hence $PV = nRT$.

37. (c) At any given pressure, graph of volume vs temperature (in °C) is a straight line and on extending to zero volume each line intercepts the temperature axis at –273.15° C.

38. (d) Given, $P_1 = 15$ atm, $P_2 = 60$ atm
$V_1 = 76\ cm^3$, $V_2 = 20.5\ cm^3$.
If the gas is an ideal gas, then according to Boyle's law, it must follow the equation,
$P_1V_1 = P_2V_2$
$P_1 \times V_1 = 15 \times 76 = 1140$
$P_2 \times V_2 = 60 \times 20.5 = 1230$
$\therefore P_1V_1 \neq P_2V_2$
∴ The gas behaves non-ideally.
The given information is not sufficient to comment on other statements.

39. (c) Number of moles, temperature and volume are same.

40. (b)

41. (c) The correct order of viscosity of the given liquids is dimethyl ether < methyl alcohol < water < glycerol.

42. (b) $\frac{P_1V_1}{T_1} = \frac{P_2V_2}{T_2}$

$$\frac{P \times 100}{T} = \frac{1.5P \times V_2}{\frac{4T}{3}}$$

$V_2 = 88.9$ mL $= 88.9$ cc

43. (c) $P_1V_1 = P_2V_2$
$V_1 = V$, $V_2 = V/10$

$$\frac{5.3 \text{ atm}}{P_2} = \frac{\frac{V}{100}}{V};\ P_2 = 5.3 \times 100 = 53 \text{ atm}$$

44. (b) $200 = \sqrt{\frac{2RT}{2\times10^{-3}}}$; $RT = 40$

$$\text{Average K.E.} = \frac{3}{2}nRT = \frac{3}{2}\times\frac{8}{2}\times40 = 240 \text{ J}$$

45. (a)

$$\begin{matrix} CH_2COO^- \\ | \\ CH_2COO^- \end{matrix} + 2H_2O \longrightarrow \begin{matrix} CH_2 \\ \| \\ CH_2 \end{matrix} + 2CO_2 + H_2 + 2OH^-$$

anode cathode

Total equivalent of
$C_2H_4 + CO_2 + H_2 = 0.2 + 0.2 + 0.2 = 0.6$
Total moles of gases

$$= \frac{0.2}{2} + \frac{0.2}{1} + \frac{0.2}{2} = 0.4$$

$$V = \frac{nRT}{P} = \frac{0.4\times0.0821\times273}{1} = 8.96\,L.$$

46. (a) $n_{N_2} = 53.57$ mole

$$\frac{P_1}{P_2} = \frac{n_1}{n_2}$$

$$n_2 = \frac{1.8}{200} = 0.009\times53.57 = 0.48213$$

n_{N_2} escape $= 53.087$

$$\text{Volume} = \frac{nRT}{P} = 1264 \text{ L}$$

47. (c) $\frac{r_{CH_3OCH_3}}{r_{CH_4}} = \sqrt{\frac{M_{CH_4}}{M_{CH_3OCH_3}}} \times \frac{p_{CH_3OCH_3}}{p_{CH_4}}$

$$= \sqrt{\frac{16}{46}} \times \frac{0.8}{0.2} = 2.36 : 1$$

48. (d) Volume of alveoli $= \frac{4}{3}\pi R^3$

$$= \frac{4}{3}\times3.14\times(5\times10^{-3})^3 \text{ mL}$$

$$= 5.23\times10^{-7} \text{ mL} = 5.23\times10^{-7} \text{ L}$$

$$n_2 = \frac{14}{100}\times n_{air} = 0.14\times\frac{PV}{RT}$$

$$= 0.14\times1\times\frac{5.23\times10^{-10}}{0.0821\times310}$$

$$= 2.87\times10^{-12}$$

Molecules $= 2.87 \times 6.02 \times 10^{23} \times 10^{-12}$
$= 1.7 \times 10^{12}$

49. (b) The graph is a straight line parallel to the axis of pressure with intercept

$$= PV_m = RT = 0.082 \text{ atm L K}^{-1}\text{mol}^{-1} \times 300\text{K}$$

$$= 24.6 \text{ atm L mol}^{-1}.$$

50. (b) P, V and R are constant
$\therefore n_1T_1 = n_2T_2, 100\times300 = n_2\times500, : n_2 = 60$
Air escaped is 40%.

51. (d) One mole of a substance contains the number of molecules which is independent of pressure.

52. (c) Molar mass ↑, 'a' increases

size of molecule ↑, 'b' increase

b(L/mol)	a (bar. L²/mol²)
$H_2 \longrightarrow 0.02661$	$CH_4 \longrightarrow 2.25$
$He \longrightarrow 0.0237$	$O_2 \longrightarrow 1.36$
$O_2 \longrightarrow 0.03183$	$H_2 \longrightarrow 0.244$
$CO_2 \longrightarrow 0.04267$	

53. (d) All of the given statements are correct for hydrogen bond.

54. (a) $V_1 = 2$ L, $T_2 = (26.1 + 273)$ K ≈ 299.1 K, $V_2 = ?$

$T_1 = (23.4 + 273)$ K $= 296.4$ K

From Charle's law,

$$\frac{V_1}{T_1} = \frac{V_2}{T_2} \Rightarrow V_2 = \frac{V_1 T_2}{T_1}$$

$$\Rightarrow V_2 = \frac{2L \times 299.1K}{296.4K} = 2L \times 1.009$$

$$= 2.018\,L$$

55. (a) Given

$P_1 = 1.5$ bar, $T_1 = 273 + 15 = 288$ K, $V_1 = V$

$P_2 = 1.0$ bar, $T_1 = 273 + 25 = 298$K, $V_2 = ?$

$$\frac{P_1 V_1}{T_1} = \frac{P_2 V_2}{T_2}$$

$$\frac{1.5 \times V}{288} = \frac{1 \times V_2}{298}$$

$V_2 = 1.55$ V i.e., volume of bubble will be almost 1.6 time to initial volume of bubble.

56. (d) By ideal gas equation,

$$P_1 V = n_1 RT$$

$$n_1 \propto P_1 \text{ and } n_2 \propto P_2$$

$$\frac{n_1}{n_2} = \frac{P_1}{P_2} \Rightarrow \frac{n_1}{n_2} = \frac{170}{570} = 0.30$$

57. (b) n, T same hence $P \propto \frac{1}{V}$,

$V_1 = 1000\ cm^3$

$V_2 = \pi (10)^2 \times 10 = 1000\pi\ cm^3$

$V_3 = \frac{4}{3}\pi (10)^3 = \frac{4}{3}\pi\, 1000\ cm^3$

∴ Pressure of the gas is minimum in (III) container, Pressure of the gas is maximum in (I),

58. (b) $\overline{KE} = \frac{3}{2}\frac{n}{N_o}RT$. Average $\overline{KE}$ is the same for all gaseous molecules

59. (d) Molar ratio of N_2 and CO is 3 : 2, i.e., 300 bar and 200 bar, respectively.

$$\frac{n_{N_2}}{n_{CO}} = \sqrt{\frac{m_{CO}}{mN_2}} \times \frac{P_{N_2}}{P_{CO}}$$

$$= \frac{300}{200} = \frac{3}{2}$$

60. (d) $U_1 \propto \sqrt{\frac{1}{m_1}}$ and $U_2 \propto \sqrt{\frac{1}{m_2}}$

$$\therefore \frac{U_1^2}{U_2^2} = \frac{m_2}{m_1} \qquad \therefore m_1 U_1^2 = m_2 U_2^2$$

6 Thermodynamics

1. Among the following the state function(s) is (are)
(i) Internal energy
(ii) Irreversible expansion work
(iii) Reversible expansion work
(iv) Molar enthalpy
(a) (ii) and (iii) (b) (i), (ii) and (iii)
(c) (i) and (iv) (d) (i) only

2. If a reaction involves only solids and liquids which of the following is true ?
(a) $\Delta H < \Delta E$ (b) $\Delta H = \Delta E$
(c) $\Delta H > \Delta E$ (d) $\Delta H = \Delta E + \Delta nRT$

3. One mole of a non-ideal gas undergoes a change of state (2.0 atm, 3.0 L, 95 K) $\rightarrow$ (4.0 atm, 5.0 L, 245 K) with a change in internal energy, ΔU = 30.0 L atm. The change in enthalpy ΔH of the process in L atm is. Tricky
(a) 40.0
(b) 42.3
(c) 44.0
(d) Not defined because pressure is not constant

4. If bond enthalpies of $N \equiv N$, $H - H$ and $N - H$ bonds are x_1, x_2 and x_3 respectively, ΔH_f^o for NH_3 will be
(a) $x_1 + 3x_2 - 6x_3$ (b) $1/2x_1 + 3/2x_2 - 3x_3$
(c) $3x_3 - 1/2x_1 - 3/2x_2$ (d) $6x_3 - x_1 - 3x_2$

5. Heat of neutralization of a strong acid HA and a weaker acid HB with KOH are – 13.7 and – 12.7 k cal mol^{-1}. When 1 mole of KOH was added to a mixture containing 1 mole each of HA and HB, the heat change was – 13.5 kcal. In what ratio is the base distributed between HA and HB. Toughnut
(a) 3 : 1 (b) 1 : 3 (c) 4 : 1 (d) 1 : 4

6. A gas expands adiabatically at constant pressure such that $T \propto V^{-1/2}$. The value of γ ($C_{p,m}/C_{v,m}$) of the gas will be :
(a) 1.30 (b) 1.50 (c) 1.70 (d) 2

7. One mole of an ideal gas at 300 K is expanded isothermally from an initial volume of 1 litre to 10 litres. The ΔE for this process is (R = 2 cal mol^{-1} K^{-1})
(a) 163.7 cal (b) zero
(c) 1381.1 cal (d) 9 lit atm

8. The latent heat of vapourization of a liquid at 500 K and 1 atm pressure is 10.0 kcal/mol. What will be the change in internal energy (ΔU) of 3 moles of liquid at the same temperature
(a) 13.0 kcal/mol (b) –13.0 kcal/mol
(c) 27.0 kcal (d) –7.0 kcal/mol

9. Equal volumes of methanoic acid and sodium hydroxide are mixed. If x is the heat of formation of water, then heat evolved on neutralization is
(a) more than x (b) equal to x
(c) less than x (d) twice x.

10. A cyclic process ABCD is shown in P–V diagram for an ideal gas. Which of the following diagram represents the same process?

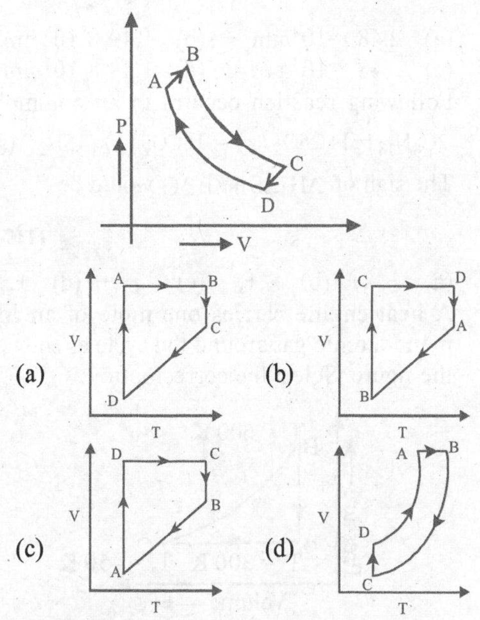

11. What is the value of change in internal energy at 1 atm in the process? Critical Thinking

$$H_2O(l, 323\ K) \longrightarrow H_2O(g, 423\ K)$$

Given : $C_{v,m}$ $(H_2O, l) = 75.0\ JK^{-1}\ mol^{-1}$;

$C_{p,m}$ $(H_2O, g) = 33.314\ JK^{-1}\ mol^{-1}$

ΔH_{vap} at 373 K = 40.7 kJ/mol

(a) 42.91 kJ/mol (b) 43086 kJ/mol
(c) 42.6 kJ/mol (d) 49.6 kJ/mol

12. Which of the following statements/relationships is **not** correct in thermodynamic changes?
(a) $\Delta U = 0$ (isothermal reversible expansion of a gas)
(b) $w = -nRT \ln \frac{V_2}{V_1}$ (isothermal reversible expansion of an ideal gas)
(c) $w = nRT \ln \frac{V_2}{V_1}$ (isothermal reversible expansion of an ideal gas)
(d) For a system of constant volume heat involved directly changes to internal energy.

13. The heat of atomization of methane and ethane are 360 kJ/mol and 620 kJ/mol, respectively. The longest wavelength of light capable of breaking the C – C bond is (Avogadro number $= 6.02 \times 10^{23}\ mol^{-1}$, $h = 6.62 \times 10^{-34}$ J s) :

Toughnut

(a) 2.48×10^4 nm (b) 1.49×10^3 nm
(c) 2.48×10^3 nm (d) 1.49×10^4 nm

14. Following reaction occurs in an automobile $2C_8H_{18}(g) + 25O_2(g) \rightarrow 16CO_2(g) + 18H_2O(g)$. The sign of $\Delta H, \Delta S$ and ΔG would be

(a) +,–,+ (b) –,+,– (c) –,+,+ (d) +,+,–

15. A heat engine carries one mole of an ideal mono-atomic gas around the cycle as shown in the figure. Select the correctoption:

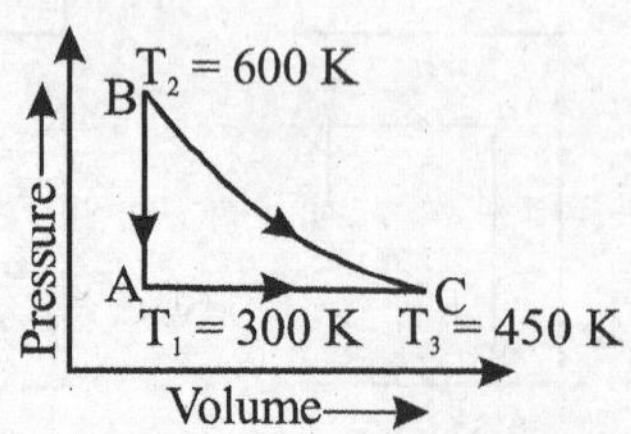

(a) $q_{AB} = 450$ R and $q_{CA} = -450$ R
(b) $q_{AB} = 450$ R and $q_{CA} = -225$ R
(c) $q_{AB} = 450$ R and $q_{CA} = -375$ R
(d) $q_{AB} = 375$ R and $q_{CA} = -450$ R

16. The standard enthalpy of formation of NH_3 is – 46.0 kJ/mol. If the enthalpy of formation of H_2 from its atoms is – 436 kJ/mol and that of N_2 is – 712 kJ/mol, the average bond enthalpy of N - H bond in NH_3 is:
(a) – 1102 kJ/mol (b) – 964 kJ/mol
(c) + 352 kJ/mol (d) + 1056 kJ/mol

17. The standard enthalpy of formation ($\Delta_f H^\circ_{298}$) for methane, CH_4 is – 74.9 kJ mol^{-1}. In order to calculate the average energy given out in the formation of a C – H bond from this it is necessary to know which one of the following?
(a) The dissociation energy of the hydrogen molecule, H_2.
(b) The first four ionisation energies of carbon.
(c) The dissociation energy of H_2 and enthalpy of sublimation of carbon (graphite).
(d) The first four ionisation energies of carbon and electron affinity of hydrogen.

18. Assuming that water vapour is an ideal gas, the internal energy change (ΔU) when 1 mol of water is vapourised at 1 bar pressure and 100°C, (given : molar enthalpy of vapourisation of water at 1 bar and 373 K = 41 kJ mol^{-1} and $R = 8.3\ J\ mol^{-1}\ K^{-1}$) will be
(a) 41.00 kJ mol^{-1} (b) 4.100 kJ mol^{-1}
(c) 3.7904 kJ mol^{-1} (d) 37.904 kJ mol^{-1}

19. An ideal gas expands against a constant external pressure of 2.0 atmosphere from 20 litre to 40 litre and absorbs 10 kJ of heat from surrounding. What is the change in internal energy of the system?
(given : 1atm-litre = 101.3 J)
(a) 4052 J (b) 5948 J
(c) 14052 J (d) 9940 J

20. For which one of the processes represented by the following equations the enthalpy (heat) change is likely to be negative
(a) $Cl^-(g) + aq \rightarrow Cl^-(aq)$
(b) $Cl(g) \rightarrow Cl^+(g) + e^-$
(c) $1/2Cl_2(g) \rightarrow Cl(g)$
(d) $Cl_2(l) \rightarrow Cl_2(g)$

21. The 'thermite reaction' involving the reaction between ferric oxide and metallic aluminium produces molten iron. Given that

$2Al + 3/2O_2 \rightarrow Al_2O_3; \Delta H_1 = -400$ kcal/ mol

$2Fe + 3/2O_2 \rightarrow Fe_2O_3; \Delta H_2 = -200$ kcal/mol.

What is ΔH for the formation of 1 mole of iron?

(a) −100 kcal (b) −200 kcal
(c) +100 kcal (d) +200 kcal

22. Given :

(I) $H_2(g) + \frac{1}{2}O_2(g) \rightarrow H_2O(l);$

$\Delta H°_{298K} = -285.9 \text{kJ mol}^{-1}$

(II) $H_2(g) + \frac{1}{2}O_2(g) \rightarrow H_2O(g);$

$\Delta H°_{298K} = -241.8 \text{kJ mol}^{-1}$

The molar enthalpy of vapourisation of water will be :

(a) 241.8 kJ mol^{-1} (b) 22.0 kJ mol^{-1}
(c) 44.1 kJ mol^{-1} (d) 527.7 kJ mol^{-1}

23. Given

Reaction	**Energy Change (in kJ)**
$Li(s) \rightarrow Li(g)$	161
$Li(g) \rightarrow Li^+(g)$	520
$\frac{1}{2}F_2(g) \rightarrow F(g)$	77
$F(g) + e^- \rightarrow F^-(g)$	(Electron gain enthalpy)
$Li^+(g) + F^-(g) \rightarrow LiF(s)$	− 1047
$Li(s) + \frac{1}{2}F_2(g) \rightarrow LiF(s)$	− 617

Based on data provided, the value of electron gain enthalpy of fluorine would be :

(a) − 300 kJ mol^{-1} (b) − 350 kJ mol^{-1}
(c) − 328 kJ mol^{-1} (d) − 228 kJ mol^{-1}

24. Fixed mass of an ideal gas contained in a 24.63 L sealed rigid vessel at 1 atm is heated from −73°C to 27°C. Calculate change in gibb's energy if entropy of gas is a function of temperature as $S = 2 + 10^{-2}T$ (J/K): (Use 1 atm L = 0.1 kJ)

(a) 1231.5 J (b) 1281.5 J
(c) 781.5 J (d) 0

25. For the reaction taking place at certain temperature $NH_2COONH_4(s) \rightleftharpoons 2NH_3(g) + CO_2(g)$, if equilibrium pressure is 3X bar then $\Delta_r G°$ would be

Tricky

(a) − RT ln 9 − 3RT ln X
(b) RT ln 4 − 3RT ln X
(c) − 3RT ln X
(d) None of these

26. Standard Gibb's free energy change for isomerization reaction

cis-2-pentene ⇌ trans-2-pentene

is − 3.67 kJ/mol at 400 K. If more trans-2-pentene is added to the reaction vessel, then

(a) more cis-2-pentene is formed
(b) equilibrium remains unaffected
(c) additional trans-2-pentene is formed
(d) equilibrium is shifted in forward direction

27. The dissolution of KCl in water is endothermic yet it dissolves in water spontaneously. Which one of the following best explains this behaviour?

(a) Endothermic processes are energetically favoured.
(b) The electrostatic bonds between the ions are not too strong.
(c) Energy changes have nothing to do with the dissolution processes.
(d) The entropy driving force causes the dissolution.

28. Hydrogen has an ionisation energy of 1311 kJ mol^{-1} and for chlorine it is 1256 kJ mol^{-1}. Hydrogen forms H^+ (aq) ions but chlorine does not form Cl^+ (aq) ions because

(a) H^+ has lower hydration enthalpy
(b) Cl^+ has lower hydration enthalpy
(c) Cl has high electron affinity
(d) Cl has high electronegativity

29. At 25°C, when 1 mole of $MgSO_4$ was dissolved in water, the heat evolved was found to be 91.2 kJ. One mole of $MgSO_4 \cdot 7H_2O$ on dissolution gives a solution of the same composition accompanied by an absorption of 13.8 kJ. The enthalpy of hydration, i.e., ΔH_h for the reaction

$MgSO_4(s) + 7H_2O(l) \longrightarrow MgSO_4.7H_2O(s)$

is:

Toughnut

(a) − 105 kJ/mol (b) − 77.4 kJ/mol
(c) 105 kJ/mol (d) None of these

30. Calculate the heat produced (in kJ) when 224 g of CaO is completely converted to $CaCO_3$ by reaction with CO_2 at 27°C in a container of fixed volume. **Critical Thinking**

Given : $\Delta H^{\circ}_f (CaCO_3, s) = -1207$ kJ/mol; $\Delta H^{\circ}_f (CaO, s) = -635$ kJ/mol, $\Delta H^{\circ}_f (CO_2, g) = -394$ kJ/mol; [Use R = 8.3 JK^{-1} mol^{-1}]

(a) 702.04 kJ (b) 721.96 kJ
(c) 712 kJ (d) 721 kJ

31. Given that:

(i) $\Delta_f H^{\circ}$ of N_2O is 82 kJ mol^{-1}

(ii) Bond energies of N ≡ N, N = N, O = O and N = O are 946, 418, 498 and 607 kJ mol^{-1} respectively,

The resonance energy of N_2O is :

(a) –88kJ (b) –66kJ (c) –62kJ (d) –44kJ

32. The difference between the reaction enthalpy change ($\Delta_r H$) and reaction internal energy change ($\Delta_r U$) for the reaction:

$$2C_6H_6(l) + 15O_2(g) \longrightarrow 12CO_2(g) + 6H_2O(l)$$

at 300 K is (R = 8.314 J $mol^{-1}K^{-1}$)

(a) 0 J mol^{-1} (b) 2490 J mol^{-1}
(c) –2490 J mol^{-1} (d) –7482 J mol^{-1}

33. From given following equations and ΔH° values, determine the enthalpy of reaction at 298 K for the reaction :

$$C_2H_4(g) + 6F_2(g) \longrightarrow 2\,CF_4(g) + 4HF(g)$$

$$H_2(g) + F_2(g) \longrightarrow 2HF(g);\ \Delta H^{\circ}_1 = -537\ kJ$$

$$C(s) + 2F_2(g) \longrightarrow CF_4(g);\ \Delta H^{\circ}_2 = -680\ kJ$$

$$2C(s) + 2H_2(g) \longrightarrow C_2H_4(g);\ \Delta H^{\circ}_3 = 52\ kJ$$

(a) – 1165 (b) – 2486
(c) + 1165 (d) + 2486

34. Substance $A_2B(g)$ can undergoes decomposition to form two set of products : **Tricky**

$A_2B(g)$ → $A_2(g) + B(g)$; ΔH° = 40 kJ / mol

$A_2B(g)$ → $A(g) + AB(g)$; Δ H° = 50 kJ / mol

If the molar ratio of A_2 (g) to A (g) is 5 : 3 in a set of product gases, then the energy involved in the decomposition of 1 mole of A_2B (g) is :

(a) 48.75 kJ/mol (b) 43.73 kJ/mol
(c) 46.25 kJ/mol (d) None of these

35. 36 mL of pure water takes 100 sec to evaporate from a vessel and heater connected to an electric source which delivers 806 watt. The ΔH_{vap} of H_2O is :

(a) 40.3 kJ/mol (b) 43.2 kJ/mol
(c) 4.03 kJ/mol (d) None of these

36. For the combustion reaction at 298 K

$$2Ag(s) + 1/2O_2(g) \longrightarrow 2Ag_2O(s)$$

Which of the following alternatives is correct? **Tricky**

(a) ΔH = ΔU
(b) ΔH > ΔU
(c) ΔH < ΔU
(d) ΔH and ΔU has no relation with each other

37. A gas undergoes change from state A to state B. In this process, the heat absorbed and work done by the gas is 5 J and 8 J, respectively. Now gas is brought back to A by another process during which 3 J of heat is evolved. In this reverse process of B to A :

(a) 10 J of the work will be done by the gas.
(b) 6 J of the work will be done by the gas.
(c) 10 J of the work will be done by the surrounding on gas.
(d) 6 J of the work will be done by the surrounding on gas.

38. The amount of energy released when 20 mL of 0.5 M NaOH are mixed with 100 mL of 0.1 M HCl is x kJ. The heat of neutralisation (in kJ mol^{-1}) is

(a) –100x (b) –50x
(c) + 100 x (d) +50x

39. Which of the following value of $\Delta_f H^{\circ}$ represent that the product is least stable ?

(a) –94.0 kcal mol^{-1} (b) –231.6 kcal mol^{-1}
(c) +21.4 kcal mol^{-1} (d) +64.8 kcal mol^{-1}

40. Enthalpy of neutralisation of HCl with NaOH is x. The heat evolved when 500 mL of 2 N HCl are mixed with 250 mL of 4N NaOH will be.

(a) 500 x (b) 100x (c) x (d) 10 x

41. The enthalpy of neutralisation of NH_4OH with HCl is –51.46 kJ mol^{-1} and the enthalpy of neutralisation of NaOH with HCl is –55.90 kJ mol^{-1}. The enthalpy of ionisation of NH_4OH is **Toughnut**

(a) – 107.36 kJ mol^{-1} (b) – 4.44 kJ mol^{-1}
(c) + 107.36 kJ mol^{-1} (d) + 4.44 kJ mol^{-1}

42. The molar heat capacity (C_p) of CD_2O is 10 cals at 1000 K. The change in entropy associated with cooling of 32 g of CD_2O vapour from 1000 K to 100 K at constant pressure will be: (D = deuterium, atomic mass = 2 u)
(a) 23.03 cal deg^{-1} (b) – 23.03 cal deg^{-1}
(c) 2.303 cal deg^{-1} (d) – 2.303 cal deg^{-1}

43. The (S°) of the following substances are:
CH_4 (g) 186.2 JK^{-1} mol^{-1}
O_2 (g) 205.2 JK^{-1} mol^{-1}
CO_2 (g) 213.6 JK^{-1} mol^{-1}
H_2O (g) 69.9.JK^{-1} mol^{-1}
The entropy change ($\Delta S°$) for the reaction $CH_4(g) + 2O_2(g) \rightarrow CO_2(g) + 2H_2O(l)$ is:

(a) – 312.5 J K^{-1} mol^{-1}
(b) – 242.8 J K^{-1} mol^{-1}
(c) – 108.1 J K^{-1} mol^{-1}
(d) – 37.6 J K^{-1} mol^{-1}

44. The enthalpy of the reaction forming PbO according to the following equation is 438 kJ. What heat energy (kJ) is released in formation of 22.3 g PbO(s)?
(Atomic masses : Pb = 207, O = 16.0)

$$2Pb(s)+O_2(g) \longrightarrow 2PbO(s)$$

(a) 21.9 (b) 28.7 (c) 14.6 (d) 34.2

45. The species which by definition has zero standard molar enthalpy of formation at 298 K is
(a) Br_2 (g) (b) Cl_2 (g)
(c) H_2O (g) (d) CH_4 (g)

46. What is Δn_{gas} for the combustion of 1 mole of benzene, when both the reactants and the products are at 298 K ?
(a) 0 (b) 1/2 (c) 3/2 (d) –3/2

47. The enthalpy of neutralisation of NH_4OH and CH_3COOH is – 10.5 kcal mol^{-1} and enthalpy of neutralisation of CH_3COOH with strong base is – 12.5 kcal mol^{-1}. The enthalpy of ionisation of NH_4OH will be
(a) 3.2 kcal mol^{-1} (b) 2.0 kcal mol^{-1}
(c) 3.0 kcal mol^{-1} (d) 4.0 kcal mol^{-1}

48. If for a given substance melting point is T_B and freezing point is T_A, then correct variation shown by, graph between entropy change and temperature is

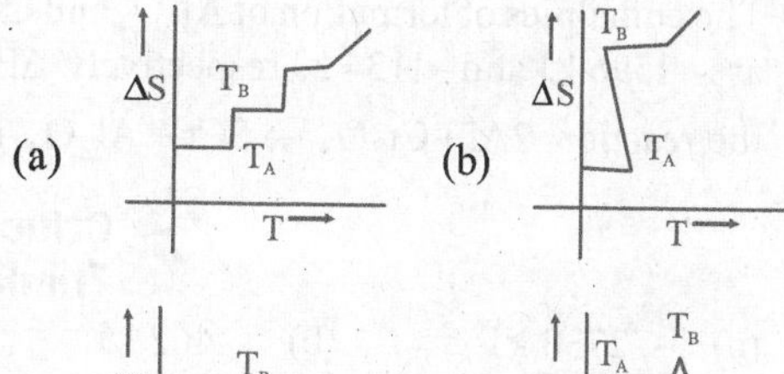

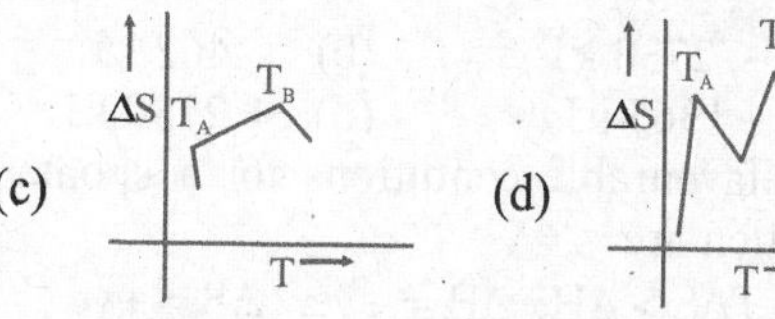

49. What is the free energy change for the conversion of 1 mole of water into steam at 373.2 K. The heat of vaporization (ΔH_v) of water of 373.2 K is 9.1 kcal mol^{-1}. The entropy change is 25.5 cal/mol deg.
(a) – 401.6 cal/mol (b) – 416.6 cal/mol
(c) 516.5 cal/mol (d) – 516.5 cal/mol

50. A certain reaction is non spontaneous at 298K. The entropy change during the reaction is 121 JK^{-1}. Is the reaction is endothermic or exothermic ? The minimum value of ΔH for the reaction is
(a) endothermic, ΔH = 36.06 kJ
(b) exothermic, ΔH = – 36.06 kJ
(c) endothermic, ΔH = 60.12 kJ
(d) exothermic, ΔH = – 60.12 kJ

51. The entropy of a sample of a certain substance increases by 0.836 J K^{-1} on adding reversibly 0.3344 J of heat at constant temperature. The temperature of the sample is:
(a) 2.5 K (b) 0.3 K
(c) 0.016 K (d) 0.4 K

52. One mole of an ideal gas is expanded isothermally and reversibly to half of its initial pressure. ΔS for the process in J K^{-1} mol^{-1} is [$\ell n\ 2 = 0.693$ and $R = 8.314$, J/(mol K)]
(a) 6.76 (b) 5.76 (c) 10.76 (d) 8.03

53. The enthalpy of hydrogenation of cyclohexene is – 119.5 kJ mol^{-1}. If resonance energy of benzene is –150.4 kJ mol^{-1}, its enthalpy of hydrogenation would be **Tricky**
(a) – 208.1 kg mol^{-1} (b) – 269.9 kg mol^{-1}
(c) – 358.5 kg mol^{-1} (d) – 508.9 kg mol^{-1}

54. The enthalpies of formation of Al_2O_3 and Cr_2O_3 are –1596 kJ and –1134 kJ respectively. ΔH for the reaction $2Al + Cr_2O_3 \rightarrow 2Cr + Al_2O_3$ is

(a) – 2730 kJ (b) – 462 kJ
(c) – 1365 kJ (d) + 2730 kJ

55. The favourable conditions for a spontaneous reaction are
(a) $T\Delta S > \Delta H$, $\Delta H = +ve$, $\Delta S = +ve$
(b) $T\Delta S > \Delta H$, $\Delta H = +ve$, $\Delta S = -ve$
(c) $T\Delta S = \Delta H$, $\Delta H = -ve$, $\Delta S = -ve$
(d) $T\Delta S = \Delta H$, $\Delta H = +ve$, $\Delta S = +ve$

56. The standard Gibb's free energy change, $\Delta G°$ is related to equilibrium constant, K_P as
(a) $K_P = -RT \ln \Delta G°$ (b) $K_P = \left[\frac{e}{RT}\right]^{\Delta G°}$
(c) $K_P = -\frac{\Delta G}{RT}$ (d) $K_P = e^{-\Delta G°/RT}$

57. Which of the following salts should cause maximum cooling when 1 mole of it is dissolved in the same volume of water ?
(a) NaCl; $\Delta H° = 5.35$ kJ mol^{-1}
(b) KNO_3; $\Delta H° = 53.5$ kJ mol^{-1}
(c) KOH; $\Delta H° = -56.0$ kJ mol^{-1}
(d) HBr; $\Delta H° = -83.3$ kJ mol^{-1}

58. The molar enthalpies of combustion of isobutane and n-butane are –2870 kJ mol^{-1} and –2875 kJ mol^{-1} respectively at 298 K and 1 atm. Calculate $\Delta H°$ for the conversion of 1 mole of n-butane to 1 mole of isobutane
(a) –8 kJ mol^{-1} (b) +8 kJ mol^{-1}
(c) –5748 kJ mol^{-1} (d) +5748 kJ mol^{-1}

59. The enthalpy of atomisation of CH_4 and C_2H_6 are 360 and 620 kcal mol^{-1} respectively. The C – C bond energy is expected to be

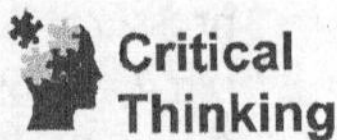

(a) 210 kcal mol^{-1} (b) 80 kcal mol^{-1}
(c) 130 kcal mol^{-1} (d) 180 kcal mol^{-1}

60. Boron can undergo the following reactions with the given enthalpy changes :

$$2B\,(s) + \frac{3}{2}O_2\,(g) \longrightarrow B_2O_3\,(s);\ \Delta H = -1260\ \text{kJ}$$

$$2B\,(s) + 3H_2\,(g) \longrightarrow B_2H_6\,(g);\ \Delta H = 30\ \text{kJ}$$

Assume no other reactions are occurring.

If in a container (operating at constant pressure) which is isolated from the surrounding, mixture of H_2 (gas) and O_2 (gas) are passed over excess of B(s), then calculate the molar ratio ($O_2 : H_2$) so that temperature of the container do not change :
(a) 15 : 3 (b) 42 : 1 (c) 1 : 42 (d) 1 : 84

61. Combustion of sucrose is used by aerobic organisms for providing energy for the life sustaining processes. If all the capturing of energy from the reaction is done through electrical process (non P–V work) then calculate maximum available energy which can be captured by combustion of 34.2 g of sucrose
Given : $\Delta H_{combustion}$ (sucrose) = – 6000 kJ mol^{-1}
$\Delta S_{combustion}$ = 180 J/Kmol and body temperature is 300 K

Toughnut

(a) 600 kJ (b) 594.6 kJ
(c) 5.4 kJ (d) 605.4 kJ

62. The factor of ΔG values is important in metallurgy. The ΔG values for the following reactions at 800°C are given as :

$$S_2(s) + 2O_2(g) \longrightarrow 2SO_2(g);\ \Delta G = -544\ \text{kJ}$$

$$2Zn(s) + S_2(s) \longrightarrow 2ZnS(s);\ \Delta G = -293\ \text{kJ}$$

$$2Zn(s) + O_2(g) \longrightarrow 2ZnO(s);\ \Delta G = -480\ \text{kJ}$$

Then ΔG for the reaction :

$$2ZnS(s) + 3O_2(g) \longrightarrow 2ZnO(s) + 2SO_2(g)$$

will be :

Tricky

(a) –357 kJ (b) –731 kJ
(c) –773 kJ (d) –229 kJ

63. 1 gram equivalent of H_2SO_4 is treated with 112 g of KOH for complete neutralization. Which of the following statementsis correct?
(a) 13.7 kcal of heat is evolved with the formation of 87 g of K_2SO_4, leaving no KOH.
(b) 27.4 kcal of heat is evolved with the formation of 87 g of K_2SO_4, leaving 4 gram equivalent of KOH.
(c) 15.7 kcal of heat is evolved with the formation of 1 gram equivalent of K_2SO_4, leaving 56 g of KOH.
(d) 13.7 kcal of heat is evolved with the formation of 87g of K_2SO_4, leaving 1 gram equivalent of KOH.

64. From the following data ΔH of the following reactions

$C(s) + \frac{1}{2}O_2(g) \longrightarrow CO(g);$ $\Delta H = -110$ kJ and

$C(s) + H_2O(g) \longrightarrow CO(g) + H_2(g);$ $\Delta H = 132$ kJ

Calculate the mole composition of the mixture of steam and oxygen on being passes over coke at 1273 K, keeping the temperature constant.

(a) 1 : 0.6 (b) 0.6 : 1 (c) 2 : 3 (d) 3 : 2

65. Gasoline has an enthalpy of combustion 24000 kJ/ gallon. When gasoline burns in an automobile engine, approximately 30% of the energy released is used to produce mechanical work. The remainder is lost as heat transfer to the engine's cooling system. As a start on estimating how much heat transfer is required, calculate what mass of water could be heated from 25°C to 75°C by the combustion of 1.0 gallon of gasoline in an automobile?

Toughnut

(Given : C (H_2O) = 4.18 J/g°C)

(a) 34.45 kg (b) 80.383 kg
(c) 22 kg (d) 224 kg

66. 2 moles of an ideal gas at 27 °C temperature is expanded reversibly from 2 L to 20 L. Find entropy change (R = 2 cal/mol K)

(a) 92.1 (b) 0 (c) 4 (d) 9.2

67. One mole of solid iron was vaporized in an oven at its boiling point of 3433 K and enthalpy of vaporization of iron is 344.3 kJ mol^{-1}. The value of entropy vaporization (in J mol^{-1}) of iron is

(a) 100 (b) 10 (c) – 100 (d) 110

68. A reaction is spontaneous at low temperature but non-spontaneous at high temperature. Which of the following is true for the reaction?

(a) $\Delta H > 0, \Delta S > 0$ (b) $\Delta H < 0, \Delta S > 0$
(c) $\Delta H > 0, \Delta S = 0$ (d) $\Delta H < 0, \Delta S < 0$

69. Which of the following pairs of processes is certain to occur in a spontaneous chemical reaction?

(a) Exothermic and increasing disorder
(b) Exothermic and decreasing disorder
(c) Endothermic and increasing disorder
(d) Endothermic and decreasing disorder

70. For the reaction

$$2C_6H_5CO_2H(s) + 15O_2(g) \longrightarrow 14CO_2(g) + 6H_2O(g)$$

$\Delta U^o = -772.7$ kJ mol^{-1} at 298 K. Calculate ΔH^o

(a) +760.3 kJ mol^{-1} (b) –760.3 kJ mol^{-1}
(c) +670.3 kJ mol^{-1} (d) –790.3 kJ mol^{-1}

71. ΔH_f^o of water is – 285.8 kJ mol^{-1}. If enthalpy of neutralisation of monoacidic strong base is -57.3 kJmol^{-1}. ΔH_f^o of OH^- ion will be

(a) – 114.25 kJ mol^{-1} (b) 114.25 kJ mol^{-1}
(c) 228.5 kJ mol^{-1} (d) – 228.5 kJ mol^{-1}

72. The heat of atomization of $PH_3(g)$ is 228 kcal mol^{-1} and that of $P_2H_4(g)$ is 335 kcal mol^{-1}. The energy of the P–P bond is

(a) 102 kcal mol^{-1} (b) 51 kcal mol^{-1}
(c) 26 kcal mol^{-1} (d) 204 kcal mol^{-1}

73. For the reaction

$$CO(g) + \frac{1}{2}O_2(g) \longrightarrow CO_2(g),\ \Delta H,\ \text{and}\ \Delta S$$

are –283 kJ and –87 JK^{-1}, respectively. It was intended to carry out this reaction at 1000, 1500, 3000 and 3500 K. At which of these temperatures would this reaction be thermodynamically spontaneous?

(a) 1500 and 3500 K
(b) 3000 and 3500 K
(c) 1000, 1500 and 3000 K
(d) 1500, 3000 and 3500 K

74. The molar entropies of HI (g) and I (g) at 298 K are 206.5, 114.6, and 180.7 J mol^{-1} K^{-1} respectively. Using the ΔG° given Below, calculate the bond energy of HI.

$$HI(g) \longrightarrow H(g) + I(g);\ \Delta G^\circ = 271.8\ \text{kJ}$$

(a) 282.4 kJ mol^{-1} (b) 298.3 kJ mol^{-1}
(c) 290.1 kJ mol^{-1} (d) 315.4 kJ mol^{-1}

75. What is the equilibrium constant if ATP hydrolysis by water produce standard free energy of –50 kJ/mol under normal body conditions ?

Tricky

(a) 2.66×10^8 (b) 5.81×10^8
(c) 1.18×10^7 (d) 1.98×10^8

76. 0.5 mole each of two ideal gases $A\left(C_{v,m}=\frac{5}{2}R\right)$ and B ($C_{v,m}$ = 3R) are taken in a container and expanded reversibly and adiabatically, during this process temperature of gaseous mixture decreased from 350 K to 250 K. Find ΔH (in cal/mol) for the process:

Toughnut

(a) – 100 R (b) – 137.5 R
(c) – 375 R (d) None of these

77. What is the free energy change 'ΔG' when 1.0 mole of water at 100 °C and 1 atm pressure is converted into steam at 100 °C and 1 atm. pressure

(a) 540 cal (b) – 9800 cal
(c) 9800 cal (d) 0 cal

78. In an irreversible process taking place at constant T and P and in which only pressure-volume work is being done, the change in Gibbs free energy (dG) and change in entropy (dS), satisfy the criteria

(a) $(\Delta S)_{V,E}>0, (\Delta G)_{T,P}<0$

(b) $(\Delta S)_{V,E}=0, (\Delta G)_{T,P}=0$

(c) $(\Delta S)_{V,E}=0, (\Delta G)_{T,P}>0$

(d) $(\Delta S)_{V,E}<0, (\Delta G)_{T,P}<0$

79. For the auto-ionization of water at 25°C,

$H_2O(l) \rightleftharpoons H^+(aq) + OH^-(aq)$

equilibrium constant is 10^{-14}.
What is ΔG° for the process?

(a) $\simeq 8 \times 10^4$ J mol^{-1} (b) $\simeq 3.5 \times 10^4$ J mol^{-1}
(c) $\simeq 2 \times 10^4$ J mol^{-1} (d) None of these

80. For which of the following process, ΔS is negative?

(a) $H_2(g) \longrightarrow 2H(g)$
(b) $2SO_3(g) \longrightarrow 2SO_2(g) + O_2(g)$
(c) $N_2(4l) \xrightarrow{\text{compressed}} N_2(2l)$
(d) C(diamond) ⟶ C(graphite)

81. What is the entropy change (in JK^{-1} mol^{-1}) when one mole of ice is converted into water at 0° C? (The enthalpy change for the conversion of ice to liquid water is 6.0 kJ mol^{-1} at 0 °C)

(a) 21.98 (b) 20.13 (c) 2.013 (d) 2.198

82. Temperature of 5 moles of a gas is decreased by 2K at constant pressure. Indicate the correct statement

Tricky

(a) Work done by gas is = 5 R
(b) Work done over the gas is = 10 R
(c) Work done by the gas = 10 R
(d) Work done = 0

83. An ideal gas occuping a volume of 2dm^3 and a pressure of 5 bar undergoes isothermal and irreversible expansion against external pressure of 1 bar. The final volume of the system and the work involved in the process is

(a) 10 dm^3, 1000 J (b) 8 dm^3, – 800 J
(c) 10 dm^3, – 800 J (d) 10 dm^3, – 1000 J

84. A piston filled with 0.04 mol of an ideal gas expands reversibly from 50.0 mL to 375 mL at a constant temperature of 37.0°C. As it does so, it absorbs 208 J of heat. The values of q and w for the process will be:
(R = 8.314 J/mol K) (ln 7.5 = 2.01)

(a) q = + 208 J, w = – 208 J
(b) q = – 208 J, w = – 208 J
(c) q = – 208 J, w = + 208 J
(d) q = + 208 J, w = + 208 J

85. The heats of neutralisation of CH_3COOH, HCOOH, HCN and H_2S are – 13.2, – 13.4, – 2.9 and – 3.8 kcal per equivalent respectively. Arrange the acids in increasing order of strength

(a) HCOOH > CH_3COOH > H_2S > HCN
(b) CH_3COOH > HCOOH > H_2S > HCN
(c) H_2S > HCOOH > CH_3COOH > HCN
(d) HCOOH > H_2S > CH_3COOH > HCN

86. The enthalpy of neutralisation of a weak acid in 1 M solution with a strong base is – 56.1 kcal mol^{-1}. If the enthalpy of ionisation of acid is 1.5 kcal mol^{-1} and enthalpy of neutralisation of the strong acid with a strong base is – 57.3 kJ eq^{-1}. What is the % ionisation of the weak acid in molar solution (assume the acid is monobasic)

Critical Thinking

(a) 25 (b) 20 (c) 15 (d) 10

87. The free energy change for the following reactions are given below,

$$C_2H_2(g)+\frac{5}{2}O_2(g)\rightarrow 2CO_2(g)+H_2O(l);\Delta G^\circ=-1234kJ$$

$$C(s)+O_2(g)\rightarrow CO_2(g);\Delta G^\circ=-394kJ$$

$$H_2(g)+\frac{1}{2}O_2(g)\rightarrow H_2O(l);\Delta G^\circ=-237kJ$$

What is the standard free energy change for the reaction $H_2(g)+2C(s)\longrightarrow C_2H_2(g)$

(a) -209 kJ (b) -2259 kJ

(c) $+2259$ kJ (d) 209 kJ

88. For the reaction

$$C_3H_8(g)+5O_2(g)\rightarrow 3CO_2(g)+4H_2O(l)$$

at constant temperature, $\Delta H-\Delta E$ is

(a) $-RT$ (b) $+RT$ (c) -3 RT (d) $+3$ RT

89. Among the following, the intensive properties are

(i) molar conductivity
(ii) electromotive force
(iii) resistance
(iv) heat capacity

(a) (i) and (ii) (b) (i), (ii) and (iii)

(c) (i) and (iv) (d) (i) only

90. What is the normal boiling point of mercury? Given : ΔH°_f (Hg, l) = 0; S°(Hg, l) = 77.4 J/K-mol ΔH°_f (Hg, g) = 60.8 kJ/mol; S°(Hg, g) = 174.4 J/K–mol

(a) 624.8 K (b) 626.8 K

(c) 636.8 K (d) None of these

91. When 1 mole of oxalic acid is treated with excess of NaOH in dilute aqueous solution, 106 kJ of heat is liberated. Predict the enthalpy of ionisation of the acid

(a) 4.3 kJ mol^{-1} (b) -4.3 kJ mol^{-1}

(c) 8.6 kJ mol^{-1} (d) -8.6 kJ mol^{-1}

92. The lattice energy of solid NaCl is 180 kcal mol^{-1} and enthalpy of solution is 1 kcal mol^{-1}. If the hydration energies of Na^+ and Cl^- ions are in the ratio 3 : 2, what is the enthalpy of hydration of sodium ion ?

(a) -107.4 kcal mol^{-1} (b) 107.4 kcal mol^{-1}

(c) 71.6 kcal mol^{-1} (d) -71.6 kcal mol^{-1}

93. Standard enthalpy of combustion of CH_4 is -890 kJ mol^{-1} and standard enthalpy of vaporisation of water is 40.5 kJ mol^{-1}. The enthalpy change of the reaction

$$CH_4(g)+2O_2(g)\longrightarrow CO_2(g)+H_2O(g)$$

(a) -809.5 kJ mol^{-1} (b) -890 kJ mol^{-1}

(c) 809 kJ mol^{-1} (d) -971 kJ mol^{-1}

94. Standard enthalpy and standard entropy changes for the oxidation of ammonia at 298 K are -382.64 kJ mol^{-1} and -45.6 JK^{-1} mol^{-1}, respectively. Standard Gibb's energy change for the same reaction at 298 K is

(a) -22.1 kJ mol^{-1} (b) -339.3 kJ mol^{-1}

(c) -439.3 kJ mol^{-1} (d) -523.2 kJ mol^{-1}

95. ΔG in $Ag_2O\rightarrow 2Ag+1/2O_2$ at a certain temperature is -10 kJ mol^{-1}. Pick the correct statement

(a) Ag_2O decomposes to Ag and O_2

(b) Ag and O_2 combines to form Ag_2O

(c) Reaction is in equilibrium

(d) Reaction does not take place

96. For complete combustion of ethanol,

$$C_2H_5OH(l)+3O_2(g)\longrightarrow 2CO_2(g)+3H_2O(l),$$

the amount of heat produced as measured in bomb calorimeter, is 1364.47 kJ mol^{-1} at 25°C. Assuming ideality the enthalpy of combustion, Δ_cH, for the reaction will be: Toughnut

(R = 8.314 kJ mol^{-1})

(a) -1366.95 kJ mol^{-1}

(b) -1361.95 kJ mol^{-1}

(c) -1460.95 kJ mol^{-1}

(d) -1350.50 kJ mol^{-1}

97. Diborane is a potential rocket fuel which undergoes combustion according to the equation

$$B_2H_6(g)+3O_2(s)\longrightarrow B_2O_3(s)+3H_2O(g)$$

Calculate the enthalpy change for the combustion of diborane. Given Critical Thinking

(i) $2B(s)+\frac{3}{2}O_2(g)\longrightarrow B_2O_3(s);$

$\Delta H=-1273$ kJ per mol

(ii) $H_2(g) + \frac{1}{2}O_2(g) \longrightarrow H_2O(l)$;

$\Delta H = -286$ kJ per mol

(iii) $H_2O(l) \longrightarrow H_2O(g)$;

$\Delta H = 44$ kJ per mol

(iv) $2B\,(s) + 3H_2\,(g) \longrightarrow B_2H_6\,(g)$;

$\Delta H = 36$ kJ per mol

(a) + 2035 kJ per mol (b) – 2035 kJ per mol
(c) + 2167 kJ per mol (d) – 2167 kJ per mol

98. What is the amount of heat (in Joules) absorbed by 18 g of water initially at room temperature heated to 100°C ? If 10 g of Cu is added to this water , than decrease in temperature (in Kelvin) of water was found to be? C (p,m) for water 75.32 J/mol K ; C (p,m) for Cu = 24.47 J/mol K.

Toughnut

(a) 5649, 369 (b) 5544, 324
(c) 5278, 342 (d) 3425, 425

99. How many molecules of ATP, undergo hydrolysis to raise the temperature of 180 kg of water which was originally at room temperature by 1°C ? C{P,m} water = 75.32 J/mol/K, ΔH{P} for ATP hydrolysis = 7 kcal/mol

Toughnut

(a) 1.5×10^{25} (b) 2.00×10^{23}
(c) 3.4×10^{25} (d) 4.0×10^{24}

100. Consider the ΔG°_f and ΔH°_f (kJ/mol) for the following oxides. Which oxide can be most easily decomposed to form the metal and oxygen gas?

(a) ZnO ($\Delta G° = -318.4$, $\Delta H° = -348.3$)
(b) Cu_2O ($\Delta G° = -146.0$, $\Delta H° = -168.8$)
(c) HgO ($\Delta G° = -58.5$, $\Delta H° = -90.8$)
(d) PbO ($\Delta G° = -187.9$, $\Delta H° = -217.3$)

Answer KEY

1	(c)	11	(c)	21	(a)	31	(a)	41	(d)	51	(d)	61	(d)	71	(d)	81	(a)	91	(c)
2	(b)	12	(c)	22	(c)	32	(d)	42	(b)	52	(b)	62	(b)	72	(b)	82	(b)	92	(a)
3	(c)	13	(b)	23	(c)	33	(b)	43	(b)	53	(a)	63	(d)	73	(c)	83	(c)	93	(a)
4	(b)	14	(b)	24	(c)	34	(b)	44	(a)	54	(b)	64	(a)	74	(b)	84	(a)	94	(b)
5	(c)	15	(c)	25	(d)	35	(a)	45	(b)	55	(a)	65	(b)	75	(a)	85	(a)	95	(a)
6	(b)	16	(b)	26	(a)	36	(c)	46	(d)	56	(d)	66	(d)	76	(c)	86	(b)	96	(a)
7	(b)	17	(c)	27	(d)	37	(d)	47	(b)	57	(b)	67	(a)	77	(d)	87	(d)	97	(b)
8	(c)	18	(d)	28	(b)	38	(a)	48	(a)	58	(a)	68	(d)	78	(a)	88	(c)	98	(a)
9	(c)	19	(b)	29	(a)	39	(d)	49	(b)	59	(b)	69	(a)	79	(a)	89	(a)	99	(a)
10	(c)	20	(a)	30	(a)	40	(c)	50	(a)	60	(d)	70	(b)	80	(c)	90	(b)	100	(c)

Hints & Solutions

1. **(c)** Internal energy and molar enthalpy are state functions. Work (reversible or irreversible) is a path function.

2. **(b)** $\Delta H = \Delta E + P\Delta V$, for solid and liquid, $\Delta V = 0$ or $\Delta H = \Delta E + \Delta nRT$, for solids and liquids $\Delta n = 0$

3. **(c)** $\Delta H = H_2 - H_1 = (E_2 + P_2V_2) - (E_1 + P_1V_1)$
$$= (E_2 - E_1) + (P_2V_2 - P_1V_1)$$
$$= 30 + (4\times5 - 2\times3) = 44\text{ L atm}$$

4. **(b)** $\frac{1}{2}N_2 + \frac{3}{2}H_2 \longrightarrow NH_3$
$\frac{1}{2}x_1 + \frac{3}{2}x_2 - 3x_3$

5. **(c)** Let x mole of KOH be neutralized by the strong acid HA. Then, moles neutralized by $HB = 1 - x$
Hence, $-13.7 \times x + (-12.7)\times(1-x) = -13.5$
$\Rightarrow \quad x = 0.8; \quad \frac{x}{1-x} = \frac{0.8}{0.2} = \frac{4}{1} \quad 4:1$

6. **(b)** $T \propto \frac{1}{\sqrt{V}}$; $TV^{1/2}$ = constant
For adiabatic process,
$TV^{\gamma-1}$ = constant
$\therefore \gamma - 1 = \frac{1}{2}, \gamma = \frac{3}{2}$

7. **(b)** For a finite change $\Delta E = C_V\Delta T$. For an isothermal process $\Delta T = 0$. Hence $\Delta E = 0$

8. **(c)** $3H_2O(l) \rightarrow 3H_2O(g)$;
$\Delta n = 3$, $\Delta E = \Delta H - \Delta nRT$
$$= 30 - 3\times\frac{2}{1000}\times500 = 27\text{ kcal}$$

9. **(c)** Neutralisation of weak acid with strong base hence < x. Extra heat is utilised to effect the ionisation of weak acid.

10. **(c)** AB : Isobaric expansion
BC : Isothermal expansion
CD : Isochoric
DA : Isothermal compression

11. **(c)** $H_2O(1, 323\text{ K}) \xrightarrow{\Delta U_1} H_2O(1, 373\text{ K})$, $\downarrow \Delta U_2$
$H_2O(g, 323\text{ K}) \xleftarrow{\Delta U_3} H_2O(1, 373\text{ K})$
$C_{v,m}(H_2O, g) = 33.314 - 8.314$
$= 25$ J/K mol
$\Delta U_2 = \Delta H_2 - \Delta n_gRT = 37.6$
$\Delta U_{total} = \Delta u_1 + \Delta u_2 + \Delta u_3$
$= C_{v,m}(l)\Delta T + \Delta V_{vap} + C_{v,m}(g)\Delta T$
$$= \frac{75\times50}{1000} + 37.6 + \frac{25\times50}{1000}$$
$= 42.6$ kJ/mol

12. **(c)** For isothermal reversible expansion.
$$w = -nRT\ln\frac{V_2}{V_1}$$

13. **(b)** In CH_4, $4 \times BE_{(C-H)} = 360$ kJ/mol
$\therefore BE_{(C-H)} = 90$ kJ/mol
In C_2H_6, $BE_{(C-C)} + 6\times BE_{(C-H)} = 620$ kJ/mol
$\therefore BE_{(C-C)} = 80$ kJ/mol
$$\therefore BE_{(C-C)} = \frac{80\times10^3}{6.023\times10^{23}}\text{ J/molecule}$$
Now, $E = \frac{hc}{\lambda}$
$$\therefore \lambda = \frac{6.626\times10^{-34}\times3\times10^8\times6.023\times10^{23}}{80\times10^3}$$
$\therefore \lambda = 1.49\times10^3$ nm

14. **(b)** For combustion reaction, ΔH is negative, $\Delta n = (16+18) - (25+2) = +7$, so ΔS is +ve, reaction is spontaneous, hence ΔG is –ve.

15. **(c)** For monoatomic gas $C_v = \frac{3}{2}R$ $C_p = \frac{5}{2}R$
At constant volume, $\Delta U = q_v = nC_{v,m}\Delta T$
At constant pressure, $\Delta H = q_p = nC_{p,m}\Delta T$

16. **(b)** Given $\frac{1}{2}N_2 + \frac{3}{2}H_2 \rightleftharpoons NH_3$;

16. (b) Given $\frac{1}{2}N_2+\frac{3}{2}H_2 \rightleftharpoons NH_3$;

$$\Delta H_f = -46.0\ kJ/mol$$

$$H+H \rightleftharpoons H_2;\ \Delta H_f = -436\ kJ/mol$$

$$N+N \rightleftharpoons N_2;\ \Delta H_f = -712\ kJ/mol$$

$$\Delta H_f(NH_3)=\frac{1}{2}\Delta H_{N-N}+\frac{3}{2}\Delta H_{H-H}-\Delta H_{N-F}$$

$$-46=\frac{1}{2}(-712)+\frac{3}{2}(-436)-\Delta H_{N-F}$$

On calculation

$$\Delta H_{N-F}=-964\ kJ/mol$$

17. (c) To calculate average enthalpy of C – H bond in methane following informations are needed

(i) dissociation energy of H_2 i.e.

$$\frac{1}{2}H_2(g) \longrightarrow H(g);\ \Delta H = x(\text{suppose})$$

(ii) Sublimation energy of C(*graphite*) to C(*g*)

$C(graphite) \longrightarrow C(g);\ \Delta H = y$ (Suppose)

Given

$$C(graphite)+2H_2(g) \longrightarrow CH_4(g);$$

$$\Delta H = 75\ kJ\ mol^{-1}$$

18. (d) Given $\Delta H = 41\ kJ\ mol^{-1} = 41000\ J\ mol^{-1}$

$T = 100°C = 273 + 100 = 373\ K$

$\Delta n = 1$

$\Delta U = \Delta H - \Delta nRT = 41000 - (1 \times 8.314 \times 373)$

$= 37898.88\ J\ mol^{-1} \simeq 37.9\ kJmol^{-1}$

19. (b) $\Delta U = q + w$

$= 10 \times 1000 - 2 \times (20) \times 101.3 = 5948\ J$

20. (a) Gaseous ions, when dissolved in water, get hydrated and heat is evolved (heat of hydration).

$Cl^-(g)+aq \rightarrow Cl^-(aq)$ is such reaction.

21. (a) I – II, gives, $2Al+Fe_2O_3 \rightarrow 2Fe+Al_2O_3$

we have $-400-(-200)=-200$; For one mole of iron the value is -100 cal

22. (c) Given

$$H_2(g) + \frac{1}{2}O_2(g) \longrightarrow H_2O(l);$$

$$\Delta H° = -285.9\ kJ\ mol^{-1} \quad ...\ (1)$$

$$H_2(g) + \frac{1}{2}O_2(g) \longrightarrow H_2O(g);$$

$$\Delta H° = -241.8\ kJ\ mol^{-1} \quad ...\ (2)$$

We have to calculate

$H_2O(l) \longrightarrow H_2O(g);\ \Delta H° = ?$

On substracting eqn. (2) from eqn. (1) we get

$H_2O(l) \longrightarrow H_2O(g);$

$\Delta H° = -241.8 - (-285.9)$

$= 44.1\ kJ\ mol^{-1}$

23. (c) Applying Hess's Law

$$\Delta_f H° = \Delta_{sub}H + \frac{1}{2}\Delta_{diss}H + I.E. + E.A + \Delta_{lattice}H$$

$$-617 = 161+520+77+E.A.+(-1047)$$

$E.A. = -617 + 289 = -328\ kJ\ mol^{-1}$

$\therefore$ electron affinity of fluorine

$= -328\ kJ\ mol^{-1}$

24. (c) At constant volume, $\frac{P_1}{T_1}=\frac{P_2}{T_2}$

$$\Rightarrow P_2 = 1 \times \frac{300}{200} = \frac{3}{2}$$

and $V_1 = 24.63$ L

for single phase

$\because\ dG = Vdp - SdT$

$$\Delta G = V\Delta P - \int(2+10^{-2}T)dT$$

$$= 1231.5 - 200 - \left(\frac{10^{-2} \times 50,000}{2}\right)$$

$= 781.5$ J

25. (d) $\Delta G° = -RT\ ln\ K_p;\ K_p = (2x)^2 X = 4x^3$

$\Delta G° = -RT\ ln\ (4X^3)$

$\Delta G° = -RT\ ln\ 4 - 3RT\ ln\ X$

26. (a) If more trans-2-pentene is added, then its concentration in right hand side will increase. But in order to maintain the K constant, concentration of cis-2-pentene will also increase. Therefore more cis-2-pentene will be formed.

27 (d)

28. (b) Hydration energy of Cl^+ is very less than H^+, hence it doesn't form Cl^+ (aq) ion.

29. **(a)** Given that

$MgSO_4(s) + nH_2O \rightarrow MgSO_4\,nH_2O;$

$\Delta_r H_1 = -91.2$ kJ/mol ...(i)

$MgSO_4.7H_2O(s) + (n-7)H_2O$

$\rightarrow MgSO_4(nH_2O)$

$\Delta_r H_2 = 13.8$ kJ/mol ...(ii)

or $\Delta H_{hyd} = \Delta_r H_1 - \Delta_r H_2$ equation (i) – (ii)

$= -91.2$ kJ/mol – 13.8 kJ/mol

$= -105$ kJ/mol

30. **(a)** $CaO(s) + CO_2(g) - CaCO_3(s)$

$\Delta_f H^\circ = \Delta H_f^\circ(CaCO_3) - \Delta H_f^\circ(CaO) - \Delta H_f^\circ(CaCO_2)$

$= -1207 - (-635) - (-394)$

$= -178$ kJ/mol

$\therefore \quad \Delta U = \Delta H - \Delta n_g RT$

$$\Delta U = -178 - \left(\frac{(-1)\times 8.3\times 300}{1000}\right)$$

$= -175.51$ kJ

$$n_{CaO} = \frac{224}{56} = 4$$

$\therefore \quad q_v = n\Delta_r U = 4 \times (-175.51)$

$= -702.04$ kJ

31. **(a)** $N_2(g) + \frac{1}{2}O_2 \rightarrow N_2O(g)$

$$N \equiv N(g) + \frac{1}{2}(O=O) \rightarrow \ddot{N} = \overset{+}{N} = \ddot{O}(g)$$

ΔH_f° = [Energy required for breaking of bonds]–[Energy released for forming of bonds]

$$= (\Delta H_{N\equiv N} + \frac{1}{2}\Delta H_{O=O} - (\Delta H_{N=N} + \Delta H_{N=O})$$

$$= (946 + \frac{1}{2}\times 498) - (418 + 607)$$

$= 170$ kJ mol^{-1}

Resonance energy $= 82 - 170$

$= -88$ kJ mol^{-1}

32. **(d)** $\Delta H = \Delta U + \Delta n_g RT$

For the reaction $\Delta n_g = 12 - 15 = -3$

$\Delta H - \Delta U = -3 \times 8.314 \times 300$

$= -7482$ J mol^{-1}

33. **(b)** $\Delta H^\circ = 2\times \Delta H_1^\circ + 2\times \Delta H_2^\circ - \Delta H_3^\circ = -2486$ kJ

34. **(b)** $\Delta_r H = \frac{5}{8}\times 40 + \frac{3}{8}\times 50 = 43.75$ kJ / mol

35. **(a)** 1 watt = 1 J/sec

Total heat supplied for 36 mL H_2O

= 806 × 100

= 80600 J

$$\Delta H_{vap} = \frac{80600}{36}\times 18$$

= 40300 J or 40.3 kJ/mol

36. **(c)** $\Delta H = \Delta U + \Delta nRT$

$\Delta n = n_P - n_R$

Now, $\Delta n = 2 - \frac{5}{2} = -\frac{1}{2}$

$\therefore \Delta H = \Delta U - \frac{1}{2}RT$

Thus, $\Delta U = \Delta H + \frac{1}{2}RT$

$\therefore \Delta U > \Delta H$

37. **(d)** $A \xrightarrow{q=+5,\ w=-8J} B$ (with return arrow from B to A)

$\Delta U_{AB} = q + w = +5 + (-8) = -3$

$q = -3, \ \Delta U_{BA} = +3$

$\Delta U_{BA} = q + w$

$\Rightarrow 3 = -3 + w \Rightarrow w = +6$ J (work done on the system).

38. **(a)** In question 0.01 gev of NaOH is being neutralised by 0.01 gev of HCl and value is x kJ, for 1 gev the value is –100x (exothermic process).

39. **(d)** As $\Delta_f H$ increases stability decreases.

40. **(c)** Enthalpy of neutralisation of HCl with NaOH is x. In question gev of HCl

$$\left(\frac{500\times 2}{1000} = 1\text{gev}\right) \text{and NaOH} \left(\frac{250\times 4}{1000} = 1\text{gev}\right)$$

hence the value x.

41. **(d)** $\underset{\text{Strong acid}}{HCl} \longrightarrow \underset{\text{(Complete ionisation)}}{H^+ + Cl^-}$ ...(i)

$\underset{\text{Weak base}}{NH_4OH \rightleftharpoons NH_4^+ + OH^-}$

$\Delta H = x$ kJ mol^{-1} ...(ii)

$H^+ + OH^- \longrightarrow H_2O$

$\Delta H = -55.90$ kJ mol^{-1} ...(iii)

(from neutralisation of strong acid and strong base)

From equation (i), (ii) and (iii)

$NH_4OH + HCl \longrightarrow NH_4^+ + Cl^- + H_2O$

$\Delta H = -51.46$ kJ mol^{-1}

$\therefore \quad x + (-55.90) = -51.46$

$x = -51.46 + 55.90$

$= 4.44$ kJ mol^{-1}

$\therefore$ Enthalpy of ionisation of NH_4OH is $\Rightarrow 4.44$ kJ mol^{-1}

42. (b) Given, $C_p = 10$ cals at 1000 K

$T_1 = 1000$ K. $T_2 = 100$ K

$m = 32$ g

$\Delta S = ?$

at constant pressure

$$\Delta S = C_p \ln \frac{T_2}{T_1}$$

$$= 2.303 \times C_p \log \frac{T_2}{T_1}$$

$$= 2.303 \times 10 \log \frac{100}{1000}$$

$= -23.03$ cal deg^{-1}

43. (b) $\Delta S° = S°_{CO_2} + 2 \times S°_{H_2O} - \left(S°_{CH_4} + 2 \times S°_{O_2}\right)$

$= (213.6 + 2 \times 69.9) - (186.2 + 2 \times 205.2)$

$= -242.8$ J K^{-1} mol^{-1}.

44. (a) $Q = \frac{1}{2} 438 \times \frac{22.3}{223} = 21.9$ kJ

45. (b) The species in its elemental form has zero standard molar enthalpy of formation at 298 K. At 298K, Cl_2 is gas while Br_2 is liquid.

46. (d) $C_6H_6\ (l) + \frac{15}{2} O_2\ (g) \rightarrow$

$6CO_2\ (g) + 3H_2O\ (l)$

$\Delta n = 6 - 15/2 = -3/2$

47. (b) $\Delta_{neu}H$ for strong base and strong acid

$= -13.7$ kcal eq^{-1}

$\Delta H_{ion}\ (CH_3COOH)$

$= -12.5 - (-13.7) = 1.2$ kcal mol^{-1}

$\Delta H_{ion}\ (NH_4OH)$

$= -10.5 - (-13.7) - \Delta H_{ion}\ (CH_3COOH)$

$= 13.7 - 10.5 - 1.2$

$= 2$ kcal mol^{-1}

48. (a) For a pure substance T_A and T_B represent the same temperature. Hence A is a correct choice.

49. (b) $\Delta G = \Delta H - T\Delta S$

$\therefore \Delta G = 9100 - 373.2 \times 25.5 = -416.6$ cal mol^{-1}

50. (a) For non spontaneous reaction

$\Delta G = +$ ve

$\Delta G = \Delta H - T\Delta S$ and

$\Delta S = 121$ JK^{-1}

For $\Delta G = +$ ve

ΔH has to be positive. Hence the reaction is endothermic.

The minimum value of ΔH can be obtained by putting $\Delta G = 0$

$\Delta H = T\Delta S = 298 \times 121$ J

$= 36.06$ kJ

51. (d) $\Delta S = \frac{\Delta H}{T}$; $T = \frac{\Delta H}{\Delta S} = \frac{0.3344}{0.836} = 0.4$K

52. (b) For isothermal process ($\Delta T = 0$)

$$\Delta S = R\ \ell n \frac{P_1}{P_2} = 8.314\ \ell n\ 2$$

$= 8.314 \times 0.693 = 5.76$

53. (a) (cyclohexene) $+ H_2 \longrightarrow$ (cyclohexane); $\Delta H = -119.5$ kJ

(benzene) $+ 3H_2 \longrightarrow$ (cyclohexane); $\Delta H = 3(-119.5)$

$= -358.5$ kJ

The resonance energy provides extra stability to the benzene molecule so it has to be overcome for hydrogenation to take place. So

$\Delta H = -358.5 - (-150.4) = -208.1$ kJ

54. (b) $2Al+\frac{3}{2}O_2 \rightarrow Al_2O_3,\ \Delta H=-1596$ kJ .....(i)

$2Cr+\frac{3}{2}O_2 \rightarrow Cr_2O_3,\ \Delta H=-1134$ kJ ....(ii)

By (i) – (ii)

$2Al+Cr_2O_3 \rightarrow 2Cr+Al_2O_3,\ \Delta H=-462$ kJ.

55. (a) $\Delta G=\Delta H\text{-}T\Delta S$, ΔH is +ve, ΔS is + ve; $T\Delta S>\Delta H$ for spontaneous process. It will make ΔG, –ve

56. (d) $\Delta G=-RT\ln K_p$ or $K_p=e^{-\Delta G/RT}$

57. (b) Dissolution of KNO_3 is endothermic, hence heat is absorbed and cooling is observed.

58. (a) Isobutane + oxygen $\rightarrow CO_2+H_2O$

$\Delta H=-2870$ kJ mol^{-1} ......(i)

n-butane + oxygen $\rightarrow CO_2+H_2O$

$\Delta H=-2878$ kJ mol^{-1} ....(ii)

(ii) – (i); n-butane – Isobutane,

$\Delta H=(-2878+2870)$
$=-8$ kJ mol^{-1}.

59. (b) Atomisation of methane

$CH_4(g) \longrightarrow C(g)+4H(g);$

$\Delta H=360$ kcal

$\therefore$ C – H bond energy $=\frac{360}{4}$

$=90$ kcal mol^{-1}

$C_2H_6(g) \longrightarrow 2C(g)+6H(g);$

$\Delta H=620$ kcal

or $H_{C-C}+6H_{C-H}=620$

$\therefore \quad H_{C-C}=620-6H_{C-H}$

$=620-6\times 90=80$ kcal mol^{-1}

60. (d) No. of moles of O_2 required to supplied 30 kJ heat to second reaction

$=\frac{30}{1260}\times\frac{3}{2}=\frac{1}{28}$

$n_{O_2} : n_{H_2} = \frac{1}{28} : 3$ or $1:84$

61. (d) No. of moles of sucrose $=\frac{34.2}{342}=0.1$

$-(\Delta G)_{T,P}=$ useful work done by the system

$-\Delta G=-\Delta H+T\Delta S$

$=+(6000\times 0.1)+\frac{180\times 0.1\times 300}{1000}$

$=605.4$ kJ

62. (b) For the reaction

$2ZnS \rightarrow 2Zn+S_2;\ \Delta G_1^\circ=293$ kJ ....(i)

$2Zn+O_2 \rightarrow 2ZnO;\ \Delta G_2^\circ=-480$ kJ ....(ii)

$S_2+2O_2 \rightarrow 2SO_2;\ \Delta G_3^\circ=-544$ kJ ....(iii)

ΔG° for the reaction

$2ZnS+3O_2 \rightarrow 2ZnO+2SO_2$

can be obtained by adding eqn. (i), (ii) and (iii)

$\Rightarrow \Delta G^\circ=293-480-544=-731$ kJ

63. (d) $\underset{\substack{98\\49}}{H_2SO_4}+\underset{\substack{112\\56}}{2KOH} \rightarrow \underset{\substack{174\\87}}{K_2SO_4}+\underset{\substack{2\text{ mole}\\1\text{ mole}}}{2H_2O}$

13.7 kcal is the heat evolved when 1 gev of strong acid is neutralised by 1 gev of strong base.

64. (a) The first reaction is exothermic and the second reaction is endothermic. If on passing the mixture of O_2 and H_2O (steam) over coke while keeping the temperature constant ΔH of both the reactions must be same.

Moles of O_2 needed to evolve 132 kJ

$=\frac{0.5\times 132}{110}=0.6$

Hence steam : O_2 ratio must be 1 : 0.6

65. (b) $q=m\times c\times \Delta T,\ m=q/(c\times \Delta T)$
$=(24\times 10^6\times 0.7)/(4.18\times 50)$
$=80383$ g or 80.383 kg

66. (d) Entropy change at constant temperature

$=2.303nR\log\frac{V_1}{V_2}=2.303\times 2\times 2\log\frac{20}{2}$

$=9.2$ cal K^{-1} mol^{-1}

67. (a) $\Delta S = \frac{\Delta H}{T} = \frac{344.3 \times 10^3}{3443} = 100 \text{ J mol}^{-1}$

68. (d) We know that $\Delta G = \Delta H - T\Delta S$
When $\Delta H < 0$ and $\Delta S < 0$ then ΔG will be negative at low temperatures (positive at high temperature) and the reaction will be spontaneous.

69. (a) Measure of disorder of a system is nothing but entropy. For a spontaneous reaction, $\Delta G < 0$. As per Gibbs Helmholtz equation,
$\Delta G = \Delta H - T\Delta S$
Thus ΔG is –ve only
when ΔH = –ve (exothermic)
and ΔS = +ve (increasing disorder)

70. (b) $\Delta H^\circ = \Delta U^\circ + \Delta nRT$

$$\Delta H^\circ = -772.7 + \frac{5 \times 8.314 \times 298}{1000}$$

$$= -760.3 \text{ kJ mol}^{-1}$$

71. (d) $H_2(g) + \frac{1}{2}O_2(g) \longrightarrow H_2O(l)$;

$\Delta H = -285.8$ kJ .... (i)

$H^+(aq) + OH^-(aq) \longrightarrow H_2O(l)$;

$\Delta H = -57.3$ kJ .... (ii)

$\frac{1}{2}H_2(g) + aq \longrightarrow H^+(aq) + e^-$; $\Delta H = 0$
(by convention) .... (iii)
(i) – (ii) – (iii) gives,

$$\frac{1}{2}H_2(g) + \frac{1}{2}O_2(g) + e^- + aq \longrightarrow OH^-(aq)$$

$\Delta H = -285.8 + 57.3 = -228.5$ kJ

72. (b) Bond dissociation energy of
$PH_3(g) = 228 \text{ kcal mol}^{-1}$

$$\text{P–H bond energy} = \frac{228}{3} = 76 \text{ kcal mol}^{-1}$$

H₂P–PH₂ (structure: each P bonded to two H atoms, P–P bond)

Bond energy of 4 (P – H) + (P –P)
$= 355 \text{ kcal mol}^{-1}$
or $4 \times 76 + (P - P) = 355 \text{ kcal mol}^{-1}$
P–P bond energy = 51 kcal mol^{-1}

73. (c) $\because \Delta G = \Delta H - T \times \Delta S$
For a spontaneous reaction, ΔG should be negative
$\Delta H = -238$ kJ, $\Delta S = -87 \text{ JK}^{-1}$
Hence, reaction will be spontaneous when $\Delta H > T \times \Delta S$. Therefore, at 1000, 1500 and 3000 K the reaction would be spontaneous.

74. (b) $\Delta S^\circ = -206.5 + 114.6 + 180.7 = 88.8$
$\Delta G^\circ = \Delta H^\circ - T\Delta S^\circ$
$\Delta H^\circ = 271.8 + 298 \times 88.8 \times 10^{-3}$
$\Delta H^\circ = 298.3 \text{ kJ mol}^{-1}$

75. (a) $\Delta G = -RT \ln K_{eq}$:
Normal body temperature = 37°C

$$\Rightarrow -50\frac{\text{kJ}}{\text{mol}} = 8.314\frac{\text{J}}{\text{K mol}} \times 310 \ln K_{eq}$$

$$\Rightarrow 19.39 = \ln K_{eq}$$

$$\Rightarrow K_{eq} = 2.6 \times 10^8$$

76. (c) $\Delta H = (n_1 C_{p,m_1} + n_2 C_{p,m_2}) \Delta T$

$$= \left(0.5 \times \frac{7}{2}R + 0.5 \times 4R\right)(-100)$$

$= -375$ R

77. (d) Condition of equilibrium, hence $\Delta G = 0$

78. (a) For spontaneous reaction, $dS > 0$ and ΔG and ΔG should be negative i.e. < 0.

79. (a) $\Delta G^\circ = -RT \ln K$
$= -8.314 \text{JK}^{-1}\text{mol}^{-1} \times 298 \text{ K} \times 2.303 \log 10^{-14}$
$= 7.98 \times 10^4 \simeq 8 \times 10^4 \text{ J mol}^{-1}$

80. (c) When gas is compressed, its entropy decreases so, ΔS is negative.

81. (a) $\Delta S = \frac{\Delta H}{T}$

$$\Delta S(\text{per mole}) = \frac{\Delta H \text{ per mole}}{T} = \frac{6000}{273}$$

$= 21.98 \text{ JK}^{-1}\text{mol}^{-1}$

82. (b) For 5 moles of gas at temperature T,
$PV_1 = 5RT$
For 5 moles of gas at temperature T – 2,
$PV_2 = 5R(T - 2)$

$\therefore\ P(V_2 - V_1) = 5R(T - 2 - T);$

$P\Delta V = -10R,$

$-P\Delta V = 10R$

When ΔV is negative, W is + ve.

83. (c) $P_1V_1 = P_2V_2$

$$\therefore \quad V = \frac{5\text{bar} \times 2\text{dm}^3}{1\text{bar}} = 10\ \text{dm}^3$$

Work, $W = -P_{ext}((V_{final} - V_{initial})$

$= -1\ \text{bar}\ (10 - 2)$

$= -1 \times 10^5\ \text{Pa} \times 8 \times 10^{-3}\text{m}^3$

$= -800$ J

84. (a) Process is isothermal reversible expansion, hence

$\Delta U = 0$, therefore $q = -w$.

Since $q = +208$ J, $W = -208$ J

85. (a) The greater the (negative value) of heat of neutralisation, the more is the strength of the acid. Hence,

$HCOOH > CH_3COOH > H_2S > HCN$

86. (b) The enthalpy of ionisation of weak acid is given by

ΔH_{ion} (HA)

$= \Delta H_N$ (weak acid / strong base) –

ΔH_N (strong acid / strong base)

$= -56.1 - (-57.3) = 1.2\ \text{kJ mol}^{-1}$

$\Delta H_{(ionisation)} = 1.5\ \text{kJ mol}^{-1}$

Hence % ionisation in 1 M solution

$$= \frac{(1.5 - 1.2)}{1.5} \times 100 = 20\%$$

87. (d) By 2× (ii) – (i) + (iii)

$H_2(g) + 2C(s) \rightarrow C_2H_2(g),\ \Delta G° = 209$ kJ

88. (c) $\Delta H = \Delta E + \Delta nRT$

$\Delta n = 3 - (1 + 5)$

$= 3 - 6 = -3$

$\Delta H - \Delta E = (-3RT)$

89. (a) Mass independent properties (molar conductivity and electromotive force) are intensive properties. Resistance and heat capacity are mass dependent, hence extensive properties.

90. (b) $Hg(l) \rightleftharpoons Hg(g)$,

$\Delta_R S° = 174.4 - 77.4 = 97$ J/K-mol

$\because\ \Delta G° = \Delta H° - T\Delta S° = 0$

$$T = \frac{\Delta H°}{\Delta S°}$$

$$= \frac{60.8 \times 1000}{97} = 626.8\ \text{K}$$

91. (c) $H^+(aq) + OH^-(aq) \rightleftharpoons H_2O\ (l);$

$\Delta H = -57.3$ kJ .... (i)

$H_2C_2O_4 + 2OH^- \rightleftharpoons$

$C_2O_4^{--} + 2H_2O;\ \Delta H = -106$ kJ .... (ii)

For the second reaction the value should have been

$2 \times (-57.3) = -114.6$ kJ

The difference

$(114.6 - 106) = 8.6\ \text{kJ mol}^{-1}$ is used to effect of the ionisation of oxalic acid.

92. (a) $\Delta H_{hyd.} = \Delta H_{sol.} - \Delta H_{lattice}$

$= 1 - 180 = -179\ \text{kcal mol}^{-1}$

Then $\Delta H_{hyd.}(Na^+) + \Delta H_{hyd.}(Cl^-) = -179$

or $\Delta H_{hyd.}(Na^+) + \frac{2}{3}\Delta H_{hyd} = -179$

or $\Delta H_{hyd.}(Na^+) = -107.4\ \text{kcal mol}^{-1}$

93. (a) $CH_4(g) + 2O_2(g) \longrightarrow$

$CO_2(g) + 2H_2O(l)$

$\Delta H = -890$ kJ .... (i)

$2H_2O(l) \longrightarrow 2H_2O(g);$

$\Delta H = 2 \times 40.5 = 81$ kJ .... (ii)

From (i) + (ii), we get

$CH_4(g) + 2O_2(g) \longrightarrow$

$CO_2(g) + 2H_2O(g)$

$\Delta H = -890 + 81 = -809$ kJ

94. (b) $\Delta G = \Delta H - T\Delta S$

$= -382.64 - (298 \times 145.6 \times 10^{-3})$

$= -339.3 \text{ kJ mol}^{-1}$

95. (a) $\Delta G = -$ve means the process is spontaneous.

96. (a) $C_2H_5OH(l) + 3O_2(g) \longrightarrow 2CO_2(g) + 3H_2O(l)$

Bomb calorimeter gives ΔU of the reaction

Given, $\Delta U = -1364.47 \text{ kJ mol}^{-1}$

$\Delta n_g = -1$

$\Delta H = \Delta U + \Delta n_g RT$

$= -1364.47 - \frac{1 \times 8.314 \times 298}{1000}$

$= -1366.93 \text{ kJ mol}^{-1}$

97. (b) For the equation

$B_2H_6(g) + 3O_2(g) \longrightarrow B_2O_3(g) + 3H_2O(g)$

Eqs. (i) + 3 (ii) + 3 (iii) – (iv)

$\Delta H = -1273 + 3(-286) + 3(44) - 36$

$= -1273 - 858 + 132 - 36$

$= -2035 \text{ kJ/mol}$

98. (a) 18gm of water at 100°C

10gm of Cu at 25°C is added.

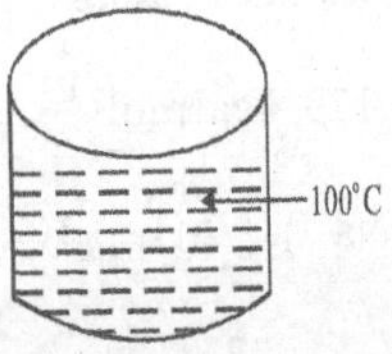

100°C

$q_p = C_{p,m} dT$

$= 75.32 \times \frac{J}{K\,mol} \times \frac{18g}{18g/mol} (373 - 298) K$

$= 75.32 \frac{J}{K} \times 75 \text{ K}$

$= 5.649 \times 10^3 \text{ J}$

If now 10g of copper is added

$C_{p,m} = 24.47 \text{ J/ mol K}$

Amount of heat gained by Cu

$= 24.47 \frac{J}{K\,mol} \times \frac{10g}{63g/mol} (373 - 298) \text{ K}$

$= 291.3 \text{ J}$

Heat lost by water = 291.30 J

$-291.30 \text{ J} = 75.32 \frac{J}{K} \times (T_2 - 373K)$

$\Rightarrow -3.947 \text{ K} = T_2 - 373 \text{ K}$

$\Rightarrow T_2 = 369.05 \text{ K}$

99. (a) $q_p = \Delta H = C_p dT$

$\Rightarrow q_p = 75.32 \frac{J}{K\,mol} \times (299 - 298) K$

$\Rightarrow q_p = 75.32 \frac{J}{K\,mol}$

For 180 kg of water, no. of moles of water

$= \frac{180 \times 10^3 g}{18g/mol} = 10^4 g \text{ moles}$

$q_p = 75.32 \frac{J}{mol} \times 10^4 \text{ moles}$

$= 753.2 \times 10^3 \text{ J} = 753.2 \text{ kJ}$

ΔH for ATP = 7 kcal / mol

$= 7 \times 4.184 \text{ kJ/mol}$

$= 29.2 \text{ kJ/mol}$

6.022×10^{23} molecules of ATP produce

= 29.2 kJ

29.2 kJ produced from

6.022×10^{23} molecules

753.2 kJ produced from

$6.022 \times 10^{23} \times \frac{75.8}{29.2}$

$= 1.5 \times 10^{25}$ molecules

100. (c) $\Delta G° = -RT \ln K$;

K ↑ stability of reactant ↓

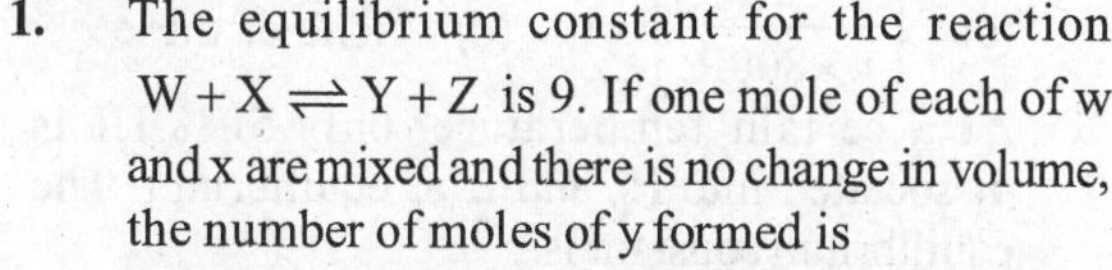

7 Equilibrium

1. The equilibrium constant for the reaction $W+X \rightleftharpoons Y+Z$ is 9. If one mole of each of w and x are mixed and there is no change in volume, the number of moles of y formed is
(a) 0.10 (b) 0.50 (c) 0.75 (d) 0.54

2. When heated, ammonium carbamate decomposes as follows:
$NH_4COONH_2(s) \rightleftharpoons 2NH_3(g)+CO_2(g)$. At a certain temperature, the equilibrium pressure of the system is 0.318 atm. K_p for the reaction is:
(a) 0.128 (b) 0.426
(c) 4.76×10^{-3} (d) None of these

3. One mole of SO_3 was placed in a one litre reaction flask at a given temperature when the reaction equilibrium was established in the reaction.
$2SO_3 \rightleftharpoons 2SO_2+O_2$. The vessel was found to contain 0.6 mole of SO_2. The value of the equilibrium constant is **Tricky**
(a) 0.36 (b) 0.675 (c) 0.45 (d) 0.54

4. For the equilibrium system
$2HX(g) \rightleftharpoons H_2(g)+X_2(g)$
the equilibrium constant is 1.0×10^{-5}. What is the concentration of HX if the equilibrium concentration of H_2 and X_2 are 1.2×10^{-3} M, and 1.2×10^{-4} M respectively.
(a) 12×10^{-4} M (b) 12×10^{-3} M
(c) 12×10^{-2} M (d) 12×10^{-1} M

5. If K_1 and K_2 are respective equilibrium constants for the two reactions
$XeF_6(g)+H_2O(g) \rightleftharpoons XeOF_4(g)+2HF(g)$
$XeO_4(g)+XeF_6(g) \rightleftharpoons XeOF_4(g)+XeO_3F_2(g)$
the equilibrium constant for the reaction
$XeO_4(g)+2HF(g) \rightleftharpoons XeO_3F_2(g)+H_2O(g)$ will be
(a) $\frac{K_1}{K_2^2}$ (b) $K_1.K_2$ (c) $\frac{K_1}{K_2}$ (d) $\frac{K_2}{K_1}$

6. If 1.0 mole of I_2 is introduced into 1.0 litre flask at 1000 K, at quilibrium ($K_c = 10^{-6}$), which one is correct
(a) $[I_2(g)] > [I^-(g)]$ (b) $[I_2(g)] < [I^-(g)]$
(c) $[I_2(g)] = [I^-(g)]$ (d) $[I_2(g)] = \frac{1}{2}[I^-(g)]$

7. In a reaction A + B $\rightleftharpoons$ C + D, the initial concentrations, of A and B were 0.9 mol. dm^{-3} each. At equilibrium the concentration of D was found to be 0.6 mol dm^{-3}. What is the value of equilibrium constant for the reaction
(a) 8 (b) 4 (c) 9 (d) 3

8. The partial pressure of CH_3OH (g), CO (g) and H_2 (g) in equilibrium mixture for the reaction, $CO(g)+2H_2(g) \rightleftharpoons CH_3OH(g)$ are 2.0, 1.0 and 0.1 atm respectively at 427°C. The value of K_p for the decomposition of CH_3OH to CO and H_2 is
(a) 10^2 atm (b) 2×10^2 atm^{-1}
(c) 50 atm^2 (d) 5×10^{-3} atm^2

9. 3.2 moles of hydrogen iodide were heated in a sealed bulb at 444°C till the equilibrium state was reached. Its degree of dissociation at this temperature was found to be 22% The number of moles of hydrogen iodide present at equilibrium are **Critical Thinking**
(a) 2.496 (b) 1.87 (c) 2 (d) 4

10. K_c for $PCl_5(g) \rightleftharpoons PCl_3(g)+Cl_2(g)$ is 0.04 at 250°C. How many moles of PCl_5 must be added to a 3 L flask to obtain a Cl_2 concentration of 0.15 M
(a) 4.2 moles (b) 2.1 moles
(c) 5.5 moles (d) 6.3 moles

11. Gaseous N_2O_4 dissociates into gaseous NO_2 according to the reaction

$$[N_2O_4(g) \rightleftharpoons 2NO_2(g)]$$

At 300 K and 1 atm pressure, the degree of dissociation of N_2O_4 is 0.2. If one mole of N_2O_4 gas is contained in a vessel, then the density of the equilibrium mixture is :
(a) 1.56 g/L (b) 6.22 g/L
(c) 3.11 g/L (d) 4.56 g/L

12. The value of K_p for the equilibrium reaction $N_2O_4(g) \rightleftharpoons 2NO_2(g)$ is 2. Toughnut

The percentage dissociation of $N_2O_4(g)$ at a pressure of 0.5 atm is
(a) 25 (b) 88 (c) 50 (d) 71

13. If $CuSO_4.5H_2O(s) \rightleftharpoons$

$$CuSO_4.3H_2O(s) + 2H_2O(g)$$

$K_P = 1.086 \times 10^{-4}$ atm^2 at 25° C. The efforescent nature of $CuSO_4.5H_2O$ can be noticed when the vapour pressure of H_2O in atmosphere is
(a) >9.72 mm (b) <7.92 mm
(c) >7.92 mm (d) <11.92 mm

14. 28g N_2 and 6.0 g of H_2 are heated over catalyst in a closed one litre flask of 450°C. The entire equilibrium mixture required 500 mL of 1.0 M H_2SO_4 for neutralisation. The value of K_c for the reaction

$$N_2(g) + 3H_2(g) \rightleftharpoons 2NH_3(g) \text{ is}$$

(a) 0.06 $mol^{-2}L^2$ (b) 0.59 $mol^{-2}L^2$
(c) 1.69 mol^2L^{-2} (d) 0.03 mol^2L^{-2}

15. A gaseous compound of molecular mass 82.1 dissociates on heating to 400 K as

$$X_2Y_4(g) \rightleftharpoons X_2(g) + 2Y_2(g)$$

The density of the equilibrium mixture at a pressure of 1 atm and temperature of 400K is $2.0gL^{-1}$. The percentage dissociation of the compound is Toughnut

(a) 12.5% (b) 48.5% (c) 90.1% (d) 25.0%

16. At 527 °C, the reaction given below has $K_c = 4$

$$NH_3(g) \rightleftharpoons \frac{1}{2}N_2(g) + \frac{3}{2}H_2(g)$$

What is the K_p for the reaction?

$$N_2(g) + 3H_2(g) \rightleftharpoons 2NH_3(g)$$

Critical Thinking

(a) $16 \times (800R)^2$ (b) $\left(\frac{800R}{4}\right)^{-2}$

(c) $\left(\frac{1}{4 \times 800R}\right)^2$ (d) None of these

17. At a certain temperature, only 50% HI is dissociated into H_2 and I_2 at equilibrium. The equilibrium constant is:
(a) 1.0 (b) 3.0 (c) 0.5 (d) 0.25

18. Some inert gas is added at constant volume to the following reaction at equilibrium

$$NH_4HS(s) \rightleftharpoons NH_3(g) + H_2S(g)$$

Predict the effect of adding the inert gas :
(a) The equilibrium shifts in the forward direction
(b) The equilibrium shifts in the backward direction
(c) The equilibrium remains unaffected
(d) The value of K_p is increased

19. The standard state Gibb's free energy change for the isomerisation reaction

cis-2-pentene $\rightleftharpoons$ trans-2-pentene

is –3.67 kJ mol^{-1} at 400 K. If more trans-2-pentene is added to the reaction vessel
(a) more cis-2-pentene is formed
(b) equilibrium shifts in the forward direction
(c) equilibrium remains unaltered
(d) more trans-2-pentene is produced

20. (1) $N_2(g) + 3H_2(g) \rightleftharpoons 2NH_3(g), K_1$
(2) $N_2(g) + O_2(g) \rightleftharpoons 2NO(g), K_2$
(3) $H_2(g) + \frac{1}{2}O_2(g) \rightleftharpoons H_2O(g), K_3$

The equation for the equilibrium constant of the reaction

Critical Thinking

$$2NH_3(g) + \frac{5}{2}O_2(g) \rightleftharpoons 2NO(g) + 3H_2O(g),$$

(K_4) in terms of K_1, K_2 and K_3 is :

(a) $\frac{K_1.K_2}{K_3}$ (b) $\frac{K_1.K_3^2}{K_2}$

(c) $K_1 K_2 K_3$ (d) $\frac{K_2.K_3^3}{K_1}$

21. On the basis of Le-Chatelier's principle, predict which of the following conditions would be unfavourable for the formation of SO_3? Given that $2SO_2 + O_2 \rightleftharpoons 2SO_3; \Delta H = -42$ kcal

Tricky

(a) Low pressure and low temperature
(b) High pressure and low temperature
(c) High temperature and low pressure
(d) High concentration of SO_2

22. The reaction, $SO_2 + Cl_2 \longrightarrow SO_2Cl_2$ is exothermic and reversible. A mixture of SO_2 (g), Cl_2 (g) and SO_2Cl_2 (l) is at equilibrium in a closed container. Now a certain quantity of extra SO_2 is introduced into the container, the volume remaining the same. Which of the following is/are true?

(a) The pressure inside the container will not change.
(b) The temperature will not change.
(c) The temperature will increase.
(d) The temperature will decrease.

23. In reaction A + 2B $\rightleftharpoons$ 2C + D, initial concentration of B was 1.5 times of [A], but at equilibrium the concentrations of A and B became equal. The equilibrium constant for the reaction is:

Critical Thinking

(a) 8 (b) 4 (c) 12 (d) 6

24. For the reaction :

$2BaO_2(s) \rightleftharpoons 2BaO(s) + O_2(g)$;

ΔH = +ve. In equilibrium condition, pressure of O_2 is dependent on

(a) mass of BaO_2
(b) mass of BaO
(c) temperature of equilibrium
(d) mass of BaO_2 and BaO both

25. For the reaction

$NH_4HS(g) \rightleftharpoons NH_3(g) + H_2S(g)$ in a closed flask, the equilibrium pressure is *P* atm. The standard free energy of the reaction would be:

Toughnut

(a) $-RT \ln p$ (b) $-RT(\ln p - \ln 2)$
(c) $-2RT \ln p$ (d) $-2RT(\ln p - \ln 2)$

26. For the reaction $H_2(g) + I_2(g) \rightleftharpoons 2HI(g)$ at 721 K, the value of equilibrium constant is 50, when equilibrium concentration of both is 5M. Value of K_p under the same conditions will be

(a) 0.02 (b) 0.2 (c) 50 (d) 50 RT

27. For a reversible gaseous reaction

$N_2 + 3H_2 \rightleftharpoons 2NH_3$ at equilibrium, if some moles of H_2 are replaced by same number of moles of T_2 (T is tritium, isotope of H and assume isotopes do not have different chemical properties) without affecting other parameter, then :

(a) the sample of ammonia obtained after sometime will be radioactive.
(b) moles of N_2 after the change will be different as compared to moles of N_2 present before the change
(c) the value of K_p or K_c will change
(d) the average molecular mass of new equilibrium will be same as that of old equilibrium

28. The ratio $\frac{K_p}{K_c}$ for the reaction

$CO(g) + \frac{1}{2}O_2(g) \rightleftharpoons CO_2(g)$ is:

(a) $\frac{1}{\sqrt{RT}}$ (b) $(RT)^{1/2}$
(c) RT (d) 1

29. Which of the following is true at chemical equilibrium?

(a) $(\Delta G)_{T,P}$ is minimum and $(\Delta S)_{U,V}$ is also minimum
(b) $(\Delta G)_{T,V}$ is minimum and $(\Delta S)_{U,V}$ is maximum
(c) $(\Delta G)_{T,V}$ is maximum and $(\Delta S)_{U,V}$ is zero
(d) $(\Delta G)_{T,P}$ is zero and $(\Delta S)_{U,V}$ is also zero

30. K_1, K_2 and K_3 are the equilibrium constants of the following reactions (I), (II) and (III) respectively:

Toughnut

(I) $N_2 + 2O_2 \rightleftharpoons 2NO_2$
(II) $2NO_2 \rightleftharpoons N_2 + 2O_2$
(III) $NO_2 \rightleftharpoons \frac{1}{2}N_2 + O_2$

The correct relation from the following is

(a) $K_1 = \frac{1}{K_2} = \frac{1}{K_3}$
(b) $K_1 = \frac{1}{K_2} = \frac{1}{(K_3)^2}$
(c) $K_1 = \sqrt{K_2} = K_3$
(d) $K_1 = \frac{1}{K_2} = K_3$

31. If K_{sp} of CaF_2 at 25°C is 1.7×10^{-10}, the combination amongst the following which gives a precipitate of CaF_2 is Toughnut

(a) 1×10^{-2} M Ca^{2+} and 1×10^{-3} M F^-
(b) 1×10^{-4} M Ca^{2+} and 1×10^{-4} M F^-
(c) 1×10^{-2} M Ca^{2+} and 1×10^{-5} M F^-
(d) 1×10^{-3} M Ca^{2+} and 1×10^{-5} M F^-

32. The solubility of PbI_2 at 25°C is 0.7 g L^{-1}. The solubility product of PbI_2 at this temperature is (molar mass of PbI_2 = 461.2 g mol^{-1})

(a) 1.40×10^{-9} (b) 0.14×10^{-9}
(c) 140×10^{-9} (d) 14.0×10^{-9}

32. I^- ions react with iodine in aqueous solution to form I_3^- ion as

$$I^-(aq) + I_2(aq) \rightleftharpoons I_3^-(aq).$$

When L of solution containing 1 mol of KI and 0.25 mol of I_2 was reacted with excess of $AgNO_3$, 0.80 mol of yellow ppt. was obtained. The stability constant of I_3^- ion is

(a) 0.20 (b) 5.0 (c) 20.0 (d) 0.05

33. Consider the expression $\Delta G = -RT\ln K_p + RT\ln Q_p$ and indicate the correct statement at equilibrium

(a) $\Delta G = 0, Q_p > K_p$ the equilibrium reaction will shift from left to right
(b) $\Delta G = 0, Q_p = K_p$ the equilibrium reaction will shift from left to right
(c) $\Delta G = \infty, Q_p < K_p$ the equilibrium reaction will shift from right to left
(d) $\Delta G < 0, Q_p > K_p$ the equilibrium reaction will shift from right to left where Q_p and K_p term refer to reaction quotient and equilibrium constant at constant pressure respectively.

34. If the synthesis of ammonia from Haber's process is carried out with exactly the same starting conditions (of partial pressure and temperature) but using D_2 (deuterium) in place of H_2. Then

(a) the equilibrium will be disturbed
(b) the composition of reaction mixture will remain same at equilibrium.
(c) Use of isotope in reaction will not produce ammonia.
(d) At equilibrium rate of forward reaction will be greater than the rate of reverse reaction

35. A solution of 0.1 M NaZ has pH = 8.90. The K_a of HZ is Critical Thinking

(a) 6.3×10^{-11} (b) 6.3×10^{-10}
(c) 1.6×10^{-5} (d) 1.6×10^{-6}

36. The following equilibrium is established when hydrogen chloride is dissolved in acetic acid.

$$HCl + CH_3COOH \rightleftharpoons Cl^- + CH_3COOH_2^+$$

The set that characterises the conjugate acid-base pairs is

(a) (HCl, CH_3COOH) and $(CH_3COOH_2^+, Cl^-)$
(b) $(HCl, CH_3COOH_2^+)$ and (CH_3COOH, Cl^-)
(c) $(CH_3COOH_2^+, HCl)$ and (Cl^-, CH_3COOH)
(d) (HCl, Cl^-) and $(CH_3COOH_2^+, CH_3COOH)$

37. A solution contains 10 mL 0.1 N NaOH and 10 mL 0.05 N H_2SO_4, pH of this solution is :

(a) less than 7 (b) 7
(c) zero (d) greater than 7

38. 8 mol of $AB_3(g)$ are introduced into a 1.0 dm^3 vessel. If it dissociates as

$2AB_3(g) \rightleftharpoons A_2(g) + 3B_2(g)$. At equilibrium, 2 mol of A_2 are found to be present. The equilibrium constant of this reaction is Critical Thinking

(a) 2 (b) 3 (c) 27 (d) 36

39. Assuming that the degree of hydrolysis is small, the pH of 0.1 M solution of sodium acetate ($K_a = 1.0 \times 10^{-5}$) will be:

(a) 5.0 (b) 6.0 (c) 8.0 (d) 9.0

40. The hydrogen ion concentration of 0.2 N CH_3COOH which is 40% dissociated is

(a) 0.08N (b) 0.12N (c) 0.80N (d) 1.2N

41. Which one of the following is the correct statement ?

(a) HCO_3^- is the conjugate base of CO_3^{2-}.
(b) NH_2^- is the conjugate acid of NH_3.
(c) H_2SO_4 is the conjugate acid of HSO_4^-.
(d) NH_3 is the conjugate base of NH_2^-.

42. Which equilibrium can be described as an acid-base reaction using the Lewis acid-base definition but not using the Bronsted-Lowry definition?
(a) $2NH_3 + H_2SO_4 \rightleftharpoons 2NH_4^+ + SO_4^{2-}$
(b) $NH_3 + CH_3COOH \rightleftharpoons NH_4^+ + CH_3COO^-$
(c) $H_2O + CH_3COOH \rightleftharpoons H_3O^+ + CH_3COO^-$
(d) $[Cu(H_2O)_4]^{2+} + 4NH_3 \rightleftharpoons [Cu(NH_3)_4]^{2+} + 4H_2O$

43. At a certain temperature the dissociation constants of formic acid and acetic acid are 1.8×10^{-4} and 1.8×10^{-6} respectively. The concentration of acetic acid solution in which the hydrogen ion has the same concentration as in 0.001 M formic acid solution is equal to **Tricky**
(a) 0.001 M (b) 0.01 M
(c) 0.1 M (d) 0.0001 M

44. One mole of O_2(g) and two moles of SO_2(g) were heated in a closed vessel of one-litre capacity at 1098 K. At equilibrium 1.6 moles of SO_3(g) were found. The equilibrium constant K_c of the reaction would be **Toughnut**
(a) 30 (b) 40 (c) 80 (d) 60

45. The increase of pressure on ice $\rightleftharpoons$ water system at constant temperature will lead to
(a) a decrease in the entropy of the system
(b) an increase in the Gibb's energy of the system
(c) no effect on the equilibrium
(d) a shift of the equilibrium in the forward direction

46. What is the decreasing order of basic strengths of
$OH^-, NH_2^-, H-C\equiv C^-$ and $CH_3-CH_2^-$
(a) $CH_3-CH_2^- > NH_2^- > H-C\equiv C^- > OH^-$
(b) $H-C\equiv C^- > CH_3CH_2^- > NH_2^- > OH^-$
(c) $OH^- > NH_2^- > H-C\equiv C^- > CH_3-CH_2^-$
(d) $NH_2^- > H-C\equiv C^- > OH^- > CH_3-CH_2^-$

47. When CO_2 dissolves in water, the following equilibrium is established
$CO_2 + 2H_2O \rightleftharpoons H_3O^+ + HCO_3^-$
for which the equilibrium constant is 3.8×10^{-7} and pH = 6.0. The ratio of $[HCO_3^-]$ to $[CO_2]$ would be
(a) 3.8×10^{-13} (b) 3.8×10^{-1}
(c) 6.0 (d) 13.4

48. The first and second dissociation constants of an acid H_2A are 1.0×10^{-5} and 5.0×10^{-10} respectively. The overall dissociation constant of the acid will be
(a) 0.2×10^5 (b) 5.0×10^{-5}
(c) 5.0×10^{15} (d) 5.0×10^{-15}

49. Values of dissociation constant, K_a are given as follows :

Acid	K_a
HCN	6.2×10^{-10}
HF	7.2×10^{-4}
HNO_2	4.0×10^{-4}

Correct order of increasing base strength of the base CN^-, F^- and NO_2^- will be :
(a) $F^- < CN^- < NO_2^-$
(b) $NO_2^- < CN^- < F^-$
(c) $F^- < NO_2^- < CN^-$
(d) $NO_2^- < F^- < CN^-$

50. Given
(i) $HCN(aq) + H_2O(l) \rightleftharpoons H_3O^+(aq) + CN^-(aq)$
$K_a = 6.2 \times 10^{-10}$
(ii) $CN^-(aq) + H_2O(l) \rightleftharpoons HCN(aq) + OH^-(aq)$
$K_b = 1.6 \times 10^{-5}$.
These equilibria show the following order of the relative base strength,
(a) $OH^- > H_2O > CN^-$ (b) $OH^- > CN^- > H_2O$
(c) $H_2O > CN^- > OH^-$ (d) $CN^- > H_2O > OH^-$

51. Which of the following statements about pH and H^+ ion concentration is incorrect?
(a) Addition of one drop of concentrated HCl in NH_4OH solution decreases pH of the solution.
(b) A solution of the mixture of one equivalent of each of CH_3COOH and NaOH has a pH of 7
(c) pH of pure neutral water is not zero
(d) A cold and concentrated H_2SO_4 has lower H^+ ion concentration than a dilute solution of H_2SO_4

52. Assuming that the buffer in the blood is $CO_2 - HCO_3^-$. Calculate the ratio of conjugate base to acid necessary to maintain blood at its proper pH of 7.4.
$K_1(H_2CO_3) = 4.5 \times 10^{-7}$

(a) 11 (b) 8 (c) 6 (d) 14

53. If the reaction between CO_2 and H_2O is
$$CO_2 + H_2O \rightleftharpoons H_2CO_3 \rightleftharpoons H^+ + HCO_3^-$$
If CO_2 escapes from the system
(a) pH will decrease
(b) H^+ concentration will decrease
(c) H_2CO_3 concentration will be altered
(d) The forward reaction is promoted

54. Calculate the pH of a solution containing 0.1 M HCO_3^- and 0.2 M CO_3^{2-}
$[K_1(H_2CO_3) = 4.2 \times 10^{-7} \times 10$ and
$K_2(HCO_3^-) = 4.8 \times 10^{-11}]$.
(a) 3.18 (b) 10.62 (c) 6.62 (d) 9.31

55. What is the pH of a 10^{-4} M OH^- solution at 330K, if K_w at 330K is $10^{-13.6}$? Tricky
(a) 4 (b) 9.0 (c) 10 (d) 9.6

56. What happens when an inert gas is added to an equilibrium keeping volume unchanged?
(a) More product will form
(b) Less product will form
(c) More reactant will form
(d) Equilibrium will remain unchanged

57. The degree of dissociation of 0.1M weak acid HA is 0.5%. If 2 mL of 1.0 M HA solution is diluted to 32 mL the degree of dissociation of acid and H_3O^+ ion concentration in the resulting solution will be respectively Toughnut
(a) 0.02 and 3.125×10^{-4}
(b) 1.25×10^{-3} and 0.02
(c) 0.02 and 1.25×10^{-3}
(d) 0.02 and 8.0×10^{-12}

58. 20 mL of 0.2 M NaOH are added to 50 mL of 0.2 M CH_3COOH ($K_a = 1.8 \times 10^{-5}$) the pH of the solution is
(a) 4.56 (b) 4.73 (c) 9.45 (d) 6.78

59. Why only As^{3+} gets precipitated as As_2S_3 and not Zn^{2+} as ZnS when H_2S is passed through an acidic solution containing As^{3+} and Zn^{2+}?
(a) Solubility product of As_2S_3 is less than that of ZnS
(b) Enough As^{3+} are present in acidic medium
(c) Zinc salt does not ionise in acidic medium
(d) Solubility product changes in presence of an acid

60. The conjugate base of hydrazoic acid is:
(a) N^{-3} (b) N_3^- (c) N_2^- (d) HN_3^-

61. NaOH is a strong base. What will be pH of 5.0×10^{-2} M NaOH solution ? (log 2 = 0.3)
(a) 14.00 (b) 13.70 (c) 13.00 (d) 12.70

62. The solubility product of Ag_2CrO_4 is 32×10^{-12}.
What is the concentration of CrO_4^{2-} ions in that solution (in g L^{-1})
(a) 2×10^{-4} (b) 8×10^{-4}
(c) 8×10^{-8} (d) 16×10^{-4}

63. What is the minimum concentration of SO_4^{2-} required to precipitate $BaSO_4$ in a solution containing 1.0×10^{-4} mole of Ba^{2+}?
K_{sp} for $BaSO_4 = 4 \times 10^{-10}$
(a) 4×10^{-10} M (b) 2×10^{-7} M
(c) 4×10^{-6} M (d) 2×10^{-3} M

64. Equimolar solutions of the following compounds are prepared separately in water. Which will have the lowest pH value ?
(a) $BeCl_2$ (b) $SrCl_2$
(c) $CaCl_2$ (d) $MgCl_2$

65. Addition of which chemical will decrease the hydrogen ion concentration of an acetic acid solution
(a) NH_4Cl (b) $Al_2(SO_4)_3$
(c) $AgNO_3$ (d) NaCN

66. What is the percentage hydrolysis of NaCN in N/80 solution when the dissociation constant for HCN is 1.3×10^{-9} and $K_w = 1.0 \times 10^{-14}$
(a) 2.48 (b) 5.26 (c) 8.2 (d) 9.6

67. The dissociation constant of 0.1 M acetic acid solution is 1.8×10^{-5}. If 1 L of this solution is mixed with 0.05 mole of HCl, what will be pH of mixture ? [log 5 = 0.7]
(a) 1.3 (b) 2.6 (c) 1.9 (d) 3.4

68. For preparing a buffer solution of pH 6 by mixing sodium acetate and acetic acid, the ratio of the concentration of salt and acid should be ($K_a = 10^{-5}$)

(a) 1 : 10 (b) 10 : 1 (c) 100 : 1 (d) 1 : 100

69. In a saturated solution of the sparingly soluble strong electrolyte $AgIO_3$ (molecular mass = 283) the equilibrium which sets is
$AgIO_3(s) \rightleftharpoons Ag^+(aq) + IO_3^-(aq)$. If the solubility product constant K_{sp} of $AgIO_3$ at a given temperature is 1.0×10^{-8}, what is the mass of $AgIO_3$ contained in 100 mL of its saturated solution? Toughnut

(a) 1.0×10^{-4} g (b) 28.3×10^{-2} g
(c) 2.83×10^{-3} g (d) 1.0×10^{-7} g.

70. At 25°C, the dissociation constant of a base, BOH, is 1.0×10^{-12}. The concentration of hydroxyl ions in 0.01 M aqueous solution of the base would be
(a) 1.0×10^{-5} mol L^{-1} (b) 1.0×10^{-6} mol L^{-1}
(c) 2.0×10^{-6} mol L^{-1} (d) 1.0×10^{-7} mol L^{-1}

71. What would be the pH of a solution obtained by mixing 5g of acetic acid and 7.5g of sodium acetate and making the volume equal to 500 mL? ($K_a = 1.75 \times 10^{-5}$, $pK_a = 4.76$)
(a) pH=4.70
(b) pH<4.70
(c) pH of solution will be equal to pH of acetic acid
(d) 4.76<pH<5.0

72. In some solutions, the concentration of H_3O^+ remains constant even when small amounts of strong acid or strong base are added to them. These solutions are known as:
(a) Ideal solutions (b) Colloidal solutions
(c) True solutions (d) Buffer solutions

73. If degree of dissociation of pure water at 100 °C is 1.8×10^{-8}, then the dissociation constant of water will be (density of H_2O = 1 g/cc) Tricky

(a) 1×10^{-12} (b) 1×10^{-14}
(c) 1.8×10^{-12} (d) 1.8×10^{-14}

74. The dissociation constant of two acids HA_1 and HA_2 are 3.14×10^{-4} and 1.96×10^{-5} respectively. The relative strength of the acids will be approximately
(a) 1 : 4 (b) 4 : 1 (c) 1 : 16 (d) 16 : 1

75. At 298K a 0.1 M CH_3COOH solution is 1.34 % ionized. The ionization constant K_a for acetic acid will be
(a) 1.82×10^{-5} (b) 18.2×10^{-5}
(c) 0.182×10^{-5} (d) None of these

76. A buffer solution is prepared by mixing 10 mL of 1.0 M CH_3COOH and 20 mL of 0.5 M CH_3COONa and then diluted to 100 mL with distilled water. If pK_a of CH_3COOH is 4.76, what is the pH of the buffer solution ? Critical Thinking

(a) 5.8 (b) 4.34 (c) 5.21 (d) 4.76

77. Which one of the following arrangements represents the correct order of the proton affinity of the given species :
(a) $I^- < F^- < HS^- < NH_2^-$
(b) $HS^- < NH_2^- < F^- < I^-$
(c) $F^- < I^- < NH_2^- < HS^-$
(d) $NH_2^- < HS^- < I^- < F^-$

78. Consider the following equilibrium
$AgCl\downarrow + 2NH_3 \rightleftharpoons [Ag(NH_3)_2]^+ + Cl^-$
White precipitate of AgCl appears on adding which of the following?
(a) NH_3 (b) aqueous NaCl
(c) aqueous HNO_3 (d) aqueous NH_4Cl

79. The degree of hydrolysis in hydrolytic equilibrium
$A^- + H_2O \rightleftharpoons HA + OH^-$
at salt concentration of 0.001 M is : ($K_a = 1 \times 10^{-5}$)
(a) 1×10^{-3} (b) 1×10^{-4}
(c) 5×10^{-4} (d) 1×10^{-6}

80. A litre of solution is saturated with AgCl. To this solution if 1.0×10^{-4} mole of solid NaCl is added, what will be the $[Ag^+]$, assuming no volume change?
(a) More (b) Less (c) Equal (d) Zero

81. If the equilibrium constant of the reaction of weak acid HA with strong base is 10^{-7}, then pOH of the aqueous solution of 0.1M NaA is Toughnut

(a) 8 (b) 10 (c) 4 (d) 5

82. How many gms of CaC_2O_4 will dissolve in one litre of saturated solution. K_{sp} of CaC_2O_4 is 2.5×10^{-9} mol^{2}lit^{-2}
(a) 0.0064 g (b) 0.0128 g
(c) 0.0032 g (d) None of these

83. Zirconium phosphate $[Zr_3(PO_4)_4]$ dissociates into three zirconium cations of charge + 4 and four phosphate anions of charge – 3. If molar solubility of zirconium phosphate is denoted by S and its solubility product by K_{sp} then which of the following relationship between S and K_{sp} is correct?

Critical Thinking

(a) $S = \{K_{sp}/(6912)^{1/7}\}$ (b) $S = \{K_{sp}/144\}^{1/7}$
(c) $S = \{K_{sp}/6912\}^{1/7}$ (d) $S = \{K_{sp}/6912\}^{7}$

84. Solid $Ba(NO_3)_2$ is gradually dissolved in a 1.0×10^{-4} M Na_2CO_3 solution. At which concentration of Ba^{2+}, precipitate of $BaCO_3$ begins to form ? (K_{sp} for $BaCO_3 = 5.1 \times 10^{-9}$)
(a) 5.1×10^{-5} M (b) 7.1×10^{-8} M
(c) 4.1×10^{-5} M (d) 8.1×10^{-7} M

85. A solution of NH_4Cl and NH_3 has pH = 8.0. Which of the following hydroxides may be precipitated when this solution is mixed with equal volume of 0.2 M of metal ion.
(a) $Ba(OH)_2 (K_{sp} = 1.1 \times 10^{-4})$
(b) $Mg(OH)_2 (K_{sp} = 3.5 \times 10^{-4})$
(c) $Fe(OH)_2 (K_{sp} = 8.1 \times 10^{-16})$
(d) $Ca(OH)_2 (K_{sp} = 2.1 \times 10^{-5})$.

86. On addition of increasing amount of $AgNO_3$ to 0.1 M each of NaCl and NaBr in a solution, what % of Br^- ion get precipitated when Cl^- ion starts precipitating. K_{sp} (AgCl) = 1.0×10^{-10}, K_{sp} (AgBr) = 1×10^{-13}
(a) 0.11 (b) 99.9 (c) 0.01 (d) 9.99

87. The percentage hydrolysis of 0.15 M solution of ammonium acetate, K_a for CH_3COOH is 1.8×10^{-5} and K_b for NH_3 is 1.8×10^{-5}
(a) 0.55 (b) 4.72 (c) 9.38 (d) 5.56

88. Which one of the following arrangements represents the correct order of solubilities of sparingly soluble salts Hg_2Cl_2, $Cr_2(SO_4)_3$, $BaSO_4$ and $CrCl_3$ respectively?

Toughnut

(a) $BaSO_4 > Hg_2Cl_2 > Cr_2(SO_4)_3 > CrCl_3$
(b) $BaSO_4 > Hg_2Cl_2 > CrCl_3 > Cr_2(SO_4)_3$
(c) $BaSO_4 > CrCl_3 > Hg_2Cl_2 > Cr_2(SO_4)_3$
(d) $Hg_2Cl_2 > BaSO_4 > CrCl_3 > Cr_2(SO_4)_3$

89. What will be the H^+ ion concentration in a solution prepared by mixing 50 mL of 0.20 M NaCl, 25 mL of 0.10 M NaOH and 25 mL of 0.30 N HCl?
(a) 0.5 M (b) 0.05 M
(c) 0.02 M (d) 0.10 M

90. The concentration of hydroxyl ion in a solution left after mixing 100 mL of 0.1 M $MgCl_2$ and 100 mL of 0.2 M NaOH
(K_{sp} of $Mg(OH)_2 = 1.2 \times 10^{-11}$) is
(a) 2.8×10^{-4} (b) 2.8×10^{-3}
(c) 2.8×10^{-2} (d) 2.8×10^{-5}

91. The 0.001M solution of $Mg(NO_3)_2$ is adjusted to pH 9, K_{sp} of $Mg(OH)_2$ is 8.9×10^{-12}. At this pH
(a) $Mg(OH)_2$ will be precipitated
(b) $Mg(OH)_2$ is not precipitated
(c) $Mg(OH)_3$ will be precipitated
(d) $Mg(OH)_3$ is not precipitated

92. The solubility (in mol L^{-1}) of AgCl ($K_{sp} = 1.0 \times 10^{-10}$) in a 0.1 M KCl solution will be
(a) 1.0×10^{-9} (b) 1.0×10^{-10}
(c) 1.0×10^{-5} (b) 1.0×10^{-11}

93. A solution is saturated with respect to $SrCO_3$ and SrF_2. The $[CO_3^{2-}]$ was found to be 1.2×10^{-3} M. The concentration of F^- in the solution would be
Given K_{sp} of $SrCO_3 = 7.0 \times 10^{-10}$ M^2,
K_{sp} of $SrF_2 = 7.9 \times 10^{-10}$ M^3,
(a) 1.3×10^{-3} M (b) 2.6×10^{-2} M
(c) 3.7×10^{-2} M (d) 5.8×10^{-7} M

94. If pK_b for fluoride ion at 25°C is 10.83, the ionisation constant of hydrofluoric acid in water at this temperature is
(a) 3.52×10^{-3} (b) 6.75×10^{-4}
(c) 5.38×10^{-2} (d) 1.74×10^{-5}

95. If K_{sp} $(PbSO_4) = 1.8 \times 10^{-8}$ and K_a $(HSO_4^-) = 1.0 \times 10^{-2}$ the equilibrium constant for the reaction.

Toughnut

$$PbSO_4(s) + H^+(aq) \rightleftharpoons HSO_4^-(aq) + Pb^{2+}(aq)$$ is
(a) 1.8×10^{-6} (b) 1.8×10^{-10}
(c) 2.8×10^{-10} (d) 1.0×10^{-2}

96. The dissociation constants of a weak acid HA and weak base BOH are 2×10^{-5} and 5×10^{-6} respectively. The equilibrium constant for the neutralisation reaction of the two is

(a) 1.0×10^{4} (b) 1.0×10^{-4}
(c) 1.0×10^{-10} (d) 2.5×10^{-1}

97. Solid $AgNO_3$ is slowly added to a solution containing each of 0.01 M NaCl and 0.001 M NaBr. What will be the concentration of Cl^- ions in solution when AgBr will just start to precipitate ? $K_{sp}(AgBr) = 3.6 \times 10^{-13}$, $K_{sp}(AgCl) = 1.8 \times 10^{-10}$.

(a) 1.8×10^{-7} (b) 3.6×10^{-10}
(c) 0.01 (d) 2×10^{-4}

98. What is the unit of K_p for the reaction?

$$CS_2(g) + 4H_2(g) \rightleftharpoons CH_4(g) + 2H_2S(g)$$

(a) atm (b) atm^{-2} (c) atm^2 (d) atm^{-1}

99. In the dissociation of PCl_5 as

$$PCl_5(g) \rightleftharpoons PCl_3(g) + Cl_2(g)$$

if the degree of dissociation is α at equilibrium pressure P, then the equilibrium constant for the reaction is

(a) $K_p = \dfrac{\alpha^2}{1+\alpha^2 P}$ (b) $K_p = \dfrac{\alpha^2 P^2}{1-\alpha^2}$

(c) $K_p = \dfrac{P^2}{1-\alpha^2}$ (d) $K_p = \dfrac{\alpha^2 P}{1-\alpha^2}$

100. The equilibrium constant for a reaction is K, and the reaction quotient is Q. For a particular reaction mixture, the ratio $\dfrac{K}{Q}$ is 0.33. This means that :

(a) The reaction mixture will equilibrate to form more reactant species
(b) the reaction mixture will equilibrate to form more product species
(c) the equilibrium ratio of reactant to product concentrations will be 3
(d) the equilibrium ratio of reactant to product concentrations will be 0.33

Answer KEY

1	(c)	11	(c)	21	(c)	31	(a)	41	(c)	51	(b)	61	(d)	71	(d)	81	(c)	91	(b)
2	(c)	12	(d)	22	(c)	32	(d)	42	(d)	52	(a)	62	(a)	72	(d)	82	(a)	92	(a)
3	(b)	13	(b)	23	(b)	33	(b)	43	(b)	53	(b)	63	(c)	73	(d)	83	(c)	93	(c)
4	(c)	14	(b)	24	(c)	34	(b)	44	(c)	54	(b)	64	(a)	74	(b)	84	(a)	94	(b)
5	(d)	15	(a)	25	(d)	35	(c)	45	(d)	55	(d)	65	(d)	75	(a)	85	(c)	95	(a)
6	(a)	16	(c)	26	(c)	36	(d)	46	(a)	56	(d)	66	(a)	76	(d)	86	(b)	96	(a)
7	(b)	17	(a)	27	(a)	37	(b)	47	(b)	57	(c)	67	(a)	77	(a)	87	(a)	97	(c)
8	(d)	18	(c)	28	(a)	38	(c)	48	(d)	58	(a)	68	(b)	78	(c)	88	(b)	98	(b)
9	(a)	19	(a)	29	(d)	39	(d)	49	(c)	59	(a)	69	(c)	79	(a)	89	(b)	99	(d)
10	(b)	20	(d)	30	(b)	40	(a)	50	(b)	60	(b)	70	(d)	80	(b)	90	(a)	100	(a)

Hints & Solutions

1. (c)

	W	+ X	⇌ Y	+ Z
Initial	1	1	0	0
At eqn.	$1-\alpha$	$1-\alpha$	α	α

$$K_{eq} = 9 = \frac{\alpha^2}{(1-\alpha)^2}$$

$\therefore \alpha = 0.75$

Moles of Y = 0.75

2. (c) $P_{total} = 3P$

$$\Rightarrow \quad P = \frac{0.318}{3} = 0.106$$

$\therefore \quad K_p = 4P^3 = 4.76 \times 10^{-3}$

3. (b)

	$2SO_3$	⇌ $2SO_2$	+ O_2
Initial	1	0	0
At equilibrium	$1-2\alpha$	2α	α

$\therefore 2\alpha = 0.6, \ \therefore \alpha = 0.3$

$[SO_3] = 1 - 2\alpha = 1 - 0.6 = 0.4$

$[SO_2] = 2\alpha = 0.6$

$[O_2] = \alpha = 0.3$

$$K_{eq} = \frac{\alpha \times (2\alpha)^2}{(1-2\alpha)^2} = \frac{0.3 \times 0.6 \times 0.6}{0.4 \times 0.4} = 0.675$$

4. (c)

	$2HX(g)$ ⇌	$H_2(g)$	+	$X_2(g)$
At eqm.		1.2×10^{-3}M		1.2×10^{-4}M

$$K_{eq} = \frac{[H_2][X_2]}{[HX]^2}$$

$$10^{-5} = \frac{1.2\times10^{-3} \times 1.2\times10^{-4}}{[HX]^2}$$

$$[HX] = \sqrt{\frac{1.2\times1.2\times10^{-7}}{10^{-5}}}$$

$= 1.2 \times 10^{-1}$

$= 12 \times 10^{-2}$ M

5. (d) $K = K_2 \times \frac{1}{K_1} = \frac{K_2}{K_1}$.

6. (a) $I_2 \rightleftharpoons 2I^-$

$1-x \qquad 2x$

$$K_c = \frac{(2x)^2}{(1-x)} = 10^{-6}$$

Soln. shows that $(1-x) > 2x$

$\therefore [I_2(g)] > [I^-(g)]$

7. (b)

	A + B ⇌		C + D	
Initial	0.9	0.9	0	0
At eqm.	0.3	0.3	0.6	0.6

$$K_c = \frac{0.6\times0.6}{0.3\times0.3} = 4$$

8. (d) $K_p = \frac{p_{CH_3OH}}{p_{CO} \times p_{H_2}} = \frac{2}{1\times(0.1)^2} = 200$;

For reverse reaction

$$\frac{1}{K_p} = \frac{1}{200} = 5\times10^{-3}\ atm^2$$

9. (a) $2HI \rightleftharpoons H_2 + I_2$. It is 22% decomposed,

$$\therefore \frac{3.20\times22}{100} = 0.704$$

(3.2–0.704) is equal to HI present at equilibrium which is = 2.496

10. (b) At equilibrium the moles of Cl_2 must be $=0.15\times3=0.45$

$$PCl_5 \rightleftharpoons PCl_3 + Cl_2$$

Eqm. Conc. $\frac{x-0.45}{3}$ $\frac{0.45}{3}$ $\frac{0.45}{3}$

$$K_c=\frac{[PCl_3][Cl_2]}{[PCl_5]}$$

$$\therefore 0.04=\frac{0.15\times0.15}{(x-0.45)/3}$$

$\therefore$ x = 2.1 moles

11. (c) $N_2O_4(g) \rightleftharpoons 2NO_2(g)$

t = 0   1   0

t = eq. $1-\alpha$ 2α

Where α = Degree of dissociation.

Mol. wt. of mixture

$$=\frac{(1-\alpha)\times M_{N_2O_4}+2\alpha\times M_{NO_2}}{(1+\alpha)}$$

$$=\frac{(1-0.2)92+2\times0.2\times46}{(1+0.2)}=76.66$$

Now, As per ideal gas equation

PV = nRT

$PM_{mixture}=dRT$

$$\therefore d=\frac{PM_{mix}}{RT}=\frac{1\times76.66}{0.0821\times300}=3.11\,g/L$$

12. (d) $N_2O_4(g) \rightleftharpoons 2NO_2(g)$

Initial moles   1   0

Moles at equil. $(1-\alpha)$ 2α

(α = degree of dissociation)

Total number of moles at equil.

$=(1-\alpha)+2\alpha$

$=(1+\alpha)$

$$p_{N_2O_4}=\frac{(1-\alpha)}{(1+\alpha)}\times P$$

$$p_{NO_2}=\frac{2\alpha}{(1+\alpha)}\times P$$

$$K_P=\frac{(p_{NO_2})^2}{p_{N_2O_4}}=\frac{\left(\frac{2\alpha}{(1+\alpha)}\times P\right)^2}{\left(\frac{1-\alpha}{1+\alpha}\right)\times P}=\frac{4\alpha^2P}{1-\alpha^2}$$

Given, $K_p=2$, $P=0.5$ atm

$$\therefore K_P=\frac{4\alpha^2P}{1-\alpha^2}$$

$$=\frac{4\alpha^2\times0.5}{1-\alpha^2}$$

$\alpha=0.707\approx0.71$

$\therefore$ Percentage dissociation

$=0.71\times100=71$

13. (b) The efflorescent salts loss water to atmosphere

$K_p=p^2_{H_2O}=1.086\times10^{-4}$;

$p_{H_2O}=1.042\times10^{-2}\,\text{atm}=7.92\,\text{mm}$

If H_2O pressure at 25°C is less than 7.92 mm. The reaction

$$CuSO_4.5H_2O(s) \longrightarrow CuSO_4.3H_2O(s)+H_2O(g)$$

will not proceed in RHS.

14. (b) Moles of $N_2=\frac{28}{28}=1$,

Moles of $H_2=\frac{6}{2}=3$

Moles of H_2SO_4 required $=\frac{500\times1}{1000}=0.5$

Moles of NH_3 neutralised by $H_2SO_4=1.0$

$$\left(2NH_3+H_2SO_4\longrightarrow(NH_4)_2SO_4\right)$$

Hence 1 mol of NH_3 by the reaction between N_2 and H_2.

$$N_2 + 3H_2 \rightleftharpoons 2NH_3$$

initial	1	3	0
at eqm	1 − 0.5	3 − 0.5 × 3	1

$$K_c = \frac{1\times 1}{0.5\times(1.5)^3} = 0.592$$

15. (a) $D = \frac{PM}{RT} = \frac{1\times 82.1}{0.0821\times 400} = 2.5\ g\ L^{-1}$

$d = 2.0\ g\ L^{-1}$ (given)

$$\alpha = \frac{D-d}{d(n-1)} = \frac{2.5-2}{2(3-1)} = 0.125 = 12.5\%$$

16. (c) If, $K_c = \frac{[N_2]^{1/2}[H_2]^{3/2}}{[NH_3]^1} = 4$

then, $\frac{[NH_3]^2}{[N_2][H_2]^3} = (4)^{-2} = \frac{1}{16}$

also $K_P = K_c \cdot (RT)^{\Delta n_g}$

$$= \frac{1}{16}\times(800R)^{-2} = \left(\frac{1}{4\times 800R}\right)^2$$

17. (a) $$HI \rightleftharpoons \frac{1}{2}H_2 + \frac{1}{2}I_2$$

at t = 0		0	0
at eq.	(1 − 0.5)	(0.5)	(0.5)

$$\left(\because \alpha = \frac{50}{100} = 0.5\right)$$

Now $K_C = \frac{(0.5)^{1/2}(0.5)^{1/2}}{0.5} = 1$

18. (c) Addition of inert gas at constant volume has no effect on equilibrium.

19. (a) Equilibrium shifted in the backward direction. (Le Chatelier's principle)

20. (d) To calculate the value of K_4 in the given equation we should apply :

eqn. (2) + eqn.(3) × 3 − eqn. (1)

hence $K_4 = \frac{K_2K_3^3}{K_1}$

21. (c) Since reaction is exothermic hence low temperature will favour forward reaction also volume is decreased by applying high pressure.

22. (c) By addition of SO_2 equilibrium will shift to RHS which is exothermic. Hence temperature will increase.

23. (b) $$A + 2B \rightleftharpoons 2C + D$$

A	2B	2C	D
a	$1.5a$	0	0
$(a-x)$	$(1.5a-2x)$	$2x$	x

Hence $K_C = \frac{(2x)^2\times x}{(a-x)(1.5a-2x)^2}$

Given, at equilibrium

$\therefore\ (a-x) = (1.5a-2x)$

$\therefore\ a = 2x$

On solving $K_C = 4$

24. (c) For the reaction

$$2BaO_2(s) \rightleftharpoons 2BaO(s) + O_2(g);$$

$\Delta H = +ve.$

At equilibrium $K_p = P_{O_2}$

Hence, the value of equilibrium constant depends only upon partial pressure of O_2. Further on increasing temperature formation of O_2 increases as this is an endothermic reaction. Hence, pressure of O_2 is dependent on temperature.

25. (d) $P_{NH_3} = P_{H_2S} = \frac{P}{2}$ atm

$$K_p = P_{NH_3}P_{H_2S} = \left(\frac{P}{2}\right)^2 = \frac{P^2}{4}$$

$$\Delta G = -RT\ln K_p = -RT\ln\left(\frac{p}{2}\right)^2$$

$$= -2RT\,[\ln p - \ln 2]$$

26. **(c)** $H_2(g) + I_2(g) \rightleftharpoons 2HI(g)$

$K_p = K_c\,(RT)^{\Delta n}$;

$\Delta n = 2 - 2 = 0; \therefore K_p = K_c$

27. **(a)** Since T_2 has similar chemical properties as H_2 so upon mixing, NH_2T and NHT_2 are formed.

28. **(a)** $K_P = K_C\,(RT)^{\Delta n_g}$

For the reaction

$$CO(g) + \frac{1}{2}O_2(g) \rightleftharpoons CO_2(g)$$

$$\Delta n_g = 1 - \left(1 + \frac{1}{2}\right) = -\frac{1}{2}$$

$$\therefore \quad K_p = \frac{K_C}{\sqrt{RT}}; \quad \frac{K_P}{K_C} = \frac{1}{\sqrt{RT}}$$

29. **(d)** The condition for the equilibrium is $(\Delta G)_{T,P} = 0$ and $(\Delta S)_{U,V} = 0$

30. **(b)** (I) $N_2 + 2O_2 \overset{K_1}{\rightleftharpoons} 2NO_2$

$$K_1 = \frac{[NO_2]^2}{[N_2][O_2]^2} \quad \text{...(i)}$$

(II) $2NO_2 \overset{K_2}{\rightleftharpoons} N_2 + 2O_2$

$$K_2 = \frac{[N_2][O_2]^2}{[NO_2]^2} \quad \text{...(ii)}$$

(III) $NO_2 \overset{K_3}{\rightleftharpoons} \frac{1}{2}N_2 + O_2$

$$K_3 = \frac{[N_2]^{1/2}\,[O_2]}{[NO_2]}$$

$$\therefore \ (K_3)^2 = \frac{[N_2][O_2]^2}{[NO_2]^2} \quad \text{...(ii)}$$

$\therefore$ from equation (i), (ii) and (iii)

$$K_1 = \frac{1}{K_2} = \frac{1}{(K_3)^2}$$

31. **(a)** When ionic product i.e. the product of the concentration of ions in the solution exceeds the value of solubility product, formation of precitpiate occurs.

$CaF_2 \rightleftharpoons Ca^{2+} + 2F^-$

Ionic product = $[Ca^{2+}]\,[F^-]^2$

when, $[Ca^{2+}] = 1\times 10^{-2}$ M

$[F^-]^2 = (1\times 10^{-3})^2$ M

$= 1\times 10^{-6}$ M

$\therefore\ [Ca^{2+}]\,[F^-]^2 = (1\times10^{-2})\,(1\times10^{-6}) = 1\times10^{-8}$

In this case,

Ionic product $(1\times 10^{-8}) >$

solubility product (1.7×10^{-10})

32. **(d)** $PbI_2 \rightleftharpoons Pb^{++} + 2I^-$

$\qquad\qquad s \qquad 2s$

$K_{sp} = s \times (2s)^2 = 4s^3$

$$= 4\times\left(\frac{0.7}{461.2}\right)^3 = 14.0\times10^{-9}$$

33. **(b)** Van't Hoff reaction isotherm is $\Delta G = \Delta G^o + RT \ln Qp$

When the reaction is in a state of equilibrium $\Delta G = 0$

Then $\Delta G^o = -RT\ln Q_p = -RT\ln K_p$

34. **(b)** The reaction mixtures starting either with H_2 or D_2 reach equilibrium with the same composition, except that D_2 and ND_3 are present instead of H_2 and NH_3.

35. **(c)** NaZ is salt of W_A/S_B

$$\therefore\ pH = \frac{1}{2}(pK_w + pK_a + \log C)$$

$8.9 \times 2 = 14 + pK_a + \log 0.1$

$17.8 = 14 + pK_a - 1$

$pK_a = 4.8,$

$K_a = \text{Antilog}\,(-4.8)$

$= 1.585 \times 10^{-5} \approx 1.6 \times 10^{-5}.$

36. (d) HCl is stronger acid than CH_3COOH and Cl^- is a stronger base than $CH_3COOH_2^+$ and is the conjugate base of HCl.

$$\underset{\text{acid}_1}{HCl} + \underset{\text{base}_2}{CH_3COOH} \rightleftharpoons \underset{\text{base}_1}{Cl^-} + \underset{\text{acid}_2}{CH_3COOH_2^+}$$

37. (b) Milliequivalents of NaOH = 10 × 0.1 = 1
Milliequivalents of H_2SO_4 = 10 × 0.05 = 0.5

$$\underset{\text{2 equ.}}{2NaOH} + \underset{\text{1 equ.}}{H_2SO_4} \longrightarrow Na_2SO_4 + H_2O$$

∴ 1 equivalent of NaOH reacts with 0.5 eq. of H_2SO_4 to give neutral (pH = 7) solution.

38. (c) $2AB_3(g) \rightleftharpoons A_2(g) + 3B_2(g)$

at $t = 0$ 8   0   0

at eq.   (8 – 2 × 2) = 4   2   3 × 2 = 6

now

$$K_C = \frac{[A_2][B_2]^3}{[AB_3]^2} = \frac{2/1 \times [6/1]^3}{[4/1]^2} = 27$$

39. (d) Sodium acetate is a salt of strong base and weak acid.

$$\therefore \ pH = 7 + \frac{1}{2}pK_a + \frac{1}{2}\log c$$

where $pk_a = -\log K_a = -\log 10^{-5} = 5$

$\log c = \log 10^{-1} = -1$

$$pH = 7 + \frac{5}{2} - \frac{1}{2} = 9.0$$

40. (a) $[H^+] = C\alpha = 0.2 \times 0.40 = 0.08$ N.

$$\left(\alpha = \frac{40}{100} = 0.40\right)$$

41. (c) HSO_4^- accepts a proton to form H_2SO_4. Thus H_2SO_4 is the conjugate acid of HSO_4^-.

$$\underset{\text{base}}{HSO_4^-} \xrightarrow{+H^+} \underset{\text{conjugate acid of } HSO_4^-}{H_2SO_4}$$

42. (d) $[Cu(H_2O)_4]^{2+} + 4NH_3 \rightleftharpoons [Cu(NH_3)_4]^{2+} + 4H_2O$ involves lose and gain of electrons. H_2O is coordinated to Cu by donating electrons (LHS). It is then removed by withdrawing electrons.

43. (b) $[H^+] = \sqrt{C \times K_a} = \sqrt{0.001 \times 1.8 \times 10^{-4}}$ for formic acid

$[H^+] = \sqrt{C_2 \times 1.8 \times 10^{-5}}$ for acetic acid

Equating and solving, $C_2 = 0.01$ M.

44. (c) $O_2 + 2SO_2 \rightleftharpoons 2SO_3$

at t = 0   1   2   0

at equil. $(1-\alpha)$ $2(1-\alpha)$ 2α

Given at equilibrium

$$2\alpha = 1.6$$

$$\alpha = 0.8$$

$$K_C = \frac{(2\alpha)^2}{(1-\alpha)(2-2\alpha)^2} = \frac{(0.8)^2}{(1-0.8)(1-0.8)^2}$$

$$= \frac{0.64}{0.008}$$

$$K_C = 80$$

45. (d) Volume of ice is greater than that of water. The direction in which the reaction will proceed can be predicted by applying Le-Chatelier's principle

$$\text{Pressure} \propto \frac{1}{\text{Volume}}$$

So equilibrium, will shift forward.

46. (a) Follow Lowry Bronsted concept for conjugate pair. Acid strength follows the order $H_2O > C_2H_2 > NH_3 > C_2H_6$ conjugate base will follow the order

$$C_2H_5^- > NH_2^- > C_2H^- > OH^-$$

47. (b) $CO_2 + 2H_2O \rightleftharpoons H_3O^+ + HCO_3^-$;

$$K_c = \frac{[H_3O^+][HCO_3^-]}{[CO_2]};$$

$$\frac{[HCO_3^-]}{[CO_2]} = \frac{3.8 \times 10^{-7}}{10^{-6}} = 3.8 \times 10^{-1}$$

48. (d) $H_2A \rightleftharpoons H^+ + HA^-$

$\therefore K_1 = 1.0 \times 10^{-5} = \frac{[H^+][HA^-]}{[H_2A]}$ (Given)

$HA^- \longrightarrow H^+ + A^{2-}$

$\therefore K_2 = 5.0\times10^{-10} = \frac{[H^+][A^{2-}]}{[HA^-]}$ (Given)

$$K = \frac{[H^+]^2[A^{2-}]}{[H_2A]} = K_1 \times K_2$$

$= (1.0 \times 10^{-5}) \times (5 \times 10^{-10}) = 5 \times 10^{-15}$

49. (c) Higher the value of K_a lower will be the value of pK_a i.e. higher will be the acidic nature. Further CN^-, F^- and NO_2^- are conjugate base of the acids HCN, HF and HNO_2 respectively hence the correct order of base strength will be

$F^- < NO_2^- < CN^-$

($\because$ stronger the acid weaker will be its conjugate base)

50. (b) The more is the value of equilibrium constant, the more is the completion of reaction or more is the concentration of products i.e. the order of relative strength would be

$OH^- > CN^- > H_2O$

51. (b) CH_3COOH is weak acid while NaOH is strong base, so one equivalent of NaOH can not be neutralized with one equivalent of CH_3COOH. Hence the solution of one equivalent of each does not have pH value as 7. Its pH will be towards basic side as NaOH is a strong base hence conc. of OH^- will be more than the conc. of H^+.

52. (a) CO_2 with H_2O forms H_2CO_3

$CO_2 + H_2O \rightleftharpoons H^+ + HCO_3^-$

$K_1 = \frac{[H^+][HCO_3^-]}{[CO_2]} = 4.5\times10^{-7}$

Again pH $= -\log[H^+] = 7.4$

$\therefore [H^+] = 4.0\times10^{-8}$

$\therefore \frac{[HCO_3^-]}{[CO_2]} = \frac{4.5\times10^{-7}}{4\times10^{-8}} = 11$

53. (b) $CO_2 + H_2O \rightleftharpoons H_2CO_3 \rightleftharpoons H^+ + HCO_3^-$

If CO_2 escapes, the equilibrium will shift to LHS and $[H^+]$ concentration will decrease

54. (b) $HCO_3^- \rightarrow H^+ + CO_3^{2-}$

$K_2 = \frac{[H^+][CO_3^{2-}]}{[HCO_3^-]} = 4.8\times10^{-11}$

$[H^+] = \frac{4.8\times10^{-11}[HCO_3^-]}{[CO_3^{2-}]}$

$= 4.8\times10^{-11}(0.1/0.2)$

pH $= -\log[H^+]$

$= -\log(4.8 \times 10^{-11} \times 0.5) = 10.62$

55. (d) Given at 330 K

$K_w = 10^{-13.6}$

i.e. $pK_w = pH + pOH$

$\because$ pOH $= -\log[OH^-]$

$13.6 = pH + pOH$

pOH $= -\log 10^{-4}$

pOH $= 4$

$\therefore$ pH $= 13.6 - 4 = 9.6$

56. (d) On adding inert gas at constant volume the total pressure of the system is increased, but the partial pressure of each reactant and product remains the same. Hence no effect on the state of equilibrium.

57. (c) $\alpha_1 = 0.005 = \sqrt{K_a \times C_1}$

Molarity of diluted solution; $2 \times 1 = 32 \times M$,

$M = \frac{1}{16}(C_2)$

$\alpha_2 = \sqrt{\frac{K_a}{C_2}} = 0.005\sqrt{16} = 0.02$

$\overset{+}{H_3O} = C_2\alpha_2 = \frac{0.02}{16} = 1.25\times10^{-3}$ M

58. (a)

NaOH	+ CH_3COOH $\rightleftharpoons$	CH_3COONa	+ H_2O
20 × 0.2	50 × 0.2	0	0
4	10	0	0
0	6	4	4

$$\text{Conc.} = \frac{\text{Millimoles}}{\text{Total Volume}}$$

$$[CH_3COOH] = \frac{6}{70}; [CH_3COONa] = \frac{4}{70}$$

$$pH = -\log 1.8\times10^{-5} + \log\frac{4/70}{6/70}, pH = 4.56$$

59. (a) K_{sp} of As_2S_3 is less than ZnS. In acidic medium ionisation of H_2S is suppresed (common ion effect) and K_{sp} of ZnS does not exceed.

60. (b) $\underset{\text{Hydrazoic acid}}{N_3H} \rightleftharpoons N_3^- + H^+$

i.e, conjugate base of hydrazoic acid is N_3^-.

61. (d) Given $[OH^-] = 5 \times 10^{-2}$

$\therefore$ $pOH = -\log 5 \times 10^{-2}$

$= -\log 5 + 2\log 10 = 1.30$

$\because$ $pH + pOH = 14$

$\because$ $pH = 14 - pOH$

$= 14 - 1.30 = 12.70$

62. (a) For ternary electrolyte; solubility

$$= \left(\frac{K_{sp}}{4}\right)^{\frac{1}{3}} = \left(\frac{32\times10^{-12}}{4}\right)^{\frac{1}{3}} = 2\times10^{-4}$$

63. (c) $[Ba^{2+}][SO_4^{2-}]$

$K_{sp} = (1.0 \times 10^{-4})[SO_4^{2-}] = 4 \times 10^{-10}$

$\therefore [SO_4^{2-}] = 4\times10^{-6}$ M

64. (a) Metal halide on hydrolysis with water form corresponding hydroxides.

The basic strength of hydroxide increases as we move down in a group. This is because of the increase in size which results in decrease of ionization energy which weakens the strength of M – O bonds in MOH and thus increases the basic strength.

$\underset{\text{Amphoteric}}{Be(OH)_2}$ $\underset{\text{Weak base}}{Mg(OH)_2}$

$\underbrace{Ca(OH)_2\ Sr(OH)_2\ Ba(OH)_2}_{\text{Strong base}}$

Hence, $Be(OH)_2$ will have lowest pH.

65. (d) $NaCN + H_2O \rightleftharpoons NaOH + HCN$ (hydrolysis)

NaOH will give OH^- ions and react with H^+ to form H_2O hence decrease the concentration of H^+ ions.

66. (a) $\alpha = \sqrt{\frac{K_w}{K_a \times c}} = \sqrt{\frac{10^{-14}}{1.3\times10^{-9}\times\frac{1}{80}}} = 2.48\%.$

67. (a) The pH of mixture is due to the HCl only, because CH_3COOH is negligibly ionised due to common ion effect. Thus,

$[H^+] = 0.05\text{ M} = 5 \times 10^{-2}$ M

$pH = -\log(5 \times 10^{-2}) = -[\log 5 + \log 10^{-2}]$

$pH = 1.3$

68. (b) $6 = -\log 10^{-5} + \log\frac{\text{salt}}{\text{acid}} = 5 + \log\frac{\text{salt}}{\text{acid}}$

$\log\frac{\text{salt}}{\text{acid}}$ must be 1. $\therefore \frac{\text{salt}}{\text{acid}} = \frac{10}{1}$ or 10 : 1.

69. (c) Let s = solubility

$AgIO_3 \rightleftharpoons \underset{s}{Ag^+} + \underset{s}{IO_3^-}$

$K_{sp} = [Ag^+][IO_3^-] = s \times s = s^2$

Given $K_{sp} = 1 \times 10^{-8}$

$\therefore s = \sqrt{K_{sp}} = \sqrt{1\times10^{-8}}$

$= 1.0 \times 10^{-4}$ mol/lit $= 1.0 \times 10^{-4} \times 283$ g/lit

($\because$ Molecular mass of $AgIO_3 = 283$)

$$= \frac{1.0\times10^{-4}\times283\times100}{1000}\text{ g/100mL}$$

$= 2.83 \times 10^{-3}$ g/100 mL

70. **(d)** Given $K_b = 1.0 \times 10^{-12}$

[BOH] = 0.01 M [OH$^-$] = ?

$$\begin{array}{llll} & BOH & \rightleftharpoons B^+ & + OH^- \\ t = 0 & c & 0 & 0 \\ t = t_{eq} & c(1-\alpha) & c\alpha & c\alpha \end{array}$$

$$K_b = \frac{c^2\alpha^2}{c(1-\alpha)} = \frac{c\alpha^2}{(1-\alpha)}$$

$$\Rightarrow 1.0 \times 10^{-12} = \frac{0.01\alpha^2}{(1-\alpha)}$$

On calculation, we get, $\alpha = 1.0 \times 10^{-5}$

Now, $[OH^-] = c\alpha = 0.01 \times 10^{-5}$

$= 1 \times 10^{-7} \text{mol L}^{-1}$

71. **(d)** $pH = pK_a + \log\frac{[salt]}{[acid]}$

$$= 4.76 + \log\frac{\frac{7.5}{500}}{\frac{5}{500}} = 4.7 + \log 1.5 = 4.87$$

Hence correct answer is $4.76 < pH < 5.0$

72. **(d)** Solutions which resist the change in the value of pH when small amount of acid or base is added to them are known as buffers.

73. **(d)** As, molarity,

$$= \frac{\text{Wt. of solute per litre of solution}}{\text{Mol. wt. of solute}}$$

Molarity of $H_2O = \frac{1000}{18}$ mol/L

$$\begin{array}{lll} H_2O & \rightleftharpoons H^+ & + OH^- \\ c(1-\alpha) & c\alpha & c\alpha \end{array}$$

Thus, $K_a = \frac{c\alpha^2}{1-\alpha} = c\alpha^2$

$$= \frac{1000}{18} \times (1.8 \times 10^{-8})^2$$

$$= 1.8 \times 10^{-14}$$

74. **(b)** $\frac{\alpha_1}{\alpha_2} = \sqrt{\frac{K_{a_1}}{K_{a_2}}} = \sqrt{\frac{3.14 \times 10^{-4}}{1.96 \times 10^{-5}}} = 4:1$

75. **(a)** $K_a = c\alpha^2 = 0.1 \times \left(\frac{1.34}{100}\right)^2 = 1.8 \times 10^{-5}$

76. **(d)** Total volume = 100 mL

$$[\text{acid}] = 10 \text{ mL} \times \frac{1.0}{100} = 0.1$$

$$[\text{salt}] = 20 \text{ mL} \times \frac{0.5}{100} = 0.1$$

$$\text{pH of acidic buffer} = pK_a + \log\frac{[salt]}{[acid]}$$

$$= 4.76 + \log\frac{0.1}{0.1} = 4.76$$

77. **(a)** The species with the greatest proton affinity will be the strongest base, and its conjugate acid will be the weakest acid. The weakest acid will have the smallest value of K_a. Since HI is a stronger acid than HF which is a stronger acid than H_2S, a partial order of proton affinity is

$I^- < F^- < HS^-$

Since NH_3 is a very weak acid, NH_2^- must be a very strong base. Therefore the correct order of proton affinity is

$I^- < F^- < HS^- < NH_2^-$

78. **(c)** $2HNO_3(aq) + [Ag(NH_3)_2]^+ + Cl^-$

$$\longrightarrow AgCl(s)\downarrow + 2NH_4 + 2NO_3^-$$

When nitric acid is added to amine solution, solution is made acidic and the complex ion dissociates and liberate silver ion to recombine with chloride ion. This is the conformatory test for silver in group 1.

79. **(a)**

$$\begin{array}{lllll} & A^- & + H_2O \rightleftharpoons & HA & + HA \\ \text{(Initial)} & 0.001M & & 0 & 0 \\ \text{(At eq.)} & (0.001-R) & & (0.001\times h) & (0.001\times h) \end{array}$$

$$K_h = \frac{(0.001\times h)(0.001\times h)}{(0.001-h)}$$

or, $K_h = 0.001 \times h^2$ (as, $0.001 - h \approx 0.001$)

$10^{-9} = 0.001 \times h^2$

$$\left(K_h = \frac{K_w}{K_a} = \frac{10^{-14}}{10^{-5}} = 10^{-9}\right)$$

$h^2 = 10^{-6}$; $\therefore h = 10^{-3}$

80. (b) The dissociation of a weak electrolyte (AgCl) is suppressed on adding the strong electrolyte having a common ion (Cl^-), hence concentration of Ag^+ ion will be less.

81. (c) Hydrolysis of a salt is reverse reaction of acid-base neutralization reaction.

$$\therefore K_h = \frac{K_w}{K_a} = \frac{10^{-14}}{10^{-7}} = 10^{-7}$$

$$[OH^-] = ch = c\times\sqrt{\frac{K_h}{c}} = \sqrt{c\times K_h}$$

$$= \sqrt{10^{-8}} = 10^{-4}$$

$\Rightarrow pOH = -\log[OH^-]$

$= -\log[10^{-4}] = 4$

82. (a) $CaC_2O_4 \rightleftharpoons Ca^{2+} + C_2O_4^{2-}$; $K_{sp} = S^2$

$\therefore S = \sqrt{K_{sp}} = \sqrt{2.5\times10^{-9}} = 5\times10^{-5}$ mol/litre

solubility $= 5\times10^{-5}\times128 = 0.0064$ g

83. (c) $\left[Zr_3(PO_4)_4\right] \rightleftharpoons \underset{3S}{3Zr^{4+}} + \underset{4S}{4PO_4^{3-}}$

$K_{sp} = (3S)^3(4S)^4$

$= 27S^3 \times 256\,S^4$

$= 6912\,S^7.$

$$\therefore \quad S = \left(\frac{K_{sp}}{6912}\right)^{1/7}$$

84. (a) Given $Na_2CO_3 = 1.0\times10^{-4}$ M

$\therefore [CO_3^{2-}] = 1.0\times10^{-4}$ M

i.e. $s = 1.0\times10^{-4}$ M

At equilibrium

$[Ba^{2+}][CO_3^{2-}] = K_{sp}$ of $BaCO_3$

$$[Ba^{2+}] = \frac{K_{sp}}{[CO_3^{2-}]} = \frac{5.1\times10^{-9}}{1.0\times10^{-4}}$$

$= 5.1\times10^{-5}$ M

85. (c) pH = 8, pOH = 6; $[OH^-] = 10^{-6}$ M;

Ionic product of $Fe(OH)_2 = 0.2\times(1\times10^{-6})^2$

$= 2\times10^{-13} > K_{sp}(= 8.1\times10^{-16})$

86. (b) To precipitate the AgCl

$[Ag^+]$ required

$$= \frac{K_{sp}(AgCl)}{[Cl^-]} = \frac{1.0\times10^{-10}}{0.1} = 1.0\times10^{-9}\text{ M}$$

$$[Br^-] \text{ left at this stage} = \frac{K_{sp}(AgBr)}{[Ag^+]}$$

$$= \frac{1.0\times10^{-13}}{1.0\times10^{-9}} = 1.0\times10^{-4}\text{ M}$$

% of remaining

$$[Br^-] = \frac{1.0\times10^{-4}}{0.1}\times100 = 0.1$$

% of Br^- to be precipitated

$= 100 - 0.1 = 99.9$

87. (a) $\alpha = \sqrt{\dfrac{K_w}{K_a\times K_b}}$

$$= \sqrt{\frac{1\times10^{-14}}{1.8\times10^{-5}\times1.8\times10^{-5}}} = 0.55\%$$

88. (b) $Cr_2(SO_4)_3 \rightleftharpoons \underset{2s}{2Cr^{3+}} + \underset{3s}{3SO_4^{2-}}$

$K_{sp} = (2s)^2(3s)^3 = 4s^2\times27s^3 = 108s^5$

$$s = \left(\frac{K_{sp}}{108}\right)^{1/5}$$

$Hg_2Cl_2 \rightleftharpoons \underset{2s}{2Hg^{2+}} + \underset{2s}{2Cl^-}$

$K_{sp} = (2s)^2\times(2s)^2 = 16s^4$

$$s = \left(\frac{K_{sp}}{16}\right)^{1/4}$$

$$BaSO_4 \rightleftharpoons \underset{s}{Ba^{2+}} + \underset{s}{SO_4^{2-}}$$

$K_{sp} = s^2$

$s = \sqrt{K_{sp}}$

$$CrCl_3 \rightleftharpoons \underset{s}{Cr^{3+}} + \underset{3s}{3Cl^-}$$

$K_{sp} = s \times (3s)^3 = 27\,s^4$

$$s = \left(\frac{K_{sp}}{27}\right)^{1/4}$$

Hence the correct order of solubilities of salts is

$$\sqrt{K_{sp}} > \left(\frac{K_{sp}}{16}\right)^{1/4} > \left(\frac{K_{sp}}{27}\right)^{1/4} > \left(\frac{K_{sp}}{108}\right)^{1/5}$$

89. **(b)** $[H^+] = \dfrac{(25 \times 0.3 - 25 \times 0.1)}{100}\text{meq} = 0.05M$

90. **(a)** $MgCl_2 + 2NaOH \rightarrow Mg(OH)_2 + 2NaCl$

100×0.1	100×0.2	0	0	Initial moles
0	0	10	20	After mixing

$$\therefore K_{sp} = \left[\frac{10}{200}\right]\left[\frac{20}{200}\right]^2 = 5 \times 10^{-4}.$$

But actually K_{sp} of

$Mg(OH)_2 = 1.2 \times 10^{-11}$

$\therefore\ 4S^3 = 1.2 \times 10^{-11}$

Find S then $[OH^-] = 2S$ which is 2.8×10^{-4}.

91. **(b)** $pH = 9;\ [H^+] = 10^{-9};[OH^-] = 10^{-5};$

$[Mg^{2+}] = 1 \times 10^{-3};$

$$[Mg^{2+}][OH^-]^2 = 1 \times 10^{-13}$$

given K_{sp} of $Mg(OH)_2 = 8.9 \times 10^{-12}$

which is more than 1×10^{-13}.

Hence $Mg(OH)_2$ will not precipitate

92. **(a)** Let solubility of AgCl = x mole/L

$$AgCl \rightleftharpoons Ag^+ + Cl^-$$

i.e., $K_{sp(AgCl)} = x \times x$

$$KCl \longrightarrow K^+ + \underset{0.1}{Cl^-}$$

$[Cl^-]$ from KCl = 0.1 m

Total $[Cl^-]$ in solution = $x + 0.1$

$K_{sp}(AgCl) = [Ag^+][Cl^-] = x(x + 0.1)$

$1.0 \times 10^{-10} = x(x + 0.1)$

$1.0 \times 10^{-10} = x^2 + 0.1x$

$1.0 \times 10^{-10} = 0.1x$ (as $x^2 << 1$)

$x = 1.0 \times 10^{-9}$ mol/L

93. **(c)** The two K_{sp} values do not differ very much. So it is a case of simultaneous equilibria, where the concentration of any species can not be neglected.

$$\frac{[Sr^{2+}][F^-]^2}{[Sr^{2+}][CO_3^{2-}]} = \frac{K_{sp_{SrF_2}}}{K_{sp_{SrCO_2}}}$$

$$= \frac{7.9 \times 10^{-10}}{7.0 \times 10^{-10}} = 1.128$$

$\therefore\ [F^-]^2 = 1.128 \times 1.2 \times 10^{-3} = 13.5 \times 10^{-4}$

$\therefore\ [F^-] = (13.5 \times 10^{-4})^{1/2}$

$= 3.674 \times 10^{-2} \approx 3.7 \times 10^{-2}$ M.

94. **(b)** $K_w = K_a \times K_b$

$K_b = 10^{-10.83} = 1.48 \times 10^{-11}$

$$\therefore\ K_a = \frac{K_w}{K_b} = \frac{10^{-14}}{1.48 \times 10^{-11}} = 6.75 \times 10^{-4}$$

95. **(a)** $PbSO_4(s) \rightleftharpoons$

$$Pb^{2+}(aq) + SO_4^{2-}(aq)\ (K_{sp})\ ...(i)$$

$HSO_4^-(aq) \rightleftharpoons$

$$H^+(aq) + SO_4^{2-}(aq)\ (K_a)\ ...(ii)$$

Subtracting equation (ii) from (i), then

$$PbSO_4(s) + H^+(aq) \rightleftharpoons Pb^{2+}(aq) + HSO_4^-(aq)\ (K_{eq})$$

$$\therefore\ K_{eq} = \frac{K_{sp}}{K_a}$$

$$= \frac{1.8\times10^{-8}}{1.0\times10^{-2}} = 1.8\times10^{-6}$$

96. (a) $HA + BOH \rightleftharpoons BA + H_2O$

$$\therefore\ K_H = \frac{K_w}{K_a \times K_b}.$$

The inverse of K_H is $K_c = \dfrac{K_a \times K_b}{K_w}$

$$= \frac{2\times10^{-5}\times5\times10^{-6}}{1\times10^{-14}} = 1.0\times 10^4$$

97. (c) $K_{sp}(AgBr) < K_{sp}(AgCl)$

Therefore, AgBr will precipitate first and at that time all Cl^- will be present.

98. (b) $K_P \longrightarrow (atm)^{\Delta n_g} = (atm)^{3-5}$
$= (atm)^{-2}$

99. (d)

	PCl_5	$\rightleftharpoons$	PCl_3	+	Cl_2
Initial	1		0		0
At eqn.	$1-\alpha$		α		α
Partial pressure =	$\frac{1-\alpha}{1+\alpha}.P$		$\frac{\alpha}{1+\alpha}.P$		$\frac{\alpha}{1+\alpha}.P$

Total mole $= 1-\alpha+\alpha+\alpha = 1+\alpha$

$$K_P = \frac{\frac{\alpha}{1+\alpha}P.\frac{\alpha}{1+\alpha}.P}{\frac{1-\alpha}{1+\alpha}P} = \frac{\alpha^2 P}{1-\alpha^2}$$

100. (a) $\because\ \dfrac{K}{Q} = 0.33$

$\Rightarrow K < Q$

i.e., reaction moves in backward direction to achieve equilibrium.

8 Redox Reactions

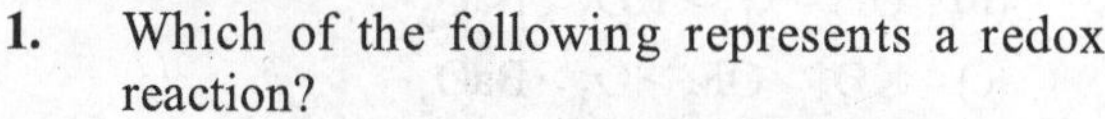

1. Which of the following represents a redox reaction?
   (a) $NaOH + HCl \longrightarrow NaCl + H_2O$
   (b) $BaCl_2 + H_2SO_4 \longrightarrow BaSO_4 + 2HCl$
   (c) $CuSO_4 + 2H_2O \longrightarrow Cu(OH)_2 + H_2SO_4$
   (d) $Zn + 2HCl \longrightarrow ZnCl_2 + H_2$
2. Which reaction involves neither oxidation nor reduction?
   (a) $CrO_4^{2-} \longrightarrow Cr_2O_7^{2-}$
   (b) $Cr \longrightarrow CrCl_3$
   (c) $Na \longrightarrow Na^+$
   (d) $2S_2O_3^{2-} \longrightarrow S_4O_6^{2-}$
3. Zn gives H_2 gas with H_2SO_4 and HCl but not with HNO_3 because **Toughnut**
   (a) Zn acts as an oxidising agent when it reacts with HNO_3
   (b) HNO_3 is weaker acid than H_2SO_4 and HCl
   (c) In electrochemical series, Zn is above hydrogen
   (d) NO_3^- is reduced in preference to hydronium ion
4. A compound of Xe and F is found to have 53.5% of Xe. What is oxidation number of Xe in this compound?
   (a) –4 (b) 0 (c) +4 (d) +6
5. Which of the following is not a disproportionation reaction?
   (a) $2PhCHO \xrightarrow{Al(OEt)_3} PhCOOCH_2Ph$
   (b) $\begin{matrix} CHO \\ | \\ COOH \end{matrix} + OH^- \longrightarrow \begin{matrix} CH_2OH \\ | \\ COO^- \end{matrix} + \begin{matrix} COO^- \\ | \\ COO^- \end{matrix}$
   (c) $NaH + H_2O \longrightarrow NaOH + H_2$
   (d) All of the above.
6. A solution contains Fe^{2+}, Fe^{3+} and I^- ions. This solution was treated with iodine at 35°C. E° for Fe^{3+}/Fe^{2+} is + 0.77 V and E° for $I_2/2I^- = 0.536$ V. The favourable redox reaction is :
   (a) I_2 will be reduced to I^-
   (b) There will be no redox reaction
   (c) I^- will be oxidised to I_2
   (d) Fe^{2+} will be oxidised to Fe^{3+}
7. Which of the following is correct code for x and y in the following reaction.

   $$2Na(s) + S(s) \longrightarrow Na_2S(s)$$
   (arrow x from Na to Na_2S above; arrow y from S to Na_2S below)

   (i) x = oxidation reaction, y = reduction reaction
   (ii) x = gain of two electrons, y = loss of two electrons,
   (iii) x = reduction reaction, y = oxidation reaction
   (iv) x = loss of two electrons, y = gain of two electrons
   (a) (i) and (ii) (b) (i) and (iv)
   (c) (ii) and (iii) (d) (iii) and (iv)
8. One mole of N_2H_4 loses 10 moles of electrons to form a new compound, y. Assuming that all nitrogen appear in the new compound, what is the oxidation state of nitrogen in y (There is no change in the oxidation state of hydrogen) **Toughnut**
   (a) –1 (b) –3 (c) +3 (d) +5
9. The oxidation state of nitrogen is correctly given for

	Compound	Oxidation state
(a)	$[Co(NH_3)_5Cl]Cl_2$	0
(b)	NH_2OH	–2
(c)	$(N_2H_5)_2SO_4$	+2
(d)	Mg_3N_2	–3

10. The equivalent mass of oxidising agent in the following reaction is **Tricky**

    $$SO_2 + 2H_2S \longrightarrow 3S + 2H_2O$$
    (a) 32 (b) 64 (c) 16 (d) 8

11. Amongst the following, identify the species with an atom in + 6 oxidation state:
(a) $[MnO_4]^-$ (b) $[Cr(CN)_6]^{3-}$
(c) Cr_2O_3 (d) CrO_2Cl_2

12. Oxidation state of sulphur in anions SO_3^{2-}, $S_2O_4^{2-}$ and $S_2O_6^{2-}$ increases in the orders:
(a) $S_2O_6^{2-} < S_2O_4^{2-} < SO_3^{2-}$
(b) $SO_6^{2-} < S_2O_4^{2-} < S_2O_6^{2-}$
(c) $S_2O_4^{2-} < SO_3^{2-} < S_2O_6^{2-}$
(d) $S_2O_4^{2-} < S_2O_6^{2-} < SO_3^{2-}$

13. In the reaction $3Mg + N_2 \rightarrow Mg_3N_2$
(a) magnesium is reduced
(b) magnesium is oxidized
(c) nitrogen is oxidized
(d) None of these

14. One gas bleaches the colour of flowers by reduction, while the other by oxidation
(a) CO and Cl_2 (b) SO_2 and Cl_2
(c) H_2S and Br_2 (d) NH_3 and SO_2

15. In the reaction :
$$C + 4HNO_3 \rightarrow CO_2 + 2H_2O + 4NO_2$$
HNO_3 act as
(a) an oxidizing agent
(b) an acid
(c) an acid as well as oxidizing agent
(d) a reducing agent.

16. Which of the following represents redox reactions?
I. $Cr_2O_7^{2-} + 2\overline{O}H \longrightarrow 2Cr_4^{2-} + H_2O$
II. $Zn + CuSO_4 \longrightarrow ZnSO_4 + Cu$
III. $2MnO_4^- + 3Mn^{2+} + 4\overline{O}H \longrightarrow 5MnO_2 + 2H_2O$
IV. $2Cu^+ \longrightarrow Cu + Cu^{2+}$
(a) I, II (b) I, III
(c) III, IV (d) II, III, IV

17. The standard reduction potentials at 298K for the following half reactions are given against each
$Zn^{2+}(aq) + 2e \rightleftharpoons Zn(s)$; –0.762 V
$Cr^{3+}(aq) + 3e \rightleftharpoons Cr(s)$; –0.740 V
$2H^+(aq) + 2e \rightleftharpoons H_2(g)$; 0.00 V
$Fe^{3+}(aq) + e \rightleftharpoons Fe^{2+}(aq)$; 0.770 V
Which is the strongest reducing agent?
(a) Zn (s) (b) Cr (s)
(c) H_2(g) (d) Fe^{3+}(aq)

18. The correct decreasing order of oxidation number of oxygen in compounds BaO_2, O_3, KO_2 and OF_2 is :
(a) $BaO_2 > KO_2 > O_3 > OF_2$
(b) $OF_2 > O_3 > KO_2 > BaO_2$
(c) $KO_2 > OF_2 > O_3 > BaO_2$
(d) $BaO_2 > O_3 > OF_2 > KO_2$

19. In the balanced chemical reaction
$$IO_3^- + aI^- + bH^+ \longrightarrow cH_2O + dI_2$$
a, b, c and d, respectively, correspond to
(a) 5, 6, 3, 3 (b) 5, 3, 6, 3
(c) 3, 5, 3, 6 (d) 5, 6, 5, 5

20. In the reaction shown below, oxidation state of the carbon in reactant and product are (*i*) and (*ii*) respectively? Is the given reaction a redox reaction?
$$Na_2CO_3(aq) + HCl(aq) \longrightarrow$$
$$Na^+(aq) + Cl^-(aq) + H_2O(l) + CO_2(g)$$
(a) (i) 6, (ii) 4, yes (b) (i) 6, (ii) 6, No
(c) (i) 4, (ii) 4, No (d) (i) 4, (ii) 4, yes

21. In the disproportionation reaction
$3\ HClO_3 \longrightarrow HClO_4 + Cl_2 + 2O_2 + H_2O$, the equivalent mass of the oxidizing agent is (molar mass of $HClO_3$ = 84.45)

(a) 16.89 (b) 32.22 (c) 84.45 (d) 28.15

22. The oxidation number of phosphorus in $Ba(H_2PO_2)_2$ is
(a) +3 (b) +2 (c) +1 (d) –1

23. How many electrons are involved in the following redox reaction?
$$Cr_2O_7^{2-} + Fe^{2+} + C_2O_4^{2-} \rightarrow Cr^{3+} + Fe^{3+} + CO_2 \text{ (Unbalanced)}$$
(a) 3 (b) 4 (c) 6 (d) 5

24. Given :
$$XNa_2HAsO_3 + YNaBrO_3 + ZHCl \rightarrow NaBr + H_3AsO_4 + NaCl$$
The values of X, Y and Z in the above redox reaction are respectively : Toughnut
(a) 2, 1, 2 (b) 2, 1, 3 (c) 3, 1, 6 (d) 3, 1, 4

25. When SO_2 is passed through acidified solution of potassium dichromate, then chromium sulphate is formed. The change in valency of chromium is
(a) +4 to +2 (b) +5 to +3
(c) +6 to +3 (d) +7 to +2

26. In which of the following transition metal complexes does the metal exhibit zero oxidation state ?
(a) $[Co(NH_3)_6]Cl_3$ (b) $[Fe(H_2O)_6]SO_4$
(c) $Ni(CO)_4$ (d) $[Fe(H_2O)_6]X_3$

27. In which of the following pairs, there is greatest difference in the oxidation number of the underlined elements ?
(a) $\underline{N}O_2$ and $\underline{N}_2O_4$ (b) $\underline{P}_2O_5$ and $\underline{P}_4O_{10}$
(c) $\underline{N}_2O$ and $\underline{N}O$ (d) $\underline{S}O_2$ and $\underline{S}O_3$

28. Which of the following statements is not correct? **Critical Thinking**
(a) The oxidation number of S in $(NH_4)_2S_2O_8$ is +6.
(b) The oxidation number of Os in OsO_4 is +8.
(c) The oxidation number of S in H_2SO_5 is +8.
(d) The oxidation number of O in KO_2 is –1/2.

29. Among NH_3, HNO_3, NaN_3 and Mg_3N_2 the number of molecules having nitrogen in negative oxidation state is
(a) 1 (b) 2 (c) 3 (d) 4

30. $$aK_2Cr_2O_7 + bKCl + cH_2SO_4 \longrightarrow xCrO_2Cl_2 + yKHSO_4 + zH_2O$$
The above equation balances when
(a) a = 2, b = 4, c = 6 and x = 2, y = 6, z = 3
(b) a = 4, b = 2, c = 6 and x = 6, y = 2, z = 3
(c) a = 6, b = 4, c = 2 and x = 6, y = 3, z = 2
(d) a = 1, b = 4, c = 6 and x = 2, y = 6, z = 3

31. The number of electrons involved in the reduction of one nitrate ion to hydrazine is
(a) 8 (b) 5 (c) 3 (d) 7

32. Which of the following do not show disproportionation reaction?
ClO_4^-, F_2, Cl_2, ClO_2^-, ClO_3^-, P_4, S_8, and ClO^-
(a) ClO_2^-, ClO_4^-, and ClO^-
(b) F_2 only
(c) F_2 and ClO_4^-
(d) ClO_4^- only

33. In the following balanced reaction,
$$X\,MnO_4^- + Y\,C_2O_4^{2-} + Z\,H^+ \rightleftharpoons X\,Mn^{2+} + 2Y\,CO_2 + \frac{Z}{2}H_2O$$
values of *X*, *Y* and *Z* respectively are **Toughnut**
(a) 2, 5, 16 (b) 8, 2, 5
(c) 5, 2, 16 (d) 5, 8, 4

34. Consider the reaction: **Tricky**
$$H_2SO_3(aq) + Sn^{4+}(aq) + H_2O(l) \rightarrow Sn^{2+}(aq) + HSO_4^-(aq) + 3H^+(aq)$$
Which of the following statements is correct?
(a) Sn^{4+} is the oxidizing agent because it undergoes oxidation
(b) Sn^{4+} is the reducing agent because it undergoes oxidation
(c) H_2SO_3 is the reducing agent because it undergoes oxidation
(d) H_2SO_3 is the reducing agent because it undergoes reduction

35. Which of the following statements is not correct about the reaction given below?
$$K_4[Fe(CN)_6] \xrightarrow{\text{Oxidation}} Fe^{3+} + CO_2 + NO_3^-$$
(a) Fe is oxidised from Fe^{2+} to Fe^{3+}.
(b) Carbon is oxidised from C^{2+} to C^{4+}.
(c) N is oxidised from N^{3-} to N^{5+}.
(d) Carbon is not oxidised.

36. Which of the following is not an intermolecular redox reaction? **Tricky**
(a) $MgCO_3 \longrightarrow MgO + CO_2$
(b) $O_2 + 2H_2 \longrightarrow 2H_2O$
(c) $K + H_2O \longrightarrow KOH + (1/2)H_2$
(d) $MnBr_3 \longrightarrow MnBr_2 + (1/2)Br_2$

37. Which of the following is not a disproportionation reaction?
(a) $P_4 + 5OH^- \longrightarrow H_2PO_2^- + PH_3$
(b) $Cl_2 + OH^- \longrightarrow Cl^- + ClO$
(c) $2H_2O_2 \longrightarrow 2H_2O + O_2$
(d) $PbO_2 + H_2O \longrightarrow PbO + H_2O_2$

38. In a balance equation $H_2SO_4 + xHI \rightarrow H_2S + yI_2 + zH_2O$, the values of x, y, z are

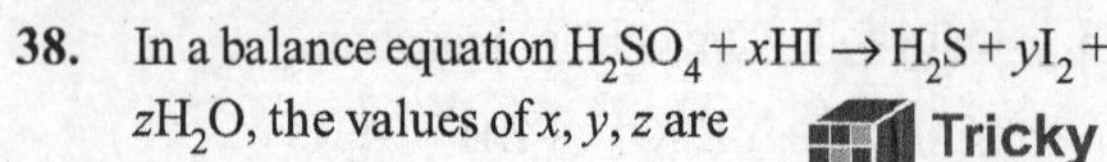

(a) $x=3, y=5, z=2$ (b) $x=4, y=8, z=5$
(c) $x=8, y=4, z=4$ (d) $x=5, y=3, z=4$

39. Given
$Fe^{3+}(aq) + e^- \rightarrow Fe^{2+}(aq); E^o = +0.77\ V$
$Al^{3+}(aq) + 3e^- \rightarrow Al(s); E^o = -1.66\ V$
$Br_2(aq) + 2e^- \rightarrow 2Br^-(aq); E^o = +1.09\ V$
Considering the electrode potentials, which of the following represents the correct order of reducing power?
(a) $Fe^{2+} < Al < Br^-$ (b) $Br^- < Fe^{2+} < Al$
(c) $Al < Br^- < Fe^{2+}$ (d) $Al < Fe^{2+} < Br^-$

40. For the reaction :
$NH_3 + OCl^- \longrightarrow N_2H_4 + Cl^-$ in basic medium, the coefficients of NH_3, OCl^- and N_2H_4 for the balanced equation are respectively
(a) 2, 2, 2 (b) 2, 2, 1 (c) 2, 1, 1 (d) 4, 4, 2

41. Which of the following act as reducing agents ?
(i) PO_4^{3-} (ii) SO_3 (iii) PO_3^{2-} (iv) NH_3
(a) (i), (ii) and (iii) (b) Only (iii)
(c) (i), (iii) and (iv) (d) (iii) and (iv)

42. In the reaction
$$Cr_2O_7^{2-} + 14H^+ + 6I^- \longrightarrow 2Cr^{3+} + 7H_2O + 3I_2$$
Which element is reduced ?
(a) I (b) O (c) H (d) Cr

43. Which of the following substances acts as an oxidising as well as a reducing agent?

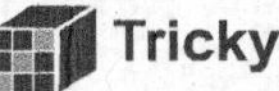

(a) Na_2O (b) $SnCl_2$
(c) Na_2O_2 (d) $NaNO_2$

44. Stronger is oxidising agent, more is
(a) standard reduction potential of that species
(b) the tendency to get itself oxidised
(c) the tendency to lose electrons by that species
(d) standard oxidation potential of that species

45. Which of the following involves transfer of five electrons ?
(a) $MnO_4^- \rightarrow Mn^{2+}$ (b) $CrO_4^{2-} \rightarrow Cr^{3+}$
(c) $MnO_4^{2-} \rightarrow MnO_2$ (d) $Cr_2O_7^{2-} \rightarrow 2Cr^{3+}$

46. The brown ring complex is formulated as $[Fe(H_2O)_5NO]SO_4$. The oxidation number of iron is
(a) 1 (b) 2 (c) 3 (d) 0

47. In which of the following reactions, there is no change in valency ? Critical Thinking
(a) $4KClO_3 \rightarrow 3KClO_4 + KCl$
(b) $SO_2 + 2H_2S \rightarrow 2H_2O + 3S$
(c) $BaO_2 + H_2SO_4 \rightarrow BaSO_4 + H_2O_2$
(d) $2BaO + O_2 \rightarrow 2BaO_2$

48. The reaction,
$2H_2O(l) \rightarrow 4H^+(aq) + O_2(g) + 4e^-$ is
(a) an oxidation reaction
(b) a reduction reaction
(c) a redox reaction
(d) a hydrolysis reaction

49. In the reaction
$2FeCl_3 + H_2S \rightarrow 2FeCl_2 + 2HCl + S$
(a) $FeCl_3$ acts as an oxidising agent
(b) Both H_2S are $FeCl_3$ are oxidised
(c) $FeCl_3$ is oxidised while H_2S is reduced
(d) H_2S acts as an oxidising agent.

50. The oxidation number of S in $Na_2S_4O_6$ is
(a) +0.5 (b) 2.5 (c) +4 (d) +6

51. In the following reaction
$4P + 3KOH + 3H_2O \longrightarrow 3KH_2PO_2 + PH_3$
(a) phosphorus is both oxidised and reduced.
(b) only phosphorus is reduced.
(c) phosphorus is not oxidised
(d) None of these

52. The standard electrode potentials of four elements A, B, C and D are –3.05, –1.66, –0.40 and +0.80. The highest chemical reactivity will be exhibited by : Tricky
(a) A (b) B (c) C (d) D

53. The oxidation number of Fe in $Na_2[Fe(CN)_5NO]$ is
(a) +2 (b) +1 (c) +3 (d) –2

54. Which of the following is the strongest reducing agent in aqueous medium?
(a) Mg (b) Na (c) Li (d) Ca

55. The oxidation state of Fe in Fe_3O_8 is
(a) 3/2 (b) 4/5 (c) 5/4 (d) 16/3

56. Which substance serves as reducing agent in the following reaction?

$$14H^+ + Cr_2O_7^{2-} + 3Ni \longrightarrow 2Cr^{3+} + 7H_2O + 3Ni^{2+}$$

(a) H_2O (b) Ni
(c) H^+ (d) $Cr_2O_7^{2-}$

57. In the reaction

$$3Br_2 + 6CO_3^{2-} + 3H_2O \longrightarrow 5Br^- + BrO_3^- + 6HCO_3^-$$

(a) bromine is oxidised and carbonate is reduced.
(b) bromine is reduced and water is oxidised
(c) bromine is neither reduced nor oxidised
(d) bromine is both reduced and oxidised

58. Which one of the following cannot function as an oxidising agent ?

(a) I^- (b) S(s)
(c) $NO_3^-(aq)$ (d) $Cr_2O_7^{2-}$

59. The oxidation number of Pt in $[Pt(C_2H_4)Cl_3]^-$ is

(a) +1 (b) +2 (c) +3 (d) +4

60. Standard electrode potentials of redox couples A^{2+}/A, B^{2+}/B, C^{2+}/C and D^{2+}/D are 0.3V, –0.5V, –0.75V and 0.9V respectively. Which of these is best oxidising agent and reducing agent respectively ?

(a) D^{2+}/D and B^{2+}/B (b) B^{2+}/B and D^{2+}/D
(c) D^{2+}/D and C^{2+}/C (d) C^{2+}/C and D^{2+}/D

Answer KEY

1	(d)	7	(b)	13	(b)	19	(a)	25	(c)	31	(d)	37	(d)	43	(d)	49	(a)	55	(d)
2	(a)	8	(c)	14	(b)	20	(c)	26	(c)	32	(c)	38	(c)	44	(a)	50	(b)	56	(b)
3	(d)	9	(d)	15	(a)	21	(a)	27	(d)	33	(a)	39	(d)	45	(a)	51	(a)	57	(d)
4	(d)	10	(c)	16	(d)	22	(c)	28	(c)	34	(c)	40	(c)	46	(a)	52	(a)	58	(a)
5	(c)	11	(d)	17	(a)	23	(a)	29	(c)	35	(d)	41	(d)	47	(c)	53	(a)	59	(b)
6	(c)	12	(c)	18	(b)	24	(c)	30	(d)	36	(a)	42	(d)	48	(d)	54	(c)	60	(c)

Hints & Solutions

1. **(d)** (a) and (b) are neutralisation reactions. The oxidation state of Cu is +2 in both reactant and product, and SO_4^{2-} ion does not change. In option (d) is a redox reaction.

$Zn \rightarrow Zn^{2+} + 2e^-$ (Oxidation)

$2H^+ + 2e^- \rightarrow H_2$ (Reduction)

2. **(a)** O. N. of Cr on both sides is + 6.

3. **(d)** Zinc gives H_2 gas with dil H_2SO_4/HCl but not with HNO_3 because in HNO_3, NO_3^- ion is reduced and give NH_4NO_3, N_2O, NO and NO_2 (based upon the concentration of HNO_3)

$[Zn + 2HNO_3 \longrightarrow Zn(NO_3)_2 + 2H] \times 4$
(nearly 6%)

$HNO_3 + 8H \longrightarrow NH_3 + 3H_2O$

$NH_3 + HNO_3 \longrightarrow NH_4NO_3$

$4Zn + 10HNO_3 \longrightarrow 4Zn(NO_3)_2 + NH_4NO_3 + 3H_2O$

Zn is on the top position of hydrogen in electrochemical series. So Zn displaces H_2 from dilute H_2SO_4 and HCl with liberation of H_2.

$Zn + H_2SO_4 \rightarrow ZnSO_4 + H_2$

4. (d) $Xe = 53.5\% \therefore F = 46.5\%$

Relative number of atoms Xe

$= \frac{53.5}{131.2} = 0.4$ and $F = \frac{46.5}{19} = 2.4$

Simple ratio Xe = 1 and F = 6 ; Molecular formula is XeF_6

O.N.of Xe is +6

5. (c) (a) and (b) are disproportionation reactions. (c) is an oxidation reaction.

$2H^- \longrightarrow H_2 + 2e^-$

6. (c) Given $Fe^{3+}/Fe^{2+} = +0.77$ V and $I_2/2I^- = 0.536$V

$2(e^- + Fe^{3+} \longrightarrow Fe^{2+})$ E° = 0.77 V

$2I^- \longrightarrow I_2 + 2e^-$ E° = – 0.536 V

$2Fe^{3+} + 2I^- \longrightarrow 2Fe^{2+} + I_2$

$E° = E°_{ox} + E°_{red}$

$= 0.77 - 0.536$

$= 0.164$ V

∴ Since value of E° is + ve reaction will take place.

7. (b) Oxidation reaction (loss of $2e^-$)

$2Na(s) + S(s) \longrightarrow Na_2S(s)$

Reduction (gain of $2e^-$) reaction

8. (c) $\overset{-4\ +4}{N_2H_4} \xrightarrow[N]{\text{loss of } 10e^-} \overset{+6}{N_2}Y;$

O.N.of N changes from – 2 to +3

9. (d)

(a) $\left[\overset{+3}{Co}\left(\overset{0}{NH_3}\right)_5\overset{-1}{Cl}\right]^{+2}\overset{-1\times 2}{Cl_2}$

$+3 + 5x + 15 - 1 = +2$

$5x + 17 = +2$

$5x = -15$

$x = -3$; Oxidation State of N = –3

(b) $NH_2OH : x + 3 - 2 = 0 \Rightarrow x = -1$

Oxidation state of N = –1

(c) $(N_2H_5)^{+1}SO_4^{2-}$:

$N_2H_5^+ : 2x + 5 = +1 \Rightarrow x = -2$

Oxidation number of N = –2

(d) $Mg_3N_2 : 2\times 3 + 2x = 0$

$2x = -6$

$x = -3$ Oxidation state of N = –3

10. (c) SO_2 is an oxidising agent

$4e^- + SO_2 \longrightarrow S$

Equivalent of $SO_2 = \frac{64}{4} = 16.$

11. (d) CrO_2Cl_2

Let O. No. of Cr = x

$\therefore\ x + 2(-2) + 2(-1) = 0$

$x - 4 - 2 = 0$

$\therefore\ x = +6$

12. (c) In SO_3^{2-}

$x + 3(-2) = -2; x = +4$

In $S_2O_4^{2-}$

$2x + 4(-2) = -2$

$2x - 8 = -2$

$2x = 6;\ \ x = +3$

In $S_2O_6^{2-}$

$2x + 6(-2) = -2$

$2x = 10;\ \ x = +5$

hence the correct order is

$S_2O_4^{2-} < SO_3^{2-} < S_2O_6^{2-}$

13. (b) In the given reaction, oxidation state of Mg is changing from 0 to +2, while in nitrogen it is changing from 0 to – 3. So oxidation of Mg and reduction of nitrogen takes place.

14. (b) SO_2 bleaches by reduction, while chlorine bleaches colour of flowers by oxidation.

15. (a) O.N. of C changes form 0 to + 4 by oxidation. Hence HNO_3 is oxidising agent.

16. (d) II, III and IV are redox reactions.

17. (a) Since oxidation potential of Zn is highest hence strongest reducing agent.

18. (b) Oxidation no. of O are + 2, 0, – 1/2 and – 1 respectively

19. (a) The balanced equation is

$12H^+ + 10e^- + 2IO_3^- \longrightarrow I_2 + 6H_2O$

$10I^- \longrightarrow 5I_2 + 10e^-$

$10I^- + 12H^+ + 2IO_3^- \longrightarrow 6H_2O + 6I_2$

or $\underset{a}{5I^-} + \underset{b}{6H^+} + IO_3^- \longrightarrow \underset{c}{3H_2O} + \underset{d}{3I_2}$

20. (c) The redox reaction involve loss or gain of electron(s) i.e. change in oxidation state. Given reaction is not a redox reaction as this reaction involves no change in oxidation state of reactant or product.

21. (a) $ClO_3^- \longrightarrow Cl_2^0$

$x - 6 = -1$ $x = 0$

$x = +5$ $x = 0$ (x = oxidation number)

$$\text{Equivalent mass} = \frac{\text{Molecular mass}}{\text{Oxidation number}}$$

$$= \frac{84.45}{5} = 16.89$$

22. (c) $Ba^{2+}(H_2PO_2)_2^{-1\times 2}$;

$H_2PO_2^- : 2 + x - 4 = -1$

$\Rightarrow x = +1$

$\therefore$ Oxidation state of P = +1

23. (a) The reaction given is

$Cr_2O_7^{2-} + Fe^{2+} + C_2O_4^{2-} \longrightarrow Cr^{3+} + Fe^{3+} + CO_2$

$Cr_2O_7^{2-} \longrightarrow 2Cr^{3+}$

On balancing

$14H^+ + Cr_2O_7^{2-} + 6e^- \longrightarrow 2Cr^{3+} + 7H_2O$ ...(i)

$Fe^{2+} \longrightarrow Fe^{3+} + e^-$ ...(ii)

$C_2O_4^{2-} \longrightarrow 2CO_2 + 2e^-$ ...(iii)

On adding all the three equations.

$Cr_2O_7^{2-} + Fe^{2+} + C_2O_4^{2-} + 14H^+ + 3e^- \longrightarrow 2Cr^{3+} + Fe^{3+} + 2CO_2 + 7H_2O$

Hence the total no. of electrons involved in the reaction = 3

24. (c) On balancing the given reaction, we find

$3Na_2HAsO_3 + NaBrO_3 + 6HCl \longrightarrow 6NaCl + 3H_3AsO_4 + NaBr$

25. (c) $K_2Cr_2O_7 + 3SO_2 + 4H_2SO_4 \rightarrow K_2SO_4 + Cr_2(SO_4)_3 + 3SO_3 + 4H_2O$

O.N.of chrominum changes from +6 to +3

26. (c) In metal carbonyls, metal always has zero O.N.

27. (d) O.N. of N in NO_2 and N_2O_4 is +4

$\therefore$ difference is zero.

O.N. of P in P_2O_5 and P_4O_{10} is +5

$\therefore$ difference is zero

O.N. of N in N_2O is +1 and in NO is +2. The difference is 1

O.N. of S in SO_2 is +4 and in SO_3 is +6. The difference is +2.

28. (c)

(a) $\left(\overset{+1}{NH_4}\right)_2 S_2O_8^{2-}$

$\therefore$ Oxidation state of S = +6

(Since $S_2O_8^{2-}$ has one peroxide bond)

(b) Oxidation state of Os = +8

(c) Oxidation state of S in H_2SO_5 = +6

(Since it has one peroxide bond)

(d) $K^{1+}O_2^{1-}$, oxidation state of O = –1/2

29. (c) Calculating the oxidation state of nitrogen in given molecules;

Oxidation state of N in NH_3 is

$x + 3 \times (+1) = 0$ or $x = -3$

Oxidation state on N in HNO_3 is

$1 + x + 3 \times (-2) = 0$ or $x = +5$

Oxidation state of N in NaN_3 is

$+1 + 3x = 0$ or $x = -\frac{1}{3}$

Oxidation state of N in Mg_3N_2 is

$3 \times 2 + 2x = 0$ or $x = -3$

Thus 3 molecules (i.e. NH_3, NaN_3 and Mg_3N_2 have nitrogen in negative oxidation states.

30. (d) The balanced equation is

$K_2Cr_2O_7 + 4KCl + 6H_2SO_4 \longrightarrow 2CrO_2Cl_2 + 6K_2SO_4 + 3H_2O$

31. (d) $\overset{+5}{N}O_3^- \longrightarrow \overset{-2}{N_2}H_4$ So, for reduction of 1 mole of NO_3^- number of electrons required is 7.

32. (c) F_2 being most electronegative element cannot exhibit any positive oxidation state. In ClO_4^- chlorine is present in its highest oxidation state i.e +7. Therefore it does not show disproportionation reaction.

33. (a) $X\,MnO_4^- + Y\,C_2O_4^{2-} + ZH^+ \rightleftharpoons X\,Mn^{+2} + 2Y\,CO_2 + \frac{Z}{2}H_2O$

First half reaction $MnO_4^- \longrightarrow Mn^{+2}$ .... (i)

On balancing

$MnO_4^- + 8H^+ + 5e^- \longrightarrow Mn^{+2} + 4H_2O$ .... (ii)

Second half reaction

$C_2O_4^{2-} \longrightarrow 2CO_2$ .... (iii)

On balancing

$C_2O_4^{2-} \longrightarrow 2CO_2 + 2e^-$ .... (iv)

On multiplying eqn. (ii) by 5 and (iv) by 2 and then adding we get

$2MnO_4^- + 5C_2O_4^{2-} + 16H^+ \longrightarrow 2Mn^{+2} + 10CO_2 + 8H_2O$

34. (c) $\overset{+4}{H_2SO_3}(aq) + Sn^{4+}(aq) + H_2O(l) \longrightarrow Sn^{2+}(aq) + \overset{+6}{HSO_4^-}(aq) + 3H^+$
Hence H_2SO_3 is the reducing agent because it undergoes oxidation.

35. (d) In CN^-, oxidation of C is +2, and it changes to +4 oxidation state in CO_2. So C is also oxidised.

36. (a) Intermolecular redox reaction and redox reactions are same.

37. (d) Disproportionation involves simultaneous oxidation and reduction of the same atom in a molecule.

38. (c) The value of *x*, *y*, *z* are 8, 4, 4 respectively hence the reaction is
$H_2SO_4 + 8HI \rightarrow H_2S + 4I_2 + 4H_2O$

39. (d) Reducing character decreases down the series. Hence the correct order is
$Al < Fe^{2+} < Br^-$

40. (c) The balanced equation :
$2NH_3 + OCl^- \longrightarrow N_2H_4 + Cl^- + H_2O$

41. (d) In (i) and (ii) both P and S are in highest oxidation state. In (iii) and (iv) ; P has oxidation state of +4 which can be oxidized to +5 state, while in case of NH_3 nitrogen has oxidation state of –3 which can be oxidised.

42. (d) $2I^- \rightarrow I_2$ is oxidation (loss of electrons) ; Cr (+6) changes to Cr (+3) by gain of electrons. Hence Cr is reduced.

43. (d) In Na_2O, $SnCl_2$ and Na_2O_2 central atom is either in lowest or highest oxidation state, so they can function either as an oxidising or a reducing agent but not both. However, the oxidation state of N in $NaNO_2$ is +3 which lies between its highest (+5) and lowest (–3) values.

44. (a) More is E°_{RP}, more is the tendency to get itself reduced or more is oxidising power.

45. (a) O.N. of Mn in MnO_4^- is +7 and in Mn^{2+} it is +2. The difference is of 5 electrons.

46. (a) $[Fe(H_2O)_5NO]SO_4$ Let O.N. of Fe be *x* then,
$1\times(x) + 5\times(0) + 1\times(+1) + 1\times(-2) = 0$
$\therefore x = +1$

47. (c) In $BaO_2 + H_2SO_4 \rightarrow BaSO_4 + H_2O_2$ all atoms are present in the same O.S. in reactants and products.

48. (d) Since there is loss of electrons, hence it is oxidation reaction.

49. (a) In $2FeCl_3 + H_2S \rightarrow 2FeCl_2 + 2HCl + S$
O.N. of S changes from –2 to 0 (hence oxidised) O.N. of Fe changes from +3 to +2 hence reduced.

50. (b) $Na_2S_4O_6 : 2 + 4x - 12 = 0$
$\therefore x = \frac{10}{4} = 2.5$
Oxidation number of S = 2.5

51. (a) $4P + 3KOH + 3H_2O \rightarrow 3KH_2PO_2 + PH_3$
O.N of P = 0, In KH_2PO_2 it is + 1, In PH_3 it is –3.
Hence P is oxidised and reduced.

52. (a) Standard electrode potential i.e. reduction potential of A is minimum (–3.05V) i.e. its oxidation potential is maximum which implies 'A' is most reactive chemically.

53. (a) $\overset{+2}{Na_2}\left[\overset{+2}{Fe}(\overset{-5}{CN})_5\overset{+1}{NO}\right]^{2-}$
Oxidation state of Fe = +2

54. (c) In an aqueous medium, Li is the strongest reducing agent, since the high negative enthalpy of hydration compensates high IE_1.

55. (d) Fe_3O_8
$3x - 16 = 0 \Rightarrow x = \frac{16}{3}$

56. (b) $(Ni \rightarrow Ni^{2+})$ Ni loses electrons hence act as reducing agent.

57. (d) $3Br_2 + 6CO_3^{2-} + 3H_2O \rightarrow 5Br^- + BrO_3^- + 6HCO_3^-$
O.N. of Br_2 changes from 0 to –1 and +5 hence it is reduced as well as oxidised.

58. (a) If an electronegative element is in its lowest possible oxidation state in a compound or in free state. It can function as a powerful reducing agent. e.g. I^-

59. (b) $\left[\overset{+2}{Pt}(\overset{0}{C_2H_4})\overset{3-}{Cl_3}\right]^-$
Oxidation number of Pt = +2

60. (c) The redox couple with maximum reduction potential will be best oxidising agent and with minimum reduction potential will be best reducing agent.

9 Hydrogen

1. When electric current is passed through an ionic hydride in molten state :
(a) hydrogen is liberated at anode
(b) hydrogen is liberated at cathode
(c) hydride ion migrates towards cathode
(d) hydride ion remains is solution

2. Among CaH_2, NH_3, NaH and B_2H_6 which are covalent hydrides?
(a) NH_3 and B_2H_6 (b) NaH and CaH_2
(c) NaH and NH_3 (d) CaH_2 and B_2H_6

3. Polyphosphates are used as water softening agents because they
(a) form soluble complexes with anionic species
(b) precipitate anionic species
(c) form soluble complexes with cationic species
(d) precipitate cationic species

4. When two ice cubes are pressed over each other, they unite to form one cube. Which of the following forces is responsible to hold them together ?
(a) Hydrogen bond formation
(b) van der Waals forces
(c) Covalent attraction
(d) Ionic interaction

5. Hydrogen molecules differs from chlorine molecule in the following respect
(a) Hydrogen molecule is non-polar but chlorine molecule is polar
(b) Hydrogen molecule is polar while chlorine molecule is non-polar
(c) Hydrogen molecule can form intermolecular hydrogen bonds but chlorine molecule does not
(d) Hydrogen molecule cannot participate in coordination bond formation but chlorine molecule can

6. The adsorption of hydrogen by metals is called
(a) Dehydrogenation (b) Hydrogenation
(c) Occlusion (d) Adsorption

7. Which of the following statements do not define the characteristic property of water "Water is a universal solvent"
(a) It can dissolve maximum number of compounds
(b) It has very low dielectric constant
(c) It has high liquid range
(d) None of these

8. A, 6 volume sample of H_2O_2

(a) will contain 6% V/V of H_2O_2
(b) will contain 6% W/V of H_2O_2
(c) would give 6 volumes of oxygen per unit volume of H_2O_2 sample at STP
(d) would give 6 volumes of oxygen per unit weight of H_2O_2 sample at STP

9. 34 g of H_2O_2 is present in 1120 mL of solution. This solution is called

Tricky

(a) 10 vol solution (b) 20 vol solution
(c) 34 vol solution (d) 32 vol solution

10. Hydrogen can be fused to form helium at
(a) high temperature and high pressure
(b) high temperature and low pressure
(c) low temperature and high pressure
(d) low temperature and low pressure

11. The oxidation states of most electronegative element in the products of reaction BaO_2 with dil. H_2SO_4 are
(a) 0 and – 1 (b) –1 and – 2
(c) – 2 and 0 (d) – 2 and + 1

12. A 5.0 mL solution of H_2O_2 liberates 1.27 g of iodine from an acidified KI solution. The percentage strength of H_2O_2 is
(a) 11.2 (b) 5.6 (c) 1.7 (d) 3.4

13. 3.4 g sample of H_2O_2 solution containing x% H_2O_2 by weight requires x mL of $KMnO_4$ solution for complete oxidation under acidic condition. The normality of $KMnO_4$ solution is

(a) 1 N (b) 2 N (c) 3 N (d) 0.5 N

14. $H_2O_2 \rightarrow 2H^+ + O_2 + 2e^-$; E° = –0.68 V. This equation represents which of the following behaviour of H_2O_2?
(a) Reducing (b) Oxidising
(c) Acidic (d) Catalytic

15. When a substance A reacts with water it produces a combustible gas B and a solution of substance C in water. When another substance D reacts with this solution of C, it also produces the same gas B on warming but D can produce gas B on reaction with dilute sulphuric acid at room temperature. A imparts a deep golden yellow colour to a smokeless flame of Bunsen burner. A, B, C and D respectively are Toughnut

(a) Na, H_2, NaOH, Zn
(b) K, H_2, KOH, Al
(c) Ca, H_2, $Ca(OH)_2$, Sn
(d) CaC_2, C_2H_2, $Ca(OH)_2$, Fe

16. HCl is added to following oxides. Which one would give H_2O_2
(a) MnO_2 (b) PbO_2 (c) BaO (d) None

17. Chemical A is used for water softening to remove temporary hardness. A reacts with sodium carbonate to generate caustic soda. When CO_2 is bubbled through a solution of A, it turns cloudy. What is the chemical formula of A?
(a) $CaCO_3$ (b) CaO
(c) $Ca(OH)_2$ (d) $Ca(HCO_3)_2$

18. Which physical property of dihydrogen is wrong?
(a) Odourless gas
(b) Tasteless gas
(c) Colourless gas
(d) Non–inflammable gas

19. Permanent hardness in water cannot be cured by :
(a) Treatment with washing soda
(b) Boiling
(c) Calgon's method
(d) Ion exchange method

20. Which of the following metal evolves hydrogen on reaction with cold dilute HNO_3 ?
(a) Mg (b) Al (c) Fe (d) Cu

21. The alum used for purifying water is
(a) Ferric alum (b) Chrome alum
(c) Potash alum (d) Ammonium alum

22. What is formed when calcium carbide reacts with heavy water?
(a) C_2D_2 (b) CaD_2
(c) Ca_2D_2O (d) CD_2

23. The hydride ion H^- is stronger base than its hydroxide ion OH^-. Which of the following reactions will occur if sodium hydride (NaH) is dissolved in water ? Tricky

(a) $H^-(aq) + H_2O \rightarrow H_3O^-$
(b) $H^-(aq) + H_2O(l) \rightarrow OH^- + H_2$
(c) $H^- + H_2O \rightarrow$ No reaction
(d) None of these

24. In context with the industrial preparation of hydrogen from water gas ($CO + H_2$), which of the following is the correct statement? Toughnut

(a) CO and H_2, are fractionally separated using differences in their densities
(b) CO is removed by absorption in aqueous Cu_2Cl_2 solution
(c) H_2 is removed through occlusion with Pd
(d) CO is oxidised to CO_2 with steam in the presence of a catalyst followed by absorption of of CO_2 in alkali

25. The dielectric constant of H_2O is 80. The electrostatic force of attraction between Na^+ and Cl^- will be
(a) reduced to 1/40 in water than in air
(b) reduced to 1/80 in water than in air
(c) will be increased to 80 in water than in air
(d) will remain unchanged.

26. Which of the following is not true?
(a) D_2O freezes at lower temperature than H_2O
(b) Reaction between H_2 and Cl_2 is much faster than D_2 and Cl_2
(c) Ordinary water gets electrolysed more rapidly than D_2O
(d) Bond dissociation energy of D_2 is greater than H_2

27. H_2O_2 is always stored in black bottles because
(a) It is highly unstable
(b) Its enthalpy of decomposition is high
(c) It undergoes auto-oxidation on prolonged standing
(d) None of these

28. Which of the following is wrong about H_2O_2? It is used
(a) as an aerating agent in production of sponge rubber
(b) as an antichlor
(c) for restoring white colour of blackened lead painting
(d) None of these

29. 100 mL of 0.01 M $KMnO_4$ oxidises 100 mL H_2O_2 in acidic medium. Volume of the same $KMnO_4$ required in alkaline medium to oxidise 100 mL of the same H_2O_2 will be (MnO_4^- changes to Mn^{2+}

in acidic medium and to MnO_2 in alkaline medium)

Critical Thinking

(a) $\frac{100}{3}$ mL (b) $\frac{500}{3}$ mL

(c) $\frac{300}{5}$ mL (d) None

30. The purity of H_2O_2 in a given sample is 85%. Calculate the weight of impure sample of H_2O_2 which requires 10 mL of M/5 $KMnO_4$ solution in titration in acidic medium

Toughnut

(a) 2 g (b) 0.2 g (c) 0.17 g (d) 0.15 g

31. Barium peroxide reacts with phosphoric acid to produce barium phosphate along with

(a) water (b) hydrogen peroxide

(c) dioxygen (d) phosphine

32. The normality and volume strength of a solution made by mixing 1.0 L each of 5.6 volume and 11.2 volume H_2O_2 solution are :

Tricky

(a) 1 N, 5.6 vol (b) 1.5 N. 5.6 vol

(c) 1.5 N, 8.4 vol (d) 1 N, 8.4 vol

33. The numbers of protons, electrons and neutrons in a molecule of heavy water are respectively :

(a) 8, 10, 11 (b) 10, 10, 10

(c) 10, 11, 10 (d) 11, 10, 10

34. 2 g of aluminium is treated separately with excess of dilute H_2SO_4 and excess of NaOH. The ratio of the volumes of hydrogen evolved is

(a) 2 : 3 (b) 1 : 1 (c) 2 : 1 (d) 1 : 2

35. Which of the following statements is correct?

(a) Hydrogen has same IP as alkali metals

(b) Hydrogen has same electronegativity as halogens

(c) Hydrogen has oxidation number of –1 and +1

(d) Hydrogen will not be liberated at anode.

36. Which of the following terms is not correct for hydrogen?

(a) Its molecule is diatomic

(b) It exists both as H^+ and H^- in different chemical compounds

(c) It is the only species which has no neutrons in the nucleus

(d) Heavy water is unstable because hydrogen is substituted by its isotope deuterium

37. Hard water is not fit for washing clothes because

(a) it contains Na_2SO_4 and KCl

(b) It gives precipitate

(c) it contains impurities

(d) it is acidic in nature

38. The amount of H_2O_2 present in 1 litre of 1.5 N H_2O_2 solution, is :

(a) 25.5 g (b) 3.0 g (c) 8.0 g (d) 2.5 g

39. 20 volume H_2O_2 solution has a strength of about

(a) 30% (b) 6% (c) 3% (d) 10%

40. When same amount of zinc is treated separately with excess of sulphuric acid and excess of sodium hydroxide solution the ratio of volumes of hydrogen evolved is

Tricky

(a) 1 : 1 (b) 1 : 2 (c) 2 : 1 (d) 9 : 4

41. Hydrogen at room temperature contains :

(a) 25% ortho + 75% para

(b) 25% para + 75% ortho

(c) 50% ortho + 50% para

(d) 60% ortho + 34% para

42. In Bosch's process, which gas is utilised for the production of hydrogen gas ?

(a) Producer gas (b) Water gas

(c) Coal gas (d) None of these

43. D_2O is preferred to H_2O, as a moderator, in nuclear reactors because

(a) D_2O slows down fast neutrons better

(b) D_2O has high specific heat

(c) D_2O is cheaper

(d) None of these

44. Consider the following statements :

1. Atomic hydrogen is obtained by passing hydrogen through an electric arc.
2. Hydrogen gas will not reduce heated aluminium oxide.
3. Finely divided palladium adsorbs large volume of hydrogen gas
4. Pure nascent hydrogen is best obtained by reacting Na with C_2H_5OH

Which of the above statements is/are correct?

(a) 1 alone (b) 2 alone

(c) 1, 2 and 3 (d) 2, 3 and 4

45. Water is :

(a) more polar than H_2S

(b) more or less identical in polarity with H_2S

(c) less polar than H_2S

(d) None of these

46. Calgon used as a water softener is

(a) $Na_2[Na_4(PO_3)_6]$ (b) $Na_4[Na_2(PO_3)_6]$

(c) $Na_4[Na_4(PO_4)_5]$ (d) $Na_4[Na_2(PO_4)_6]$

47. In which of the following reactions, H_2O_2 is acting as a reducing agent
(a) $H_2O_2 + SO_2 \rightarrow H_2SO_4$
(b) $2KI + H_2O_2 \rightarrow 2KOH + I_2$
(c) $PbS + 4H_2O_2 \rightarrow PbSO_4 + 4H_2O$
(d) $Ag_2O + H_2O_2 \rightarrow 2Ag + H_2O + O_2$

48. Which of the following statements are correct?
(i) Hydrogen peroxide is industrially prepared by the auto-oxidation of 2-alkyl anthraquinols
(ii) One millilitre of 30% H_2O_2 means that solution will give 100 V of oxygen at STP
(iii) Dihedral angle of H_2O_2 in gas phase is 90.2° and in solid phase dihedral angle is 111.5°
(a) (i), (ii) and (iii) (b) (i) and (iii)
(c) (ii) and (iii) (d) (i) and (ii)

49. When a sample of hard water is passed through the layer of sodium zeolite resulting which of the following ions will not be present in the resulting sample of water obtained?
(a) Mg^{2+} and Ca^{2+} (b) Ca^{2+} and Na^+
(c) Mg^{2+} and CO_3^{2-} (d) CO_3^{2-} and Cl^-

50. Which statement is wrong?
(a) Ordinary hydrogen is an equilibrium mixture of ortho and para hydrogen
(b) In ortho hydrogen spin of two nuclei is in same direction
(c) Ortho and para forms do not resemble in their chemical properties
(d) In para hydrogen spin of two nuclei is in opposite direction.

51. An orange coloured solution acidified with H_2SO_4 and treated with a substance 'X' gives a blue coloured solution of CrO_5. The substance 'X' is

Toughnut

(a) H_2O_2 (b) H_2O
(c) dil HCl (d) conc HCl

52. Hydrogen peroxide acts both as an oxidising and as a reducing agent depending upon the nature of the reacting species. In which of the following cases H_2O_2 acts as a reducing agent in acid medium?
(a) MnO_4^- (b) $Cr_2O_7^{2-}$
(c) SO_3^{2-} (d) KI

53. Pick out the correct statement
(a) By decreasing the temperature pure para-hydrogen can be obtained
(b) By increasing the temperature pure ortho-hydrogen can be obtained
(c) By decreasing the temperature pure ortho-hydrogen can be obtained
(d) By increasing the temperature pure para-hydrogen can be obtained

54. An inorganic compound gives off O_2 when heated, turns an acidic solution of KI violet and reduces acidified $KMnO_4$. The compound is

(a) SO_3 (b) KNO_3
(c) H_2O_2 (d) All of these

55. The reaction of H_2O_2 with sulphur is an example of........ reaction
(a) Addition (b) Oxidation
(c) Reduction (d) Redox

56. Which substance does not speed up decomposition of H_2O_2?
(a) Glycerol (b) Pt
(c) Gold (d) MnO_2

57. When zeolite (hydrated sodium aluminium silicate) is treated with hard water, the sodium ions are exchanged with
(a) H^+ ions (b) Ca^{2+} ions
(c) SO_4^{2-} ions (d) OH^- ions

58. Saline hydrides react explosively with water, such fires can be extinguished by
(a) water (b) carbon dioxide
(c) sand (d) None of these

59. 30 volume hydrogen peroxide means
(a) 30% of H_2O_2 solution
(b) 30 cm^3 solution contains 1g of H_2O_2
(c) 1 cm^3 of solution liberates 30 cm^3 of O_2 at STP
(d) 30 cm^3 of solution contains 1 mole of H_2O_2

60. Hydrogen peroxide in its reaction with KIO_4 and NH_2OH respectively, is acting as a
(a) Reducing agent, oxidising agent
(b) Reducing agent, reducing agent
(c) Oxidising agent, oxidising agent
(d) Oxidising agent, reducing agent

Answer KEY

1	(a)	7	(b)	13	(b)	19	(b)	25	(b)	31	(b)	37	(b)	43	(d)	49	(a)	55	(d)
2	(a)	8	(c)	14	(a)	20	(a)	26	(a)	32	(c)	38	(a)	44	(c)	50	(c)	56	(a)
3	(c)	9	(a)	15	(a)	21	(c)	27	(c)	33	(b)	39	(b)	45	(a)	51	(a)	57	(b)
4	(a)	10	(a)	16	(d)	22	(a)	28	(d)	34	(b)	40	(a)	46	(a)	52	(a)	58	(c)
5	(d)	11	(b)	17	(c)	23	(b)	29	(b)	35	(c)	41	(b)	47	(d)	53	(a)	59	(c)
6	(c)	12	(d)	18	(d)	24	(d)	30	(b)	36	(d)	42	(b)	48	(d)	54	(c)	60	(a)

Hints & Solutions

1. **(a)** At anode : $2H^- \longrightarrow H_2\uparrow + 2e^-$
2. **(a)** N and B belong to *p*-block and since the electronegativity difference between these elements and H is very less, they form covalent hydrides.
3. **(c)** Polyphosphates form soluble complexes with Ca^{2+} and Mg^{2+} and are removed.
4. **(a)** Two ice cubes stick to each other due to H-bonding
5. **(d)** Chlorine has lone pair which it can donate to form coordinate bond, while hydrogen cannot.
6. **(c)** Occlusion is the phenomenon of adsorption of hydrogen by metal.
7. **(b)** Water has high dielectric constant i.e., 78.39 C^2/Nm^2, high liquid range and can dissolve maximum number of compounds. That is why it is used as universal solvent.
8. **(c)** 6 volume H_2O_2 would give 6 volumes of O_2 per unit volume of H_2O_2
9. **(a)** $N = \dfrac{W_2 \times 1000}{E.w_2 \times VmL}$

$$= \frac{34 \times 1000}{17 \times 1120} = \frac{200}{12}$$

1 N of $H_2O_2 = 5.6$ vol strength

$\dfrac{200}{12}$ N of $H_2O_2 = 5.6 \times \dfrac{200}{12} = 10$

vol strength

10. **(a)** A fusion reaction between hydrogen nuclei is difficult because positively charged nuclei repel each other. However, at very high temperatures of the order of 10^6 to 10^7 K, the nuclei may have sufficient energy to overcome the repulsive forces and thus fuse. This is why, fusion reactions are also called thermonuclear reactions.
11. **(b)** $BaO_2 + H_2SO_4 \rightarrow BaSO_4 + H_2O_2$.
oxygen has common O.S. as –2 and in peroxides as –1.
12. **(d)** $H_2O_2 \equiv KI \equiv I_2$

mEq of $H_2O_2 = 5 \times N$

$$\text{mEq of } I_2 = \frac{\text{Weight}}{\text{E.w}} \times 1000$$

$$\frac{1.27}{127} \times 1000 = 10 \text{ mEq}$$

$I_2 = \text{M.w} = 254$

$$\text{E.w} = \frac{\text{M.w}}{2} = \frac{254}{2} = 127$$

$\therefore$ mEq of H_2O_2 = mEq of I_2

$5 \times N = 10 \quad \therefore N = 2$

1 N of H_2O_2 = 1.7% of H_2O_2

2 N of H_2O_2 = 3.4% H_2O_2

13. (b) 100 g of H_2O_2 sample solution contains = x g of H_2O_2

3.4 g of H_2O_2 sample solution contains = $\frac{x}{100} \times 3.4$

Weight of $H_2O_2 = \frac{3.4\ x}{100}$

Eq. of $H_2O_2 = \frac{3.4\ x}{100} \times \frac{1}{17}$

mEq of H_2O_2

$= \frac{3.4x}{100 \times 17} \times 1000 = \frac{34\ x}{17} = 2x$

mEq of $KMnO_4 = x \times N$

$x \times N = 2x$

N = 2

14. (a) As H_2O_2 is loosing electrons, so it is acting as reducing agent.

15. (a) $\underset{'A'}{2Na} + 2H_2O \rightarrow \underset{'C'}{2NaOH} + \underset{'B'}{H_2} \uparrow$

$\underset{'D'}{Zn} + \underset{'C'}{2NaOH} \rightarrow Na_2ZnO_2 + \underset{'B'}{H_2} \uparrow$

$\underset{'D'}{Zn} + \text{dil. } H_2SO_4 \rightarrow ZnSO_4 + \underset{'B'}{H_2} \uparrow$

Na produces golden yellow colour with smokeless flame of Bunsen burner.

16. (d) MnO_2, PbO_2 and BaO will not give H_2O_2 with HCl. MnO_2 and PbO_2 will give Cl_2 and BaO will react with HCl to give $BaCl_2$ and water.

17. (c) $Ca(OH)_2$ is used for the softening of temporary hard water.

$$Ca(OH)_2(aq) \xrightarrow{CO_2} \underset{\text{cloudiness}}{CaCO_3(s)} + H_2O(l)$$

18. (d) H_2 is a highly inflammable gas.

19. (b) Only temporary hardness which is due to HCO_3^- (bicarbonate) ions is removed by boiling.

20. (a) $Mg + 2HNO_3(\text{dil.}) \rightarrow Mg(NO_3)_2 + H_2$

(Mg and Mn give H_2 with dil HNO_3)

21. (c) Potash alum is used for purifying water.

22. (a) $CaC_2 + 2D_2O \rightarrow C_2D_2 + Ca(OD)_2$

23. (b) $H^-(aq) + H_2O(l) \rightarrow OH^- + H_2$. Since H^- is a strong base it will abstract H^+ to form H_2O.

24. (d) On the industrial scale hydrogen is prepared from water gas according to following reaction sequence

$$\underbrace{CO + H_2}_{\text{water gas}} + \underbrace{H_2O}_{\text{(steam)}} \xrightarrow{\text{catalyst}} CO_2 + 2H_2$$

$$\xrightarrow[\text{(alkali)}]{2NaOH} Na_2CO_3 + H_2O$$

From the above it is clear that CO is first oxidised to CO_2 which is then absorbed in NaOH.

25. (b) Electrostatic forces of attraction are reduced to 1/80th in water.

26. (a) D_2O actually has higher freezing point (3.8°C) than water H_2O (0°C)

27. (c) H_2O_2 is unstable liquid and decomposes into water and oxygen either on standing or on heating.

28. (d) H_2O_2 show all these properties.

29. (b) $MnO_4^- + 5e^- \rightarrow Mn^{2+}$ (acidic)

$MnO_4^- + 3e^- \rightarrow MnO_2$ (basic)

100 mL $H_2O_2 \equiv 100 \times 5N$

$MnO_4^- \equiv V \times 3N\ MnO_2$

$N = \frac{500}{3}$ mL

30. (b) mEq of $MnO_4^- \equiv$ mEq of H_2O_2

$10 \times \frac{M}{5} \times 5 \equiv$ mEq of H_2O_2

$\Rightarrow 10$ mEq $\equiv$ mEq of H_2O_2

Weight of $H_2O_2 = 10 \times 10^{-3} \times \frac{34}{2} \times 0.17$

$= 0.2$ g

Weight of impure H_2O_2

$= \frac{100}{85} \times 0.17 = 0.2$ g

31. (b) $3BaO_2 + 2H_3PO_4 \rightarrow Ba_3(PO_4)_2 + 3H_2O_2$

Barium peroxide, Phosphoric acid, Barium phosphate, Hydrogen peroxide

32. (c) 1 N = 5.6 vol

⇒ Normalities of two solutions are 1 N and 2 N

Normality of mixture

$$= \frac{1\times1+1\times2}{1+1} = \frac{3}{2} = 1.5\text{ N}$$

Volume strength

$$= \frac{3}{2}\times5.6 = \frac{16.8}{2} = 8.4\text{ vol}$$

33. (b) Heavy water is D_2O hence

number of electrons = 2 + 8 = 10

number of protons = 10

Atomic mass of D_2O = 4 + 16 = 20

hence number of neutron

= Atomic mass – number of protons

= 20 – 10 = 10

34. (b) $2Al + 3H_2SO_4 \rightarrow Al_2(SO_4)_3 + 3H_2$

$2Al + 2NaOH + 2H_2O \rightarrow 2NaAlO_2 + 3H_2$

The ratio of volumes of hydrogen evolved is 1:1.

35. (c) In metal hydrides the O.S. of hydrogen –1 otherwise it is +1.

36. (d) Heavy water is stable.

37. (b) $2RCOONa + Mg^{2+} \rightarrow \underset{\text{Insoluble}}{(RCOO)_2Mg} + 2Na^+$

$2RCOONa + Ca^{2+} \rightarrow \underset{\text{Insoluble}}{(RCOO)_2Ca} + 2Na^+$

38. (a) Molecular weight of H_2O_2 = 34

Equivalent weight of H_2O_2 = 17

∴ 1 L of 1 N H_2O_2 has = 17 g of H_2O

∴ 1 L of 1.5 N H_2O_2 has

= 1.5 × 17 = 25.5 g of H_2O_2

39. (b) ∵ 22.4 litre O_2 at N.T.P. obtained by 68 gm of H_2O_2

∴ 1 litre O_2 at N.T.P. obtained by

$\frac{68}{22.4}$ gm of H_2O_2

∴ 20 litre O_2 at N.T.P. obtained by

$\frac{68}{22.4}\times20$ gm of H_2O_2 = 60.71 gm of H_2O_2

∴ 1000 ml O_2 at N.T.P. obtained by = 60.71 gm of H_2O_2

∴ 100 ml O_2 at N.T.P. obtained by

$$= \frac{60.71}{1000}\times100 = 6.071\%$$

40. (a) $Zn + H_2SO_4 \rightarrow ZnSO_4 + H_2$

$Zn + 2NaOH \rightarrow Na_2ZnO_2 + H_2$

∴ Ratio of volumes of H_2 evolved is 1 : 1

41. (b) It is a fact.

42. (b) $\underbrace{CO + H_2}_{\text{water gas}} + H_2O \xrightarrow{\text{catalyst}} CO_2 + 2H_2$

43. (d) H_2O absorbs neutrons more than D_2O and this decreases the number of neutrons for the fission process.

44. (c) Pure hydrogen is evolved by reacting absolute alcohol and Na

$$C_2H_5OH + Na \rightarrow C_2H_5ONa + \frac{1}{2}H_2$$

other statements are correct.

45. (a) Polarity of bond depends on difference in electronegativity of the two concerned atoms. H_2O is more polar than H_2S because oxygen (in O–H) is more electronegative than sulphur (in S–H).

46. (a) The complex salt of metaphosphoric acid sodium hexametaphosphate $(NaPO_3)_6$, is known as calgon. It is represented as $Na_2[Na_4(PO_3)_6]$

47. (d) SO_2 changes to H_2SO_4 (O.N. changes from +4 to +6 oxidation)

$2KI \rightarrow I_2$ (O.S. changes from –1 to 0 oxidation)

$PbS \rightarrow PbSO_4$ (O.S. changes from –2 to +6 oxidation)

$Ag_2O \rightarrow 2Ag$ (O.S. changes from +1 to 0 reduction)

48. (d) Dihedral angle of H_2O_2 in gas phase is 111.5° and in solid phase it is 90.2°

49. (a) As Na^+ ions in sodium zeolite are replaced by Ca^{2+} and Mg^{2+} ions present in hard water, these two ions will not be present.

50. (c) Ortho and para forms of hydrogen resemble in their chemical properties.

51. (a) $K_2Cr_2O_7$ an orange coloured solution, on acidification with dil H_2SO_4 is oxidised by H_2O_2 to blue peroxide of chromium, CrO_5, which is unstable.

$$K_2Cr_2O_7 + H_2SO_4 \rightarrow K_2SO_4 + H_2Cr_2O_7$$

$$[H_2O_2 \rightarrow H_2O + [O]] \times 4$$

$$H_2Cr_2O_7 + 4[O] \rightarrow 2CrO_5 + H_2O$$

$$K_2Cr_2O_7 + H_2SO_4 + 4H_2O_2 \rightarrow 2CrO_5 + K_2SO_4 + 5H_2O$$

52. (a) H_2O_2 acts as a reducing agent only in presence of strong oxidising agents (i.e., MnO_4^-) in acidic as well as alkaline medium.

$$2KMnO_4 + 3H_2SO_4 + 5H_2O_2 \longrightarrow K_2SO_4 + 2MnSO_4 + 8H_2O + 5O_2$$

53. (a) Pure para form can be obtained at low temperature (20K) but pure ortho form is very difficult to obtain.

54. (c) $2H_2O_2 \xrightarrow{\Delta} 2H_2O + O_2$

$2KI + H_2O_2 \rightarrow 2KOH + I_2$

$$2KMnO_4 + 3H_2SO_4 + 5H_2O_2 \rightarrow K_2SO_4 + 2MnSO_4 + 8H_2O + 5O_2$$

55. (d)

$$H_2S + H_2O_2 \longrightarrow 2H_2O + S$$

(oxidation: $H_2S \rightarrow S$; reduction: $H_2O_2 \rightarrow H_2O$)

In this reaction H_2S is oxidised to sulphur and H_2O_2 is reduced to H_2O, hence this reaction show oxidation-reduction both i.e., redox reaction.

56. (a) Glycerol, phosphoric acid or acetanilide is added to H_2O_2 to check its decomposition.

57. (b) Na zeolite + $CaCl_2 \rightarrow$ Ca zeolite + 2NaCl

58. (c) Fire due to action of water on saline hydrides cannot be extinguished with water or CO_2. These hydrides can reduce CO_2 at high temperature to produce O_2.

59. (c) 30 vol of H_2O_2 means one volume of H_2O_2 on decom-position will give 30 volume of oxygen.

60. (a) $KIO_4 + H_2O_2 \rightarrow KIO_3 + H_2O + O_2$

Thus H_2O_2 is acting as a reducing agent

$2NH_2OH + H_2O_2 \rightarrow N_2 + 4H_2O$

Here H_2O_2 is acting as an oxidising agent

10 The s-Block Elements

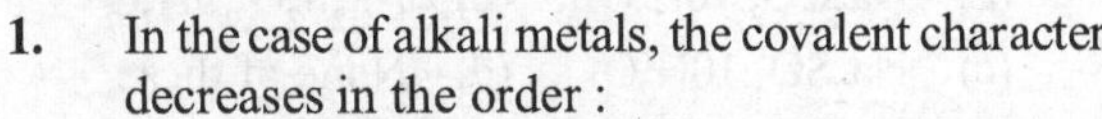

1. In the case of alkali metals, the covalent character decreases in the order :
(a) $MF > MCl > MBr > MI$
(b) $MF > MCl > MI > MBr$
(c) $MI > MBr > MCl > MF$
(d) $MCl > MI > MBr > MF$

2. The mobility of metal ions in aqueous medium (Li^+, Na^+, K^+, Rb^+) in the electric field, follows the order
(a) $Li^+ > Na^+ > K^+ > Rb^+$
(b) $Rb^+ > Na^+ = K^+ > Li^+$
(c) $Li^+ < Na^+ < K^+ < Rb^+$
(d) $Na^+ = K^+ > Rb^+ > Li^+$

3. The main oxides formed on combustion of Li, Na and K in excess of air are, respectively:

(a) Li_2O_2, Na_2O_2 and KO_2
(b) Li_2O, Na_2O_2 and KO_2
(c) Li_2O, Na_2O and KO_2
(d) LiO_2, Na_2O_2 and K_2O

4. Which of the following oxides of potassium is not known ?
(a) K_2O (b) K_2O_4 (c) KO_3 (d) K_2O_3

5. Nitrogen dioxide cannot be prepared by heating
(a) KNO_3 (b) $Pb(NO_3)_2$
(c) $Cu(NO_3)_2$ (d) $AgNO_3$

6. Incorrect statement is:
(a) $NaHCO_3$ and $KHCO_3$ have same crystal structure
(b) On heating $LiNO_3$ decomposes into Li_2O and NO_2
(c) Among alkali metals, Li metal impart red colour to flame
(d) Li_2SO_4 does not form alum

7. Which is not correctly matched Toughnut
(i) Basic strength $Cs_2O < Rb_2O < K_2O < Na_2O < Li_2O$ of oxides.
(ii) Stability of $Na_2O_2 < K_2O_2 < Rb_2O_2 < Cs_2O_2$ peroxides.
(iii) Stability of $LiHCO_3 < NaHCO_3 < KHCO_3 < RbHCO_3 < CsHCO_3$.
(iv) Melting point $NaF < NaCl < NaBr < NaI$
(a) (i) and (iv) (b) (i) and (iii)
(c) (i) and (ii) (d) (ii) and (iii)

8. The solubility order for alkali metal fluoride in water is :
(a) $LiF < RbF < KF < NaF$
(b) $RbF < KF < NaF < LiF$
(c) $LiF > NaF > KF > RbF$
(d) $LiF < NaF < KF < RbF$

9. Based on lattice energy and other considerations, which one of the following alkali metal chloride is expected to have the highest melting point?
(a) NaCl (b) KCl (c) LiCl (d) RbCl

10. In crystals which one of the following ionic compounds would you expect maximum distance between centres of cations and anions?
(a) LiF (b) CsF (c) CsI (d) LiI

11. Which of the following statements about Na_2O_2 is not correct? Critical Thinking
(a) It is diamagnetic in nature
(b) It is derivative of H_2O_2
(c) Na_2O_2 oxidises Cr^{3+} to CrO_4^{2-} in acid medium.
(d) It is the super oxide of sodium

12. The correct order of thermal stability of hydroxides is:
(a) $Ba(OH)_2 < Ca(OH)_2 < Sr(OH)_2 < Mg(OH)_2$
(b) $Mg(OH)_2 < Sr(OH)_2 < Ca(OH)_2 < Ba(OH)_2$
(c) $Mg(OH)_2 < Ca(OH)_2 < Sr(OH)_2 < Ba(OH)_2$
(d) $Ba(OH)_2 < Sr(OH)_2 < Ca(OH)_2 < Mg(OH)_2$

13. The reducing power of a metal depends on various factors. Suggest the factor which makes Li metal the strongest reducing agent in aqueous solution:
(a) Sublimation enthalpy
(b) Ionisation enthalpy
(c) Hydration enthalpy
(d) Electron-gain enthalpy

14. Select the correct statements :
   I. Cs^+ is more highly hydrated than the other alkali metal ions
   II. Among the alkali metals Li, Na, K and Rb, lithium has the highest melting point
   III. Among the alkali metals, only lithium forms a stable nitride by direct combination with nitrogen
   (a) I, II and III (b) I and II
   (c) I and III (d) II and III
15. In the following sequence of reaction, identify the compounds (A), (B) (C) and (D):

$$\underset{\text{Solution}}{Na_2CO_3} \xrightarrow{SO_2} (A) \xrightarrow{Na_2CO_3} (B) \xrightarrow[\text{Heat}]{S} (C) \xrightarrow{AgNO_3} (D)$$

   (a) $Na_2SO_3, NaHSO_3, Na_2S, Ag_2S$
   (b) $NaHSO_3, Na_2SO_3, Na_2S_2O_3, Ag_2S$
   (c) $NaHSO_3, Na_2SO_4, Na_2S, Ag_2O$
   (d) $Na_2SO_3, Na_2SO_4, Na_2S_2O_3, Ag$
16. Which of the following does not illustrate the anomalous properties of lithium?
   (a) The melting point and boiling point of Li are comparatively high
   (b) Li is much softer than the other group I metals
   (c) Li forms a nitride Li_3N unlike group I metals
   (d) The ion of Li and its compounds are more heavily hydrated than those of the rest of the group
17. The order solubility in water of alkaline earth metal sulphates down the group is Be > Mg > Ca > Sr > Ba. This is due to increases in
   (a) ionization energy
   (b) melting point
   (c) coordination number
   (d) All of these
18. A chloride dissolves appreciably in cold water. When placed on a Pt wire in bunsen flame, no distinctive colour is noted. Which cation could be present?
   (a) Be^{2+} (b) Ba^{2+} (c) Pb^{2+} (d) Ca^{2+}
19. An unknown inorganic compound (X) loses its water of crystallization on heating and its aqueous solution gives the following reactions :

   (a) It gives a white turbidity with dilute HCl solution
   (b) It decolourises a solution of iodine in potassium iodide
   (c) It gives a white precipitate with silver nitrate solution which turns black on standing.

   Identify the compound (X)
   (a) $Na_2CO_3.10H_2O$ (b) $Na_2S_2O_3.5H_2O$
   (c) $Na_2SO_4.10H_2O$ (d) None of these
20. A metal M readily forms its sulphate MSO_4 which is water-soluble. It forms its oxide MO which becomes inert on heating. It forms an insoluble hyroxide $M(OH)_2$ which is soluble in NaOH solution. Then M is
   (a) Mg (b) Ba (c) Ca (d) Be
21. Melting point of calcium halides decreases in the order
   (a) $CaF_2 > CaCl_2 > CaBr_2 > CaI_2$
   (b) $CaI_2 > CaBr_2 > CaCl_2 > CaF_2$
   (c) $CaBr_2 > CaI_2 > CaF_2 > CaCl_2$
   (d) $CaCl_2 > CaBr_2 > CaI_2 > CaF_2$
22. Which of the following is not correct ?

   (a) $2Li_2O \xrightarrow[673K]{heat} Li_2O_2 + 2Li$
   (b) $2K_2O \xrightarrow[673K]{heat} K_2O_2 + 2K$
   (c) $2Na_2O \xrightarrow[673K]{heat} Na_2O_2 + 2Na$
   (d) $2Rb_2O \xrightarrow[673K]{heat} Rb_2O_2 + 2Rb$
23. The alkali metals form salt-like hydrides by the direct synthesis at elevated temperature. The thermal stability of these hydrides decreases in which of the following orders ?
   (a) CsH > RbH > KH > NaH > LiH
   (b) KH > NaH > LiH > CsH > RbH
   (c) NaH > LiH > KH > RbH > CsH
   (d) LiH > NaH > KH > RbH > CsH
24. Which of the following is not correct about the solution when moderate amount of sodium metal is dissolved in liquid ammonia at low temperature?
   (a) Na^+ ions are produced in solution
   (b) We get a blue coloured solution
   (c) the solution acts as a good conductor of electricity
   (d) Liquid ammonia remains diamagnetic

25. The order of solubility of lithium halides in non polar solvents follows the order
   (a) LiI > LiBr > LiCl > LiF
   (b) LiF > LiI > LiBr > LiCl
   (c) LiCl > LiF > LiI > LiBr
   (d) LiBr > LiCl > LiF > LiI

26. Which of the following is most stable?
   (a) Na_3N (b) Li_3N
   (c) K_3N (d) Rb_3N

27. $ZnCl_2 + NaHCO_3 \xrightarrow{Heat} (A) \xrightarrow{Heat}$
   $(B) + (C)\ H_2O\ (B) + NaOH \rightarrow (D)$
   Identify the compound (D) present in the solution. Toughnut
   (a) $ZnCO_3$ (b) $Zn(OH)_2$
   (c) ZnO   (d) Na_2ZnO_2

28. Which of the following has correct increasing basic strength?
   (a) MgO<BeO<CaO<BaO
   (b) BeO<MgO<CaO<BaO
   (c) BaO<CaO<MgO<BeO
   (d) CaO<BaO<BeO<MgO

29. Which of the following statement is false ?
   (a) Strontium decomposes water readily than beryllium.
   (b) Barium carbonate melts at a higher temperature than calcium carbonate.
   (c) Barium hydroxide is more soluble in water than magnesium hydroxide.
   (d) Beryllium hydroxide is more basic than barium hydroxide.

30. Which of the following is the best method for preparation of BeF_2?
   (a) Reaction of Be with F_2
   (b) Thermal decomposition of $(NH_4)_2BeF_4$
   (c) Reaction of Be with HF
   (d) All of the above are equally effective

31. Aqueous solution of a group 2 element is precipitated by adding Na_2CO_3, then this precipitate is tested on flame, no light in visible region is observed , this element can be Tricky
   (a) Ba   (b) Mg   (c) Ca   (d) Sr

32. Which of the following are found in biological fluids Na^+, Mg^{2+}, Ca^{2+}, K^+, Sr^{2+}, Li^+ and Ba^{2+}?
   (a) Mg^{2+}, Ca^{2+}, and Sr^{2+}
   (b) $Na^{2+}$and K^+
   (c) Na^+, K^+, $Mg^{2+}$and Ca^{2+}
   (d) Sr^+, Li and Ba^{2+}

33. Metallic magnesium is prepared by
   (a) reduction of MgO by coke.
   (b) electrolysis of aqueous solution of $Mg(NO_3)_2$.
   (c) displacement of Mg by iron from $MgSO_4$ solution.
   (d) electrolysis of molten $MgCl_2$.

34. A compound (A) is used in preparation of washing soda to recover ammonia in Solvay's process. When CO_2 is bubbled through an aqueous solution of (A), the solution turns milky. It is used in white washing due to disinfectant nature. What is the chemical formula of A?

   (a) $Ca(HCO_3)_2$ (b) CaO
   (c) $Ca(OH)_2$ (d) $CaCO_3$

35. BeF_2 is soluble in water whereas fluorides of other alkaline earth metals are insoluble because of
   (a) ionic nature of BeF_2.
   (b) covalent nature of BeF_2.
   (c) greater hydration energy of Be^{2+} ion as compared to its lattice energy.
   (d) none of these.

36. Which category of salts of alkaline earth metals is not found in solid sate, but found in solution state?
   (a) Carbonates   (b) Bicarbonates
   (c) Hydroxides   (d) Sulphates

37. Sodium carbonate solution in water is alkaline due to
   (a) hydrolysis of Na^+
   (b) hydrolysis of CO_3^{2-}
   (c) hydrolysis of both Na^+ and CO_3^{2-} ions
   (d) None of these

38. When a crystal of caustic soda is exposed to air, a liquid layer is deposited because :
   (a) Crystal loses water
   (b) Crystal absorbs moisture and CO_2
   (c) Crystal melts
   (d) Crystal sublimes

39. Which of the following statements is correct for $CsBr_3$?
(a) It is a covalent compound.
(b) It contains Cs^{3+} and Br^- ions.
(c) It contains Cs^+ and Br_3^- ions
(d) It contains Cs^+, and Br^- and lattice Br_2 molecule.

40. The pair of compounds which cannot exits together in solution is

Tricky

(a) $NaHCO_3$ and NaOH
(b) Na_2CO_3 and $NaHCO_3$
(c) Na_2CO_3 and NaOH
(d) $NaHCO_3$ and NaCl

41. Na_2SO_3 and $NaHCO_3$ may be distinguished by treating their aqueous solution with :
(a) MgO (b) $MgSO_4$
(c) litmus solution. (d) dil. acid

42. Substance which absorbs CO_2 and violently reacts with H_2O with sound is :
(a) H_2SO_4 (b) $CaCO_3$
(c) ZnO (d) CaO

43. When sulphur is heated with NaOH (aq). The compounds formed are
(a) $Na_2S + H_2O$
(b) $Na_2SO_3 + H_2O$
(c) $Na_2S + Na_2S_2O_3 + H_2O$
(d) $Na_2S_2O_3 + H_2O$

44. The metal X is prepared by the electrolysis of fused chloride. It reacts with hydrogen to form a colourless solid from which hydrogen is released on treatment with water. The metal is :
(a) Ca (b) Al (c) Zn (d) Cu

45. Acidified solution of sodium thiosulphate is unstable because in thiosulphate

(a) the sulphur atoms are at unstable oxidation state of +2
(b) the two sulphur atoms are at different oxidation states of + 6 and – 2
(c) the S-S bond are unstable bonds.
(d) sulphur is in zero oxidation state.

46. Two metals (A) and (B) belong to the same group of the periodic table. Metal (A) forms an insoluble oxide but a soluble sulphate, metal (B) forms a soluble oxide but an insoluble sulphate. Both metals (A) and (B) form hydroxides which are soluble in alkalis. (A) and (B) are
(a) Ba and Mg (b) Na and K
(c) Mg and Ba (d) K and Rb

47. Among the given statements, the incorrect one is
(a) Be differs much from other alkali metals than Li does from other alkali metals.
(b) Be generally forms covalent compounds.
(c) Be forms a very strong complex, $[Be(H_2O)_4]^{2+}$.
(d) Be usually has more than four water of crystallisation associated with it.

48. Calcium imide on hydrolysis gives gas (B) which on oxidation by bleaching powder gives gas (C). Gas (C) on reaction with magnesium give compound (D) which on hydrolysis gives again gas (B). Identify (B), (C) and (D).

Toughnut

(a) NH_3, N_2, Mg_3N_2
(b) N_2, NH_3, MgNH
(c) N_2, N_2O_5, $Mg(NO_3)_2$
(d) NH_3, NO_2, $Mg(NO_2)_2$

49. Which one is the correct statement with reference to solubility of $MgSO_4$ in water?
(a) SO_4^{2-} ion mainly contributes towards hydration energy.
(b) Sizes of Mg^{2+} and SO_4^{2-} are similar.
(c) Hydration energy of $MgSO_4$ is higher in comparison to its lattice energy.
(d) Ionic potential (charge/radius ratio) of Mg^{2+} is very low.

50. A solid compound 'X' on heating gives CO_2 gas and a residue. The residue mixed with water forms 'Y'. On passing an excess of CO_2 through 'Y' in water, a clear solution 'Z', is obtained. On boiling 'Z', compound 'X' is reformed. The compound 'X' is

Critical Thinking

(a) $Ca(HCO_3)_2$ (b) $CaCO_3$
(c) Na_2CO_3 (d) K_2CO_3

51. Amongst the following hydroxides, the one which has the lowest value of K_{sp} is :
(a) $Ca(OH)_2$ (b) $Mg(OH)_2$
(c) $Be(OH)_2$ (d) $Ba(OH)_2$

52. For which one of the following minerals, the composition given is incorrect?
(a) Soda ash — (Na_2CO_3)
(b) Carnallite — ($KCl . MgCl_2 . 6H_2O$)
(c) Borax — ($Na_2B_4O_7 . 7H_2O$)
(d) Glauber's salt — ($Na_2SO_4 . 10H_2O$)

53. The difference of number of water molecules in gypsum and plaster of Paris is

Critical Thinking

(a) $\frac{5}{1}$ (b) 2 (c) 1/2 (d) $1\frac{1}{2}$

54. Which is not obtained when metal carbides react with H_2O?

(a) $Al_4C_3 + H_2O \longrightarrow CH \equiv CH$

(b) $CaC_2 + H_2O \longrightarrow CH \equiv CH$

(c) $Mg_4C_3 + H_2O \longrightarrow CH_3C \equiv CH$

(d) $Be_2C + H_2O \longrightarrow CH_4$

55. $CaCl_2$ is preferred over NaCl for clearing ice on roads particularly in very cold countries. This is because:

(a) $CaCl_2$ is less soluble in H_2O than NaCl

(b) $CaCl_2$ is hygroscopic but NaCl is not

(c) Eutectic mixture of $CaCl_2/H_2O$ freezes at −55°C while that of $NaCl/H_2O$ freezes at −18°C

(d) NaCl makes the road slipperty but $CaCl_2$ does not

56. Which of the following statement about the sulphate of alkali metal is correct?

(a) Except Li_2SO_4 all sulphate of other alkali metals are soluble in water

(b) All sulphate salts of alkali metals except lithium sulphate forms alum.

(c) All sulphate salts of alkali metals except lithium sulphate do not decompose at high temperature

(d) All of the above

57. Mg_2C_3 reacts with water forming propyne gas. C_3^{4-} ions has:

(a) two sigma and two pi bonds

(b) three sigma and one pi bond

(c) two sigma and one pi bond

(d) two sigma and three pi bonds

58. If NaOH is added to an aqueous solution of Zn^{2+} ions, a white precipitate appears and on adding excess NaOH, the precipitate dissolves. In this solution zinc exists in the :

(a) both in cationic and anionic parts

(b) there is no zinc left in the solution

(c) cationic part

(d) anionic part.

59. Which is incorrect statement?

(a) The heats of hydration of the dipositive alkaline earth metal decrease with an increase in their ionic size

(b) $NaNO_3$ forms Na_2O_2 on heating

(c) Hydration of alkali metal ion is less than that of IIA group

(d) Alkaline earth metal ions, because of their much larger charge to size ratio, exert a much stronger electrostatic attraction on the oxygen of water molecule surrounding them

60. Which of the following does not form an oxide on heating?

(a) $ZnCO_3$ (b) $CaCO_3$

(c) Li_2CO_3 (d) Na_2CO_3

Answer KEY

1	(c)	7	(a)	13	(c)	19	(b)	25	(a)	31	(b)	37	(b)	43	(c)	49	(c)	55	(c)
2	(c)	8	(d)	14	(d)	20	(d)	26	(b)	32	(c)	38	(b)	44	(a)	50	(b)	56	(d)
3	(b)	9	(a)	15	(b)	21	(a)	27	(d)	33	(d)	39	(c)	45	(b)	51	(c)	57	(a)
4	(b)	10	(c)	16	(b)	22	(a)	28	(b)	34	(c)	40	(a)	46	(c)	52	(c)	58	(d)
5	(a)	11	(d)	17	(c)	23	(d)	29	(d)	35	(c)	41	(b)	47	(d)	53	(d)	59	(b)
6	(a)	12	(d)	18	(a)	24	(d)	30	(b)	36	(b)	42	(d)	48	(a)	54	(a)	60	(d)

Hints & Solutions

1. (c) As the size of halide ion decreases, $I^- > Br^- > Cl^- > F^-$, tendency to get polarised by cation M^+ decreases and hence covalent character decreases.
2. (c) Down the group with decrease in positive charge tendency to get hydrated decreases and hence the size of hydrated ion decreases. Thus mobility increases.
3. (b) On heating with excess of air Li, Na and K forms following oxides

$$4Li + O_2 \longrightarrow \underset{\text{Lithium monoxide}}{2Li_2O}$$

$$4Na + O_2 \xrightarrow{575K} \underset{\text{Sodium peroxide}}{Na_2O_2}$$

$$K + O_2 \longrightarrow \underset{\text{Potassium superoxide}}{KO_2}$$

4. (b) O_4^{2-} ion is not possible and K_2O_4 is unknown.
5. (a) Only nitrates of heavy metals and lithium decompose on heating to produce NO_2.
6. (a) The crystal structures of $NaHCO_3$ and $KHCO_3$ both show hydrogen bonding but are different. In $NaHCO_3$, the HCO_3^- ions are linked into an infinite chain, while in $KHCO_3$, a dimeric anion is formed.
7. (a) Basic strength of the oxides increases in the order
$Li_2O < Na_2O < K_2O < Rb_2O < Cs_2O$. The increase in basic strength is due to the decrease in I.E. and increase in electropositive character.
The melting points of the halides decrease in the order NaF > NaCl > NaBr > NaI, as the size of the halide ion increases. The decrease in melting point is due to increase in the covalent character with increase in the size of anion according to Fajan's rule.
8. (d) Higher the lattice enthalpy lower will be solubility i.e.,

$$\text{lattice enthalpy} \propto \frac{1}{\text{Solubility}}$$

Since the lattice enthalpy of alkali metals follow the order
Li > Na > K > Rb
Hence the correct order of solubility is
LiF < NaF < KF < RbF
9. (a) NaCl due it to its largest lattice energy has the highest melting point.
10. (c) As Cs^+ ion has larger size than Li^+ and I^- has larger size than F^-, therefore maximum distance between centres of cations and anions is in CsI.
11. (d) Na_2O_2 is peroxide of sodium not super oxide. The formula of sodium superoxide is NaO_2.
12. (d) Stability of ionic compounds decreases with decrease in lattice enthalpy. Thus stability of alkaline earth metal hydroxides decreases with decrease in lattice enthalpy as the size of alkali earth metal cations increases down the group.
13. (c) Due to high hydration energy of Li^+ cation, the standard reduction potential of Li^+ is more negative among all alkali metal cations hence Li act as strong reducing agent in water.
14. (d) Amongst alkali metal ions, Li ions are highly hydrated.
15. (b)

$$Na_2CO_3 + 2SO_2 + H_2O \rightarrow \underset{(A)}{2NaHSO_3} + CO_2$$

$$\underset{(A)}{2NaHSO_3} + Na_2CO_3 \rightarrow \underset{(B)}{2Na_2SO_3} + H_2O + CO_2$$

$$\underset{(B)}{2Na_2SO_3} + S \xrightarrow{Heat} \underset{(C)}{2Na_2S_2O_3}$$

$$Na_2S_2O_3 + AgNO_3 \rightarrow \underset{(D)}{Ag_2S} + H_2SO_4$$

16. (b) Actually Li is harder than other alkali metals.
17. (c) In crystalline sulphates, the coordination number of the ions increases from Be to Ba and so also the lattice energy hence solubility in H_2O decreases.

18. (a) Due to very high I.E. of Be, its valence electrons require high energy to remove (excite) which is not available in bunsen flame hence Be^{2+} ions do not impart colour to the flame.

19. (b) $X = Na_2S_2O_3.5H_2O$

$$Na_2S_2O_3 + 2HCl \longrightarrow 2NaCl + H_2O + S + SO_2$$

$$KI_3 + 2Na_2S_2O_3 \longrightarrow KI + 2NaI + Na_2S_4O_6$$

$$2AgNO_3 + Na_2S_2O_3 \longrightarrow Ag_2S_2O_3 + 2NaNO_3$$

$$Ag_2S_2O_3 + H_2O \longrightarrow Ag_2S + H_2SO_4$$

20. (d) Beryllium shows anomalous properties due to its small size.

21. (a) As the size of the anion increases, the covalent character increases and hence the m.p. decrease.

22. (a) Lithium does not form peroxide.

23. (d) The stability of alkali metal hydrides decreases from Li to Cs. It is due to the fact that M–H bonds become weaker with increase in size of alkali metals as we move down the group from Li to Cs. Thus the order of stability of hydrides is

$LiH > NaH > KH > RbH > CsH$

24. (d) Liquid ammonia remains paramagnetic

25. (a) As the size of the anion increases from F^- to I^-, the covalent character increase and hence the solubility in non -polar solvent increases

$LiI > LiBr > LiCl > LiF$

26. (b) Only Li_3N is stable, others are not formed at all.

27. (d) $ZnCl_2 + 2NaHCO_3 \xrightarrow{\Delta} \underset{(A)}{Zn(HCO_3)_2} + 2NaCl$

$$\underset{(A)}{Zn(HCO_3)_2} \xrightarrow{\Delta} \underset{(B)}{ZnCO_3} + H_2O + \underset{(C)}{CO_2 \uparrow}$$

$$\underset{(B)}{ZnCO_3} + 2NaOH \rightarrow \underset{(D)}{Na_2ZnO_2} + 2H_2O$$

28. (b) The basic character of oxides increases down the group.

29. (d) $Be(OH)_2$ is amphoteric, but the hydroxides of other alkaline earth metals are basic. The basic strength increases gradually.

30. (b) Thermal decomposition of $(NH_4)_2BeF_4$ is the best method for preparation of BeF_2.

31. (b) Electrons in Mg due to its small size are tightly bound so they cannot be excited by the flame.

32. (c) Monovalent sodium and potassium ions and divalent magnesium and calcium ions are found in large proportions in biological fluids.

33. (d) $\underset{(Molten)}{MgCl_2} \xrightarrow{Electrolysis} \underset{Cation}{Mg^{+2}} + \underset{Anion}{2Cl^-}$

Anode $\rightarrow 2Cl^- \rightarrow 2Cl + 2e^-, Cl + Cl \rightarrow Cl_2$

Cathode $\rightarrow Mg^{+2} + 2e^- \rightarrow Mg$

34. (c) $NH_3 \uparrow \xrightarrow[\text{Solvay's process}]{NH_4Cl} \underset{(A)}{Ca(OH)_2} \xrightarrow{CO_2} CaCO_3$

35. (c) Be^{2+} being small in size is heavily hydrated and heat of hydration exceeds the lattice energy. Hence BeF_2 is soluble in water.

36. (b) The bicarbonates of alkaline earth metals exist in solution only.

37. (b) Due to anionic hydrolysis

$$CO_3^{2-} + H_2O \rightleftharpoons HCO_3^- + \overline{O}H$$

38. (b) It is hygroscopic and deliquescent. It absorbs moisture and CO_2 to form Na_2CO_3

$$2NaOH + CO_2 \longrightarrow Na_2CO_3 + H_2O$$

39. (c) $CsBr_3$ may be represented as $Cs^+Br_3^-$

40. (a) $NaHCO_3 + NaOH \rightarrow Na_2CO_3 + H_2O$

($NaHCO_3$ is an acidic salt)

41. (b) $2MgSO_4 + 2Na_2CO_3 + H_2O \longrightarrow MgCO_3.Mg(OH)_2 + 2Na_2SO_4 + CO_2$

Basic magnesium carbonate

42. (d) $CaO + CO_2 \longrightarrow CaCO_3$

$CaO + H_2O \longrightarrow Ca(OH)_2$

hissing sound and $\Delta H = -ve$

43. (c) $4S + 6NaOH \longrightarrow Na_2S_2O_3 + 2Na_2S + 3H_2O$

44. (a) $Ca + H_2 \longrightarrow CaH_2 + 2H_2O$

$\longrightarrow Ca(OH)_2 + H_2$

45. (b) $Na_2S_2O_3 + 2HCl \longrightarrow$

(Disproportionation)

$2NaCl + SO_2 + S + H_2O$

46. (c) MgO is insoluble, whereas $MgSO_4$ is soluble, whereas BaO is soluble, but $BaSO_4$ is insoluble.

47. (d) Hydrated Be^{2+} ion exists as $[Be(H_2O)_4]^{2+}$.

48. (a) $Ca(NH) + 2H_2O \longrightarrow Ca(OH)_2 + \underset{(B)}{NH_3(g)}$

$\underset{(B)}{2NH_3} + 3CaOCl_2 \longrightarrow \underset{(C)}{N_2(g)} + 3CaCl_2 + 3H_2O$

$\underset{(C)}{N_2(g)} + 3Mg \longrightarrow \underset{(D)}{Mg_3N_2}$

$\underset{(D)}{Mg_3N_2} + 6H_2O \longrightarrow 3Mg(OH)_2 + \underset{(B)}{NH_3}$

49. (c) $MgSO_4$ is the only alkaline earth metal sulphate which is soluble in water and for solubility hydration energy should be greater than lattice energy.

50. (b)

$$\underset{\text{Solid}}{X} \xrightarrow{\text{heat}} CO_2 + \text{Residue} \xrightarrow{H_2O} Y \xrightarrow{\text{excess } CO_2} \underset{\text{Clear solution}}{Z} \xrightarrow{\text{boil}} X$$

The given properties coincide with $CaCO_3$

$$\underset{'X'}{CaCO_3} \xrightarrow{\text{heat}} CO_2 + \underset{\text{Residue}}{CaO} \xrightarrow{H_2O} \underset{'Y'}{Ca(OH)_2} \xrightarrow{\text{excess } CO_2} \underset{'Z'}{Ca(HCO_3)_2} \xrightarrow{\text{boil}} CaCO_3$$

51. (c) $Be(OH)_2$ is insoluble in water and thus has lower value of K_{sp}.

52. (c) Barax is $Na_2B_4O_7 . 10H_2O$.

53. (d) Gypsum is $CaSO_4 . 2H_2O$, where as plaster of paris is $CaSO_4 \frac{1}{2} H_2O$.

No. of water $= \left(2 - \frac{1}{2}\right) \longrightarrow \frac{3}{2}$ or $1\frac{1}{2}$ molecules

54. (a) $Al_4C_3 + 12H_2O \longrightarrow 4Al(OH)_3 + 3CH_4$

55. (c) Due to much lower freezing point of eutectic mixture of $CaCl_2/H_2O$.

56. (d) All are facts

57. (a) The structure of C_3^{4-} ion is $\overset{-2}{C} = C = \overset{-2}{C}$ i.e, $2\sigma\ 2\pi$ bonds.

58. (d) $Zn^{2+} + 2NaOH \longrightarrow Zn(OH)_2 + 2Na^+$

$Zn(OH)_2 + 2NaOH \longrightarrow Na_2ZnO_2 + 2H_2O$

$Na_2ZnO_2 \longrightarrow 2Na^+ + (ZnO_2)^{2-}$

59. (b) $NaNO_3 \xrightarrow{\Delta} NaNO_2 + O_2$

60. (d) Na_2CO_3 does not decompose to form Na_2O.

11 The p-Block Elements (Group 13 and Group 14)

1. Which of the following hydroxide is acidic?
   (a) $Al(OH)_3$ (b) $Ca(OH)_2$
   (c) $Tl(OH)_3$ (d) $B(OH)_3$

2. Al is more reactive than Fe but Al is less easily corroded than Fe because
   (a) it is a noble metal
   (b) oxygen forms a protective oxide layer
   (c) iron undergoes reaction easily with water
   (d) Fe form mono and divalent ions.

3. Often a ground glass stopper gets stuck in the neck of a glass bottle containing NaOH solution. This is due to :
   (a) The presence of dirt particles in between.
   (b) The formation of solid silicate in between by the reaction of SiO_2 of glass with NaOH.
   (c) The formation of Na_2CO_3 in between by the reaction of CO_2 of air and NaOH.
   (d) Glass contains a boron compound which forms a precipitate with the NaOH solution.

4. $(Si_2O_5)_n^{2n-}$ anion is obtained when:

   Toughnut

   (a) no oxygen of a SiO_4^{4-} tetrahedron is shared with another SiO_4^{4-} tetrahedron
   (b) one oxygen of a SiO_4^{4-} tetrahedron is shared with another SiO_4^{4-} tetrahedron
   (c) two oxygen of a SiO_4^{4-} tetrahedron are shared with another SiO_4^{4-} tetrahedron
   (d) three oxygen of a SiO_4^{4-} tetrahedron are shared with another SiO_4^{4-} tetrahedron

5. Boric acid is a weak monobasic acid and acts as Lewis acid
   (a) By donating H^+
   (b) By accepting OH^-
   (c) By donating lone pair of electrons
   (d) By accepting lone pair of electrons.

6. 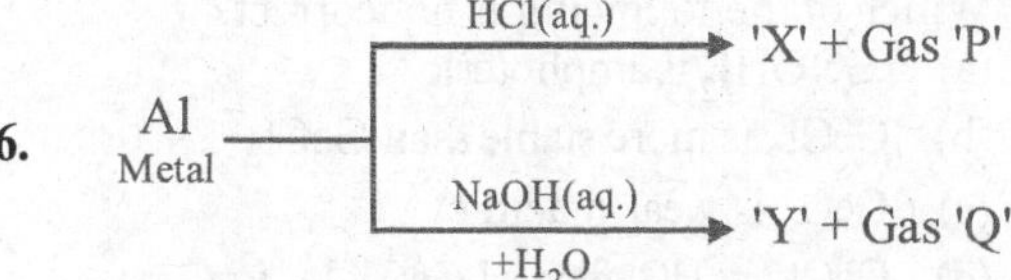

   The incorrect statement regarding above reactions is:

   Critical Thinking

   (a) Al shows amphoteric character
   (b) Gas 'P' and 'Q' are different
   (c) Both X and Y are water soluble
   (d) Gas Q is inflammable

7. Ge(II)compounds are powerful reducing agents whereas Pb(IV)compounds are strong oxidants. It is because
   (a) Pb is more electropositive than Ge
   (b) ionization potential of lead is less than that of Ge
   (c) ionic radii of Pb^{2+} and Pb^{4+} are larger than those of Ge^{2+} and Ge^{4+}
   (d) of more pronounced inert pair effect in lead than in Ge

8. Which of the following conceivable structures for CCl_4 will have a zero dipole moment?
   (a) Square planar
   (b) Square pyramid (carbon at apex)
   (c) Irregular tetrahedron
   (d) None of these

9. Least thermally stable is –

   (a) CCl_4 (b) $SiCl_4$ (c) $GeCl_4$ (d) $GeBr_4$

10. To a piece of charcoal, sulphuric acid is added. Then:
    (a) there is no reaction
    (b) water gas is formed
    (c) SO_2 and CO_2 are evolved
    (d) CO and SO_2 are evolved

11. The correct statement with respect to CO is
    (a) it combines with H_2O to give carbonic acid
    (b) it reacts with haemoglobin in RBC
    (c) it is powerful oxidising agent
    (d) it is used to prepare aerated drinks

12. CO_2 and N_2 are non-supporters of combustion. However for putting out fires CO_2 is preferred over N_2 because CO_2
(a) does not burn
(b) forms non-combustible products with burning substances
(c) is denser than nitrogen
(d) is a more reactive gas

13. Which of the following is not correct?
(a) $Ge(OH)_2$ is amphoteric
(b) $GeCl_2$ is more stable than $GeCl_4$
(c) GeO_2 is weakly acidic
(d) $GeCl_4$ in HCl forms $[GeCl_2]^{2-}$ ion

14. Which of the following is similar to graphite?
(a) B (b) BN (c) B_2H_6 (d) B_4C

15. In the following sets of reactants which two sets best exhibit the amphoteric characters of $Al_2O_3.xH_2O$? Toughnut

Set 1: $Al_2O_3.xH_2O$ (s) and OH^- (aq)
Set 2: $Al_2O_3.xH_2O$ (s) and H_2O (l)
Set 3: $Al_2O_3.xH_2O$ (s) and H^+ (aq)
Set 4: $Al_2O_3.xH_2O$ (s) and NH_3 (aq)
(a) 1 and 2 (b) 1 and 3
(c) 2 and 4 (d) 3 and 4

16. The gas evolved on heating CaF_2 and SiO_2 with concentrated H_2SO_4, on hydrolysis gives a white gelatinous precipitate. The precipitate is:
(a) hydrofluosilicic acid
(b) silica gel
(c) silicic acid
(d) calciumfluorosilicate

17. On adding ammonium hydroxide solution to $Al_2(SO_4)_3$(aq): Tricky
(a) A precipitate is formed which does not dissolve in excess of ammonium hydroxide
(b) A precipitate is formed which dissolves in excess of ammonia solution
(c) No precipitate is formed
(d) None of these

18. Which of the following statements are correct?
(i) Aluminium forms $[AlF_6]^{3-}$ ion while boron forms only $[BF_4]^-$ ion due to presence of *d*-orbitals in aluminium.
(ii) The first member of a group differs from the heavier members in its ability to form pπ-pπ multiple bonds to itself and to other second row elements. While heavier member forms dπ-pπ bonds.
(iii) *d*-orbitals contribute more to the overall stability of molecules than pπ-pπ bonding of second row elements.
(a) (i) (ii) (iii) (b) (i) (iii)
(c) (i) (ii) (d) (ii) (iii)

19. An aqueous solution of $FeSO_4$, $Al_2(SO_4)_3$ and chrome alum is heated with excess of Na_2O_2 and filtered. The materials obtained are :

(a) a colourless filtrate and a green residue
(b) a yellow filtrate and a green residue
(c) a yellow filtrate and a brown residue
(d) a green filtrate and a brown residue

20. Which statement is not true about potash alum?
(a) On heating it melts and loses its water of crystallization.
(b) It's aqueous solution is basic in nature.
(c) It is used in dyeing industries.
(d) It's empirical formula is $KAl(SO_4)_2.12H_2O$.

21. Anhydrous aluminium chloride (Al_2Cl_6) is covalent compound and soluble in water giving:
(a) Al^{3+} and Cl^- ions
(b) $[Al(H_2O)_6]^{3+}$ and Cl^- ions
(c) $[AlCl_2(H_2O)_4]^+$ and $[AlCl_4(H_2O)_2]^-$ ions
(d) none of the above

22. Borax is converted into crystalline boron by the following steps: Critical Thinking

$$\text{Borax} \xrightarrow{X} H_3BO_3 \xrightarrow{\Delta} B_2O_3 \xrightarrow[\Delta]{Y} B$$

X and Y are respectively:
(a) HCl, Mg (b) HCl, C
(c) C, Al (d) HCl, Al

23. On controlled hydrolysis and condensation, R_3SiCl yields
(a) $R_3Si-O-SiR_3$
(b) $\text{(}R_3Si-O-SiR_3\text{)}_n$
(c) R_3SiOH
(d)

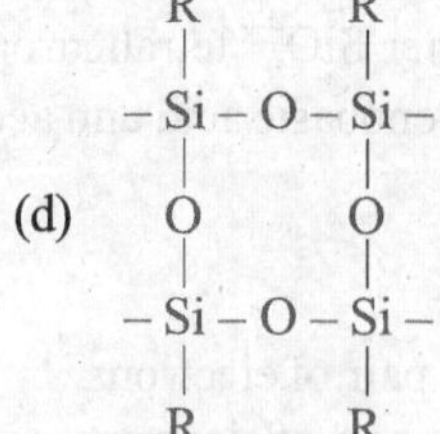

24. $B(OH)_3 + NaOH \rightarrow NaBO_2 + Na[B(OH)_4] + H_2O$
How can this reaction is made to proceed in forward direction?
(a) addition of cis 1, 2-diol
(b) addition of borax
(c) addition of trans 1, 2-diol
(d) addition of Na_2HPO_4

25. Anhydrous $AlCl_3$ cannot be obtained from which of the following reactions? Critical Thinking
(a) Heating $AlCl_3.6H_2O$
(b) By passing dry HCl over hot aluminium powder
(c) By passing dry Cl_2 over hot aluminium powder
(d) By passing dry Cl_2 over a hot mixture of alumina and coke

26. The role of fluorspar (CaF_2) which is added in small quantities in the electrolytic reduction of alumina dissolved in fused cryolite (Na_3AlF_6) is
(a) as a catalyst
(b) to make the fused mixture very conducting
(c) to increase the temperature of the melt.
(d) to decrease the rate of oxidation of carbon at the anode.

27. The correct order of increasing C – O bond length of CO, CO_2 and CO_3^{2-} is:
(a) $CO_3^{2-} < CO_2 < CO$ (b) $CO_2 < CO_3^{2-} < CO$
(c) $CO < CO_3^{2-} < CO_2$ (d) $CO < CO_2 < CO_3^{2-}$

28. In silicon dioxide
(a) there are double bonds between silicon and oxygen atoms
(b) silicon atom is bonded to two oxygen atoms
(c) each silicon atom is surrounded by two oxygen atoms and each oxygen atom is bonded to two silicon atoms
(d) each silicon atom is surrounded by four oxygen atoms and each oxygen atom is bonded to two silicon atoms.

29. Example of a three-dimensional silicate is:
(a) Zeolites (b) Ultramarines
(c) Feldspars (d) Beryls

30. Identify the incorrect statement : Tricky
(a) In $(Si_3O_9)^{6-}$, tetrahedral SiO_4 units share two oxygen atoms.
(b) Trialkylchlorosilane on hydrolysis gives R_3SiOH.
(c) $SiCl_4$ undergoes hydrolysis to give H_4SiO_4.
(d) $(Si_3O_9)^{6-}$ has cyclic structure.

31. Aluminium chloride exists as dimer, Al_2Cl_6 in solid state as well as in solution of non-polar solvents such as benzene. When dissolved in water, it gives
(a) $[Al(OH)_6]^{3-} + 3HCl$
(b) $[Al(H_2O)_6]^{3+} + 3Cl^-$
(c) $Al^{3+} + 3Cl^-$
(d) $Al_2O_3 + 6HCl$

32. When Al is added to NaOH solution
(a) No action takes place
(b) $NaAlO_2$ is formed and H_2 is evolved
(c) $Al(OH)_3$ is formed and H_2 is evolved
(d) Na_2AlO_2 is formed and H_2 is evolved

33. Which of the following statement(s) is / are incorrect for CO_2?
(i) In laboratory CO_2 is prepared by the action of dilute HCl on calcium carbonate
(ii) Carbon dioxide is a poisonous gas
(iii) Increase in carbon dioxide content in atmosphere lead to increase in green house effect.
(iv) CO_2 as dry ice is used as a refrigerant for ice cream and frozen food.
(a) (i) and (ii) (b) Only (ii)
(c) (i), (ii) and (iii) (d) (ii) and (iii)

34. A group 14 element is oxidised to form corresponding oxide which is gaseous in nature, when dissolved in water pH of the water decreases further addition of group 2 hydroxides leads to precipitation. This oxide can be

(a) GeO_2 (b) CO (c) CO_2 (d) SnO_2

35. PbF_4, $PbCl_4$ exist but $PbBr_4$ and PbI_4 do not exist because of
(a) large size of Br^- and I^-
(b) strong oxidising character of Pb^{4+}
(c) strong reducing character of Pb^{4+}
(d) low electronegativity of Br^- and I^-.

36. Which one of the following is a correct set for SiO_2?
(a) Linear acidic
(b) Linear,basic
(c) Tetrahedral, acidic
(d) Angular, basic.

37. Which is not the use of orthoboric acid?
(a) As an antiseptic and eye wash.
(b) In glass industry.
(c) In glazes for pottery.
(d) In borax - bead test.

38. The factor responsible for weak acidic nature of B–F bonds in BF_3 is Tricky
(a) large electronegativity of fluorine
(b) three centred two electron bonds in BF_3
(c) $p\pi$ - $d\pi$ back bonding
(d) $p\pi$ - $p\pi$ back bonding

39. Which of the following has the minimum heat of dissociation:
(a) $(CH_3)_3N: \rightarrow BF_3$
(b) $(CH_3)_3N: \rightarrow B(CH_3)_2F$
(c) $(CH_3)_3N: \rightarrow B(CH_3)_3$
(d) $(CH_3)_3N: \rightarrow B(CH_3)F_2$

40. Lead is not affected by dil. HCl in cold because
(a) Pb is less electronegative than H
(b) PbO film is formed which resists chemical attack by acid
(c) $PbCl_2$ protective coating gets formed on Pb surface
(d) PbO_2 film is always present on Pb surface, which resist chemical attack

41. The catenation tendency of C, Si and Ge is in the order Ge < Si < C. The bond energies (in kJ mol^{-1}) of C – C, Si – Si and Ge–Ge bonds are respectively Critical Thinking
(a) 348, 297, 260 (b) 297, 348, 260
(c) 348, 260, 297 (d) 260, 297, 348

42. The melting pt. of group 13 follows the order
(a) B > Al > Ga > In > Tl
(b) B > Al < Ga > In > Tl
(c) B > Al > Tl > In > Ga
(d) B > Al < Ga < In < Tl

43. AlI_3, when reacts with CCl_4, gives
(a) $AlCl_3$ (b) CI_4
(c) Al_4C_3 (d) both (a) and (b)

44. Which of the following properties of aluminium makes it useful for food packaging ?
(a) Good electrical conductivity
(b) Good thermal conductivity
(c) Low density
(d) Non toxicity

45. A metal, M forms chlorides in its +2 and +4 oxidation states. Which of the following statements about these chlorides is correct?
(a) MCl_2 is more ionic than MCl_4
(b) MCl_2 is more easily hydrolysed than MCl_4
(c) MCl_2 is more volatile than MCl_4
(d) MCl_2 is more soluble in anhydrous ethanol than MCl_4

46. Among the following substituted silanes the one which will give rise to cross linked silicone polymer on hydrolysis is Critical Thinking
(a) R_4Si (b) R_2SiCl_2
(c) $RSiCl_3$ (d) R_3SiCl

47. Which of the following statements is false?
(a) Water gas is a mixture of hydrogen and carbon monoxide
(b) Producer gas is a mixture of CO and nitrogen
(c) Water gas is a mixture of water vapour and hydrogen
(d) Natural gas consists of methane, ethane and gaseous hydrocarbons.

48. Lead pipes are not suitable for drinking water because
(a) lead forms basic lead carbonate
(b) lead reacts with water containing air to form $Pb(OH)_2$
(c) a layer of lead dioxide is deposited over pipes
(d) lead reacts with air to form litharge

49. Choose the correct sequence for the geometry of the given molecules
Borazone, Borazole, $B_3O_6^{3-}$
['P' stands for planar and 'N' stands for non-planar] Toughnut
(a) NP, NP, NP (b) P, P, NP
(c) NP, P, NP (d) NP, P, P

50. Which out of the following compounds does not exist?
(a) BF_3 (b) $TlCl_3$
(c) $TlCl_5$ (d) Both (b) and (c)

51. Anhydrous $AlCl_3$ is prepared from

(a) conc. HCl and Al metal
(b) aluminium and Cl_2
(c) dry HCl gas + heated Al metal
(d) dil. HCl and Al metal

52. Boric acid is polymeric due to
(a) its acidic nature
(b) the presence of hydrogen bonds
(c) its monobasic nature
(d) its geometry

53. Al_2O_3 can be converted to anhydrous $AlCl_3$ by heating

Toughnut

(a) Al_2O_3 with NaCl in solid state
(b) a mixture of Al_2O_3 and carbon in dry Cl_2 gas
(c) Al_2O_3 with Cl_2 gas
(d) Al_2O_3 with HCl gas

54. In which of the following, a salt of the type KMO_2 is obtained?
(a) $B_2H_6 + KOH(aq) \rightarrow$
(b) $Al + KOH(aq) \rightarrow$
(c) Both
(d) None

55. Bauxite ore is generally contaminated with impurity of oxides of two elements X and Y. Which of the following statement is correct?

(a) X is a non-metal and belongs to the third period while Y is a metal and belongs to the fourth period.
(b) One of two oxides has three-dimensional polymeric structure.
(c) Both (a) and (b) are correct.
(d) None of the above.

56. A solid element (symbol Y) conducts electricity and forms two chlorides YCl_n (colourless volatile liquid) and YCl_{n-2} (a colourless solid). To which one of the following groups of the periodic table does Y belong?
(a) 13 (b) 14 (c) 15 (d) 16

57. The structure and hybridization of $Si(CH_3)_4$ is
(a) Bent, *sp* (b) Trigonal, sp^2
(c) Octahedral, d^2sp^3 (d) Tetrahedral, sp^3

58. When steam reacts with red hot coke to form CO_2 and hydrogen :
(a) Water acts as an oxidising agent.
(b) Water acts as a reducing agent.
(c) Carbon acts as an oxidising agent.
(d) There is no oxidation or reduction.

59. Which of the following is/are not correctly matched?
(i) GeO_2 – Acidic
(ii) PbO_2– Amphoteric
(iii) CO – Neutral
(iv) SiO_2 – Amphoteric
(a) (i) and (iv) (b) (iv) only
(c) (ii) only (d) (iii) only

60. The shape of gaseous $SnCl_2$ is
(a) tetrahedral (b) linear
(c) angular (d) T-shaped

Answer KEY

1	(d)	7	(d)	13	(b)	19	(c)	25	(a)	31	(b)	37	(d)	43	(d)	49	(d)	55	(c)
2	(b)	8	(d)	14	(b)	20	(b)	26	(b)	32	(b)	38	(d)	44	(c)	50	(c)	56	(b)
3	(b)	9	(d)	15	(b)	21	(c)	27	(d)	33	(b)	39	(c)	45	(a)	51	(c)	57	(d)
4	(d)	10	(c)	16	(d)	22	(d)	28	(d)	34	(c)	40	(c)	46	(c)	52	(b)	58	(a)
5	(b)	11	(b)	17	(a)	23	(a)	29	(c)	35	(b)	41	(a)	47	(a)	53	(b)	59	(b)
6	(b)	12	(c)	18	(c)	24	(a)	30	(b)	36	(c)	42	(c)	48	(b)	54	(c)	60	(c)

Hints & Solutions

1. (d) $B(OH)_3$ is acid because it can take OH^- ions.
H_3BO_3 or $B(OH)_3 + OH^- \rightarrow B(OH)_4^-$
2. (b) The protective oxide layer prevents Al from further corrosion.
3. (b) $SiO_2 + 2NaOH \rightarrow \underset{\text{Sodium silicate}}{Na_2SiO_3} + H_2O$
4. (d)

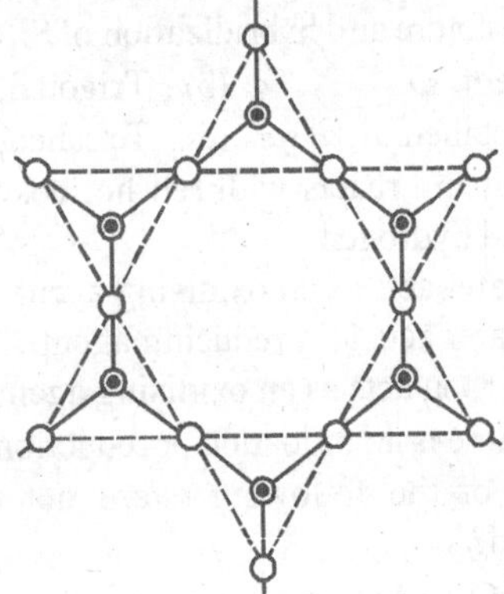

Total No. oxygen atoms per silicon atom
$= \frac{1}{2} + \frac{1}{2} + \frac{1}{2} + 1 = 2.5$
$\therefore$ Formula $Si_2O_5^{2-}$.
5. (b) $B(OH)_3 + 2H_2O \rightleftharpoons [B(OH)_4]^- + H_3O^+$
6. (b)

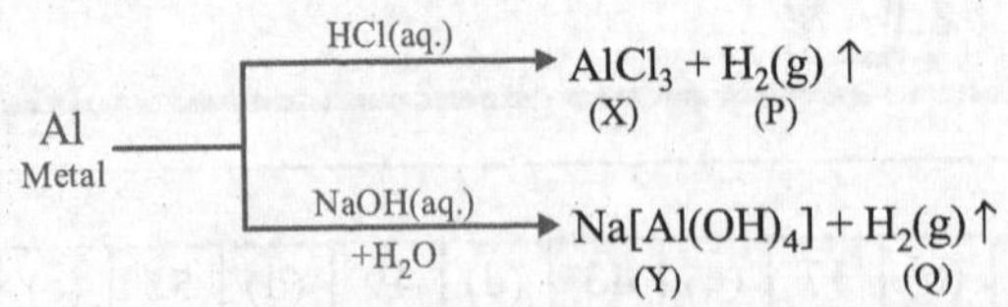

7. (d) Ge(II) tends to acquire Ge (IV) state by loss of electrons. Hence it is reducing in nature. Pb (IV) tends to acquire Pb (II) O.S. by gain of electrons. Hence it is oxidising in nature. This is due to inert pair effect.
8. (d) CCl_4 is tetrahedral in nature.
9. (d) The thermal stability of tetrahalides decreases in order
$CX_4 > SiX_4 > GeX_4 > SnX_4$ and in terms of same metal with different halides is in order of $MF_4 > MCl_4 > MBr_4 > MI_4$.
10. (c) Charcoal is a pure form of carbon, its reaction with hot conc. H_2SO_4 is as follows:
$C + 2H_2SO_4 \longrightarrow 2H_2O + 2SO_2 + CO_2$
11. (b) CO react with haemoglobin, forms carboxy haemoglobin and stopes the supply of O_2
12. (c) CO_2 being more dense covers the igniting material more effectively than N_2.
13. (b) Ge^{4+} is more stable than Ge^{2+}. Hence $GeCl_4$ is more stable than $GeCl_2$
14. (b) Graphite and boron nitride have similar structure.
15. (b) Aluminium oxide is amphoteric oxide because it shows the properties of the both acidic and basic oxides. It reacts with both acids and bases to form salt and water.
$Al_2O_3.xH_2O + 2NaOH \longrightarrow NaAlO_2 + H_2O$
$Al_2O_3.xH_2O + HCl \longrightarrow AlCl_3 + H_2O$
16. (d) $2CaF_2 + SiO_2 + H_2SO_4 \longrightarrow SiF_4 + H_2O + CaSO_4 \xrightarrow{\text{hydrolysis}} CaSiF_4$
17. (a) $Al_2(SO_4)_3 + 6NH_4OH \longrightarrow 2Al(OH)_3 + 3(NH_4)_2SO_4$
$Al(OH)_3 + NaOH \longrightarrow \underset{\text{Soluble complex}}{Na^+[Al(OH)_4]^-}$
It is insoluble in NH_4OH.
18. (c) *d*-orbitals are of higher energy than the *p*-orbitals, they contribute less to the overall stability of molecules than $p\pi$-$p\pi$ bonding of the second row elements.
19. (c) The filtrate is yellow due to CrO_4^{2-} ion and residue is brown due to $Fe(OH)_3$.

20. (b) Alum form acidic solution due to hydrolysis of Al^{3+}.

21. (c) $AlCl_3.6H_2O \xrightarrow{\text{dissociation}} [AlCl_2(H_2O)_4]^+ + [AlCl_4(H_2O)_2]^-$

22. (d) $\text{Borax} \xrightarrow{HCl} H_3BO_3$

$$B_2O_3 \xrightarrow[\Delta]{Al} B(\text{crystalline})$$

$$B_2O_3 \xrightarrow[\Delta]{Mg} B(\text{Amorphous})$$

23. (a) R_3SiCl on hydrolysis forms only a dimer. $R_3SiOH + HOSiR_3 \rightarrow R_3Si-O-SiR_3$.

24. (a) H_3BO_3 acts as weak monobasic Lewis acid. $B(OH)_3 + NaOH \rightarrow Na[B(OH)_4]$
On addition of cis-1, 2-diol in H_3BO_3 solution, acidic strength of H_3BO_3 increases due to chelation effect.

25. (a) $2Al + 6HCl \xrightarrow{\Delta, \text{air}} 2AlCl_3 + 3H_2$

$$2Al + 3Cl_2 \longrightarrow 2AlCl_3$$

$$Al_2O_3 + 3C + 3Cl_2 \xrightarrow{1000\,°C} \underset{\text{vapours}}{2AlCl_3} + 3CO$$

$\downarrow$ Cooling

Anhydrous $AlCl_3$

$$AlCl_3.6H_2O \xrightarrow{\Delta} Al(OH)_3 + 3HCl + 3H_2O$$

Thus $AlCl_3$ cannot be obtained by this method.

26. (b) CaF_2 when added to fused cryolite, lowers the m.p. and increases the conductivity.

27. (d) Structures of CO, CO_2 and CO_3^{2-} are :

$C \equiv \ddot{O}$ $\quad O=C=O$

[Resonance structures of CO_3^{2-}: three structures, each with C bonded to one O by a double bond and to two O^-]

Compound	CO	CO_2	CO_3
Bond order	3	2	1.33

$$\text{Bond order} \propto \frac{1}{\text{Bond length}}$$

Hence, the decreasing (C—O) bond length is :

$CO < CO_2 < CO_3^{2-}$

28. (d) In SiO_2 (quartz), each of O-atom is shared between two SiO_4^{4-} tetrahedra.

[Structure: two SiO_4 tetrahedra sharing an O atom]

29. (c) The feldspars are most abundant aluminosilicate minerals in the Earth surface. The silicon atoms and aluminium atoms occupy the centres of interlinked tetrahedra of SiO_4^{4-} and AlO_4^{5-}. These tetrahedra connect at each corner to other tetrahedra forming an intricate, three dimesional, negatively charged framework. The sodium cations sit within the voids in this structure.

30. (b) The hydrolysis of Trialkylchlorosilane R_3SiCl yields dimer :

$$R_3Si-O-SiR_3$$

31. (b) $Al_2Cl_6 + 12H_2O \rightleftharpoons 2[Al(H_2O)_6]^{3+} + 6Cl^-$

32. (b) $Al + NaOH + H_2O \rightarrow NaAlO_2 + H_2 \uparrow$.

33. (b) Carbon dioxide is not a poisonous gas.

34. (c) CO_2 forms carbonic acid H_2CO_3, when dissolved in water, CO is neutral, whereas other two GeO_2 and SnO_2 are solids.

35. (b) F and Cl are more oxidising in nature and can achieve Pb in (IV) O.S. but Br_2 and I_2 can not achieve Pb in (IV) O.S. secondly Pb^{4+} is strong in oxidising nature and in its presence, Br^- and I^- can not exist.

36. (c) SiO_2 being oxide of non metal is acidic in nature and silicon is bonded to O atoms tetrahedrally. It has giant structure.

37. (d) Borax on heating gives B_2O_3 and $NaBO_2$ which is glassy mass and used for borax-bead test.

38. (d) It is $p\pi - p\pi$ back bonding involving B and F. The smaller atoms show more back bonding.

39. (c) Due to + I effect of methyl groups the Lewis character of $B(CH_3)_3$ decreases and coordination becomes weaker.

40. (c) Pb with dil. HCl forms protective coating of $PbCl_2$.

41. (a) The linking of identical atoms with each other to form long chains is called catenation. However, this property decreases from carbon to lead. Decrease of this property is associated with M-M bond energy which decreases from carbon to lead.

42. (c) Due to structural changes, melting point, increases from Ga to Tl and Ga has the lowest melting point.

43. (d) AlI_3, on reaction with CCl_4, gives the $AlCl_3$

$$4AlI_3 + 3CCl_4 \longrightarrow 4AlCl_3 + 3CI_4$$

44. (c) Due to the low density of aluminium, it is useful for food packaging.

45. (a) Metal atom in the lower oxidation state forms the ionic bond and in the higher oxidation state the covalent bond because higher oxidation state means small size and great polarizing power and hence greater the covalent character. Hence MCl_2 is more ionic than MCl_4.

46. (c) The cross linked polymers will be formed by $RSiCl_3$

$$nRSiCl_3 \xrightarrow[-3nHCl]{3nH_2O} nR-\underset{\underset{OH}{|}}{\overset{\overset{OH}{|}}{Si}}-OH \longrightarrow$$

```
   |    |
   O    O
   |    |
R-Si-O-Si
   |    |
   O    O
   |    |
R-Si-O-Si-R
   |    |
   O    O
   |    |
```

(Cross linked polymer)

47. (a) Water gas is $CO + H_2$

48. (b) $2Pb + 2H_2O + O_2 \rightarrow 2Pb(OH)_2$

49. (d) Borazone : A crystalline form of boron nitride which has diamond like structure. (Non-planar)
Borazole : Inorganic benzene : $B_3N_3H_6$ (planar)

$B_3O_6^{3-}$:

```
        O⁻
        |
        B
      /   \
     O     O
     |     |
     B     B
   /   \ /   \
 ⁻O     O     O⁻
```

planar

50. (c) Because Tl^{+5} does not exist

51. (c) $2Al + 3Cl_2 \xrightarrow{\Delta} 2AlCl_3$ (anhydrous)

52. (b) In Boric acid each B atom is sp^2 hybridized and contains BO_3^{3-} units which are held together by hydrogen bonds.

53. (b) Al_2O_3 can be converted to anhydrous $AlCl_3$ by heating a mixture of Al_2O_3 and carbon in dry Cl_2

$$Al_2O_3 + 3C + 3Cl_2 \xrightarrow{1000°C} \underset{\text{vapours}}{2AlCl_3} + 3CO$$

vapours $\xrightarrow{\text{cooled}}$ Solid anhydrous aluminium chloride

54. (c) $B_2H_6 + 2KOH + 2H_2O \rightarrow \underset{\text{Potassium metaborate}}{2KBO_2} + 6H_2$

$$2Al + 2KOH + 2H_2O \rightarrow \underset{\text{Potassium meta aluminate}}{2KAlO_2} + 3H_2$$

55. (c) Two oxides present in bauxite as an impurity are SiO_2 and Fe_2O_3. Si belongs to the third period and Fe to the fourth period. SiO_2 has a three dimensional structure.

56. (b) $SnCl_4$ is colourless volatile liquid and $SnCl_2$ is colourless solid Sn conducts electricity and it belongs to 14 group.

57. (d) Hybridisation $= \frac{1}{2}(4+4+0-0) = 4$, sp^3 tetrahedral.

58. (a) $\underset{\text{Steam}}{H_2O} + C \longrightarrow CO_2 + H_2$

H_2O oxidises $C \longrightarrow CO_2$, hence H_2O acts as oxidising agent.

59. (b) SiO_2 is acidic oxide.

60. (c) Shape of $SnCl_2$ is angular due to sp^2 hybridisation and having the following structure

```
   Cl
  /
Sn   ,
  \
   Cl
```

12 Organic Chemistry – Some Basic Principles & Techniques

1. The correct statement about the compounds A, B, and C

$$\text{(A)}: \begin{array}{c} COOCH_3 \\ H-\!\!\!+\!\!\!-OH \\ H-\!\!\!+\!\!\!-OH \\ COOH \end{array} \quad \text{(B)}: \begin{array}{c} COOH \\ H-\!\!\!+\!\!\!-OH \\ H-\!\!\!+\!\!\!-OH \\ COOCH_3 \end{array} \quad \text{(C)}: \begin{array}{c} COOH \\ H-\!\!\!+\!\!\!-OH \\ HO-\!\!\!+\!\!\!-H \\ COOCH_3 \end{array}$$

is

Tricky

(a) A and B are identical
(b) A and B are diastereomers
(c) A and C are enantiomers
(d) A and B are enantiomers

2. $C_6H_5CHO + HCN \longrightarrow C_6H_5-\underset{OH}{\overset{H}{C}}-CN$

The product would be
(a) a racemate
(b) optically active
(c) a meso compound
(d) a mixture of diastereomers

3. Arrange the following (w, x, y, z) in decreasing order of their boiling points:

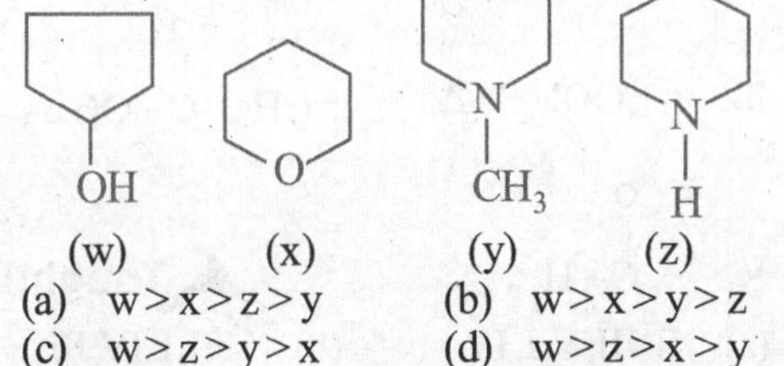

(w) (x) (y) (z)

(a) w > x > z > y (b) w > x > y > z
(c) w > z > y > x (d) w > z > x > y

4. How many degrees of unsaturation are there in the following compound?

Critical Thinking

(a) 6 (b) 7 (c) 10 (d) 11

5. Arrange in the order of increasing acidity.

(I) cyclohexyl–OH (II) cyclohexyl–NH_2 (III) phenyl–OH

(a) III < I < II (b) I < III < II
(c) III < II < I (d) II < I < III

6. Which of the following is optically active ?

(a) $CH_3-\underset{OH}{\overset{H}{C}}-\underset{OH}{\overset{H}{C}}-C_2H_5$

(b) $CH_3-CH(OH)-CH_3$

(c) $CH_3-\underset{O}{\overset{\|}{C}}-\underset{Br}{\overset{Br}{C}}-COOH$

(d) $CH_3-CH_2-\overset{O}{\overset{\|}{C}}-OH$

7. Keto-enol tautomerism is observed in :

(a) $H_5C_6-\overset{O}{\overset{\|}{C}}-CH_2-\overset{O}{\overset{\|}{C}}-C_6H_5$

(b) $H_5C_6-\overset{O}{\overset{\|}{C}}-CH_3$

(c) $H_5C_6-\overset{O}{\overset{\|}{C}}-H$

(d) Both (a) and (b)

8. The change in the state of hybridization of the asterisked carbon in the following reaction.

$CH_3\overset{*}{C}N \xrightarrow{H_2O} CH_3\overset{*}{C}ONH_2$ is

(a) sp^3 to sp^2 (b) sp^3 to sp
(c) sp to sp^2 (d) sp^2 to sp^3

9. In the reaction shown below the six membered ring is generated by shifting which bond

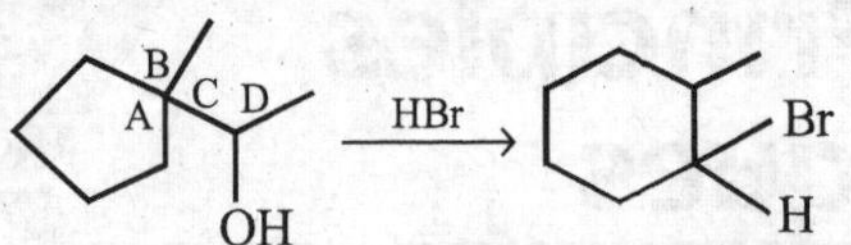

(a) A (b) B (c) C (d) D

10. Which among the following is substitution reaction ?

(a) $>C=O + RMgX \longrightarrow >C<^{OMgX}_{R}$

(b) Diene + maleic anhydride $\longrightarrow$ product

(c) $R_3COH \xrightarrow[\text{Conc.HCl}]{\text{Anhy.ZnCl}_2}$ Turbidity

(d) All of these

11. In allene (C_3H_4), the type(s) of hybridization of the carbon atoms is (are):

(a) sp and sp^3 (b) sp^2 and sp

(c) only sp^2 (d) sp^2 and sp^3

12. The correct IUPAC name of the following compound

is:

(a) 4 - methyl - 3 - ethylhexane

(b) 3 - ethyl - 4 - methylhexane

(c) 3, 4 - ethylmethylhexane

(d) 4 - ethyl - 3 - methylhexane

13. Which molecule will be most reactive for S_N1 reaction?

(a) Cl, CH_3, CH_3 (b) CH_3, Br

(c) CH_3, Cl, CH_3, CH_3 (d) CH_3, I

14. Arrange the following in decreasing order of solubility in water

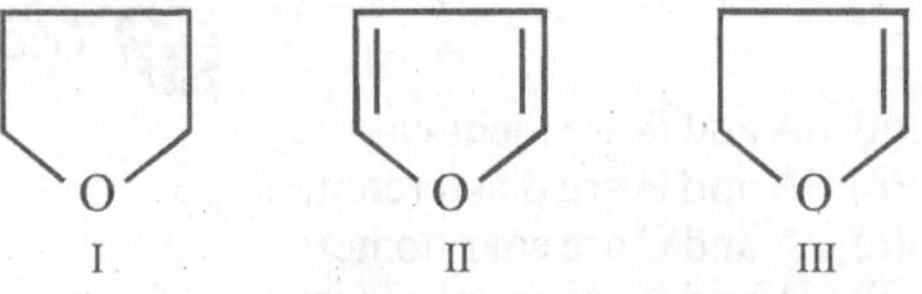

(a) I > III > II

(b) III > II > I

(c) II > III > I

(d) All are equally soluble

15. The correct priorities for the substituents shown below, according to the E-Z sequence rule is

I. $-CN$ II. $-CBr(CH_3)_2$

III. $-COOH$ IV. $-CH_2-\overset{O}{\overset{||}{C}}-OCH_3$

V. 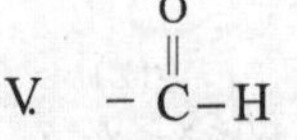

Toughnut

(a) II, III, V, I, IV (b) V, II, I, IV, III

(c) III, IV, I, II, V (d) II, V, I, IV, III

16. The alcohol product(s) of the reduction of 2-methyl-3-pentanone with $LiAlH_4$ is (are)

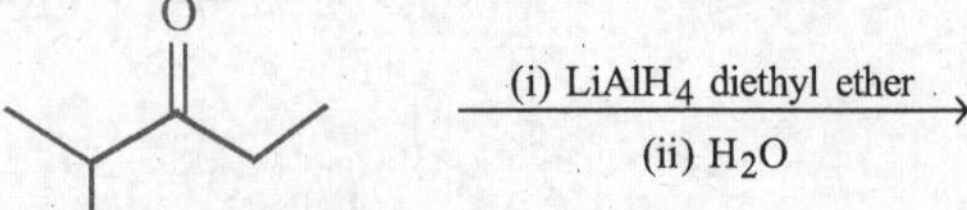

(a) a single enantiomer

(b) racemic mixture

(c) two diastereoisomers

(d) two structural isomers

17. Identify which of the strucutre below are meso structures?

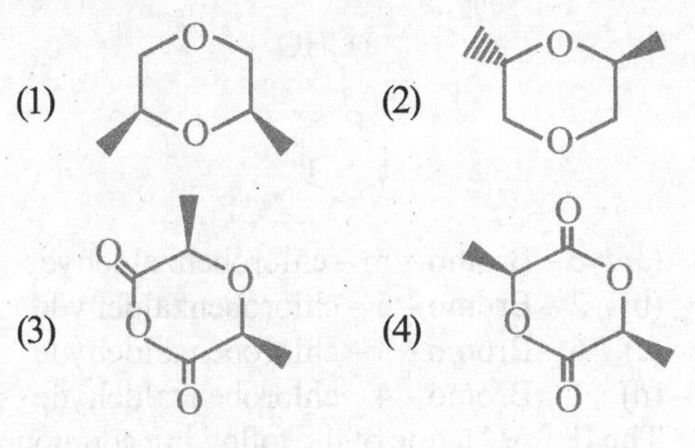

(5)

(a) 1 and 3 (b) 1, 3 and 5
(c) 1, 3 and 4 (d) 2 and 5

18. Select the most stable carbocation amongst the following

(a)

(b)

(c)

(d)

19. Which of the following reactions is elimination reaction?

(a) cyclohexanone$=O + Ph_3P = CH_2 \longrightarrow$ methylenecyclohexane$= CH_2 + Ph_3PO$

(b) $C_2H_5Cl + KOH \rightarrow C_2H_5OH + KCl$

(c) $C_2H_5-\overset{H}{\underset{\overset{+}{N}(CH_3)_3}{C}}-CH_3 \xrightarrow{\text{Alc KOH}}$

$CH_3CH=CH-CH_3 + \ddot{N}(CH_3)_3$

(d) None of them

20. In the following groups Tricky

–OAc (I) –OMe (II)

$-OSO_2Me$ (III) $-OSO_2CF_3$ (IV)

the order of leaving group ability is

(a) I > II > III > IV (b) IV > III > I > II
(c) III > II > I > IV (d) II > III > IV > I

21. A mixture containing the following four compounds is extracted with 1M HCl. The compound that goes to aqueous layer is :

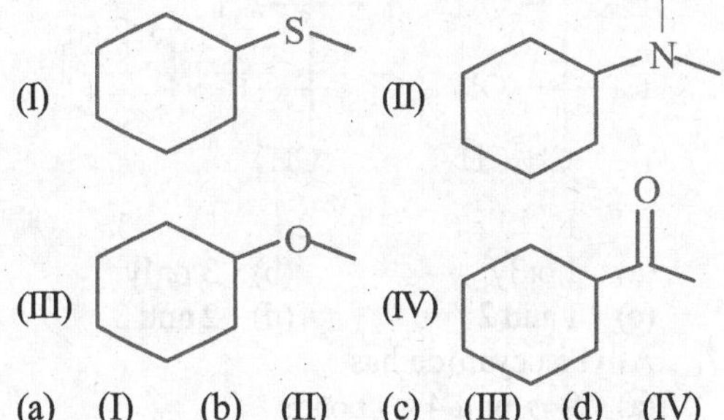

(a) (I) (b) (II) (c) (III) (d) (IV)

22. Most stable carbocation among the following is:

(a) CH_3

(b)

(c) $\overset{+}{C}H_2$

(d) $\overset{+}{C}H_3$

23. Which of the following cannot be represented by resonance structures?

(a) Dimethyl ether (b) Nitrate anion
(c) Carboxylate anion (d) Toluene

24. The IUPAC name of the compound

$H_2C-CH-CH_3$ (with O bridging) is

(a) 1, 2-Propoxide (b) Propylene oxide
(c) 1, 2-Oxo propane (d) 1, 2-Epoxy propane

25. $CH_3-\overset{\overset{O}{||}}{C}\underset{a}{-}O\underset{b}{-}CH_3$;

The correct relation between the bond lengths *a* and *b* is:

(a) $a = b$
(b) $b > a$
(c) $b < a$
(d) Impossible to predict

26. Which of the following compounds are meso forms?

1: CH_3; H–C–OH; H–C–OH; CH_2CH_3

2: CH_3; H–C–Cl; Cl–C–H; CH_3

3: CH_3, CH_3 (cyclohexane)

(a) 1 only (b) 3 only
(c) 1 and 2 (d) 2 and 3

27. Allyl isocyanide has

(a) 9 σ and 4 π - bonds
(b) 8 σ and 5 π - bonds
(c) 9 σ , 3 π and 2 non- bonded electrons
(d) 8 σ, 3 π and 4 non - bonded electrons

28. Which one of the following is a free-radical substitution reaction? **Tricky**

(a) $CH_3CHO + HCN \longrightarrow CH_3CH(OH)CN$

(b) $C_6H_5CH_3 + Cl_2 \xrightarrow{\text{Boiling}} C_6H_5CH_2Cl$

(c) $C_6H_6 + CH_3Cl \xrightarrow{\text{Anhy. AlCl}_3} C_6H_5CH_3$

(d) $C_6H_5CH_2Cl + AgNO_2 \longrightarrow C_6H_5CH_2NO_2$

29. (I) $\longleftrightarrow$ (II) $\longleftrightarrow$ (III)

The most stable canonical structure among the given structure is :

(a) I (b) II
(c) III (d) all are equally stable

30. Which of the following is the IUPAC name of the compound

CHO, Br, Cl

(a) o – Bromo – m – chlorobenzaldehyde
(b) 2 – Bromo – 5 – chlorobenzaldehyde
(c) 6 – Bromo – 3 – chlorobenzaldehyde
(d) 1 – Bromo – 4 – chlorobenzaldehyde

31. The IUPAC name of the following compound is

$\overset{CH_3}{\underset{H}{}}\!\!>C=C<\!\!\overset{H}{\underset{C\equiv C-CH_2CH_3}{}}$

(a) (E)-2-hepten-4-yne
(b) (Z)-5-hepten-3-yne
(c) (E)-5-hepten-3-yne
(d) (Z)-2-hepten-4-yne

32. Dipole moment is shown by

(a) 1, 2-dichlorobenzene
(b) *trans* 2, 3-dichloro-2-butene
(c) 1, 4-chlorobenzene
(d) *trans*-1, 2-dinitroethene

33. The optically inactive compound from the following is :

(a) 2 - chloropropanal
(b) 2 - chlorobutane
(c) 2 - chloropentane
(d) 2 - chloro - 2- methylbutane

34. Tautomerism is exhibited by–

(1) $C_6H_5-CH=CH-OH$

(2) O= =O

(3) =O, O

(4) 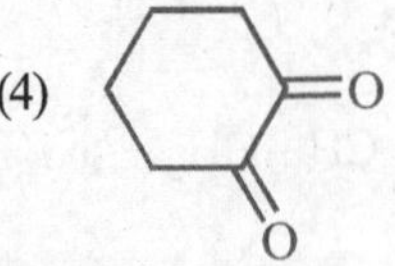

(a) (1), (3) and (4) (b) (2), (3), and (4)
(c) (1), (2) and (4) (d) None of these

35. Which of the following represents the correct order of stability of the given carbocations ?

(a) $F_3\overset{+}{C} > F_3C-\overset{|}{\underset{|}{\overset{+}{C}}} > \overset{+}{C}H_3$

(b) $H_3\overset{+}{C} > F_3C-\overset{|}{\underset{|}{\overset{+}{C}}} > F_3\overset{+}{C}$

(c) $F_3C-\overset{|}{\underset{|}{\overset{+}{C}}} > F_3\overset{+}{C} > H_3\overset{+}{C}$

(d) $F_3C-\overset{|}{\underset{|}{\overset{+}{C}}} > H_3\overset{+}{C} > F_3\overset{+}{C}$

36. Which of the following behaves both as a nucleophile and as an electrophile?

(a) $CH_3C\equiv N$ (b) CH_3OH
(c) $CH_2=CHCH_3$ (d) CH_3NH_2

37. What is the decreasing order of strength of the bases OH^-, NH_2^-, $HC\equiv C^-$ and $CH_3CH_2^-$?

(a) $CH_3CH_2^- > NH_2^- > HC\equiv C^- > OH^-$
(b) $HC\equiv C^- > CH_3CH_2^- > NH_2^- > OH^-$
(c) $OH^- > NH_2^- > HC\equiv C^- > CH_3CH_2^-$
(d) $NH_2^- > HC\equiv C^- > OH^- > CH_3CH_2^-$

38. Pick up the correct statement regarding the following resonating structures of the anilinium ion

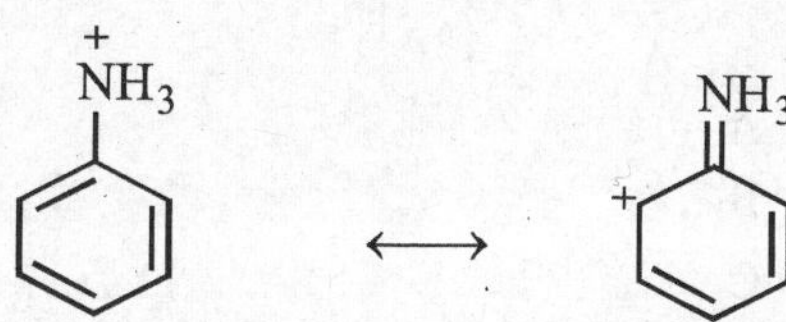

(a) Structure II is not acceptable because carbonium ions are less stable than ammonium ions
(b) II is not acceptable because it is non-aromatic
(c) II is not acceptable because here nitrogen has 10 valence electrons
(d) II is an acceptable canonical structure.

39. Which of the following numberings is correct ?

Toughnut

A (F at 1, Br at 3; ring numbered 1, 2, 3, 4, 5)
B (Br at 1, F at 3; ring numbered 1, 2, 3, 4, 5)
C (Br at 3, F at 5; ring numbered 1, 2, 3, 4, 5)
D (Br at 2, F at 5; ring numbered 1, 2, 3, 4, 5)

(a) A (b) B (c) C (d) D

40. The two compounds given below are

HO–⟨cyclohexane⟩–COOH and HO–⟨cyclohexane⟩–COOH

(a) identical (b) enantiomers
(c) diastereo isomers (d) regiomers

41. The number of structural isomers for C_6H_{14} is:

Toughnut

(a) 4 (b) 3 (c) 6 (d) 5

42. Which of the following pairs of compounds are positional isomers?

(a) $CH_3-CH_2-CH_2-\underset{\underset{O}{\|}}{C}-CH_3$ and $CH_3-CH_2-\underset{\underset{O}{\|}}{C}-CH_2-CH_3$

(b) $CH_3-CH_2-CH_2-CH_2-CHO$ and $CH_3-CH_2-CH_2-\overset{\overset{O}{\|}}{C}-CH_3$

(c) $CH_3-CH_2-CH_2-\underset{\underset{O}{\|}}{C}-CH_3$ and $CH_3-\underset{\underset{CH_3}{|}}{CH}-CH_2-CHO$

(d) $CH_3-CH_2-\underset{\underset{O}{\|}}{C}-CH_2-CH_3$ and $\begin{matrix}CH_3\\CH_3\end{matrix}\!\!>CH-CH_2-CHO$

43. The correct nucleophilicity order is

(a) $CH_3^- < NH_2^- < HO^- < F^-$

(b) $CH_3^- \simeq NH_2^- > HO^- \simeq F^-$

(c) $CH_3^- > NH_2^- > HO^- > F^-$

(d) $NH_2^- > F^- > HO^- > CH_3^-$

44.

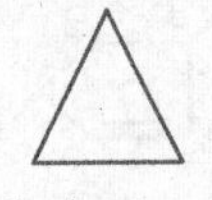
Cyclopropane (I)

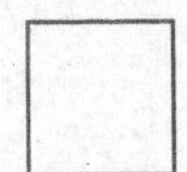
Cyclobutane (II)

Cyclopentane (III)

The correct order of heats of combustion of above compounds is:

(a) I > II > III (b) II > I > III

(c) III > II > I (d) III > I > II

45. Which of the following optically active compounds racemizes in dil. KOH/ CH_3OH solution? **Critical Thinking**

(a)

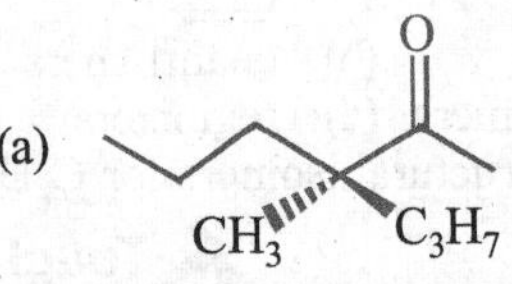

(b)

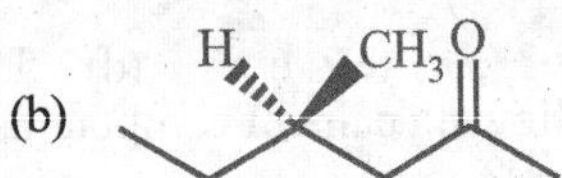

(c)

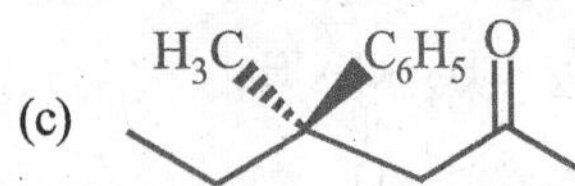

(d)

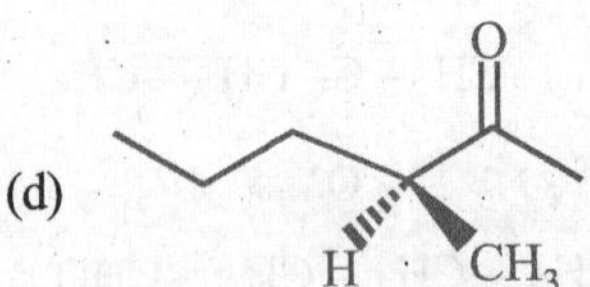

46. The correct number of chiral centres in

C_2H_5 C_2H_5 (decalin ring with two ethyl substituents)

(a) 1 (b) 3 (c) 4 (d) 2

47. An optically active compound (A) is treated with NaI/acetone (B) is obtained

$$\begin{array}{c} CH_3 \\ H - \!\!\!\!+\!\!\!\!- I \\ C_2H_5 \\ (A) \end{array} \xrightarrow[\text{acetone}]{\text{NaI}} \begin{array}{c} CH_3 \\ I - \!\!\!\!+\!\!\!\!- H \\ C_2H_5 \\ (B) \end{array}$$

What is true about (B)? **Tricky**

(a) optically active

(b) optically inactive and racemic form

(c) optically inactive

(d) None of these

48. Which of the following is correct set of physical properties of the geometrical isomers ? **Toughnut**

I: $CH_3(H)C=C(H)Cl$ — CH_3 and H on left carbon; H (up) and Cl (down) on right carbon

II: $CH_3(H)C=C(Cl)H$ — CH_3 and H on left carbon; Cl (up) and H (down) on right carbon

	Dipole moment	B.P.	M.P.	Stability
(a)	I > II	I > II	II > I	I > II
(b)	II > I	II > I	II > I	II > I
(c)	I > II	I > II	I > II	I > II
(d)	II > I	II > I	I > II	I > II

49. Which of the following resonance structure is lowest in energy?

(A) $H_3C-C(=\ddot{O}:)-\ddot{O}-H$ (H–C(H)(H)–C with double-bonded :O: and single-bonded $\ddot{O}$–H)

(B) H–C(H)(H)–C(–$:\ddot{O}:^-$)$=\overset{+}{\ddot{O}}$–H

(C) H–C(H)(H)–$\overset{+}{C}$(–$:\ddot{O}:^-$)–$\ddot{O}$–H

(a) A

(b) B

(c) C

(d) All have same energy

50. A solution of (+) - 2 - chloro - 2 - phenylethane in toluene racemises slowly in the presence of small amounts of $SbCl_5$ due to the formation of

(a) Carbanion (b) Carbene

(c) Free radical (d) Carbocation

51. In the following compounds

(I) 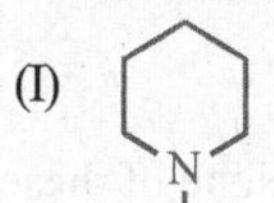 (II)

(III) 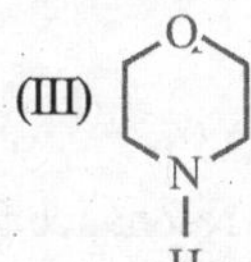 (IV)

the order of basicity is :

(a) IV > I > III > II (b) III > I > IV > II
(c) II > I > III > IV (d) I > III > II > IV

52. Which one of the following compounds has non zero dipole moment?

(a) 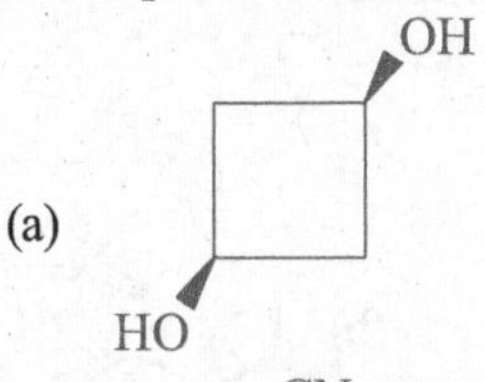

(b) 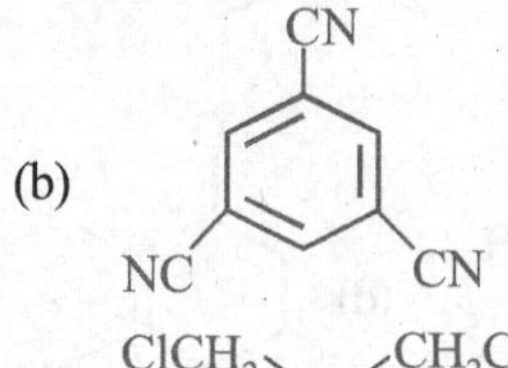

(c) $ClCH_2$ CH_2Cl / $ClCH_2$ CH_2Cl

(d) HN NH

53. Dipole moment of which ketone is maximum? Tricky

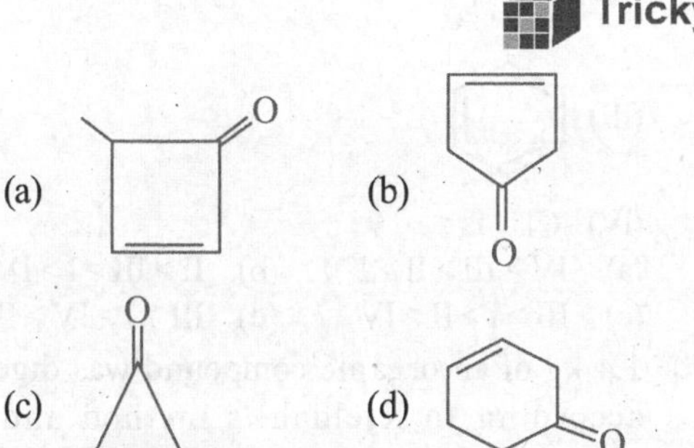

54. Which of the following orders is correct for heat of hydrogenation of these compounds?

(I) CH_3, CH_3 (II) CH_2, CH_3 (III) CH_3, CH_3

(a) I > II > III (b) III > II > I
(c) II > III > I (d) III > I > II

55. Which one of the following acids does not exhibit optical isomerism?

(a) Lactic acid (b) Tartaric acid
(c) Maleic acid (d) α-amino acids

56. How many cyclic structures are possible for C_4H_6? Critical Thinking

(a) 3 (b) 5 (c) 6 (d) 4

57. Maleic acid and fumaric acids are

(a) Chain isomers
(b) Functional isomers
(c) Tautomers
(d) Geometrical isomers

58. One among the following compounds will not give effervescence with sodium carbonate:

(a) $C_6H_5CO_2H$

(b) SO_3H (benzenesulphonic acid)

(c) C_6H_5OH

(d) OH, NO_2, NO_2, NO_2 (picric acid)

59. Which of the following alkenes is the most stable? Critical Thinking

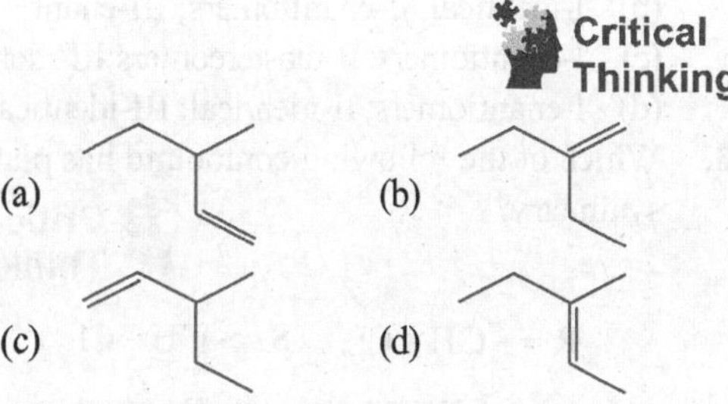

60. Which of the following statements is not correct?

(a) Carbocation posses sextet of electrons.
(b) The order of carbocation stability is :
$\overset{+}{C}H_3 > (CH_3)_2\overset{+}{C}H > (CH_3)_3\overset{+}{C}$
(c) Carbocations have trigonal planar shape
(d) Carbocations are formed by heterolytic cleavage

61. The enol form of acetone after treatment with D_2O gives:

(a) $CH_3-\overset{OD}{\overset{|}{C}}=CH_2$ (b) $CH_3-\overset{O}{\overset{||}{C}}-CD_3$

(c) $CH_2=\overset{OH}{\overset{|}{C}}-CH_2D$ (d) $CD_2=\overset{OD}{\overset{|}{C}}-CD_3$

62. Indicate whether each of the following pairs are identical or not? **Toughnut**

(I) [Newman projection: F (top); Br, H; H, CH_3; H (bottom)] and [wedge-dash structure: H, $C-CH_2Br$, H_3C, F]

(II) [wedge-dash structure: H, Cl, CH_3, CH_2CH_3, H, Cl] and [wedge-dash structure: Cl, H, H, Cl, CH_3, CH_2CH_3]

(III) [Fischer projection: CH_3; H—Br; CH_3—H; Br] and [Fischer projection: CH_3; Br—H; H—CH_3; Br]

(a) I-enantiomers; II-diastereomers; III-enantiomers
(b) I-identical; II-enantiomers; III-enantiomers
(c) I-enantiomers; II-diastereomers; III-identical
(d) I-enantiomers; II-identical; III-identical

63. Which of the following compound has plane of symmetry? **Critical Thinking**

$R=-\underset{CH_3}{\underset{|}{CH}}-Cl$, $S=-\underset{Br}{\underset{|}{CH}}-Cl$

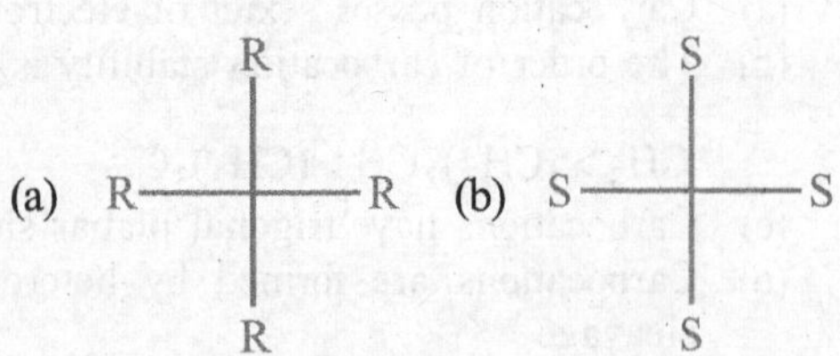

(c) [Fischer projection: R (top); S—S; R (bottom)] (d) None of these

64. Which of the following is not a resonance form of the enolate ion formed in the following acid base equilibrium

[2-methylcyclohexane-1,3-dione] + NaOH ⇌

(a) (b)

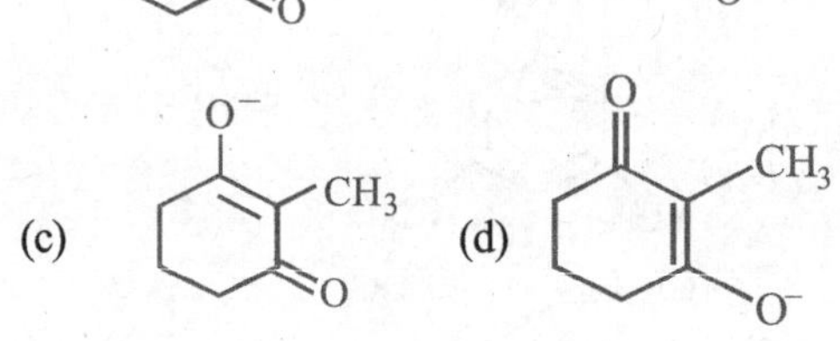

65. The order of stability of the following carbocations :

(I) $CH_2=CH-\overset{+}{C}H_2$

(II) $CH_3-CH_2-\overset{+}{C}H_2$

(III) $C_6H_5\overset{+}{C}H_2$

(IV) $\overset{+}{C}H_3$ is :

(a) IV > III > II > I (b) II > III > I > IV
(c) III > I > II > IV (d) III > I > IV > II

66. 1.4 kg of an organic compound was digested according to Kjeldahl's method and the ammonia evolved was absorbed in 60 mL of M/10 H_2SO_4 solution. The excess sulphuric acid required 20 mL of M/10 NaOH solution for neutralization. The percentage of nitrogen in the compound is: **Toughnut**

(a) 10 (b) 3 (c) 24 (d) 5

67. Which of the following would react most readily with nucleophiles?

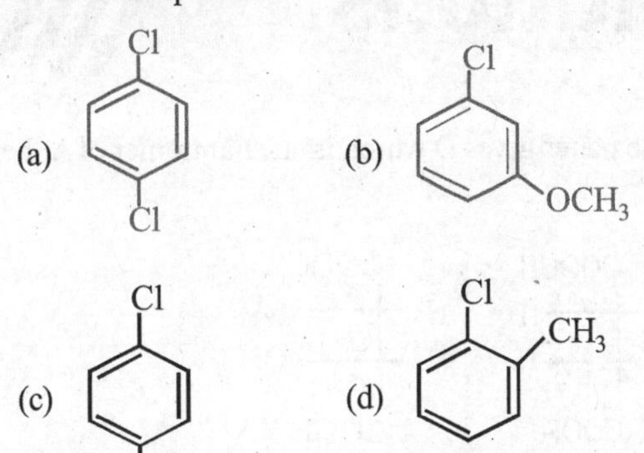

(c) Cl, NO_2 (d) Cl, CH_3

68. The shape of transition state is –

$$Nu^- + H\!\rhd\!\overset{R}{\underset{R}{C}}\!-X \longrightarrow [\text{Transition state}] \longrightarrow Nu\!\rhd\!\overset{R}{\underset{R}{C}}\!-H + X^-$$

(a) Triangular planar
(b) Square pyramidal
(c) Trigonal bipyramidal
(d) Tetrahedral

69. Which one of the following is most stable?

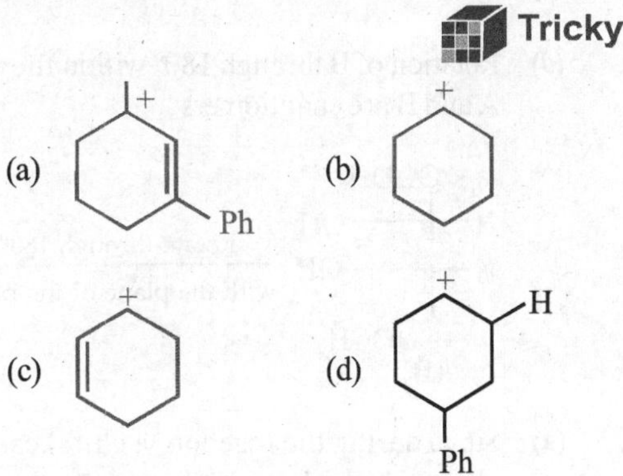

(c) + (d) + H Ph

70. In nucleophilic substitution reaction, order of halogens as incoming (attacking) nucleophile is: $I^- > Br^- > Cl^-$

The order of halogens as departing nucleophile should be :

(a) $Br^- > I^- > Cl^-$ (b) $I^- > Br^- > Cl^-$
(c) $Cl^- > Br^- > I^-$ (d) $Cl^- > I^- > Br^-$

Answer KEYs

1	(d)	8	(c)	15	(a)	22	(a)	29	(c)	36	(a)	43	(c)	50	(d)	57	(d)	64	(b)
2	(a)	9	(a)	16	(b)	23	(a)	30	(b)	37	(a)	44	(c)	51	(d)	58	(c)	65	(c)
3	(d)	10	(c)	17	(a)	24	(d)	31	(a)	38	(c)	45	(d)	52	(a)	59	(d)	66	(a)
4	(d)	11	(b)	18	(b)	25	(b)	32	(a)	39	(d)	46	(d)	53	(c)	60	(b)	67	(c)
5	(d)	12	(b)	19	(c)	26	(b)	33	(d)	40	(a)	47	(b)	54	(c)	61	(a)	68	(c)
6	(a)	13	(d)	20	(b)	27	(c)	34	(a)	41	(d)	48	(c)	55	(c)	62	(a)	69	(a)
7	(d)	14	(a)	21	(b)	28	(b)	35	(b)	42	(a)	49	(a)	56	(b)	63	(d)	70	(b)

Hints & Solutions

1. (d) Rotation of B through 180° within the plane of the paper gives D which is an enantiomer of A, hence A and B are enantiomers

(B): COOH on top; H—C—OH; H—C—OH; $COOCH_3$ at bottom $\xrightarrow[\text{with the plane of the paper}]{\text{Rotate through 180°}}$ (D): $COOCH_3$ on top; OH—C—H; HO—C—H; COOH at bottom ⋮ (A): $COOCH_3$ on top; H—C—OH; H—C—OH; COOH at bottom

2. (a) Since during the reaction, a chiral carbon is created and further since the CN^- ion can attack the planar aldehyde group both from the top and the bottom face of the aldehyde group with equal ease, therefore, a 50:50 mixture of the two enantiomers, i.e. a racemic mixture is obtained.

3. (d) B.P. $\propto$ extent of H bonding $\propto$ surface area of molecule.

4. (d) Degree of unsaturation = 11

5. (d) Order of Acidic strength $\propto$ stability of conjugate base.

Cyclohexyl–NH_2 < Cyclohexyl–OH < Phenyl–OH

i.e., Cyclohexyl–$\bar{N}H$ < Cyclohexyl–O^- < Phenyl–O^-

Resonance stabilized {phenoxide ion}

6. (a) $CH_3-\overset{H}{\underset{OH}{C^*}}-\overset{H}{\underset{OH}{C^*}}-C_2H_5$

7. (d) Keto-enol tautomerism is shown by carbonyl compounds having α-hydrogen atom.

(a) $H_5C_6-\overset{O}{\overset{\|}{C}}-\overset{\alpha}{C}H_2-\overset{O}{\overset{\|}{C}}-C_6H_5 \rightleftharpoons H_5C_6-\overset{O}{\overset{\|}{C}}-\underset{H}{C}=\overset{OH}{C}-C_6H_5$

enol form

(b) $H_5C_6-\overset{O}{\overset{\|}{C}}-\overset{\alpha}{C}H_3 \rightleftharpoons C_6H_5-\overset{OH}{C}=CH_2$

Keto form enol form

(c) $C_6H_5-\overset{O}{\overset{\|}{C}}-H$

no α–H atom

8. (c) $\overset{sp}{H_3C-C}\equiv N \xrightarrow{H_2O} H_3C-\overset{sp^2}{\underset{\underset{O}{\|}}{C}}-NH_2$

9. (a) Five membered ring is being converted into six membered ring. Hence the sixth bond must come from within the ring.

10. (c) $R_3COH \xrightarrow[\text{Conc. HCl}]{\text{Anhy. } ZnCl_2} R_3C.Cl$

It is substitution reaction. $R_3C.Cl$ being insoluble in H_2O give turbid solution.

11. (b) $H_2\overset{sp^2}{C}=\overset{sp}{C}=\overset{sp^2}{C}H_2$

12. (b) $CH_3CH_2-\underset{\underset{\underset{CH_3}{|}}{\underset{CH_2}{|}}}{CH}-\overset{\overset{CH_3}{|}}{CH}-CH_2CH_3$

3-Ethyl-4-methyl hexane

13. (d) Order of stability of carbocations is 3° > 2° > 1° and I is the best leaving group among halogens.

14. (a) Higher the electron density on O, stronger is the H-bond with water and thus more is the solubility. Thus solubility of the three ethers follow the order

I > III > II

I: electron pair localised; III: electron pair delocalised; II: electron pair more delocalised

15. (a) (i) $-\underset{\underset{N}{|}}{\overset{\overset{N}{|}}{C}}-\underset{\underset{C}{|}}{\overset{\overset{C}{|}}{N}}$, (ii) $-\underset{\underset{Br}{|}}{\overset{\overset{CH_3}{|}}{C}}-CH_3$, (iii) $-\underset{\underset{C-O}{|}}{\overset{\overset{OH}{|}}{C}}-\underset{\underset{C}{|}}{O}$, (iv) $-\underset{\underset{H}{|}}{\overset{\overset{H}{|}}{C}}-\underset{\underset{O-C}{|}}{\overset{\overset{O-C}{|}}{C}}-OCH_3$

(v) $-\underset{\underset{H}{|}}{\overset{\overset{O-C}{|}}{C}}-\underset{\underset{C}{|}}{O}$

Arrange (NNN), (BrCC), (OOO), (CHH), (OOH) in increasing atomic number. The order is ii, iii, v, i, iv.

16. (b) The resulting compound is $C_3H_7CH(OH)C_2H_5$ which is optically inactive and reaction leads the racemisation

17. (a) Meso compounds are those having 2 or more chiral centers along with symmetry.

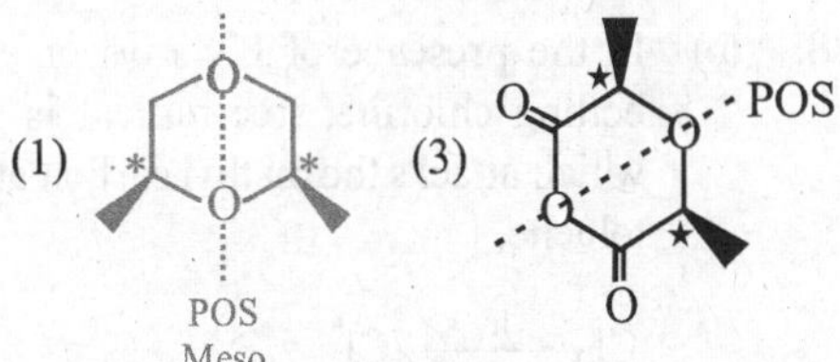

18. (b) Structure (b) is a 3° carbocation, while (a) is 2° and (c) and (d) are 1° carbocations; thus (b) is the most stable.

19. (c) The reaction (c) is Hoffmann elimination

20. (b) The more the electrons withdrawing groups attached to - O atom, the more is the ease of leaving.

21. (b) When the given mixture is shaken with 1 M HCl, amine get protonated and becomes cation ($RNH_2^{\oplus}$), which does not dissolve in organic solvent but usually dissolve in H_2O due to its charge. So, shaking with aqueous HCl will pull amines into the aqueous phase and leave all other compounds in organic layer.

22. (a) Stability of carbocation $\propto$ no. of α–H present on carbocation.

23. (a) Ethers due to absence of delocalized pair of electrons do not show resonance.

24. (d) 1, 2-Epoxy propane is the correct IUPAC name of given compound.

25. (b) $CH_3 - \overset{\overset{\delta^-}{O}}{\overset{\|}{C}} = \overset{\delta^+}{\ddot{O}} - CH_3$

due to resonance this bond has partial double bond characters.

26. (b) (structure: 1,3-dimethylcyclohexane with Me wedges, chiral centres marked *, dashed plane of symmetry)

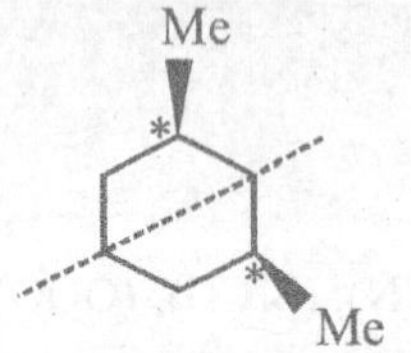

Meso $\rightarrow$ Compounds having chiral centres and have plane or centre of symmetry are known as meso compound.

27. (c) Allyl isocyanide has the structure

$H_2C = CH - CH_2 - N \overset{\rightarrow}{=} C$

it has 9σ and 3π bonds along with 2 non-bonded electrons

28. (b) In the presence of UV rays or energy, by boiling chlorine, free radical is generated which attacks the methyl carbon atom of the toluene.

$$Cl_2 \xrightarrow{h\nu} 2Cl^{\bullet}$$

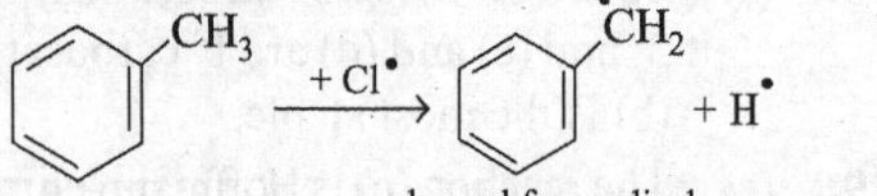

$$C_6H_5CH_3 \xrightarrow{+Cl^{\bullet}} C_6H_5\dot{C}H_2 + H^{\bullet}$$

benzyl free radical

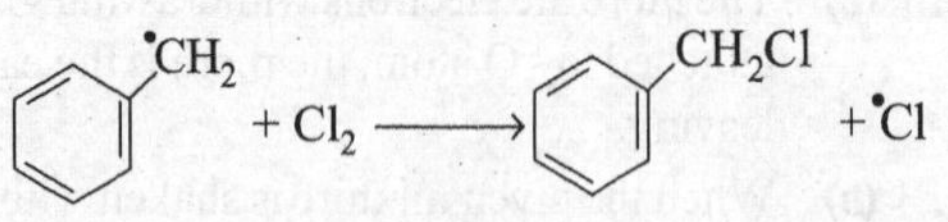

$$C_6H_5\dot{C}H_2 + Cl_2 \longrightarrow C_6H_5CH_2Cl + \dot{C}l$$

29. (c) 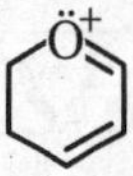 (pyrylium ion, $\ddot{O}^+$ in ring) It has complete octet for all atoms.

30. (b) The compound is a derivative of benzaldehyde. Start numbering C-atoms of benzene nucleus from C-atom bearing – CHO group.

31. (a) $\overset{1}{CH_3}(H)\overset{2}{C} = \overset{3}{C}(H) - \overset{4}{C} \equiv \overset{5}{C} - \overset{6}{CH_2}\overset{7}{CH_3}$

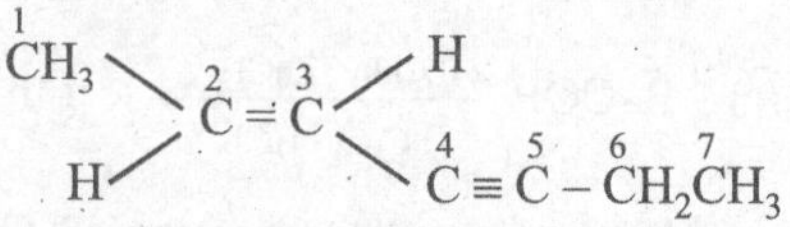

(E) – 2 – hepten – 4 – yne

32. (a) In 1, 2-dichlorobenzene the two dipoles are at 60° (i.e. unsymmetric).

Thus possesses dipole moment.

33. (d) The optically inactive compound must contains achiral carbon atom(s). Option (d) contains achiral carbon atom

$CH_3 - \underset{H}{\overset{Cl}{C}} - CHO$ 2- chloropropanol

$CH_3 - CH_2 - \underset{H}{\overset{Cl}{C}} - CH_3$ 2-chlorobutane (chiral)

$CH_3 - CH_2 - CH_2 - \underset{H}{\overset{Cl}{C}} - CH_3$ 2-chloropentane (chiral)

$CH_3 - CH_2 - \underset{Cl}{\overset{CH_3}{C}} - CH_3$ 2-chloro-2-methylbutane (achiral)

34. (a)

(a) 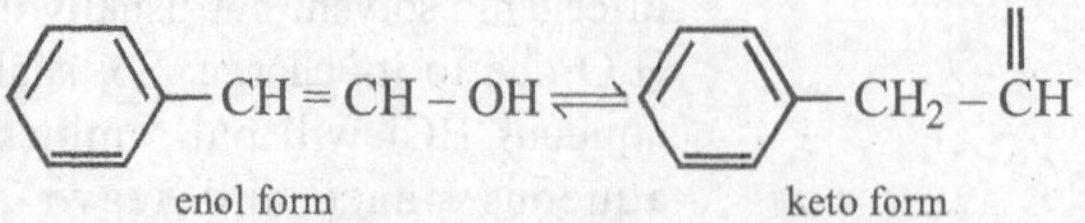 $C_6H_5 - CH = CH - OH \rightleftharpoons C_6H_5 - CH_2 - \overset{\overset{O}{\|}}{CH}$

enol form keto form

(b) 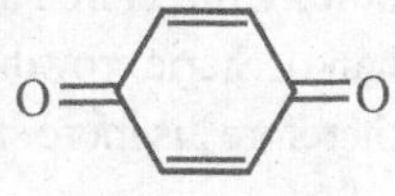 O=⟨cyclohexadiene⟩=O cannot tautomerise

(c) 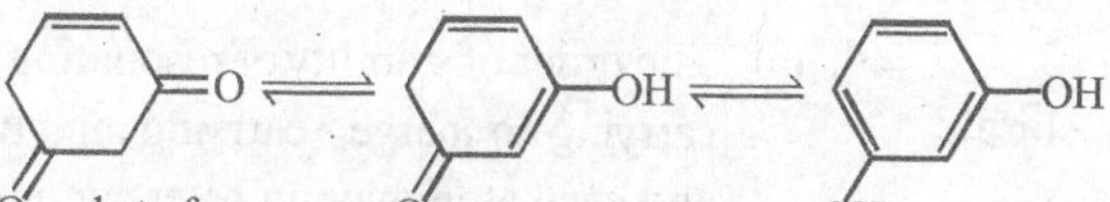 keto form ⇌ ⇌ enol form

(d) 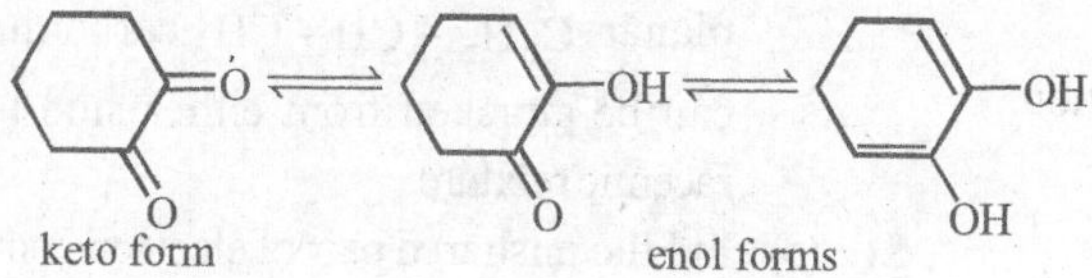 keto form ⇌ ⇌ enol forms

35. (b) – I group destablises carbocation and since inductive effect decreases with increasing length of carbon chain. Therefore (b) is the correct option.

36. (a) Due to the presence of a lone pair of electrons on N, $CH_3C \equiv N$: acts as a nucleophile. Further due to greater electronegativity of N than C, the C atom of – C ≡ N carries a positive charge and hence behaves as an electrophile.

37. (a) Stronger the acid, weaker the conjugate base. Since acid character follows the order

$H_2O > HC \equiv CH > NH_3 > CH_3 - CH_3$

(Acid character),

the basic character of their conjugate bases decreases in the reverse order, i.e.,

$CH_3\ CH_2^- > NH_2^- > HC \equiv C^- > OH^-$

(Basic character)

38. (c) N has no d orbital, hence it can't accommodate more than 8 electrons in its valence shell.

39. (d) 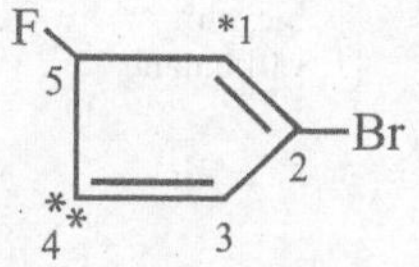 F, *1, 5, —Br, 2, 3, **4

The numbering of C-atom starts from $\overset{*}{C}$ or $\overset{**}{C}$. But numbering from $\overset{*}{C}$ give minimum locant (2) to Br which is correct.

40. (a) The given compounds are identical.

41. (d)

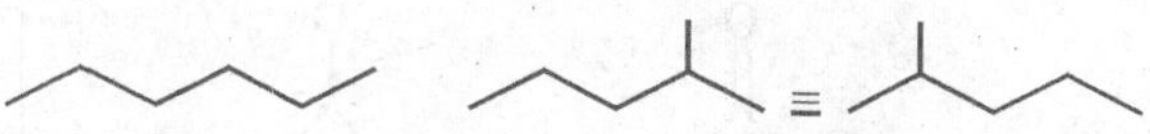

(1) (2) are same (3) (4) (5)

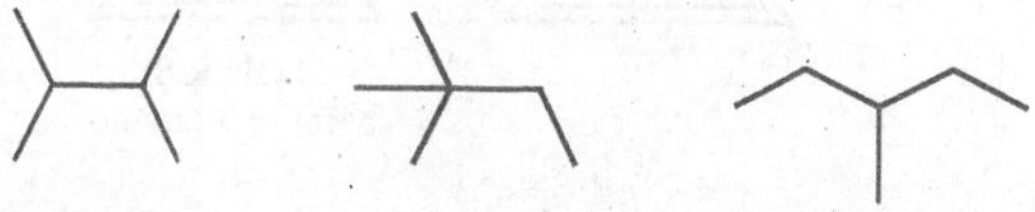

Total 5 structural isomers.

42. (a) Pentan-2-one and pentan-3-one are position isomers.

(b), (c), (d) contain different compounds aldehyde and ketones.

43. (c) Nucleophilicity increases with the decrease in electronegativity of the central atom. Since electronegativity follows the order: F > O > N > C; nucleophilicity of the concerned group will follow the reverse order i.e.,

$CH_3^- > NH_2^- > OH^- > F^-$

44. (c) More the no. of carbons more will be the heat of combustion.

45. (d) The compounds containing $-\overset{O}{\overset{||}{C}}-$ group adjacent to asymmetric carbon atom carrying a hydrogen easily racemise.

46. (d) The molecule contains two similar chiral centres.

47. (b) The reaction proceeds through planar $H-\overset{CH_3}{\overset{|}{C^+}}-C_2H_5$ leading to racemisation.

48. (c) In compounds

$\underset{H}{\overset{CH_3}{>}}C=C\underset{Cl}{\overset{H}{<}}$ (I) $\qquad \underset{H}{\overset{CH_3}{>}}C=C\underset{H}{\overset{Cl}{<}}$ (II)

I has more dipole moment than II, hence its boiling point will be higher. Melting point depends on symmetry, therefore I has higher melting point than II. Steric crowding in II is more than in I therefore I is more stable than II.

49. (a) The order of stability of resonating structures carrying no charge > carrying minimum charge and each atom having octet complete.

50. (d) $SbCl_5$ pulls Cl^- to form $SbCl_6^-$ leaving behind planar $C_6H_5-\overset{+}{C}H-CH_3$ carbonium ion. It can be attacked from either side leading to racemic mixture.

51. (d) In I the unshared pair of electrons on N always available for protonation in III due to presence of electronegative O atom the electron density on N is decreased . In II and IV resonance suppresses the basic character.

52. (a) Cis molecule has non-zero dipole moment.

53. (c)

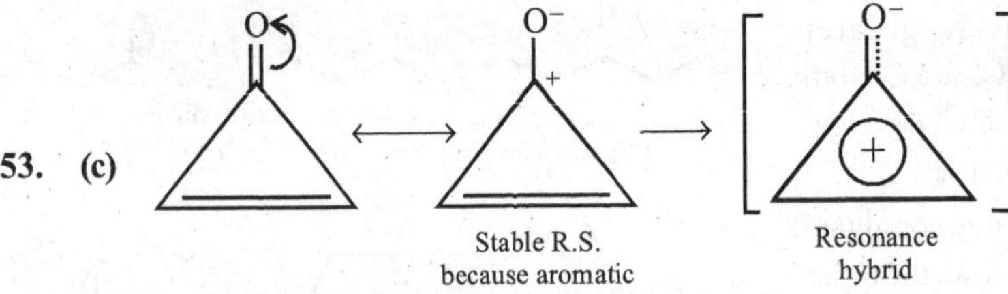

so molecule has high dipole moment.

54. (c) Heat of hydrogenation $\propto \dfrac{1}{\text{Stability of alkene}}$.

55. (c) Optical isomers contain an asymmetric (chiral) carbon atom (a carbon atom attached to four different atoms or groups). Therefore all except maleic acid exhibit optical isomerism.

56. (b) Five cyclic structures are possible for C_4H_6. These are as following:

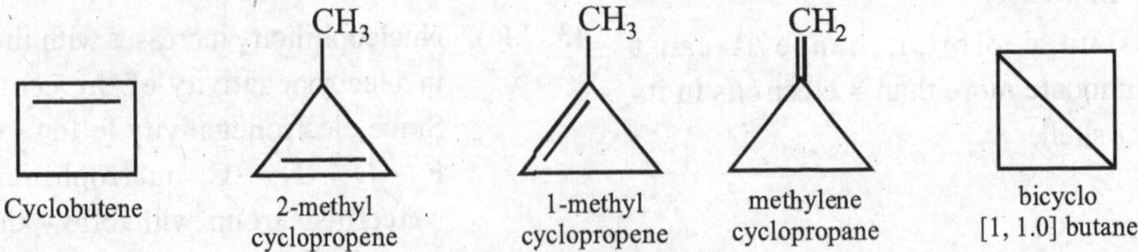

57. (d) Maleic acid and fumaric acids are geometrical isomers.

$$\begin{array}{c} H-C-COOH \\ \| \\ H-C-COOH \end{array} \qquad \begin{array}{c} H-C-COOH \\ \| \\ HOOC-C-H \end{array}$$

Maleic acid $\qquad$ Fumaric acid

58. **(c)** Phenol is less acidic than H_2CO_3 so reaction does not favours in forward direction.

59. **(d)** Stability of Alkene ∝ No. of H-atoms or more hyperconjugation.

60. **(b)** The order of stability of carbocations is :

$$(CH_3)_3\overset{+}{C} > (CH_3)_2\overset{+}{C}H > \overset{+}{C}H_3$$

61. **(a)** The enol form of acetone is

$$CH_3-\overset{OH}{\overset{|}{C}}=CH_2 \text{ with } D_2O \text{ it gives}$$

$$CH_3-\overset{OD}{\overset{|}{C}}=CH_2$$

62. **(a)** (I)

F (top), Br, H, H, CH_3, H (Newman projection) ≡ H–C(–Me)(a F, b CH_2Br, c Me) & H, CH_2Br (b), Me (c), F (a)

R-Configuration　　　　S-Configuration

Enantiomer

(II)

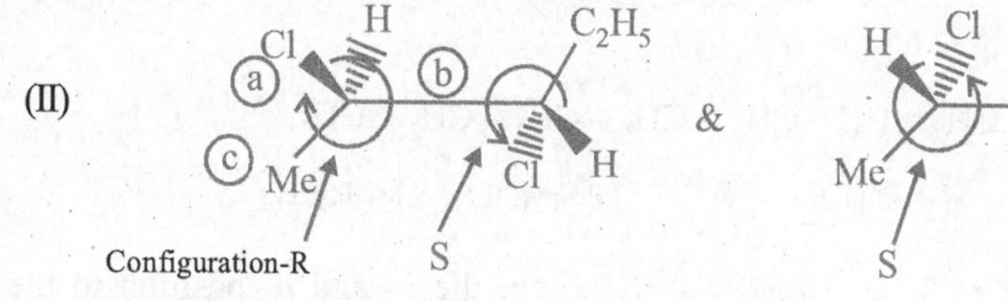

Configuration-R　　　S-Configuration

are diastereoisomers.

(III)

Me: H—Br, Me—H, Br　|　Me: Br—H, H—Me, Br

Enantiomer

63. **(d)** Consider R(top), R—R, R(bottom); H_3C, Cl, CH_3, CH_3, Cl, Cl, Cl, CH_3

no plane can pass in such a way that molecule has symmetry.

Similarly for S(top), S—S, S(bottom) and R(top), S—S, R(bottom)

64. (b) (a),(c) and (d) are resonating structures.

65. (c) Higher stability of allyl and aryl substituted methyl carbocation is due to dispersal of positive charge due to resonance

$CH_2 = CH - \overset{+}{C}H_2 \longleftrightarrow \overset{+}{C}H_2 - CH = CH_2$

Resonating structures of allyl carbocation

Resonating structures of benzyl carbocation

Hence the correct order of stability will be

$C_6H_5\overset{+}{C}H_2 > CH_2 = CH - \overset{+}{C}H_2 > CH_3 - CH_2 - \overset{+}{C}H_2 > \overset{+}{C}H_3$

Benzyl, III   Allyl, I   Propyl, II   Methly, IV

66. (a) Milli equivalents of H_2SO_4

$= 60 \times \frac{M \times 2}{10} = 12$

Milli equivalents of NaOH $= 20 \times \frac{M}{10} = 2$

Milli equivalent of $NH_3 = 12 - 2 = 10$

% of nitrogen

$= \frac{1.4 \times (N \times V) NH_3}{(\text{Wt. of organic compound})}$

$\frac{1.4 \times 10}{1.4} = 10$

67. (c) Aryl halides do not undergo nucleophilic substitution under ordinary conditions. The low reactivity of halogen atom in aryl halides is due to resonance. However, aryl halides can be made to undergo nucleophilic substitution either under drastic condition (high temperature, pressure or very strong nucleophiles) or by activating the nuclear halogen by introducing electron withdrawing group e.g. NO_2, –CHO, CN etc. in the *o*- and *p*- position to the nuclear halogen. Hence, *p*-nitrochlorobenzene would react most readily with nucleophiles.

68. (c) Shape of transition state is as follows:

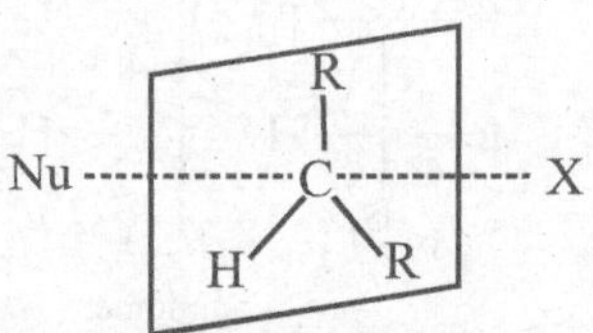

Trigonal bipyramidal

69. (a) 3° carbocations are most stable.

70. (b) Since the leaving group breaks away as a base, it is easier to displace weaker bases as compared to stronger bases. Thus less basic the substituent, the more easily it is displaced.

Since the basic strength of the given groups is in order.

$I^- < Br^- < Cl^-$

Thus the order of halogen leaving groups is

$I^- > Br^- > Cl^-$

13 Hydrocarbons

1. Which of the following would not give 2-phenylbutane as the major product in a Friedel-Crafts alkylation reaction on benzene ring?
   (a) 1-butene + HF
   (b) 2-butanol + H_2SO_4
   (c) Butanoyl chloride + $AlCl_3$ then Zn, HCl
   (d) Butyl chloride + $AlCl_3$

2. Which of the following compounds are antiaromatic

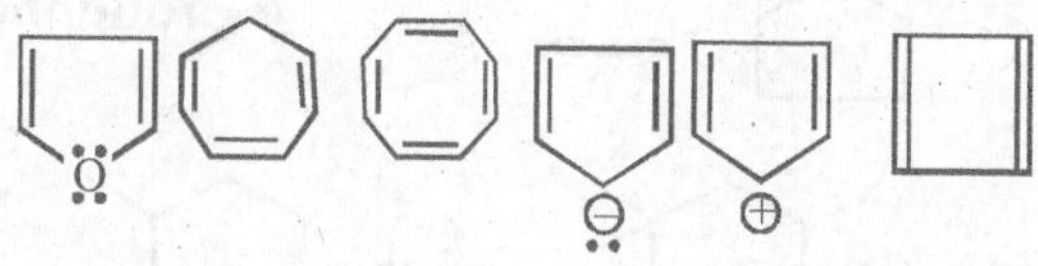

(I) (II) (III) (IV) (V) (VI)

   (a) (I) and (V) (b) (II) and (V)
   (c) (I) and (IV) (d) (V) and (VI)

3. Which one of the following reactions is expected to readily give a hydrocarbon product in good yields ? **Toughnut**

   (a) $RCOOK \xrightarrow[\text{oxidation}]{\text{Electrolytic}}$
   (b) $RCOO^-Ag^+ \xrightarrow{Br_2}$
   (c) $CH_3CH_3 \xrightarrow[h\upsilon]{Cl_2}$
   (d) $(CH_3)_3CCl \xrightarrow{C_2H_5OH}$

4. (A) $\xrightarrow{H_2(1\ \text{mole})/Pt}$

   Double bond equivalent (degree of unsaturation) of (A) is:
   (a) 1 (b) 2 (c) 3 (d) 4

5. An alkene having molecular formula C_7H_{14} was subjected to ozonolysis in the presence of zinc dust. An equimolar amount of the following two compounds was obtained

   $(CH_3)(CH_3)C=O$ and $(CH_3)(CH_3CH_2)C=O$

   The IUPAC name of the alkene is
   (a) 3, 4-dimethyl-3-pentene
   (b) 3, 4-dimethyl-2-pentene
   (c) 2, 3-dimethyl-3-pentene
   (d) 2, 3-dimethyl-2-pentene

6. Which of the following will yield a mixture of 2-chlorobutene and 3-chlorobutene on treatment with HCl ? **Critical Thinking**
   (a) $CH_2=C=CH-CH_3$
   (b) $H_2C=C(CH_3)-CH=CH_2$
   (c) $CH_2=CH-CH=CH_2$
   (d) $HC\equiv C-CH=CH_2$

7. Which of the following statements is incorrect regarding dehydrohalogenation of alkenes?
   (a) During the reaction hydrogen atom is eliminated from the β - carbon atom.
   (b) Rate of reaction for same alkyl group; Iodine > Bromine > Chlorine
   (c) Rate of reaction; $(CH_3)_3C- > (CH_3)_2CH- > CH_3CH_2-$
   (d) Only nature of halogen atom determine rate of the reaction.

8. Which of the following will form alkynide?
   (a) cyclohexyl$-C\equiv C-CH_3$
   (b) cyclohexyl$-C\equiv C-H$
   (c) cyclohexyl$-C\equiv C-$cyclohexyl
   (d) $CH_3-C\equiv C-CH_3$

9. $CH_3-\underset{\underset{}{}}{\overset{\overset{\displaystyle CH_3}{|}}{CH}}-C\equiv CH \xrightarrow{\text{excess HBr}}$

The product of the above reaction is:

(a) $CH_3-\overset{\overset{\displaystyle CH_3}{|}}{CH}-\overset{\overset{\displaystyle Br}{|}}{C}=\overset{\overset{\displaystyle Br}{|}}{CH_2}$

(b) $CH_3-\overset{\overset{\displaystyle CH_3}{|}}{CH}-\overset{\overset{\displaystyle Br}{|}}{C}=CH_2$

(c) $CH_3-\overset{\overset{\displaystyle CH_3}{|}}{CH}-\underset{\underset{\displaystyle Br}{|}}{\overset{\overset{\displaystyle Br}{|}}{C}}-CH_3$

(d) $CH_3-\overset{\overset{\displaystyle CH_3}{|}}{CH}-CH_2-\underset{\underset{\displaystyle Br}{|}}{\overset{\overset{\displaystyle Br}{|}}{CH}}$

10. What is the end product of the following sequences of operations? Critical Thinking

$CaC_2 \xrightarrow{H_2O} A \xrightarrow[Hg^{2+}]{dil.H_2SO_4} B \xrightarrow[H_2]{Ni} C$

(a) Methyl alcohol (b) Acetaldehyde
(c) C_2H_5OH (d) C_2H_4

11. [Structure: cyclohexane ring bearing CH_3 and H^a on one carbon, H^b on adjacent carbon, OH and H^d on lower carbons, and a CH_2-H^c substituent] $+ Br^{\bullet} \longrightarrow$

$Br^{\bullet}$ will abstract which of the hydrogen most readily? Toughnut

(a) a (b) b (c) c (d) d

12. Consider the following reaction

$H_3C-\underset{\underset{\displaystyle D}{|}}{CH}-\underset{\underset{\displaystyle CH_3}{|}}{CH}-CH_3+\overset{\bullet}{Br}\longrightarrow 'X'+HBr$

Identify the structure of the major product 'X'

(a) $H_3C-\underset{\underset{\displaystyle D}{|}}{CH}-\underset{\underset{\displaystyle CH_3}{|}}{CH}-\overset{\bullet}{CH_2}$

(b) $H_3C-\underset{\underset{\displaystyle D}{|}}{CH}-\underset{\underset{\displaystyle CH_3}{|}}{\overset{\bullet}{C}}-CH_3$

(c) $H_3C-\underset{\underset{\displaystyle D}{|}}{\overset{\bullet}{C}}-\underset{\underset{\displaystyle CH_3}{|}}{CH}-CH_3$

(d) $H_3C-\overset{\bullet}{CH}-\underset{\underset{\displaystyle CH_3}{|}}{CH}-CH_3$

13. Predict the product (A) of the following reaction

[Structure: isopropylidenecyclopentane] $\underset{\Delta}{\overset{H^+}{\rightleftharpoons}}$ A Toughnut

(a) [Structure] (b) [Structure]
(c) [Structure] (d) [Structure]

14. Consider the following sequence of reactions

$CH_3CH=CH_2 \xrightarrow[700K]{Cl_2} A \xrightarrow[420K,12atm]{Na_2CO_3}$

$B \xrightarrow[(ii)\ NaOH]{(i)\ HOCl} C$

Compound 'C' is

(a) $\underset{\underset{\displaystyle CH_2OH}{|}}{\overset{\overset{\displaystyle CH_2OH}{|}}{CHOH}}$

(b) $CH_3\underset{\underset{\displaystyle OH}{|}}{CH}COONa$

(c) $HOCH_2-CH=CH_2$

(d) $CH_3\underset{\underset{\displaystyle OH}{|}}{CH}COCl$

15. Choose the correct alkyne and reagents for the preparation of

(a) , $HgSO_4$, H_2SO_4, H_2O

(b) , $HgSO_4$, H_2SO_4, H_2O

(c) , BH_3, H_2O_2, NaOH

(d) , BH_3, H_2O_2, NaOH

16. The major product of the following reaction

$$H_3C-C\equiv CH \xrightarrow[\text{ether}]{CH_3CH_2MgCl} \xrightarrow[H_3O^+]{} $$

(a)

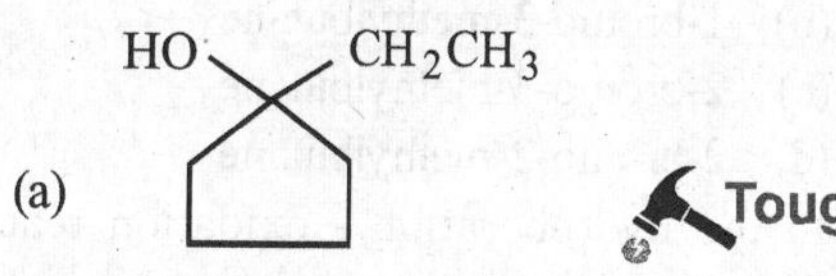

Toughnut

(b) $CHCH_3$

(c) HO CH_2 $C\equiv CH$

(d) HO C $\equiv$C-CH_3

17. A compound (X) (C_5H_8) reacts with ammonical $AgNO_3$ to give a white precipitate, and on oxidation with hot alkaline $KMnO_4$ gives the acid, $(CH_3)_2CHCOOH$, therefore X is –

(a) $CH_2=CH-CH=CH-CH_3$

(b) $CH_3-CH=CH-CH_2-CH_3$

(c) $(CH_3)_2CH-C\equiv CH$

(d) $(CH_3)_2C=C=CH_2$

18. The major product obtained in the following reaction

$\xrightarrow[FeBr_3]{Br_2}$?

(a) Br

(b) Br

(c) Br Br

(d) Br

19. $\xrightarrow{O_3}$ Toughnut

Which of the following products cannot be obtained in ozonolysis of o-xylene?

(a) $\begin{array}{c} CHO \\ | \\ CHO \end{array}$

(b) $CH_3-\overset{O}{\overset{\|}{C}}-\overset{O}{\overset{\|}{C}}-H$

(c) $CH_3-\overset{O}{\overset{\|}{C}}-\overset{O}{\overset{\|}{C}}-CH_3$

(d) $CH_3-\overset{O}{\overset{\|}{C}}-\overset{O}{\overset{\|}{C}}-CHO$

20. Compare rate of dehydration of (I), (II) and (III) by conc. H_2SO_4. **Tricky**

I. OH

II. OH

III. OH

(a) (I) > (III) > (II) (b) (I) > (II) > (III)
(c) (II) > (I) > (III) (d) (II) > (III) > (I)

21. In which step of the following reaction, sp^3 carbon changes to sp^2 carbon ?

$+ E^+ \xrightarrow{\text{Step 1}}$ (H, E, +) $\xrightarrow{\text{Step 2}}$ (E) $+ H^+$

(a) Step 1 (b) Step 2
(c) Both (d) None

22. Toluene reacts with excess of Cl_2 in presence of sun light to give a product which on hydrolysis by reaction with NaOH gives **Toughnut**

(a) COOH (b) COONa

(c) COCl, Na (d) None of these

23. The major product formed in the reaction is:

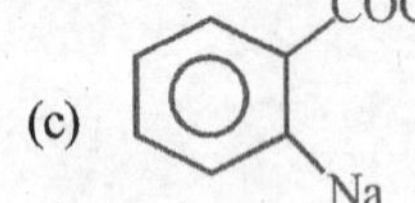

Critical Thinking

$\xrightarrow[\text{(mononitration)}]{\text{Conc. } HNO_3/\text{conc. } H_2SO_4}$

(a) O, C—O, O_2N

(b) O_2N—, O, C—O—

(c) O, C—O—, O_2N

(d) O, C—O—, —NO_2

24. The major product obtained in the photo catalysed bromination of 2-methylbutane is:
(a) 1-bromo-2-methylbutane
(b) 1-bromo-3-methylbutane
(c) 2-bromo-3-methylbutane
(d) 2-bromo-2-methylbutane

25. In the hydroboration - oxidation reaction of propene with diborane, H_2O_2 and NaOH, the organic compound formed is:
(a) CH_3CH_2OH (b) $CH_3CHOHCH_3$
(c) $CH_3CH_2CH_2OH$ (d) $(CH_3)_3COH$

26. In the given reaction,

$C{\equiv}CCH_3$ $\xrightarrow{H^+/Hg^{2+}}$ A the product 'A' is

(a) OH, $HC{=}C{-}CH_3$

(b) $HO{-}C{=}CHCH_3$

(c) $O{=}C{-}CH_2CH_3$ (attached to benzene ring)

(d) $CH_2{-}\overset{O}{\overset{\|}{C}}{-}CH_3$ (attached to benzene ring)

27. The product of the reaction between ethyl benzene and N-bromosuccinamide is

(a) CH_2-CH_2-Br (attached to benzene ring)

(b) CH_2-CH_3 (attached to benzene ring, with Br at meta position)

(c) CH_2-CH_3 (attached to benzene ring, with Br at para position)

(d) $\overset{Br}{\overset{|}{CH}}-CH_3$ (attached to benzene ring)

28. In preparation of alkene from alcohol using Al_2O_3 which is effective factor ?

(a) Porosity of Al_2O_3
(b) Temperature
(c) Concentration
(d) Surface area of Al_2O_3

29. Which one of the following compounds would have the highest heat of hydrogenation?

(a) $CH_2=CH_2$
(b) $CH_3-CH_2-CH=CH_2$
(c) $CH_3CH=CHCH_3$
(d) $(CH_3)_2C=C(CH_3)_2$

30. Which of the following types of reaction occur when a reactant has got a double bond ?

(i) Addition
(ii) Photolysis
(iii) Nucleophilic substitution
(iv) Polymerization

(a) (i) and (iv) (b) (i) and (ii)
(c) (i), (ii) and (iv) (d) (i), (ii), (iii) and (iv)

31. HOCl reacts on 3-methyl-2-pentene, the main product will be :

(a) $CH_3-CH=\underset{Cl}{\underset{|}{C}}-CH(OH)-CH_3$

(b) $CH_3-\overset{OH}{\overset{|}{CH}}-\underset{Cl}{\underset{|}{\overset{CH_3}{\overset{|}{C}}}}-CH_2-CH_3$

(c) $CH_3-\overset{Cl}{\overset{|}{CH}}-\underset{OH}{\underset{|}{\overset{CH_3}{\overset{|}{C}}}}-CH_2-CH_3$

(d) None of these

32. The major product of the following reaction sequence is: **Tricky**

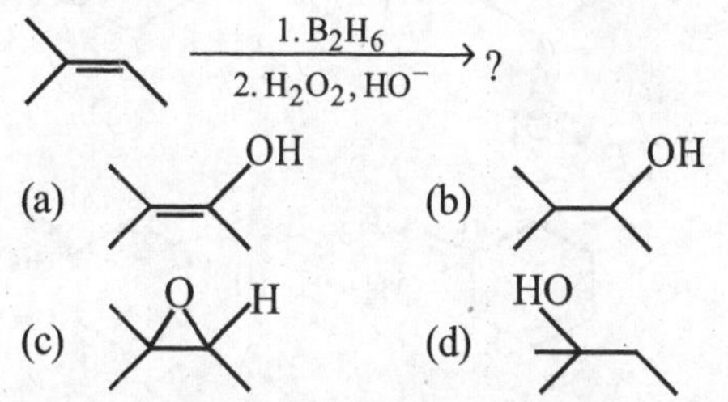

33. All the hydrocarbons shown are very weak acids. One, however, is far more acidic than the others. Which one is the strongest acid?

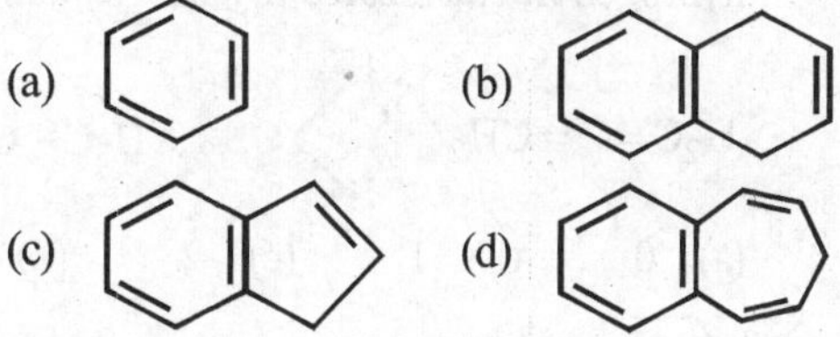

34. Which one amongst the following carbocations is most stable?

(a) $C_6H_5-\overset{+}{C}H-C_6H_5$

(b) (cyclohexenyl cation)

(c) $CH_3-\underset{CH_3}{\underset{|}{\overset{+}{C}}}-CH_3$

(d)

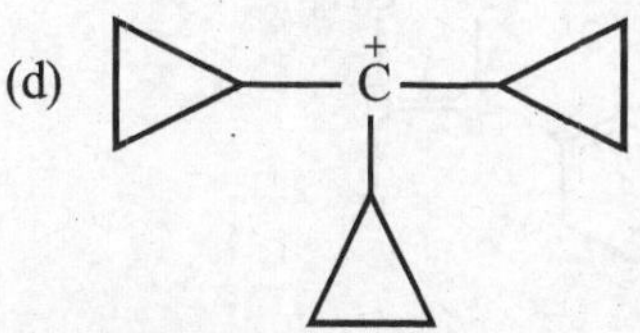

35. Toluene on treatment with CrO_3 and $(CH_3CO)_2O$ followed by hydrolysis with dil.HCl gives
(a) benzaldehyde
(b) benzoic acid
(c) phenol
(d) phenylacetaldehyde

36. What is the correct product of reaction?

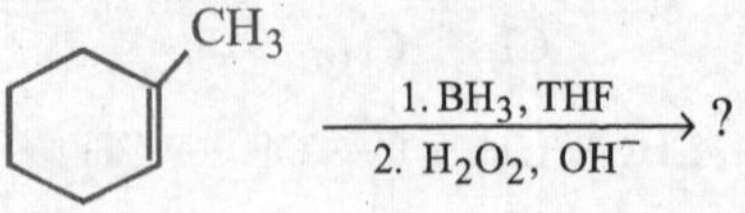

$$\text{1-methylcyclohexene} \xrightarrow[\text{2. } H_2O_2,\ OH^-]{\text{1. } BH_3,\ THF} ?$$

Toughnut

(a) 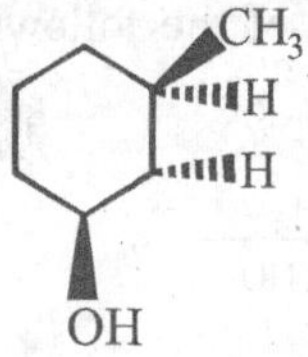 CH_3, H, H, OH (b) 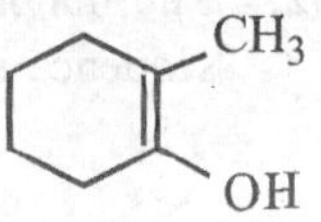 CH_3, OH (c) 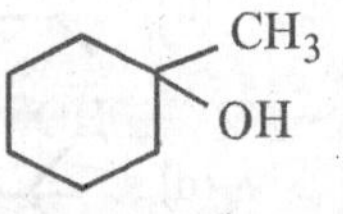 CH_3, OH (d) 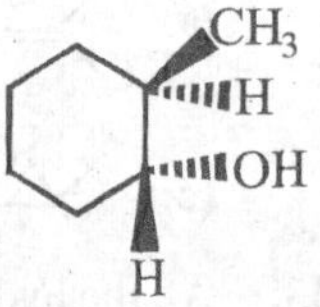 CH_3, H, OH, H

37. Of the three isomeric C_3H_4 hydrocarbons shown below how many can exist with all carbon and hydrogen nuclei located in a single plane

$H_2C=C=CH_2$, cyclopropene ($CH=CH$, CH_2), $H_3C-C\equiv CH$

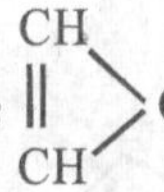

(a) 0 (b) 1 (c) 2 (d) 3

38. 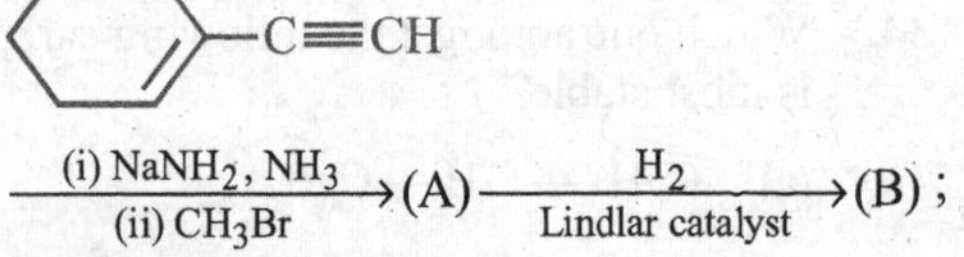 (cyclohexenyl)$-C\equiv CH$ $\xrightarrow[\text{(ii) } CH_3Br]{\text{(i) } NaNH_2,\ NH_3}$ (A) $\xrightarrow[\text{Lindlar catalyst}]{H_2}$ (B);

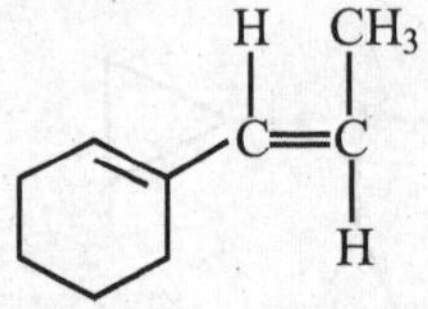

Product (B) is:

(a) 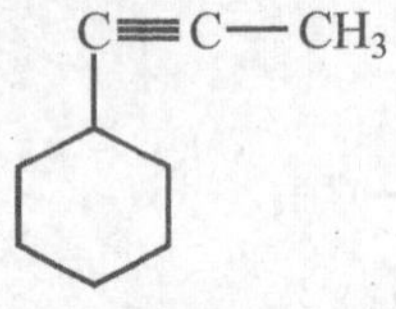 $C\equiv C-CH_3$ (on cyclohexane)

(b) H, CH_3, $C=C$, H (on cyclohexene)

(c) H, $-C=C$, H, CH_3 (on cyclohexene)

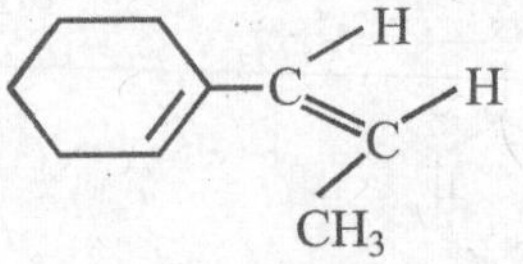

(d) 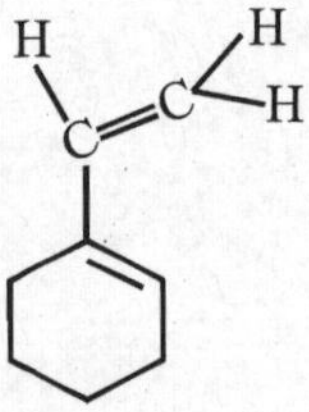 H, H, $C=C$, H (on cyclohexene)

39. Among the following free radical bromination reactions, select those in which 2° halide is the major product

Toughnut

(P) 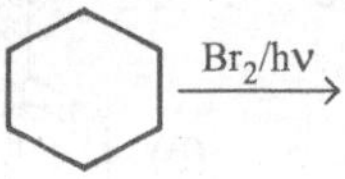 cyclohexane $\xrightarrow{Br_2/h\nu}$

(Q) 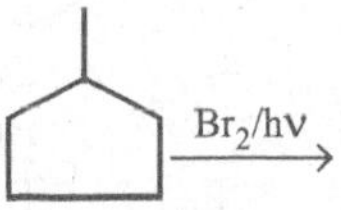 methylcyclopentane $\xrightarrow{Br_2/h\nu}$

(R) 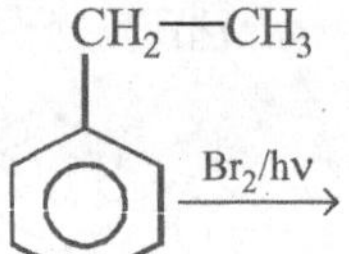 CH_2-CH_3 (ethylbenzene) $\xrightarrow{Br_2/h\nu}$

(S) 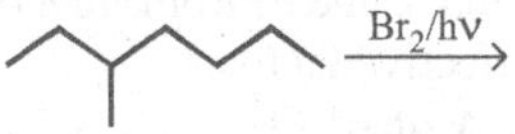 $\xrightarrow{Br_2/h\nu}$

(T) 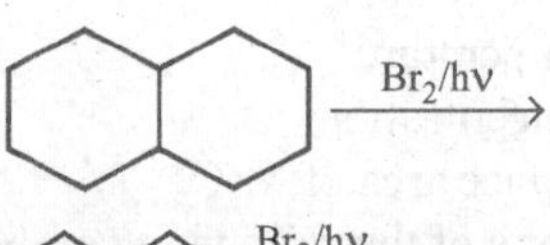 $\xrightarrow{Br_2/h\nu}$

(U) 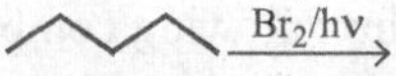 $\xrightarrow{Br_2/h\nu}$

(a) P, Q, R, S (b) P, R, U
(c) P, R, S, T (d) P, Q, R, S, T

40. The gas liberated by the electrolysis of dipotassium succinate solution is:
(a) Ethane (b) Ethyne
(c) Ethene (d) Propene

41. In the presence of peroxide, HCl and HI do not give anti-Markownikoff's addition of alkenes because:

Critical Thinking

(a) One of the steps is endothermic in HCl and HI
(b) Both HCl and HI are strong acids
(c) HCl is oxidizing and the HI is reducing
(d) All the steps are exothermic is HCl and HI

42. Consider the following sequence of reactions

$$CH_3CH=CH_2 \xrightarrow[700K]{Cl_2} A \xrightarrow[420K, 12\ atm]{Na_2CO_3} B$$

Compound 'B' is

(a) $CH_2OH-CH=CH_2$ (drawn vertically: CH_2OH | CH = CH_2)

(b) $CH_3CH(OH)COONa$ (drawn as $CH_3CHCOONa$ with OH below CH)

(c) $HOCH_2-CH=CH_2$

(d) $CH_3CH(OH)COCl$ (drawn as $CH_3CHCOCl$ with OH below CH)

43. How many monochlorobutanes will be obtained on chlorination of n-butane?
(a) 5 (b) 2 (c) 3 (d) 4

44. The cycloalkane having the lowest heat of combustion per CH_2 group **Critical Thinking**

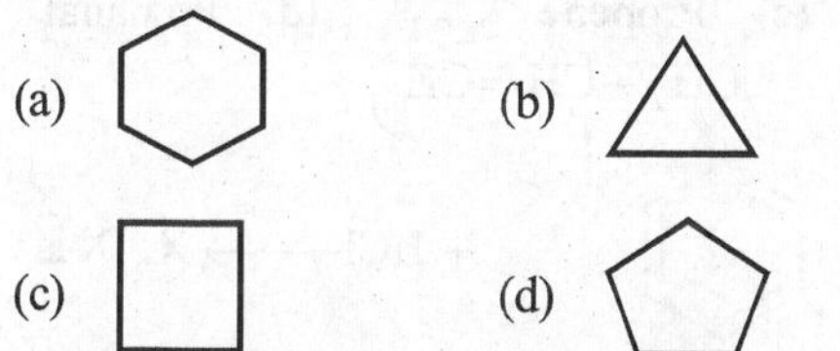

45. Which one of the following conformations of cyclohexane is chiral?
(a) Boat (b) Twist boat
(c) Rigid (d) Chair

46. Which one of the following compounds gives acetone $(CH_3)_2C=O$ as one of the product of its ozonolysis?

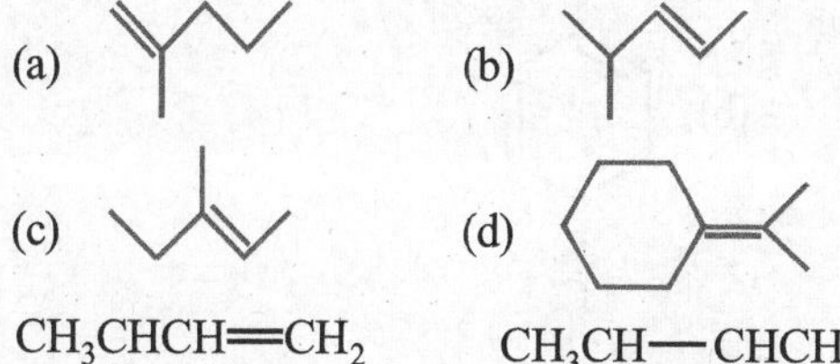

47.

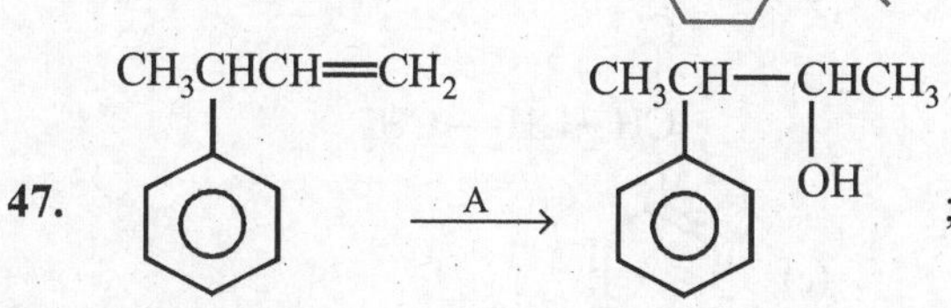

Reagent A may be:

(a) H_2O/H^+
(b) BH_3, $THF/H_2O_2-OH^-$
(c) $Hg(OCOCH_3)_2$, $H_2O/NaBH_4 . NaOH$
(d) All are possible

48. In reaction sequence

$$CH_2=CH_2 \xrightarrow[\text{acid}]{\text{Hypochlorous}} M \xrightarrow{R} \begin{matrix} CH_2OH \\ | \\ CH_2OH \end{matrix}$$

molecule 'M' and reagent 'R' respectively are
(a) CH_3CH_2Cl and NaOH
(b) CH_3CH_2OH and H_2SO_4
(c) $CH_2(Cl)CH_2OH$ and aqueous $NaHCO_3$
(d) H_2C-CH_2 (bridged by O) and heat

49. What is the major product expected from the following reaction?

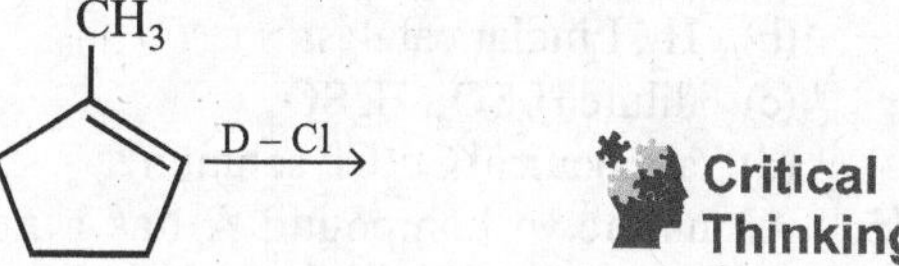

Critical Thinking

Where D is an isotope of hydrogen

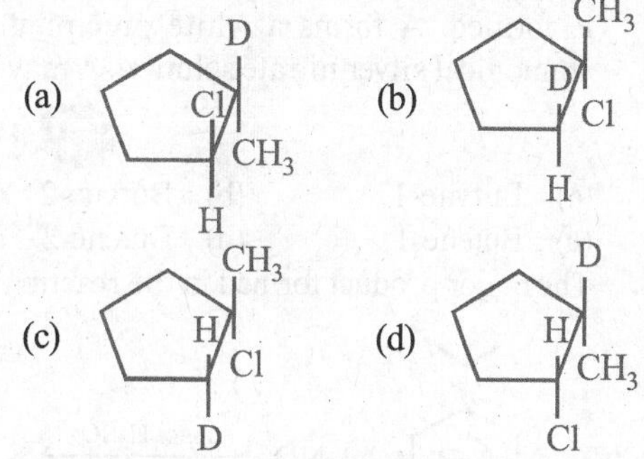

50. The reagent needed for converting

$$Ph-C\equiv C-Ph \longrightarrow \begin{matrix} Ph \\ H \end{matrix}\!\!>C=C<\!\!\begin{matrix} H \\ Ph \end{matrix}$$

(a) Cat. Hydrogenation
(b) H_2/ Lindlar Cat.
(c) Li/NH_3
(d) $LiAlH_4$

51. Addition of HI to double bond of propene yields isopropyl iodide and not n-propyl iodide as the major product, because addition proceeds through
(a) a more stable carbonium ion
(b) a more stable carbanion
(c) a more stable free radical
(d) homolysis

52. The principal organic product formed in the reaction,

$CH_2 = CH(CH_2)_8COOH + HBr \xrightarrow{\text{peroxide}}$ is

(a) $CH_3CHBr(CH_2)_8COOH$
(b) $CH_2 = CH(CH_2)_8COBr$
(c) $CH_2BrCH_2(CH_2)_8COOH$
(d) $CH_2 = CH(CH_2)_7CHBrCOOH$

53. Acid catalyzed hydration of alkenes except ethene leads to the formation of
(a) mixture of secondary and tertiary alcohols
(b) mixture of primary and secondary alcohols
(c) secondary or tertiary alcohol
(d) primary alcohol

54. Identify the reagent from the following list which can easily distinguish between 1-butyne and 2-butyne
(a) bromine, CCl_4
(b) H_2, Lindlar catalyst
(c) dilute H_2SO_4, $HgSO_4$
(d) ammonical Cu_2Cl_2 solution

55. An unknown compound A has a molecular formula C_4H_6, when A is treated with an excess of Br_2, a new substance B with formula $C_4H_6Br_2$ is formed. A forms a white precipitate with ammonical silver nitrate solution. A may be

(a) Butyne-1 (b) Butyne-2
(c) Butene-1 (d) Butene-2

56. The major product formed in the reaction is:

(4-isopropyltoluene) + $HNO_3 \xrightarrow{\text{conc. } H_2SO_4}$

(a) NO_2 ortho to isopropyl group (b) NO_2 ortho to methyl group

(c) NO_2 replacing the methyl group (para to isopropyl) (d) NO_2 on the isopropyl carbon

57. (Benzene ring)$-CH_2 - CH = CH_2$

On mercuration-demercuration produces the major product:

(a) (Benzene ring)$-CH_2 - CH(OH) - CH_3$
(b) (Benzene ring)$-CH_2 - CH_2 - CH_2 - OH$
(c) (Benzene ring)$-CH_2 - CH(OH) - CH_2(OH)$
(d) (Benzene ring)$-CH_2 - COOH$

58. Which one of the following class of compounds is obtained by polymerization of acetylene?
(a) Poly-yne (b) Poly-ene
(c) Poly-ester (d) Poly-amine

59. The hydration of propyne results in formation of
(a) Acetone (b) Propanol-1
(c) Propene (d) Propanal

60. (Benzene ring)$-CH_2 - CH = CH_2$ + HCl ⟶ X, X is:

(a) (Benzene ring)$-CH_2 - CH(Cl) - CH_3$
(b) (Benzene ring)$-CH_2 - CH_2 - CH_2Cl$
(c) (Benzene ring)$-CH(Cl) - CH_2 - CH_3$

(d) [Structure: 2-chloro allylbenzene — $CH_2-CH=CH_2$ on benzene ring with Cl]

61. [Methylcyclopentane] $\xrightarrow{Br_2/hv}$ Major (X) $\xrightarrow[KOH/\Delta]{Alcoholic}$ Major (Y) $\xrightarrow[Peroxide]{H-Br}$ Major (Z) z is:

(a) [Bromocyclohexane] (b) [1-bromo-3-methylcyclopentane]

(c) [1-bromo-2-methylcyclopentane] (d) [Bromomethylcyclopentane]

62. 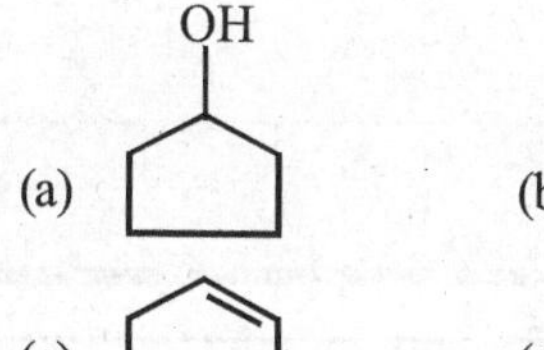 $\xrightarrow[Zn]{O_3}$ A $\xrightarrow[Or\ LiAlH_4]{H_2Ni}$ B $\xrightarrow[\Delta]{H+}$ (C);

Product (C) of the reaction is: **Toughnut**

(a) [Cyclopentanol] (b) [Cyclopentanone]

(c) [Cyclopentene] (d) [Cyclopentadiene]

63. Which of the following is correct? **Critical Thinking**

(a) [Cyclopentene] $\xrightarrow{C_6H_5COOOH}$ [Cyclopentene oxide, cis H, H]

(b) 

[Cyclopentene] $\xrightarrow[(ii)\ H_2O_2]{(i)\ O_3}$ $OHC.CH_2.CH_2.CHO$

(c) [Cyclopentene] $\xrightarrow{KMnO_4}$ [cis-Cyclopentane-1,2-diol]

(d) [Cyclopentene] $\xrightarrow[KMnO_4]{Alk.}$ $HO-\overset{O}{\overset{\|}{C}}-(CH_2)_3-\overset{O}{\overset{\|}{C}}-O-H$

64. Which of the following change is correct

(a) $>C=C<$ $\xrightarrow[wet\ CH_3COOH]{I_2,\ CH_3COOAg}$ [$>C(OH)-C(OH)<$, both OH on same side]

(b) $>C=C<$ $\xrightarrow[dry\ CH_3COOH]{I_2,\ CH_3COOAg}$ [$>C(OH)-C(OH)<$, OH on opposite sides]

(c) Both (a) and (b) one correct

(d) Neither (a) nor (b)

65. Choose the correct product of the following reaction **Toughnut**

[1,2-Dimethylcyclopentene] $\xrightarrow[(2)\ H_3O^+]{(1)\ C_6H_5COOOH}$?

(a) [Cyclopentane with OH and CH_3 on one carbon, CH_3 and H on adjacent carbon]

(b) [trans-1,2-Dimethylcyclopentane-1,2-diol]

(c) [1,2-Dimethylcyclopentane-1,2-diol, OH and CH_3 variant]

(d) [Cyclopentane with OH and CH_3 on one carbon, H and CH_3 on adjacent carbon]

66. The two compounds A and B obtained from 1-butyne can be distinguished by **Tricky**

$$B \xleftarrow[\text{(ii) } H_2O_2/OH^-]{\text{(i) } BH_3} CH_3CH_2C\equiv CH \xrightarrow{H^+/Hg^{2+}} A$$

(a) $NaHSO_3$ (b) litmus solution
(c) iodoform test (d) 2,4-DNP

67.

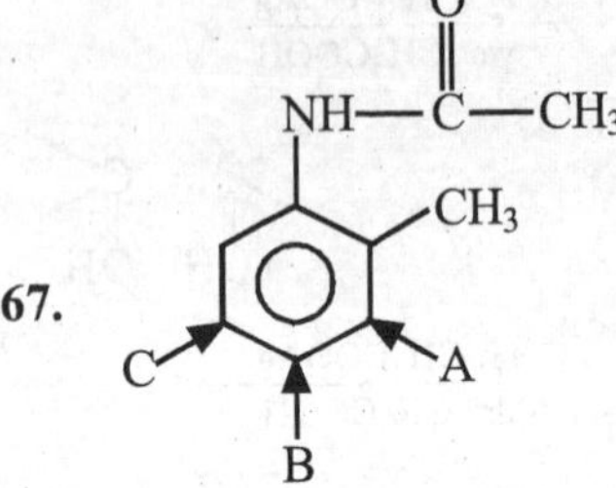

Identify the position where electrophilic aromatic substitution (EAS) is most favourable.

(a) A (b) B
(c) C (d) A and C

68. Arrange the following in decreasing order of reactivity towards EAS (electrophilic aromatic substitution)

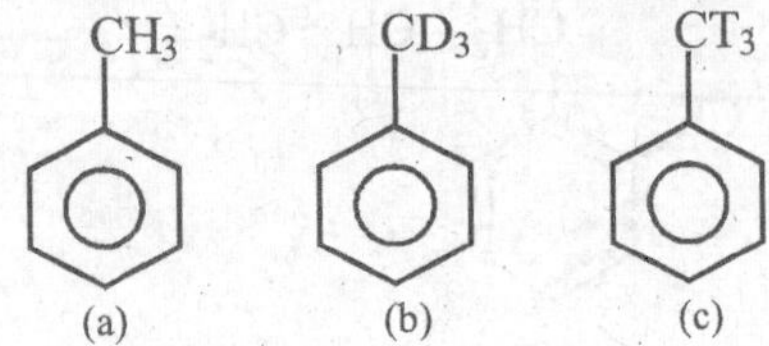

(a) a > b > c (b) c > b > a
(c) a > c > b (d) c > a > b

69. Which of the following is an aromatic species?

(a) [cation with NO_2, H] (b) [cation with CH_3, H]

(c) [N–H ring] (d) All of the three

70. Cl–◇–Br $\xrightarrow{\text{Na/Dry ether}}$ (A) 97%

Product (A) of above reaction is:

(a) (b)

(c) (d)

Answer KEYs

1	(c)	8	(b)	15	(b)	22	(b)	29	(a)	36	(d)	43	(b)	50	(c)	57	(a)	64	(c)
2	(d)	9	(c)	16	(d)	23	(d)	30	(a)	37	(a)	44	(a)	51	(a)	58	(b)	65	(c)
3	(a)	10	(c)	17	(c)	24	(d)	31	(c)	38	(c)	45	(b)	52	(c)	59	(a)	66	(c)
4	(c)	11	(a)	18	(a)	25	(c)	32	(b)	39	(b)	46	(d)	53	(c)	60	(c)	67	(b)
5	(d)	12	(b)	19	(d)	26	(c)	33	(c)	40	(c)	47	(c)	54	(d)	61	(c)	68	(a)
6	(a)	13	(d)	20	(b)	27	(d)	34	(d)	41	(a)	48	(c)	55	(a)	62	(c)	69	(c)
7	(d)	14	(a)	21	(b)	28	(d)	35	(a)	42	(a)	49	(b, c)	56	(b)	63	(a)	70	(b)

Hints & Solutions

1. (c) The Friedal-crafts acylation reaction will give propyl phenyl ketone which further on Clemmenson's reduction will give butyl benzene

$$C_6H_6 + CH_3CH_2CH_2COCl \xrightarrow{AlCl_3} C_6H_5COCH_2CH_2CH_3 \xrightarrow{Zn\text{-}Hg/HCl} \underset{\text{Butyl benzene}}{C_6H_5CH_2CH_2CH_2CH_3}$$

2. (d) For a compound to be aromatic it must have $(4n + 2)\ \pi e^-$ and system should be planar where n is an integer. (V) and (VI) both have 4 e^- and planar structure. Hence are Antiaromatic.

3. (a) $2RCOOK \xrightarrow[\text{oxidation}]{\text{Electrolytic}} \underset{\text{Anode}}{2RCOO^-} + \underset{\text{Cathode}}{2K^+}$

At anode $2RCOO^- \rightarrow 2RCOO^\bullet + 2e^-$

$2RCOO^\bullet \rightarrow R{-}R + 2CO_2$

At cathode $2K^+ + 2e^- \rightarrow 2K$

$2K + H_2O \rightarrow 2KOH + H_2 \uparrow$

4. (c) Molecular formula

(A) $\xrightarrow[Pt]{H_2}$ $\Rightarrow$ DBE of (A) = 3

5. (d) $(CH_3)_2C{=}O + O{=}C(CH_3)(C_2H_5) \longleftarrow \underset{\text{2,3-Dimethyl-2-pentene}}{(CH_3)_2C{=}C(CH_3)CH_2CH_3}$

6. (a) $CH_2{=}C{=}CH{-}CH_3 \xrightarrow{H^+} \left[CH_2{=}\overset{+}{C}{-}CH_2{-}CH_3 + CH_2{=}CH{-}\overset{+}{C}H{-}CH_3\right]$

$$\xrightarrow{Cl^-} \underset{\text{2–Chlorobutene}}{CH_2{=}C(Cl){-}CH_2CH_3} + \underset{\text{3–Chlorobutene}}{CH_2{=}CH{-}CH(Cl)CH_3}$$

7. (d) Nature of halogen atom and the alkyl group both determine rate of reaction.
8. (b) Only 1-Alkynes form alkynides

9. (c) $CH_3{-}CH(CH_3){-}C{\equiv}CH \xrightarrow{H^+Br^-} CH_3{-}CH(CH_3){-}C(Br){=}CH_2 \xrightarrow{H^+Br^-} CH_3{-}CH(CH_3){-}CBr_2{-}CH_3$

10. (c) $CaC_2 \xrightarrow{H_2O} \underset{\substack{\text{Acetylene}\\(A)}}{HC \equiv CH} \xrightarrow[Hg^{+2}]{Dil.H_2SO_4} \underset{\substack{\text{Acetaldehyde}\\(B)}}{CH_3CHO} \xrightarrow[H_2]{Ni} \underset{(C)}{C_2H_5OH}$

11. (a) Bromine is more selective

∴ abstract that hydrogen which forms stable free- radical

∴ 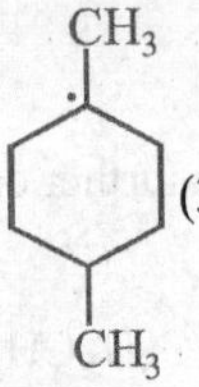 (3° free-radical) is most stable.

12. (b) $Br^{\bullet}$ is less reactive and more selective and so the most stable free radical (3°) will be the major product.

13. (d) 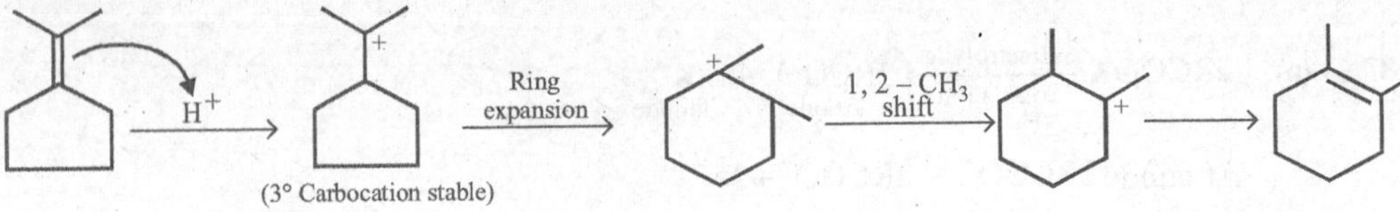 $\xrightarrow{H^+}$ (3° Carbocation stable) $\xrightarrow{\text{Ring expansion}}$ $\xrightarrow{1,2-CH_3 \text{ shift}}$ $\longrightarrow$

14. (a) 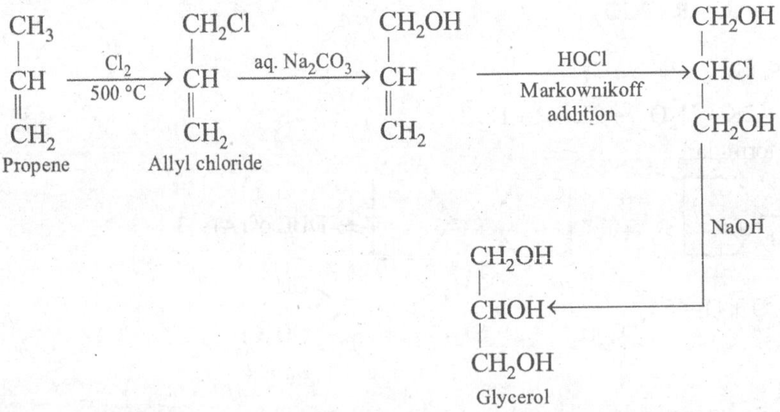 $CH_3-CH=CH_2$ (Propene) $\xrightarrow[500\,°C]{Cl_2}$ $CH_2Cl-CH=CH_2$ (Allyl chloride) $\xrightarrow{aq.\ Na_2CO_3}$ $CH_2OH-CH=CH_2$ $\xrightarrow[\text{Markownikoff addition}]{HOCl}$ $CH_2OH-CHCl-CH_2OH$ $\xrightarrow{NaOH}$ $CH_2OH-CHOH-CH_2OH$ (Glycerol)

15. (b) 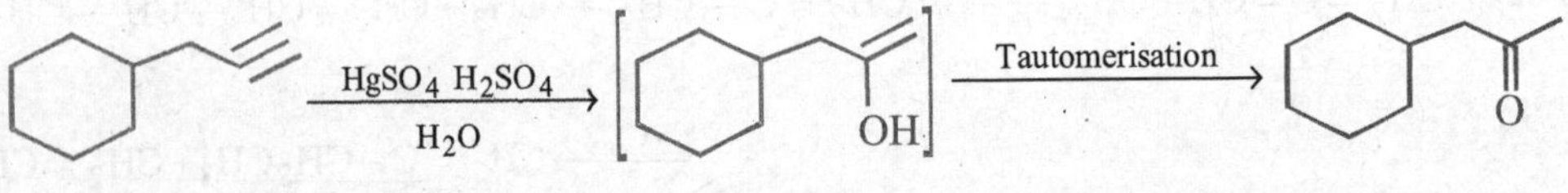 $\xrightarrow[H_2O]{HgSO_4\ H_2SO_4}$ [OH] $\xrightarrow{\text{Tautomerisation}}$ O

16. (d) $H_3C-C\equiv C-\boxed{H + CH_3CH_2}\ MgCl \longrightarrow H_3C-C\equiv C-MgCl + CH_3CH_3$

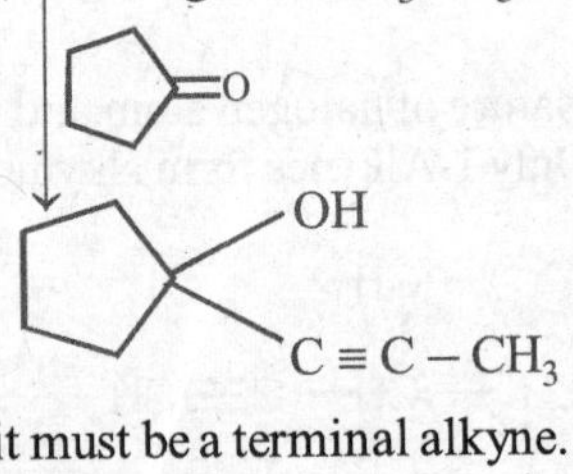

$\downarrow$ (cyclopentanone, =O)

OH, $C\equiv C-CH_3$

17. (c) Compound X reacts with ammonical $AgNO_3$ solution, so it must be a terminal alkyne.

Formation of $(CH_3)_2CH\ COOH$ on oxidation of X with hot alkaline $KMnO_4$ further confirms that X is $(CH_3)_2\ CHC\equiv CH$.

18. (a) 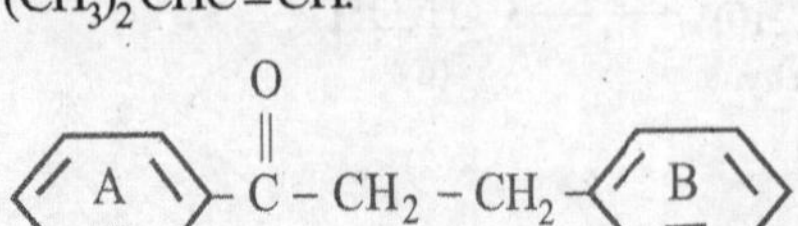 (A) $-\overset{O}{\overset{\|}{C}}-CH_2-CH_2-$ (B)

The ring B has o/p directing and (A) has m-directing group. Hence the product will be

$C_6H_5-\overset{O}{\overset{||}{C}}-CH_2-CH_2-C_6H_4-Br$

19. (d) (o-xylene resonance structures) (o-xylene) which on ozonolysis give a, b and c.

20. (b) Formation of carbocation is r.d.s.

(stability order): 7 R.S. > 8 R.S. > Vinylic (least stable)

R.S. = Resonating Structure

21. (b) Step 1 involves the conversion of sp^2 carbon to sp^3 carbon (carbon having electrophile E) while step 2 involves the conversion of sp^3 carbon to sp^2 carbon.

22. (b)

$$C_6H_5CH_3 \xrightarrow[h\nu]{3Cl_2} C_6H_5CCl_3 \xrightarrow{HOH} C_6H_5COOH \xrightarrow[-H_2O]{NaOH} C_6H_5COONa$$

23. (d) $C_6H_5-\overset{O}{\overset{||}{C}}-O-C_6H_5$ ⟵ because of steric factor para is major.

(– M) (+ M)

24. (d) The order of substitution in different alkanes is

$3° > 2° > 1°$

Thus the bromination of 2-methyl butane mainly gives 2-Bromo - 2 - methyl butane

$$\underset{\text{2-methyl butane}}{CH_3-CH_2-\overset{CH_3}{\overset{|}{C}H}-CH_3} \xrightarrow{Br_2} \underset{\substack{\text{1-Bromo-2methyl butane}\\ \text{(minor)}}}{CH_3-CH_2-\overset{CH_3}{\overset{|}{C}H}-CH_2Br} + \underset{\substack{\text{2-Bromo-2-methyl butane}\\ \text{(major)}}}{CH_3-CH_2-\overset{CH_3}{\overset{|}{\underset{Br}{\underset{|}{C}}}}-CH_3}$$

25. (c) $CH_3CH=CH_2 \xrightarrow[H_2O_2,NaOH]{B_2H_6} CH_3CH_2CH_2OH$

26. (c)

$$C_6H_5-C\equiv C-CH_3 \xrightarrow{H^+/Hg^{2+}} C_6H_5-C(OH)=\overset{H}{\overset{|}{C}}-CH_3 \xrightarrow{\text{tautomerisation}} C_6H_5-C(=O)-CH_2CH_3$$

27. (d)

$$C_6H_5CH_2CH_3 + \text{N-bromosuccinamide} \longrightarrow C_6H_5-\overset{Br}{\overset{|}{C}H}-CH_3$$

N – bromosuccinamide

28. (d) The amount of alcohol absorbed depends upon the surface area of Al_2O_3

29. (a) The heat of hydrogenation of an alkene depends upon its stability. Higher the stability, lower the heat of hydrogenation. Since $CH_2=CH_2$ has no substituent, it is the least stable alkene and hence has the highest heat of hydrogenation

30. (a) Addition reaction occurs on a double bond.
The compound containing double bonds are also undergo polymerisation. So, the correct option are both (i) and (iv) e.g.

$$CH_2 = CH_2 + HOCl \longrightarrow CH_2OH - CH_2Cl$$
(addition reaction)

$$nCH_2 = CH_2 \xrightarrow{\text{Polymerisation}} \left[CH_2 - CH_2\right]_n$$
Polyethene

31. (c) The reaction follows Markownikoff rule

$$CH_3CH_2-\underset{CH_3}{C}=\underset{H}{C}-CH_3 + \overset{\delta-\ \delta+}{HO\ Cl} \longrightarrow CH_3CH_2-\overset{CH_3}{\underset{OH}{C}}-\overset{H}{\underset{Cl}{C}}-CH_3$$

3-Methyl-2-pentene

32. (b) Hydroboration oxidation reaction.

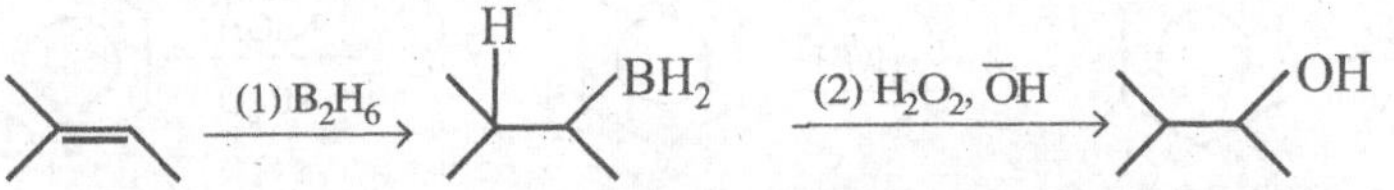

$$\xrightarrow{(1)\ B_2H_6} \quad \xrightarrow{(2)\ H_2O_2,\ \bar{O}H}$$

33. (c) 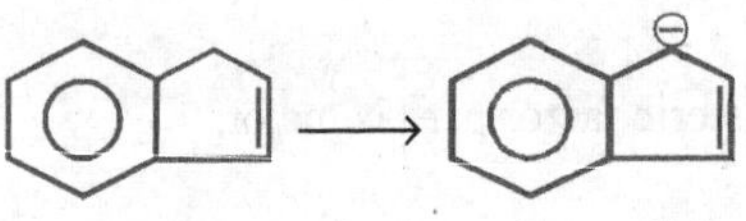 $\longrightarrow$ (Aromatic)

34. (d) very stable carbocation.

35. (a) $C_6H_5CH_3 \xrightarrow[(CH_3CO)_2O]{CrO_3} C_6H_5CH(OCOCH_3)_2 \xrightarrow[-(CH_3CO)_2O]{H^+} C_6H_5CHO$

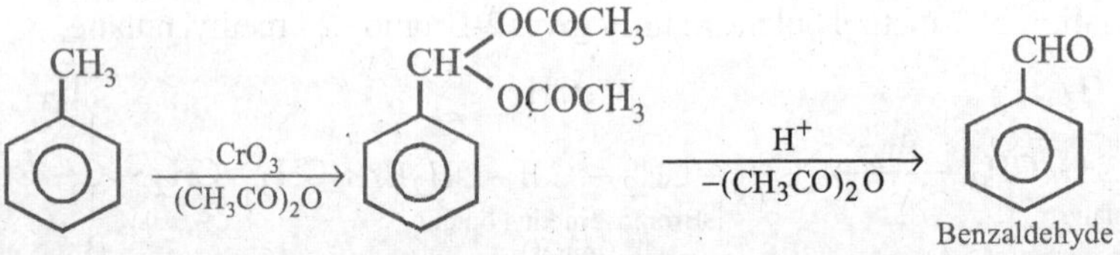

Benzaldehyde

36. (d) The reaction is hydroboration which is addition of H_2O to alkene (anti Makownikoff's's rule).

37. (a) In $H_2C = C = CH_2$ hybridisation of C atoms is sp^2 and sp. The molecule is monoplanar but H–atoms lie in different planes. See allene structure.

38. (c) $C_6H_9-C\equiv C-H \xrightarrow{\overset{+}{Na}\overset{-}{N}H_2} C_6H_9-C\equiv \overset{-}{C}\overset{+}{Na} \xrightarrow{CH_3-Br} C_6H_9-C\equiv C-CH_3 \xrightarrow{H_2/Pd/BaSO_4} C_6H_9(H)C=C(H)CH_3$

cis

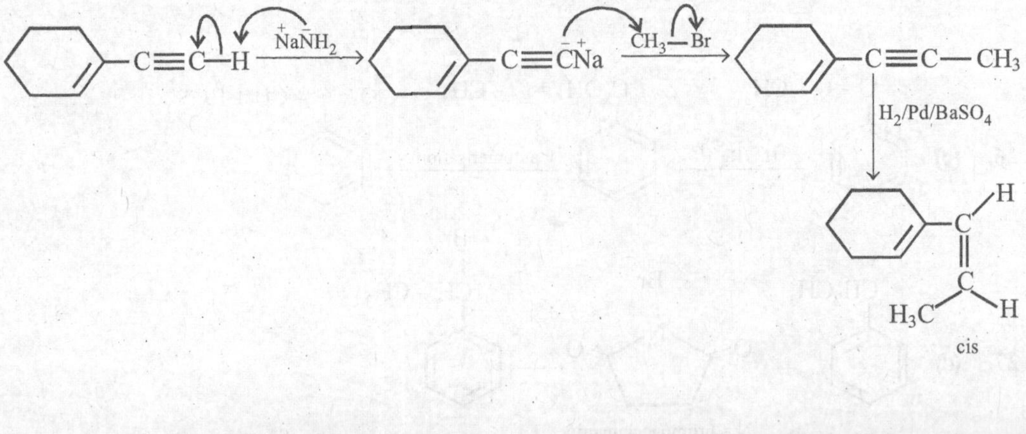

39. (b) Bromine is more selective

∴ it will form 3° halide if there is presence of 3° hydrogen.

∴ Q, S, T, form 3° halide as major and P, R, U form 2° halide as major.

40. (c) Ethene is obtained by electrolysis of dipotassium succinate as follows

$$\begin{matrix} CH_2COOK \\ | \\ CH_2COOK \end{matrix} \xrightarrow{\text{ionization}} \begin{matrix} CH_2COO^- \\ | \\ CH_2COO^- \end{matrix} + 2K^+$$

Pot. Succinate

$$2H_2O \xrightleftharpoons{\text{ionization}} 2OH^- + 2H^+$$

At anode :

$$\begin{matrix} CH_2COO^- \\ | \\ CH_2COO^- \end{matrix} \xrightarrow{-2e^-} \left[\begin{matrix} CH_2COO \\ | \\ CH_2COO \end{matrix}\right] \longrightarrow \begin{matrix} CH_2 \\ \| \\ CH_2 \end{matrix} + 2CO_2$$

Unstable

At cathode :

$$2H^+ + 2e^- \longrightarrow [2H] \longrightarrow H_2$$

41. (a) Anti-Markownikoff addition is possible only in case of HBr and not in HCl and HI. In HBr both the chain initiation and propagation steps are exothermic, while in HCl, first step is exothermic, and second step is endothermic and in HI, no step is exothermic. Hence HCl and HI do not undergo anti-Markownikoff's addition.

42. (a)

$$\begin{matrix} CH_3 \\ | \\ CH \\ \| \\ CH_2 \end{matrix} \xrightarrow[500\,°C]{Cl_2} \begin{matrix} CH_2Cl \\ | \\ CH \\ \| \\ CH_2 \end{matrix} \xrightarrow{aqNa_2CO_3} \begin{matrix} CH_2OH \\ | \\ CH \\ \| \\ CH_2 \end{matrix}$$

Propene Allyl chloride

43. (b) n-Butane, $\overset{1}{CH_3}\ \overset{2}{CH_2}\ \overset{2}{CH_2}\ \overset{1}{CH_3}$ has two different hydrogen atoms, marked by1 and 2.

44. (a) More is the stability, lower is the heat of combustion (See Baeyer's strain theory). Cyclohexane is more stable.

45. (b) Chiral conformation will not have plane of symmetry. Since twist boat does not have plane of symmetry it is chiral.

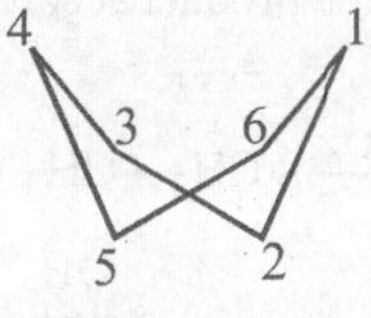

4 1 3 6 5 2

46. (d) $\xrightarrow{O_3}$ + =O

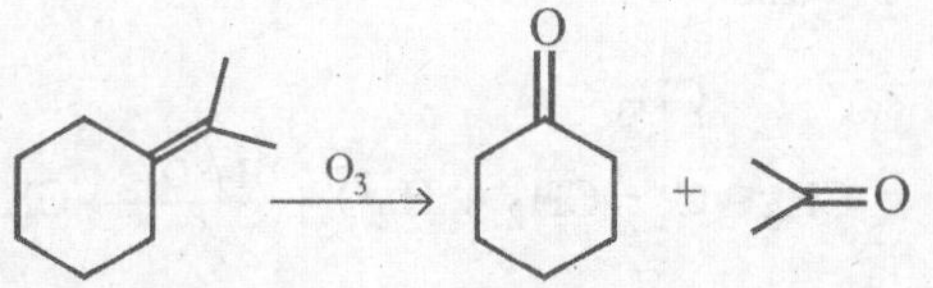

47. (c) Oxymercuration - demercuration reaction addition of H_2O molecule is according to Markovnikoff's rule.

48. **(c)** $CH_2 = CH_2 + HOCl \longrightarrow \underset{CH_2Cl}{\overset{CH_2OH}{|}} \xrightarrow[R]{NaHCO_3} \underset{\underset{(M)}{CH_2OH}}{\overset{CH_2OH}{|}}$

49. **(b, c)** 1-methylcyclopentene $\xrightarrow{D-Cl}$ carbocation (CH_3, +, D)

Formation of above species is more favourable. The stability order of carbocations is

$$-\overset{|}{C}+ < -\overset{|}{C}+ < -\overset{|}{\underset{|}{C}}+$$
$$1° \qquad 2° \qquad 3°$$

Now, carbocation formed is sp^2 hybridised that is triangular planar as shown below

H, H, H, H, H, H, Cl^-, +, CH_3, D, Cl^-

Either the $Cl^{\ominus}$ attack from above or below formation of the plane of the molecule, so there is probability of resultant product as shown below i.e., both product will form.

H, Cl, CH_3, D + H, CH_3, Cl, D

≡

D, CH_3, Cl, H

50. **(c)** With Lithium in liquid ammonia, trans-alkene is almost an exclusive product.

$$Ph-C\equiv C-Ph \xrightarrow[\text{Birch reduction}]{\text{Li in liq. } NH_3} \underset{H}{\overset{Ph}{>}}C=C\underset{Ph}{\overset{H}{<}}$$

51. **(a)** All are examples of electrophilic addition reactions and involve the formation of stable carbocation leading to the formation of addition product according to Markovnikov's rule

52. **(c)** Anti Markovnikov's addition

53. **(c)** $CH_3-CH=CH_2 + H_2O \xrightarrow{H_2SO_4} CH_3-\underset{\underset{\text{2° alcohol}}{OH}}{\underset{|}{CH}}-CH_3$

$$CH_3-\overset{\overset{CH_3}{|}}{C}=CH_2 + H_2O \xrightarrow{H_2SO_4} CH_3-\overset{\overset{CH_3}{|}}{\underset{\underset{\text{3° alcohol}}{OH}}{\underset{|}{C}}}-CH_3$$

Addition follows Markovnikov's rule.

54. **(d)** Ammonical cuprous chloride gives precipitate (red) only with terminal alkynes, i.e. alkynes having acidic hydrogen atom, such as butyne-1 ($CH \equiv CCH_2CH_3$)

55. **(a)** Conversion of C_4H_6 to $C_4H_6Br_2$ indicates that the compound is either butyne-1 or butyne-2. However, white precipitate with ammonical silver nitrate solution indicates that it is a terminal alkyne, i.e. butyne-1 and not butyne-2

56. **(b)** Hyperconjugation $\rightarrow$ more α H, more reactive o, p site.

57. **(a)**

$$C_6H_5-CH_2-CH=CH_2 \xrightarrow[\text{(ii) NaBH}_4]{\text{(i) Hg(OAC)}_2} C_6H_5-CH_2-\underset{\underset{OH}{|}}{CH}-CH_3$$

58. **(b)** Poly-ene. e.g

$$\underset{\text{acetylene}}{3\,CH \equiv CH} \longrightarrow \underset{\text{Benzene}}{C_6H_6} \qquad \underset{\text{Propyne}}{3CH_3-C\equiv CH} \longrightarrow \underset{\text{1, 3, 5-Trimethyl benzene}}{C_6H_3(CH_3)_3}$$

59. **(a)**

$$\underset{\text{Propyne}}{CH_3-C\equiv CH} + H_2O \xrightarrow[HgSO_4]{H_2SO_4} \left[\underset{\substack{\text{2-Propene}\\\text{(unstable)}}}{CH_3-\overset{\overset{OH}{|}}{C}=CH_2}\right] \longleftrightarrow \underset{\substack{\text{2-Propanone}\\\text{(acetone)}}}{CH_3-\overset{\overset{O}{\|}}{C}-CH_3}$$

60. **(c)**

$$C_6H_5-CH_2-CH=CH_2 \xrightarrow{H^+} C_6H_5-\underset{H}{CH}-\overset{+}{CH}-CH_3$$

3° allylic carbocation

$$\downarrow$$

$$C_6H_5-\underset{Cl}{CH}-CH_2-CH_3 \xleftarrow{Cl^+} C_6H_5-\overset{+}{CH}-CH_2-CH_3$$

(Since tertiary carbocation is more stable)

61. **(c)** X= 1-bromo-1-methylcyclopentane Y= 1-methylcyclopentene Z= 1-bromo-2-methylcyclopentane

62. (c) Bicyclopentylidene $\xrightarrow[Zn]{O_3}$ cyclopentanone $\xrightarrow{LiAlH_4}$ cyclopentanol $\xrightarrow[\Delta, H_2O]{H^+}$ cyclopentene

63. (a) Peracids give epoxides (alkene oxide) with alkene. In (b), (c), and (d) the products are wrong.

64. (c) Both (a) and (b) are correct. Wet CH_3COOH gives cis addition and dry CH_3COOH gives trans addition products.

65. (c) Peracids gives trans hydroxylated products with alkenes.

66. (c) $\underset{\text{(an aldehyde)}}{CH_3CH_2CH_2CHO} \xleftarrow[\text{(ii)} H_2O_2/OH^-]{\text{(i)} BH_3} CH_3CH_2C \equiv CH \xrightarrow{H^+/Hg^{2+}} \underset{\text{(A ketone with } -COCH_3 \text{ group)}}{CH_3CH_2COCH_3}$

67. (b) $-NH-\overset{O}{\overset{||}{C}}-CH_3$ (+ M group) (More activating group)

68. (a) $-CH_3 > -CD_3 > -CT_3$

Order of hyperconjugation on the basis of bond energy. Strong bond means less hyperconjugation.

69. (c) An aromatic compound have cyclic clouds of delocalised (4n + 2) π electrons above and below the plane of the molecule. Among the given three compounds, only compound (c) satisfies these conditions.

Carbon bearing $-NO_2$ (or $-CH_3$) group is sp^3 hybridised and does not have a p orbital with the result delocalisation of π electrons over the complete ring is interrupted.

4π electrons and 2 non-bonding electrons (present in p orbital of N) form a cyclic cloud of 6π electrons

70. (b) It is an example of intramolecular Wurtz reaction

$Na \longrightarrow Na^+ + e^-$

14 Environmental Chemistry

1. Which of the following statement is false?
   (a) London smog is oxidising in nature
   (b) Photochemical smog causes irritation in eyes
   (c) London smog is a mixture of smoke and fog
   (d) Photochemical smog results in the formation of PAN
2. Which of the following acts as a sink for CO?
   (a) Plants
   (b) Haemoglobin
   (c) Microorganisms present in the soil
   (d) Oceans
3. Global warming is due to increase of:
   (a) methane and nitrous oxide in atmosphere.
   (b) methane and CO_2 in atmosphere.
   (c) methane and O_3 in atmosphere.
   (d) methane and CO in atmosphere.
4. Which of the following statements about the depletion of ozone layer is correct?

Critical Thinking

   (a) The problem of ozone depletion is less serious at poles because NO_2 solidifies and is not available for consuming $ClO^•$ radicals.
   (b) The problem of ozone depletion is more serious at poles because ice crystals in the clouds over poles act as catalyst for photochemical reactions involving the decomposition of ozone of $Cl^•$ and $ClO^•$ radicals.
   (c) Freons, chlorofluorocarbons, are inert. Chemically, they do not react with ozone in stratosphere.
   (d) Oxides of nitrogen also do not react with ozone in stratosphere.
5. Excess nitrate in drinking water can cause
   (a) methemoglobinemia
   (b) kidney damage
   (c) liver damage
   (d) laxative effect
6. Soil is polluted by
   I. pesticides
   II. synthetic fertilizers
   III. green manure
   Choose the correct option.
   (a) I and III (b) I and II
   (c) II and III (d) I, II and III
7. Identify the pollutant gases largely responsible for the discoloured and lustreless nature of marble of the Taj Mahal?
   (a) O_3 and CO_2 (b) CO_2 and NO2
   (c) SO_2 and NO_2 (d) SO_2 and O_3
8. Which of the following is/are the hazardous pollutant(s) present in automobile exhaust gases?
   (i) N_2 (ii) CO
   (iii) CH_4 (iv) Oxides of nitrogen
   (a) (ii) and (iii) (b) (i) and (ii)
   (c) (ii) and (iv) (d) (i) and (iii)
9. Which of the following is the major cause of global warming?
   (a) re-radiation of U.V. rays by CO_2 and H_2O.
   (b) re-radiation of I.R. rays by CO_2 and H_2O.
   (c) re-radiation of I.R. rays by O_2 and N_2.
   (d) re-radiation of U.V. rays by O_2 and N_2.
10. The false statement among the followings :
   (a) The average residence time of NO is one month.
   (b) Limestone acts as a sink for SO_x.
   (c) SO_x can be removed from flue gases by passing through a solution of citrate ions.
   (d) Ammonia acts as a sink for NO_x.
11. Phosphate fertilizers when added to water leads to
   (a) increased growth of decomposers.
   (b) reduced algal growth.
   (c) increased algal growth.
   (d) nutrient enrichment (eutrophication).

12. Phosphate pollution is caused by
(a) sewage and agricultural fertilizers
(b) weathering of phosphate rocks only
(c) agricultural fertilizers only
(d) phosphate rocks and sewage

13. When rain is accompanied by a thunderstorm, the collected rain water will have a pH value

Critical Thinking

(a) slightly lower than that of rain water without thunderstorm.
(b) slightly higher than that when the thunderstorm is not there.
(c) uninfluenced by occurrence of thunderstorm.
(d) which depends upon the amount of dust in air.

14. Minamata disease of Japan is due to pollution of
(a) Aresenic (b) Lead
(c) Cyanide (d) Mercury

15. Which causes death of fishes in water bodies polluted by sewage?

Tricky

(a) Foul smell
(b) Pathogens
(c) Clogging of gills by silt
(d) Decrease in D.O.

16. Photochemical smog consists of excessive amount of X, in addition to aldehydes, ketones, peroxyacetyl nitrate (PAN), and so forth X is :
(a) CO (b) CH_4 (c) O_3 (d) CO_2

17. Water sample is reported to be highly polluted if BOD (Biological Oxygen Demand) value of sample becomes
(a) more than 17 ppm. (b) equal to 10 ppm.
(c) equal to 5 ppm. (d) less than 5 ppm.

18. The region which is greatly affected by air pollution is
(a) thermosphere (b) stratosphere
(c) troposphere (d) mesosphere

19. Which of the following is most abundant hydrocarbon pollutant?
(a) Butane (b) Ethane
(c) Methane (d) Propane

20. The aromatic compounds present as particulates are
(a) polycyclic aromatic hydrocarbons
(b) benzene
(c) toluene
(d) nitrobenzene

21. The substance having the largest concentration in acid rain?
(a) H_2CO_3 (b) HNO_3
(c) HCl (d) H_2SO_4

22. BOD is connected with
(a) microbes and organic matter.
(b) organic matter.
(c) microbes.
(d) None of the above

23. What is the concentration of dissolved oxygen in cold water?
(a) 5 ppm (b) 10 ppm
(c) 200, 000 ppm (d) 100 ppm

24. Green chemistry means such reactions which :
(a) Produce colour during reactions.
(b) Reduce the use and production of hazardous chemicals.
(c) Are related to the depletion of ozone layer.
(d) Study the reactions in plants.

25. Use of which of the following solvent in dry cleaning will result in less harm to ground water?
(a) $Cl_2C=CCl_2$ (b) Liquid CO_2
(c) H_2O_2 (d) None of these

26. The statement which is not true
(a) NO_2 does not play any role in photochemical smog
(b) SO_3 is more harmful air pollutant than SO_2
(c) SO_2 dos not affect larynx (voice box)
(d) NO is more toxic to living tissues than NO_2

27. Synthesis of ethanal commercially from which of the following reagent is the part of green chemistry?

Toughnut

(a) CH_3CH_2OH (b) $CH_2=CH_2$
(c) $HC\equiv CH$ (d) All of these

28. Green house gases can be arranged in 'Global Warming Potential' sequence as
(a) $N_2O > CFC > CH_4 > CO_2$
(b) $CFC > N_2O > CH_4 > CO_2$
(c) $CFC > CO_2 > N_2O > CH_4$
(d) $CO_2 > CFC > N_2O > CH_4$

29. Addition of phosphate fertilisers to water bodies causes:
(a) Increase in amount of dissolved oxygen in water
(b) Deposition of calcium phosphate
(c) Increase in fish population
(d) Enhanced growth of algae

30. Which forms the part of hazy fumes of photochemical smog?
(a) SO_2 (b) Nitrogen dioxide
(c) PAN formation (d) Aldehydes

31. Which one of the following depletes ozone layer?
(a) CO (b) NO and freons
(c) SO_2 (d) CO_2

32. Which of the following chemical, harmful to ozone, is released by chlorofluorocarbon?
(a) Sulphur dioxide (b) Fluorine
(c) Chlorine (d) Nitrogen dioxide

33. In Antarctica ozone depletion is due to the formation of following compound
(a) acrolein
(b) peroxyacetyl nitrate
(c) SO_2 and SO_3
(d) chlorine nitrate

34. Ozone is an important constituent of stratosphere because it
(a) destroys bacteria which are harmful to human life.
(b) prevents the formation of smog over large cities.
(c) removes poisonous gases of the atmosphere by reacting with them.
(d) absorbs ultraviolet radiation which is harmful to human life.

35. The statement which is not correct about control of particulate pollution: Toughnut
(a) In electrostatic precipitator, the particulates are made to acquire positive charge which are then attracted by the negative electrode and removed.
(b) Gravity settling chamber removes larger particles from the air.
(c) Cyclone collector removes fine particles in the diameter range 5-20 microns.
(d) Wet scrubbers are used to wash away all types of particulates.

36. Green house gases Critical Thinking
(a) allow shorter wavelength to enter earth's atmosphere while doesn't allow longer wavelength to leave the earth's atmosphere.
(b) allow longer wavelength to enter earth atmosphere while doesn't allow shorter wavelength to leave the surface
(c) do not have wavelength specific character.
(d) show wavelength specific behaviour near the earth while far from earth these have wavelength independent behaviour.

37. The greenhouse effect is because of the
(a) presence of gases, which in general are strong infrared absorbers, in the atmosphere
(b) presence of CO_2 only in the atmosphere
(c) pressure of O_3 and CH_4 in the atmosphere
(d) N_2O and chlorofluorohydrocarbons in the atmosphere

38. Carbon monoxide (CO) is harmful to man because
(a) it forms carbolic acid.
(b) it generates excess CO_2.
(c) it is carcinogenic.
(d) it competes with O_2 for haemoglobin.

39. Which of the following can control the photochemical smog?
(A) Use of catalytic converters in automobiles.
(B) Plantation of trees like pinus, pyrus vitis etc.
(C) Using less sulphur containing fossil fuels.
(a) A and C (b) B
(c) A and B (d) A, B and C

40. High concentration of which of the following in atmosphere leads to stiffness of flower buds which eventually fall off from plants?
(a) NO_2 (b) SO_2 (c) CFC (d) Smog

41. Identify the wrong statement in the following:

(a) Chlorofluorocarbons are responsible for ozone layer depletion.
(b) Greenhouse effect is responsible for global warming.
(c) Acid rain is mostly because of oxides of nitrogen and sulphur.
(d) Ozone layer does not permit infrared radiation from the sun to reach the earth.

42. Ozone hole refers to
(a) increase in concentration of ozone.
(b) hole in ozone layer.
(c) reduction in thickness of ozone layer in troposphere.
(d) reduction in thickness of ozone layer in stratsophere.

43. Which one of the following statements is not true? **Tricky**
(a) Dissolved oxygen (DO) in cold water can reach a concentration upto 10 ppm.
(b) Clean water would have a BOD value of 5 ppm.
(c) Fluoride deficiency in drinking water is harmful. Soluble fluoride is often used to bring its concentration upto 1 ppm.
(d) When the pH of rain water is higher than 6.5, it is called acid rain.

44. The smog is essentially caused by the presence of
(a) oxides of sulphur and nitrogen
(b) O_2 and N_2
(c) O_2 and O_3
(d) O_3 and N_2

45. Presence of which of the following fuel gas in the exhaust fumes shows incomplete combustion of fuel?
(a) Sulphur dioxide
(b) Carbon monoxide and water vapour
(c) Carbon monoxide
(d) Nitrogen dioxide

46. Which of the following statements about polar stratosphere clouds (PSCs) is not correct?

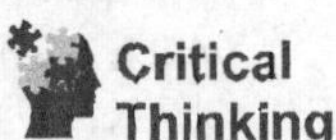

(a) PSCs do not react with chlorine nitrate and HCl.
(b) Type I clouds are formed at about –77ºC and contain solid $HNO_3 . 3H_2O$.
(c) Type II clouds are formed at about –85ºC and contain some ice.
(d) A tight whirlpool of wind called Polar Vortex is formed which surrounds Antarctica.

47. The irritant red haze in the traffic and congested places is due to presence of which of the following?
(i) Oxides of sulphur
(ii) Oxides of nitrogen
(iii) Carbon dioxide
(iv) Mists, smoke and dust
(v) Smog
(a) (i), (iv) and (v) (b) (iii) only
(c) (ii) only (d) (ii) and (v)

48. Which one of the following substances used in dry cleaning is a better strategy to control environmental pollution?
(a) Sulphur dioxide
(b) Carbon dioxide
(c) Nitrogen dioxide
(d) Tetrachloroethylene

49. In a coal fired power plant, electrostatic precipitators are installed to control emission of
(a) SO_2 (b) NO_x (c) SPM (d) CO

50. Which among the following statements is *false*?
(a) Oil slick in sea water increases D.O. value
(b) The main reason for river water pollution is industrial and domestic sewage discharge
(c) Surface water contains a lot of organic matter mineral nutrients and radioactive materials
(d) Oil spill in sea water causes heavy damage to fishery

51. Lichens do not like to grow in cities
(a) because of absence of the right type of algae and fungi
(b) because of lack of moisture
(c) because of SO_2 pollution
(d) because natural habitat is missing

52. Which of the following statement(s) is/are correct?
(i) Classical smog is a mixture of smoke, fog and sulphur dioxide.
(ii) Classical smog is also called oxidising smog
(iii) Hydrocarbons, NO_2 and PAN are components of photochemical smog.
(a) (i) and (iii) (b) (i) and (ii)
(c) (iii) only (d) (i), (ii) and (iii)

53. Which one of the following pairs is mismatched
(a) Fossil fuel burning - release of CO_2
(b) Nuclear power - radioactive wastes
(c) Solar energy - Greenhouse effect
(d) Biomass burning - release of CO_2

54. In almost all Indian metropolitan cities like Delhi, the major atmospheric pollutant(s) is/are
(a) suspended particulate matter (SPM)
(b) oxides of sulphur
(c) carbon dioxide and carbon monoxide
(d) oxides of nitrogen

55. Which of the following statements are not correct? **Critical Thinking**

(i) F^- ion concentration above 2 ppm causes brown mottling in teeth.
(ii) Excessive F^- (over 10 ppm) causes harmful effect to bones and teeth.
(iii) Excessive lead in drinking water causes disease methemoglobinemia
(iv) Deficiency of sulphate in drinking water causes laxative effect.

(a) (ii) and (iv) (b) (ii) and (iii)
(c) (ii), (iii) and (iv) (d) (iii) and (iv)

56. Which of the following statements is incorrect?
(a) Smoke particulates consist of solid or mixture of solid and liquid particles formed during combustion of organic matter.
(b) Herbicides and insecticides that miss their target and travel through air form mists.
(c) Organic solvents, metals and metallic oxides form fume particles
(d) None of these

57. Which of the following practices will not come under green chemistry?
(a) If possible, making use of soap made of vegetable oils instead of using synthetic detergents
(b) Using H_2O_2 for bleaching purpose instead of using chlorine based bleaching agents
(c) Using bicycle for travelling small distances instead of using petrol/diesel based vehicles
(d) Using plastic cans for storing substances

58. Which of the following statement(s) is/are true about waste recycling? **Toughnut**

(i) Clothes can be made from recycled plastic waste.
(ii) Fuel that has high octane rating and contains no lead can be obtained from plastic waste.
(iii) Technology has now been developed to produce electricity from the garbage.

(a) (ii) only (b) (ii) and (iii)
(c) (iii) only (d) All of these

59. Which of the following is/are formed when ozone reacts with the unburnt hydrocarbons in polluted air? **Critical Thinking**

(i) Formaldehyde
(ii) Acrolein
(iii) Peroxyacetyl nitrate
(iv) Formic acid

(a) (i) and (iv) (b) (ii) only
(c) (iii) only (d) (i), (ii) and (iii)

60. Which one of the following statement is **not** true?
(a) pH of drinking water should be between 5.5–9.5.
(b) Concentration of DO below 6 ppm is good for the growth of fish.
(c) Clean water would have a BOD value of less than 5 ppm.
(d) Oxides of sulphur, nitrogen and carbon are the most widespread air pollutant.

Answer KEYs

1	(a)	7	(c)	13	(a)	19	(c)	25	(b)	31	(b)	37	(a)	43	(d)	49	(c)	55	(d)
2	(c)	8	(c)	14	(d)	20	(a)	26	(b)	32	(c)	38	(d)	44	(a)	50	(a)	56	(d)
3	(b)	9	(b)	15	(d)	21	(d)	27	(b)	33	(a)	39	(c)	45	(c)	51	(c)	57	(d)
4	(c)	10	(a)	16	(c)	22	(a)	28	(b)	34	(d)	40	(b)	46	(a)	52	(a)	58	(d)
5	(a)	11	(d)	17	(a)	23	(b)	29	(d)	35	(a)	41	(d)	47	(c)	53	(c)	59	(d)
6	(b)	12	(a)	18	(c)	24	(b)	30	(b)	36	(a)	42	(d)	48	(d)	54	(a)	60	(b)

Hints & Solutions

1. (a) London smog is reducing in nature.
2. (c) CO is converted into CO_2 by microorganism present in soil.
3. (b) The main gas responsible for the global warming is carbon dioxide. Other contributors include methane released from landfills and agriculture.
4. (c) Freons and chlorofluorocarbon find their way to stratosphere through without being destroyed in the troposphere due to their low reactivity. In the stratosphere the Cl and Br atoms are liberated from the parent compounds by the action of ultraviolet light e.g.

$$CF_2Cl_2 + h\nu \longrightarrow CF_2Cl + Cl$$

$$CFCl_3 + h\nu \longrightarrow CFCl_2 + Cl$$

The Cl atoms can destroy ozone molecules through a variety of catalytic cycles.

5. (a) Excessive concentration of nitrate in drinking water is harmful and can cause methemoglobinemia (blue baby syndrome).
6. (b) Pesticides and synthetic fertilizers pollute the soil.
7. (c) SO_2 and NO_2 are responsible for the discoloured and lustureless nature of marble.
8. (c) CO and oxides of Nitrogen are poisnous gases present in automobile exhaust gases.
9. (b)
10. (a) The average residence time of NO is 4 days.
11. (d) Addition of phosphate fertilizers to water leads to nutrient enrichment (eutrophication)
12. (a) Phosphate pollution is caused by sewage and agricultural fertilizers.
13. (a) Normal rain water has pH 5.6. Thunderstorm results in the formation of NO and HNO_3 which lowers the pH.
14. (d) Minamata is caused by Hg poisoning.
15. (d) Decrease in D.O causes death of fish
16. (c) Chemical pollutants in photochemical smog are nitrogen oxides (NO and NO_2), volatile organic compounds, ozone (O_3) and peroxyacetyl nitrate.
17. (a) Water sample is reported to be highly polluted if BOD value of sample is more than 17 ppm.
18. (c) Air pollution greatly affect the troposphere.
19. (c) Most abundant hydrocarbon pollutant is methane.
20. (a) PAH (Poly Aromatic Hydrocarbon)
21. (d) Acid rain contains $H_2SO_4 > HNO_3 > HCl$.
22. (a) BOD is connected with microbes and organic matter
23. (b) In cold water, dissolved oxygen can reach a concentration upto 10 ppm, whereas oxygen in air is about 200, 000 ppm.
24. (b) Green chemistry may be defined as the programme of developing new chemical products and chemical processes or making improvements in the already existing compounds and processes so as to make less harmful to human health and environment. This means the same as to reduce the use and production of hazardous chemicals.

i.e. correct answer is option (b).

25. (b) Replacement of earlier used tetra-chloroethene as solvent for dry cleaning by liquid CO_2 results in less harm to ground water.
26. (b) SO_3 is more harmful pollutant than SO_2.

27. (b) Ethanal is commerically prepared by one step oxidation of ethene in the presence of ionic catalyst in aqueous medium with an yield of 90%.

$$CH_2 = CH_2 + O_2 \xrightarrow[\text{Pb(II)|Cu(II) (in water)}]{\text{Catalyst}} CH_3CHO$$

28. (b) $CFC > N_2O > CH_4 > CO_2$
29. (d)
30. (b) NO_2 forms a part of photochemical smog.
31. (b) NO and freons are responsible for ozone layer depletion.
32. (c) $CF_2Cl_2 \xrightarrow{hv} CF_2Cl + \dot{C}l$

$$\dot{C}l + O_3 \longrightarrow Cl\dot{O} + O_2$$

33. (a) In antarctica, ozone depletion is due to formation of acrolein.
34. (d) Ozone absorbs U.V. radiations harmful to human life.
35. (a) Particulates acquire negative charge and are attracted by the positive electrode.
36. (a) Radiation coming from sun or outerspace have high energy or short wavelength, which are allowed to enter by green house gases. However, radiation emitted by earth is in infrared region, having long wavelength, are reflected back by the envelope of green house gases.
37. (a) Green house gases such as CO_2, ozone, methane, the chlorofluorocarbon compounds and water vapour form a thick cover around the earth which prevents the IR rays emitted by the earth to escape. It gradually leads to increase in temperature of atmosphere.
38. (d) CO is highly toxic and impairs respiration. CO combines with haemoglobin of blood and reduces its O_2 carrying capacity.
39. (c) Usually catalytic converters are used in the automobiles, which prevent the release of nitrogen oxide and hydrocarbons to the atmosphere. Certain plants e.g., Pinus, Juniparus. Quercus, Pyrus and Vitis can metabolise nitrogen oxide and therefore, their plantation could help in this matter.
40. (b) High concentration of SO_2 leads to stiffness of flower buds.
41. (d) Ozone layer acts as a shield and does not allow ultraviolet radiation from sun to reach earth. It does not prevent infra-red radiation from sun to reach earth, thus option (d) is wrong statement and so it is the correct answer.
42. (d) Ozone hole is reduction in ozone layer in stratosphere.
43. (d) Acid rain is the rain water containing sulphuric acid and nitric acid which are formed from the oxides of sulphur and nitrogen present in the air as pollutants and rain water has a pH range of 4-5.
44. (a) Smog is caused by oxides of sulphur and nitrogen.
45. (c) Presence of CO in the exhaust fumes shows incomplete combustion.
46. (a) PSCs react with chlorine nitrate and HCl to give HOCl and Cl_2.
47. (c) The irritant red haze in the traffic and congested places is due to presence of oxides of nitrogen.
48. (d) Tetrachloroethylene is widely used for dry-cleaning fabrics and metal degreasing operations.
49. (c) SPM $\rightarrow$ Suspended particulates matter.
50. (a) Oil slick in sea water decreases D.O value.
51. (c) Because lichens are very sensitive to sulphur dioxide and in cities the amount of SO_2 is high, so they do not grow in cities.
52. (a) Classical smog is also called reducing smog.
53. (c) Solar energy is not responsible for green house effect instead it is a source of energy for the plants and animals.
54. (a)

55. **(d)** For statement (iii), Methemoglobinemia (blue baby syndrome) is caused due to excess of nitrate in drinking water.
For statement (iv), Excessive sulphate (> 500 ppm) in drinking water causes laxative effect, otherwise at moderate levels it is harmless.

56. **(d)** 57. **(d)**

58. **(d)** All the given statements are true for about waste recycling.

59. **(d)** $3CH_4 + 2O_3 \rightarrow 3CH_2{=}O + 3H_2O$
Formaldehyde
$CH_2{=}CHCH{=}O$; CH_3COONO_2
Acrolein Peroxyacetyl nitrate (PAN)

60. **(b)** The ideal value of DO for growth of fishes is 8 mg/L. 7 mg/L is desirable range, below this value fishes get susceptible to disease. A value of 2 mg/L or below is lethal for fishes.

15 The Solid State

1. How many unit cells are present in a cube-shaped ideal crystal of NaCl of mass 1.00 g ?
[Atomic masses : Na = 23, Cl = 35.5]
(a) 5.14×10^{21} unit cells
(b) 1.28×10^{21} unit cells
(c) 1.71×10^{21} unit cells
(d) 2.57×10^{21} unit cells

2. In a solid 'AB' having the NaCl structure, '*A*' atoms occupy the corners of the cubic unit cell. If all the face-centered atoms along one of the axes are removed, then the resultant stoichiometry of the solid is
Critical Thinking
(a) AB_2 (b) A_2B (c) A_4B_3 (d) A_3B_4

3. Edge length of a cube is 400 pm. Its body diagonal would be
(a) 500 pm (b) 566 pm
(c) 600 pm (d) 693 pm

4. The second order Bragg diffraction of X-rays with = 1.00 Å from a set of parallel planes in a metal occurs at an angle 60°. The distance between the scattering planes in the crystal is
(a) 0.575 Å (b) 1.00 Å
(c) 2.00 Å (d) 1.15 Å

5. The radii of Na^+ and Cl^- ions are 95 pm and 181 pm respectively. The edge length of NaCl unit cell is
Toughnut
(a) 276 pm (b) 138 pm
(c) 552 pm (d) 415 pm

6. CsBr has *bcc* structure with edge length 4.3. The shortest interionic distance in between Cs^+ and Br^- is
(a) 3.72 (b) 1.86 (c) 7.44 (d) 4.3

7. Potassium crystallizes with a
(a) body-centred cubic lattice
(b) face-centred cubic lattice
(c) simple cubic lattice
(d) orthorhombic lattice

8. First three nearest neighbour distances for body centred cubic lattice are respectively:
Critical Thinking
(a) $\sqrt{2}a, a, \sqrt{3}a$ (b) $\frac{a}{\sqrt{2}}, a, \sqrt{3}a$
(c) $\frac{\sqrt{3}a}{2}, a, \sqrt{2}a$ (d) $\frac{\sqrt{3}a}{2}, a, \sqrt{3}a$

9. A crystalline solid
(a) changes abruptly from solid to liquid when heated.
(b) has no definite melting point.
(c) undergoes deformation of its geometry easily.
(d) has an irregular 3-dimensional arrangements.

10. Which of the following is not a crystalline solid?
(a) KCl (b) CsCl
(c) Glass (d) Rhombic S

11. Which one of the following forms a molecular solid when solidified?
(a) Silicon carbide (b) Calcium fluoride
(c) Rock salt (d) Methane

12. The intermetallic compound LiAg crystallizes in a cubic lattice in which both lithium and silver atoms have coordination number of eight. To what crystal class does the unit cell belong?
(a) Simple cubic (b) Face-centred cubic
(c) Body-centred cubic (d) None of these

13. Which set of following characteristics for ZnS crystal is correct?
(a) Coordination number (4 : 4); *ccp*; Zn^{2+} ion in the alternate tetrahedral voids.
(b) Coordination number (6 : 6); *hcp*; Zn^{2+} ion in all tetrahedral voids.
(c) Coordination number (6 : 4); *hcp*; Zn^{2+} ion in all octahedral voids.
(d) Coordination number (4 : 4); *ccp*; Zn^{2+} ion in all tetrahedral voids.

14. The ionic radii of X^+ and Y^- ions are 146 and 216 pm respectively. The probable type of structure shown by it is
(a) CsCl type (b) Rock salt type
(c) Zinc blende type (d) CaF_2 type

15. In CsCl type structure, the co-ordination number of Cs^+ and Cl^- respectively are
(a) 6, 6 (b) 6, 8 (c) 8, 8 (d) 8, 6

16. Copper crystallises in fcc with a unit length of 361pm. What is the radius of copper atom ?
(a) 157pm (b) 128pm
(c) 108pm (d) 181pm

17. In a face centred cubic lattice, atoms of A form the corner points and atoms of B form the face centred points. If two atoms of A are missing from the corner points, the formula of the ionic compound is :
(a) AB_3 (b) AB_4 (c) A_2B_5 (d) AB_2

18. Which one of the following statements about packing in solids is incorrect ?
(a) Coordination number in bcc mode of packing is 8.
(b) Coordination number in hcp mode of packing is 12.
(c) Void space in hcp mode of packing is 32%.
(d) Void space is ccp mode of packing is 26%.

19. Among the following the incorrect statement is

Critical Thinking

(a) Density of crystals remains unaffected due to Frenkel defect.
(b) In BCC unit cell the void space is 32%.
(c) Density of crystals decreases due to Schottky defect.
(d) Electrical conductivity of semiconductors and metals increases with increase in temperature.

20. In a face centered cubic lattice atoms A are at the corner points and atoms B at the face centered points. If atom B is missing from one of the face centered points, the formula of the ionic compound is:
(a) AB_2 (b) A_5B_2 (c) A_2B_3 (d) A_2B_5

21. A cubic crystal has
(a) thirteen planes of symmetry and nine axis of symmetry.
(b) thirteen axis of symmetry and nine planes of symmetry.
(c) thirteen axis and thirteen planes of symmetry.
(d) nine planes and nine axis of symmetry.

22. The appearance of colour in solid alkali meta halides is generally due to:
(a) Schottky defect (b) Frenkel defect
(c) Interstitial position (d) F-centres

23. A solid AB crystallises as NaCl structure and the radius of the cation is 0.100nm. The maximum radius of the anion can be:
(a) 0.137 nm (b) 0.241 nm
(c) 0.274 nm (d) 0.482 nm

24. In a monoclinic unit cell, the relation of sides and angles are respectively:
(a) $a = b \neq c$ and $\alpha = \beta = \gamma = 90°$
(b) $a \neq b \neq c$ and $\alpha = \beta = \gamma = 90°$
(c) $a \neq b \neq c$ and $\beta = \gamma = 90° \neq \alpha$
(d) $a \neq b \neq c$ and $\alpha \neq \beta \neq \gamma \neq 90°$

25. A solid is made of two elements X and Z. The atoms Z are in *ccp* arrangement while the atoms X occupy all the tetrahedral sites. What is the formula of the compound?

Tricky

(a) XZ (b) XZ_2
(c) X_2Z (d) X_2Z_3

26. Among the following which is the best description of water in the solid phase?
(a) Covalent solid (b) Molecular solid
(c) Ionic solid (d) Network solid

27. CsBr crystallises in a body centered cubic lattice. The unit cell length is 436.6 pm. Given that the atomic mass of Cs = 133 and that of Br = 80 amu and Avogadro number being 6.02×10^{23} mol^{-1}, the density of CsBr is
(a) 0.425 g/cm^3 (b) 8.5 g/cm^3
(c) 4.25 g/cm^3 (d) 82.5 g/cm^3

28. Compound that will show the highest lattice energy
(a) KF (b) NaF (c) CsF (d) RbF

29. The edge length of unit cell of a metal having molecular weight 75 g/mol is 5Å which crystallizes in cubic lattice. If the density is 2g/cc then find the radius of metal atom. ($N_A = 6 \times 10^{23}$). Give the answer in pm.

Toughnut

(a) 217 pm (b) 210 pm
(c) 220 pm (d) 205 pm

30. Ammonium chloride crystallizes in a body centred cubic lattice with edge length of unit cell of 390 pm. If the size of chloride ion is 180 pm, the size of ammonium ion would be
(a) 174 pm (b) 158 pm
(c) 142 pm (d) 126 pm

31. The cubic unit cell of a metal (molar mass $= 63.55 g\ mol^{-1}$) has an edge length of 362 pm. Its density is $8.92 g\ cm^{-3}$.
The type of unit cell is
(a) primitive (b) face centered
(c) body centered (d) end centered

32. A solid has a '*bcc*' structure. If the distance of nearest approach between two atoms is 1.73Å, the edge length of the cell is
(a) 314.20 pm (b) 1.41 pm
(c) 200 pm (d) 216 pm

33. When molten zinc is converted into solid state, it acquires *hcp* structure. The number of nearest neighbours of Zn will be
(a) 6 (b) 12 (c) 8 (d) 4

34. The radius of a calcium ion is 94 pm and of the oxide ion is 146 pm. The possible crystal structure of calcium oxide will be
(a) tetrahedral (b) trigonal
(c) octahedral (d) pyramidal

35. An alloy of copper, silver and gold is found to have cubic lattice in which Cu atoms constitute *ccp*. If Ag atoms are located at the edge centres and Au atom is present at body centre, the alloy will have the formula

Tricky

(a) CuAgAu (b) Cu_4Ag_4Au
(c) Cu_4Ag_3Au (d) Cu_4Ag_6Au

36. A solid has a structure in which 'W' atoms are located at the corners of a cubic lattice 'O' atoms at the centre of edges and Na atoms at the centre of the cube. The formula for the compound is
(a) Na_2WO_3 (b) Na_2WO_2
(c) $NaWO_2$ (d) $NaWO_3$

37. If calcium crystallizes in bcc arrangement and the radius of Ca atom is 96 pm, then the volume of unit cell of Ca is
(a) $10.9 \times 10^{-36} m^3$ (b) $10.9 \times 10^{-30} m^3$
(c) $21.8 \times 10^{-30} m^3$ (d) $21.8 \times 10^{-36} m^3$

38. How many unit cells are present in 5.0 gm of crystal AB (formula mass of AB = 40) having rock salt type structure? (N_A = Avogadro's no.)
(a) N_A (b) $\frac{N_A}{10}$
(c) $4N_A$ (d) none of these

39. AB; crystallizes in a body centred cubic lattice with edge length '*a*' equal to 387 pm. The distance between two oppositely charged ions in the lattice is
(a) 335 pm (b) 250 pm
(c) 200 pm (d) 300 pm

40. The number of unit cells in 58.8 g of NaCl is nearly
(a) 6×10^{20} (b) 3×10^{22}
(c) 1.5×10^{23} (d) 0.5×10^{24}

41. An element having an atomic radius of 0.14 nm crystallizes in an *fcc* unit cell. What is the length of a side of the cell ?
(a) 0.56 nm (b) 0.24 nm
(c) 0.96 nm (d) 0.4 nm

42. Which of the following has maximum value of cation/anion ratio?

(a) KCl (b) NaCl
(c) CaF_2 (d) $MgCl_2$

43. Which of the following statements is correct for ionic solids ?
(a) Ionic solids are soluble in CCl_4, C_6H_6, etc.
(b) Under the electric field cations and anions acquire translatory motion in opposite directions.
(c) Structural units have strong electrostatic force of attraction.
(d) Structural units have dipole-dipole interactions.

44. Total volume of atoms present in a face-centred cubic unit cell of a metal is (*r* is atomic radius)
(a) $\frac{12}{3}\pi r^3$ (b) $\frac{16}{3}\pi r^3$
(c) $\frac{20}{3}\pi r^3$ (d) $\frac{24}{3}\pi r^3$

45. An element occuring in the bcc structure has 12.08×10^{23} unit cells. The total number of atoms of the element in these cells will be
(a) 24.16×10^{23} (b) 36.18×10^{23}
(c) 6.04×10^{23} (d) 12.08×10^{23}

46. In an antifluorite structure, cations occupy
(a) octahedral voids (b) centre of cube
(c) tetrahedral voids (d) corners of cube

47. The number of atoms in 100 g of an *fcc* crystal with density, $d = 10\ g/cm^3$ and cell edge equal to 100 pm, is equal to

Toughnut

(a) 1×10^{25} (b) 2×10^{25}
(c) 3×10^{25} (d) 4×10^{25}

48. Crystal defect indicated in the diagram below is

$Na^+Cl^-Na^+Cl^-Na^+Cl^-$
Cl^- □ Cl^-Na^+ □ Na^+
Na^+Cl^- □ $Cl^-Na^+Cl^-$
$Cl^-Na^+Cl^-Na^+$ □ Na^+

(a) Interstitial defect
(b) Schottky defect
(c) Frenkel defect
(d) Frenkel and Schottky defects

49. In the laboratory, sodium chloride is made by burning sodium in the atmosphere of chlorine. The salt obtained is yellow in colour. The cause of yellow colour is: **Tricky**

(a) Presence of Na^+ ions in the crystal lattice
(b) Presence of Cl^- ions in the crystal lattice
(c) Presence of electrons in the crystal lattice
(d) Presence of face centred cubic crystal lattice.

50. Each of the following solids show, the Frenkel defect except

(a) ZnS (b) AgBr (c) AgI (d) KCl

51. Schottky defect in crystals is observed when

(a) an ion leaves its normal site and occupies an interstitial site.
(b) unequal number of cations and anions are missing from the lattice.
(c) density of the crystal increases.
(d) equal number of cations and anions are missing from the lattice.

52. In the solid state, MgO has the same structure as that of sodium chloride. The number of oxygens surrounding each magnesium in MgO is

(a) 6 (b) 1 (c) 2 (d) 4

53. Which is incorrect statement ?

(a) In NaCl structure, tetrahedral voids are unoccupied.
(b) In ZnS structure, octahedral voids are unoccupied.
(c) In CaF_2 structure, all tetrahedral voids are occupied.
(d) In Na_2O structure, all tetrahedral voids are unoccupied.

54. The r_+ / r_- ratio of ZnS is 0.402. Pick out the incorrect statements from the following?

(a) ZnS is 4 : 4 coordination compound.
(b) ZnS does not crystallize in rock salt type lattice because r_+ / r_- is too small to avoid overlapping of S^{2-} ions.
(c) Zn^{2+} ion is too small to fit precisely into the octahedral voids of S^{2-} ions.
(d) Zn^{2+} ion is too large to fit into the octahedral voids of S^{2-} ions.

55. For *fcc*, if AB is just like the rock salt like structure then, A^+ and B^- are located at –

(a) A^+ - Tetrahedral voids ; B^- - Corner
(b) A^+ - Corner and faces ; B^- - Octahedral voids
(c) A^+ - Octahedral voids ; B^- - Corner and faces
(d) A^+ - Corner and faces ; B^- - Octahedral and tetrahedral voids

56. A metal has a *fcc* lattice. The edge length of the unit cell is 404 pm. The density of the metal is 2.72 g cm^{-3}. The molar mass of the metal is : (N_A Avogadro's constant = 6.02×10^{23} mol^{-1})

(a) 30 g mol^{-1} (b) 27 g mol^{-1}
(c) 20 g mol^{-1} (d) 40 g mol^{-1}

57. What is the energy gap between valence band and conduction band in crystal of insulators ?

(a) Both the bands are overlapped with each other
(b) Very small
(c) Infinite
(d) Very large

58. The addition of arsenic to germanium makes the latter a

(a) metallic conductor
(b) intrinsic semiconductor
(c) mixed conductor
(d) extrinsic semiconductor

59. A mineral having the formula AB_2 crystallizes in ccp lattice with A atoms occupying the lattice points. Pick out the correct statement of the following

(a) 100% occupancy of tetrahedral voids, C.N. of B = 4
(b) 100% occupancy of octahedral voids, C.N. of B = 4
(c) 50% occupancy of tetrahedral voids, C.N. of A = 4
(d) 100% occupancy of octahedral voids, C.N. of A = 4

60. Superconductors are derived from compounds of
(a) *p*-block elements (b) lanthanides
(c) actinides (d) transition elements

61. If NaCl is doped with 10^{-4} mol % of $SrCl_2$, the concentration of cation vacancies will be ($N_A = 6.02 \times 10^{23}\ mol^{-1}$)
(a) $6.02 \times 10^{16}\ mol^{-1}$ (b) $6.02 \times 10^{17}\ mol^{-1}$
(c) $6.02 \times 10^{14}\ mol^{-1}$ (d) $6.02 \times 10^{15}\ mol^{-1}$

62. Predict coordination number of the cation in crystals of the following compounds:

Tricky

1. MgO : $r_c = 0.65$ Å; $r_a = 1.40$ Å,
2. MgS : $r_c = 0.65$ Å; $r_a = 1.84$ Å,

(a) 6, 4 (b) 4, 6 (c) 3, 4 (d) 6, 8

63. An element (atomic mass = 100 g / mol) having *bcc* structure has unit cell edge 400 pm. Then, density of the element is
(a) 10.376 g/cm³ (b) 5.188 g/cm³
(c) 7.289 g/cm³ (d) 2.144 g/cm³⁶

64. MgO crystallizes in a cubic type crystal system. The ionic radii for Mg^{2+} and O^{2-} are 0.066 and 0.140 nm respectively. One can conclude that the Mg^{2+} ions occupy
(a) a cubic hole in a simple cubic structure.
(b) every tetrahedral hole in a close packed structure.
(c) an octahedral hole in a cubic close packed structure.
(d) every other tetrahedral hole in a close packed structure.

65. A crystal lattice with alternate positive and negative ions has radius ratio 0.524. Its co-ordination number is
(a) 4 (b) 3 (c) 6 (d) 12

66. The correct statement regarding F– centre is
(a) electrons are held in the voids of crystals.
(b) F – centre produces colour to the crystals.
(c) conductivity of the crystal increases due to F – centre.
(d) All of these

67. The number of octahedral void(s) per atom present in a cubic close-packed structure is
(a) 1 (b) 3 (c) 2 (d) 4

68. If we mix a pentavalent impurity in a crystal lattice of germanium, what type of semiconductor formation will occur?
(a) *p*-type (b) *n*-type
(c) Both (a) and (b) (d) Neither (a) nor (b)

69. A NaCl crystal is changed into CsCl type structure by:

Critical Thinking

(a) Increasing temperature and decreasing pressure.
(b) Increasing pressure and decreasing temperature.
(c) Increasing both temperature and pressure
(d) None of these.

70. To get a *n*- type semiconductor, the impurity to be added to silicon should have which of the following number of valence electrons
(a) 1 (b) 2 (c) 3 (d) 5

71. The empty space in the body centred cubic lattice is
(a) 68% (b) 52.4% (c) 47.6% (d) 32%

72. A solid with high electrical and thermal conductivity is
(a) Si (b) Li (c) NaCl (d) Ice

73. How many effective Na^+ and Cl^- ions are present respectively in a unit cell of NaCl solid (Rock salt structure) if all ions along line connecting opposite face centres are absent?
(a) 3, 3 (b) $\frac{7}{2}, 4$ (c) $\frac{7}{2}, \frac{7}{2}$ (d) $4, \frac{7}{2}$

74. In a normal spinel type structure, the oxide ions are arranged in *ccp* whereas 1/8 tetrahedral holes are occupied by Zn^{2+} ions and 50% of octahedral holes are occupied by Fe^{3+} ions. The formula of the compound is

Tricky

(a) $Zn_2Fe_2O_4$ (b) $ZnFe_2O_3$
(c) $ZnFe_2O_4$ (d) $ZnFe_2O_2$

75. Non stoichiometric defects are formed by
(a) *s*- block elements
(b) *p*-block elements
(c) either *s*-block elements or *d*-block elements
(d) only *d*-block elements.

76. Which of the following metal oxides is anti-ferromagnetic in nature?
(a) MnO_2 (b) TiO_2 (c) VO_2 (d) CrO_2

77. A molecule A_2B (Mwt. = 166.4) occupies triclinic lattice with a 5 Å, b = 8 Å, and c = 4 Å. If the density of AB_2 is 5.2 g cm⁻³, the number of molecules present in one unit cell is

Toughnut

(a) 2 (b) 3 (c) 4 (d) 5

78. Doping of silicon (Si) with boron (B) leads to :
(a) *n*-type semiconductor
(b) *p*-type semiconductor
(c) Metal
(d) Insulator

79. Non-stoichiometric compound have the properties of:
(a) Electrical conductance
(b) Isolation
(c) Insulation
(d) None of these

80. Which one of the following statements is correct?

(a) NaCl is a paramagnetic salt
(b) $CuSO_4$ is a diamagnetic salt
(c) MnO is an example of ferromagnetic substance
(d) Ferrimagnetic substance like $ZnFe_2O_4$ becomes paramagnetic on heating

Answer KEYs

1	(d)	9	(a)	17	(b)	25	(c)	33	(b)	41	(d)	49	(c)	57	(d)	65	(c)	73	(a)
2	(d)	10	(c)	18	(c)	26	(b)	34	(c)	42	(c)	50	(d)	58	(d)	66	(d)	74	(c)
3	(d)	11	(d)	19	(d)	27	(b)	35	(c)	43	(c)	51	(d)	59	(a)	67	(a)	75	(d)
4	(d)	12	(c)	20	(d)	28	(b)	36	(d)	44	(b)	52	(a)	60	(a)	68	(b)	76	(a)
5	(c)	13	(a)	21	(b)	29	(a)	37	(b)	45	(a)	53	(d)	61	(b)	69	(b)	77	(b)
6	(a)	14	(b)	22	(d)	30	(b)	38	(d)	46	(c)	54	(d)	62	(a)	70	(d)	78	(b)
7	(a)	15	(c)	23	(b)	31	(b)	39	(a)	47	(d)	55	(c)	63	(b)	71	(d)	79	(d)
8	(c)	16	(b)	24	(c)	32	(c)	40	(c)	48	(b)	56	(b)	64	(c)	72	(b)	80	(d)

Hints & Solutions

1. (d) Since in NaCl type of structure 4 formula units form a cell.

58.5 gm. of NaCl = 6.023×10^{23} atoms

$$1 \text{ gm of NaCl} = \frac{6.023\times10^{23}}{58.5} \text{atoms}$$

4 atoms constitute 1 unit cell

$$\therefore \frac{6.023\times10^{23}}{58.5} \text{ atoms constitute}$$

$$= \frac{6.023\times10^{23}}{58.5\times4}$$

$= 2.57\times 10^{21}$ unit cells.

2. (d) Effective number of 'A' atoms

$$= \left(8\times\frac{1}{8}\right)+\left(4\times\frac{1}{2}\right)=3$$

Effective number of 'B' atoms

$$= \left(12\times\frac{1}{4}\right)+1=4$$

$\therefore$ Formula of the solid $= A_3B_4$.

3. (d) Body diagonal (d) of a cubic crystal of edge length (*a*) is given by,

$d = a\sqrt{3}$

putting $a = 400$ pm, we get

$d = \sqrt{3}\times400$ pm $= 692.8$ pm ≈ 693 pm.

4. (d) Order of Bragg diffraction (n) = 2; Wavelength (λ) = 1Å and angle (θ) = 60°.

We know from the Bragg's equation

$n\lambda = 2d\sin\theta$

or $2\times1 = 2d\sin 60°$

$$\Rightarrow 2\times1 = 2.d.\frac{\sqrt{3}}{2} \Rightarrow d = \frac{2}{\sqrt{3}} = 1.15\text{Å}$$

(where d = Distance between the scattering planes)

5. (c) In a *fcc* lattice, the distance between the cation and anion is equal to the sum of their radii, which is equal to half of the edge length of unit cell,

i.e. $r^+ + r^- = \frac{a}{2}$ (where a = edge length)

$r^+ = 95$ pm, $r^- = 181$ pm
Edge length
$= 2r^+ + 2r^- = (2 \times 95 + 2 \times 181)$ pm
$= (190 + 362)$ pm $= 552$ pm.

6. (a) For *bcc* structure, atomic radius, $r = \frac{\sqrt{3}}{4}a$

$= \frac{\sqrt{3}}{4} \times 4.3 = 1.86$

Since, r = half the distance between two nearest neighbouring atoms.
∴ Shortest interionic distance
$= 2 \times 1.86 = 3.72$

7. (a) Potassium crystallises in *bcc* lattice.

8. (c)

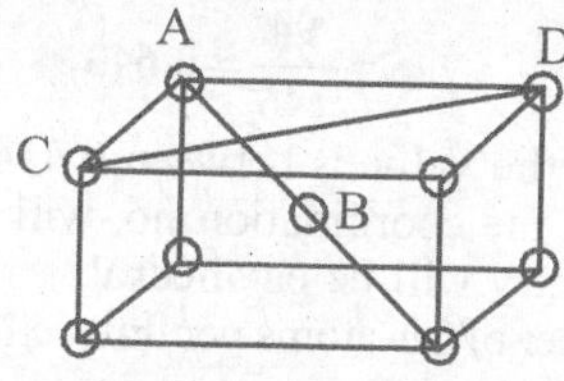

$AB = \frac{\sqrt{3}a}{2}$ (nearest)
$AC = a$ (next - nearest)
$CD = \sqrt{2}a$ (next - next - nearest)

9. (a) In crystalline solid there is perfect arrangement of the constituent particles only at 0 K. They have sharp M.P.

10. (c) Glass is an amorphous solid.

11. (d)

12. (c) A body-centred cubic system consists of all eight corners plus one atom at the centre of cube.

13. (a) ZnS has cubic close packed (*ccp*) structure. The S^{2-} ions are present at the corners of the cube and at the centre of each face. Zinc ions occupy half of the tetrahedral sites. Each zinc ion is surrounded by four sulphide ions which are disposed towards the corners of a regular tetrahedron. Similarly, S^{2-} ion is surrounded by four Zn^{2+} ions.

14. (b) The radius ratio $\frac{r_+}{r_-} = \frac{146}{216} = 0.675$

It lies between 0.414 – 0.732. Hence, it exhibits rock salt type structure.

15. (c) The coordination number is 8 : 8 for Cs^+ and Cl^-.

16. (b) For fcc,

$r = \frac{\sqrt{2}a}{4} = \frac{a}{2\sqrt{2}} = 0.3535a$

given $a = 361$ pm
$r = 0.3535 \times 361 = 128$ pm

17. (b) A form corner points and two atoms of A are missing from corner

∴ Atoms at corner (A) $= 6 \times \frac{1}{8} = \frac{3}{4}$

Atoms at face centre (B) $= 6 \times \frac{1}{2} = 3$

∴ $A_{3/4}B_3$ i.e., AB_4

18. (c) The *hcp* arrangement of atoms occupies 74% of the available space and thus has 26% vacant space.

19. (d) The electrical resistance of metals depends upon temperature. Electrical resistance decreases with decrease in temperature and becomes zero near the absolute temperature. Material in this state is said to possess super conductivity.

20. (d) Number of atoms $A = \frac{1}{8} \times 8 = 1$

Number of atoms $B = \frac{1}{2} \times 5 = \frac{5}{2}$

(Since atom *B* is missing from one of the face centered point)
∴ Formula of ionic compound
$= AB_{5/2} = A_2B_5$

21. (b)

22. (d) The appearance of colour in alkali metal halides is generally due to F-centres. F centers in NaCl produce a band of optical absorption towards the blue end of visible spectrum, thus the colour appears yellow under transmitted light. On the other hand, KCl with F-centers appears magenta, and KBr appears blue.

23. (b) Solid AB crystallizes as NaCl structure, so it has coordination number 6 and its r^+/r^- ranges from 0.414 – 0.732.
For maximum radius of anion, we have to take the lower limit of the range 0.414– 0.732.

So, $\frac{r^+}{r^-} = 0.414$

$\Rightarrow r^- = \frac{0.100}{0.414}$ nm $= 0.241$ nm

24. (c)

25. (c) Let the number of Z atoms in the *ccp* arrangement = 100
Thus the number of tetrahedral sites = 200
Since all the tetrahedral sites are occupied by X atoms, the number of X atoms = 200
Hence ratio of X : Z = 2 : 1
Thus the formula is X_2Z

26. (b) Molecular solid is the best description of water in the solid phase. For example, ice is hydrogen bonded molecular solid.

27. (b) For body centred cubic lattice $Z = 2$
Atomic mass of unit cell = 133 + 80
= 213 a.m.u
Volume of cell = $(436.6 \times 10^{-10})^3$ cm^3
Density,

$$\rho = \frac{ZM}{a^3 N_A}$$

$$= \frac{2\times 213}{(436.6\times 10^{-10})^3 \times 6.02\times 10^{23}}$$

= 8.50 g/cm^3

28. (b) The amount of energy released when cation and anions are brought from infinity to their respective sites in the crystal lattice to form one mole of the ionic compounds is called the lattice energy. Na^+ being smallest in size have highest lattice energy.

29. (a) $\rho = \dfrac{ZM}{N_A V}$

$$Z = \frac{\rho N_A V}{M} = \frac{2\times 6\times 10^{23}\times (5\times 10^{-8})^3}{75}$$

$Z = 2$, which represents *bcc* structure

$$\therefore \quad r = \frac{\sqrt{3}}{4}a = \frac{\sqrt{3}}{4}\times 5 = 2.165 Å = 216.5 \text{ pm}$$

$\approx$ 217 pm

30. (b) For bcc structure

Interionic distance $= r^+ + r^- = \dfrac{\sqrt{3}}{2}a$

(a = edge length)

$$r^+_{NH_4^+} + r^-_{Cl^-} = \frac{\sqrt{3}}{2}a$$

$$r^+_{NH_4^+} + 180 = \frac{\sqrt{3}}{2}\times 390$$

$$r^+_{NH_4^+} = 338\text{pm} - 180\text{pm} = 158 \text{ pm}$$

31. (b) $\rho = \dfrac{ZM}{N_A V}$

$$Z = \frac{\rho N_A V}{M}$$

$$= \frac{8.92\times 6.02\times 10^{23}\times (362)^3\times 10^{-30}}{63.55} = 4$$

$\therefore$ It has *fcc* unit cell

32. (c) For bcc structure

$$d = \frac{\sqrt{3}\,a}{2}$$

where d = distance between two atoms
a = edge length

$$1.73 = \frac{\sqrt{3}}{2}a$$

$$a = \frac{2\times 1.73}{\sqrt{3}} = 2Å = 200\text{pm}$$

33. (b) *hcp* is a closed packed arrangement in which the unit cell is hexagonal and coordination number is 12.

34. (c) As per formula,

$$\text{radius ratio} = \frac{\text{radius of cation}}{\text{radius of anion}}$$

$$= \frac{94}{146} = 0.643$$

Since the value is between 0.414 – 0.732 hence the coordination no. will be 6 and geometry will be octahedral.

35. (c) Number of Cu-atoms per unit cell

$$= \frac{1}{8}\times 8 + \frac{1}{2}\times 6 = 4.$$

Number of Ag-atoms per unit cell

$$= \frac{1}{4}\times 12 = 3$$

Number of Au-atoms per unit cell = 1 (at body centre)
Formula : Cu_4Ag_3Au

36. (d) In a unit cell, W atoms at the corner

$$= \frac{1}{8}\times 8 = 1$$

O-atoms at the centre of edges $= \dfrac{1}{4}\times 12 = 3$

Na-atoms at the centre of the cube = 1
W : O : Na = 1 : 3 : 1
Hence, formula = $NaWO_3$

37. (b) For bcc lattice,

$$\sqrt{3}a = 4R \Rightarrow a = \frac{4\times 96}{\sqrt{3}} \text{pm} = 221.7 \text{ pm}$$

(where R is the radius of Ca atom)
Colume of unit cell = $a^3 = (211.7\times 10^{-12})^3$ m^3
= 10.9×10^{-30} m^3

38. (d) 4 formula units of AB are present in a unit cell

$$\therefore \quad \text{no. of unit cell} = \frac{4}{40}\times\frac{N_A}{4}$$

$\Rightarrow 0.025 N_A$

39. (a) For *bcc* lattice body diagonal = $a\sqrt{3}$.
The distance between the two oppositely charged ions

$$=\frac{a}{2}\sqrt{3}=\frac{387\times1.732}{2}=335 \text{ pm}$$

40. (c) Number of unit cells

$$=\frac{N_A\times \text{Wt. of substance}}{\text{M. Wt.}\times Z}$$

$$=\frac{6.023\times10^{23}\times58.8}{58.8\times4}=1.5\times10^{23}$$

41. (d) For a *fcc* unit cell

$$r=\frac{\sqrt{2}a}{4}$$

$$a=\frac{4r}{\sqrt{2}}=2\sqrt{2}\times0.14=0.39\approx0.4 \text{ nm.}$$

42. (c) Among the three options KCl, NaCl and $MgCl_2$, the size of anion is same. So larger the cation, larger will be the cation/anion ratio i.e., KCl will have larger cation/anion ratio among the three. So, we left with two options KCl and CaF_2. Among these two CaF_2 will have maximum value of cation/anion ratio because decrease in ionic radii of anion from Cl^- to F^- does not overcome the effect of decrease in ionic radii of cation from K^+ to Ca^{2+}.

43. (c)

44. (b) The face centered cubic unit cell contains 4 atom
∴ Total volume of atoms

$$=4\times\frac{4}{3}\pi r^3=\frac{16}{3}\pi r^3$$

45. (a) There are two atoms in a *bcc* unit cell.
So, number of atoms in 12.08×10^{23} unit cells
$=2\times12.08\times10^{23}=24.16\times10^{23}$ atoms.

46. (c) In antifluorite structure the oxide ions (anions) form a face-centered cubic array and the metal ions (cations) occupy all the tetrahedral voids. Example : Na_2O

47. (d) $M=\frac{\rho\times a^3\times N_A\times10^{-30}}{Z}$

$$=\frac{10\times(100)^3\times6.02\times10^{23}\times10^{-30}}{4}=15.05$$

∴ Number of atoms in 100 g

$$=\frac{6.02\times10^{23}}{15.05}\times100=4\times10^{25}$$

48. (b) When equal number of cations (Na^+) and anions (Cl^-) are missing from their regular lattice positions, we have schottky defect.

49. (c) Due to presence of F-centres.

50. (d) In KCl, co-ordination number of cation and anion is 6 and 6 respectively. KCl is highly ionic so Schottky defect is common.
Note : Schottky defect is common in compounds having high coordination number while Frenkel defect is common in compounds with low coordination number.

51. (d) If in an ionic crystal of the type A^+, B^-, equal number of cations and anions are missing from their lattice sites so that the electrical neutrality is maintained. The defect is called Schottky defect.

52. (a) Since MgO has a rock salt structure. In this structure each cation is surrounded by six anions and vice versa.

53. (d) In Na_2O all tetrahedral void are occupied.

54. (d) Zn^{2+} ion is too small to fit into the octahedral voids of S^{2-} ions.

55. (c) AB is just like NaCl. Thus twelve A^+ are at edges and 1 within body of *fcc* i.e. in octahedral voids and six B^- at faces and 8 at corner.

56. (b) Density is given by

$$d=\frac{Z\times M}{N_A a^3}$$; where Z = number of formula units present in unit cell, which is 4 for *fcc*
a = edge length of unit cell. M = Molecular mass

$$2.72=\frac{4\times M}{6.02\times10^{23}\times(404\times10^{-10})^3}$$

$$(\because 1\text{pm}=10^{-10}\text{cm})$$

$$M=\frac{2.72\times6.02\times(404)^3}{4\times10^7}$$

$=26.99=27$ gm mole^{-1}

57. (d) When insulators (non metal atoms) interact to form a solid, their atomic orbitals mix to form two bunch of orbitals, separated by a large band gap. Electrons cannot therefore be promoted to an empty level, where they could move freely.

58. (d) Extrinsic semiconductor.

59. (a) Number of A cations per unit cell = 4
Number of B anions per unit cell must be $=2\times4=8$
Number of tetrahedral voids per unit cell $=2\times4=8$

Number of octahedral voids per unit all $= 1 \times 4 = 4$.
Hence, occupancy of tetrahedral voids $= 100\%$

60. **(a)** Superconductors are derived from compound of p block elements.

61. **(b)** Since each Sr^{++} ion provides one cation vacancy, hence
Concentration of cation vacancies
= mole % of $SrCl_2$ added
$= 10^{-4}$ mole %
$= \frac{10^{-4}}{100} \times 6.023 \times 10^{23} = 6.023 \times 10^{17}$

62. **(a)** For MgO, $\frac{r_c}{r_a} = \frac{0.65}{1.4} = 0.47$
Mg^{2+} is in octahedral void so C.N. = 6
For MgS, $\frac{r_c}{r_a} = \frac{0.65}{1.84} = 0.353$
Mg^{2+} is in tetrahedral void so C.N. = 4

63. **(b)** $\rho = \frac{Z \times M}{N_A \times a^3}$

$= \frac{2 \times 100}{6.023 \times 10^{23} \times (400 \times 10^{-10})^3}$

$= 5.188\ g/cm^3$

64. **(c)** $\frac{r_c}{r_a} = \frac{0.066}{0.140} = 0.47$
So, cation (Mg^{2+}) fitted in octahedral holes of O^{2-} are present at fcc lattice point.

65. **(c)** $\frac{r^+}{r^-} = 0.524$. It is in between 0.414 – 0.732.

Hence C. No. = 6.

66. **(d)** All the given statements are correct about F-centres.

67. **(a)** Number of octahedral voids in *ccp*, is equal to effective number of atoms.
In *ccp*, effective number of atoms are 4 so, 4 octahedral voids.
So, 1 octahedral voids per atom.

68. **(b)** n-type, since electron is set free.

69. **(b)** Low co-ordination compound can be changed into high co-ordination compound by increasing pressure and decreasing temperature.

$$\underset{(6,6)}{NaCl} \underset{Temp.}{\overset{Pr.}{\rightleftharpoons}} \underset{(8,8)}{CsCl}$$

70. **(d)** For *n*-type, impurity added to silicon should have more than 4 valence electrons.

71. **(d)** Packing fraction of *bcc* = 68%
Empty space = 100 – 68 = 32%

72. **(b)** Out of the given substances, only Li has high electrical and thermal conductivity as Li is a metallic solid.

73. **(a)** No. of $Na^+ = 4 - 1 = 3$

No. of $Cl^- = 4 - \left(2 \times \frac{1}{2}\right) = 3$

74. **(c)** Number of O-atoms per unit cell

$= \frac{1}{8} \times 8 + \frac{1}{2} \times 6 = 4$.

Number of octahedral holes per unit cell $= 1 \times 4 = 4$
Number of Fe^{3+} ions per unit cell

$= \frac{50 \times 4}{100} = 2$

Number of tetrahedral voids per unit cell $= 2 \times 4 = 8$.

Number of Zn^{2+} ions per unit cell $= \frac{1}{8} \times 8 = 1$

Hence, formula : $ZnFe_2O_4$

75. **(d)** In non stoichiometric defects ratio of cation and anion is not same that is represented by chemical ideal formula, this is due to variable oxidation nos of cation. Transition elements show variable oxidation nos.

76. **(a)** MnO_2

77. **(b)** Volume of unit cell $= a \times b \times c$
$= 5 \times 10^{-8} \times 8 \times 10^{-8} \times 4 \times 10^{-8}$
$= 1.6 \times 10^{-22}\ cm^3$
Mass of unit cell $= 1.6 \times 10^{-22} \times 5.2$
$= 8.32 \times 10^{-22}$ g
Number of molecules in one unit cell

$= \frac{8.32 \times 10^{-22}\ g}{166.4\ g\ mol^{-1}} = 3$

78. **(b)**

79. **(d)** Non-stoichiometric compounds are chemical compounds with an elemental composition that cannot be represented by a ratio of well-defined natural numbers, and are therefore in violation of the law of definite proportions. They are most often solids that contain random defects, resulting in the deficiency of one element. So, none of given option is correct.

80. **(d)** Ferrimagnetic substance become paramagnetic on heating. This is due to randomisation of spins on heating.

16 Solutions

1. For preparing 0.1 N solution of a compound from its impure sample (percentage purity of which is known) weight of the substance required will be
   (a) less than the theoretical weight.
   (b) more than the theoretical weight.
   (c) same as the theoretical weight.
   (d) None of these.
2. A solution is prepared by dissolving 10 g NaOH in 1250 mL of a solvent of density 0.8 g/mL. The molality of the solution in mol kg^{-1} is
   (a) 0.25 (b) 0.2 (c) 0.008 (d) 0.0064
3. Two solutions of a substance (non electrolyte) are mixed in the following manner. 480 mL of 1.5 M first solution + 520 mL of 1.2 M second solution. What is the molarity of the final mixture?
   (a) 2.70 M (b) 1.344 M
   (c) 1.50 M (d) 1.20 M
4. Formation of a solution from two components can be considered as
   (i) Pure solvent $\rightarrow$ separated solvent molecules, ΔH_1
   (ii) Pure solute $\rightarrow$ separated solute molecules, ΔH_2
   (iii) Separated solvent and solute molecules $\rightarrow$ Solution, ΔH_3
   Solution so formed will be ideal if
   (a) $\Delta H_{soln} = \Delta H_3 - \Delta H_1 - \Delta H_2$
   (b) $\Delta H_{soln} = \Delta H_1 + \Delta H_2 + \Delta H_3$
   (c) $\Delta H_{soln} = \Delta H_1 + \Delta H_2 - \Delta H_3$
   (d) $\Delta H_{soln} = \Delta H_1 - \Delta H_2 - \Delta H_3$
5. Which of the following liquid pairs shows a positive deviation from Raoult's law ?
   (a) Water - Nitric acid
   (b) Benzene - Methanol
   (c) Water - Hydrochloric acid
   (d) Acetone - Chloroform
6. Which will form maximum boiling point azeotrope?
   (a) $HNO_3 + H_2O$ solution
   (b) $C_2H_5OH + H_2O$ solution
   (c) $C_6H_6 + C_6H_5CH_3$ solution
   (d) None of these
7. A solution of acetone in ethanol
   (a) shows a positive deviation from Raoult's law
   (b) behaves like a non ideal solution
   (c) obeys Raoult's law
   (d) shows a negative deviation from Raoult's law
8. An aqueous solution is 1.00 molal in KI. Which change will cause the vapour pressure of the solution to increase? **Tricky**
   (a) Addition of NaCl
   (b) Addition of Na_2SO_4
   (c) Addition of 1.00 molal KI
   (d) Addition of water
9. A binary liquid solution is prepared by mixing *n*-heptane and ethanol. Which one of the following statements is correct regarding the behaviour of the solution?
   (a) The solution is non-ideal, showing -ve deviation from Raoult's Law.
   (b) The solution is non-ideal, showing +ve deviation from Raoult's Law.
   (c) *n*-heptane shows +ve deviation while ethanol shows –ve deviation from Raoult's Law.
   (d) The solution formed is an ideal solution.
10. P_A and P_B are the vapour pressure of pure liquid components, A and B, respectively of an ideal binary solution. If X_A represents the mole fraction of component A, the total pressure of the solution will be. **Tricky**
    (a) $P_A + X_A(P_B - P_A)$ (b) $P_A + X_A(P_A - P_B)$
    (c) $P_B + X_A(P_B - P_A)$ (d) $P_B + X_A(P_A - P_B)$

11. Vapour pressure of chloroform ($CHCl_3$) and dichloromethane (CH_2Cl_2) at 25°C are 200 mm Hg and 41.5 mm Hg respectively. Vapour pressure of the solution obtained by mixing 25.5 g of $CHCl_3$ and 40 g of CH_2Cl_2 at the same temperature will be: (Molecular mass of $CHCl_3$ = 119.5 u and molecular mass of CH_2Cl_2 = 85 u).
(a) 90.92 mm Hg (b) 115.0 mm Hg
(c) 147.9 mm Hg (d) 285.5 mm Hg

12. The vapour pressure of a solution of the liquids A (P° = 80 mm Hg and x_A = 0.4) and B (P° = 120 mm Hg and x_B = 0.6) is found to be 100 mm Hg. It shows that the solution exhibits
(a) positive deviation from ideal behaviour
(b) negative deviation from ideal behaviour
(c) ideal behaviour
(d) positive deviation for lower conc. and negative for higher conc.

13. Vapour pressure (in torr) of an ideal solution of two liquids *A* and *B* is given by : $P = 52X_A + 114$ where X_A is the mole fraction of *A* in the mixture. The vapour pressure (in torr) of equimolar mixture of the two liquids will be : Tricky
(a) 166 (b) 83 (c) 140 (d) 280

14. The vapour pressure (at the standard boiling point of water) of an aqueous solution containing 28% by mass of a non-volatile normal solute (molecular mass = 28) will be
(a) 152 torr (b) 608 torr
(c) 760 torr (d) 547 torr

15. Lowering of vapour pressure of 1.00 *m* solution of a non-volatile solute in a hypothetical solvent of molar mass 40 g at its normal boiling point, is :
(a) 29.23 torr (b) 30.4 torr
(c) 35.00 torr (d) 40.00 torr

16. Equal masses of a solute are dissolved in equal amount of two solvents *A* and *B*, respective molecular masses being M_A and M_B. The relative lowering of vapour pressure of solution in solvent *A* is twice that of the solution in solvent *B*. If the solutions are dilute, M_A and M_B are related as Toughnut
(a) $M_A = M_B$ (b) $2M_A = M_B$
(c) $M_A = 2M_B$ (d) $M_A = 4M_B$

17. At 20°C and 1.00 atm partial pressure of H_2, 18 mL of H_2 (STP) dissolves in 1 L of water. If 2 L of water is exposed to a gaseous mixture having a total pressure of 1425 torr (excluding the vapour pressure of water) and containing 80% H_2 by volume, the volume of H_2(STP) dissolved is Toughnut
(a) 27 mL (b) 54 mL
(c) 33.75 mL (d) 67.50 mL

18. At a particular temperature, the vapour pressures of two liquids A and B are respectively 120 and 180 mm of mercury. If 2 moles of A and 3 moles of B are mixed to form an ideal solution, the vapour pressure of the solution at the same temperature will be (in mm of mercury)
(a) 156 (b) 145 (c) 150 (d) 108

19. Which one of the following gases has the lowest value of Henry's law constant?
(a) N_2 (b) He (c) H_2 (d) CO_2

20. Which one of the following binary mixtures forms an azeotrope with minimum boiling point type?
(a) acetone-ethanol
(b) H_2O-HNO_3
(c) benzene-toluene
(d) *n*-hexane-*n*-heptane

21. The vapour pressure of pure benzene and toluene at a particular temperature are 100 mm and 50 mm respectively. Then the mole fraction of benzene in vapour phase in contact with equimolar solution of benzene and toluene is

(a) 0.67 (b) 0.75 (c) 0.33 (d) 0.50

22. 12 g of urea is dissolved in 1 litre of water and 68.4 g of sucrose is dissolved in 1 litre of water. The lowering of vapour pressure of first case is
(a) equal to second
(b) greater than second
(c) less than second
(d) double that of second

23. Which of the following statement is correct if the intermolecular forces in liquids A, B and C are in the order A < B < C?
(a) B evaporates more readily than A
(b) B evaporates less readily than C
(c) A and B evaporates at the same rate
(d) A evaporates more readily than C

24. At 80° C, the vapour pressure of pure liquid 'A' is 520 mm Hg and that of pure liquid 'B' is 1000 mm Hg. If a mixture solution of 'A' and 'B' boils at 80° C and 1 atm pressure, the amount of 'A' in the mixture is (1 atm = 760 mm Hg)

Toughnut

(a) 52 mol percent (b) 34 mol percent
(c) 48 mol percent (d) 50 mol percent

25. A solution at 20°C is composed of 1.5 mol of benzene and 3.5 mol of toluene. If the vapour pressure of pure benzene and pure toluene at this temperature are 74.7 torr and 22.3 torr, respectively, then the total vapour pressure of the solution and the benzene mole fraction in equilibrium with it will be, respectively:

Toughnut

(a) 35.8 torr and 0.280 (b) 38.0 torr and 0.589
(c) 30.5 torr and 0.389 (d) 30.5 torr and 0.480

26. Choose the correct statement with respect to the vapour pressure of a liquid among the following:
(a) Increases linearly with increasing temperature
(b) Increases non-linearly with increasing temperature
(c) Decreases linearly with increasing temperature
(d) Decreases non-linearly with increasing temperature

27. For an ideal solution of two components A and B, which of the following is true?
(a) $\Delta H_{mixing} < 0$ (zero)
(b) $\Delta H_{mixing} > 0$ (zero)
(c) A – B interaction is stronger than A – A and B – B interactions
(d) A– A, B – B and A – B interactions are identical.

28. Vapour pressure of pure benzene is 119 torr and that of toluene is 37.0 torr at the same temperature. Mole fraction of toluene in vapour phase which is in equilibrium with a solution of benzene and toluene having a mole fraction of toluene 0.50, will be:
(a) 0.137 (b) 0.237 (c) 0.435 (d) 0.205

29. How many grams of methyl alcohol should be added to 10 litre tank of water to prevent its freezing at 268 K? (K_f for water is 1.86 K kg mol^{-1})
(a) 880.07 g (b) 899.04 g
(c) 886.02 g (d) 868.06 g

30. Henry's law constant K of CO_2 in water at 25°C is 3.0×10^{-2} mol L^{-1}atm^{-1}. The mass of CO_2 present in 100 L of soft drink bottled with a partial pressure of CO_2 of 4 atm at the same temperature is
(a) 5.28 g (b) 12.0 g (c) 428 g (d) 528 g

31. Vapour pressure of solution containing 2 mol of liquid A ($P_A^o = 80$ torr) and 3 mol of liquid B ($P_B^o = 100$ torr) is 87 torr. We can conclude that

Critical Thinking

(a) there is negative deviation from Raoult's law
(b) boiling point is higher than that expected for ideal solution
(c) molecular attractions between unlike molecules are stronger than those between like molecules
(d) All of these statements are correct

32. At 15°C and 1 atmosphere partial pressure of hydrogen, 20 mL of hydrogen measured at STP dissolves in 1 L of water. If water at 15°C is exposed to a gaseous mixture having a total pressure of 1500 mm of Hg (excluding the vapour pressure of water) and containing 80% hydrogen by volumne, then the volume of hydrogen measured at STP that will dissolve in 1 L of water is
(a) 20.0 mL (b) 31.6 mL
(c) 36.1 mL (d) 26.3 mL

33. The vapour pressure of two pure liquids A and B that form an ideal solution, are 400 and 800 mm of Hg respectively at a temperature t°C. The mole fraction of A in a solution of A and B whose boiling point is t°C will be
(a) 0.4 (b) 0.8 (c) 0.1 (d) 0.2

34. The total vapour pressure of a 4 mole % solution of NH_3 in water at 293 K is 50.0 torr. The vapour pressure of pure water is 17.0 torr at this temperature. Applying Henry's and Raoult's laws, the total vapour pressure for a 5 mole % solution is

Toughnut

(a) 58.25 torr (b) 33 torr
(c) 42.1 torr (d) 52.25 torr

35. Which one of the following aqueous solutions will exihibit highest boiling point?
(a) 0.015 M urea (b) 0.01 M KNO_3
(c) 0.01 M Na_2SO_4 (d) 0.015 M glucose

36. The elevation in boiling point of a solution of 13.44 g of $CuCl_2$ in 1 kg of water using the following information will be (Molecular weight of $CuCl_2$ = 134.4 g and K_b = 0.52 K kg mol^{-1})
(a) 0.16 (b) 0.05 (c) 0.1 (d) 0.2

37. The temperature at which 10% aqueous solution (*w/v*) of glucose will exhibit the osmotic pressure of 14 bar is –
(a) 307.3°C (b) 307.3 K
(c) 34 K (d) –34°C

38. KBr is 80% ionized in solution. The freezing point of 0.4 molal solution of KBr is :
$K_f(H_2O)$ = 1.86 K kg/mole
(a) 274.339 K (b) –1.339 K
(c) 257.3 K (d) –1.339°C

39. A solution of urea (mol. mass 56 g mol^{-1}) boils at 100.18°C at the atmospheric pressure. If K_f and K_b for water are 1.86 and 0.512 K kg mol^{-1} respectively, the above solution will freeze at
Toughnut
(a) 0.654 °C (b) – 0.654 °C
(c) 6.54 °C (d) – 6.54 °C

40. The solution containing 4.0 g of a polyvinyl chloride polymer in 1 litre of dioxane was found to have an osmotic pressure 6.0×10^{-4} atmosphere at 300K, the value of R used is 0.082 litre atmosphere $mole^{-1}K^{-1}$. The molecular mass of the polymer was found to be
Toughnut
(a) 3.0×10^2 (b) 1.6×10^5
(c) 5.6×10^4 (d) 6.4×10^2

41. If the various terms in the given below expressions have usual meanings, the van't Hoff factor (*i*) cannot be calculated by which one of the following expressions
(a) $\pi V = \sqrt{i}nRT$
(b) $\Delta T_f = iK_f.m$
(c) $\Delta T_b = iK_b.m$
(d) $\dfrac{P^o_{solvent} - P_{solution}}{P^o_{solvent}} = i\left(\dfrac{n}{N+n}\right)$

42. A 0.001 molal solution of $[Pt(NH_3)_4Cl_4]$ in water had a freezing point depression of 0.0054°C. If K_f for water is 1.80, the correct formulation for the above molecule is
(a) $[Pt(NH_3)_4Cl_3]Cl$ (b) $[Pt(NH_3)_4Cl_2]Cl_2$
(c) $[Pt(NH_3)_4Cl]Cl_3$ (d) $[Pt(NH_3)_4Cl_4]$

43. Which of the following has the lowest freezing point?
(a) 0.1 m sucrose (b) 0.1 m urea
(c) 0.1 m ethanol (d) 0.1 m glucose

44. The van't Hoff factor *i* for a compound which undergoes dissociation in one solvent and association in other solvent is respectively :
(a) less than one and greater than one.
(b) less than one and less than one.
(c) greater than one and less than one.
(d) greater than one and greater than one.

45. If 0.1 M solution of glucose and 0.1 M solution of urea are placed on two sides of the semipermeable membrane to equal heights, then it will be correct to say that
(a) there will be no net movement across the membrane
(b) glucose will flow towards urea solution
(c) urea will flow towards glucose solution
(d) water will flow from urea solution to glucose

46. Determination of the molar mass of acetic acid in benzene using freezing point depression is affected by :
(a) Partial ionization (b) Dissociation
(c) Complex formation (d) Association

47. The observed osmotic pressure for a 0.10 M solution of $Fe(NH_4)_2(SO_4)_2$ at 25°C is 10.8 atm. The experimental (observed) value of van't Hoff factor will be :
(R = 0.082 L atm k^{-1} mol^{-1})

(a) 4.42 (b) 4.00
(c) 3.42 (d) 5.42

48. 12 g of a nonvolatile solute dissolved in 108 g of water produces the relative lowering of vapour pressure of 0.1. The molecular mass of the solute is :
(a) 80 (b) 60 (c) 20 (d) 40

49. A molecule M associates in a given solvent according to the equation M $\rightleftharpoons$ $(M)_n$. For a certain concentration of M, the van't Hoff factor was found to be 0.9 and the fraction of associated molecules was 0.2. The value of *n* is:

(a) 3 (b) 5 (c) 2 (d) 4

50. A solution containing 0.85 g of $ZnCl_2$ in 125.0 g of water freezes at –0.23°C. The apparent degree of dissociation of the salt is (K_f for water = 1.86K kg mol^{-1}, atomic mass: Zn = 65.3 and Cl = 35.5)

Toughnut

(a) 1.36% (b) 73.5% (c) 7.35% (d) 2.47%

51. 0.5 molal aqueous solution of a weak acid (HX) is 20% ionised. If K_f for water is 1.86 K kg mol^{-1}, the lowering in freezing point of the solution is

(a) 0.56 K (b) 1.12 K
(c) – 0.56 K (d) – 1.12 K

52. 200 mL of an aqueous solution of a protein contains its 1.26 g. The osmotic pressure of this solution at 300 K is found to be 2.57×10^{-3} bar. The molar mass of protein will be
(R = 0.083 L bar mol^{-1} K^{-1})

(a) 51022 g mol^{-1} (b) 122044 g mol^{-1}
(c) 31011 g mol^{-1} (d) 61038 g mol^{-1}

53. Freezing point of an aqueous solution is (–0.186)°C. Elevation of boiling point of the same solution is K_b = 0.512°C, K_f = 1.86°C, find the increase in boiling point.

Toughnut

(a) 0.186 °C (b) 0.0512 °C
(c) 0.092 °C (d) 0.2372 °C

54. The relationship between osmotic pressure at 273 K when 10 g glucose (P_1), 10 g urea (P_2), and 10 g sucrose (P_3) are dissolved in 250 mL of water is

(a) $P_1 > P_2 > P_3$ (b) $P_3 > P_1 > P_2$
(c) $P_2 > P_1 > P_3$ (d) $P_2 > P_3 > P_1$

55. Which observation(s) reflect(s) colligative properties?

Critical Thinking

(i) A 0.5 m NaBr solution has a higher vapour pressure than a 0.5 m $BaCl_2$ solution at the same temperature
(ii) Pure water freezes at the higher temperature than pure methanol
(iii) A 0.1 m NaOH solution freezes at a lower temperature than pure water

Choose the correct answer from the codes given below

(a) (i), (ii) and (iii) (b) (i) and (ii)
(c) (ii) and (iii) (d) (i) and (iii)

56. The difference between the boiling point and freezing point of an aqueous solution containing sucrose (molecular wt = 342 g $mole^{-1}$) in 100 g of water is 105°C. If K_f and K_b of water are 1.86 and 0.51 K kg mol^{-1} respectively, the weight of sucrose in the solution is about

Tricky

(a) 34.2 g (b) 342 g (c) 7.2 g (d) 72 g

57. A solution containing 1.8 g of a compound (empirical formula CH_2O) in 40 g of water is observed to freeze at –0.465° C. The molecular formula of the compound is
(K_f of water = 1.86 kg K mol^{-1})

(a) $C_2H_4O_2$ (b) $C_3H_6O_3$
(c) $C_4H_8O_4$ (d) $C_6H_{12}O_6$

58. Freezing point of an aqueous solution is – 0.186°C. If the values of K_b and K_f of water are respectively 0.52 K kg mol^{-1} and 1.86 K kg mol^{-1}, then the elevation of boiling point of the solution in K is

(a) 0.52 (b) 1.04 (c) 1.34 (d) 0.052

59. If the elevation in boiling point of a solution of 10 g of solute (mol. wt. = 100) in 100 g of water is ΔT_b, the ebullioscopic constant of water is

(a) 10 (b) 10 ΔT_b (c) ΔT_b (d) $\frac{\Delta T_b}{10}$

60. Which one of the following statements is incorrect?

Critical Thinking

(a) The correct order of osmotic pressure for 0.01 M aqueous solution of each compound is $BaCl_2 > KCl > CH_3COOH >$ Sucrose
(b) The osmotic pressure (π) of a solution is given by the equation π = MRT, where M is the molarity of the solution
(c) Raoult's law states that the vapour pressure of a component over a solution is proportional to its mole fraction
(d) Two sucrose solutions of same molality prepared in different solvents will have the same freezing point depression

61. π_1, π_2, π_3 and π_4 atm are the osmotic pressures of 5% (mass/volume) solutions of urea, fructose, sucrose and KCl respectively at certain temperature. The correct order of their magnitudes is :

(a) $\pi_1 > \pi_4 > \pi_2 > \pi_3$ (b) $\pi_1 > \pi_4 > \pi_2 > \pi_3$
(c) $\pi_4 > \pi_1 > \pi_2 > \pi_3$ (d) $\pi_4 > \pi_1 > \pi_3 > \pi_2$

62. Which of the following has been arranged in the increasing order of freezing point?
(a) $0.025M\ KNO_3 < 0.1M\ NH_2CSNH_2 < 0.05M\ BaCl_2 < 0.1M\ NaCl$
(b) $0.1M\ NaCl < 0.05M\ BaCl_2 < 0.1M\ NH_2CSNH_2 < 0.025M\ KNO_3$
(c) $0.1M\ NH_2CSNH_2 < 0.1\ M\ NaCl < 0.05M\ BaCl_2 < 0.025M\ KNO_3$
(d) $0.025M\ KNO_3 < 0.05M\ BaCl_2 < 0.1M\ NaCl < 0.1M\ NH_2CSNH_2$

63. 0.010M solution an acid *HA* freezes at –0.0205°C. If K_f for water is 1.860 K kg mol^{-1}, the ionization constant of the conjugate base of the acid will be (assume 0.010 M = 0.010 m) **Toughnut**
(a) 1.1×10^{-4} (b) 1.1×10^{-3}
(c) 9.0×10^{-11} (d) 9.0×10^{-12}

64. 1.0 M aqueous solution of $CrCl_3.6H_2O$ freezes at –5.58°C. Assuming complete ionization of the hydrated complex, which of the following isomers conforms to the observation (K_f for water = 1.86 K kg mol^{-1}) ?
(a) $[Cr(H_2O)_3Cl_3].3H_2O$
(b) $[Cr(H_2O)_4Cl_4]Cl.2H_2O$
(c) $[Cr(H_2O)_5Cl]Cl_2.H_2O$
(d) $[Cr(H_2O)_6]Cl_3$

65. Assuming degree of ionization to be unity in each case, which of the following equimolal solutions would freeze at the lowest temperature ?
(a) $[Pt(NH_3)_6]Cl_4$ (b) $[Pt(NH_3)_5Cl]Cl_3$
(c) $[Pt(NH_3)_4Cl_2]Cl_2$ (d) $[Pt(NH_3)_3Cl_3]Cl$

66. Which of the following aqueous solutions should have the highest osmotic pressure?
(a) 0.011 M $AlCl_3$ at 50 °C
(b) 0.03 M NaCl at 25 °C
(c) 0.012 M $(NH_4)_2SO_4$ at 25 °C
(d) 0.03 M NaCl at 50 °C

67. At 760 torr pressure and 20°C temperature, 1 L of water dissolves 0.04 gm of pure oxygen or 0.02 gm of pure nitrogen. Assuming that dry air is composed of 20% oxygen and 80% nitrogen (by volume), the masses (in g/L) of oxygen and nitrogen dissolved by 1 L of water at 20°C exposed to air at a total pressure of 760 torr are respectively: **Toughnut**
(a) 0.008, 0.016 (b) 0.016, 0.008
(c) 0.16, 0.08 (d) 0.04, 0.02

68. If sodium sulphate is considered to be completely dissociated into cations and anions in aqueous solution, the change in freezing point of water (ΔT_f), when 0.01 mol of sodium sulphate is dissolved in 1 kg of water, is (K_f = 1.86 K kg mol^{-1})
(a) 0.372 K (b) 0.0558 K
(c) 0.0744 K (d) 0.0186 K

69. The osmotic pressure (at 27° C) of an aqueous solution (200mL) containing 6 g of a protein is 2×10^{-3} atm. If R = 0.080 L atm mol^{-1}K^{-1}, the molecular weight of protein is
(a) 7.2×10^5 (b) 3.6×10^5
(c) 1.8×10^5 (d) 1.0×10^5

70. At temperature 327 °C and concentration C, osmotic pressure of a solution is P, the same solution at concentration C/2 and at temperature 427 °C shows osmotic pressure of 2 atm, value of P will be **Tricky**
(a) $\frac{12}{7}$ (b) $\frac{24}{7}$ (c) $\frac{6}{5}$ (d) $\frac{5}{6}$

71. In countries nearer to polar region, the roads are sprinkled with $CaCl_2$. This is
(a) to minimise the snow present on roads
(b) to minimise pollution
(c) to minimise the accumulation of dust on the road
(d) to minimise the wear and tear of the roads

72. If α is the degree of dissociation of Na_2SO_4, the vant Hoff's factor (*i*) used for calculating the molecular mass is
(a) $1+\alpha$ (b) $1-\alpha$ (c) $1+2\alpha$ (d) $1-2\alpha$

73. Which has the minimum freezing point ?
(a) One molal NaCl aq. solution
(b) One molal $CaCl_2$ aq. solution
(c) One molal KCl aq. solution
(d) One molal urea aq. solution

74. On mixing 3 g of non - volatile solute in 200 mL of water, its boiling point (100°) becomes 100.52°C. If K_b for water is 0.6 K/m then molecular wt. of solute is : **Toughnut**
(a) 10.5 g mol^{-1} (b) 12.6 g mol^{-1}
(c) 15.7 g mol^{-1} (d) 17.3 g mol^{-1}

75. Benzene freezes at 5.50°C. If the freezing point of 2.48 g of phosphorous in 100 g benzene is 4.48°C, the atomicity of phosphorus in benzene is (K_f (benzene) = 5.12 K kg mol^{-1}): Toughnut

(a) 1 (b) 3 (c) 4 (d) 8

76. A solute X has a solubility of 10.0 g in 100 g of water at 25°C. If a solution containing 4.0 g of solute in 50 g of water is subjected to evaporation at constant temperature of 25°C, the osmotic pressure of the solution

(a) remains unchanged throughout
(b) increases continuously
(c) increases to constant value
(d) decreases to constant value

77. At 17°C, the osmotic pressure of sugar solution is 580 torr. The solution is diluted and the temperature is raised to 57°C, when the osmotic pressure is found to be 165 torr. The extent of dilution is:

(a) 2 times (b) 3 times
(c) 4 times (d) 5 times

78. Which of the following terms has units of temperature? Critical Thinking

I : K_f II : mK_f III : iK_f IV : $\frac{mK_f}{i}$

(a) I, II (b) I, III (c) II, III (d) II, IV

79. 0.5 m solution of a complex of iron and cyanide ions has the depression of f.pt. to be 3.72 K (K_f for water = 1.86 K molal^{-1}). The formula of the complex is:

(a) $K_4[Fe(CN)_6]$ (b) $K_2[Fe(CN)_4]$
(c) $K_3[Fe(CN)_6]$ (d) $Fe(CN)_4$

80. A living cell contains a solution which is isotonic with 0.2 M glucose solution. What osmotic pressure develops when the cell is placed in 0.05 M $BaCl_2$ solution at 300 K? Toughnut

(a) 1.23 atm (b) 3.69 atm
(c) 6.15 atm (d) None of these

Answer KEYs

1	(b)	9	(b)	17	(b)	25	(b)	33	(c)	41	(a)	49	(c)	57	(d)	65	(a)	73	(b)
2	(a)	10	(d)	18	(a)	26	(b)	34	(a)	42	(b)	50	(b)	58	(d)	66	(d)	74	(d)
3	(b)	11	(a)	19	(d)	27	(d)	35	(c)	43	(c)	51	(b)	59	(c)	67	(a)	75	(c)
4	(b)	12	(b)	20	(a)	28	(b)	36	(a)	44	(c)	52	(d)	60	(d)	68	(b)	76	(c)
5	(b)	13	(c)	21	(a)	29	(d)	37	(b)	45	(a)	53	(b)	61	(c)	69	(b)	77	(c)
6	(a)	14	(b)	22	(a)	30	(d)	38	(d)	46	(d)	54	(c)	62	(b)	70	(b)	78	(d)
7	(a)	15	(a)	23	(d)	31	(d)	39	(b)	47	(a)	55	(d)	63	(c)	71	(a)	79	(c)
8	(d)	16	(c)	24	(d)	32	(b)	40	(b)	48	(c)	56	(d)	64	(c)	72	(c)	80	(a)

Hints & Solutions

1. **(b)** More than theoretical weight since impurity will not contribute.

2. **(a)** Given w = 10 g, Mol. mass = 40 g

Weight of solvent = 1250 × 0.8 g = 10000 g = 1 kg

$$\therefore \text{molality} = \frac{10}{40 \times 1} = 0.25\,\text{m}$$

3. **(b)** From the molarity equation.

$$M_1V_1 + M_2V_2 = MV$$

$$M = \frac{M_1V_1 + M_2V_2}{V} \text{ where } V = V_1 + V_2$$

$$M = \frac{480 \times 1.5 + 520 \times 1.2}{480 + 520} = 1.344\text{ M}$$

4. (b) For an ideal solution, $\Delta H_{mixing} = 0$

$\Delta H = \Delta H_1 + \Delta H_2 + \Delta H_3$

(According to Hess's law)

i.e., for ideal solutions there is no change in magnitude of the attractive forces in the two components present.

5. (b) Positive deviations are shown by such solutions in which solvent-solvent and solute-solute interactions are stronger than the solute-solvent interactions. In such solution, the interactions among molecules becomes weaker. Therefore their escaping tendency increases which results in the increase in their partial vapour pressures. In pure methanol there exists intermolecular H–bonding.

---O – H--- O—H --- O—H---
　　|　　　|　　　|
　CH_3　　CH_3　　CH_3

On adding benzene, its molecules come between methanol molecules there by breaking H-bonds which weaken intermolecular forces. This results in increase in vapour pressure.

6. (a) The solutions (liquid mixture) which boils at constant temperature and can distil as such without any change in composition are called azeotropes.

Solution of HNO_3 and H_2O will form maximum boiling point azeotrope. Maximum boiling azeotropes show negative deviation from Raoult's law.

	Composition (%)	Boiling Point
HNO_3	68.0	359 K
H_2O	32.0	373 K

Boiling point of the azeotrope of these two solutions is 393.5 K.

7. (a) A solution of acetone in ethanol shows positive deviation from Raoult's law. It is because ethanol molecules are strongly hydrogen bonded. When acetone is added, these molecules break the hydrogen bonds and ethanol becomes more volatile. Therefore its vapour pressure is increased.

8. (d) When the aqueous solution of one molal KI is diluted with water, concentration decreases, therefore the vapour pressure of the resulting solution increases.

9. (b) For this solution intermolecular interactions between heptane and ethanol are weaker than heptane -heptane and ethanol -ethanol interactions hence the solution of heptane and ethanol is non-ideal and shows positive deviation from Raoult's law.

10. (d) $P = P_A X_A + P_B X_B$

$= P_A X_A + P_B (1 - X_A)$

$\Rightarrow P_A X_A + P_B - P_B X_A$

$\Rightarrow P_B + X_A (P_A - P_B)$

11. (a) $n_{CHCl_3} = \frac{25.5}{119.5} = 0.213$

$n_{CH_2Cl_2} = \frac{40}{85} = 0.47$

$P_T = P_A^o X_A + P_B^o X_B$

$= 200 \times \frac{0.213}{0.683} + 41.5 \times \frac{0.47}{0.683}$

$= 62.37 + 28.55 = 90.92$

12. (b) $P_{total} = P_A^o \times X_A + P_B^o \times X_B$

$= 80.0 \times 0.4 + 120.0 \times 0.6 = 104$ mm Hg

The observed P_{total} is 100 mm Hg which is less than 104 mm Hg. Hence the solution shows negative deviation.

13. (c) Total V.P.,

$P = P_A^o X_A + P_B^o X_B = P_A^o X_A + P_B^o (1 - X_A)$

$= (P_A^o - P_B^o) X_A + P_B^o$

Thus, $P_B^o = 114$ torr ; $P_A^o - P_B^o = 52$

or $P_A^o = 166$ torr

Hence $P = 166 \times \frac{1}{2} + 114 \times \frac{1}{2} = 140$ torr

14. (b) Moles of solute $= \frac{28}{28} = 1$;

moles of water $= \frac{100 - 28}{18} = 4$

V.P. of solution

$= P_{H_2O}^o \times X_{H_2O} = 760 \times \frac{4}{5} = 608$ torr

15. (a) $P^o_{solvent} = 760$ torr (at the b. pt.);

$n_{solute} = 1$

$n_{solvent} = \frac{1000}{40} = 25$;

$\Delta P = 760 \times X_{solute} = \frac{760}{26}$ torr $= 29.23$ torr

16. **(c)** For dilute solution,

$$\frac{\Delta P}{P^\circ} \Rightarrow \frac{n_{solute}}{n_{solvent}} \text{ For solution in A,}$$

$$\frac{\Delta P_A}{P_A^\circ} = \frac{W/M}{W_A/M_A} = \frac{W}{M} \times \frac{M_A}{W_A} \quad \text{....(i)}$$

For solution in B, $\frac{\Delta P_B}{P_A^\circ} = \frac{W}{M} \times \frac{M_B}{W_B}$ ....(ii)

From (i) and (ii),

$$\frac{\Delta P_A/P_A^\circ}{\Delta P_B/P_B^\circ} = 2 = \frac{M_A W_B}{M_B W_A} = \frac{M_A}{M_B} \ (W_A = W_B)$$

$M_A = 2M_B$

17. **(b)** $V_{dissolved} = KP_{H2}$

P_{H2} in the mixture $= X_{H2}P = \frac{80 \times 1425}{100} = 1140$

torr $= \frac{1140}{760} = 1.5$ atm

Hence, $V_{dissolved}$ in 2L of water $= 2 \times K \times P_{H2}$

$= 2 \times 18 \times 1.5 = 54$ mL

18. **(a)** Vapour pressure of solution $= P_A + P_B$

$$= P_A^\circ X_A + P_B^\circ X_B = \frac{120 \times 2}{5} + \frac{180 \times 3}{5}$$

$= 156$ mm Hg

19. **(d)** According to Henry's law the mass of a gas dissolved per unit volume of solvent is proportional to the pressure of the gas at constant temperature m = Kp i.e. as the solubility increases, value of Henry's law constant decreases. Since CO_2 is most soluble in water among the given set of gases. Therefore CO_2 has the lowest value of Henry's law constant.

20. **(a)** Minimum boiling azeotrope is formed by solution showing positive deviation. e.g. acetone – ethanol.

21. **(a)** Total vapour pressure = vapour pressure of pure benzene + vapour pressure of toluene

$= 100 + 50 = 150$ mm

We know,

$P^\circ_{C_6H_6} = P \times X_{C6H6}$

$100 = 150 \times X_{C6H6}$

$$X_{C6H6} = \frac{100}{150} = 0.67$$

22. **(a)** Moles of urea $= \frac{12}{60} = 0.2$

Moles of sucrose $= \frac{68.4}{342} = 0.2$

Both are non electrolyte hence lowering of V.P. will be same.

23. **(d)** Lesser the intermolecular forces, the more the volatile character.

24. **(d)** At 1 atmospheric pressure the boiling point of mixture is 80°C.

At boiling point the vapour pressure of mixture, $P_T = 1$ atmosphere $= 760$ mm Hg. Using the relation,

$P_T = P_A^\circ x_A + P_B^\circ x_B$, we get

$P_T = 520x_A + 1000(1 - x_A)$

$\{\because P_A^o = 520$ mm Hg,

$P_B^\circ = 1000$ mm Hg; $x_A + x_B = 1$

or $760 = 520x_A + 1000 - 1000x_A$ or $480x_A = 240$

or $x_A = \frac{240}{480} = = \frac{1}{2}$ or 50 mol percent

25. **(b)** Total V.P. of solution $= \overset{\circ}{P_A} x_A + \overset{\circ}{P_B} x_B$

Given, $\overset{\circ}{P_A} = 74.7$ torr, $\overset{\circ}{P_B} = 22.3$ torr

$n_{benzene} = 1.5$ mol, $n_{toluene} = 3.5$ mol

$n_{solution} = 1.5 + 3.5 = 5$ mol

$$x_A = \frac{n_{benzene}}{n_{solution}} = \frac{1.5}{5}$$

Total V.P. of solution

$$= \left(\frac{1.5}{5} \times 74.7 + \frac{3.5}{5} \times 22.3\right) \text{torr}$$

$= (22.4 + 15.6)$ torr $= 38$ torr

Mole fraction of benzene in vapour form =

$$\frac{22.4}{38} = 0.589$$

26. **(b)** If we plot a graph between the vapour pressure and the temperature. We would get a curve that rises faster as T increases, giving a curved line.

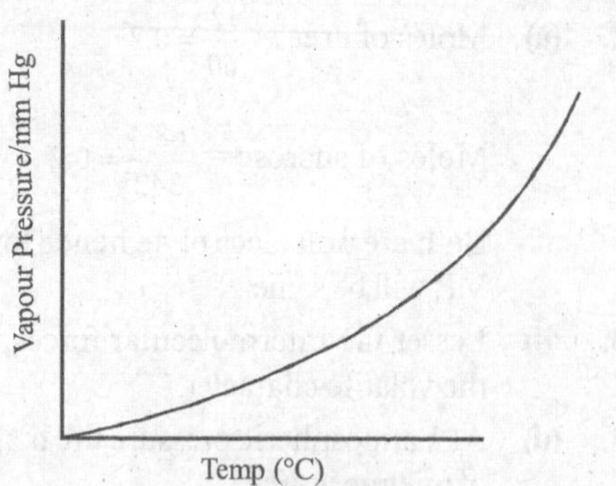

27. (d) Solutions in which solute - solute and solvent-solvent interactions are almost similar to solute solvent interactions are known as ideal solution.

28. (b) $P_A = P_A^o \times x_A = \text{total pressure} \times y_A$

$P_B = P_B^o \times x_B = \text{total pressure} \times y_B$

where x and y represents mole fraction in liquid and vapour phase respectively.

$$\frac{P_B^o x_B}{P_A^o x_A} = \frac{y_B}{y_A}; \frac{P_B^o(1-x_A)}{P_A^o x_A} = \frac{1-y_A}{y_A}$$

on putting values $\frac{119(1-0.50)}{37\times 0.50} = \frac{1-y_A}{y_A}$

on solving $y_A = 0.237$

29. (d) $\Delta T_f = K_f m$

where m = molality

$$273-268 = 1.86 \times \frac{w}{M \times v}$$

$$5 = 1.86 \times \frac{w}{32\times 10}$$

$$w = \frac{5\times 32\times 10}{1.86}$$

$= 860.2 \approx 868.06$ g

30. (d) Moles of CO_2 dissolved

$= K \times P_{CO_2} \times V_{bottle} = 3\times 10^{-2} \times 4 \times 100 = 12$;

Mass of $CO_2 = 12 \times 44 = 528$ g

31. (d) For ideal solution vapour pressure of solution

$= P_A^o X_A + P_B^o X_B$

$= 80\times\frac{2}{5} + 100\times\frac{3}{5} = 92$ torr

Since observed vapour pressure of solution < ideal vapour pressure, the solution shows negative deviation.

32. (b) From the given data

$P_{H2} = 1500 \times 0.80$

$= 1200$ mm of Hg $= \frac{1200}{760}$ atmosphere $= 1.58$ atmosphere

If 's are volumes of gas dissolved by same volume of liquid, then from Henry's law

$$\frac{V_2}{V_1} = \frac{P_2}{P_1}; \text{ or } \frac{V_2}{20} = \frac{1.58}{1.0}$$

or $V_2 = 31.60$ mL

33. (c) V.P. of solution at t°C = 760 mm

[at b.p., V.P. of solution =atompheric pressure]

Thus $= P_A^o.x_A + P_B^o.x_B$

or $P = P_A^o.x_A + P_B^o.(1-x_A)$ [$\because x_A + x_B = 1$]

or $760 = 400X_A + 800(1-X_A)$ [$\because P = 760$ mm of Hg]

or $-800 + 760 = -400\,x_A$

or $-40 = -400\,x_A$

or $x_A = \frac{40}{400} = 0.1$

Thus mole fraction of in solution is 0.1

34. (a) The given data are

$P_{water} = 17.0$ torr;

P_{total} (4 mole % solution)

$= P_{NH_3} + P_{water} = 50.0$ torr

$X_{NH_3} = 0.04$ and $X_{water} = 0.96$

Now according to Raoult's law;

$P_{water} = X_{water}\, P^o_{water}$

$= 0.96 \times 17.0$ torr $= 16.32$ torr

Now Henry's law constant for ammonia is

$$K_H(NH_3) = \frac{P_{NH_3}}{X_{NH_3}} = \frac{33.68 \text{ torr}}{0.04} = 842 \text{ torr}$$

Hence, for 5 mole % solution, we have

$P_{NH_3} = K_H(NH_3) X_{NH_3}$

$= (842 \text{ torr})(0.05) = 16.15$ torr

Thus, P_{total} (5 mole % solution)

$= P_{NH_3} + P_{water} = 42.1 + 16.15 = 58.25$ torr

35. (c) $\Delta T_b = K_b \times m$

Elevation in boiling point is a colligative property, which depends upon the no. of particles (concentration of solution). Thus greater the number of particles, greater is the elevation in boiling point and hence greater will be its boiling point.

$Na_2SO_4 \rightleftharpoons 2Na^+ + SO_4^{2-}$

Since Na_2SO_4 has maximum number of particles hence it has maximum boiling point.

36. (a) (i) $i = \dfrac{\text{No. of particles after ionisation}}{\text{No. of particles before ionisation}}$

(ii) $\Delta T_b = i \times K_b \times m$

$CuCl_2$	$\longrightarrow$	Cu^{2+}	+	$2Cl^-$
1		0		0
$(1-\alpha)$		α		2

$i = \dfrac{1+2\alpha}{1}, = 1 + 2\alpha$

Assuming 100% ionization

So, $i = 1 + 2 = 3$

$\Delta T_b = 3 \times 0.52 \times 0.1 = 0.156 \approx 0.16$

$\left(m = \dfrac{13.44}{134.4} = 0.1\right)$

37. (b) $\pi = \dfrac{nRT}{V}$ or $T = \dfrac{\pi V}{nR}$

10% (w/v) aqueous solution of glucose means 10 g glucose is present in 100 mL of H_2O

$= \dfrac{14 \times 100 \times 10^{-3} \times 180}{0.082 \times 10} = 307.3$ K

38. (d) $\Delta T_f = iK_f m$

KBr	$\longrightarrow$	K^+	+	Br^-
1		0		0
$(1-\alpha)$		α		α

$i = 1 + \alpha$

Here $\alpha = 0.8$ (80% ionization)

$= 1.8$

$\Delta T_f = (1.8)(1.86)(0.4) = 1.339$

$T_f = -1.339$ °C

39. (b) As $\Delta T_f = K_f m$

$\Delta T_b = K_b.m$

Hence, we have $m = \dfrac{\Delta T_f}{K_f} = \dfrac{\Delta T_b}{K_b}$

or $\Delta T_f = \Delta T_b \dfrac{K_f}{K_b}$

$\Rightarrow [\Delta T_b = 100.18 - 100 = 0.18°C]$

$= 0.18 \times \dfrac{1.86}{0.512} = 0.654°C$

As the freezing point of pure water is 0°C,

$\Delta T_f = 0 - T_f$

$0.654 = 0 - T_f$

$\therefore T_f = -0.654$

Thus the freezing point of solution will be -0.654°C.

40. (b) $\pi V = \dfrac{w}{m} RT$

$\therefore 6 \times 10^{-4} \times 1 = \dfrac{4}{m} \times 0.0821 \times 300;$

$m = 1.64 \times 10^5$.

41. (a) Van't Hoff equation is

$\pi V = inRT$

For depression in freezing point.

$\Delta T_f = i \times K_f \times m$

For elevation in boiling point.

$\Delta T_b = i \times K_b \times m$

For lowering of vapour pressure,

$$\frac{P^\circ_{solvent} - P_{solution}}{P^\circ_{solvent}} = i\left(\frac{n}{N+n}\right)$$

42. (b) $\Delta T_f = i.K_f.m$; $0.0054 = i \times 1.8 \times 0.001$

$i = 3$ so it is $[Pt(NH_3)_4Cl_2]Cl_2$.

43. (c) From the relation, $\Delta T_f = iK_f m$, it is evident that $\Delta T_f \propto i$. Since i is maximum for ethanol (others are non electrolytes), ΔT_f (depression in freezing point) will be maximum for C_2H_5OH and hence freezing point will be minimum for ethanol solution.

44. (c) If compound dissociates in solvent $i > 1$ and on association $i < 1$.

45. (a) As both the solutions are isotonic hence there is no net movement of the solvent through the semipermeable membrane between two solutions.

46. (d) Acetic acid contain carboxylic group – COOH which can form H-bonding so acetic acid dimerises.

$$CH_3-C(=O\cdots H-O)(-O-H\cdots O=)C-CH_3$$

47. (a) Given $\pi_{ob} = 10.8$ atm

$\pi_{nor} = CRT = 0.10 \times 0.0821 \times 298 = 2.446$

now experimental value of (i)

$$= \frac{\text{Observed osmotic pressure}}{\text{Normal osmotic pressure}}$$

$$= \frac{10.8}{2.446} = 4.42$$

48. (c) $\frac{P^o - P_s}{P^o} = \frac{n}{N} = \frac{w}{m} \times \frac{M}{W}$

$$0.1 = \frac{12}{m} \times \frac{18}{108}$$

$$m = \frac{12 \times 18}{0.1 \times 108} = 20$$

49. (c) van't Hoff factor (i) and the degree of association are related as below :

$$i = 1 - \alpha\left(1 - \frac{1}{n}\right)$$

$$0.9 = 1 - 0.2\left(1 - \frac{1}{n}\right)$$

On solving,

$$\left(1 - \frac{1}{n}\right) = \frac{1}{2}$$

$$\frac{1}{n} = 1 - \frac{1}{2} = \frac{1}{2}$$

$$\therefore n = 2$$

50. (b) $\text{Mol.wt.} = \frac{k_f \times w \times 1000}{\Delta T_f \times W}$

$$= \frac{1.86 \times 0.85 \times 1000}{0.23 \times 125} \approx 55 \text{ g}$$

Where

$w = 0.85$g

$W = 125$ g

$\Delta T_f = 0°C - (-23°C) = 23°C$

Now, $i = \frac{M_{normal}}{M_{observed}} = \frac{136.3}{55} = 2.47$

$$\underset{1-\alpha}{ZnCl_2} \rightleftharpoons \underset{\alpha}{Zn^{++}} + \underset{2\alpha}{2Cl^-}$$

Van't Hoff factor (i)

$$= \frac{1 - \alpha + \alpha + 2\alpha}{1} = 2.47$$

$$\therefore \quad \alpha = 0.735 = 73.5\%$$

51. (b) As $\Delta T_f = K_f . m$

For,

	HX	⇌	H^+	+	X^-
t = 0	1		0		0
t = eq	(1 – 0.20)		0.20		0.20

Total no. of moles $= 1 - 0.20 + 0.20 + 0.20$

$= 1 + 0.20 = 1.2$

$\therefore \Delta T_f = 1.2 \times 1.86 \times 0.5 = 1.1160 \approx 1.12$ K

52. (d) Osmotic pressure $(\pi) = CRT$

$$(\pi) = \frac{\text{wt} \times 1000}{\text{Molecular mass} \times V} RT$$

$$= 2.57 \times 10^{-3} = \frac{1.26 \times 1000}{\text{Mol.mass} \times 200} \times 0.083 \times 300$$

Molecular mass = 61038 g

53. (b) $\Delta T_b = K_b \frac{W_B}{M_B \times W_A} \times 1000$;

$$\Delta T_f = K_f \frac{W_B}{M_B \times W_A} \times 1000;$$

$$\frac{\Delta T_b}{\Delta T_f} = \frac{K_b}{K_f} = \frac{\Delta T_b}{-0.186} = \frac{0.512}{1.86} = 0.0512°C.$$

54. (c) Moles of glucose $\frac{10}{180} = 0.05$,

urea $= \frac{10}{60} = 0.16$, Sucrose $= \frac{10}{342} = 0.029$

Hence osmotic presure $P_2 > P_1 > P_3$ $(\pi \propto C)$

55. (d) Colligative properties depends upon the no. of particles. Since methanol is non electrolyte hence cannot be considered.

56. **(d)** $(100+\Delta T_b)-(0-\Delta T_f)=105$

$\Delta T_b+\Delta T_f=5$

$m(K_b+K_f)=5$

$m=\frac{5}{2.37}$ i.e., $\frac{5}{2.37}$ moles in 1000 g water

(or) $\frac{5}{2.37\times10}$ moles in 100 g water

$\therefore$ Wt. of sucrose $=\frac{5}{2.37\times10}\times342=72g$

57. **(d)** $\Delta T_f=K_f\times m$

$M=\frac{1000\times K_f\times w_2(\text{solute})}{\Delta T_f\times w_1(\text{solvent})}$

$=\frac{1000\times1.86\times1.8}{0.465\times40}\Rightarrow M=180$

Molecular formula = (empirical formula)$_n$

$n=\frac{\text{Molecular mass}}{\text{Empirical formula mass}}=\frac{180}{30}=6$

Molecular formula $=(CH_2O)_6=C_6H_{12}O_6$.

58. **(d)** $\Delta T_f=i.K_f.m;\ \Delta T_b=i.K_b.m$

$\frac{\Delta T_f}{\Delta T_b}=\frac{K_f}{K_b}$

$\Delta T_f=0-(-0.186°C)=0.186°C$

$\frac{0.186}{\Delta T_b}=\frac{1.86}{0.52}\Rightarrow\Delta T_b=\frac{0.52\times0.186}{1.86}=0.052$

59. **(c)** $\Delta T_b=\frac{K_b\times w\times1000}{M\times W}$;

$\therefore K_b=\frac{\Delta T_b\times100\times100}{10\times1000}=\Delta T_b$

60. **(d)** $\Delta T_f=K_f\times m\times i$. Since K_f has different values for different solvents, hence even if m is the same ΔT_f will be different

61. **(c)** Moles of urea $=\frac{5}{60}$;

moles of fructose $=\frac{5}{180}$;

$C_{urea}=\frac{0.08\times1000}{100}\Rightarrow0.833$

moles of sucrose $=\frac{5}{342}$;

$C_{sucrose}=0.0146\times\frac{1000}{100}=0.146$

moles of KCl(effective)

$=2\times\frac{5}{74.5}=\frac{5}{37.25};\ \frac{0.134\times1000}{100}=1.34$

$\pi\propto C$

this order is $\pi_4>\pi_1>\pi_2>\pi_3$

62. **(b)** Greater is the effective molarity (i × C), higher the ΔT_f value and lower the freezing point.

(i) 0.1 M NaCl $=i\times C=2\times0.1=0.2$

(ii) 0.05 M $BaCl_2=i\times C=3\times0.05=0.15$

(iii) 0.1 M $NH_2CSNH_2=i\times C=1\times0.1=0.1$

(iv) 0.025 M $KNO_3=i\times C=2\times0.025=0.50$

Thus, order is (i) < (ii) < (iii) < (iv)

63. **(c)** $\Delta T_f(\text{normal})=K_fm=1.86\times0.01=0.0186$;

$i=\frac{\Delta T_{f(obs)}}{\Delta T_{f(nor)}}=\frac{0.0205}{0.0186}=1.10=1+\alpha$;

$\alpha=0.1$

$K_a=\frac{C\alpha^2}{1-\alpha}=\frac{0.01\times0.1^2}{1-0.1}=\frac{1}{9}\times10^{-3}$;

$K_b=\frac{K_w}{K_a}=1.0\times10^{-14}\times9\times10^3=9\times10^{-11}$

64. **(c)** $i=\frac{5.58}{1.86}=3$;

$[Cr(H_2O)_5Cl]Cl_2.H_2O\longrightarrow[Cr(H_2O)_5Cl]^{2+}+2Cl^-$

65. **(a)** $[Pt(NH_3)_6]Cl_4\rightarrow[Pt(NH_3)_6]^{4+}+4Cl^-$;

$i=1+4=5$ (maximum)

66. **(d)** $\pi\propto iCT$; iCT is maximum for 0.03 M NaCl at 50 °C.

67. **(a)** Mass of oxygen $=\frac{20}{100}\times0.04=0.008\ g/L$

Mass of nitrogen

$=\frac{80}{100}\times0.02=0.016\ g/L$

68. **(b)** Sodium sulphate dissociates as

$Na_2SO_4(s) \longrightarrow 2Na^+ + SO_4^{--}$

hence van't hoff factor $i = 3$

Now $\Delta T_f = i k_f .m$

$= 3 \times 1.86 \times 0.01 = 0.0558\,K$

69. **(b)** Using the relation,

$$M = \frac{w.R.T}{\pi V},$$

we get $M = \frac{6 \times 0.080 \times 300}{2 \times 10^{-3} \times 0.2} = 3.6 \times 10^5$

70. **(b)** $\pi V = CRT$

$$\frac{\pi_1}{\pi_2} = \frac{C_1T_1}{C_2T_2}$$

$\pi_1 = P, \pi_2 = 2$ atm. $C_1 = C$, $C_2 = \frac{C}{2}$

$T_1 = 600\,K$, $T_2 = 700\,K$

$$\frac{P}{2} = \frac{2 \times C \times 600}{C \times 700}$$

$$P = \frac{24}{7}.$$

71. **(a)** $CaCl_2$ acts as a non-volatile solute and results in depression in freezing point. Thus, snow is reduced and prevents blocking of roads in the polar region.

72. **(c)** $Na_2SO_4 \longrightarrow 2Na^+ + SO_4^{2-}$

	Na_2SO_4	$2Na^+$	SO_4^{2-}
Mol. before dissociation	1	0	0
Mol. after dissociation	$1-\alpha$	2α	1α

$i = 1 - \alpha + 2\alpha + \alpha = 1 + 2\alpha$

73. **(b)** Among the given options, $CaCl_2$ solution will produce maximum (three) ions per molecule, so it will show minimum freezing point.

74. **(d)** $\Delta T_b = K_b \frac{w}{M} \times \frac{1000}{W}$

$0.52 = 0.6 \times \frac{3}{m} \times \frac{1000}{200}$ $(W = 200 \times 1))$

$m = \frac{1.8 \times 5}{0.52} = 17.3\,g\,mol^{-1}$

75. **(c)** Molecular mass of phosphorous

$$= \frac{1000 \times K_f \times W_{solute}}{\Delta T_f \times W_{benzene}}$$

$$= \frac{1000 \times 5.12 \times 2.48}{1.02 \times 100} = 124.5 = (P)_x$$

or, $x = \frac{124.5}{31} = 4$

76. **(c)** The given solution is unsaturated (saturated one contains 5 g of solute in 50 g of water). On evaporation, concentration of solution increases to saturation and thereafter becomes constant due to excess solid separating out. Hence, osmotic pressure first increases and then becomes constant.

77. **(c)** $\pi_1 = \frac{nRT_1}{V_1}$, $\pi_2 = \frac{nRT_2}{V_2}$

Hence, $\frac{\pi_1}{\pi_2} = \frac{580}{165} = \frac{V_2T_1}{V_1T_2} = \frac{V_2 \times 290}{V_1 \times 330}$

$\Rightarrow \frac{V_2}{V_1} = 4$

78. **(d)** $mK_f = \Delta T_f$ has the units of temperature. Van't Hoff factor, i, is dimensionless quantity.

79. **(c)** $\Delta T_f(\text{normal}) = K_f m = 1.86 \times 0.5 = 0.93°$

Assuming 100% ionization, i = n = No. of ions per molecule

$$= \frac{\text{Observed}\,\Delta T_f}{\text{Normal}\,\Delta T_f} = \frac{3.72}{0.93} = 4$$

$K_3[Fe(CN)_6] \longrightarrow 3K^+ + [Fe(CN)_6]^{3-}$

no. of ions = 4

80. **(a)** For non electrolyte solute (glucose) i = 1

For $BaCl_2 \rightleftharpoons Ba^{2+} + 2Cl^-$

i = 3 (100 % ionized)

i = (number of ions after ionization × α) + (1 – α)

= (3 × 1) + (1 – 1) = 3

$\pi = (iC_1 - i_2C_2)RT$

$\Rightarrow (1 \times 0.2 - 3 \times 0.05)\, 0.0821 \times 300$

$\Rightarrow 1.23$ atm

17 Electrochemistry

1. Which of the following statement is correct?
   (a) Cathode is –ve terminal in both, glavanic and electrolytic cells.
   (b) Anode is +ve terminal in both, galvanic and electrolytic cells.
   (c) Cathode and anode are –ve terminal in electrolytic and galvanic cell.
   (d) Cathode and anode are +ve terminal in electrolytic and galvanic cell.

2. Based on the cell notation for a spontaneous reaction, at the anode :
   $Ag\,(s)\,|\,AgCl\,(s)\,|\,Cl^-\,(aq)\,||\,Br^-\,(aq)\,|\,Br_2\,(l)\,|\,C\,(s)$
   (a) AgCl gets reduced
   (b) Ag gets oxidized
   (c) Br^- gets oxidized
   (d) Br_2 gets reduced

3. For the electrochemical cell

   $Pt\,(s)\,|\,\underset{1\,atm}{H_2(g)}\,|\,H^+(1M)\,|$, Cu (s) which one of the following statements is true?
   (a) H^+ ions are formed at anode and Cu is deposited at cathode.
   (b) H_2 is liberated at cathode and Cu is deposited at anode.
   (c) Oxidation occurs at cathode.
   (d) Reduction occurs at anode.

4. The standard reduction potentials at 298K for the following half reactions are given against each

   $Zn^{2+}(aq) + 2e^- \rightleftharpoons Zn(s)\,;\ -0.762\,V$

   $Cr^{3+}(aq) + 3e^- \rightleftharpoons Cr\,(s);\ -0.740\,V$

   $2H^+(aq) + 2e^- \rightleftharpoons H_2\,(g)\,;\ 0.00\,V$

   $Fe^{3+}(aq) + e^- \rightleftharpoons Fe^{2+}(aq)\,;\ 0.770\,V$

   Which is the strongest reducing agent?

   Tricky

   (a) Zn (*s*) (b) Cr (*s*)
   (c) $H_2(g)$ (d) $Fe^{3+}(aq)$

5. Zn gives H_2 gas with H_2SO_4 and HCl but not with HNO_3 because
   Tricky
   (a) Zn acts as oxidizing when reacts with HNO_3.
   (b) HNO_3 is weaker acid than H_2SO_4 and HCl.
   (c) In electrochemical series Zn is above hydrogen.
   (d) NO_3^- is reduced in preference to hydronium.

6. A smuggler could not carry gold by depositing iron on the gold surface since
   (a) gold is denser
   (b) iron rusts
   (c) gold has higher reduction potential than iron
   (d) gold has lower reduction potential than iron

7. Standard electrode potential for Sn^{4+}/Sn^{2+} couple is + 0.15 V and that for the Cr^{3+}/Cr couple is –0.74V. These two couples in their standard state are connected to make a cell. The cell potential will be:
   (a) +1.19 V (b) +0.89 V
   (c) +0.18 V (d) +1.83 V

8. A solution contains Fe^{2+}, Fe^{3+} and I^- ions. This solution was treated with iodine at 35°C. E° for Fe^{3+}/Fe^{2+} is + 0.77 V and E° for $I_2/2I^- = 0.536$ V. The favourable redox reaction is :
   (a) I_2 will be reduced to I^-
   (b) There will be no redox reaction
   (c) I^- will be oxidised to I_2
   (d) Fe^{2+} will be oxidised to Fe^{3+}

9. Consider the following four electrodes:
   Critical Thinking
   $P = Cu^{2+}$ (0.0001 M)/Cu(*s*)
   $Q = Cu^{2+}$ (0.1 M)/Cu(*s*)
   $R = Cu^{2+}$ (0.01 M)/Cu(*s*)
   $S = Cu^{2+}$ (0.001 M)/Cu(*s*)
   If the standard reduction potential of Cu^{2+}/Cu is +0.34 V, the reduction potentials in volts of the above electrodes follow the order.
   (a) P > S > R > Q (b) S > R > Q > P
   (c) R > S > Q > P (d) Q > R > S > P

10. Following cell has EMF 0.7995V.
$Pt | H_2 (1 atm) | HNO_3 (1M) \| AgNO_3 (1M) | Ag$
If we add enough KCl to the Ag cell so that the final Cl^- is 1M. Now the measured emf of the cell is 0.222V. The K_{sp} of AgCl would be –

Toughnut

(a) $1 \times 10^{-9.8}$ (b) $1 \times 10^{-19.6}$
(c) 2×10^{-10} (d) 2.64×10^{-14}

11. Standard cell voltage for the cell $Pb | Pb^{2+} \| Sn^{2+} | Sn$ is – 0.01 V. If the cell is to exhibit $E_{cell} = 0$, the value of $[Sn^{2+}]/[Pb^{2+}]$ should be antilog of –
(a) +0.3 (b) 0.5 (c) 1.5 (d) –0.5

12. An electrochemical cell is set up as:
$Pt; H_2 (1atm)|HCl (0.1 M) \| CH_3COOH (0.1 M)| H_2$ (1atm); Pt. The e.m.f of this cell will not be zero, because

Tricky

(a) the temperature is constant
(b) e.m.f depends on molarities of acids used
(c) acids used in two compartments are different
(d) pH of 0.1 M HCl and 0.1 M CH_3COOH is not same

13. A variable, opposite external potential (E_{ext}) is applied to the cell
$Zn|Zn^{2+} (1 M) \| Cu^{2+} (1 M) | Cu$, of potential 1.1 V. When $E_{ext} < 1.1$ V and $E_{ext} > 1.1$ V, respectively electrons flow from :
(a) anode to cathode in both cases
(b) cathode to anode and anode to cathode
(c) anode to cathode and cathode to anode
(d) cathode to anode in both cases

14. The standard electrode potentials $\left(E^o_{M^+/M}\right)$ of four metals A, B, C and D are – 1.2 V, 0.6 V, 0.85 V and – 0.76 V, respectively. The sequence of deposition of metals on applying potential is:
(a) A, C, B, D (b) B, D, C, A
(c) C, B, D, A (d) D, A, B, C

15. A current of 10.0 A flows for 2.00 h through an electrolytic cell containing a molten salt of metal X. This results in the decomposition of 0.250 mol of metal X at the cathode. The oxidation state of X in the molten salt is: (F = 96,500 C)

Toughnut

(a) 1+ (b) 2+ (c) 3+ (d) 4+

16. Which of the following will form a cell with the highest voltage?
(a) $1 M Ag^+, 1 M Co^{2+}$
(b) $2 M Ag^+, 2 M Co^{2+}$
(c) $0.1 M Ag^+, 2 M Co^{2+}$
(d) $2 M Ag^+, 0.1 M Co^{2+}$

17. The standard EMF for the cell reaction,
$Zn + Cu^{2+} \longrightarrow Cu + Zn^{2+}$ is 1.1 volt at 25°C. The EMF for the cell reaction, when 0.1 M Cu^{2+} and 0.1 M Zn^{2+} solutions are used, at 25°C is
(a) 1.10V (b) 0.10V
(c) –1.10V (d) –0.110V

18. Consider the following reactions
(i) $Cd^{2+} (aq) + 2e^- \longrightarrow Cd(s), E^\circ = -0.40$ V
(ii) $Ag^+ (aq) + e^- \longrightarrow Ag(s), E^\circ = 0.80$ V
For the galvanic cell involving the above reactions. Which of the following is not correct ?

(a) Molar concentration of the cation in the cathodic compartment changes faster than that of the cation in the anodic compartment.
(b) E_{cell} increase when Cd^{2+} solution is diluted.
(c) Twice as many electrons pass through the cadmium electrode as through silver electrode.
(d) E_{cell} decreases when Ag^+ solution is diluted.

19. Given that: $E^\circ_{Ag^+/Ag} = 0.80V$ and $[Ag^+] = 10^{-3} M$;
$E^\circ_{Hg_2^{2+}/Hg} = 0.785 V$ and $[Hg_2^{2+}] = 10^{-1} M$
which is true for the cell reaction
$$2Hg(l) + 2Ag^+(aq) \rightarrow 2Ag(s) + Hg_2^{2+}(aq) ?$$
(a) The forward reaction is spontaneous
(b) The backward reaction is spontaneous
(c) $E_{cell} = 0.163\ V$
(d) $E_{cell} = 1.585\ V$

20. Use the following standard electrode potentials, calculate ΔG° in kJ/mol for the indicated reaction:

$$5Ce^{4+}(aq) + Mn^{2+}(aq) + 4H_2O(l) \longrightarrow 5Ce^{3+}(aq) + MnO_4^-(aq) + 8H^+(aq)$$

$$MnO_4^-(aq) + 8H^+(aq) + 5e^- \longrightarrow Mn^{2+}(aq) + 4H_2O(l); E^\circ = +1.51\ V$$

$$Ce^{4+}(aq) + e^- \longrightarrow Ce^{3+}(aq); E^\circ = +1.61\ V$$

(a) −9.65 (b) −24.3 (c) −48.25 (d) −35.2

21. If 0.01 M solution of an electrolyte has a resistance of 40 ohms in a cell having a cell constant of 0.4 cm^{-1}, then its molar conductance in $ohm^{-1}\ cm^2\ mol^{-1}$ is **Critical Thinking**

(a) 10^2 (b) 10^4 (c) 10 (d) 10^3

22. When a concentrated solution of an electrolyte is diluted
(a) its specific conductance increases.
(b) its equivalent conductivity decreases.
(c) its specific conductivity decreases and equivalent conductivity increases.
(d) both specific and equivalent conductivity increase.

23. The limiting equivalent conductivity of NaCl, KCl and KBr are 126.5, 150.0 and 151.5 S cm^2 eq^{-1}, respectively. The limiting equivalent ionic conductivity for Br^- is 78 S $cm^2 eq^{-1}$. The limiting equivalent ionic conductivity for Na^+ ions would be :

(a) 128 (b) 125 (c) 49 (d) 50

24. The ionic conductivity of Ba^{2+} and Cl^- at infinite dilution are 127 and 76 $ohm^{-1}\ cm^2\ eq^{-1}$ respectively. The equivalent conductivity of $BaCl_2$ at infinity dilution (in $ohm^{-1} cm^2\ eq^{-1}$) would be :

(a) 203 (b) 279 (c) 101.5 (d) 139.5

25. Aluminium oxide may be electrolysed at 1000°C to furnish aluminium metal (At. Mass = 27 amu; 1 Faraday = 96,500 Coulombs). The cathode reaction is– $Al^{3+} + 3e^- \rightarrow Al$ **Critical Thinking**

To prepare 5.12 kg of aluminium metal by this method we require electricity of

(a) 5.49×10^1 C (b) 5.49×10^4 C
(c) 1.83×10^7 C (d) 5.49×10^7 C

26. When electric current is passed through acidified water, 112 mL of hydrogen gas at STP collected at the cathode in 965 seconds. The current passed in amperes is **Tricky**

(a) 1.0 (b) 0.5 (c) 0.1 (d) 2.0

27. On passing current through two cells, connected in series containing solution of $AgNO_3$ and $CuSO_4$, 0.18 g of Ag is deposited. The amount of the Cu deposited is:

(a) 0.529 g (b) 10.623 g
(c) 0.0529 g (d) 1.2708 g

28. If x is the specific resistance of the solution and N is the normality of the solution, the equivalent conductivity of the solution is given by

(a) $\frac{1000x}{N}$ (b) $\frac{1000}{Nx}$
(c) $\frac{1000N}{x}$ (d) $\frac{Nx}{1000}$

29. For an electrolyte solution of 0.05 mol L^{-1}, the conductivity has been found to be 0.0110 S cm^{-1}. The molar conductivity is
(a) 0.055 S $cm^2\ mol^{-1}$
(b) 550 S $cm^2\ mol^{-1}$
(c) 0.22 S $cm^2\ mol^{-1}$
(d) 220 S $cm^2\ mol^{-1}$

30. The highest electrical conductivity of the following aqueous solutions is of
(a) 0.1 M difluoroacetic acid
(b) 0.1 M fluoroacetic acid
(c) 0.1 M chloroacetic acid
(d) 0.1 M acetic acid

31. A current of 2.0 A passed for 5 hours through a molten metal salt deposits 22.2 g of metal (At wt. = 177). The oxidation state of the metal in the metal salt is **Toughnut**

(a) +1 (b) +2 (c) +3 (d) +4

32. The specific conductivity of N/10 KCl solution at 20°C is 0.212 $ohm^{-1}\ cm^{-1}$ and the resistance of the cell containing this solution at 20°C is 55 ohm. The cell constant is

(a) 4.616 cm^{-1} (b) 11.66 cm^{-1}
(c) 2.173 cm^{-1} (d) 3.324 cm^{-1}

33. The equivalent conductance at infinite dilution of a weak acid such as HF
(a) can be determined by extrapolation of measurements of dilute solutions of HCl, HBr and HI
(b) can be determined by measurement of very dilute HF solutions
(c) can be determined from measurements of dilute solutions of NaF, NaCl and HCl
(d) is an undefined quantity

34. Conductance of 0.1 M KCl (conductivity = X $Ohm^{-1}cm^{-1}$) filled in a conductivity cell is Y Ohm^{-1}. If the conductance of 0.1 M NaOH filled in the same cell is Z Ohm^{-1}, the molar conductance of NaOH will be **Toughnut**
(a) $10^3 \frac{XZ}{Y}$ (b) $10^4 \frac{XZ}{Y}$
(c) $10 \frac{XZ}{Y}$ (d) $0.1 \frac{XZ}{Y}$

35. In electrolytic reduction of a nitroarene with 50% current efficiency 20.50 g of the compound is reduced by 2×96500 C of electric charge. The molar mass of the compound is
(a) 123.0 g (b) 61.5 g
(c) 10.2 g (d) 20.5 g

36. Electrolysis of NaCl solution with inert electrodes for certain period of time gave 600 cm^3 of 1.0 M NaOH in the electrolytic cell. During the same period 31.80 g of copper was deposited in a copper voltmeter in series with the electrolytic cell. What is the percent current efficiency in the electrolytic cell ? (At. wt. of Cu = 63. 6)
(a) 40 (b) 25 (c) 60 (d) 50

37. Given the ionic conductance of $\begin{matrix} COO^- \\ | \\ COO^- \end{matrix}$, K^+, and Na^+ are 74, 50, and 73 $cm^2\ ohm^{-1}\ eq^{-1}$, respectively. The equivalent conductance at infinite dilution of the salt $\begin{matrix} COONa \\ | \\ COOK \end{matrix}$ is **Toughnut**
(a) 197 $cm^2\ ohm^{-1}\ eq^{-1}$
(b) 172 $cm^2\ ohm^{-1}\ eq^{-1}$
(c) 135.5 $cm^2\ ohm^{-1}\ eq^{-1}$
(d) 160.5 $cm^2\ ohm^{-1}\ eq^{-1}$

38. On passing a current of 1.0 ampere for 16 min and 5 sec through one litre solution of $CuCl_2$, all copper of the solution was deposited at cathode. The strength of $CuCl_2$ solution was (Molar mass of Cu= 63.5; Faraday constant = 96500 $Cmol^{-1}$)

(a) 0.01 N (b) 0.01 M
(c) 0.02 M (d) 0.2 N

39. Electrode potential of the half cell Pt (s) | Hg (l) | Hg_2Cl_2(s) | Cl^- (aq) can be increased by :
(a) Increasing [Cl^-]
(b) Decreasing [Cl^-]
(c) Increasing Hg_2Cl_2 (s)
(d) Decreasing Hg (l)

40. Identify the correct statement :
(a) Corrosion of iron can be minimized by forming a contact with another metal with a higher reduction potential
(b) Iron corrodes in oxygen free water
(c) Corrosion of iron can be minimized by forming an impermeable barrier at its surface
(d) Iron corrodes more rapidly in salt water because its electrochemical potential is higher

41. During the charging of lead storage battery, the reaction at anode is represented by :
(a) $Pb^{2+} + SO_4^{2-} \longrightarrow PbSO_4$
(b) $PbSO_4 + 2H_2O \longrightarrow PbO_2 + SO_4^{2-} + 4H^+ + 2e^-$
(c) $Pb \longrightarrow Pb^{2+} + 2e^-$
(d) $Pb^{2+} + 2e^- \longrightarrow Pb$

42. Which colourless gas evolves, when NH_4Cl reacts with zinc in a dry cell battery
(a) NH_4 (b) N_2 (c) H_2 (d) Cl_2

43. Given
$Fe^{3+}(aq) + e^- \rightarrow Fe^{2+}(aq)$; E° = + 0.77 V
$Al^{3+}(aq) + 3e^- \rightarrow Al(s)$; E° = – 1.66 V
$Br_2(aq) + 2e^- \rightarrow 2Br^-$; E° = + 1.09 V
Considering the electrode potentials, which of the following represents the correct order of reducing power?
(a) $Fe^{2+} < Al < Br^-$ (b) $Br^- < Fe^{2+} < Al$
(c) $Al < Br^- < Fe^{2+}$ (d) $Al < Fe^{2+} < Br^-$

44. How many electrons would be required to deposit 6.35 g of copper at the cathode during the electrolysis of an aqueous solution of copper sulphate? (Atomic mass of copper = 63.5 u, N_A = Avogadro's constant):

(a) $\frac{N_A}{20}$ (b) $\frac{N_A}{10}$ (c) $\frac{N_A}{5}$ (d) $\frac{N_A}{2}$

45. Given :

$E^o_{\frac{1}{2}Cl_2/Cl^-} = 1.36\,V$, $E^o_{Cr^{3+}/Cr} = -0.74\,V$,

$E^o_{Cr_2O_7^{2-}/Cr^{3+}} = 1.33\,V$, $E^o_{MnO_4^-/Mn^{2+}} = 1.51\,V$

The correct order of reducing power of the species (Cr, Cr^{3+}, Mn^{2+} and Cl^-) will be :

(a) $Mn^{2+} < Cl^- < Cr^{3+} < Cr$
(b) $Mn^{2+} < Cr^{3+} < Cl^- < Cr$
(c) $Cr^{3+} < Cl^- < Mn^{2+} < Cr$
(d) $Cr^{3+} < Cl^- < Cr < Mn^{2+}$

46. The oxidation potential of a hydrogen electrode at pH = 10 and p_{H_2} = 1 atm is

(a) –0.59 V (b) 0.00 V
(c) +0.59 V (d) 0.059 V

47. If the E^o_{cell} for a given reaction has a negative value, then which of the following gives the correct relationships for the values of ΔG^o and K_{eq}?

(a) $\Delta G^o > 0; K_{eq} > 1$ (b) $\Delta G^o < 0; K_{eq} > 1$
(c) $\Delta G^o < 0; K_{eq} < 1$ (d) $\Delta G^o > 0; K_{eq} < 1$

48. The EMF of the cell

Tl/Tl^+ (0.001M) || Cu^{2+} (0.01M) /Cu is 0.83. The cell EMF can be increased by

(a) Increasing the concentration of Tl^+ ions.
(b) Increasing the concentration of Cu^{2+} ions.
(c) Increasing the concentration of Tl^+ and Cu^{2+} ions.
(d) None of these

49. Small quantities of solutions of compounds TX, TY and TZ are put into separate test tubes containing X, Y and Z solution. TX does not react with any of these. TY reacts with both X and Z. TZ reacts with X. The decreasing order of state of oxidation of the anions X^-, Y^-, Z^- is

Tricky

(a) Y^-, Z^-, X^- (b) Z^-, X^-, Y^-
(c) Y^-, X^-, Z^- (d) X^-, Z^-, Y^-

50. An unknown metal M displaces nickel from nickel (II) sulphate solution but does not displace manganese from manganese sulphate solution. Which order represents the correct order of reducing power?

(a) Mn > Ni > M (b) Ni > Mn > M
(c) Mn > M > Ni (d) M > Ni > Mn

51. What is the e.m.f for the given cell?

$Cr \mid Cr^{3+}(1.0M) \parallel Co^{2+}(1.0M) \mid Co$

($E°$ for $Cr^{3+}/Cr = -0.74$ volt and $E°$ for $Co^{2+}/Co = -0.28$ volt)

(a) – 0.46 volt (b) – 1.02 volt
(c) + 0.46 volt (d) 1.66 volt

52. The standard electrode potential ($E°$) for OCl^-/Cl^- and $Cl^-/\frac{1}{2}Cl_2$ respectively are 0.94 V and –1.36 V. The $E°$ value for $OCl^-/\frac{1}{2}Cl_2$ will be

(a) –0.42 V (b) –2.20 V
(c) 0.52 V (d) 1.04 V

53. For the reduction of silver ions with copper metal, the standard cell potential was found to be + 0.46 V at 25°C. The value of standard Gibbs energy, $\Delta G°$ will be (F = 96500 C mol^{-1})

(a) – 89.0 kJ (b) – 89.0 J
(c) – 44.5 kJ (d) – 98.0 kJ

54. In a cell that utilises the reaction

$Zn(s) + 2H^+(aq) \rightarrow Zn^{2+}(aq) + H_2(g)$

addition of H_2SO_4 to cathode compartment, will

(a) increase the E and shift equilibrium to the right
(b) lower the E and shift equilibrium to the right
(c) lower the E and shift equlibrium to the left
(d) increase the E and shift equilibrium to the left

55. For a cell reaction involving two electron change, the standard EMF of the cell is 0.295 V at 2°C. The equilibrium constant of the reaction at 25°C will be:

(a) 29.5×10^{-2} (b) 10
(c) 1×10^{10} (d) 2.95×10^{-10}

56. On the basis of the information available from the reaction $\frac{4}{3}Al + O_2 \rightarrow \frac{2}{3}Al_2O_3, \Delta G = -827$ kJ mol^{-1} of O_2 the minimum e.m.f required to carry out an electrolysis of Al_2O_3 is ($F = 96500$ C mol^{-1})

Critical Thinking

(a) 8.56V (b) 2.14V (c) 4.28V (d) 6.42V

57. The e.m.f. of a Daniell cell at 298 K is E_1.

$$Zn\left|\begin{matrix} ZnSO_4 \\ (0.01\ M) \end{matrix}\right|\left|\begin{matrix} CuSO_4 \\ (1.0\ M) \end{matrix}\right|Cu$$

When the concentration of $ZnSO_4$ is 1.0 M and that of $CuSO_4$ is 0.01 M, the e.m.f. changed to E_2. What is the relationship between E_1 and E_2?

(a) $E_2 = 0 \neq E_1$ (b) $E_1 > E_2$

(c) $E_1 < E_2$ (d) $E_1 = E_2$

58. Mark the false statement?

(a) A salt bridge is used to eliminate liquid junction potential

(b) The Gibbs free energy change, ΔG is related with electromotive force E as $\Delta G = -nFE$

(c) Nernst equation for single electrode potential is $E = E^o - \frac{RT}{nF}\log a_{M^{n+}}$

(d) The efficiency of a hydrogen-oxygen fuel cell is 23%

59. The electrode potential $E_{(Zn^{2+}/Zn)}$ of a zinc electrode at 25°C with an aqueous solution of 0.1 M $ZnSO_4$ is [$E^o_{(Zn^{2+}/Zn)} = -0.76$ V. Assume $\frac{2.303RT}{F} = 0.06$ at 298 K].

(a) +0.73 (b) −0.79 (c) −0.82 (d) −0.70

60. The reduction potential (in volt) of a hydrogen electrode set up with a 2×10^{-2} M aqueous solution of a weak mono basic acid ($K_a = 5 \times 10^{-5}$) at one atmosphere and 25°C is

(a) +0.09 (b) +0.18 (c) −0.09 (d) −0.18

61. A solution of copper sulphate ($CuSO_4$) is electrolysed for 10 minutes with a current of 1.5 amperes. The mass of copper deposited at the cathode (at. mass of Cu = 63u) is :

(a) 0.3892g (b) 0.2938g

(c) 0.2398g (d) 0.3928g

62. A battery is constructed of Cr and $Na_2Cr_2O_7$. The unbalanced chemical equation when such a battery discharges is following:

$$Na_2Cr_2O_7 + Cr + H^+ \rightarrow Cr^{3+} + H_2O + Na^+$$

If one Faraday of electricity is passed through the battery during the charging, the number of moles of Cr^{3+} removed from the solution is

Tricky

(a) $\frac{4}{3}$ (b) $\frac{1}{3}$ (c) $\frac{3}{3}$ (d) $\frac{2}{3}$

63. The standard potentials of Ag^+/Ag, $Hg_2^{2+}/2Hg$, Cu^{2+}/Cu and Mg^{2+}/Mg electrodes are 0.80, 0.79, 0.34 and −2.37 V, respectively. An aqueous solution which contains one mole per litre of the salts of each of the four metals is electrolyzed. With increasing voltage, the correct sequence of deposition of the metals at the cathode is

(a) Ag, Hg, Cu, Mg (b) Cu, Hg, Ag only

(c) Ag, Hg, Cu only (d) Mg, Cu, Hg, Ag

64. When electric current is passed through a cell having an electrolytic solution, the cations move towards the cathode and anions towards the anode. If anode is pulled out from the solution

(a) the cations and anions will move towards the cathode

(b) the cations will continue to move towards cathode and anions will stop moving

(c) both the cations and anions will stop moving

(d) the cations and anions will start moving randomly.

65. At 298K the standard free energy of formation of H_2O (l) is −237.20 kJ/mol while that of its ionisation into H^+ ion and hydroxyl ions is 80 kJ/mol, then the emf of the following cell at 298 K will be

Toughnut

[Take Faraday constant F = 96500 C]

H_2(g, 1 bar) | H^+ (1M) || OH^- (1M) | O_2(g, 1 bar)

(a) 0.40 V (b) 0.81 V

(c) 1.23 V (d) −0.40 V

66. A concentration cell is a galvanic cell in which

(a) decrease in free energy in a spontaneous chemical process appears as electrical energy

(b) decrease in free energy in a spontaneous physical process appears as electrical energy

(c) decrease in free energy in a spontaneous physical or chemical process appears as electrical energy

(d) a non-spontaneous physical or chemical process produces electrical energy.

67. In the silver plating of copper, $K[Ag(CN)_2]$ is used instead of $AgNO_3$. The reason is
(a) a thin layer of Ag is formed on Cu
(b) more voltage is required
(c) Ag^+ ions are completely removed from solution
(d) less availability of Ag^+ ions, as Cu cannot displace Ag from $[Ag(CN)_2]^-$ ion

68. 1.0 L each of a buffer containing 1 mole NH_3 and 1 mol of NH_4^+ were placed in the cathodic and anodic half-cells and 965 C of electricity was passed. If anodic and cathodic half cells reactions involve oxidation and reduction of water only as

$$2H_2O \longrightarrow 4H^+ + O_2 - 4e^-;$$
$$2H_2O + 2e^- \longrightarrow H_2 + 2OH^-$$

Then pH of 
(a) cathodic solution will increase
(b) anodic solution will decrease
(c) both the solutions will remain practically constant
(d) both the solutions will increase

69. The emf of a particular voltaic cell with the cell reaction $Hg_2^{2+} + H_2 \rightleftharpoons 2Hg + 2H^+$ is 0.65 V. The maximum electrical work of this cell when 0.5 g of H_2 is consumed.
(a) -3.12×10^4 J (b) -1.25×10^5 J
(c) 25.0×10^6 J (d) None of these

70. A cell reaction would be spontaneous if the cell potential and $\Delta_r G$ are respectively :
(a) positive and negative
(b) negative, negative
(c) zero, zero
(d) positive, zero

71. The Nernst equation E = E° – RT/nF ln Q indicates that the Q will be equal to equilibrium constant K_c when :
(a) E = E° (b) RT / nF = 1
(c) E = zero (d) E° = 1

72. Thermodynamic efficiency of a cell is given by:
(a) $\frac{\Delta H}{\Delta G}$ (b) $\frac{nFE}{\Delta G}$ (c) $\frac{-nFE}{\Delta H}$ (d) nFE°

73. A graph was plotted between molar conductivity of various electrolytes (NaCl, HCl and NH_4OH) and $\sqrt{c}$ (in mol L^{-1}). Correct set is : Tricky

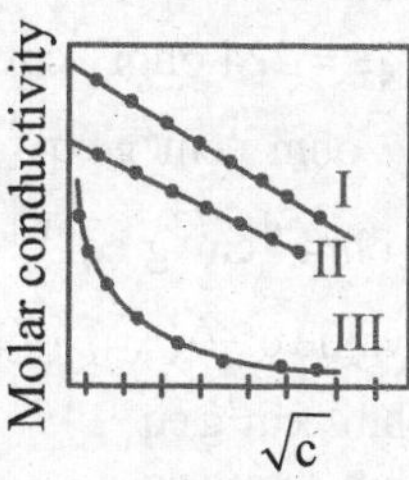

(a) I (NaCl), II (HCl), III, (NH_4OH)
(b) I (HCl), II (NaCl), III, (NH_4OH)
(c) I (NH_4OH), II (NaCl), III, (HCl)
(d) I (NH_4OH), II (HCl), III, (NaCl)

74. HNO_3 (aq) is titrated with NaOH(aq) conductometrically, graphical representation of the titration is :

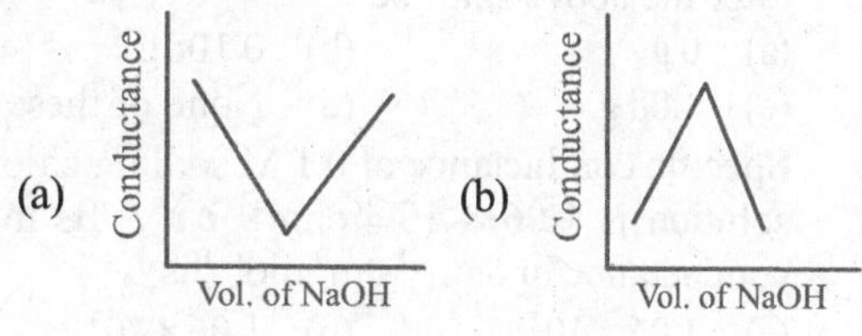

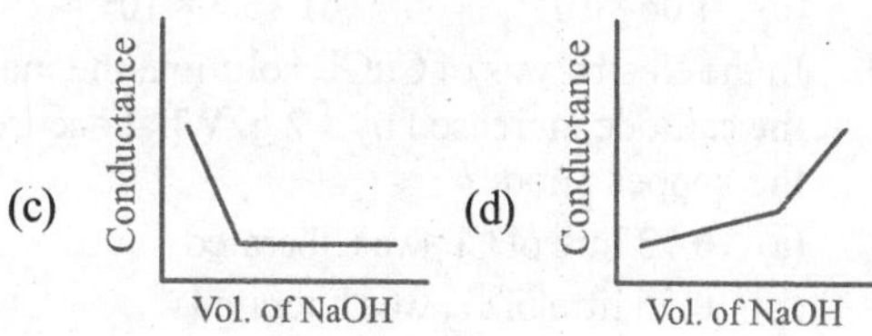

75. In the electrolysis of water, one faraday of electrical energy would liberate
(a) one mole of oxygen
(b) one gram atom of oxygen
(c) 8 g oxygen
(d) 22.4 lit. of oxygen

76. Resistance of a conductivity cell filled with a solution of an electrolyte of concentration 0.1 M is 100 Ω. The conductivity of this solution is 1.29 S m^{-1}. Resistance of the same cell when filled with 0.02 M of the same solution is 520 Ω. The molar conductivity of 0.02 M solution of electrolyte will be Critical Thinking
(a) 1.24×10^{-4} S m^2 mol^{-1}
(b) 12.4×10^{-4} S m^2 mol^{-1}
(c) 124×10^{-4} S m^2 mol^{-1}
(d) 1240×10^{-4} S m^2 mol^{-1}

77. $\Lambda_{ClCH_2COONa} = 224\ ohm^{-1}cm^2g\ eq^{-1}$,

$\Lambda_{NaCl} = 38.2\ ohm^{-1}cm^2g\ eq^{-1}$,

$\Lambda_{HCl} = 203\ ohm^{-1}cm^2g\ eq^{-1}$,

What is the value of Λ_{ClCH_2COOH}

(a) $288.5\ ohm^{-1}cm^2g\ eq^{-1}$
(b) $289.5\ ohm^{-1}cm^2g\ eq^{-1}$
(c) $388.8\ ohm^{-1}cm^2g\ eq^{-1}$
(d) $59.5\ ohm^{-1}cm^2g\ eq^{-1}$

78. 1.08 g of pure silver was converted into silver nitrate and its solution was taken in a beaker. It was electrolysed using platinum cathode and silver anode. 0.01 Faraday of electricity was passed using 0.15 volt above the decomposition potential of silver. The silver content of the beaker after the above shall be

(a) 0 g (b) 0.108 g
(c) 1.08 g (d) None of these

79. Specific conductance of 0.1 M sodium chloride solution is $1.06 \times 10^{-2}\ ohm^{-1}\ cm^{-1}$. Its molar conductance in $ohm^{-1}\ cm^2\ mol^{-1}$ is

(a) 1.06×10^2 (b) 1.06×10^3
(c) 1.06×10^4 (d) 5.3×10^2

80. In the electrolysis of $CuCl_2$ solution, the mass of the cathode increased by 3.2 g. What occured at the copper anode?

(a) 0.12 litre of Cl_2 was liberated
(b) 0.56 litre of O_2 was liberated
(c) 0.1 mol Cu^{2+} passed into the solution.
(d) 0.05 mol of Cu^{2+} passed into the solution.

81. Given

$E^\circ_{Cu^{2+}/Cu} = 0.34V$, $E^\circ_{Cu^{2+}/Cu} = 0.15V$

Standard electrode potential for the half cell Cu^+/Cu is

(a) 0.38 V (b) 0.53 V (c) 0.19 V (d) 0.49 V

82. At 298 K, the standard reduction potentials are 1.51 V for $MnO_4^- | Mn^{2+}$, 1.36 V for $Cl_2 | Cl^-$, 1.07 V for $Br_2 | Br^-$, and 0.54 V for $I_2 | I^-$. At pH = 3, permanganate is expected to oxidize : $\left(\frac{RT}{F} = 0.059V\right)$ Toughnut

(a) Cl^-, Br^- and I^- (b) Br^- and I^-
(c) Cl^- and Br^- (d) I^- only

83. Electrode potentials (E°) are given below :

$Cu^+/Cu = +0.52\ V$,
$Fe^{3+}/Fe^{2+} = +0.77\ V$,
$\frac{1}{2}I_2(s)/I^- = +0.54\ V$,
$Ag^+/Ag = +0.88V$.

Based on the above potentials, strongest oxidizing agent will be :

(a) Cu^+ (b) Fe^{3+} (c) Ag^+ (d) I_2

84. Which of the following is a highly corrosive salt?

(a) Hg_2Cl_2 (b) $HgCl_2$
(c) $FeCl_2$ (d) $PbCl_2$

85. The Zn acts as sacrificial or cathodic protection to prevent rusting of iron because :

(a) E°_{OP} of Zn < E°_{OP} of Fe
(b) E°_{OP} of Zn > E°_{OP} of Fe
(c) E°_{OP} of Zn = E°_{OP} of Fe
(d) Zn is cheaper than iron

86. Among the following cells:

(I) Leclanche cell
(II) Nickel-Cadmium cell
(III) Lead storage battery
(IV) Mercury cell

primary cells are

(a) I and II (b) I and III
(c) II and III (d) I and IV

87. How long a current of 3 amperes has to be passed through a solution of $AgNO_3$ to coat a metal surface of 80 cm² and 0.005 mm thick layer. Density of Ag is $10.5\ g\ cm^{-3}$

(a) 125.1 seconds (b) 12.5 seconds
(c) 155.2 seconds (d) 200 seconds

88. Al_2O_3 is reduced by electrolysis at low potentials and high currents. If 4.0×10^4 amperes of current is passed through molten Al_2O_3 for 6 hours, what mass of aluminium is produced? (Assume 100% current efficiency. At. mass of Al = 27 g mol⁻¹)

(a) 8.1×10^4 g (b) 2.4×10^5 g
(c) 1.3×10^4 g (d) 9.0×10^3 g

89. The resistance of 1 N solution of acetic acid is 250 ohm, when measured in a cell of cell constant $1.15\ cm^{-1}$. The equivalent conductance (in $ohm^{-1}\ cm^2\ equiv^{-1}$) of 1 N acetic acid will be

(a) 4.6 (b) 9.2 (c) 18.4 (d) 0.023

90. Specific conductance of 0.1 *M* HA is $3.75 \times 10^{-4}\ ohm^{-1}\ cm^{-1}$. If $\lambda^\infty(HA) = 250\ ohm^{-1}\ cm^2\ mol^{-1}$, the dissociation constant K_a of HA is :

(a) 1.0×10^{-5} (b) 2.25×10^{-4}
(c) 2.25×10^{-5} (d) 2.25×10^{-13}

91. Equivalent conductance at infinite dilution, λ° of NH_4Cl, NaOH and NaCl are 128.0, 217.8 and 109.3 $ohm^{-1}\ cm^2\ eq^{-1}$ respectively. The equivalent conductance of 0.01 N NH_4OH is 9.30 $ohm^{-1}\ cm^2\ eq^{-1}$, then the degree of ionization of NH_4OH at this temperature would be

(a) 0.04 (b) 0.1 (c) 0.39 (d) 0.62

92. I_2 (s) $|I^-$ (0.1 M) half cell is connected to a H^+ (aq) $|H_2$ (1 bar) $|$ Pt half cell and e.m.f. is found to be 0.7714 V. if $E^\circ_{I_2|I^-} = 0.535$ V, find the pH of $H^+ | H_2$ half-cell. **Toughnut**

(a) 1 (b) 3 (c) 5 (d) 7

93. Consider the following standard electrode potentials and calculate the equilibrium constant at 25° C for the indicated disproportion nation reaction :

$$3Mn^{2+}(aq) \longrightarrow Mn(s) + 2Mn^{3+}(aq)$$

$$Mn^{3+}(aq) + e^- \longrightarrow Mn^{2+}(aq); E^\circ = 1.51\ V$$

$$Mn^{2+}(aq) + 2e^- \longrightarrow Mn(s); \quad E^\circ = -1.185\ V$$

(a) 1.2×10^{-43} (b) 2.4×10^{-73}
(c) 6.3×10^{-92} (d) 1.5×10^{-62}

94. In a hydrogen-oxygen fuel cell, combustion of hydrogen occurs to

(a) produce high purity water
(b) create potential difference between two electrodes
(c) generate heat
(d) remove adsorbed oxygen from elctrode surfaces

95. The standard reduction potentials at 25°C of Li^+/Li, Ba^{2+}/Ba, Na^+/Na and Mg^{2+}/Mg are $-3.03, -2.73, -2.71$ and -2.37 volt respectively. Which one of the following is the strongest oxidising agent?

(a) Na^+ (b) Li^+ (c) Ba^{2+} (d) Mg^{2+}

96. The $E^\circ_{M^{3+}/M^{2+}}$ values for Cr, Mn, Fe and Co are $-0.41, +1.57, +0.77$ and $+1.97$V respectively. For which one of these metals the change in oxidation state from +2 to +3 is easiest?

(a) Fe (b) Mn (c) Cr (d) Co

97. The solution of $CuSO_4$ in which copper rod is immersed is diluted to 10 times. The reduction electrode potential **Toughnut**

(a) Increases by 30 mV
(b) Decreases by 30 mV
(c) Increases by 59 mV
(d) Decreases by 59 mV

98. For a given reaction: $M^{(x+n)} + ne^- \longrightarrow M^{x+}$, E^-_{red} is known along with $M^{(x+n)}$ and M^{x+} ion concentrations. Then

(a) n can be evaluated
(b) x can be evaluated
(c) (x + n) can be evaluated
(d) n, x, (x + n) can be evaluated

99. The emf of the cell Pt Cl_2 (g) (P_1 atm)/ Cl^- (aq) (1M)/ Cl_2 (g) (P_2 atm) Pt will be positive when

(a) $P_1 = P_2$ (b) $P_1 < P_2$
(c) $P_1 > P_2$ (d) None of these

100. Equivalent conductivity can be expressed in terms of specific conductance (κ) and concentration (N) in gram equivalent per dm^{-3} as :

(a) $\kappa \times N$ (b) $\dfrac{\kappa \times 1000}{N}$

(c) $\dfrac{\kappa \times N}{1000}$ (d) $\kappa \times N \times 1000$

Answer KEYs

1	(c)	11	(a)	21	(d)	31	(c)	41	(b)	51	(c)	61	(b)	71	(c)	81	(b)	91	(a)
2	(b)	12	(d)	22	(c)	32	(b)	42	(c)	52	(a)	62	(c)	72	(c)	82	(b)	92	(b)
3	(a)	13	(c)	23	(d)	33	(c)	43	(d)	53	(a)	63	(c)	73	(b)	83	(c)	93	(c)
4	(a)	14	(c)	24	(b)	34	(b)	44	(c)	54	(a)	64	(d)	74	(a)	84	(b)	94	(b)
5	(d)	15	(c)	25	(d)	35	(a)	45	(a)	55	(c)	65	(a)	75	(c)	85	(b)	95	(d)
6	(c)	16	(d)	26	(a)	36	(c)	46	(c)	56	(b)	66	(b)	76	(a)	86	(d)	96	(c)
7	(b)	17	(a)	27	(c)	37	(c)	47	(d)	57	(b)	67	(d)	77	(c)	87	(a)	97	(b)
8	(c)	18	(c)	28	(b)	38	(a)	48	(b)	58	(c)	68	(c)	78	(a)	88	(a)	98	(a)
9	(d)	19	(b)	29	(d)	39	(b)	49	(a)	59	(b)	69	(a)	79	(a)	89	(a)	99	(b)
10	(a)	20	(c)	30	(a)	40	(c)	50	(c)	60	(d)	70	(a)	80	(d)	90	(c)	100	(b)

Hints & Solutions

1. (c) Cathode and anode are (–) ve terminals in electrolytic and galvanic cell.

2. (b) Ag becomes oxidized and Br^- becomes reduced.

3. (a) $A : H_2(g) \rightleftharpoons 2H^+(aq) + 2e^-$

   $C : Cu^{2+}(aq) + 2e^- \rightleftharpoons Cu(s)$

4. (a) Since oxidation potential of Zn is highest hence strongest reducing agent.

5. (d) The reduction potential of NO_3^- ion is more than H^+ ion.

6. (c) Gold having higher E^o_{Red} and oxidises $Fe \longrightarrow Fe^{2+}$.

7. (b) Given $E_{Sn^{4+}/Sn^{2+}} = +0.15\,V$

   $E_{Cr^{3+}/Cr} = -0.74\,V$

   $E^o_{cell} = E^o_{cathode} - E^o_{anode}$
   $= 0.15 - (-0.74) = +0.89\,V$

8. (c) Given $Fe^{3+}/Fe^{2+} = +0.77\,V$
   and $I_2/2I^- = 0.536 V$
   $2\,(e^- + Fe^{3+} \longrightarrow Fe^{2+})\quad E^\circ = 0.77\,V$
   $2I^- \longrightarrow I_2 + 2e^- \quad E^\circ = -0.536\,V$
   $2Fe^{3+} + 2I^- \longrightarrow 2Fe^{2+} + I_2$
   $E^\circ = E^\circ_{ox} + E^\circ_{red}$
   $= 0.77 - 0.536 = 0.164\,V$
   So, reaction will take place.

9. (d) $E_{red} = E^o_{red} + \frac{0.591}{n}\log[M^{n+}]$

   Lower the concentration of M^{n+}, lower is the reduction potential.
   Hence order of reduction potential is :
   $Q > R > S > P$

10. (a) $2Ag^+ + H_2 \longrightarrow 2H^+ + 2Ag$

   $$E = E^\circ - \frac{0.0591}{2}\log\frac{[H^+]^2}{P_{H_2} \times [Ag^+]^2}$$

   $$0.222 = 0.7995 - \frac{0.0591}{2}\log\frac{1}{[Ag^+]^2}$$

   $[Ag^+] = 10^{-9.8}$
   $K_{sp} = [Ag^+][Cl^-] = (10^{-9.8}) \times (1) = 10^{-9.8}$

11. (a) Apply Nernst equation to the reaction
   $Pb + Sn^{2+} \rightarrow Pb^{2+} + Sn$

   $$E_{cell} = E^\circ + \frac{0.059}{2}\log\frac{[Sn^{2+}]}{[Pb^{2+}]}$$

   or $\log\frac{[Sn^{2+}]}{[Pb^{2+}]} = \frac{0.01 \times 2}{0.059} = 0.3 \quad (\because E_{cell} = 0)$

   or $\frac{[Sn^{2+}]}{[Pb^{2+}]} = \text{antilog}\ (0.3)$

12. (d) For a concentration cell having different concentrations of ions.

   $$E = -\frac{0.0591}{n}\log\frac{c_1}{c_2}$$

   If all the concentrations are identical then obviously the cell voltage is zero. But as the pH of 0.1 M HCl (strong acid) and pH of 0.1M CH_3COOH is (weak acid) not same, therefore the cell voltage is not zero.

13. (c) EMF of galvanic cell = 1.1 volt
   If E_{ext} < EMF then electron flows steadily from anode to cathode while If E_{ext} > EMF then electron flows from cathode to anode as polarity is changed.

14. (c) As the value of reduction potential decreases the reducing power increases i.e.
   C < B < D < A
   (0.85). (0.6). (–0.76) (–1.2)

15. (c) According to Faraday's first law of electrolysis

   $$W = \frac{E \times i \times t}{96500}$$

   Where E = equivalent weight

   $$= \frac{\text{mol. mass of metal (M)}}{\text{oxidation state of metal } (x)}$$

   Substituting the value in the formula

   $$W = \frac{M}{x} \times \frac{i \times t}{96500}$$

   or $x = \frac{M}{W} \times \frac{i \times t}{96500} = \frac{10 \times 2 \times 60 \times 60}{96500 \times 0.250} = 3$

   $\left[\text{Given : no. of moles} = \frac{M}{W} = 0.250\right]$

   Hence oxidation state of metal is (+3)

16. (d) Oxidation potential of Co is more than Ag, hence cell reaction will be

$$Co + 2Ag^+ \longrightarrow Co^{2+} + 2Ag$$

$E_{cell} = E^o_{cell} - \frac{RT \ln}{nF} \frac{[Co^{2+}]}{[Ag^+]^2}$, the lesser the value of the factor $\frac{[Co^{2+}]}{[Ag^+]^2}$ greater will be value of E_{cell}.

17. (a) Since concentration of ions is the same hence $E_{cell} = E^o_{cell}$

18. (c) $Cd(s) + 2\,Ag^+(aq) \longrightarrow Cd^{2+}(aq) + Ag\,(s)$,

$$E_{cell} = E^o_{cell} - \frac{0.059}{2} \log \frac{[Cd^{2+}]}{[Ag^+]^2}$$

19. (b) $E_{cell} = E^o_{cell} - \frac{0.0592}{2} \log \frac{[Hg_2^{2+}(aq)]}{[Ag^+(aq)]^2}$

$$= (0.80 - 0.785) - \frac{0.0592}{2} \log \frac{10^{-1}}{(10^{-3})^2} = -0.133\text{V}$$

hence backward reaction is spontaneous.

20. (c) $E^o_{cell} = E^o_{RP(RHS)} - E^o_{RP\,(LHS)}$

$= 1.61 - 1.51 \Rightarrow 0.10\text{ V}$

$\Delta G° = -nFE° \Rightarrow -5 \times 96500 \times 0.10\text{ J}$

$\Delta G° = -48.25\text{ kJ mol}^{-1}$

21. (d) Molarity = 0.01 M ; Resistance = 40 ohm;

Cell constant $\frac{l}{A} = 0.4\text{cm}^{-1}$.

Specific conductivity (κ)

$$= \frac{\text{cell constant}}{\text{resistance}} = \frac{0.4}{40} = 0.01\text{ ohm}^{-1}\text{ cm}^{-1}$$

Molar conductance $(\wedge_m) = \frac{1000\kappa}{\text{Molarity}}$

$$= \frac{1000 \times 0.01}{0.01} = 10^3\text{ ohm}^{-1}\text{ cm}^2\text{ mol}^{-1}$$

22. (c) On dilution number of ions decrease in unit volume hence specific conductance decreases. But separation between ions also increase hence equivalent conductance increases.

23. (d) $\Lambda^\infty_m(NaBr) = \Lambda^\infty_m(NaCl) + \Lambda^\infty_m KBr - \Lambda^\infty_m(KCl)$

$\lambda^\infty_m(Na^+) + \lambda^\infty_m(Br^-) = 126.5 + 151.5 - 150$

$\lambda^\infty_m(Na^+) = 50\text{ S cm}^2\text{ eq}^{-1}$.

24. (b) Equivalent conductivity of $BaCl_2$

$\Lambda^\infty_m(BaCl_2) = \lambda^\infty_m(Ba^{2+}) + 2\lambda^\infty_m(Cl^-)$

$= 127 + 2 \times 76 = 279\text{ ohm}^{-1}\text{ cm}^2\text{ eq}^{-1}$.

25. (d) 1 mole of $e^- = 1F = 96500\text{ C}$

27g of Al is deposited by 3×96500 C

5120 g of Al will be deposited by

$$= \frac{3 \times 96500 \times 5120}{27} = 5.49 \times 10^7\text{ C}$$

26. (a) $2H^+ + 2e^- \longrightarrow H_2$

$E_H(\text{Eq. wt}) = \frac{2}{2} = 1$

$$1\text{ g} = \frac{22400}{2} = 11200\text{ mL (STP)}$$

Total charge passed $= \frac{96500 \times 112}{11200} = 965$

$Q = It = 965$

$I = \frac{965}{965} = 1\text{ amp.}$

27. (c) Using Faraday's second law of electrolysis,

$$\frac{\text{Weight of Cu deposited}}{\text{Weight of Ag deposited}} = \frac{\text{Equ. wt. of Cu}}{\text{Equ. wt. of Ag}}$$

$$\Rightarrow \frac{w_{Cu}}{0.18} = \frac{63.5}{2} \times \frac{1}{108}$$

$$\Rightarrow w_{Cu} = \frac{63.5 \times 18}{2 \times 108 \times 100} = 0.0529\text{ g.}$$

28. (b) Specific resistance = x

$\therefore$ Specific conductance (or conductivity)

$$= \kappa = \frac{1}{x}.$$

$$\therefore \Lambda_{eq} = \frac{\kappa \times 1000}{N} = \frac{1000}{xN}$$

29. (d) $\Lambda_m = \frac{\kappa \times 1000}{M} = \frac{0.0110 \times 1000}{0.05}$

$= 220\text{ S cm}^2\text{ mol}^{-1}$

30. (a) Difluoroacetic acid being strongest acid will furnish maximum number of ions showing highest electrical conductivity.

31. (c) $m = \frac{E.wt \times Q}{96500}$;

$\therefore E.wt = \frac{m \times 96500}{Q} = \frac{22.2 \times 96500}{2 \times 5 \times 60 \times 60} = 60.3$

Oxidation state $= \frac{\text{At wt.}}{\text{Eq. wt.}} = \frac{177}{60.3} = 3$

32. (b) Cell constant

$= \kappa \times R$

$= 0.212\ ohm^{-1}\ cm^{-1} \times 55\ ohm$

$= 11.66\ cm^{-1}$

33. (c) Kohlrausch Law, $\Lambda_{eq}(NaF) + \Lambda_{eq}(HCl) - \Lambda_{eq}(NaCl) = \Lambda_{eq}(HF)$

34. (b) Conductivity (X) = conductance (c) × cell constant

$\therefore$ Cell constant $= \frac{X}{Y}$

Conductivity of NaOH $= \frac{X}{Y}.Z$

Molar conductance of NaOH

$= \frac{X}{Y}.Z \times \frac{1000}{0.1} = \frac{XZ}{Y} 10^4$

35. (a) $ArNO_2 + 6H^+ + 6e^- \longrightarrow ArNH_2 + 2H_2O$

$W = \frac{E.wt \times Q}{96500}; 20.50 = \frac{M}{6} \times \frac{2 \times 96500}{96500} \times \frac{50}{100}$

$\therefore M = 123.0\ g$

36. (c) Moles of NaOH formed $= \frac{600 \times 1}{100} = 0.6$

or geq of NaOH = 0.6 geq of Cu deposited

$= \frac{31.8}{63.6/2} = 1.0$

$\therefore$ Current efficiency $= \frac{0.6 \times 100}{1} = 60\%$

37. (c) Total charge = 2

Number of equivalent of ion

$= \frac{\text{Charge on the ion}}{\text{Total charge}}$

$\therefore$ Eq of $\begin{pmatrix} COO^- \\ | \\ COO^- \end{pmatrix} = \frac{2}{2} = 1$

Eq of $Na^+ = \frac{1}{2}$, Eq of $K^+ = \frac{1}{2}$

$\therefore \lambda^\circ_{eq} \begin{pmatrix} COONa \\ | \\ COOK \end{pmatrix}$

$= \lambda^\circ_{eq} \begin{pmatrix} COO^- \\ | \\ COO^- \end{pmatrix} + \frac{1}{2}\lambda^\circ Na^+ + \frac{1}{2}\lambda^\circ K^+$

$= 74 + \frac{50}{2} + \frac{73}{2} = 135.5\ ohm^{-1}\ cm^2\ eq^{-1}$

38. (a) By Faraday's 1st law of electrolysis,

$\frac{W}{E} = \frac{Q}{96500}$ (where Q = it = charge of ion)

We know that no. of gram equivalent

$= \frac{W}{E} = \frac{it}{96500} = \frac{1 \times 965}{96500} = \frac{1}{100}$

(where i = 1 A, t = 16 × 60 + 5 = 965 sec.)

Since, we know that

Normality

$= \frac{\text{No. of gram equivalent}}{\text{Volume (in litre)}} = \frac{\frac{1}{100}}{1} = 0.01\ N$

39. (b) $E = E^\circ - \frac{0.0591}{2} \log \frac{1}{[Cl^-]^2}$; as $[Cl^-] \downarrow, E \uparrow$

40. (c) Corrosion of iron can be minimized by forming an impermeable barrier at its surface.

41. (b) During charging, the lead storage battery behaves like an electrolytic cell. So, at anode the reaction is

$PbSO_4 + 2H_2O \longrightarrow PbO_2 + 4H^+ + SO_4^{2-} + 2e^-$

42. (c) $2NH_4Cl + Zn \rightarrow 2NH_3 + ZnCl_2 + H_2 \uparrow$.

43. (d) Reducing character decreases down the series. Hence the correct order is

$Al < Fe^{2+} < Br^-$

44. (c) $Cu \longrightarrow Cu^{++} + 2e^-$

i.e, to deposit 1 mole of Cu at cathode from $Cu^{2+}\ SO_4^{2-}$ solution = 2 moles of electrons are required

i.e, To deposit

$6.35g = \frac{6.35}{63.5} \times 2 = \frac{2}{10} = \frac{1}{5}$ moles

Thus total no. of electrons required $= \frac{N_A}{5}$

45. (a) Lower the value of reduction potential higher will be reducing power hence the correct order will be
$Mn^{2+} < Cl^- < Cr^{3+} < Cr$

46. (c) $E_{oxidation} = 0.059\,pH = 0.059 \times 10 = 0.59\,V$

47. (d) Standard Gibbs free energy is given as
$\Delta G° = -nE°F$
If $E°_{cell} < 0$ i.e. – ve
$\Delta G° > 0$
Further $\Delta G° = -RT\,\ell n\,K_{eq}$

48. (b) The oxidation potential
$\propto \frac{1}{\text{Concentration of ions}}$ and reduction potential $\propto$ concentration of ions. The cell voltage can be increased by decreasing the concentration of ions around anode or by increasing the concentration of ions around cathode

49. (a) Oxidising tendency $\propto \frac{1}{\text{Electrode potential}}$
$TX \longrightarrow$ No reaction
$TY \longrightarrow X, Z$
$TZ \longrightarrow X$
$\Rightarrow$ order of electrode potential is
$TY < TZ < TX$
$\Rightarrow$ Order of oxidation of the anion is
$Y^- > Z^- > X^-$

50. (c) Oxidation potential of M is more than Ni and less than Mn. Hence reducing power Mn > M > Ni

51. (c) $E°_{Cr^{3+}/Cr} = -0.74\,V$, $E°_{Co^{2+}/Co} = -0.28\,V$
The given cell reaction is
$Cr\,|\,Cr^{3+}(1.0\,M)\,||\,Co^{2+}(1.0\,M)\,|\,Co$
$\therefore$ Cr is anode and Co is cathode
$E°_{cell} = E°_C - E°_A = -0.28 - (-0.74)$
$= -0.28 + 0.74 = +0.46\,V$

52. (a) $OCl^- \longrightarrow Cl^-;\ E° = 0.94\,V$ (I)
$Cl^- \longrightarrow \frac{1}{2}Cl_2 + e^-;\ E° = 1.36\,V$ (II)
Add (I) + (II) $OCl^- \longrightarrow \frac{1}{2}Cl_2;\ E° = 0.94 - 1.36$
$= -0.42\,V$

53. (a) $Cu + 2Ag^+(aq) \longrightarrow Cu^{2+}(aq) + 2Ag(s)$
Here, $n = 2$, $E°_{cell} = +0.46\,V$
$\Delta G° = -nE°F$
$= \frac{-2 \times 0.46 \times 96500}{1000}\,kJ \simeq -89\,kJ$

54. (a) $Zn(s) + 2H^+(aq) \rightleftharpoons Zn^{2+}(aq) + H_2(g)$
$E_{cell} = E°_{cell} - \frac{0.059}{2}\log\frac{[Zn^{2+}][H_2]}{[H^+]^2}$
Addition of H_2SO_4 will increase $[H^+]$ and E_{cell} will also increase and the equilibrium will shift towards RHS

55. (c) Using the relation,
$E°_{cell} = \frac{2.303RT}{nF}\log K_{eq} = \frac{0.0591}{n}\log K_{eq}$
$\therefore 0.295\,V = \frac{0.0591}{2}\log K_{eq}$
or $\log K_{eq} = \frac{2 \times 0.295}{0.0591} = 10$
or $K_{eq} = 1 \times 10^{10}$

56. (b) $\Delta G = -nEF$
For 1 mole of Al, $n = 3$
$\therefore$ for $\frac{4}{3}$ mole of Al, $n = 3 \times \frac{4}{3} = 4$
According to question,
$-827 \times 1000 = -4 \times E \times 96500$
$E = \frac{827 \times 1000}{4 \times 96500} = 2.14V$
$\therefore$ minimum e.m.f. required = 2.14 V

57. (b) Cell reaction is, $Zn + Cu^{2+} \rightarrow Zn^{2+} + Cu$
$E_{cell} = E°_{cell} - \frac{RT}{nF}\ell n\frac{[Zn^{2+}]}{[Cu^{2+}]}$
Greater the factor $\left[\frac{(Zn^{2+})}{(Cu^{2+})}\right]$, less is the EMF
Hence $E_1 > E_2$

58. (c) Correct Nernst equation is
$E = E° + \frac{2.303RT}{nF}\log a_{M^{n+}}$.

59. (b) For $Zn^{2+} \longrightarrow Zn$
$E_{Zn^{2+}/Zn} = E°_{Zn^{2+}/Zn} - \frac{2.303RT}{nF}\log\frac{[Zn]}{[Zn^{2+}]}$

$$= -0.76 - \frac{0.06}{2}\log\frac{1}{[0.1]} = -0.76 - 0.03$$

$$E_{Zn^{2+}/Zn} = -0.79V$$

60. **(d)** $2H^+ + 2e^- \longrightarrow H_2$

$E_{cell} = 0.06 \log [H^+] = 0.06 \times \log\sqrt{K_a C}$

$= 0.06 \times \log 10^{-3} = -0.18$ V

61. **(b)** W = Zit

where Z = Electrochemical equivalent

Eq. wt. of copper $= \frac{63}{2} = 31.5$

$Z = \frac{31.5}{96500}$

$W = Zit = \frac{31.5}{96500} \times 1.5 \times 10 \times 60 = 0.2938g$

62. **(c)** Reduction half reaction :

$Cr_2O_7^{2-} + 6e^- + 14H^+ \longrightarrow 2Cr^{3+} + 7H_2O$

Oxidation half reaction :

$Cr \longrightarrow Cr^{3+} + 3e^-$

Overall reaction :

$Cr_2O_7^{2-} + Cr + 14H^+ + 3e^- \longrightarrow 3Cr^{3+} + 7H_2O$

3F of electricity = 3 moles of Cr^{3+}

1F of electricity $= \frac{3}{3}$ moles of Cr^{3+}

63. **(c)** More the reduction potential, more is the deposition of metals at cathode. Cation having E^0 value less than – 0.83V (reduction potential of H_2O) will not deposit from aqueous solution. Hence correct order of A deposition of the metal at the cathode is

Ag > Hg > Cu

64. **(d)** On pulling out one of the electrodes the electric field applied to the solution disappears and hence the ions start moving randomly.

65. **(a)** Cell reaction

cathode :

$H_2O(l) + \frac{1}{2}O_2(g) + 2e^- \longrightarrow 2OH^-(aq)$

anode : $H_2(g) \longrightarrow 2H^+(aq) + 2e^-$

$H_2O(l) + \frac{1}{2}O_2(g) + H_2(g) \longrightarrow 2H^+(aq) + 2OH^-(aq)$

Also we have

$H_2(g) + \frac{1}{2}O_2(g) \longrightarrow H_2O(l)$;

$\Delta G_f^\circ = -237.2$ kJ / mole

$H_2O(l) \longrightarrow H^+(aq) + OH^-(aq)$;

$\Delta G^\circ = 80$ kJ / mol

Hence for cell reaction

$\Delta G^\circ = -237.2 + (2 \times 80) = -77.20$ kJ / mol

$\therefore\ E^\circ = -\frac{\Delta G^\circ}{nF} = \frac{77200}{2 \times 96500} = -0.40V$

66. **(b)** In concentration cell the spontaneous process is physical in nature involving transfer of matter from higher concentration to lower concentration in indirect manner.

67. **(d)** In the silver plating of copper, $K[Ag(CN)_2]$ is used instead of $AgNO_3$. Copper being more electropositive readily precipitate silver from their salt solution

$Cu + 2AgNO_3 \longrightarrow Cu(NO_3)_2 + Ag$

whereas in $K[Ag(CN)_2]$ solution a complex anion $[Ag(CN)_2]^-$ is formed and hence Ag^+ are less available in the solution and therefore copper cannot displace Ag from its complex ion.

68. **(c)** Due to buffer action the pH will remain practically constant.

69. **(a)** $W_{max} = -n.FE$;

$W_{max} = -2 \times 96500 \times 0.65 = -1.25 \times 10^5$ J

0.5g $H_2 = 0.25$ mole

Hence W_{max}

$= -1.25 \times 10^5 \times 0.25 = -3.12 \times 10^4$ J

70. **(a)** For spontaneous

$\Delta_r G^\circ = -nFE^\circ < 0$

$\therefore E^\circ > 0$

71. **(c)** When $Q = K_C$, $E = 0$

72. **(c)** Thermodynamic efficiency of a cell is given by the ratio of Gibbs function change to enthalpy change in the overall cell reaction.

73. **(b)** Ionic molar conductivity of H^+ is very high and NH_4OH is a weak electrolyte.

74. **(a)** Molar conductivity of H^+ and OH^- are very high as compare to other ions.

Initially conductance of solution sharply decreases due to consumption of free H^+. After complete neutralization further slightly increases due to presence of OH^-.

75. (c) According to the definition 1 F or 96500 C is the charge carried by 1 mol of electrons when water is electrolysed

$2H_2O \longrightarrow 4H^+ + O_2 + 4e^-$

So, 4 Faraday of electricity liberate = 32 g of O_2.

Thus 1 Faraday of electricity liberate

$= \frac{32}{4}$ g of O_2 = 8 g of O_2

76. (a) Conductivity $(\kappa) = \frac{\ell}{R.a.}$

$\kappa = \frac{l}{R.a}$; cell constant. $\left(\frac{l}{a}\right) = 1.29 \times 100 = 129$

Again conductivity of 0.02M solution

$\kappa = \frac{1}{520} \times 129$

$\Lambda_m = \frac{\kappa \times 1000}{M} = \frac{129}{520} \times \frac{1000}{0.02}$

$= 1.24 \times 10^{-4}$ S m^{2}mol^{-1}

77. (c) $ClCH_2COONa + HCl \rightarrow ClCH_2COOH + NaCl$

$\Lambda_{ClCH_2COONa} + \Lambda_{HCl} = \Lambda_{ClCH_2COOH} + \Lambda_{NaCl}$

$224 + 203 = \Lambda_{ClCH_2COOH} + 38.2$

$\Lambda_{ClCH_2COOH} = 427 - 38.2$

$= 388.8$ ohm^{-1}cm^{2}g eq^{-1}

78. (a) $\underset{1F}{Ag^+ + e^-} \longrightarrow \underset{108g}{Ag}$

1 F = 1 mole of electrons = 96500 C

0.01F = 1.08 g Ag; Ag left = 1.08 – 1.08 = 0 g

79. (a) $\Lambda_m = \frac{\kappa \times 1000}{M} = \frac{1.06 \times 10^{-2} \times 1000}{0.1}$

$= 1.06 \times 10^2$

80. (d) The amount. of copper deposited at cathode by reduction of Cu^{2+} ions is $\frac{3.2}{63} = 0.05$ moles.

The same amount 0.05 mole of Cu^{2+} must pass into solution from anode by oxidation

81. (b) $Cu^{2+} + e^- \longrightarrow Cu^+$;

$E_1^o = 0.15V; \Delta G_1^o = -n_1 E_1^o F$

$Cu^{2+} + 2e \longrightarrow Cu$;

$E_2^o = 0.34V; \Delta G_2^o = -n_2 E_2^o F$

On subracting eq.(i) from eq. (ii) we get

$Cu^+ + e^- \longrightarrow Cu$; $\Delta G^o = \Delta G_2^o - \Delta G_1^o$

$-nE^o F = -(n_2 E^o F - n_1 E_1^o F)$

$E^o = \frac{n_2 E_2^o F - n_1 E_1^o F}{nF}$

$= \frac{2 \times 0.34 - 0.15}{1} = 0.53V$

82. (b) $MnO_4^- + 8H^+ + 5e^- \longrightarrow Mn^{2+} + 4H_2O$

$E = 1.51 - \frac{0.059}{5} \log \frac{[Mn^{2+}]}{[MnO_4^-][H^+]^8}$

Taking Mn^{2+} and MnO_4^- in standard state i.e. 1 M,

$E = 1.51 - \frac{0.059}{5} \times 8 \log \frac{1}{[H^+]}$

$= 1.51 - \frac{0.059}{5} \times 8 \times 3 = 1.2268$ V

Hence at this pH, MnO_4^- will oxidise only Br^- and I^- as SRP of Cl_2/Cl^- is 1.36 V which is greater than that for MnO_4^-/Mn^{2+}.

83. (c) Higher the value of reduction potential stronger will be the oxidising hence based on the given values Ag^+ will be strongest oxidizing agent.

84. (b) $E^o_{Hg^{2+}|Hg} > E^o_{Hg_2^{2+}|Hg} > E^o_{Fe^{2+}|Fe} > E^o_{Pb^{2+}|Pb}$.

85. (b) E^o_{OP} of Zn > E^o_{OP} of Fe.

86. (d) Primary cells are those cells, in which the reaction occurs only once and after use over a period of time, it becomes dead and cannot be reused again. e.g., Leclanche cell and mercury cell.

87. **(a)** Volume = Area × thickness

Mass = Volume × density

∴ Mass of Ag to be deposited

$$= \frac{80 \times 0.005}{10} \times 10.5 = 0.42 \text{ g}$$

$$\text{Amount deposited} = \frac{i \times t \times E.wt}{96500}$$

$$\therefore 0.42 = \frac{108 \times 3 \times t}{98500} \quad \therefore t = 125.1 \text{ seconds}$$

88. **(a)** $\because Q = i \times t$

$\therefore Q = 4.0 \times 10^4 \times 6 \times 60 \times 60 \, C = 8.64 \times 10^8 C$

Now since 96500 C liberates 9 g of Al

8.64×10^8 C liberates

$$\frac{9}{96500} \times 8.64 \times 10^8 \text{ g Al} = 8.1 \times 10^4 \text{ g of Al}$$

89. **(a)** $\kappa = \frac{\text{Cell constant}}{R} = \frac{1.15}{250}$

$$\Lambda_{eq} = \frac{1.15}{250} \times \frac{1000}{1} = 4.6$$

90. **(c)** $\lambda_m = \frac{1000\kappa}{0.1} = \frac{1000 \times 3.75 \times 10^{-4}}{0.1} = 3.75;$

$$\alpha = \frac{\lambda_m}{\lambda_m^\infty} = \frac{3.75}{250} = 1.5 \times 10^{-2};$$

$K_a = C\alpha^2 = 0.1 \times (1.5 \times 10^{-2})^2 = 2.25 \times 10^{-5}$

91. **(a)** $\Lambda_{eq}^\infty(NH_4OH) = \Lambda_{eq}^\infty(NH_4Cl) + \Lambda_{eq}^\infty(NaOH) - \Lambda_{eq}^\infty(NaCl)$

$= 129.8 + 217.8 - 109.3 = 238.3 \text{ ohm}^{-1}\text{cm}^2\text{eq}^{-1}$

$$\alpha = \frac{\Lambda_{eq}}{\Lambda_{eq}^\infty} = \frac{9.30}{238.3} = 0.04$$

92. **(b)** The cell reaction is

$$H_2(g) + I_2(s) \rightleftharpoons 2H^+(aq) + 2I^-(aq)$$

$$0.7714 = 0.535 - \frac{0.0591}{2}\log\frac{[H^+]^2[I^-]^2}{P_{H_2}}$$

pH = 3

93. **(c)** $2Mn^{2+} \longrightarrow 2Mn^{3+} + 2e^-, \Delta G_1^\circ$

$Mn^{2+} + 2e^- \rightarrow Mn, \quad \Delta G_2^\circ$

$3Mn^{2+}(aq) \rightarrow Mn(s) + 2Mn^{3+}(aq)$

$-2 \times F \times E_3^\circ = -2 \times F \times [-1.51] - 2 \times F \times (-1.185)$

$E_3^\circ = -2.695$

$$E_3^\circ = +\frac{0.0591}{2}\log K_{eq};\ K_{eq} = 6.3 \times 10^{-92}.$$

94. **(b)** In $H_2 - O_2$ fuel cell, the combustion of H_2 occurs to create potential difference between the two electrodes

95. **(d)** Higher the reduction potential, stronger is the oxidising agent.

96. **(c)** The given values show that Cr has highest negative value of $E^\circ_{M^{3+}/M^{2+}}$ i.e., Cr is the strongest reducing agent among given metals therefore its oxidation will be easiest.

97. **(b)** $(E_{Cu^{2+}|Cu})_1 = E^0_{Cu^{2+}|Cu} + \frac{0.059}{2}\log[Cu^{2+}]_1$

$[Cu^{2+}(aq) + 2e^- \longrightarrow Cu(s)]$ and

$$(E_{Cu^{2+}|Cu})_2 = E^0_{Cu^{2+}|Cu} + \frac{0.059}{2}\log[Cu^{2+}]_2$$

$$\Rightarrow \Delta E_{Cu^{2+}|Cu} = -\frac{0.059}{2}\log\left(\frac{C_2}{C_1}\right)$$

$$= -\frac{0.059}{2}\log 10 = -0.030 \text{ V} = -30 \text{ mV}$$

98. **(a)** $E_{red} = E_{red}^\circ + \frac{0.059}{n}\log\frac{[M^{(x+n)}]}{[M^{x+}]}$

99. **(b)** The cell reaction is

$$Cl_2(g)(P_2 \text{ atm}) \longrightarrow Cl_2(aq)(P_1 \text{atm})$$

$$E_{cell} = -\frac{0.0592}{2}\log\frac{P_1}{P_2} = \frac{0.0592}{2}\log\frac{P_2}{P_1}$$

E_{cell} will be positive when $P_2 > P_1$.

100. **(b)** $\Lambda_{eq} = \frac{\kappa \times 1000}{N}$

18 Chemical Kinetics

1. For the reaction
$N_2O_5(g) \longrightarrow 2NO_2(g) + 1/2\ O_2(g)$ the value of rate of disappearance of N_2O_5 is given as 6.25×10^{-3} mol $L^{-1}s^{-1}$. The rate of formation of NO_2 and O_2 is given respectively as :
(a) 6.25×10^{-3} mol $L^{-1}s^{-1}$ and 6.25×10^{-3} mol $L^{-1}s^{-1}$
(b) 1.25×10^{-2} mol $L^{-1}s^{-1}$ and 3.125×10^{-3} mol $L^{-1}s^{-1}$
(c) 6.25×10^{-3} mol $L^{-1}s^{-1}$ and 3.125×10^{-3} mol $L^{-1}s^{-1}$
(d) 1.25×10^{-2} mol $L^{-1}s^{-1}$ and 6.25×10^{-3} mol $L^{-1}s^{-1}$

2. The reactions rate $N_2(g) + 3H_2(g) \rightarrow 2NH_3(g)$ was measured $\frac{d[NH_3]}{dt} = 2\times10^{-4}$ mol sec^{-1}. The rates of reactions expressed in terms of N_2 and H_2 are

	Rate in terms of N_2. ($mol\ L^{-1}sec^{-1}$)	Rate in terms of H_2 ($mol\ L^{-1}sec^{-1}$)
(a)	2×10^{-4}	2×10^{-4}
(b)	3×10^{-4}	1×10^{-4}
(c)	1×10^{-4}	3×10^{-4}
(d)	2×10^{-1}	2×10^{-3}

3. For the reaction $A + B \longrightarrow C + D$. The variation of the concentration of the products is given by the curve **Tricky**

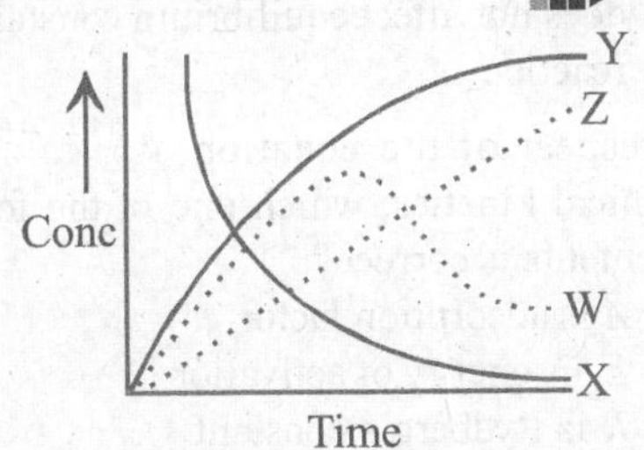

(a) Z (b) Y (c) W (d) X

4. For the reaction $H_2(g) + Br_2(g) \rightarrow 2HBr(g)$, the experimental data suggest, rate = $k[H_2][Br_2]^{1/2}$. The molecularity and order of the reaction are respectively
(a) $2, \frac{3}{2}$ (b) $\frac{3}{2}, \frac{3}{2}$ (c) 1, 1 (d) $1, \frac{1}{2}$

5. For a chemical reaction $t_{1/2}$ is 2.5 hours at room temperature. How much of the reactant will be left after 7.5 hours if initial weight of reactant was 160 g?

(a) 10 g (b) 40 g (c) 80 g (d) 20 g

6. The rate law for a reaction between the substances A and B is given by rate = $k[A]^n[B]^m$. On doubling the concentration of A and halving the concentration of B, the ratio of the new rate to the earlier rate of the reaction will be as
(a) $(m + n)$ (b) $(n - m)$
(c) $2^{(n+m)}$ (d) $\frac{1}{2^{(m+n)}}$

7. The order of a reaction, with respect to one of the reacting component Y, is zero. It implies that:
(a) the reaction is going on at a constant rate
(b) the rate of reaction does not vary with temperature
(c) the reaction rate is independent of the concentration of Y
(d) the rate of formation of the activated complex is zero

8. For a first order reaction, a plot of log (a – x) against time is a straight line with a negative slope equal to
(a) $\frac{-k}{2.303}$ (b) $-2.303\,k$
(c) $\frac{2.303}{k}$ (d) $-\frac{E_a}{2.303\,R}$

9. A following mechanism has been proposed for a reaction
$2A + B \rightarrow D + E$
$A + B \rightarrow C + D$ (slow)
$A + C \rightarrow E$ (fast)
The rate law expression for the reaction is:
(a) $r = k[A]^2[B]$ (b) $r = k[A][B]$
(c) $r = k[A]^2$ (d) $r = k[A][C]$

10. The rate law for the reaction below is given by the expression k [A][B]
$A + B \rightarrow$ Product
If the concentration of B is increased from 0.1 to 0.3 mole, keeping the value of A at 0.1 mole, the rate constant will be:
(a) 3k (b) 9k (c) k/3 (d) k

11. The half life of a radioactive element is 20 min. The time interval between the stages of its 33% and 67% decay is **Toughnut**
(a) 40 min (b) 20 min (c) 30 min (d) 25 min

12. Which of the following reactions is **not** of the first order?
(a) Inversion of sucrose in the presence of acid
(b) Acid-catalyzed hydrolysis of ethyl acetate
(c) Hydrolysis of tertiary butyl halide using alkali
(d) Oxidation of I^- ion by $S_2O_8^{2-}$ ion

13. Rate constant $k = 1.2 \times 10^3$ mol$^-$ 1 L s^{-1} and $E_a = 2.0 \times 10^2$ kJ mol^{-1}. When $T \rightarrow \infty$:
(a) $A = 2.0 \times 10^2$ kJ mol^{-1}
(b) $A = 1.2 \times 10^3$ mol^{-1} L s^{-1}
(c) $A = 1.2 \times 10^3$ mol L^{-1} s^{-1}
(d) $A = 2.4 \times 10^3$ kJ mol^{-1} s^{-1}

14. A catalyst lowers the activation enegy of a reaction from 20 kJ mol^{-1} to 10 kJ mol^{-1}. The temperature at which the uncatalyzed reaction will have the same rate as that of the catalyzed at 27 °C is
(a) – 123 °C (b) 327 °C
(c) 32.7 °C (d) +23 °C

15. For a reaction, the rate constant is expressed as $k = Ae^{-40000/T}$. The energy of the activation is
(a) 40000 cal (b) 88000 cal
(c) 80000 cal (d) 8000 cal

16. The activation energy of the reaction, $A + B \rightarrow C + D + 38$ kcal is 20 kcal. What would be the activation energy of the following reaction.
$C + D \rightarrow A + B$
(a) 20 kcal (b) –20 kcal
(c) 18 kcal (d) 58 kcal

17. The half-life of a reaction is inversely proportional to the square of the initial concentration of the reactant. Then the order of the reaction is
(a) 0 (b) 1 (c) 2 (d) 3

18. For a reaction $A + 2B \rightarrow C$, the amount of C formed by starting the reaction with 5 moles of A and 8 moles of B is
(a) 5 moles (b) 8 moles
(c) 16 moles (d) 4 moles

19. The rate law for the reaction $2X + Y \rightarrow Z$ is Rate = k[X][Y]. The correct statement with regard to this relation is **Critical Thinking**
(a) the unit of k is s^{-1}
(b) the rate of the reaction is independent of [X] and [Y]
(c) for this reaction $t_{1/2}$ is independent of initial concentrations of reactant
(d) the rate of formation of Z is half of the rate of disappearance of X

20. In a reaction, $2A \rightarrow$ products, the concentration of A decreases from 0.50 M to 0.38 M in 10 min. What is the rate of reaction (in M s^{-1}) during this interval?
(a) 0.012 (b) 0.024
(c) 2×10^{-3} (d) 2×10^{-4}

21. A chemical reaction is catalyzed by a catalyst X. Hence X
(a) reduces enthalpy of the reaction
(b) decreases rate constant of the reaction
(c) increases activation energy of the reaction
(d) does not affect equilibrium constant of the reaction

22. In respect of the equation $k = Ae^{-E_a/RT}$ in chemical kinetics, which one of the following statements is correct ?
(a) A is adsorption factor
(b) E_a is energy of activation
(c) R is Rydberg's constant
(d) k is equilibrium constant

23. The minimum energy a molecule should possess in order to enter into a fruitful collision is known as
(a) reaction energy (b) collision energy
(c) activation energy (d) threshold energy

24. For a reaction, activation energy $(E_a) = 0$ and rate constant $(k) = 3.2\times10^{6}\,s^{-1}$ at 300 K. What is the value of the rate constant at 310 K
(a) $3.2\times10^{-12}\,s^{-1}$ (b) $3.2\times10^{6}\,s^{-1}$
(c) $6.4\times10^{12}\,s^{-1}$ (d) $6.4\times10^{6}\,s^{-1}$

25. In the Arrhenius plot of $\ln k$ Vs $\frac{1}{T}$, a linear plot is obtained with a slope of -2×10^{4} K. The energy of activation of the reaction (in kJ mole^{-1}) is (R value is 8.3 J K^{-1} mol^{-1})

Toughnut

(a) 83 (b) 166 (c) 249 (d) 332

26. The rate of reaction is doubled for every 10°C rise in temperature. The increase in reaction rate as a result of temperature rise from 10°C to 100°C is
(a) 112 (b) 512 (c) 400 (d) 614

27. Consider the reaction $A \rightarrow 2B + C$, $\Delta H = -15$ kcal. The energy of activation of backward reaction is 20 kcal mol^{-1}. In presence of catalyst the energy of activation of forward reaction is 3 kcal mol^{-1}. At 400 K the catalyst causes the rate of the reaction to increase by the number of times equal to

(a) $e^{3.5}$ (b) $e^{2.5}$
(c) $e^{-2.5}$ (d) $e^{2.303}$

28. A reaction takes place in various steps. The rate constant for first, second, third and fifth steps are k_1, k_2, k_3 and k_5 respectively. The overall rate constant is given by $k = \frac{k_2}{k_3}\left(\frac{k_1}{k_5}\right)^{1/2}$

If activation energy are 40, 60, 50 and 10 kJ/mol respectively, the overall energy of activation (kJ/mol) is :

Toughnut

(a) 10 (b) 20
(c) 25 (d) none of these

29. Consider the two hypothetical reactions given below :
I $aA \rightarrow$ Products, $k = x\,\text{mol}^{-1}\,\text{L}\,\text{min}^{-1}$
II $bB \rightarrow$ Products, $k = y\,\text{min}^{-1}$
The half-lives of both the reactions are the same, equal to 1 hr when molar concentration of the reactant is 1.0 M in each case. If these reactions are started at the same time taking 1M of the reactant in each case, the ratio $[A]/[B]$ after 3 hr will be :
(a) 0.5 (b) 4 (c) 1 (d) 2

30. Consider the consecutive reactions :

$$A \xrightarrow{k=2\times10^{-5}s^{-1}} B \xrightarrow{k=8\times10^{-6}s^{-1}} C \xrightarrow{k=3\times10^{-3}s^{-1}} D$$

The rate determining step of the reaction is :
(a) $A \rightarrow B$ (b) $C \rightarrow D$
(c) $B \rightarrow C$ (d) $A \rightarrow D$

31. A + 2B → C, the rate equation for this reaction is given as Rate = K[A][B].
If the concentration of A is kept the same but that of B is doubled what will happen to the rate itself ?
(a) halved (b) the same
(c) doubled (d) quadrupled

32. In the reaction of formation of sulphur trioxide by contact process $2SO_2 + O_2 \rightleftharpoons 2SO_3$ the rate of reaction was measured as $\frac{d[O_2]}{dt} = -2.5\times10^{-4}\,\text{mol L}^{-1}\text{s}^{-1}$. The rate of reaction is terms of $[SO_2]$ in mol L^{-1} s^{-1} will be:
(a) -1.25×10^{-4} (b) -2.50×10^{-4}
(c) -3.75×10^{-4} (d) -5.00×10^{-4}

33. For the reaction, $2N_2O_5 \rightarrow 4NO_2 + O_2$, the rate equation can be expressed in two ways $-\frac{d[N_2O_5]}{dt} = k[N_2O_5]$ and $+\frac{d[NO_2]}{dt} = k'[N_2O_5]$
k and k′ are related as:
(a) $k = k'$ (b) $2k = k'$
(c) $k = 2k'$ (d) $k = 4k'$

34. The reaction $2N_2O_5 \rightleftharpoons 2N_2O_4 + O_2$ is
(a) bimolecular and of second order
(b) unimolecular and of first order
(c) bimolecular and of first order
(d) bimolecular and of zero order

35. For the reaction system : $2NO(g) + O_2(g) \rightarrow 2NO_2(g)$ volume is suddenly reduced to half its value by increasing the pressure on it. If the reaction is of first order with respect to O_2 and second order with respect to NO, the rate of reaction will
(a) diminish to one-eighth of its initial value
(b) increase to eight times of its initial value
(c) increase to four times of its initial value
(d) diminish to one-fourth of its initial value

36. Units of rate constant of first and zero order reactions in terms of molarity M unit are respectively
(a) sec^{-1}, $M.sec^{-1}$ (b) sec^{-1}, M
(c) $M.sec^{-1}$, sec^{-1} (d) M, sec^{-1}

37. The plot of concentration of the reactant Vs time for a reaction is a straight line with a negative slope. The reaction follows a rate equation of

(a) zero order (b) first order
(c) second order (d) third order

38. The differential rate law for the reaction $H_2(g) + I_2(g) \rightarrow 2HI(g)$ is

(a) $-\frac{d[H_2]}{dt} = -\frac{d[I_2]}{dt} = -\frac{d[HI]}{dt}$

(b) $\frac{d[H_2]}{dt} = \frac{d[I_2]}{dt} = \frac{1}{2}\frac{d[HI]}{dt}$

(c) $\frac{1}{2}\frac{d[H_2]}{dt} = \frac{1}{2}\frac{d[I_2]}{dt} = -\frac{d[HI]}{dt}$

(d) $-2\frac{d[H_2]}{dt} = -2\frac{d[I_2]}{dt} = \frac{d[HI]}{dt}$

39. Which of the following will react at the highest rate?

Critical Thinking

(a) 1 mole of A and 1 mole of B in a 1-L vessel
(b) 2 mole of A and 2 mole of B in a 2-L vessel
(c) 3 mole of A and 3 mole of B in a 3-L vessel
(d) All would react at the same rate

40. In the reaction, $A + 2B \longrightarrow 6C + 2D$, If the initial rate $-\frac{d[A]}{dt}$ at t = 0 is 2.6×10^{-2} M sec^{-1}, what will be the value of $-\frac{d[B]}{dt}$ at t = 0?
(a) 8.5×10^{-2} M sec^{-1} (b) 2.5×10^{-2} M sec^{-1}
(c) 5.2×10^{-2} M sec^{-1} (d) 7.5×10^{-2} M sec^{-1}

41. The reaction $L \longrightarrow M$ is started with 10.0 g of L. After 30 and 90 minutes 5.0 g and 1.25 g of L respectively are left. The order of the reaction is
(a) 0 (b) 1 (c) 2 (d) 3

42. A first order reaction is half-completed in 45 minutes. How long does it need for 99.9% of the reaction to be completed?
(a) 20 hours (b) 10 hours
(c) $7\frac{1}{2}$ hours (d) 5 hours

43. A reaction which is of first order w.r.t. reactant A, has a rate constant 6 min^{-1}. If we start with [A] = 0.5 mol L^{-1}, when would [A] reach the value of 0.05 mol L^{-1}
(a) 0.384 min (b) 0.15 min
(c) 3 min (d) 3.84 min

44. The time taken for 90% of a first order reaction to complete is approximately
(a) 1.1 times that of half-life
(b) 2.2 times that of half-life
(c) 3.3 times that of half-life
(d) 4.4 times that of half-life

45. The rate constant of a reaction is 0.0693 min^{-1}. Starting with 10 mol, the rate of the reaction after 10 min is
(a) 0.0693 mol min^{-1}
(b) 0.0693×2 mol min^{-1}
(c) 0.0693×5 mol min^{-1}
(d) $0.0693 \times (5)^2$ mol min^{-1}

46. The plot that represents the zero order reaction is :

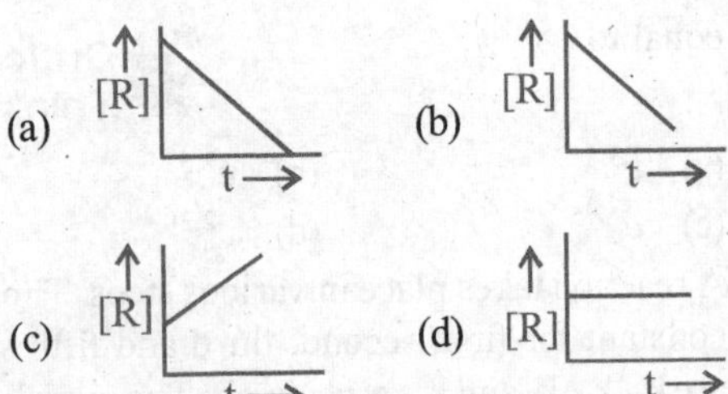

47. Which of the following reactions is **not** of the first order?

Critical Thinking

(a) Inversion of sucrose in the presence of acid
(b) Acid-catalyzed hydrolysis of ethyl acetate
(c) Hydrolysis of tertiary butyl halide using alkali
(d) Oxidation of I^- ion by $S_2O_8^{2-}$ ion

48. A first order reaction is 50% completed in 20 minutes at 27°C and in 5 minutes at 47°C. The energy of activation of the reaction is :

Toughnut

(a) 43.85 kJ/mol (b) 55.14 kJ/mol
(c) 11.97 kJ/mol (d) 6.65 kJ/mol

49. For the first order reaction $A \rightarrow B + C$ is carried out at 27°C. If 3.8×10^{-16} % of the reactant molecules exists in the activated state, the E_a (activation energy) of the reaction is:
(a) 12 kJ/mol (b) 831.4 kJ/mol
(c) 100 kJ/mol (d) 88.57 kJ/mol

50. A catalyst lowers the activation energy of a certain reaction from 83.314 to 75 kJ mol^{-1} at 500 K. What will be the rate of reaction as compared to uncatalysed reaction? Assume other things being equal.
(a) Double (b) 28 times
(c) 7.38 times (d) 7.38×10^3 times

51. Rate of a reaction can be expressed by Arrhenius equation as: $k = Ae^{-E_a/RT}$
In this equation, E_a represents
(a) the total energy of the reacting molecules at a temperature, T
(b) the fraction of molecules with energy greater than the activation energy of the reaction
(c) the energy below which all the colliding molecules will react
(d) the energy below which colliding molecules will not react

52. The reason for almost doubling the rate of reaction on increasing the temperature of the reaction system by 10°C is
(a) The value of threshold energy increases
(b) Collision frequency increases
(c) The fraction of the molecule having energy equal to threshold energy or more increases
(d) Activation energy decreases

53. In Arrhenius plot, intercept is equal to
(a) $\frac{-E_a}{R}$ (b) $\ell n A$
(c) $\ell n K$ (d) $\log_{10} A$

54. A reaction rate constant is given by

$k = 1.2 \times 10^{14} e^{-25000/RT} sec^{-1}$. It means
(a) log k versus log T will give a straight line with a slope as –25000
(b) log k versus T will give a straight line with slope as 25000
(c) log k versus 1/T will give a straight line with slope as –25000/R
(d) log k versus 1/T will give a straight line

55. The velocity of a reaction is doubled for every 10°C rise in temp. If the temp. is raised to 50°C from 0 °C the reaction velocity increases by about
(a) 12 times (b) 16 times
(c) 32 times (d) 50 times

56. Which of the following statements is incorrect?
(a) Activation energy for the forward reaction is equals to activation energy for the reverse reaction
(b) For a reversible reaction, an increase in temperature increases the reaction rate for both the forward and the backward reaction
(c) The larger the initial reactant concentration for a second order reaction, the shorter is its half-life.
(d) When Δt is infinitesimally small, the average rate equals the instantaneous rate

57. Consider the following statements:

I. Increase in concentration of reactant increases the rate of a zero order reaction.
II. Rate constant k is equal to collision frequency A if $E_a = 0$.
III. Rate constant k is equal to collision frequency A if $E_a = \infty$.
IV. ln k Vs T is a straight line.
V. ln k Vs 1/T is a straight line.
Correct statements are
(a) I and IV (b) II and V
(c) III and IV (d) II and III

58. The time required for 10% completion of a first order reaction at 298 K is equal to that required for its 25% completion at 308 K. If the pre-exponential factor for the reaction is $3.56 \times 10^9 s^{-1}$, the rate constant at 318 K is:

(a) 18.39 kcal mol^{-1} (b) 20 kcal mol^{-1}
(c) 16 kcal mol^{-1} (d) 21.5 kcal mol^{-1}

59. For the reaction, $3A+2B \rightarrow C+D$, the differential rate law can be written as:

(a) $\frac{1}{3}\frac{d[A]}{dt}=\frac{d[C]}{dt}=k[A]^n[B]^m$

(b) $-\frac{d[A]}{dt}=\frac{d[C]}{dt}=k[A]^n[B]^m$

(c) $+\frac{1}{3}\frac{d[A]}{dt}=-\frac{d[C]}{dt}=k[A]^n[B]^m$

(d) $-\frac{1}{3}\frac{d[A]}{dt}=\frac{d[C]}{dt}=k[A]^n[B]^m$

60. The instantaneous rate of disappearance of MnO_4^- ion in the following reaction is $4.56\times10^{-3}\,Ms^{-1}$ $2MnO_4^- + 10I^- + 16H^+ \rightarrow 2Mn^{2+} + 5I_2 + 8H_2O$

The rate of appearance I_2 is :

(a) $4.56\times10^{-4}\,Ms^{-1}$ (b) $1.14\times10^{-2}\,Ms^{-1}$

(c) $1.14\times10^{-3}\,Ms^{-1}$ (d) $5.7\times10^{-3}\,Ms^{-1}$

61. The rate constant of a zero order reaction is $2.0\times10^{-2}\,mol\,L^{-1}\,s^{-1}$. If the concentration of the reactant after 25 seconds is 0.5 M. What is the initial concentration?

(a) 0.5 M (b) 1.25 M

(c) 12.5 M (d) 1.0 M

62. The reaction $A \rightarrow B$ follows first order kinetics. The time taken for 0.8 mole of A to produce 0.6 mole of B is 1 hour. What is the time taken for conversion of 0.9 mole of A to produce 0.675 mole of B? Toughnut

(a) 2 hours (b) 1 hour

(c) 0.5 hour (d) 0.25 hour

63. The rate of a first order reaction is $1.5\times10^{-2}\,mol\,L^{-1}\,min^{-1}$ at 0.5 M concentration of the reactant. The half life of the reaction is

(a) 0.383 min (b) 23.1 min

(c) 8.73 min (d) 7.53 min

64. For a first order reaction $A \longrightarrow B$ the reaction rate at reactant concentration of 0.01 M is found to be $2.0\times10^{-5}\,mol\,L^{-1}\,s^{-1}$. The half life period of the reaction is

(a) 30 s (b) 220 s (c) 300 s (d) 347 s

65. A reaction proceeds by first order, 75% of this reaction was completed in 32 min. The time required for 50% completion is Critical Thinking

(a) 8 min (b) 16 min

(c) 20 min (d) 24 min

66. Consider the reaction : $N_2(g)+3H_2(g)\rightarrow 2NH_3(g)$

The equality relationship between $\frac{d[NH_3]}{dt}$ and $-\frac{d[H_2]}{dt}$ is

(a) $+\frac{d[NH_3]}{dt}=-\frac{2}{3}\frac{d[H_2]}{dt}$

(b) $+\frac{d[NH_3]}{dt}=-\frac{3}{2}\frac{d[H_2]}{dt}$

(c) $\frac{d[NH_3]}{dt}=-\frac{d[H_2]}{dt}$

(d) $\frac{d[NH_3]}{dt}=-\frac{1}{3}\frac{d[H_2]}{dt}$

67. The rate of a reaction increases four-fold when the concentration of reactant is increased 16 times. If the rate of reaction is $4\times10^{-6}\,mol\,L^{-1}\,s^{-1}$ when the concentration of the reactant is $4\times10^{-4}\,mol\,L^{-1}$. The rate constant of the reaction will be

(a) $2\times10^{-4}\,mol^{1/2}\,L^{-1/2}\,s^{-1}$

(b) $1\times10^{-2}\,s^{-1}$

(c) $2\times10^{-4}\,mol^{-1/2}\,L^{1/2}\,s^{-1}$

(d) $25\,mol^{-1}\,L\,min^{-1}$

68. In the reaction $A \rightarrow B + C$, rate constant is 0.001 Ms^{-1}. If we start with 1 M of A then conc. of A and B after 10 minuter are respectively.

(a) 0.5 M, 0.5 M (b) 0.6 M, 0.4 M

(c) 0.4 M, 0.6 M (d) 0.6 M 0.5 M

69. The half life for the virus inactivation if in the beginning 1.5% of the virus is inactivated per minute is (Given: The reaction is of first order)

(a) 76 min (b) 66 min (c) 56 min (d) 46 min

70. The reaction of A_2 and B_2 follows the equation

$$A_2(g)+B_2(g)\rightarrow 2AB(g)$$

The following data were observed

$[A_2]_0$	$[B_2]_0$	Initial rate of appearance of AB(g) (in Ms^{-1})
0.10	0.10	2.5×10^{-4}
0.20	0.10	5×10^{-4}
0.20	0.20	10×10^{-4}

The value of rate constant for the above reaction is:

(a) 2.5×10^{-4} (b) 2.5×10^{-2}
(c) 1.25×10^{-2} (d) None of these

71. In the presence of an acid, the initial concentration of cane sugar was reduced from 0.20 to 0.10 M in 5 hours and from 0.2 to 0.05 M in 10 hours. The reaction is of: Tricky

(a) Zero order (b) First order
(c) Second order (d) Third order

72. The activation energies of the forward and backward reactions in the case of a chemical reaction are 30.5 and 45.4 kJ/mol respectively. The reaction is :

(a) exothermic
(b) endothermic
(c) neither exothermic nor endothermic
(d) independent of temperature

73. A homogeneous catalytic reaction takes place through the three alternative plots A, B and C shown in the given figure. Which one of the following indicates the relative ease with which the reaction can take place?

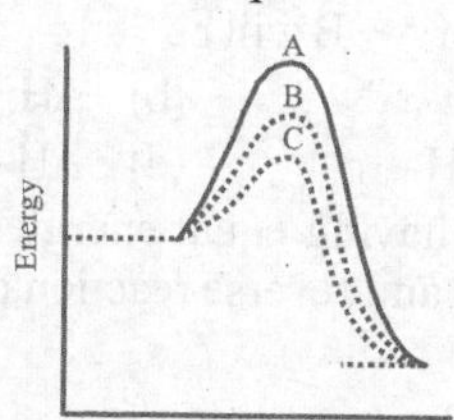

(a) A>B>C (b) C>B>A
(c) A>C>B (d) A=B=C

74. A catalyst is a substance which :

(a) is always in the same phase as in the reaction
(b) alters the equilibrium in a reaction
(c) does not participate in the reaction but alters the rate of reaction
(d) participates in the reaction and provides an easier pathway for the same

75. For an exothermic reaction, the energy of activation of the reactants is

(a) equal to the energy of activation of products
(b) less than the energy of activation of products
(c) greater than the energy of activation of products
(d) sometimes greater and sometimes less than that of the products

76. When a biochemical reaction is carried out in laboratory in the absence of enzyme then rate of reaction obtained is 10^{-6} times, then activation energy of reaction in the presence of enzyme is

(a) $\frac{6}{RT}$
(b) different from E_a obtained in laboratory
(c) P is required
(d) can't say anything

77. A chemical reaction was carried out at 300 K and 280 K. The rate constants were found to be k_1 and k_2 respectively. then

(a) $k_2=4k_1$ (b) $k_2=2k_1$
(c) $k_2=0.25\,k_1$ (d) $k_2=0.5\,k_1$

78. In a reaction at 27°C, 10^{-3}% reactant molecules manage to cross over the barrier of transition state. The energy of these molecules in excess of the average value will be ($R=2$ cal K^{-1} mol^{-1}) :

(a) 6.91 *kcal* mol^{-1} (b) 3.00 *kcal* mol^{-1}
(c) 4.15 *kcal* mol^{-1} (d) 5.10 *kcal* mol^{-1}

79. The activation energy for a simple chemical reaction $A\rightarrow B$ is E_a in forward direction. The activation energy for reverse reaction

(a) is always double of E_a
(b) is negative of E_a
(c) is always less than E_a
(d) can be less than or more than E_a

80. The slope in Arrhenius plot, is equal to:

(a) $-\frac{E_a}{2.303R}$ (b) $\frac{E_a}{R}$
(c) $-\frac{R}{2.303E_a}$ (d) None of these

81. A radioactive isotope having a half - life period of 3 days was received after 12 days. If 3g of the isotope is left in the container, what would be the initial mass of the isotope?

(a) 12g (b) 36g (c) 48g (d) 24g

82. In a chemical reaction A is converted into B. The rates of reaction, starting with initial concentrations of A as 2×10^{-3} M and 1×10^{-3} M, are equal to 2.40×10^{-4} Ms^{-1} and 0.60×10^{-4} Ms^{-1} respectively. The order of reaction with respect to reactant A will be

(a) 0 (b) 1.5 (c) 1 (d) 2

83. For a reaction $A \rightarrow$ Products, a plot of log $t_{1/2}$ versus log a_0 is shown in the figure. If the initial concentration of A is represented by a_0, the order of the reaction is

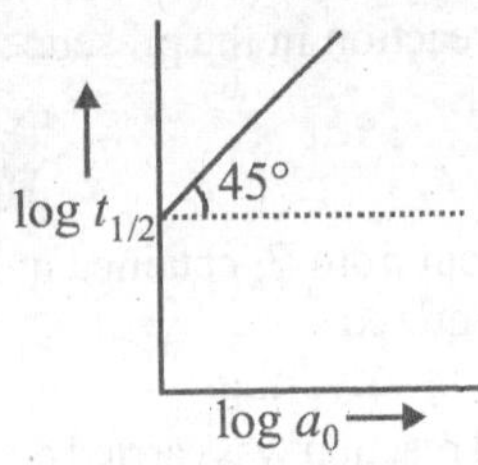

(a) one (b) zero (c) two (d) three

84. The rate constant of a reaction with a virus is $3.3 \times 10^{-4}\ s^{-1}$. Time required for the virus to become 75% inactivated is

(a) 35 min (b) 70 min
(c) 105 min (d) 17.5 min

85. A Geigger Muller counter is used to study the radioactive process. In the absence of radioactive substance A, it counts 3 disintegration per second (dps). At the start in the presence of A, it records 23 dps; and after 10 min 13 dps,

(i) What does it count after 20 min
(ii) What is the half life of A? Toughnut

(a) 8 dps, 10 min (b) 5 dps, 10 min
(c) 5 dps, 20 min (d) 5 dps, 5 min

86. The half life period for catalytic decomposition of AB_3 at 50 mm Hg is 4 hrs and at 100 mm Hg it is 2 hrs. The order of reaction is

(a) 1 (b) 3 (c) 2 (d) 0

87. The rate equation for a reaction,

$N_2O \longrightarrow N_2 + 1/2O_2$

is Rate = $k[N_2O]^0 = k$. If the initial concentration of the reactant is a mol Lit^{-1}, the half-life period of the reaction is

(a) $t_{\frac{1}{2}} = \frac{a}{2k}$ (b) $-t_{\frac{1}{2}} = ka$

(c) $t_{\frac{1}{2}} = \frac{a}{k}$ (d) $t_{\frac{1}{2}} = \frac{k}{a}$

88. The rate constant for a first order reaction whose half life is 480 sec, is :

(a) $1.44 \times 10^{-3}\ sec^{-1}$ (b) $1.44 \times sec^{-1}$
(c) $0.72 \times 10^{-3}\ sec^{-1}$ (d) $2.88 \times 10^{-3}\ sec^{-1}$

89. The hypothetical reaction Toughnut

$A_2 + B_2 \longrightarrow 2AB$; follows the following mechanism $A_2 \xrightarrow{\text{Fast}} A + A$,

$A + B_2 \xrightarrow{\text{Slow}} AB + B$, $A + B \xrightarrow{\text{Fast}} AB$.

The order of the overall reaction is

(a) 0 (b) 1 (c) 2 (d) 3/2

90. If half-life of a substance is 5 yrs, then the total amount of substance left after 15 years, when initial amount is 64 grams is

(a) 16 g (b) 2 g (c) 32 g (d) 8 g

91. In a reversible reaction the energy of activation of the forward reaction is 50 kcal. The energy of activation for the reverse reaction will be

(a) $<$ 50 kcal
(b) either greater than or less than 50 kcal
(c) 50 kcal
(d) $>$ 50 kcal

92. Activation energy of a chemical reaction can be determined by Critical Thinking

(a) evaluating rate constant at standard temperature
(b) evaluating velocities of reaction at two different temperatures
(c) evaluating rate constants at two different temperatures
(d) changing concentration of reactants

93. For the exothermic reaction $A + B \rightarrow C + D, \Delta H$ is the heat of reaction and E_a is the energy of activation. The energy of activation for the formation of A + B will be

(a) E_a (b) ΔH
(c) $E_a + \Delta H$ (d) $\Delta H - E_a$

94. A reaction having equal energies of activation for forward and reverse reaction has :

(a) $\Delta G = 0$
(b) $\Delta H = 0$
(c) $\Delta H = \Delta G = \Delta S = 0$
(d) $\Delta S = 0$

95. The rate coefficient (k) for a particular reactions is $1.3 \times 10^{-4} M^{-1} s^{-1}$ at 100°C, and $1.3 \times 10^{-3} M^{-1} s^{-1}$ at 150°C. What is the energy of activation (E_a) (in kJ) for this reaction? (R = molar gas constant = $8.314\ JK^{-1} mol^{-1}$)

(a) 16 (b) 60 (c) 99 (d) 132

96. The reaction X $\rightarrow$ Y is an exothermic reaction. Activation energy of the reaction for X into Y is 150 kJ mol^{-1}. Enthalpy of reaction is 135 kJ mol^{-1}. The activation energy for the reverse reaction, Y $\rightarrow$ X will be:
(a) 280 kJ mol^{-1} (b) 285 kJ mol^{-1}
(c) 270 kJ mol^{-1} (d) 15 kJ mol^{-1}

97. The activation energy for a reaction which doubles the rate when the temperature is raised from 298 K to 308 K is

Critical Thinking

(a) 59.2 kJ mol^{-1} (b) 39.2 kJ mol^{-1}
(c) 52.9 kJ mol^{-1} (d) 29.5 kJ mol^{-1}

98. $t_{1/4}$ can be taken as the time taken for the concentration of a reactant to drop to $\frac{3}{4}$ of its initial value. If the rate constant for a first order reaction is k, the $t_{1/4}$ can be written as

Toughnut

(a) $0.75/k$ (b) $0.69/k$ (c) $0.29/k$ (d) $0.10/k$

99. $A \rightarrow B, \Delta H = -10 kJ\ mol^{-1}, E_{a(f)} = 50\ kJ\ mol^{-1}$, then E_a of $B \rightarrow A$ will be
(a) 40 kJ mol^{-1} (b) 50kJ mol^{-1}
(c) – 50kJ mol^{-1} (d) 60 kJ mol^{-1}

100. For a certain reaction, rate = $k \times [H^+]^n$. If pH of reaction changes from two to one, the rate becomes 100 times of its value at pH = 2, the order of reaction is –
(a) 1 (b) 2 (c) 0 (d) 3

Answer KEYs

1	(b)	11	(b)	21	(d)	31	(c)	41	(b)	51	(d)	61	(d)	71	(b)	81	(c)	91	(b)
2	(c)	12	(d)	22	(b)	32	(d)	42	(c)	52	(b)	62	(b)	72	(a)	82	(d)	92	(c)
3	(b)	13	(b)	23	(d)	33	(b)	43	(a)	53	(b)	63	(b)	73	(b)	83	(b)	93	(c)
4	(a)	14	(b)	24	(b)	34	(c)	44	(c)	54	(c)	64	(d)	74	(c)	84	(b)	94	(b)
5	(d)	15	(c)	25	(b)	35	(b)	45	(c)	55	(c)	65	(b)	75	(b)	85	(a)	95	(b)
6	(c)	16	(d)	26	(b)	36	(a)	46	(d)	56	(a)	66	(a)	76	(b)	86	(c)	96	(b)
7	(c)	17	(d)	27	(b)	37	(a)	47	(d)	57	(b)	67	(a)	77	(c)	87	(a)	97	(c)
8	(a)	18	(d)	28	(c)	38	(d)	48	(b)	58	(a)	68	(c)	78	(a)	88	(a)	98	(c)
9	(b)	19	(d)	29	(d)	39	(d)	49	(c)	59	(d)	69	(d)	79	(d)	89	(d)	99	(d)
10	(d)	20	(d)	30	(c)	40	(c)	50	(c)	60	(b)	70	(c)	80	(a)	90	(d)	100	(b)

Hints & Solutions

1. **(b)** $N_2O_5(g) \longrightarrow 2NO_2(g) + 1/2\, O_2(g)$

$$-\frac{d}{dt}[N_2O_5] = +\frac{1}{2}\frac{d}{dt}[NO_2] = 2\frac{d}{dt}[O_2]$$

$$\frac{d}{dt}[NO_2] = 1.25 \times 10^{-2}\ mol\ L^{-1}s^{-1}\ \text{and}$$

$$\frac{d}{dt}[O_2] = 3.125 \times 10^{-3}\ mol\ L^{-1}s^{-1}$$

2. **(c)** $N_2 + 3H_2 \rightleftharpoons 2NH_3$; Rate is given by any of the expressions

$$\frac{-d[N_2]}{dt} = -\frac{1d[H_2]}{3dt} = \frac{1}{2}\frac{d(NH_3]}{dt}$$

Rate of disappearance of $N_2 = \frac{1}{2}$ the rate of formation of $NH_3 = 1 \times 10^{-4}$
Rate of disappearance of H_2 = 3/2 the rate of formation of $NH_3 = 3 \times 10^{-4}$

3. **(b)** The curve Y shows the increase in concentration of products with time.

4. **(a)** The order of reaction is $\frac{3}{2}$ and molecularity is 2.

5. (d) Using the relation
$[A] = [A]_0(1/2)^n$ [n = number of half-lives]
$T = n \times t_{1/2}$
Here, $n = \frac{7.5}{2.5} = 3$
$\therefore [A] = 160 \times \left(\frac{1}{2}\right)^3 = 160 \times \frac{1}{8} = 20\,g$

6. (c) $Rate_1 = k[A]^n[B]^m$; $Rate_2 = k[2A]^n[½B]^m$
$\therefore \frac{Rate_2}{Rate_1} = \frac{k[2A]^n[½B]^m}{k[A]^n[B]^m}$
$= [2]^n[½]^m = 2^n.2^{-m}$
$= 2^{n-m}$

7. (c) Let us consider a reaction,
$xX + yY \longrightarrow aA + bB$
rate $= [X]^x[Y]^y$
It is given that order of reaction w.r.t. component Y is zero.
Hence, rate $= [X]^x$
i.e., rate becomes independent of the concentration of Y.

8. (a) $t = \frac{2.303}{k}\log\frac{a}{a-x}$
or $t = \frac{2.303}{k}\log a - \frac{2.303}{k}\log(a-x)$
$\log(a-x) = \frac{-k}{2.303}t + \log a$

9. (b) From slow reaction
Rate = k [A] [B]

10. (d) Rate constant is independent of concentration.

11. (b) Change in 67% to 33% is almost half the concentration change.
$67\% \xrightarrow[\text{change}]{\text{Half}} 33.5(\approx 33\%)$
So time interval between the stages of its 33% and 67% decay is same as $t_{1/2} = 20$ min.

12. (d) It is a second order reaction, first order both w.r.t $S_2O_8^{2-}$ and I^-.
$\therefore r = k[S_2O_8^{2-}][I^-]$
All other options are of first order reaction.

13. (b) $k = Ae^{-Ea/RT}$
If $T \longrightarrow \infty$, $k = A$

14. (b) $\frac{E'_a}{T_1} = \frac{E'_a}{T_2} = \frac{10}{300} = \frac{20}{T_2}$
$\therefore T_2 = 600\,K = 327°C$

15. (c) $k = Ae^{-E_a/RT}$
$\therefore \frac{-E_a}{R} = -40000$
$\therefore E_a = 40000 \times 2 = 80000$ cal

16. (d) $E_{a_2} = 58$

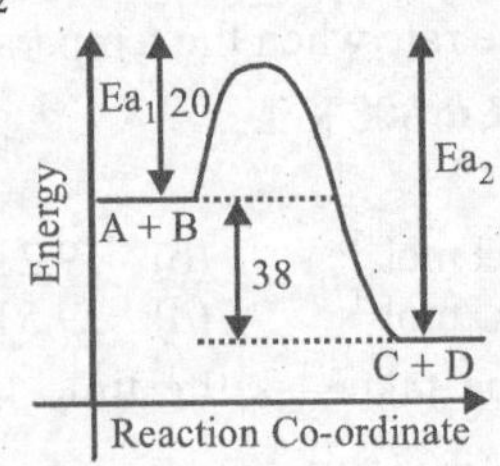

17. (d) $t_{1/2} \propto \frac{1}{a^2}$
We know that $t_{1/2} \propto \frac{1}{a^{n-1}}$
i.e. $n = 3$
Thus reaction is of 3rd order.

18. (d) $A + 2B \longrightarrow C$
1 mole of A reacts with 2 moles of B to give 1 mol of C.
$\therefore$ 5 moles of A would react with 10 moles of B to give 5 moles of C.
But, only 8 moles of B are available
$\therefore$ B acts as a limiting reagent.
2 moles of B gives 1 mole of C
$\therefore$ 8 moles of B will give $1/2 \times 8 = 4$ moles of C.

19. (d) The given reaction is : $2X + Y \longrightarrow Z$
$-\frac{d[X]}{2dt} = \frac{d[Z]}{dt}$
$\therefore$ Rate of formation of Z is half of the rate of disappearance of X.

20. (d) Rate of reaction $= \frac{d[A]}{dt}$
Given, $[A]_{initial} = 0.50$ M
$[A]_{final} = 0.38$ M
$dt = 10$ min $= 600$ sec
$d[A] = 0.12$
Rate $= \frac{0.12}{600} = 2 \times 10^{-4}\,M\,s^{-1}$.

21. (d) A catalyst affects equally both forward and backward reactions, therefore it does not affect equilibrium constant of reaction.

22. (b) In equation $k = Ae^{-E_a/RT}$; A = Frequency factor
k = velocity constant, R = gas constant and E_a = energy of activation

23. (d) The definition of threshold energy.
24. (b) When $E_a = 0$, rate constant is independent of temperature.
25. (b) $k = Ae^{-E_a/RT}$

$\ln k = \ln A - E_a/RT$

For $\ln k$ Vs $1/T$

$\ln A$ = intercept

$-E_a/R$ = slope = -2×10^4 K

$\therefore E_a = 8.3 \times 2 \times 10^4$ J mol^{-1}

$= 16.6 \times 10^4$ J mol^{-1} or 166 kJ mol^{-1}

26. (b) As the rate of reaction get doubled for every 10°C rise in temperature. Hence the increase in reaction rate as a result of temperature rise from 10°C to 100°C is equal to $= 2^9 = 512$
27. (b) $E_a(f) - E_a(b) = \Delta H = -15\ kcal$

$\Rightarrow E_a(f) = -15 + 20 = 5\ kcal$

$$\frac{k\ (\text{catalyst})}{k} = e^{\frac{E_a - E_{a(\text{catalyst})}}{RT}}$$

$$= e^{\frac{(5-3)\times 10^3}{2\times 400}} = e^{2.5}$$

28. (c) $k = A.e^{-E_a/(RT)}$

$\therefore$ Effective overall energy of activation

$$E_a = E_a(2) - E_a(3) + \frac{1}{2}E_a(1) - \frac{1}{2}E_a(5)$$

$$= 60 - 50 + \frac{1}{2}\times 40 - \frac{1}{2}\times 10 = 25 \text{ kJ/mol}$$

29. (d) Units of k indicate that reaction I is of second order and reaction II is first order. For I reaction, $t_{1/2} \propto 1/a$, first $t_{1/2} = 1$ hr, second $t_{1/2} = 2$ hr

$$[A] = 1\text{M} \xrightarrow{1\text{hr}} 0.5\text{M} \xrightarrow{2\text{hr}} 0.25\text{M}$$

$$[B] = 1\text{M} \xrightarrow{1\text{hr}} 0.5\text{M} \xrightarrow{1\text{hr}}$$

$$0.25\text{M} \xrightarrow{1\text{hr}} 0.125\text{M}$$

$$\frac{[A]}{[B]} = \frac{0.25\text{M}}{0.125\text{M}} = 2$$

30. (c) The slowest step determines the rate.
31. (c) Rate = k [A][B] = R

R' = k[A][2B]

$$\frac{R}{R'} = \frac{k[A][B]}{k[A][2B]} = \frac{k[A][B]}{2k[A][B]}$$

$\Rightarrow$ 2R = R' i.e., rate become doubles.

32. (d) From rate law

$$-\frac{1}{2}\frac{dSO_2}{dt} = -\frac{dO_2}{dt} = \frac{1}{2}\frac{dSO_3}{dt}$$

$$\therefore -\frac{dSO_2}{dt} = -2\times\frac{dO_2}{dt}$$

$= -2 \times 2.5 \times 10^{-4}$

$= -5 \times 10^{-4}$ mol L^{-1} s^{-1}

33. (b) Rate of disappearance of reactant = Rate of appearance of products

$$-\frac{1}{2}\frac{d[N_2O_5]}{dt} = \frac{1}{4}\frac{d[NO_2]}{dt}$$

$$\frac{1}{2}k[N_2O_5] = \frac{1}{4}k'[N_2O_5]$$

$$\frac{k}{2} = \frac{k'}{4}$$

$\therefore k' = 2k$

34. (c) It is bimolecular first order reaction since Rate $\propto [N_2O_5]$
35. (b) $r = k[O_2][NO]^2$. When the volume is reduced to 1/2, the conc. will double

$\therefore$ New rate $= k[2O_2][2NO]^2 = 8k[O_2][NO]^2$

The new rate increases to eight times of its initial.

36. (a) For a zero order reaction.

rate $= k[A]^0$ i.e., rate $= k$

hence unit of k = M.sec^{-1}

For a first order reaction.

rate $= k[A]$

k = M.sec^{-1}/M = sec^{-1}

37. (a) Plots of conc. [A] Vs time, t

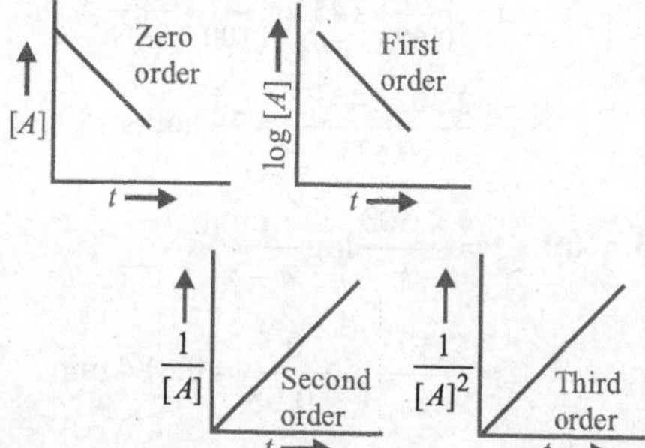

38. (d) rate of appearance of HI $= \frac{1}{2}\frac{d[HI]}{dt}$

rate of disappearance of $H_2 = \frac{-d[H_2]}{dt}$

rate of disappearance of $I_2 = \frac{-d[I_2]}{dt}$

hence $\frac{-d[H_2]}{dt} = -\frac{d[I_2]}{dt} = \frac{1}{2}\frac{d[HI]}{dt}$

or $-\frac{2d[H_2]}{dt} = -\frac{2d[I_2]}{dt} = \frac{d[HI]}{dt}$

39. (d) Since all have same concentration of reactants, all would react at same time.

40. (c) Rate of reaction

$= \frac{\text{Rate of disappearance / appearance}}{\text{Stoichiometric coefficient}}$

$= -\frac{d[A]}{dt} = \frac{1}{2}\frac{d[B]}{dt} = 2.6\times10^{-2}$

$= -\frac{1}{2}\frac{d[B]}{dt}$

$5.2\times10^{-2} = -\frac{d[B]}{dt}$

41. (b) After every 30 minutes the amount is reduced to $\frac{1}{2}$ therefore $t_{1/2}$ is 30 minutes. In 90 minutes the amount is reduced to $\frac{1}{8}$ i.e. $\frac{1}{2^n}$. Here $n = 3$. True for 1st order reaction.

42. (c) $k = \frac{0.693}{45}$ further

$t = \frac{2.303}{0.693}\times 45 \log\frac{100}{100-99.9}$

$= \frac{2.303\times45\times3}{0.693} = 7\frac{1}{2}$ hours.

43. (a) $t = \frac{2.303}{k}\log\frac{a}{a-x}$

$= \frac{2.303}{6}\log\frac{0.5}{0.05} = 0.384$ min.

44. (c) $t_{90\%} = \frac{2.303}{k}\log\frac{100}{100-90}$ (I)

$t_{50\%} = \frac{2.303}{k}\log\frac{100}{100-50}$ (II)

Dividing $\frac{t_{90\%}}{t_{50\%}} = \frac{\log 10}{\log 2}$

$\therefore t_{90\%} = 3.3 t_{50\%}$

45. (c) $t_{1/2} = \frac{0.693}{k} = \frac{0.693}{0.0693} = 10$ min

Reactant after 10 min = 5 mol

Rate $\left(\frac{dx}{dt}\right) = k[A] = 0.0693 \times 5$ mol min^{-1}

46. (d) For a zero order, rate of reaction does not change with time.

47. (d) It is a second order reaction, first order both w.r.t $S_2O_8^{2-}$ and I^-.

$\therefore r = k[S_2O_8^{2-}][I^-]$

All other options are of first order reaction.

48. (b) $k_{1(300)} = \frac{0.693}{20}$; $k_{2(320)} = \frac{0.693}{5}$

$\ln\frac{k_{2(320)}}{k_{1(300)}} = \frac{E_a}{R}\left[\frac{1}{T_1} - \frac{1}{T_2}\right]$

$E_a = \frac{2.303\, RT_1T_2}{(T_2 - T_1)}\log\frac{k_2}{k_1}$

$= \frac{2.303\times8.314}{20\times1000}\times300\times320 \log 4$

$= 55.14$ kJ/mol

49. (c) $e^{-\frac{E_a}{RT}} = 3.8\times\frac{10^{-16}}{100}$

$-\frac{E_a}{RT} = \ln 3.8\times10^{-18}$

$E_a = 100$ kJ/mol

50. (c) $\frac{k_2}{k_1} = \frac{Ae^{-E_{a_2}/RT}}{Ae^{-E_{a_1}/RT}} = e^{(E_{a_1} - E_{a_2})/RT}$

$2.303\log\frac{k_2}{k_1} = \frac{E_{a_1} - E_{a}}{RT}$

$= \frac{(83.314-75)\times10^3}{8.314\times500} = 2$

$\log k_2 = \frac{2}{2.303} = 0.868$

Taking Antilog

$k_2 = 7.38$

51. (d) In Arrhenius equation $k = Ae^{-E_a/RT}$, E_a is the energy of activation, which is required by the colliding molecules to react resulting in the formation of products.

52. (b) When the temperature is increased, energy in form of heat is supplied which increases

the kinetic energy of the reacting molecules. this will increase the number of collisions and ultimately the rate of reaction will be enhanced.

53. (b) According to Arrhenius equation,

$k = A.e^{-E_a/RT}$

where k = rate constant

or $\ln k = \ln A - \frac{E_a}{RT}$

A = frequency factor

E_a = Energy of activation.

54. (c) $k = 1.2 \times 10^{14} e^{-25000/RT}$ $\sec^{-1}$ or

$\log k = \log 1.2 \times 10^{14} - \frac{25000}{R}.\frac{1}{T}$

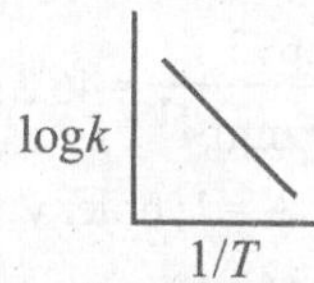

Equation of straight line

slope $= -\frac{25000}{R}$

55. (c) There are 5 tens hence $(2)^5 = 32$.

56. (a) E_a(F.R.) $\neq E_a$(B.R.) E_a can be calculated.

57. (b) According to Arrhenius equation,

$k = Ae^{-Ea/RT}$

$\therefore$ when $E_a = 0, k = A$

Also $\ln k$ vs $1/T$ is a straight line with slope $= -E_a/R$.

$\therefore$ Statements (ii) and (v) are correct.

58. (a) Let the initial concentration $(A) = 100$

Final concentration at 298 K = 100 – 10 = 90

Final concentration at 308 K = 100 – 25 = 75

Substituting the values in the 1st order rate reaction

$$t = \frac{2.303}{k_{298}} \log \frac{100}{90} \quad ...(i)$$

$$t = \frac{2.303}{k_{308}} \log \frac{100}{75} \quad ...(ii)$$

From (i) and (ii) $\frac{k_{308}}{k_{208}} = 2.73$

Substituting the value in the following relation

$$E_a = \frac{2.303\, R \times T_1 \times T_2}{T_2 - T_1} \log \frac{k_2}{k_1}$$

$$= \frac{2.303 \times 8.314 \times 298 \times 308}{308 - 298} \log 2.73$$

$E_a = 76.62$ kJ mol^{-1} = 18.39 kcal mol^{-1}

59. (d) For the reaction

$3A + 2B \longrightarrow C + D$

Rate of disappearance of A = Rate of appearance of C reaction

$$= -\frac{1}{3}\frac{d[A]}{dt} = \frac{d[C]}{dt} = k[A]^n[B]^m$$

60. (b) Given $-\frac{dMnO_4^-}{dt} = 4.56 \times 10^{-3}$ Ms^{-1}

From the reaction given,

$$-\frac{1}{2}\frac{dMnO_4^-}{dt} = \frac{4.56 \times 10^{-3}}{2} Ms^{-1}$$

$$-\frac{1}{2}\frac{dMnO_4^-}{dt} = \frac{1}{5}\frac{dI_2}{dt}$$

$$\therefore \quad -\frac{5}{2}\frac{dMnO_4^-}{dt} = \frac{dI_2}{dt}$$

On substituting the given value

$$\therefore \quad \frac{dI_2}{dt} = \frac{4.56 \times 10^{-3} \times 5}{2} = 1.14 \times 10^{-2} \text{ M/s}$$

61. (d) For a zero order reaction

Rate constant = k = $\frac{a - x}{t}$

$2 \times 10^{-2} = \frac{a - 0.5}{25}$

$a - 0.5 = 0.5$

$a = 1.0$ M

62. (b) A $\rightarrow$ B For a first order reaction

Given $a = 0.8$ mol, $(a - x) = 0.8 - 0.6 = 0.2$

$k = \frac{2.303}{1}\log\frac{0.8}{0.2}$ or $k = 2.303 \log 4$

again $a = 0.9$, $a - x = 0.9 - 0.675 = 0.225$

$k = \frac{2.303}{t}\log\frac{0.9}{0.225}$

$2.303 \log 4 = \frac{2.303}{t}\log 4$

Hence t = 1 hour

63. (b) For a first order reaction, A $\rightarrow$ Products

$$r = k[A] \text{ or } k = \frac{r}{[A]}$$

$$\Rightarrow k = \frac{1.5\times10^{-2}}{0.5} = 3\times10^{-2}$$

Further, $t_{1/2} = \frac{0.693}{k} = \frac{0.693}{3\times10^{-2}} = 23.1$

64. (d) Given [A] = 0.01 M

Rate = 2.0×10^{-5} mol L^{-1} s^{-1}

For a first order reaction

Rate = k[A]

$$k = \frac{2.0\times10^{-5}}{[0.01]} = 2\times10^{-3}$$

$$t_{1/2} = \frac{0.693}{2\times10^{-3}} = 347 \text{ sec}$$

65. (b) Given: 75% reaction gets completed in 32 min

Thus, $k = \frac{2.303}{t}\log\frac{a}{(a-x)}$

$$= \frac{2.303}{32}\log\frac{100}{(100-75)}$$

$$= \frac{2.303}{32}\log 4 = 0.0433 \text{ min}^{-1}$$

Now we can use this value of k to get the value of time required for 50% completion of reaction

$$t = \frac{2.303}{k}\log\frac{a}{(a-x)} = \frac{2.303}{0.0433}\log\frac{100}{50}$$

$$= \frac{2.303}{0.0433}\log 2 = 16 \text{ min}$$

66. (a) If we write rate of reaction in terms of concentration of NH_3 and H_2, then

$$\text{Rate of reaction} = \frac{1}{2}\frac{d[NH_3]}{dt} = -\frac{1}{3}\frac{d[H_2]}{dt}$$

$$\text{So, } \frac{d[NH_3]}{dt} = -\frac{2}{3}\frac{d[H_2]}{dt}$$

67. (a) Rate $\propto \sqrt{\text{Concentration}}$

$$= k\sqrt{\text{Concentration}}$$

$$k = \frac{\text{Rate}}{(\text{Concentration})^{1/2}}$$

$$= \frac{4\times10^{-6}}{(4\times10^{-4})^{1/2}} = \frac{4\times10^{-6}}{2\times10^{-2}}$$

$$= 2\times10^{-4} \text{ mol}^{1/2} \text{ L}^{-1/2} \text{ s}^{-1}$$

68. (c) $[A]_t = [A] - kt = 1 - 0.001\times10\times60$

$= 0.4$ M

$[B]_t = 0.001\times10\times60 = 0.6$ M

69. (d) For the first order reaction for small finite change

$$k_1 = \frac{1}{[A]}\frac{\Delta[A]}{\Delta t} \Rightarrow \frac{\Delta[A]/[A]}{\Delta t} = 1.5\% \text{ min}^{-1}$$

$= 0.015$ min^{-1}

$$t_{1/2} = \frac{0.693}{0.015 \text{ min}^{-1}} = 46.2 \text{ min} \approx 46 \text{ min}$$

70. (c) Order w.r.t. A = 1; order w.r.t. B = 1

$$\text{Rate} = \frac{1}{2}\frac{d(AB)}{dt} = k_r[A][B];$$

$$\frac{1}{2}\times(2.5\times10^4) = k_r(0.1)(0.1)$$

$k_r = 1.25\times10^{-2}$

71. (b) $k_1 = \frac{1}{5}\ln\frac{0.2}{0.1}; k_2 = \frac{1}{10}\ln\frac{0.2}{0.05}; k_1 = k_2.$

72. (a) Exothermic because of activation energy $E_b > E_f$

73. (b) More activations energy, slow reaction rate.

74. (c) A catalyst does not take part in chemical reaction but it alters the rate of reaction.

75. (b) Activation energy of reactant is less than the energy of activation of products.

76. (b) The presence of enzyme (catalyst) increases the speed of reaction by lowering the energy barrier, i.e., a new path is followed with lower activation energy.

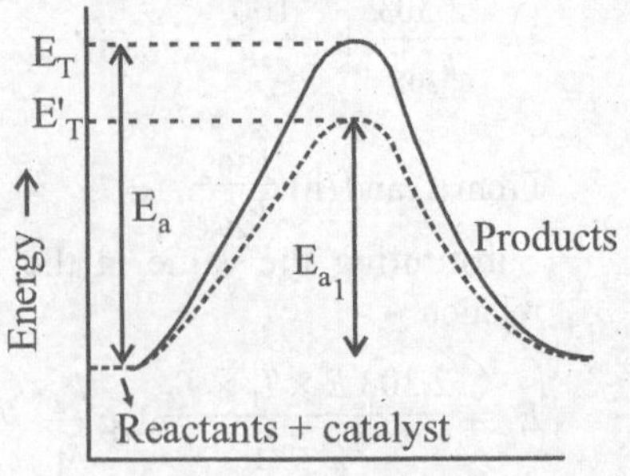

Here E_T is the threshold energy.
E_a and E_{a1} is energy of activation of reaction in absence and presence of catalyst respectively.

77. (c) The rate constant doubles for 10° C rise in temperature.
For 20° C rise, the rate constant will be 4 times
$\therefore k_1 = 4k_2$ or $k_2 = 0.25\, k_1$

78. (a) $e^{-E_a/RT} = 10^{-3}\% = 10^{-5}$;
$E_a = 2.303 \times 2 \times 300 \times 5$ cal
$= 6.91\, k\, cal\, mol^{-1}$

79. (d) The activation energy of reverse reaction will depend upon whether the forward reaction is exothermic or endothermic.
As $\Delta H = E_a$ (forward reaction) $- E_a$ (backward reaction)
For exothermic reaction
$\Delta H = -ve$
$\therefore -\Delta H = E_a(f) - E_a(b)$
or $E_a(f) = E_a(b) - \Delta H$
$\therefore E_a(f) < E_a(b)$
for endothermic reaction
$\Delta H = +ve$
$\therefore \Delta H = E_a(f) - E_a(b)$ or $E_a(f) = \Delta H + E_a(b)$
$\therefore E_a(f) > E_a(b)$.

80. (a) Arrhenius equation is given by
$k = Ae^{-E_a/RT}$
Taking log on both sides, we get
$\log k = \log A - \dfrac{E_a}{2.303RT}$
Arrhenius plot a graph between $\log k$ and $\dfrac{1}{T}$ whose slope is $\dfrac{-E_a}{2.303R}$.

81. (c) Given $t_{1/2} = 3$
Total time T = 12
No. of half lives $(n) = \dfrac{12}{3} = 4$

$$\left(\frac{1}{2}\right)^n = \frac{N}{N_o} \quad \therefore \quad \left(\frac{1}{2}\right)^4 = \frac{3}{N}$$

$$\frac{3}{N} = \frac{1}{16}$$

N = 48 g

82. (d) $A \longrightarrow B$

Initial concentration	Rate of reaction
2×10^{-3} M	2.40×10^{-4} Ms^{-1}
1×10^{-3} M	0.60×10^{-4} Ms^{-1}

rate of reaction
$r = k[A]^x$
where x = order of reaction
hence
$2.40 \times 10^{-4} = k\,[2 \times 10^{-3}]^x$ ......(i)
$0.60 \times 10^{-4} = k\,[1 \times 10^{-3}]^x$ ......(ii)
On dividing eqn.(i) from eqn. (ii) we get
$4 = (2)^x$
$\therefore x = 2$
i.e. order of reaction = 2

83. (b) Plot given is for zero order reaction.

84. (b) $t_{1/2} = \dfrac{0.693}{k} = 2100\,s = 35$ min
$t_{75\%} = 2t_{1/2} = 2 \times 35 = 70$ min

85. (a) In the absence of A, 3 dps is zero error, hence
Initial count = 23 – 3 = 20 dps
After 10 min = 13 – 3 = 10 dps
After 20 min = 5 dps
(recorded = 5 + 3 = 8 dps)
(50% fall in 10 min, $T_{50} = 10$ min)

86. (c) $t_{1/2} \propto \dfrac{1}{(p)^{n-1}}$ where n is the order of reaction

$$\frac{2}{4} = \left(\frac{50}{100}\right)^{n-1} \quad \text{or} \quad \frac{1}{2} = \left(\frac{1}{2}\right)^{n-1}$$

$n - 1 = 1;\ n = 2$

87. (a) For a zero order reaction
$t_{1/2} = \dfrac{a}{2k}$

88. (a) $k = \dfrac{0.693}{t_{1/2}} = \dfrac{0.693}{480} = 1.44 \times 10^{-3} s^{-1}$

89. (d) $A_2 + B_2 \longrightarrow 2AB$;
$A_2 \longrightarrow A + A$ (Fast);
$A + B_2 \longrightarrow AB + B$ (Slow)
Rate law = $k[A][B_2]$ put value of [A] from Ist reaction since A is intermediate
$\sqrt{k[A_2]} = A$

$\therefore$ Rate law equation $= K\sqrt{k[A_2]}[B_2]$

$\therefore$ Order $= \frac{1}{2} + 1 = \frac{3}{2}$

90. **(d)** $t_{1/2} = 5$ years, $T = 15$ years hence total number of half life periods $= \frac{15}{5} = 3$

$\therefore$ Amount left $= \frac{64}{(2)^3} = 8g$

91. **(b)** $\Delta H = E_a(f) - E_a(b)$

Thus energy of activation for reverse reaction depend upon whether reaction is exothermic or endothermic.

If reaction is exothermic,

$\Delta H = -ve,\ E_a(b) > E_a(f)$

If reaction is endothermic,

$\Delta H = +ve\ \ E_a(b) < E_a(f)$

92. **(c)** We know that the activation energy of chemical reaction is given by formula

$= \frac{k_2}{k_1} = \frac{E_a}{2.303R}\left[\frac{T_2 - T_1}{T_1 T_2}\right]$, where k_1 is the rate constant at temperature T_1 and k_2 is the rate constant at temperature T_2 and E_a is the activation energy. Therefore activation energy of chemical reaction is determined by evaluating rate constant at two different temperatures.

93. **(c)** For the exothermic reaction the energy of products is always less than the reactants. If E_a is the energy of activation for the forward reaction, the energy of activation for backward reaction is $E_a + \Delta H$

94. **(b)** $\Delta H = E_a(f) - E_a(b) = 0$

95. **(b)** According to Arrhenius equation

$$\log\frac{k_2}{k_1} = \frac{E_a}{2.303R}\left(\frac{1}{T_1} - \frac{1}{T_2}\right)$$

$$\log\frac{1.3\times10^{-3}}{1.3\times10^{-4}} = \frac{E_a}{2.303\times8.314}\left[\frac{1}{373} - \frac{1}{423}\right]$$

$$1 = \frac{E_a}{2.303\times8.314}\left[\frac{1}{373} - \frac{1}{423}\right]$$

$E_a = 60$ kJ / mol

96. **(b)** $X \longrightarrow Y;\ \Delta H = -135$ kJ/mol,

$E_a = 150$ kJ/mol

For an exothermic reaction

$E_{a(F.R.)} = \Delta H + E'_{a(B.R.)}$

$150 = -135 + E'_{a(B.R.)}$

$E'_{a(B.R.)} = 285$ kJ/mol

97. **(c)** Activation energy can be calculated from the equation.

$$\frac{\log K_2}{\log K_1} = \frac{-E_a}{2.303\,R}\left(\frac{1}{T_2} - \frac{1}{T_1}\right)$$

Given $\frac{\log K_2}{\log K_1} = 2$ $T_2 = 308$ K; $T_1 = 298$ K

$$\therefore\ \log 2 = \frac{-E_a}{2.303\times8.314}\left(\frac{1}{308} - \frac{1}{298}\right)$$

$E_a = 52.9$ kJ mol^{-1}

98. **(c)** $t_{1/4} = \frac{2.303}{k}\log\frac{1}{3/4} = \frac{2.303}{k}\log\frac{4}{3}$

$= \frac{2.303}{k}(\log 4 - \log 3) = \frac{2.303}{k}(2\log 2 - \log 3)$

$= \frac{2.303}{k}(2\times0.301 - 0.4771) = \frac{0.29}{k}$

99. **(d)** $A \longrightarrow B,\ \Delta H = -10$ kJ mol^{-1}

It is an exothermic reaction.

$E_{a(b)} = E_{a(f)} - (\Delta H)$

$= 50 - (-10) = 60$ kJ

100. **(b)** pH = 2 ; $r_1 = k \times (10^{-2})^n$ {Q $[H^+] = 10^{-pH}$}

pH = 1 ; $r_2 = k \times (10^{-1})^n$

Given $r_2 = 100\,r_1$

$$\Rightarrow \left(\frac{10^{-1}}{10^{-2}}\right)^n = 100$$

$10^n = 100$

$\therefore\ n = 2$

19 Surface Chemistry

1. The volume of gases NH_3, CO_2 and H_2 adsorbed by one gram of charcoal at 300 K are in order of:
(a) $H_2 > CO_2 > NH_3$ (b) $NH_3 > H_2 > CO_2$
(c) $NH_3 > CO_2 > H_2$ (d) $CO_2 > NH_3 > H_2$

2. A plot of $\log\left(\frac{x}{m}\right)$ against log P for the adsorption of a gas on a solid gives a straight line with slope equal to : Tricky
(a) $\frac{1}{n}$ (b) n (c) log K (d) K

3. The heat evolved in chemisorption lies in the range (in kJ/mol) of :
(a) 80 to 240 (b) 20 to 40
(c) 40 to 80 (d) 20 to 100

4. Identify the correct statement regarding enzymes________.
(a) Enzymes are specific biological catalysts that cannot be poisoned
(b) Enzymes are normally heterogeneous catalysts that are very specific in their action
(c) Enzymes are specific biological catalysts that can normally function at very high temperatures ($T \approx 1000K$)
(d) Enzymes are specific biological catalysts that possess well-defined active sites

5. Catalytic poisons act by :
(a) making the products chemically inactive.
(b) increasing the rate of the backward reaction.
(c) chemical combination with any one of the reactants.
(d) preferential adsorption on the catalyst surface.

6. The size of particles in suspension, true solution and colloidal solution varies in the order:
(a) suspension > colloidal > true solution
(b) true solution > suspension > colloidal
(c) suspension > colloidal = true solution
(d) none of these

7. Gold number of a lyophilic sol is such a property that:
(a) The larger its value, the greater is the peptizing power.
(b) The lower its value, the greater is the peptizing power.
(c) The lower its value, the greater is the protecting power.
(d) The larger its value, the greater is the protecting power.

8. Among the following, correct statement is :

(a) Brownian movement is more pronounced for smaller particles than for bigger–particles.
(b) Sols of metal sulphides are lyophilic.
(c) Hardy Schulze law states that bigger the size of the ons, the greater is its coagulating power.
(d) One would expect charcoal to adsorb chlorine more than hydrogen sulphide.

9. Under the influence of an electric field, the particles in a sol migrate towards cathode. The coagulation of the same sol is studied using NaCl, Na_2SO_4 and Na_3PO_4 solutions. Their coagulating values will be in the order
(a) $NaCl > Na_2SO_4 > Na_3PO_4$
(b) $Na_2SO_4 > Na_3PO_4 > NaCl$
(c) $Na_3PO_4 > Na_2SO_4 > NaCl$
(d) $Na_2SO_4 > NaCl > Na_3PO_4$

10. 3.6 gram of oxygen is adsorbed on 1.2 g of metal powder. What volume of oxygen adsorbed per gram of the adsorbent at 1 atm and 273 K?

(a) $0.19\,Lg^{-1}$ (b) $1\,Lg^{-1}$
(c) $2.1\,L\,g^{-1}$ (d) None of these

11. For the graph below, select correct order of temperature?

Tricky

Extent of adsorption (x/m) — Pressure (P) — P_s — T_1, T_2, T_3

(a) $T_1 > T_2 > T_3$ (b) $T_2 > T_3 > T_1$
(c) $T_3 > T_2 > T_1$ (d) $T_1 = T_2 = T_3$

12. 100 mL of 0.6 M acetic acid is shaken with 2 g activated carbon. The final concentration of the solution after adsorption is 0.5 M. What is the amount of acetic acid adsorbed per gram of carbon?

Toughnut

(a) 0.6 g (b) 0.3 g
(c) 1.2 g (d) None of these

13. The efficiency of an enzyme in catalysing a reaction is due to its capacity
(a) to form a strong enzyme-substrate complex
(b) to decrease the bond energies of substrate molecule
(c) to change the shape of the substrate molecule
(d) to lower the activation energy of the reaction

14. A chemical reaction is catalyzed by a catalyst X. Hence X
(a) reduces enthalpy of the reaction
(b) decreases rate constant of the reaction
(c) increases activation energy of the reaction
(d) does not affect equilibrium constant of the reaction

15. The gold numbers of protective colloids A, B, C and D are 0.04, 0.004, 10 and 40 respectively. The protective powers of A, B, C and D are in the order :
(a) A > B > C > D (b) B > A > C > D
(c) D > C > A > B (d) D > C > B > A

16. The density of gold is 19 g/cm^3. If 1.9×10^{-4} g of gold is dispersed in one litre of water to give a sol having spherical gold particles of radius 10 nm, then the number of gold particles per mm^3 of the sol will be :

Toughnut

(a) 1.9×10^{12} (b) 6.3×10^{14}
(c) 6.3×10^{10} (d) 2.4×10^{6}

17. A thixotropic gel is characterised by
(a) reversible transformation of gel into sol on heating
(b) irreversible transformation of gel into sol by stirring
(c) reversible transformation of gel into sol by stirring
(d) swelling on placing in the dispersion medium

18. In an experiment, addition of 4.0 ml of 0.005 *M* $BaCl_2$ to 16.0 ml of arsenius sulphide sol just causes the complete coagulation in 2 hrs. The flocculating value of the effective ion is

Critical Thinking

(a) Cl^-, 1.0 (b) Cl^-, 2.0
(c) Ba^{2+}, 1.0 (d) Ba^{2+}, 0.5

19. Choose the correct statement
(a) *Latex* is an example of solid-in-liquid sol
(b) *Latex* is a negative sol
(c) *Latex* is a positive sol
(d) None of the above is correct

20. Arrange the following electrolytes in the increasing order of coagulating power for ferric hydroxide sol

Na_3PO_4	KCl	K_2SO_4	NaCl
I	II	III	IV

(a) I < II < III < IV (b) II = IV < III < I
(c) II = IV < I < III (d) II = III < IV = I

21. Although, nitrogen does not adsorb on a surface at room temperature, it adsorbs on the surface at 83 K. Which one of the following statements is correct ?
(a) At 83 K, there is formation of monolayer
(b) At 83 K, nitrogen is adsorbed as atoms
(c) At 83 K, nitrogen molecules are held by chemical bonds
(d) At 83 K, there is formation of multimolecular layers

22. The following statements relate to the adsorption of gases on a solid surface. Identify the incorrect statement among them :
(a) Enthalpy of adsorption is negative
(b) Energy appears as heat
(c) On adsorption, the residual forces on the surface are increased
(d) Entropy of adsorption is negative

23. If x is the mass of the gas adsorbed on mass m of the adsorbent at pressure p, Freundlich adsorption isotherm gives a straight line on plotting
(a) x/m vs p (b) x/m vs $1/p$
(c) $\log x/m$ vs $\log p$ (d) $\log x/m$ vs p

24. The correct statement for both the processes of physisorption and chemisorption is
(a) both are endothermic
(b) chemisorption is endothermic but physisorption is exothermic
(c) both are exothermic
(d) physisorption is endothermic but chemisorption is exothermic.

25. Which one of the sols acts as protective colloid?
(a) As_2S_3 (b) Gelatin
(c) Au (d) $Fe(OH)_3$

26. Given below are a few electrolytes, indicate which one among them will bring about the coagulation of a gold sol quickest and in the least of concentration? Tricky
(a) NaCl (b) $MgSO_4$
(c) $Al_2(SO_4)_3$ (d) $K_4[Fe(CN)_6]$

27. Which of the following has minimum flocculation value for positively charged sol?
(a) Cl^- (b) SO_4^{2-}
(c) PO_4^{3-} (d) $[Fe(CN)_6]^{4-}$

28. Butter is a colloid formed when
(a) fat is dispersed in water
(b) fat globules are dispersed in water
(c) water is dispersed in fat
(d) None of the these

29. Cloud or fog is a colloidal system in which the dispersed phase and the dispersion medium are
(a) Gas, Liquid (b) Liquid, gas
(c) Liquid, liquid (d) Solid, Liquid

30. Which of the following constitutes irreversible colloidal system in water as dispersion medium?
(a) Clay (b) Platinum
(c) $Fe(OH)_3$ (d) All of these

31. At the high pressure, Langmuir adsorption isotherm takes the form
(a) $\frac{x}{m}=\frac{ap}{1+bp}$ (b) $\frac{x}{m}=\frac{a}{b}$
(c) $\frac{x}{m}=ap$ (d) $\frac{m}{x}=\frac{b}{a}+\frac{1}{ap}$

32. Methylene blue, from its aqueous solution, is adsorbed on activated charcoal at 25°C. For this process, which of the following statement is correct ? Critical Thinking
(a) The adsorption requires activation at 25°C
(b) The adsorption is accompanied by a decrease in enthalpy
(c) The adsorption increases with increase of temperature
(d) The adsorption is irreversible

33. In an experiment, 200 mL of 0.5 M oxalic acid is shaken with 10 g of activated charcoal and filtered. The concentration of the filtrate is reduced to 0.4 M. The amount of adsorption $\left(\frac{x}{m}\right)$ is Toughnut
(a) 0.9 (b) 1.8 (c) 0.18 (d) 0.09

34. Equal volume each of two sols of AgI, one obtained by adding $AgNO_3$ to slight excess of KI and another obtained by adding KI to slight excess of $AgNO_3$, are mixed together. Then:
(a) the two sols will stabilize each other
(b) the sol particles will acquire more electric charge
(c) the sols will coagulate each other mutually
(d) a true solution will be obtained

35. Under the influence of an electric field, the particles in a sol migrate towards cathode. The coagulation of the same sol is studied using NaCl, Na_2SO_4 and Na_3PO_4 solutions. Their coagulating values will be in maximum for:
(a) NaCl (b) Na_2SO_4
(c) Na_3PO_4 (d) same for all

36. According to Langmuir adsorption isotherm, the amount of gas adsorbed at very high pressures
(a) reaches a constant limiting value
(b) goes on increasing with pressure
(c) goes on decreasing with pressure
(d) increases first and then decreases with pressure

37. Under ambient conditions, which among the following surfactants will form micelles in aqueous solution at lowest molar concentration? Tricky
(a) $CH_3-(CH_2)_8-CO\bar{O}\ Na^+$
(b) $CH_3(CH_2)_{11}\overset{+}{N}(CH_3)_3\overset{-}{Br}$
(c) $CH_3-(CH_2)_{13}-OSO_3^-\ Na^+$
(d) $CH_3(CH_2)_{15}\overset{+}{N}(CH_3)_3\overset{-}{Br}$

38. The migration of dispersion medium under the influence of an electric potential is called :
(a) Cataphoresis (b) Electro-osmosis
(c) Electrophoresis (d) Sedimentation

39. Flocculation value of $BaCl_2$ is much less than that of KCl for sol A and flocculation value of Na_2SO_4 is much less than that of NaBr for sol B. The correct statement among the following is :

(a) Both the sols A and B are negatively charged.
(b) Sol A is positively charged and Sol B is negatively charged.
(c) Both the sols A and B are positively charged.
(d) Sol A is negatively charged and sol B is positively charged.

40. Which of the following statements regarding difference between adsorption and absorption is incorrect?
(a) Adsorption is a surface whereas absorption is a bulk phenomena.
(b) Water vapours are absorbed by anhydrous $CaCl_2$ but adsorbed by silica gel.
(c) Adsorption and absorption take place individually. They can not occur simultaneously.
(d) All of the above statements are correct.

41. In Langmuir's model of adsorption of a gas on a solid surface

Critical Thinking

(a) the mass of gas striking a given area of surface is proportional to the pressure of the gas
(b) the mass of gas striking a given area of surface is independent of the pressure of the gas
(c) the rate of dissociation of adsorbed molecules from the surface does not depend on the surface covered
(d) the adsorption at a single site on the surface may involve multiple molecules at the same time.

42. Which one of the following statements is not correct in respect of lyophilic sols?
(a) There is a considerable interaction between the dispersed phase and dispersion medium
(b) These are quite stable and are not easily coagulated
(c) They carry charge
(d) The particles are hydrated

43. Lyophilic sols are more stable than lyophobic sols because:
(a) the colloidal particles have positive charge
(b) the colloidal particles have negative charge
(c) the colloidal particles are solvated
(d) there is strong electrostatic repulsion between the colloidal particles

44. Select incorrect statement:
(a) Micelles are associated colloids
(b) The electrical charge on a colloid particle is indicated by electrophoresis
(c) Formation of micelles takes place above Kraft temperature
(d) Formation of micelles takes place below CMC

45. Point out the false statement
(a) Brownian movement and Tyndall effect are shown by colloidal systems.
(b) Gold number is a measure of the protective power of a lyophillic colloid
(c) The colloidal solution of a liquid in liquid is called gel
(d) Hardy - Schulze rule is related with coagulation.

46. Which gas will be adsorbed on a solid to greater extent?
(a) Having non - polar molecule
(b) Having highest critical temperature
(c) Having lowest critical temperature
(d) Having lowest critical pressure

47. Which of the following adsorption isotherms represents the absorption of a gas by a solid involving multilayers of formation? (P_s = saturation pressure)

(a) 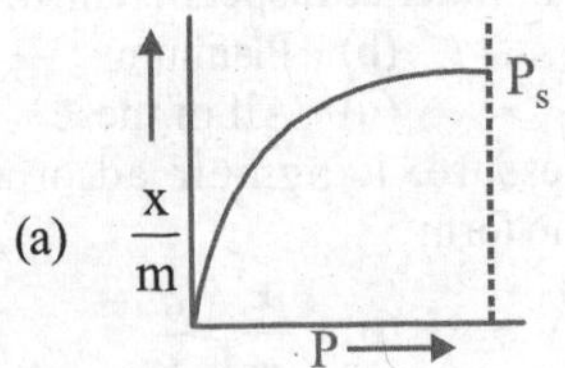

(b) 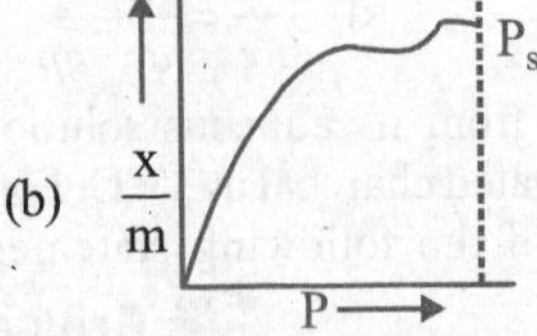

(c)

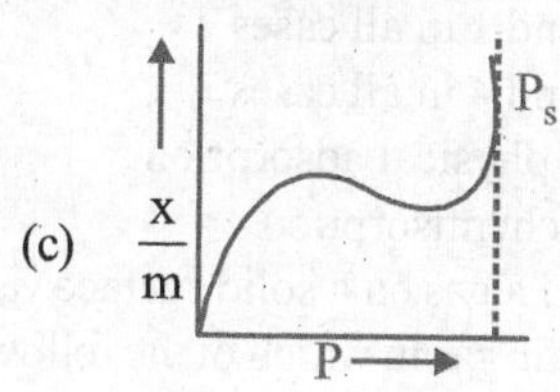

(d)

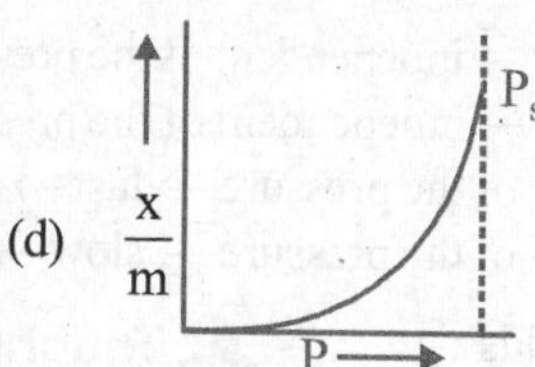

48. A detergent ($C_{12}H_{25}SO_4^-Na^+$) solution becomes a colloidal sol at a concentration of 10^{-3} M. On an average 10^{13} colloidal particles are present in 1 mm^3. What is the average number of ions which are contained by one colloidal particle (micelle)? [Given: $N_A = 6 \times 10^{23} mol^{-1}$]

Toughnut

(a) 6×10^7 (b) 10
(c) 60 (d) None of these

49. Which of the following is most powerful to coagulate the negative colloid?
(a) $ZnSO_4$ (b) Na_3PO_4
(c) $AlCl_3$ (d) $K_4[Fe(CN)_6]$

50. Coagulation value of the electrolytes $AlCl_3$ and NaCl for As_2S_3 sol are 0.093 and 52 respectively. How many times $AlCl_3$ has greater coagulating power than NaCl ?
(a) 930 (b) 520
(c) 560 (d) None of these

51. The protecting power of lyophilic colloidal sol is expressed in terms of:
(a) coagulation value
(b) gold number
(c) critical miscelle concentration
(d) oxidation number

52. Which of the following statements is correct for the spontaneous adsorption of a gas?
(a) ΔS is negative and, therefore, ΔH should be highly positive
(b) ΔS is negative and therefore, ΔH should be highly negative
(c) ΔS is positive and, therefore, ΔH should be negative
(d) ΔS is positive and, therefore, ΔH should also be highly positive

53. Which property of colloids is not dependent on the charge on colloidal particles?
(a) Coagulation (b) Electrophoresis
(c) Electro - osmosis (d) Tyndall effect

54. Smoke is an example of:
(a) Solid dispersed in solid
(b) Gas dispersed in liquid
(c) Solid dispersed in gas
(d) Gas dispersed in solid

55. On addition of 1 mL of 10% NaCl solution to 10 mL gold sol in the presence of 0.025 g of starch, the coagulation is just prevented. Starch has gold number

(a) 2.5 (b) 25 (c) 0.25 (d) 0.025

56. Fog is a colloidal solution of
(a) solid particles dispersed in gas
(b) solid particles dispersed in a liquid
(c) liquid particles dispersed in gas
(d) gaseous particles dispersed in a liquid

57. Select incorrect statement:
(a) Soap and detergent lower the interfacial surface tension between oil and water
(b) Basic principle of peptization is reverse of coagulation
(c) Soap and detergent used as emulsifiers
(d) Lyophilic sols need stabilizing agent

58. Which of the following is most effective in causing the coagulation of ferric hydroxide sol?
(a) KCl (b) KNO_3
(c) K_2SO_4 (d) $K_3[Fe(CN)_6]$

59. Aluminium hydroxide forms a positively charged sol. Which of the following ionic substances should be most effective in coagulating the sol?
(a) NaCl (b) $CaCl_2$
(c) $Fe_2(SO_4)_3$ (d) K_3PO_4

60. Colloidal gold is prepared by
(a) Mechanical dispersion
(b) Peptisation
(c) Bredig's Arc method
(d) Hydrolysis

61. Colloid of which one of the following can be prepared by electrical dispersion method as well as reduction method ? Tricky
(a) Sulphur
(b) Ferric hydroxide
(c) Arsenious sulphide
(d) Gold

62. The cause of Brownian movement is
(a) Heat changes in liquid state
(b) Convectional currents
(c) The impact of molecules of the dispersion medium on the colloidal particles.
(d) Attractive forces between the colloidal particles and molecules of dispersion medium.

63. During adsorption Tricky
(a) $T\Delta S$ is positive
(b) $\Delta H - T\Delta S$ is negative
(c) ΔH is positive
(d) $T\Delta S$ and ΔG become zero

64. Which of the following gas molecules have maximum value of enthalpy of physisorption?
(a) C_2H_6 (b) Ne (c) H_2O (d) H_2

65. Adsorption is accompanied by the evolution of heat. So according to Le-Chatelier principle the amount of substance adsorbed should
(a) increase with decrease in temperature
(b) increase with increase in temperature
(c) decrease with decrease in temperature
(d) decrease with increase in temperature

66. In Freundlich adsorption isotherm, the value of $1/n$ is : Tricky
(a) between 0 and 1 in all cases
(b) between 2 and 4 in all cases
(c) 1 in case of physical adsorption
(d) 1 in case of chemisorption

67. The adsorption of a gas on a solid surface varies with pressure of the gas in which of the following manner.
(a) Fast → slow → independent of the pressure
(b) Slow → fast → independent of the pressure
(c) Independent of the pressure → fast → slow
(d) Independent of the pressure → slow → fast

68. Associated colloids :
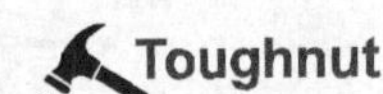

(a) raise both the surface tension and viscosity of water
(b) lower both the surface tension and viscosity of water
(c) lower the surface tension and raise the viscosity of water
(d) have greater concentration at the surface layer than the bulk of the solution

69. Hydrolysis of urea is an example of
(a) homogeneous catalysis
(b) heterogeneous catalysis
(c) biochemical catalysis
(d) zeolite catalysis

70. When $KClO_3$ is heated, it decomposes into $KCl + O_2$. If some MnO_2 is added, the reaction goes much faster because
(a) MnO_2 decomposes to give O_2
(b) MnO_2 provides heat by reacting
(c) better contact is provided by MnO_2
(d) MnO_2 acts as a catalyst

Answer KEYs

1	(c)	8	(a)	15	(b)	22	(c)	29	(b)	36	(a)	43	(c)	50	(c)	57	(d)	64	(c)
2	(a)	9	(a)	16	(d)	23	(c)	30	(d)	37	(d)	44	(d)	51	(b)	58	(d)	65	(a)
3	(a)	10	(c)	17	(c)	24	(c)	31	(b)	38	(b)	45	(c)	52	(b)	59	(d)	66	(a)
4	(d)	11	(c)	18	(c)	25	(b)	32	(b)	39	(b)	46	(b)	53	(d)	60	(c)	67	(a)
5	(d)	12	(b)	19	(b)	26	(c)	33	(c)	40	(c)	47	(a)	54	(c)	61	(d)	68	(c)
6	(a)	13	(d)	20	(b)	27	(d)	34	(c)	41	(a)	48	(c)	55	(b)	62	(c)	69	(c)
7	(c)	14	(d)	21	(d)	28	(c)	35	(a)	42	(c)	49	(c)	56	(c)	63	(b)	70	(d)

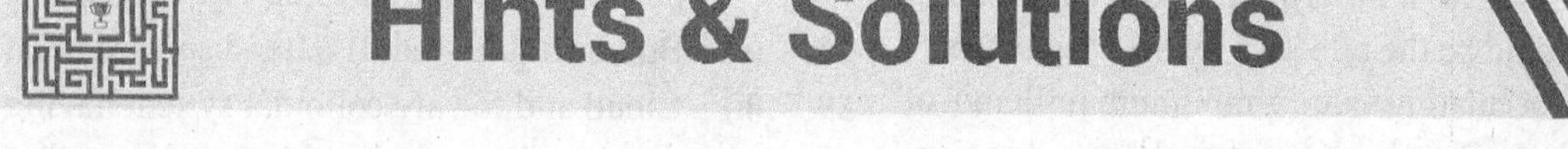

Hints & Solutions

1. (c) Van der Waal force ↑ adsorption ↑

2. (a) $\frac{x}{m} = KP^{\frac{1}{n}}$

$\log \frac{x}{m} = \log K + \frac{1}{n} \times \log P$

Slope $= \frac{1}{n}$

3. (a) ΔH : 80 to 240 kJ mol^{-1}

4. (d) Enzymes are specific biological catalysts possessing well - defined active sites.

5. (d) The catalytic poisons decrease the activity of the catalyst because they are preferentially adsorbed on the surface of catalyst.

6. (a)

Solution	Particle size
True solution	Size < 1 nm
Colloidal solution	1 nm < size < 1000 nm
Suspension	Size > 1000 nm

7. (c) The reciprocal of gold number is directly proportional to the protecting power.

8. (a) Brownian movement is the random motion of particles suspended in a fluid (a liquid or gas) resulting from their collision with the fast moving atom or molecules in the liquid or gaseous state of the matter. That means smaller particles are responsible for the Brownian movement than for bigger particles.

9. (a) Since the sol particles migrate towards cathode, they are positively charged. Hence, anions would be effective in coagulation. Greater is the valency of effective ion, smaller will be its coagulating value.

10. (c) Mass of O_2 per gram of adsorbent

$= \frac{3.6}{1.2} = 3$

No. of moles of O_2 per gram of adsorbent $= \frac{3}{32}$

Volume of O_2 per gram of adsorbent

$= \frac{3}{32} \times \frac{0.0821 \times 273}{1} = 2.10$

11. (c) Temp ↑ Extent of adsorption ↓

12. (b) Mass of acetic acid adsorbed by 2 g charcoal

$= 100 \times 10^{-3} \times (0.6 - 0.5) \times 60$

$\Rightarrow 0.6; \frac{x}{m} = \frac{0.6}{2} \Rightarrow 0.3$

13. (d) Efficiency of catalysing property of a catalyst is inversely proportional to activation energy.

14. (d) A catalyst affects equally both forward and backward reactions, therefore it does not affect equilibrium constant of reaction.

15. (b) Smaller is the 'gold number' of protective colloid, greater is its protective power.

16. (d) Volume of gold dispersed in 1 L water

$= \frac{\text{Mass}}{\text{Density}} = \frac{1.9 \times 10^{-4}\,\text{g}}{19\text{gm cm}^{-3}} = 1 \times 10^{-5}\,\text{cm}^3$

Radius of gold sol particle

$= 10\,\text{nm} = 10 \times 10^{-7}\,\text{cm} = 10^{-6}\,\text{cm}$

Volume of gol sol particle $= \frac{4}{3}\pi r^3$

$= \frac{4}{3} \times \frac{22}{7} \times (10^{-6})^3 = 4.19 \times 10^{-18}\,\text{cm}^3$

∴ No. of gold sol particles in $1 \times 10^{-5}\,\text{cm}^3$

$= \frac{1 \times 10^{-5}}{4.19 \times 10^{-18}} = 2.38 \times 10^{12}$

∴ No. of gold sol particles in one mm^3

$= \frac{2.38 \times 10^{12}}{10^6} = 2.38 \times 10^6$

17. (c) Thixotropic gel changes into sol on mechanical agitation, which on standing sets to gel.

18. **(c)** As_2S_3 sol is negatively charged owing to preferential adsorption of S^{2-} ions. Cation would be the effective ion in coagulation. Flocculating value = minimum milli mol of the effective ion per litre of sol =

$$\frac{4\times0.005\times10^3}{4+16}=1.0$$

19. **(b)** *Latex* is a negative sol.
Latex is an *emulsion* in which the rubber particles forming emulsion are stabilized by various proteins.

20. **(b)** Ferric hydroxide sol is positively charged and hence anions would be effective in causing coagulation. Greater the valence of the effective ion, more will be its coagulating power.

21. **(d)** At 83 K physical absorption takes place effectively so multimolecular layers formed.

22. **(c)** Adsorption is a phenomena of attracting and retaining the molecules of a substance on the surface of a liquid or a solid resulting into a higher concentration of the molecules on the surface.
After adsorption there is a decrease in the residual forces due to bond formation ΔG, ΔH & ΔS all are negative in the case of adsorption.

23. **(c)** Freundlich adsorption isotherm gives straight line on plotting log x/m vs log p as show below.

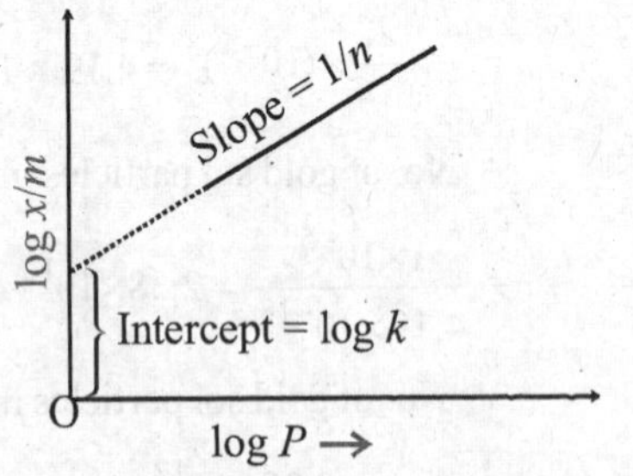

24. **(c)** Both are exothermic.

25. **(b)** Gelatin act as protective colloids.

26. **(c)** Gold sol. have negative charge. So $Al_2(SO_4)_3$ is most effective for coagulation.

27. **(d)** For positive charge colloids coagulating power ↑coagulation value ↓.$[Fe(CN)_6]^{4-}$.

28. **(c)** Butter (liquid - solid) water dispersed in fat

29. **(b)** Cloud and fog are colloidal system having dispersed phase liquid and dispersion medium gas

30. **(d)** All are irreversible colloidal systems.

31. **(b)** Langmuir adsorption isotherm, is

$$\frac{x}{m}=\frac{ap}{1+bp}$$

At high pressure, 1 + bp = bp

$$\text{or}\quad \frac{x}{m}=\frac{ap}{bp}=\frac{a}{b}$$

32. **(b)** The adsorption of methylene blue on activated charcoal is an example of physiosorption which is exothermic, multilayer and does not have energy barrier.

33. **(c)** Mass of oxalic acid adsorbed by 10 g charcoal

$$=200\times10^{-3}(0.5-0.4)\times90=1.8\text{ g}$$

$$\frac{x}{m}=\frac{1.8}{10}=0.18$$

34. **(c)** The sols obtained in the two cases will be oppositely charged to coagulate each other.

35. **(a)** The sol particles migrate towards cathode. So they are positively charged. Hence, anions would be effective in coagulation. Greater is the valence of effective ion, smaller will be its coagulating value.

36. **(a)** According to Langmuir adsorption isotherm, the amount of gas adsorbed at very high pressure reaches a constant limiting volume.

37. **(d)** Greater the surface area, greater the van der waal forces of attraction and therefore at lesser concentration micelle formation will take place. In case of $CH_3(CH_2)_{15}\overset{\oplus}{N}(CH_3)_3Br^-$ due to greater chain length, greater will be van der waal forces.

38. (b) The motion of a liquid through a membrane under the influence of an applied electric field is known as electro-osmosis.

39. (b) In first case the given compounds have same anion but different cations having different charge hence they will precipitate negatively charged sol i.e. 'A'.
In second case the given compounds have similar cation but different anion with different charge. Hence they will precipitate positively charged sol. i.e. 'B'.

40. (c) Both adsorption and absorption can take place simultaneously. The term sorption is used to describe both the processes.

41. (a) According to Langmuir's model of adsorption of a gas on a solid surface, the mass of gas adsorbed (x) per gram of the adsorbent (m) is directly proportional to the pressure of the gas (p) at constant temperature.

42. (c) In lyophilic sol no charge present.

43. (c) D.P. particles have great affinity towards D.M. in lyophilic colloids.

44. (d) Formation of micelles takes place below CMC.

45. (c) Liquid - liquid system is known as emulsion.

46. (b) Critical temp. ↑ Attractive force ↑
Extent of adsorption ↑.

47. (a) 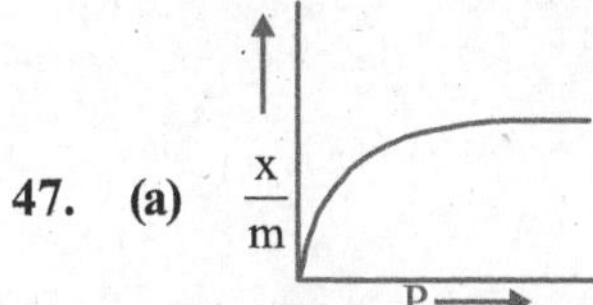

48. (c) No. of sodium lauryl sulphate molecules
$(CH_3(CH_2)_{11}SO_4^-Na^+)$ in 1 litre solution
$= 10^{-3} \times 6 \times 10^{23} = 6 \times 10^{20}$
No. of sodium lauryle sulphate molecules per mm$^3 = 6 \times 10^{14}$
No. of colloidal particles per mm$^3 = 10^{13}$
No. of molecules per colloidal particle

$$= \frac{6\times10^{14}}{10^{13}} = 60$$

49. (c) According to Hardy-Schulze rule "The amount of electrolyte required to coagulate a fixed amount of a sol depends upon the sign of charge and valency of the flocculating ion."
Thus, the coagulating power vary in the order.
$Al^{3+} > Zn^{2+} > Na^+$

50. (c) $\frac{\text{Coagulation power of } AlCl_3}{\text{Coagulation power of NaCl}}$

$$= \frac{\text{Coagulation value of NaCl}}{\text{Coagulation value of } AlCl_3}$$

$$= \frac{52}{0.093} = 559.13 \approx 560$$

51. (b) The lyophobic sols are less stable than lyophilic sols. The lyophilic sols are thus used to protect the lyophobic sols. This property of lyophilic sols is known as protective action of lyophilic sols which is represented by gold number.

52. (b) For adsorption $\Delta S < 0$ and for a spontaneous change $\Delta G = -$ ve
hence ΔH should be highly negative which is clear from the equation
$\Delta G = \Delta H - T\Delta S$
$= -\Delta H - T(-\Delta S) = -\Delta H + T\Delta S$
So if ΔH is highly negative ΔG will also be (– ve)

53. (d) Tyndall effect is optical property, whereas other properties are electrical properties. Hence dependent on the charge on colloids

54. (c) Smoke is an example of solid particles dispersed in gas.

55. (b) Gold Number is the minimum amount of lyophilic colloide in miligrams (starch in given case) which prevents the flocculation of 10 ml gold sol by the addition of 1ml of 10% NaCl solution.
∴ Gold number = miligram of starch
$= 0.025 \times 10^3$ mg
$= 25$

56. (c) Fog has liquid as dispersed phase and gas as dispersion medium.

57. (d) No stabilising agent required for the preparation of lyophilic sol.

58. (d) $Fe(OH)_3$ is positive sol. $K_3[Fe(CN)_6]$ will provide $[Fe(CN)_6]^{3-}$ for coagulation having highest magnitude of –ve charge among given options.

59. (d) $Al(OH)_3$ is positively charged sol. K_3PO_4 has greater negative charge. Hence, it is most effective in coagulation

60. (c) Colloidal gold is prepared by Bredig's arc method.

61. (d) Gold by Bredig's method (Dispersion method) and by reduction method

$AuCl_3$ + Tannic acid $\rightarrow$ Gold sol

62. (c) It is due to impact of molecules of dispersion medium on the colloidal particles.

63. (b) $\Delta G = \Delta H - T\Delta S$. Adsorption is spontaneous process ΔG should be negative.

Hence $\Delta H - T\Delta S$ should be negative.

64. (c) The more the liquefiable nature of a gas, the more is the enthalpy of adsorption. Water is more liquefiable.

65. (a) The extent of adsorption increases with decrease in temperature.

66. (a) According to Freundlich adsorption isotherm

$$\frac{x}{m} = kp^{\frac{1}{n}}$$

at low pressure, $\frac{1}{n} = 1$

$$\therefore \frac{x}{m} \propto p^1$$

at high pressure, $\frac{1}{n} = 0$

$$\frac{x}{m} \propto p^{\circ}$$

i.e., the value of n varies between 0 to 1

67. (a) Adsorption of a gas on solid is represented by following equilibria,

Gas (Adsorbate) + Solid (Adsorbent) $\rightleftharpoons$ Gas adsorbed on solid + Heat

Initially adsorption increases with increase in pressure at a particular temperature then got slow. After attaining equilibrium, adsorption becomes independent of pressure.

68. (c) Associate colloids decrease surface tension and increase viscosity.

69. (c) Hydrolysis of urea can be represented as follows

$$H_2N-\underset{\underset{O}{\|}}{C}-NH_2 + H_2O \xrightarrow[\text{(enzyme)}]{\text{Urease}} 2NH_3 + CO_2$$

Since it involves biological catalyst (enzyme), so it is an example of biochemical catalysis.

70. (d) $2KClO_3 \xrightarrow{MnO_2} 2KCl + 3O_2$

20 General Principles and Processes of Isolation of Elements

1. Which one of the following is not a sulphide ore?
(a) Magnetite (b) Iron pyrites
(c) Copper glance (d) Sphalerite

2. Which of the following pair is incorrectly matched?
(a) Magnetite – Fe_3O_4
(b) Copper glance – Cu_2S
(c) Calamine – $ZnCO_3$
(d) Zincite – ZnS

3. The impurities associated with mineral used in metallurgy are called collectively?
(a) Slag (b) Flux (c) Gangue (d) Ore

4. Galena is an ore of
(a) Pb (b) Hg (c) Zn (d) Cu

5. The process of converting hydrated alumina into anhydrous alumina is called
(a) roasting (b) smelting
(c) dressing (d) calcination

6. Which of the following statements regarding metallurgy of iron is incorrect?

(a) Reaction $Fe_3O_4 + 4CO \longrightarrow 3Fe + 4CO_2$ belongs to lower temperature range (500 – 800K) of the blast furnace.
(b) Reaction $FeO + CO \longrightarrow Fe + CO_2$ belongs to higher temperature range (900 – 1500K) of the blast furnace.
(c) The iron obtained from blast furnace is cast iron with 3% carbon.
(d) For reduction of iron oxide to occur ΔG of the couple of following reactions should be negative

$$FeO(s) \longrightarrow Fe(s) + \frac{1}{2}O_2(g)$$

$$C(s) + \frac{1}{2}O_2(g) \longrightarrow CO(g)$$

7. The most electropositive metals are isolated from their ores by
(a) high temperature reduction with carbon
(b) self reduction
(c) thermal decomposition
(d) electrolysis of fused ionic salts

8. Which of the following condition favours the reduction of a metal oxide to metal?

(a) $\Delta H = +ve, T\Delta S = +ve$ at low temperature
(b) $\Delta H = +ve, T\Delta S = -ve$ at any temperature
(c) $\Delta H = -ve, T\Delta S = -ve$ at high temperature
(d) $\Delta H = -ve, T\Delta S = +ve$ at any temperature

9. Which of the following statements regarding electrolytic refining of copper is incorrect?

(a) In this process anode is made up of impure copper and pure copper strips are taken as cathode.
(b) Acidic or basic solution of copper sulphate is used as electrolyte
(c) Antimony, tellurium, silver and gold are some of the metals deposits as anode mud during this process
(d) Zinc can be also refined by electrolytic refining method.

10. If an impurity in a metal has a greater affinity for oxygen and is more easily oxidised than the metal, then the purification of metal may be carried out by
(a) Poling (b) Zone refining
(c) Electrolytic refining (d) Cupellation

11. van Arkel method of purification of metals involves converting the metal to a
(a) volatile stable compound
(b) volatile unstable compound
(c) non volatile stable compound
(d) None of the above

12. In order to refine "blister copper" it is melted in a furnace and is stirred with green logs of wood. The purpose is
(a) to expel the dissolved gases in blister copper.
(b) to bring the impurities to surface and oxidize them.
(c) to increase the carbon content of copper.
(d) to reduce the metallic oxide impurities with hydrocarbon gases liberated from the wood.

13. The value of $\Delta_f G^o$ for formation of Cr_2O_3 is -540 kJmol^{-1} and that of Al_2O_3 is -827 kJ mol^{-1} What is the value of $\Delta_r G^\circ$ for the reaction?

$$\frac{4}{3}Al(s)+\frac{2}{3}Cr_2O_3(s)\rightarrow\frac{2}{3}Al_2O_3(s)+\frac{4}{3}Cr(s).$$

Toughnut

(a) -574 kJ mol^{-1} (b) -287kJ mol^{-1}
(c) $+574$ kJ mol^{-1} (d) $+287$kJ mol^{-1}

14. Which of the following statement is not correct about Ellingham diagram?
(a) Δ G increases with an increase in temperature
(b) It consists of plots of $\Delta_f G^o$ Vs T for formation of oxides
(c) A coupling reaction can be well expressed by this diagram
(d) It express the kinetics of the reduction process

15. Iron is obtained on large scale from Fe_2O_3 by
(a) Reduction with CO (b) Reduction with Al
(c) Calcination (d) Passing H_2

16. Aluminothermic process is used for the extraction of metals, whose oxides are
(a) fusible
(b) not easily reduced by carbon
(c) not easily reduced by hydrogen
(d) strongly basic.

17. When $CuSO_4$ is electrolysed using platinum electrodes,

Critical Thinking

(a) copper is liberated at cathode, sulphur at anode
(b) copper is liberated at cathode, oxygen at anode
(c) sulphur is liberated at cathode, oxygen at anode
(d) oxygen is liberated at cathode, copper at anode

18. Which one of the following is a mineral of iron ?
(a) Malachite (b) Cassiterite
(c) Pyrolusite (d) Magnetite

19. In the extraction of copper from its sulphide ore, the metal finally obtained by the reduction of cuprous oxide with

Toughnut

(a) iron (II) sulphide (b) carbon monoxide
(c) copper (I) sulphide (d) sulphur dioxide

20. Calamine is an ore of :
(a) zinc (b) aluminium
(c) iron (d) copper

21. Which one of the following ores is known as Malachite:
(a) Cu_2O (b) Cu_2S
(c) $CuFeS_2$ (d) $Cu(OH)_2.CuCO_3$

22. In the isolation of metals, calcination process usually results in :
(a) metal hydroxide (b) metal sulphide
(c) metal oxide (d) metal carbonate

23. Cryolite is
(a) Na_3AlF_6 and used in the electrolysis of alumina for decreasing electrical conductivity
(b) Na_3AlF_6 and used in the electrolysis of alumina for lowering the melting point of alumina
(c) Na_3AlF_6 and used in the electrolytic purification of alumina
(d) Na_3AlF_6 and used in the electrolysis of alumina

24. Froth floatation process is used for the metallurgy of
(a) chloride ores (b) amalgams
(c) oxide ores (d) sulphide ores

25. Which of the following metal is leached by cyanide process
(a) Ag (b) Na (c) Al (d) Cu

26. Which of the following statements, about the advantage of roasting of sulphide ore before reduction is not true? Tricky

(a) The ΔG_f^o of the sulphide is greater than those for CS_2 and H_2S.

(b) The ΔG_f^o is negative for roasting of sulphide ore to oxide.

(c) Roasting of the sulphide to the oxide is thermodynamically feasible.

(d) Carbon and hydrogen are suitable reducing agents for reduction of metal sulphides.

27. Sulphide ores of metals are usually concentrated by froth flotation process. Which one of the following sulphide ores offer an exception and is concentrated by chemical leaching?

Toughnut

(a) Galena (b) Copper pyrite
(c) Sphalerite (d) Argentite

28. Pb and Sn are extracted from their chief ore by
(a) carbon reduction and self reduction.
(b) self reduction and carbon reduction.
(c) electrolysis and self reduction.
(d) self reduction and electrolysis.

29. Which of the following elements is present as the impurity to the maximum extent in the pig iron?
(a) Manganese (b) Carbon
(c) Silicon (d) Phosphorus

30. Which of the following pairs of metals is purified by van Arkel method ?
(a) Ga and In (b) Zr and Ti
(c) Ag and Au (d) Ni and Fe

31. The following reactions take place in the blast furnace in the preparation of impure iron. Identify the reaction pertaining to the formation of the slag. Toughnut
(a) $Fe_2O_3(s) + 3\,CO(g) \rightarrow 2\,Fe\,(l) + 3\,CO_2\,(g)$
(b) $CaCO_3\,(s) \rightarrow CaO\,(s) + CO_2\,(g)$
(c) $CaO\,(s) + SiO_2(s) \rightarrow CaSiO_3(s)$
(d) $2C(s) + O_2\,(g) \rightarrow 2\,CO(g)$

32. Aluminium is extracted from alumina (Al_2O_3) by electrolysis of a molten mixture of
(a) $Al_2O_3 + HF + NaAlF_4$
(b) $Al_2O_3 + CaF_2 + NaAlF_4$
(c) $Al_2O_3 + Na_3AlF_6 + CaF_2$
(d) $Al_2O_3 + KF + Na_3AlF_6$

33. Correct match is:
(a) Bayer's method - Na_2CO_3
(b) Matte - 98% Cu_2S+2% FeS
(c) van Arkel method - AgI
(d) Thomas slag - Raw material for cement industry

34. Consider the following metallurgical processes:
(I) Heating impure metal CO and distilling the resulting volatile carbonyl (b.p. 43°C) and finally decomposition at 150°–200°C to get the pure metal
(II) Heating the sulphide ore in air until a part is converted to oxide and then furher heating in the absence of air to let the oxide react with unchanged metal sulphide
(III) Electrolysis of the molten electrolyte containing approximately equal amounts of the metal chloride and NaCl to obtained the metal

The processes used for obtaning magnesium, nickel and copper are respectively:

(a) (I), (II) and (III) (b) (II), (III) and (I)
(c) (III), (I) and (II) (d) (II), (I) and (III)

35. When copper pyrites is roasted in excess of air, a mixture of CuO + FeO is formed. FeO is present as impurity. This can be removed as slag during reduction of CuO. The flux added to form slag is
(a) SiO_2, which is an acidic flux
(b) Limestone, which is a basic flux
(c) SiO_2, which is the basic flux
(d) CaO, which is a basic flux

36. One of the processes used for concentration of ores is *Froth floatation process*. This process is generally used for concentration of *sulphide ores*. Sometimes in this process we add NaCN as a depressant. NaCN is generally added in case of ZnS and PbS minerals. What is the purpose of

addition of NaCN during the process of Froth floatation ?

(a) NaCN causes reduction by precipitation

(b) A soluble complex is formed by reaction between NaCN and ZnS while PbS forms froth

(c) A soluble complex is formed by reaction between NaCN and PbS while ZnS forms froth

(d) A precipitate of $Pb(CN)_2$ is produced while ZnS remain unaffected.

37. Which of the following statement(s) is/are correct?

(i) Cast iron is used in the manufacture of railway sleepers

(ii) Wrought iron is used in the manufacture of anchors, bolts, chains etc.

(iii) Nickel steel is used in making pendulums.

(a) Only (i) (b) (i) and (ii)

(c) (i), (ii) and (iii) (d) Only (iii)

38. A coupled reaction takes place as follow–

$A + B \longrightarrow C + D \qquad \Delta G^o = +x\,kj$

$D + E \longrightarrow F \qquad \Delta G^o = -y\,kj$

for the spontaneity of reaction

$A + B + E \longrightarrow C + F$, which of the following is correct?

Tricky

(a) $2x = y$ (b) $x < y$

(c) $x > y$ (d) $x = (y) \times T\Delta S$

39. The form of iron obtained from blast furnace is:

(a) Steel (b) Cast Iron

(c) Pig Iron (d) Wrought Iron

40. Calcination is the process in which :

(a) ore is heated above its melting point to expel H_2O or CO_2 or SO_2

(b) ore is heated below its melting point to expel volatile impurities

(c) ore is heated above its melting point to remove S, As and Sb as SO_2, As_2O_3 and Sb_2O_3 respectively

(d) ore is heated below its melting point to expel H_2O or CO_2

41. In Goldschmidt aluminothermic process which of the following reducing agents is used :

(a) calcium (b) coke

(c) Al-powder (d) sodium

42. Pyrolusite is a/an

(a) oxide ore (b) sulphide ore

(c) carbide ore (d) Not an ore

43. Among the following statements the incorrect one is

(a) Calamine and siderite are carbonates

(b) Argentite and cuprite are oxides

(c) Zinc blende and iron pyrites are sulphides

(d) Malachite and azurite are ores of copper.

44. During smelting an additional substance is added which combines with impurities to form a fusible product. It is known as

(a) slag (b) mud (c) gangue (d) flux

45. Which of the following processes involve the roasting process?

(a) $ZnCO_3 \rightarrow ZnO + CO_2$

(b) $Fe_2O_3 + 3C \rightarrow 2Fe + 3CO$

(c) $2PbS + 3O_2 \rightarrow 2PbO + 2SO_2$

(d) $Al_2O_3 . 2H_2O \rightarrow Al_2O_3 + 2H_2O$

46. Before introducing FeO in blast furnace , it is converted to Fe_2O_3 by roasting so that

(a) it may not be removed as slag with silica

(b) it may not evaporate in the furnace

(c) presence of it may increase the m.pt. of charge

(d) None of these

47. In electrorefining of copper some gold is deposited as

(a) anode mud (b) cathode mud

(c) electrolyte (d) None of these

48. In the extraction of copper from its sulphide ore, the metal is finally obtained by the reduction of cuprous oxide with :

(a) Copper (I) sulphide (Cu_2S)

(b) Sulphur dioxide (SO_2)

(c) Iron sulphide (FeS)

(d) Carbon monoxide (CO)

49. $XCl_2\,(excess) + YCl_2 \longrightarrow XCl_4 + Y\downarrow$;

$YO \xrightarrow[>400^\circ]{\Delta} \frac{1}{2}O_2 + Y$, Ore of Y would be:

(a) Siderite (b) Cinnabar

(c) Malachite (d) Hornsilver

50. Refractory materials are generally used in furnaces because
(a) they possess great structural strength
(b) they can withstand high temperature
(c) they are chemically inert
(d) they do not require replacement

51. In electro-refining of metal the impure metal is made the anode and a strip of pure metal, the cathode, during the electrolysis of an aqueous solution of a complex metal salt. This method cannot be used for refining of
(a) silver (b) copper
(c) aluminium (d) sodium

52. When the sample of copper with zinc impurity is to be purified by electrolysis, the appropriate electrodes are

	cathode	anode
(a)	pure zinc	pure copper
(b)	impure sample	pure copper
(c)	impure zinc	impure sample
(d)	pure copper	impure sample.

53. In the extraction of aluminium by Hall-Heroult process, purified Al_2O_3 is mixed with CaF_2 to
(i) lower the melting point of Al_2O_3.
(ii) increase the conductivity of molten mixture.
(iii) reduce Al^{3+} into Al(s).
(iv) acts as catalyst.

(a) (i) and (ii) (b) (i), (ii) and (iii)
(c) (iii) and (iv) (d) (ii), (iii) and (iv)

54. In Hall-Heroult process how much carbon anode is burnt away to produce each 1kg of aluminium?

(a) 0.3 kg (b) 0.5 kg (c) 1 kg (d) 0.1 kg

55. Which of the oxide groups among the following cannot be reduced by carbon?
(a) Cu_2O, SnO_2 (b) CaO, K_2O
(c) PbO, Fe_2O_4 (b) Fe_2O_3, ZnO

56. Extraction of zinc from zinc blende is achieved by
(a) roasting followed by self-reduction
(b) electrolytic reduction
(c) roasting followed by reduction with carbon
(d) roasting followed by reduction with another metal

57. The substance used as froth stabilisers in froth-floatation process is
(a) Potassium ethyl xanthate
(b) Aniline
(c) Sodium cyanide
(d) Copper sulphate

58. When a metal is to be extracted from its ore and the gangue associated with the ore is silica, then
(a) an acidic flux is needed
(b) a basic flux is needed
(c) both acidic and basic fluxes are needed
(d) Neither of them is needed

59. Which one of the following ores is not concentrated by froth floatation process?
(a) Copper pyrites (b) Pyrargyrite
(c) Pyrolusite (d) Zinc blende

60. Which of the following reactions is an example for calcination process ?

Tricky

(a) $2Ag + 2HCl + (O) \rightarrow 2AgCl + H_2O$
(b) $2Zn + O_2 \rightarrow 2ZnO$
(c) $2ZnS + 3O_2 \rightarrow 2ZnO + 2SO_2$
(d) $MgCO_3 \rightarrow MgO + CO_2$

61. Heating of ore in the absence of air below its melting point is called
(a) leaching (b) roasting
(c) smelting (d) calcination

62. Heating pyrites to remove sulphur is called
(a) smelting (b) calcination
(c) liquation (d) roasting

63. Complexes formed in the following methods are

Toughnut

I. Mond's process for purification of nickel.
II. Removal of lead poisoning from the body.
III. Cyanide process for extraction of silver.
IV. Froth flotation process for separation of ZnS from galena ore by using depressant.

	I	II	III	IV
(a)	$Ni(CO)_4$	$[Pb(EDTA]^{2-}$	$[Ag(CN)_2]^-$	$[Zn(CN)_2]$
(b)	$Ni(CO)_4$	$[Pb(EDTA]^{2-}$	$[Ag(CN)_2]^-$	$[Zn(CN)_4]^{2-}$
(c)	$Ni(CO)_6$	$[Pb(EDTA)]^{4-}$	$[Ag(CN)_2]^-$	$[Zn(CN)_6]^{4-}$
(d)	$Ni(CO)_4$	$[Pb(EDTA)]^{2-}$	$[Ag(CN)_4]^{3-}$	$Zn(CN)_4]^{2-}$

64. The method of zone refining of metals is based on the principle of:
(a) Greater solubility of the impurity in the molten state than in the solid
(b) Greater mobility of the pure metal than that of the impurity
(c) Higher melting point of the impurity than that of the pure metal
(d) Greater noble character of the solid metal than that of the impurity

65. $\Delta_f G^-$ vs T plot in the Ellingham diagram slopes downward for the reaction
(a) $Mg + \frac{1}{2}O_2 \rightarrow MgO$

(b) $2Ag + \frac{1}{2}O_2 \rightarrow Ag_2O$
(c) $C + \frac{1}{2}O_2 \rightarrow CO$
(d) $CO + \frac{1}{2}O_2 \rightarrow CO_2$

66. Purification of silicon element used in semiconductors is done by
(a) zone refining (b) heating
(c) froth floatation (d) heating in vacuum

67. In the electrolysis of alumina to obtain aluminium metal, cryolite is added mainly to
(a) lower the melting point of alumina
(b) dissolve alumina in molten cryolite
(c) remove the impurities of alumina
(d) increase the electrical conductivity

68. Cassiterite is concentrated by
(a) levigation
(b) electromagnetic separation
(c) floatation
(d) liquefaction

69. Calcination is used in metallurgy for removal of?

(a) Water and sulphide
(b) Water and CO_2
(c) CO_2 and H_2S
(d) H_2O and H_2S

70. Which of the following metals is extracted by the electrometallurgical method ?
(a) Cu (b) Fe (c) Na (d) Ag

Answer KEYs

1	(a)	8	(d)	15	(a)	22	(c)	29	(b)	36	(b)	43	(b)	50	(b)	57	(b)	64	(a)
2	(d)	9	(b)	16	(b)	23	(b)	30	(b)	37	(c)	44	(d)	51	(d)	58	(b)	65	(c)
3	(c)	10	(d)	17	(b)	24	(d)	31	(c)	38	(b)	45	(c)	52	(d)	59	(c)	66	(a)
4	(a)	11	(a)	18	(d)	25	(a)	32	(c)	39	(b)	46	(a)	53	(a)	60	(d)	67	(a)
5	(d)	12	(d)	19	(a)	26	(d)	33	(b)	40	(d)	47	(a)	54	(b)	61	(d)	68	(b)
6	(c)	13	(b)	20	(a)	27	(d)	34	(c)	41	(c)	48	(a)	55	(b)	62	(d)	69	(b)
7	(d)	14	(d)	21	(d)	28	(b)	35	(a)	42	(a)	49	(b)	56	(c)	63	(b)	70	(c)

Hints & Solutions

1. (a) The formula of magnetite is Fe_3O_4.
2. (d) Zincite is ZnO.
3. (c) Impurities associated with minerals are called gangue or matrix.
4. (a) Galena is an ore of lead. It is PbS.
5. (d) $Al_2O_3.2H_2O \rightarrow Al_2O_3 + 2H_2O$ is calcination.
6. (c) The iron obtained from blast furnace is pig iron with 4% carbon and impurities like S, P, Mn etc., in small amount.
7. (d) Most electropositive metals are obtained by electrolysis of their fused ionic salts.
8. (d) $\Delta H = -ve$, $T\Delta S = +ve$ at any temperature
9. (b) During electrolytic refining of copper electrolyte used is acidified solution of copper sulphate.
10. (d) Cupellation can be used in such a case.
11. (a) $Ti + 2I_2 \xrightarrow{523K} \underset{\text{Volatile stable compound}}{TiI_4}$

$$\xrightarrow{1700K} \underset{\text{Pure metal}}{Ti} + 2I_2$$

12. (d) $2Cu_2S + 3O_2 \rightarrow 2Cu_2O + 2SO_2$

$$3Cu_2O + \underset{\text{(From green logs of wood)}}{CH_4} \rightarrow 6Cu + 2H_2O + CO$$

13. (b) The two equation are:

$$\frac{4}{3}Al(s) + O_2(g) \rightarrow \frac{2}{3}Al_2O_3(s),$$

$$\Delta_f G° = -827 kJ\ mol^{-1} \quad ...(1)$$

$$\frac{4}{3}Cr(s) + O_2(g) \rightarrow \frac{2}{3}Cr_2O_3(s),$$

$$\Delta_f G° = -540 kJ\ mol^{-1} \quad ...(2)$$

Subtracting equation (ii) from equation (i) we have,

$$\frac{4}{3}Al(s) + \frac{2}{3}Cr_2O_3(s), \rightarrow \frac{2}{3}Al_2O_3(s) + \frac{4}{3}Cr(s),$$

$$\Delta_r G° = -287 kJ\ mol^{-1}$$

14. (d) Ellingham diagrams are based on thermodynamic concepts. It does not tell anything about the kinetics of the reduction process.
15. (a) $Fe_2O_3 + 3CO \rightarrow 2Fe + 3CO_2$ (in blast furnace)
16. (b) When reduction by carbon is not satisfactory in case of metals having high m.pt., aluminothermic process is used.
17. (b) $CuSO_4 \rightleftharpoons Cu^{2+} + SO_4^{2-}$

$$H_2O \rightleftharpoons H^+ + OH^-$$

At cathode : $Cu^{2+} + 2e^- \rightarrow Cu$

At anode : $4OH^- \rightarrow 2H_2O + O_2 + 4e^-$

18. (d) Magnetite – Fe_3O_4
Malachite – $CuCO_3 \cdot Cu(OH)_2$
Cassiterite – SnO_2.
Pyrolusite – MnO_2
19. (a) Highly electropositive metals like Al, K, Na etc. are extracted by the electrolytic reduction.
zone refining method is used for obtaining metals of high purity e.g. Ge
Froth flotation process is suitable for sulphide ores
Cyanide process is used for the extraction of gold.
20. (a) $ZnCO_3$ = calamine.

21. (d) Malachite is a copper carbonate hydroxide mineral with the formula $Cu(OH)_2.CuCO_3$.

22. (c) Calcination is the process of converting an ore into its oxide by heating it strongly below its melting point either in absence or limited supply of air. It is usually done to convert metal carbonates and hydroxides to their respective oxides.

$$\underset{\text{Limestone}}{CaCO_3} \xrightarrow{\Delta} \underset{\text{Calcium oxide}}{CaO} + CO_2 \uparrow$$

$$\underset{\text{Malachite}}{CuCO_3.Cu(OH)_2} \xrightarrow{\Delta} 2CuO + H_2O \uparrow + CO_2 \uparrow$$

23. (b) Na_3AlF_6 is cryolite and used in the electrolysis of alumina to lower the melting point and increase electrical conductivity.

24. (d) Froth floatation process is used for the concentration of sulphide ores.

25. (a) Ag is leached by cyanide process.

26. (d) The sulphide ore is roasted to oxide before reduction because the ΔG_f^o of most of the sulphides are greater than those of CS_2 and H_2S, therefore neither C nor H can reduce metal sulphide to metal. Further, the standard free energies of formation of oxide are much less than those of SO_2. Hence oxidation of metal sulphides to metal oxide is thermodynamically favourable.

27. (d) Leaching is the selective dissolution of the desired mineral leaving behind the impurities in a suitable dissolving agent e.g.,

Argentitie or Silver glance, Ag_2S is an ore of silver. Silver is extracted from argentite by the mac-Arthur and Forest process (leaching process).

$$Ag_2S + 4NaCN \rightarrow 2Na[Ag(CN)_2] + Na_2S$$

$$4Au + 8KCN + 2H_2O + O_2 \rightarrow 4K[Au(CN)_2] + 4KOH$$

28. (b) PbO & $PbSO_4$ get reduced by PbS itself which is already present in mixture so because the reduction took place by itself, hence is known as self reduction.

$$2PbO + PbS \xrightarrow{\Delta} 3Pb + SO_2 \uparrow$$

$$PbSO_4 + PbS \xrightarrow{\Delta} 2Pb + 2SO_2 \uparrow$$

29. (b) Pig iron or cast iron contains 3 – 5% carbon and varying amounts of Mn, Si, P and S which makes the iron hard and brittle.

30. (b) Zr and Ti are purified by van Arkel method.

$$Zr(s) + 2I_2(g) \xrightarrow{870K} ZrI_4(g)$$

$$ZrI_4(g) \xrightarrow[\text{Tugsten filament}]{2075K} Zr(s) + 2I_2(g)$$

$$Ti(s) + 2I_2(s) \xrightarrow{523K} TiI_4(g)$$

$$TiI_4 \xrightarrow{1700\ K} \underset{\text{Pure titanium}}{Ti(s)} + 2I_2(g)$$

31. (c) In blast furnace at about 1270 K, calcium carbonate is almost completely decomposed to give CaO which acts as a flux and combines with SiO_2 present as impurity (gangue) in the ore to form calcium silicate (fusible slag)

$$\underset{\text{(basic flux)}}{CaO(s)} + \underset{\text{(acidic flux)}}{SiO_2(s)} \longrightarrow \underset{\text{(slag)}}{CaSiO_3(s)}$$

32. (c) Fused alumina (Al_2O_3) is a bad conductor of electricity. Therefore, cryolite (Na_3AlF_6) and fluorspar (CaF_2) are added to purified alumina which not only make alumina a good conductor of electricity but also reduce the melting point of the mixture to around 1140 K.

33. (b) Roasted mass obtained from roasting step is called matte. (98% Cu_2S + 2% FeS)

34. (c) (I) $\underset{\text{(impure)}}{Ni} + 4CO \xrightarrow{50°C} \underset{\text{(volite)}}{[Ni(CO)_4]} \xrightarrow{230°C} \underset{\text{(pure)}}{Ni} + 4CO \uparrow$

(II) $Cu_2S + \frac{3}{2}O_2 \xrightarrow{\Delta} Cu_2O + SO_2$

$$Cu_2S + 2Cu_2O \xrightarrow[\text{temp.}]{\text{high}} 6Cu + SO_2$$

(III) $MgCl_2(s) \xrightarrow{\text{electrolysis}} Mg^{2+}(l) + 2Cl^-(l)$

At cathode :

$Mg^{2+} 2e^- \longrightarrow Mg(s)$

At anode :

$2Cl^-(l) \longrightarrow Cl_2(g) + 2e^-$

35. (a) $FeO + SiO_2 \rightarrow FeSiO_3$.

36. (b) $NaCN + ZnS \longrightarrow \underset{\text{Soluble complex}}{Na_2[Zn(CN)_4]}$

A layer of this zinc complex is formed on the surface of ZnS and due to this ZnS is prevented from the froth formation while PbS form froth. (i.e., NaCN is added as depressant for ZnS)

37. (c) All three statements are correct

38. (b) For a spontaneous reaction, ΔG^o must be negative and it can be possible only in this case when $x < y$

39. (b) The iron obtained from blast furnace is in the form of cast iron.

40. (d) Calcination is a process of heating a substance to a high temperature but below the melting or fusion point, causing loss of moisture, reduction or oxidation and dissociation into simpler substances.

41. (c) Reduction by powdered aluminium is known as Gold-Schmidt aluminothermic process. This process is employed in cases where metals have very high m.p. and are to be extracted from their oxides.

42. (a) MnO_2 is pyrolusite (oxide ore).

43. (b) Cuprite is Cu_2O and Argentite is Ag_2S.

44. (d) Flux is added which combines with impurities to form slag.

45. (c) Roasting involves heating the ore strongly in presence of excess of air (O_2).

46. (a) FeO is capable forming slag with SiO_2

$SiO_2 + FeO \rightarrow FeSiO_3$

47. (a) Gold is deposited as anode mud.

48. (a) Cuprous oxide formed during roasting of cuprous sulphide is mixed with few amount of cuprous sulphide and heated in a reverberatory furnace to get metallic copper.

$$2Cu_2O + Cu_2S \rightarrow 6Cu + SO_2$$

49. (b) $\underset{(XCl_2)}{SnCl_2} + \underset{(YCl_2)}{HgCl_2} \longrightarrow \underset{(XCl_4)}{SnCl_4} + \underset{(Y)}{Hg}$

$HgO + \xrightarrow[>400°C]{\Delta} Hg + \frac{1}{2}O_2$

HgS : Cinnabar

50. (b) Refractories are the substances which can withstand very high temperature without melting or becoming soft.

51. (d) Na reacts vigorously with water (exothermic process)

52. (d) Pure metal always deposits at cathode.

53. (a) In the extraction of aluminium by Hall-Heroult process purified Al_2O_3 is mixed with CaF_2 to lower the melting point by Al_2O_3 and increase the conductivity of molten mixture.

54. (b) For each kg of Al produced, about 0.5 kg of carbon anode is burnt away.

55. (b) Ca and K are strong reducing agents, hence their oxides can not be reduced with carbon.

56. (c) Extraction of Zn from zinc blends is achieved by roasting followed by reduction with carbon. Reactions involved are :

$2ZnS + 3O_2 \xrightarrow{\text{heat}} 2ZnO + 2SO_2$ (Roasting)

$ZnO + C(\text{Coke}) \xrightarrow{\text{heat}} Zn + CO$ (Reduction)

57. (b) Froth stabilizers cresols, aniline are added to stabilize the froth and enhance the non-wettability of the mineral particles.

58. (b) Since silica is acidic impurity the flux must be basic.

$$CaO + SiO_2 \rightarrow CaSiO_3$$

59. (c) Pyrolusite is MnO_2. Hence not concentrated by froth floatation process.

60. (d) Decomposition of carbonates and hydrated oxides.

61. (d) It is definition of calcination.

62. (d) Pyrites are sulphur ores and are converted into oxide by roasting.

63. (b)

(I) Mond's process

$$\underset{\text{Impure}}{Ni + 4CO} \xrightarrow{50°C} \underset{\substack{\text{Volatile}\\\text{complex}}}{[Ni(CO)_4]}$$

$$\xrightarrow{200°C} \underset{\text{Pure}}{Ni} + 4CO \uparrow$$

(II) Removal of lead poisoning

$$Pb^{2+} + EDTA^{4-} \rightleftharpoons [Pb(EDTA)]^{2-}$$

(III) Cyanide process for extraction of silver

$$2Ag + 4NaCN + H_2O + \frac{1}{2}O_2 \rightleftharpoons$$

$$2Na[Ag(CN)_2] + 2NaOH$$

(IV) Separation of ZnS from galena by using depressant in froth flotation process

64. (a) Zone refining is based on the difference in solubility of impurities in molten and solid state of the metal. This method is used for obtaining metals of very high purity.

65. (c) $C(s) + \frac{1}{2}O_2(g) \rightarrow CO(g)$; ΔS increases.

Hence, as the temperature increases, $T\Delta S$ increases and hence ΔG ($\Delta H - T\Delta S$) decreases. In other words, the slope of the curve for formation of CO decreases. However, for all other oxides, it increases.

66. (a) Silicon is purified by zone refining. Metals of high purity are always purified by zone refining.

67. (a) Fused alumina (Al_2O_3) is a bad conductor of electricity. Therefore, cryolite (Na_3AlF_6) and fluorspar (CaF_2) are added to purified alumina which not only make alumina a good conductor of electricity but also reduce the melting point of the mixture to around 1140 K.

68. (b) Cassiterite contains the magnetic impurities of $FeSO_4$ and concentrated by electromagnetic separation.

69. (b) Calcination is used for removal of volatile impurities and decompose carbonates.

70. (c) Na is obtained by elctrolytic reduction being electropositive in nature.

21 The p-Block Elements (Group 15, 16, 17 & 18)

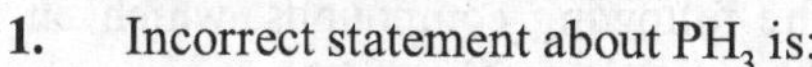

1. Incorrect statement about PH_3 is:
   (a) It is produced by hydrolysis of Ca_3P_2.
   (b) It gives black ppt. (Cu_3P_2) with $CuSO_4$ solution.
   (c) Spontaneously burns in presence of P_2H_4.
   (d) It does not react with B_2H_6.
2. The number of P–OH bonds and the oxidation state of phosphorus atom in pyrophosphoric acid ($H_4P_2O_7$) respectively are : Toughnut
   (a) four and four (b) five and four
   (c) five and five (d) four and five
3. P_2H_4 can be removed from phosphine containing traces of it
   (a) by passing impure PH_3 gas through a freezing mixture.
   (b) by passing the impure PH_3 gas through HI and then its treatment with KOH (aq).
   (c) by both (a) and (b).
   (d) by none of these.
4. $NH_4ClO_4 + HNO_3(dil.) \longrightarrow HClO_4 + [X]$
   $[X] \xrightarrow{\Delta} Y(g)$
   [X] and [Y] are respectively –
   (a) NH_4NO_3 & N_2O (b) NH_4NO_2 & N_2
   (c) HNO_4 & O_2 (d) None of these
5. For H_3PO_3 and H_3PO_4, the correct choice is

   (a) H_3PO_3 is dibasic and reducing
   (b) H_3PO_4 is diabasic and non-reducing
   (c) H_3PO_4 is tribasic and reducing
   (d) H_3PO_4 is tribasic and non reducing
6. One mole of calcium phosphide on reaction with excess of water gives
   (a) one mole of phosphine.
   (b) two moles of phosphoric acid.
   (c) two moles of phosphine.
   (d) one mole of phosphorus oxide.
7. In nitrogen family, the H-M-H bond angle in the hydrides gradually becomes closer to 90° on going from N to Sb. This shows that gradually
   (a) the basic strength of the hydrides increases.
   (b) almost pure p-orbitals are used for M-H bonding.
   (c) the bond energies of M-H bonds increases.
   (d) the bond pairs of electrons become nearer to the central atom.
8. Brown colour in HNO_3 can be removed by
   (a) adding Mg powder.
   (b) boiling the acid.
   (c) passing NH_3 through acid.
   (d) passing air through warm acid.
9. In nitroprusside ion, the iron and NO exist as Fe^{2+} and NO^+ rather than Fe^{3+} and NO. These forms can be differentiated by
   (a) estimating the concentration of iron.
   (b) measuring the concentration of CN^-.
   (c) measuring the solid state magnetic moment.
   (d) thermally decomposing the compound.
10. A deep brown gas is formed by mixing two colourless gases which are
    (a) NO_2 and O_2 (b) N_2O and NO
    (c) NO and O_2 (d) NH_3 and HCl
11. Which of the following compounds does not exist?
    (a) $AsCl_5$ (b) $SbCl_3$ (c) $BiCl_5$ (d) $SbCl_5$
12. The brown ring test for NO_2^- and NO_3^- is due to the formation of complex ion with a formula
    (a) $[Fe(H_2O)_6]^{2+}$ (b) $[Fe(NO)(CN)_5]^{2+}$
    (c) $[Fe(H_2O)_5NO]^{2+}$ (d) $[Fe(H_2O)(NO)_5]^{2+}$
13. Which of the following is a cyclic phosphate ? Toughnut
    (a) $H_3P_3O_{10}$ (b) $H_6P_4O_{13}$
    (c) $H_5P_5O_{15}$ (d) $H_7P_5O_{16}$
14. An element (X) forms compounds of the formula XCl_3, X_2O_5 and Ca_3X_2 but does not form XCl_5. Which of the following is the element X ?
    (a) B (b) Al (c) N (d) P
15. P_4O_{10} is not used to dry NH_3 gas because

    (a) P_4O_{10} reacts with moisture in NH_3.
    (b) P_4O_{10} is not a drying agent.
    (c) P_4O_{10} is acidic and NH_3 is basic.
    (d) P_4O_{10} is basic and NH_3 is acidic.

16. Pick out the wrong statement.
   (a) Nitrogen has the ability to form pπ-pπ bonds with itself.
   (b) Bismuth forms metallic bonds in elemental state.
   (c) Catenation tendency is higher in nitrogen when compared with other elements of the same group.
   (d) Nitrogen has higher first ionisation enthalpy when compared with other elements of the same group.

17. Which of the following is incorrect for white and red phosphorus ?
   (a) They are both soluble in CS_2.
   (b) They can be oxidised by heating in air.
   (c) They consist of the same kind of atoms.
   (d) They can be converted into one another.

18. BCl_3 is a planar molecule whereas NCl_3 is pyramidal because
   (a) BCl_3 has no lone pair of electrons but NCl_3 has a lone pair of electrons.
   (b) B—Cl bond is more polar than N—Cl bond.
   (c) nitrogen atom is smaller than boron atom.
   (d) N—Cl bond is more covalent than B—Cl bond.

19. A metal X on heating in nitrogen gas gives Y. Y on treatment with H_2O gives a colourless gas which when passed through $CuSO_4$ solution gives a blue colour. Y is
   (a) $Mg(NO_3)_2$ (b) Mg_3N_2
   (c) NH_3 (d) MgO

20. Which of the following oxy-acids has the maximum number of hydrogens directly attached to phosphorus?
   (a) $H_4P_2O_7$ (b) H_3PO_2
   (c) H_3PO_3 (d) H_3PO_4

21. Number of sigma bonds in P_4O_{10} is
   (a) 6 (b) 7 (c) 17 (d) 16

22. Which of the following compound has a P–P bond?
   (a) $H_4P_2O_5$ (b) $(HPO_3)_3$
   (c) $H_4P_2O_6$ (d) $H_4P_2O_7$

23. Which one of the following does not have a pyramidal shape?
   (a) $(CH_3)_3N$ (b) $(SiH_3)_3N$
   (c) $P(CH_3)_3$ (d) $P(SiH_3)_3$

24. Consider the following sequence of reaction.
$$Na + NH_3(g) \longrightarrow [X] \xrightarrow{N_2O} [Y] \xrightarrow{Heat} \underset{\text{Gas Pure}}{[Z]}$$
   Identify (Z) gas : **Toughnut**
   (a) N_2 (b) NH_3 (c) O_2 (d) H_2

25. Among the following compounds, which on heating do not produce N_2?
   (a) $(NH_4)_2Cr_2O_7$ (b) $NH_4Cl + NaNO_2$
   (c) $NH_4Cl + CaO$ (d) $Ba(N_3)_2$

26. $A + H_2O \longrightarrow B + HCl$

   $B + H_2O \longrightarrow C + HCl$
   Compound (A), (B) and (C) will be respectively :
   (a) $PCl_5, POCl_3, H_3PO_3$
   (b) $PCl_5, POCl_3, H_3PO_4$
   (c) $SOCl_2, POCl_3, H_3PO_3$
   (d) $PCl_3, POCl_3, H_3PO_4$

27. A white precipitate is obtained on hydrolysis of
   (a) PCl_5 (b) NCl_3 (c) $BiCl_3$ (d) $AsCl_3$

28. Superphosphate of lime is obtained from the reaction of :
   (a) Bones with gypsum.
   (b) Calcium phosphate with sulphuric acid.
   (c) Calcium phosphate with HCl.
   (d) Calcium carbonate with phosphoric acid.

29. Phosphine is not obtained by which of the following reaction?
   (a) White P is heated with NaOH.
   (b) Red P is heated with NaOH.
   (c) Ca_3P_2 reacts with water.
   (d) Phosphorus trioxide is boiled with water.

30. Which of the following statements is not true ?
   (a) NO_2 can be prepared by heating $Pb(NO_3)_2$.
   (b) NO_2 is red - brown gas.
   (c) NO_2 is diamagnetic.
   (d) NO_2 readily dimerises to N_2O_4.

31. The deep blue colour produced on adding excess of ammonia to copper sulphate is due to presence of
   (a) Cu^{2+} (b) $[Cu(NH_3)_4]^{2+}$
   (c) $[Cu(NH_3)_6]^{2+}$ (d) $[Cu(NH_3)_2]^{2+}$

32. Conc. HNO_3 is heated with P_2O_5 to form :
   (a) NO_2 (b) NO (c) N_2O_5 (d) N_2O

33. From the following information
$X + H_2SO_4 \rightarrow Y$ (a colourless and irritating gas)
$Y + K_2Cr_2O_7 + H_2SO_4 \rightarrow$ (green coloured solution)
Identify the pair X and Y.
(a) $\bar{C}r$, HCl (b) SO_3^{2-}, SO_2
(c) S^{2-}, H_2S (d) CO_3^{2-}, CO_2

34. H_2SO_4 and H_2SO_3 can be distinguished by the addition of
(a) magnesium powder
(b) $NaHSO_4$ solution
(c) $FeCl_3$ solution
(d) litmus solution

35. Which of the following is the key step in the manufacture of sulphuric acid?
(a) Burning of sulphur or sulphide ores in air to generate SO_2.
(b) Conversion of SO_2 to SO_3 by the reaction with oxygen in presence of catalyst.
(c) Absorption of SO_3 in H_2SO_4 to give oleum.
(d) Both (b) and (c).

36. $FeCl_3$ solution on reaction with SO_2 changes to
(a) $FeCl_2$ (b) $Fe_2(SO_4)_3$
(c) $Fe_2(SO_3)_3$ (d) $FeSO_4$

37. Hydrolysis of one mole of peroxodisulphuric acid produces

Critical Thinking

(a) two moles of sulphuric acid.
(b) two moles of peroxomonosulphuric acid.
(c) one mole of sulphuric acid and one mole of peroxomonosulphuric acid.
(d) one mole of sulphuric acid, one mole of peroxomonosulphuric acid and one mole of hydrogen peroxide.

38. Concentrated H_2SO_4 is not used to prepare HBr from KBr because it
(a) oxidizes HBr.
(b) reduces HBr.
(c) causes disproportionation of HBr.
(d) reacts too slowly with KBr.

39. Which reaction represents the oxidizing behaviour of H_2SO_4?
(a) $2PCl_5 + H_2SO_4 \rightarrow 2POCl_3 + 2HCl + SO_2Cl_2$
(b) $2NaOH + H_2SO_4 \rightarrow Na_2SO_4 + 2H_2O$
(c) $NaCl + H_2SO_4 \rightarrow NaHSO_4 + HCl$
(d) $2HI + H_2SO_4 \rightarrow I_2 + SO_2 + 2H_2O$

40. Which of the following is not correctly matched?
(a) SF_4 – gas (b) SeF_4 – liquid
(c) TeF_4 – solid (d) SF_6 – solid

41. Which of the following on thermal decomposition gives oxygen gas ?
(a) Ag_2O (b) Pb_3O_4
(c) PbO_2 (d) All of these

42. When sugar is treated with conc. sulphuric acid, the sugar is charred. In this process, sugar is
(a) reduced (b) oxidized
(c) sulphonated (d) dehydrated

43. The number of S—S bonds in SO_3, $S_2O_3^{2-}$ $S_2O_6^{2-}$ and $S_2O_8^{2-}$ respectively are

Toughnut

(a) 1, 0, 0, 1 (b) 1, 0, 1, 0
(c) 0, 1, 1, 0 (d) 0, 1, 0, 1

44. Correct order of O – O bond length (increasing) in O_2, H_2O_2 and O_3 is
(a) $H_2O_2 < O_3 < O_2$ (b) $O_2 < O_3 < H_2O_2$
(c) $O_3 < O_2 < H_2O_2$ (d) $O_3 < H_2O_2 < O_2$

45. Iron sulphide is heated in air to form A, an oxide of sulphur. A is dissolved in water to give an acid. The basicity of this acid is
(a) 2 (b) 3 (c) 1 (d) zero

46. The shape of the molecule SF_3Cl_3 is
(a) trigonal bipyramidal
(b) cubic
(c) octahedral
(d) tetrahedral

47. What are the products formed when chlorine is passed through aqueous hypo solution?
(a) $Na_2SO_3 + HCl + S$
(b) $Na_2SO_3 + SO_3 + HCl$
(c) $Na_2SO_4 + HCl + S$
(d) $Na_2SO_4 + HCl + SO_2$

48. Which of the following is incorrect?

(a) M.p of monoclinic sulphur > m.p. of rhombic sulphur.
(b) Specific gravity of rhombic sulphur > specific gravity of monoclinic sulphur.
(c) Monoclinic sulphur is stable below 369 K.
(d) Both rhombic sulphur and monoclinic sulphur have S_8 molecules.

49. Which of the following form of the sulphur shows paramagnetic behaviour ?
(a) S_8 (b) S_6
(c) S_2 (d) All of these

50. Which of the following has pπ- dπ bonding ?
(a) NO_3^- (b) SO_3^{2-}
(c) BO_3^{3-} (d) CO_3^{2-}

51. Sulphuric acid reacts with PCl_5 to give
(a) thionyl chloride
(b) sulphur monochloride
(c) sulphuryl chloride
(d) sulphur tetrachloride

52. By which of the following SO_2 is formed?
(a) Reaction of dil. H_2SO_4 with O_2
(b) Hydrolysis of dil. H_2SO_4
(c) Reaction of conc. H_2SO_4 with Cu
(d) None of these

53. The acid which has a peroxy linkage is
(a) Sulphurous acid (b) Pyrosulphuric acid
(c) Dithionic acid (d) Caro's acid

54. Excess of KI reacts with $CuSO_4$ solution and then $Na_2S_2O_3$ solution is added to it. Which of the statements is incorrect for this reaction ?
(a) $Na_2S_2O_3$ is oxidised
(b) CuI_2 is formed
(c) Cu_2I_2 is formed
(d) Evolved I_2 is reduced

55. Of the interhalogen AX_3 compounds, ClF_3 is most reactive but BrF_3 has higher conductance in liquid state. This is because **Critical Thinking**
(a) BrF_3 has higher molecular mass.
(b) ClF_3 is more volatile.
(c) BrF_3 dissociates into BrF_2^+ and BrF_4^- most easily.
(d) Electrical conductance does not depend on concentration.

56. Which of the following is the best description for the behaviour of bromine in the reaction given below ?

$H_2O + Br_2 \rightarrow HOBr + HBr$

(a) Proton acceptor only
(b) Both oxidized and reduced
(c) Oxidized only
(d) Reduced only

57. Oxidation of thiosulphate with iodine gives
(a) sulphate ion (b) sulphite ion
(c) tetrathionate ion (d) sulphide ion

58. Which of the following is correct about the reaction?

$3NaClO \xrightarrow{heat} NaClO_3 + 2NaCl$

(a) It is disproportionation reaction
(b) Oxidation number of Cl decreases as well as increases in this reaction
(c) This reaction is used for the manufacture of halates
(d) All of these

59. The reaction of $KMnO_4$ and HCl results in
(a) oxidation of Mn in $KMnO_4$ and production of Cl_2.
(b) reduction of Mn in $KMnO_4$ and production of H_2.
(c) oxidation of Mn in $KMnO_4$ and production of H_2.
(d) reduction of Mn in $KMnO_4$ and production of Cl_2.

60. The correct order of reactivity of halogens with alkalies is
(a) $F > Cl > Br > I$ (b) $F < Cl > Br < I$
(c) $F < Cl < Br < I$ (d) $F < Cl < Br > I$

61. One mole of fluorine is reacted with two moles of hot and concentrated KOH. The products formed are KF, H_2O and O_2. The molar ratio of KF, H_2O and O_2 respectively is **Tricky**
(a) 1 : 1 : 2 (b) 2 : 1 : 0.5
(c) 1 : 2 : 1 (d) 2 : 1 : 2

62. Chlorine cannot displace
(a) fluorine from NaF
(b) iodine from NaI
(c) bromine from NaBr
(d) None of these

63. When chlorine reacts with cold and dilute solution of sodium hydroxide, the products obtained are
(a) $Cl^- + ClO^-$ (b) $Cl^- + ClO_2^-$
(c) $Cl^- + ClO_3^-$ (d) $Cl^- + ClO_4^-$

64. A greenish yellow gas reacts with an alkali metal hydroxide to form a halate which can be used in fire works safety matches. The gas and halate respectively are
(a) Br_2, $KBrO_3$ (b) Cl_2, $KClO_3$
(c) I_2, $NaIO_3$ (d) Cl_2, $NaClO_3$

65. In the reaction

$HNO_3 + HF \longrightarrow H_2\overset{+}{N}O_3 + F^-$ base is

(a) HF (b) HNO_3
(c) HF and HNO_3 (d) None of these

66. The aqueous solution of which of the following has maximum pH? Toughnut

(a) NaClO (b) $NaClO_2$
(c) $NaClO_3$ (d) $NaClO_4$

67. Which one of the following oxides of chlorine is obtained by passing dry chlorine over silver chlorate at 90°C?

(a) Cl_2O (b) ClO_3 (c) ClO_2 (d) ClO_4

68. Identify the incorrect statement among the following

(a) Br_2 reacts with hot and strong NaOH solution to give NaBr and H_2O.
(b) Ozone reacts with SO_2 to give SO_3.
(c) Silicon reacts with NaOH(aq) in the presence of air to give Na_2SiO_3 and H_2O.
(d) Cl_2 reacts with excess of NH_3 to give N_2 and HCl.

69. The least number of oxyacids are formed by:

(a) Chlorine (b) Nitrogen
(c) Fluorine (d) Sulphur

70. Chlorine water on standing loses its colour and forms:

(a) HCl only (b) HCl and $HClO_2$
(c) HCl and HOCl (d) HOCl and $HOCl_2$

71. Shapes of certain interhalogen compounds are stated below. Which one of them is not correctly stated?

(a) IF_7 : pentagonal bipyramidal
(b) BrF_5 : trigonal bipyramidal
(c) BrF_3 : planar T-shaped
(d) ICl_3 : planar dimeric

72. Electron gain enthalpy with negative sign of fluorine is less than that of chlorine due to:

(a) High ionization enthalpy of fluorine
(b) Smaller size of chlorine atom
(c) Smaller size of fluorine atom
(d) Bigger size of 2*p* orbital of fluorine

73. Which of the following xenon-oxo compounds may not be obtained by hydrolysis of xenon fluorides?

(a) XeO_2F_2 (b) $XeOF_4$
(c) XeO_3 (d) XeO_4

74. Trigonal bipyramidal geometry is shown by:

(a) XeO_3 (b) XeO_3F_2
(c) $FXeOSO_2F$ (d) $[XeF_8]^{2-}$

75. XeO_4 molecule is tetrahedral having: Critical Thinking

(a) Two pπ – dπ bonds
(b) One pπ – dπ bonds
(c) Four pπ – dπ bonds
(d) Three pπ – dπ bonds

76. The compound of xenon with zero dipole moment is

(a) XeO_3 (b) XeF_4
(c) $XeOF_4$ (d) XeO_2

77. XeF_6 dissolves in anhydrous HF to give a good conducting solution which contains:

(a) H^+ and XeF_7^- ion
(b) HF_2^- and XeF_5^+ ions
(c) $HXeF_6^+$ and F^- ions
(d) none of these

78. Incorrect statement regarding following reactions is: Toughnut

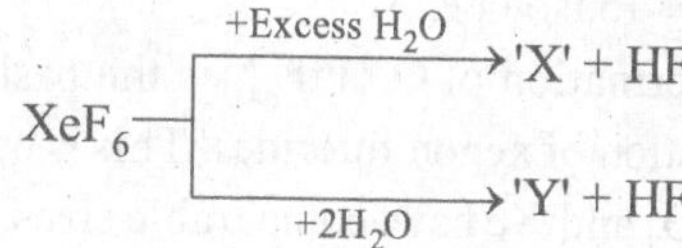

(a) 'X' is explosive
(b) 'Y' is an oxyacid of xenon
(c) Both are example of non-redox reaction
(d) XeF_6 can undergo partial hydrolysis.

79. In the clathrates of xenon with water, the nature of bonding between xenon and water molecule is

(a) covalent
(b) hydrogen bonding
(c) co-ordinate
(d) dipole-induced dipole interaction

80. XeF_4 on partial hydrolysis produces

(a) XeF_4 (b) $XeOF_2$
(c) $XeOF_4$ (d) XeO_3

81. Which one of the following is correct pair with respect to molecular formula of xenon compound and hybridization state of xenon in it?

(a) XeF_4, sp^3 (b) XeF_2, sp
(c) XeF_2, sp^3d (d) XeF_4, sp^2

82. Which of the following noble gas is least polarisable?

(a) He (b) Xe (c) Ar (d) Ne

83. What are the products formed in the reaction of xenon hexafluoride with silicon dioxide?
(a) $XeSiO_4 + HF$ (b) $XeF_2 + SiF_4$
(c) $XeOF_4 + SiF_4$ (d) $XeO_3 + SiF_2$

84. Which statement about noble gases is not correct?
(a) Xe forms XeF_6.
(b) Ar is used in electric bulbs.
(c) Kr is obtained during radioactive disintegration.
(d) He has the lowest b.p. among all the noble gases.

85. Density of nitrogen gas prepared from air is slightly greater than that of nitrogen prepared by chemical reaction from a compound of nitrogen due to the presence of
(a) argon
(b) carbon dioxide
(c) some N_3 molecules analogous to O_3.
(d) greater amount of N_2 molecules derived from N-15 isotope.

86. The formation of $O_2^+[PtF_6]^-$ is the basis for the formation of xenon fluorides. This is because
(a) O_2 and Xe have comparable sizes.
(b) both O_2 and Xe are gases.
(c) O_2 and Xe have comparable ionisation energies.
(d) Both (a) and (c).

87. The reason for not storing XeF_6 in a glass or a quartz vessel is that **Toughnut**
(a) it forms an explosive having the formula XeO_2F_2.
(b) it forms an explosive having the formula $XeOF_4$.
(c) it forms XeO_2 which is explosive substance.
(d) it forms XeO_6^{4-} which is explosive in nature.

88. SbF_5 reacts with XeF_4 to form an adduct. The shapes of cation and anion in the adduct are respectively: **Critical Thinking**
(a) square planar, trigonal bipyramidal
(b) T-shaped, octahedral
(c) square pyramidal, octahedral
(d) square planar, octahedral

89. Out of (i) XeO_3 (ii) $XeOF_4$ and (iii) XeF_6, the molecules having same number of lone pairs on Xe are
(a) (i) and (ii) only (b) (i) and (iii) only
(c) (ii) and (iii) only (d) (i), (ii) and (iii)

90. The geometry of $XeOF_4$ by VSEPR theory is :
(a) pentagonal planar
(b) octahedral
(c) square pyramidal
(d) trigonal bipyramidal

Answer KEYs

1	(d)	10	(c)	19	(b)	28	(b)	37	(c)	46	(c)	55	(c)	64	(b)	73	(d)	82	(a)
2	(d)	11	(c)	20	(b)	29	(b)	38	(a)	47	(c)	56	(b)	65	(b)	74	(b)	83	(c)
3	(c)	12	(c)	21	(d)	30	(c)	39	(d)	48	(c)	57	(c)	66	(a)	75	(c)	84	(c)
4	(a)	13	(c)	22	(c)	31	(b)	40	(d)	49	(c)	58	(d)	67	(c)	76	(b)	85	(a)
5	(a)	14	(c)	23	(b)	32	(c)	41	(d)	50	(b)	59	(d)	68	(d)	77	(b)	86	(d)
6	(c)	15	(c)	24	(a)	33	(b)	42	(d)	51	(c)	60	(a)	69	(c)	78	(b)	87	(b)
7	(b)	16	(c)	25	(c)	34	(c)	43	(c)	52	(c)	61	(b)	70	(c)	79	(d)	88	(b)
8	(d)	17	(a)	26	(b)	35	(b)	44	(b)	53	(d)	62	(a)	71	(b)	80	(b)	89	(d)
9	(c)	18	(a)	27	(c)	36	(a)	45	(a)	54	(b)	63	(a)	72	(c)	81	(c)	90	(c)

Hints & Solutions

1. (d) PH_3 (Lewis base) can react with B_2H_6 (Lewis acid).

2. (d)

O O
|| ||
HO–P–O–P–OH
| |
OH OH

Pyrophosphoric acid ($H_4P_2O_7$)

Oxidation State :
Each P atom is bound to one oxygen = –1
2OH = –1 × 2 = –2
1 Oxygen = – 2
Total = –5
P = +5.

3. (c) Both the methods (a) and (b) can be used. When passed through freezing mixture the P_2H_4 present condenses and pure PH_3 is obtained.
When passed through HI, PH_3 is absorbed forming $PH_4I.PH_4I$ when treated with KOH (aq) yields pure phosphine.

$$PH_3 + HI \longrightarrow PH_4I$$
$$PH_4I + KOH(aq) \longrightarrow KI + H_2O + PH_3 \uparrow$$

4. (a) $NH_4ClO_4 + HNO_3 \rightarrow HClO_4 + \underset{(X)}{NH_4NO_3}$

$$NH_4NO_3 \xrightarrow{\Delta} \underset{(Y)}{N_2O} + 2H_2O$$

5. (a) The structure of H_3PO_3 is as given below.

H
|
O=P–OH
|
OH

There are only two –OH groups and hence it is dibasic. The oxidation number of P in this acid is +3, whereas P may have +5 oxidation state also. Therefore, H_3PO_3 can be oxidised which means H_3PO_3 is a reducing agent.

6. (c) $\underset{1\text{ mole}}{Ca_3P_2} + 6H_2O \rightarrow 3Ca(OH)_2 + \underset{2\text{ moles}}{2PH_3}$

7. (b) With the decrease in the electronegativity of central atom the bond angle decreases

8. (d) Nitric acid decomposes to give NO_2 which is brown

$$4HNO_3 \rightarrow 4NO_2 + 2H_2O + O_2$$

When air is passed through acid the reaction proceeds towards left hand side and brown colour diminishes.

9. (c) The nitroprusside ion is $[Fe(CN)_6NO^+]^{2-}$. The magnetic moment measurements reveal the presence of 4 unpaired electrons in Fe which must be then in Fe^{2+} ($3d^6$) and not Fe^{3+}($3d^5$)

10. (c) $2NO + O_2 \rightarrow 2NO_2$ brown

11. (c) Due to inert pair effect the lower O.S. is more common at the bottom of group

12. (c) $[Fe(H_2O)_5NO]^{2+}$ ion is formed

13. (c) $H_5P_5O_{15}$ or $(HPO_3)_5$. It is metaphosphoric acid which is a cyclic phosphate.

14. (c) Nitrogen can form NCl_3, N_2O_5 Ca_3N_2 and not NCl_5 since it has no d atomic orbitals in valence shell

15. (c) $\underset{\text{Acidic}}{P_4O_{10}} + 2H_2O \longrightarrow \underset{\text{meta phosphoric acid}}{4HPO_3}$

$$\xrightarrow[\text{basic}]{NH_3} \underset{\text{meta ammonium phosphate}}{NH_4PO_3}$$

16. (c) Catenation tendency is higher in phosphorus when compared with other elements of same group.

17. (a) Both white and red phosphorus are not soluble in CS_2 only white phosphorus is soluble in CS_2.

18. (a) In BCl_3 there is sp^2 hybridisation (planar) whereas in NCl_3 hybridisation is sp^3 hence shape is pyramidal

19. (b) $3Mg + N_2 \xrightarrow{\Delta} Mg_3N_2 \xrightarrow{6H_2O} 3Mg(OH)_2 + 2NH_3$;

$CuSO_4 + 4NH_3 \rightarrow \underset{\text{deep blue colour}}{[Cu(NH_3)_4]SO_4}$

20. (b)

(a) $H_4P_2O_7 \Rightarrow O{=}P(OH)_2{-}O{-}P(OH)_2{=}O$

Pyrophosphoric acid

(b) $H_3PO_2 \Rightarrow$ HO–P(=O)(H)–H

Hypophosphorous acid

(c) $H_3PO_3 \Rightarrow$ HO – P(=O)(H) – OH

Phosphorous acid

(d) $H_3PO_4 \Rightarrow$ HO – P(=O)(OH) – OH

orthophosphoric acid

21. (d) [Structure of P_4O_{10}: four P atoms each bearing a P=O, linked through bridging O atoms]

22. (c) $H_4P_2O_6$ has P–P linkage

HO – P(=O)(OH) — P(=O)(OH) – OH

23. (b) In case of $N(SiH_3)_3$, N atom is sp^2 hybridised, the lone pair is present in $2p$ orbital and it is transferred to empty d orbital of Si forming $d\pi - p\pi$ bond. Hence nitrogen with sp^2 hybridization has trigonal planar shape.

24. (a) $Na + NH_3(g) \xrightarrow{\Delta} \underset{[X]}{NaNH_2} \xrightarrow{N_2O} \underset{[Y]}{NaN_3} \xrightarrow{\Delta} \underset{[Z]}{N_2} \uparrow$

25. (c)

(a) $(NH_4)_2Cr_2O_7 \xrightarrow{\Delta} Cr_2O_7 + N_2 + H_2O$

(b) $NH_4Cl + NaNO_2 \xrightarrow{\Delta} NH_4NO_2 + NaCl$

(c) $2NH_4Cl + CaO \xrightarrow{\Delta} CaCl_2 + 2NH_3 \uparrow + H_2O$

(d) $NH_4NO_2 \xrightarrow{\Delta} N_2 \uparrow + 2H_2O$

$Ba(N_3)_2 \xrightarrow{\Delta} Ba + 3N_2 \downarrow$

26. (b) $\underset{(A)}{PCl_5} + H_2O \longrightarrow \underset{(B)}{POCl_3} + 2HCl$

$\underset{(B)}{POCl_3} + 3H_2O \longrightarrow \underset{(C)}{H_3PO_4} + 3HCl$

27. (c) $BiCl_3 + H_2O \rightarrow BiOCl + 2HCl.BiOCl$

gives white ppt. which is used as white pigment, under the name of **pearl white.**

28. (b) $Ca_3(PO_4)_2 + 2H_2SO_4 + 5H_2O \rightarrow$

$Ca(H_2PO_4)_2.H_2O + 2CaSO_4.2H_2O$

29. (b) Red P does not react with NaOH to give PH_3.

30. (c) NO_2 is paramagnetic and readily dimerises to N_2O_4 which is diamagnetic.

31. (b) $CuSO_4 + 4NH_3 \rightarrow [Cu(NH_3)_4]SO_4$

Blue complex due to $Cu(NH_3)_4^{2+}$

32. (c) $P_2O_5 + 2HNO_3 \rightarrow N_2O_5 + 2HPO_3$

33. (b) $\underset{(X)}{SO_3^{2-}} + 2H^{\oplus} \rightarrow SO_2 + H_2O$

$\underset{(Y)}{3SO_2} + K_2Cr_2O_7 + H_2SO_4 \rightarrow$

$K_2SO_4 + \underset{\text{Green coloured solution}}{Cr_2(SO_4)_3} + H_2O$

34. (c) $FeCl_3$ acts as oxidant and H_2SO_3 as reductant.

35. (b) The key step in the manufacture of H_2SO_4 is catalytic oxidation of SO_2 with O_2 to give SO_3 in presence of V_2O_5.

36. (a) $2FeCl_3 + SO_2 + 2H_2O \rightarrow 2FeCl_2 + H_2SO_4 + 2HCl$

37. (c) HO–S(=O)(=O)–O–O–S(=O)(=O)–OH + H_2O ⟶
peroxodisulphuric acid

HO–S(=O)(=O)–O–OH + HO–S(=O)(=O)–OH
peroxomonosulphuric acid   sulphuric acid

38. (a) $2HBr + H_2SO_4 \rightarrow Br_2 + 2H_2O + SO_2$

HBr is oxidised to Br_2

39. (d) $2HI + H_2SO_4 \rightarrow I_2 + SO_2 + 2H_2O$ (HI is oxidised to I_2) (H_2SO_4 is reduced to SO_2)

40. (d) All hexafluorides of group 16 elements are gaseous in nature.

41. (d) $2Ag_2O\,(s) \longrightarrow 4Ag\,(s) + O_2\,(g)$

$2Pb_3O_4\,(s) \longrightarrow 6PbO\,(s) + O_2\,(g)$

$2PbO_2\,(s) \longrightarrow 2PbO\,(s) + O_2\,(g)$

42. (d) $C_{12}H_{22}O_{11} \xrightarrow[\Delta]{H_2SO_4} \underset{\text{black}}{12C} + 11H_2O$

43. (c)

(SO_3): O=S(=O)=O ; $(S_2O_3^{2-})$: S=S(–O⁻)(=O)–O⁻ ; $(S_2O_6^{2-})$: O=S(=O)(–O⁻)–S(=O)(=O)–O⁻ ;

$(S_2O_8^{2-})$: O=S(=O)(–O⁻)–O–O–S(=O)(=O)–O⁻

Hence (c) is the correct option.

44. (b) $H_2O_2 > O_3 > O_2$

B. O. is 1, 1.5 and 2 respectively.

H–O–O–H; O₃ (O–O=O with O at apex); O=O

45. (a) $4FeS + 7O_2 \rightarrow 2Fe_2O_3 + 4SO_2$

$SO_2 + H_2O \rightarrow H_2SO_3 \rightleftharpoons 2H^+ + SO_3^{2-}$

It is dibasic

46. (c) Hybridisation is sp^3d^2, hence octahedral

47. (c) $Na_2S_2O_3 + Cl_2 + H_2O \longrightarrow 2HCl + S + Na_2SO_4$

48. (c) Monoclinic sulphur is stable above 369 K.

49. (c) S_2 is paramagnetic. It contains two unpaired electrons in the antibonding π^* orbital

50. (b) $p\pi - d\pi$ bonding is present in SO_3^{2-}, N, B, C have no vacant d atomic arbitals.

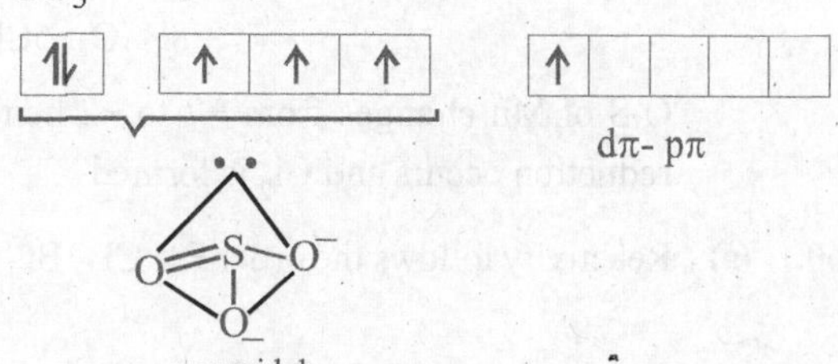

51. (c) $HO.SO_2OH + 2PCl_5 \longrightarrow ClSO_2Cl + 2POCl_3 + 2HCl$
Sulphuryl chloride

52. (c) $Cu + 2H_2SO_4 \text{(conc)} \longrightarrow CuSO_4 + SO_2 + 2H_2O$

53. (d) Caro's acid is H_2SO_5 which contains one S – O – O – H peroxy linkage. It is also known as permonosulphuric acids.

$$H-O-O-\overset{\overset{\large O}{||}}{\underset{\underset{\large O}{||}}{S}}-OH$$

Caro's acid

54. (b) $4\overset{-1}{K}I + 2CuSO_4 \rightarrow \overset{0}{I_2} + Cu_2I_2 + 2K_2SO_4$

$\overset{0}{I_2} + 2Na_2\overset{2+}{S_2}O_3 \rightarrow Na_2\overset{-2.5}{S_4}O_6 + 2Na\overset{-1}{I}$

In this CuI_2 is not formed.

55. (c) In liquid state BrF_3 dissociates into BrF_2^+ and BrF_4^- ions most easily.

56. (b) $H_2O + \overset{0}{Br_2} \longrightarrow HO\overset{+1}{Br} + H\overset{-1}{Br}$

Thus here oxidation number of Br increases from 0 to +1 and also decreases from 0 to –1. Thus it is oxidised as well as reduced.

57. (c) $2Na_2S_2O_3 + I_2 \rightarrow \underset{\text{Sodium tetrathionate}}{Na_2S_4O_6} + 2NaI$

58. (d) $Na\overset{+1}{Cl}O \rightarrow Na\overset{+5}{Cl}O_3 + 2Na\overset{-1}{Cl}$.

All statements are correct as evident from the reaction.

59. (d) $2K\overset{+7}{Mn}O_4 + 16HCl \rightarrow 2\overset{+2}{Mn}Cl_2 + 2KCl + 8H_2O + 5Cl_2$

O.S of Mn changes from +7 to +2 hence reduction occurs and Cl_2 is formed.

60. (a) Reactivity follows the order F > Cl > Br > I

61. (b) $2F_2 + 4KOH \rightarrow 4KF + O_2 + 2H_2O$ for 1 mole of F_2 the molar ratio.

F_2	KOH	KF	O_2	H_2O
1	2	2	$\frac{1}{2}$	1

62. (a) Chlorine cannot displace F from NaF. The reactivity follows the order F > Cl > Br > I

63. (a) $2NaOH + Cl_2 \rightarrow NaCl + NaOCl + H_2O$

hence Cl^- and OCl^-

64. (b) $3Cl_2 + 6KOH \rightarrow KClO_3 + 5KCl + 3H_2O$

$KClO_3$ is used in fire works and safety matches and Cl_2 is greenish yellow gas.

65. (b) In $HNO_3 + HF \rightarrow \overset{+}{H_2NO_3} + F^-$ HNO_3 accepts H^+ hence it is a base.

66. (a) NaClO is a salt of strong base and weak acid hence on hydrolysis the solution will be most basic, therefore maximum pH.

67. (c) Pure ClO_2 is obtained by passing dry Cl_2 over $AgClO_3$ at 90°C.

$$2AgClO_3 + Cl_2\text{(dry)} \xrightarrow{90°C} 2AgCl + 2ClO_2 + O_2$$

68. (d) Chlorine reacts with excess of ammonia to produce ammonium chloride and nitrogen.
$3Cl_2 + 8NH_3 \text{(excess)} \longrightarrow 6NH_4Cl + N_2$

69. (c) Flourine is the most electronegative element & has least tendency to form double bonds.

70. (c) Chlorinated water is yellow in colour on standing reaction occurs
$Cl_2 + H_2O \rightarrow HCl + HOCl$

71. (b) The molecular geometry of BrF_5 is square pyramidal with asymmetric charge distribution on the central atom.

:F: :F: Br :F: :F: :F:

72. (c) The electron gain enthalpy order for halogens is

Cl > F > Br > I

Due to small size of fluorine the extra electron to be added feels more electron-electron repulsion. Therefore fluorine has less value for electron affinity than chlorine.

73. (d) $XeF_6 + H_2O \xrightarrow{\text{Partial}} XeOF_4 + 2HF$

$XeF_6 + 2H_2O \xrightarrow{\text{Partial}} XeO_2F_2 + 4HF$

$XeF_6 + 3H_2O \xrightarrow{\text{Complete}} XeO_3 + 6HF$

74. (b) The hybridization of XeO_3F_2 is sp^3d and its structure is trigonal bipyramidal in which oxygen atoms are situated on the plane and the fluoride atoms are on the top and bottom.

75. (c) Xenon undergo sp^3 hybridization.

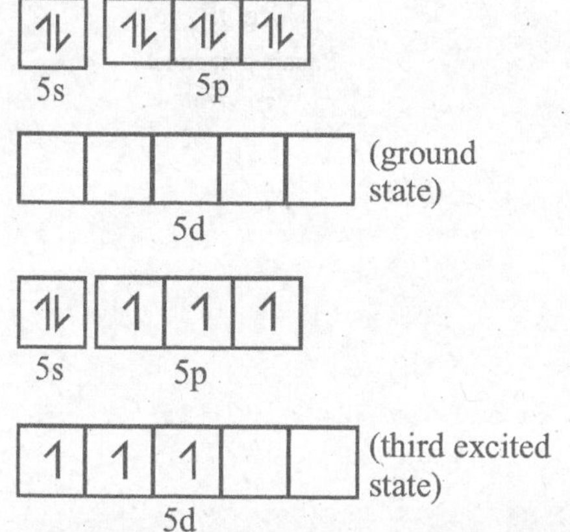

In the fourth excited state xenon atom, has 8 unpaired electrons

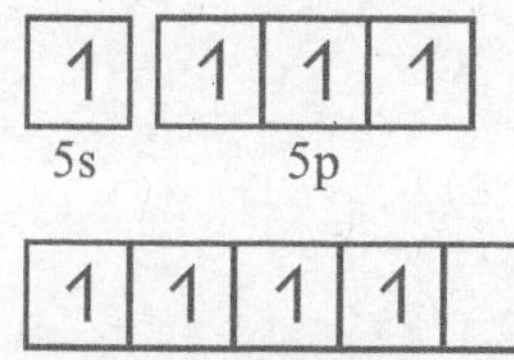

One s and three p orbital undergo sp^3 hybridization. Four sp^3 hybrid orbitals form four σ bonds with oxygen atoms. They are $\sigma sp^3 - p$. Four $p\pi - d\pi$ bonds are also formed with oxygen atoms by the unpaired electrons.

76. (b) XeF_4 has zero dipole moment. It has square planar structure due to which the bond moments of Xe – F cancel each other.

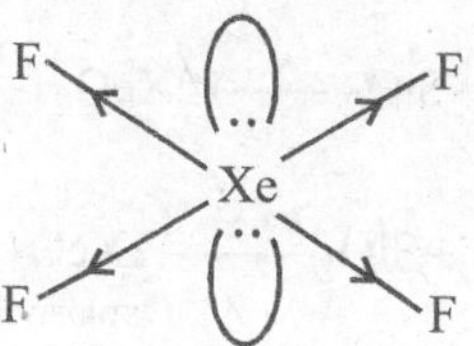

77. (b) $HF + XeF_6 \longrightarrow XeF_5^+ + HF_2^-$

78. (b)

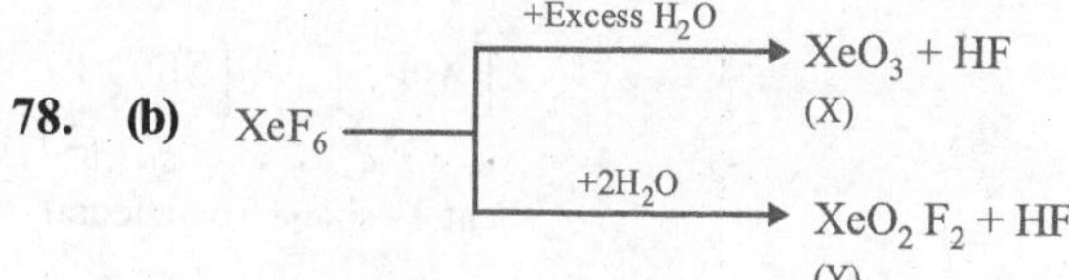

Y is not an oxyacid of xenon.

79. (d) In clathrates the forces are dipole - induced dipole interaction

80. (b) $XeF_4 + H_2O \rightarrow 2HF + XeOF_2$

81. (c) Hybridisation in each case is XeF_4, sp^3d^2, XeF_2, sp^3d.

82. (a) The smaller the size the least is the polarisability.

83. (c) $2XeF_6 + SiO_2 \rightarrow SiF_4 + 2XeOF_4$

84. (c) Helium is obtained during radioactive decay.

85. (a) Air contains about 1% inert gases, mainly Ar. (At.wt 40) The atomic wt. of N_2 is 28.

86. (d)

(i) The first ionization energy of xenon (1, 170 kJ mol^{-}) is quite close to that of dioxygen (1,180 kJ mol^{-1}).

(ii) The molecular diameters of xenon and dioxygen are almost identical.

Based on the above similarities Barlett (who prepared $O_2^+[PtF_6]^-$ compound) suggested that since oxygen combines with PtF_6, so xenon should also form similar compound with PtF_6.

87. (b) $2XeF_6 + SiO_2 \longrightarrow 2XeOF_4 + SiF_4$
(glass) (explosive nature)

$2XeOF_4 + SiO_2 \longrightarrow 2XeO_2F_2 + SiF_4$

$2XeO_2F_2 + SiO_2 \longrightarrow 2XeO_3 + SiF_4$
(explosive)

88. (b) $XeF_4 + SbF_5 \longrightarrow [XeF_3]^+[SbF_6]^- \longrightarrow$

$[XeF_3]^+ + [SbF_6]^-$
sp^3d sp^3d^2
bent T–shape octahedral

89. (d) XeO_3^-

– 1 lP

$XeOF_4^-$

– 1 lP

XeF_6^-

– 1 lP

90. (c) In $XeOF_4$, Xe is sp^3d^2 hybridised

$XeOF_4$
Square pyramidal
(sp^3d^2)

22 The d-and f-Block Elements

1. Which of the following is not a condition for complex formation?
   (a) Small atomic size
   (b) High nuclear charge
   (c) Variable oxidation states
   (d) Availability of vacant *d* orbitals
2. Zn gives H_2 gas with H_2SO_4 and HCl but not with HNO_3 because
   (a) Zn acts as an oxidising agent when react with HNO_3.
   (b) HNO_3 is weaker acid than H_2SO_4 and HCl.
   (c) In electrochemical series Zn is above hydrogen.
   (d) NO_3^- ion is reduced in preference to hydroniumion.
3. Which one of the following species is stable in aqueous solution ? Toughnut
   (a) Cr^{2+} (b) MnO_4^{2-}
   (c) MnO_4^{3-} (d) Cu^+
4. Which of the following is not correct about transition metals?
   (a) Their melting and boiling points are high
   (b) Their compounds are generally coloured
   (c) They can form ionic or covalent compounds
   (d) They do not exhibit variable valency
5. Zinc and mercury do not show variable valency like *d*-block element because
   (a) they are soft.
   (b) their *d*-shells are complete.
   (c) they have only two electrons in the outermost shell.
   (d) their *d*-shells are incomplete.
6. Which one of the following is diamagnetic ion?
   (a) Co^{2+} (b) Cu^{2+} (c) Mn^{2+} (d) Sc^{3+}
7. Which group contains coloured ions out of
   1. Cu^{2+} 2. Ti^{4+} 3. Co^{2+} 4. Fe^{2+}
   (a) 1, 2, 3, 4 (b) 1, 3, 4
   (c) 2, 3 (d) 1, 2
8. A compound of a metal ion M^{x+} (Z = 24) has a spin only magnetic moment of $\sqrt{15}$ Bohr Magnetons. The number of unpaired electrons in the compound are
   (a) 2 (b) 4 (c) 5 (d) 3
9. Cuprous ion is colourless while cupric ion is coloured because
   (a) both have half filled *p*-and *d*-orbitals.
   (b) cuprous ion has incomplete *d*-orbital and cupric ion has a complete *d*-orbital.
   (c) both have unpaired electrons in the *d*-orbitals.
   (d) cuprous ion has complete *d*-orbital and cupric ion has an incomplete *d*-orbital.
10. Among the following, the compound that is both paramagnetic and coloured, is
    (a) $KMnO_4$ (b) CuF_2
    (c) $K_2Cr_2O_7$ (d) All are coloured
11. Which of the following oxides of manganese is amphoteric?
    (a) MnO_2 (b) Mn_2O_3
    (c) Mn_2O_7 (d) MnO
12. In the laboratory, manganese (II) salt is oxidised to permanganate ion in aqueous solution by
    (a) hydrogen peroxide (b) conc. nitric acid
    (c) peroxy disulphate (d) dichromate
13. When Cu reacts with $AgNO_3$ solution, the reaction takes place is
    (a) oxidation of Cu (b) reduction of Cu
    (c) oxidation of Ag^+ (d) reduction of NO_3^-
14. $CuSO_4$ reacts with KCN solution and forms Toughnut
    (a) Cu(CN) (b) $Cu(CN)_2$
    (c) $K_3[Cu(CN)_4]$ (d) $K_4[Cu(CN)_6]$
15. Which of the following compounds gives red precipitate with $AgNO_3$?
    (a) KI (b) K_2CrO_4
    (c) NaBr (d) $NaNO_3$
16. When horn silver ore is dissolved in excess of sodium cyanide solution compound formed is
    (a) AgCN (b) $Na[Ag(CN)_2]$
    (c) $Na_2[Ag(CN)_3]$ (d) $Na_4[Ag(CN)_5]$

17. In the equation

$$4M + 8CN^- + 2H_2O + O_2 \rightarrow$$

$$4\left[M(CN)_2\right]^- + 4OH^-$$

Identify the metal M. It is

(a) copper (b) iron (c) gold (c) zinc

18. Mercury is the only metal which is liquid at 0°C. This is due to its

(a) very high ionisation energy and weak metallic bond.
(b) low ionisation potential.
(c) high atomic weight.
(d) high vapour pressure.

19. If $KMnO_4$ is reduced by oxalic acid in an acidic medium then oxidation number of Mn changes from

(a) 4 to 2 (b) 6 to 4
(c) +7 to +2 (d) 7 to 4

20. In the preparation of $KMnO_4$ pyrolusite (MnO_2) is first converted to potassium manganate (K_2MnO_4). In this conversion, the oxidation state of manganese changes from

(a) +1 to +3 (b) +2 to +4
(c) +3 to +5 (d) +4 to +6

21. In an alkaline condition $KMnO_4$ reacts as follows:

$$2KMnO_4 + 2KOH \rightarrow 2K_2MnO_4 + H_2O + O$$

Its equivalent weight is

(a) 31.6 (b) 52.7 (c) 49 (d) 158

22. In which of the following compounds manganese has oxidation number equal to that of iodine in KIO_4?

(a) Potassium manganate
(b) Potassium permanganate
(c) Manganous chloride
(d) Manganese chloride

23. The colour of $CuCr_2O_7$ solution in water is green because

(a) $Cr_2O_7^{2-}$ ions are green.
(b) Cu^{2+} ions are green.
(c) Both $Cr_2O_7^{2-}$ and Cu^{2+} ions are green.
(d) Cu^{2+} ions are blue and $Cr_2O_7^{2-}$ ions are yellow.

24. The correct order of magnetic moments (spin only values in B.M.) among is

(a) $[Fe(CN)_6]^{4-} > [MnCl_4]^{2-} > [CoCl_4]^{2-}$
(b) $[MnCl_4]^{2-} > [Fe(CN)_6]^{4-} > [CoCl_4]^{2-}$
(c) $[MnCl_4]^{2-} > [CoCl_4]^{2-} > [Fe(CN)_6]^{4-}$
(d) $[Fe(CN)_6]^{4-} > [CoCl_4]^{2-} > [MnCl_4]^{2-}$

(Atomic nos. : Mn = 25, Fe = 26, Co = 27)

25. Out of $[TiF_6]^{2-}$, $[CoF_6]^{3-}$, Cu_2Cl_2 and $[NiCl_4]^{2-}$ (Z of Ti = 22, Co = 27, Cu = 29, Ni = 28), the colourless species are:

(a) Cu_2Cl_2 and $[NiCl_4]^{2-}$
(b) $[TiF_6]^{2-}$ and Cu_2Cl_2
(c) $[CoF_6]^{3-}$ and $[NiCl_4]^{2-}$
(d) $[TiF_6]^{2-}$ and $[CoF_6]^{3-}$

26. The basic character of the transition metal monoxides follows the order (Atomic Nos., Ti = 22, V = 23, Cr = 24, Fe = 26)

(a) TiO > VO > CrO > FeO
(b) VO > CrO > TiO > FeO
(c) CrO > VO > FeO > TiO
(d) TiO > FeO > VO > CrO

27. Which of the following statement is incorrect ?

(a) Across a period from Sc to Cu the densities increase with increasing atomic number.
(b) The melting point of transition elements rise to a maximum from Sc to Cr and then decreases from Fe to Zn.
(c) Transition elements have high enthalpies of atomization and in 3*d* series increase regularly from Sc to Cu.
(d) On going down a group from 3*d* to 6d series the stability of higher oxidation state increases with increasing atomic number.

28. On the basis of data given below,

$$E^\ominus_{Sc^{3+}/Sc^{2+}} = -0.37,\ E^\ominus_{Mn^{3+}/Mn^{2+}} = +1.57$$

$$E^\ominus_{Cr^{2+}/Cr} = -0.90,\ E^\ominus_{Cu^{2+}/Cu} = 0.34$$

Which of the following statements is incorrect?

(a) Sc^{3+} has good stability due of $[Ar]3d^04s^0$ configuration.
(b) Mn^{3+} is more stable than Mn^{2+}.
(c) Cr^{2+} is reducing in nature.
(d) Copper does not give H_2 on reaction with dil. H_2SO_4.

29. A compound of iron exists as a dimer in vapour state. It is hygroscopic in nature and dissolves in water giving brown acidic solution. The compound is

(a) Fe_3O_4 (b) $FeSO_4$ (c) $FeCl_3$ (f) FeO_2

30. A metal M and its compound can give the following observable changes in a consequence of reactions

Tricky

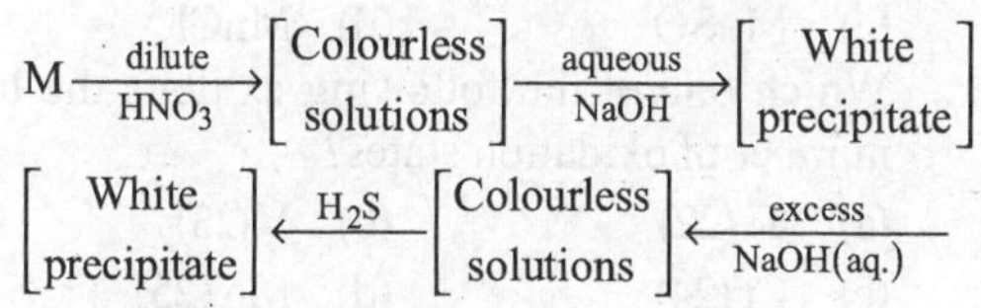

(a) Mg (b) Pb (c) Zn (d) Sn

31. $(X) + K_2CO_3 + Air \xrightarrow{heat} (Y)$

$(Y) + Cl_2 \longrightarrow (Z)$ Pink

Which of the following is correct?

(a) X = black, MnO_2, Y = Blue, K_2CrO_4, Z = $KMnO_4$

(b) X = green, Cr_2O_3, Y = Yellow, K_2CrO_4, Z = $K_2Cr_2O_7$

(c) X = black, MnO_2, Y = green, K_2MnO_4, Z = $KMnO_4$

(d) X = black, Bi_2O_3, Y = colourless $KBiO_2$, Z = $KBiO_3$

32. $CuSO_4(aq.) \xrightarrow{H_2S\uparrow} M\downarrow \xrightarrow[\text{of KCN}]{\text{Excess}} N + O$

Then final products N and O are respectively.

(a) $[Cu(CN)_4]^{3-}$, $(CN)_2$ (b) $CuCN$, $(CN)_2$

(c) $[Cu(CN)_4]^{2-}$, $(CN)_2$ (d) $Cu(CN)_2$, K_2S

33. Which of the following statements is false ?

Toughnut

(a) $Na_2Cr_2O_7$ is less soluble than $K_2Cr_2O_7$.

(b) $Na_2Cr_2O_7$ is primary standard in volumetry.

(c) CrO_4^{2-} is tetrahedral in shape.

(d) $Cr_2O_7^{2-}$ has a Cr–O–Cr bond.

34. A red solid is insoluble in water. However it becomes soluble if some KI is added to water. Heating the red solid in a test tube results in liberation of some violet coloured fumes and droplets of a metal appear on the cooler parts of the test tube. The red solid is

(a) HgI_2 (b) HgO

(c) Pb_3O_4 (d) $(NH_4)_2Cr_2O_7$

35. Of Cr (VI) as $Cr_2O_7^{2-}$ and CrO_4^{2-}, which is better oxidising agent?

(a) CrO_4^{2-}, basic medium

(b) $Cr_2O_7^{2-}$, basic medium

(c) $Cr_2O_7^{2-}$, acidic medium

(d) CrO_3, basic medium

36. A blue colouration is not obtained when

(a) ammonium hydroxide dissolves in copper sulphate.

(b) copper sulphate solution reacts with $K_4[Fe(CN)_6]$.

(c) ferric chloride reacts with sod. ferrocyanide.

(d) anhydrous $CuSO_4$ is dissolved in water.

37. Which of the statements is not true?

(a) On passing H_2S through acidified $K_2Cr_2O_7$ solution, a milky colour is observed.

(b) $Na_2Cr_2O_7$ is preferred over $K_2Cr_2O_7$ in volumetric analysis.

(c) $K_2Cr_2O_7$ solution in acidic medium is orange.

(d) $K_2Cr_2O_7$ solution becomes yellow on increasing the pH beyond 7.

38. The blue complex formed on addition of conc. NH_4OH solution to a Cu^{2+} salt solution has the structure?

(a) $[Cu(NH_4)_4]^{2+}$ (b) $[Cu(NH_3)_2]^{2+}$

(c) $[Cu(NH_3)_4]^{2+}$ (d) $[Cu(NH_4)_2]^{2+}$

39. Identify the incorrect statement

(a) $CuSO_4$ reacts with KCl in aqueous solution to give Cu_2Cl_2.

(b) $CuSO_4$ reacts with KI in aqueous solution to give Cu_2I_2.

(c) $CuSO_4$ react with NaOH and glucose in aqueous medium to give Cu_2O.

(d) $CuSO_4$ on strong heating gives CuO.

40. The black compound formed during the reaction between sodium thiosulphate and silver nitrate is

(a) silver thiosulphate ($Ag_2S_2O_3$)

(b) silver sulphide (Ag_2S)

(c) silver sulphate (Ag_2SO_4)

(d) silver sulphite (Ag_2SO_3)

41. When $KMnO_4$ is added to oxalic acid, the decolourisation is slow in the beginning but becomes instantaneous after sometime because

(a) Mn^{2+} acts as autocatalyst.
(b) CO_2 is formed as the product.
(c) Reaction is exothermic.
(d) MnO_4^- catalyses the reaction.

42. Number of electrons transfered in each case when $KMnO_4$ acts as an oxidising agent to give MnO_2, Mn^{2+}, $Mn(OH)_3$ and MnO_4^{2-} are respectively

(a) 3,5,4 and 1 (b) 4,3,1 and 5
(c) 1,3,4 and 5 (d) 5,4,3 and 1

43. Chloro compound of Vanadium has only spin magnetic moment of 1.73 BM. This Vanadium chloride has the formula: Toughnut

(a) VCl_2 (b) VCl_4 (c) VCl_3 (d) VCl_5

44. Which pair of elements with the given atomic numbers is expected to have similar properties?

(a) 40, 72 (b) 20, 36 (c) 10, 28 (d) 11, 12

45. When concentrated HCl is added to an aqueous solution of $CoCl_2$, its colour changes from reddish pink to deep blue. Which complex ion gives blue colour in this reaction? Toughnut

(a) $[CoCl_4]^{2-}$ (b) $[CoCl_6]^{3-}$
(c) $[CoCl_6]^{4-}$ (d) $[Co(H_2O)_6]^{2+}$

46. Which of the following is not formed when H_2S reacts with acidic $K_2Cr_2O_7$ solution?

(a) $CrSO_4$ (b) $Cr_2(SO_4)_3$
(c) K_2SO_4 (d) S

47. Potassium dichromate when heated with concentrated sulphuric acid and a soluble chloride, gives brown-red vapours of:

(a) CrO_3 (b) $CrCl_3$
(c) CrO_2Cl_2 (d) Cr_2O_3

48. When a small amount of $KMnO_4$ is added to concentrated H_2SO_4, a green oily compound is obtained which is highly explosive in nature. Compound may be:

(a) $MnSO_4$ (b) Mn_2O_7
(c) MnO_2 (d) Mn_2O_3

49. Identify incorrect statement:

(a) Cu_2O is colourless.
(b) Copper (I) compounds are colourless except when colour results from charge transfer.
(c) Copper (I) compounds are diamagnetic.
(d) Cu_2S is black.

50. Which of the following compounds has colour but no unpaired electrons?

(a) $KMnO_4$ (b) K_2MnO_4
(c) $MnSO_4$ (d) $MnCl_2$

51. Which one of the following exhibits the large number of oxidation states?

(a) Ti (22) (b) V (23)
(c) Cr (24) (d) Mn (25)

52. The element with which of the following outer electron configuration may exhibit the largest number of oxidation states in its compounds:

(a) $3d^5 4s^2$ (b) $3d^8 4s^2$
(c) $3d^7 4s^2$ (d) $3d^6 4s^2$

53. Because of lanthanoid contraction, which of the following pairs of elements have nearly same atomic radii ? (Numbers in the parenthesis are atomic numbers).

(a) Zr (40) and Nb (41) (b) Zr (40) and Hf (72)
(c) Zr (40) and Ta (73) (d) Ti (22) and Zr (40)

54. Which is not the correct statement?
(At. nos. Ce = 58, Lu = 71, La = 57, Yb = 70)

(a) Colour of Yb^{3+} ion is pink.
(b) La^{3+} is diamagnetic.
(c) Ce^{4+} has f^0 configuration.
(d) Lu^{3+} had f^{14} configuration.

55. Magnetic moment of Gd^{3+} ion (Z = 64) is

(a) 3.62 BM (b) 9.72 BM
(c) 7.9 BM (d) 10.60 BM

56. Which of the following forms stable + 4 oxidation state?

(a) La (Z = 57) (b) Eu (Z = 63)
(c) Ce (Z = 58) (d) Gd (Z = 64)

57. The number of unpaired electrons in Gadolinium [Z = 64] is

(a) 3 (b) 8 (c) 6 (d) 2

58. Which one of the following elements shows maximum number of different oxidation states in its compounds? Toughnut

(a) Eu (b) La (c) Gd (d) Am

59. The radius of La^{3+} (Atomic number of La = 57) is 1.06Å. Which one of the following given values will be closest to the radius of Lu^{3+} (Atomic number of Lu = 71)?
(a) 1.40 Å (b) 1.06 Å (c) 0.85 Å (d) 1.60 Å

60. The electronic configuration of gadolinium (At. No. 64) is
(a) $[Xe]4f^8 5d^1 6s^2$ (b) $[Xe]4f^7 5d^1 6s^2$
(c) $[Xe]4f^3 5d^5 6s^2$ (d) $[Xe]4f^6 5d^2 6s^2$

61. In which of the following lanthanoids oxidation state +2 is most stable?
(a) Ce (b) Eu (c) Tb (d) Dy

62. Which of the following lanthanoid ions is diamagnetic?
(At nos. Ce = 58, Sm = 62, Eu = 63, Yb = 70)
(a) Sm^{2+} (b) Eu^{2+} (c) Yb^{2+} (d) Ce^{2+}

63. Lanthanoid contraction can be observed in
(a) At (b) Gd (c) Ac (d) Lw

64. Lanthanoides and actinoides resemble in
(a) Electronic configuration
(b) Oxidation state
(c) Ionization energy
(d) Formation of complexes

65. The Ce (Z = 58) belongs to IIIrd group of periodic table. If it furnish one α particle to form an element 'X', then X belongs to
(a) IIIrd group (b) IInd group
(c) I[st] group (d) zero group

66. In aqueous solutions Eu^{2+} acts as
(a) an oxidising agent.
(b) a reducing agent.
(c) can act either of these.
(d) can act as redox agent.

67. Arrange (I) Ce^{3+}, (II) La^{3+}, (III) Pm^{3+} and (IV) Yb^{3+} in increasing order of their ionic radii.
(a) IV < III < I < II (b) I < IV < III < II
(c) IV < III < II < I (d) III < II < I < IV

68. Which of the following statement is **not correct**?

(a) $La(OH)_3$ is less basic than $Lu(OH)_3$.
(b) In lanthanide series ionic radius of Ln^{3+} ions decreases.
(c) La is actually an element of transition series rather than lanthanide series.
(d) Atomic radii of Zr and Hf are same because of lanthanide contraction.

69. The reason for the stability of Gd^{3+} ion is
(a) 4*f* subshell — half filled.
(b) 4*f* subshell — completely filled.
(c) Possesses the general electronic configuration of noble gases.
(d) 4*f*-subshell empty.

70. Knowing that the chemistry of lanthanoids(Ln) is dominated by its + 3 oxidation state, which of the following statements is incorrect?
(a) The ionic size of Ln (III) decrease in general with increasing atomic number.
(b) Ln (III) compounds are generally colourless.
(c) Ln (III) hydroxide are mainly basic in character.
(d) Because of the large size of the Ln (III) ions the bonding in its compounds is predominantly ionic in character.

Answer KEYs

1	(c)	8	(d)	15	(b)	22	(b)	29	(c)	36	(b)	43	(b)	50	(a)	57	(b)	64	(b)
2	(d)	9	(d)	16	(b)	23	(d)	30	(c)	37	(b)	44	(a)	51	(d)	58	(d)	65	(b)
3	(b)	10	(b)	17	(c)	24	(c)	31	(c)	38	(c)	45	(a)	52	(a)	59	(c)	66	(b)
4	(d)	11	(a)	18	(a)	25	(b)	32	(a)	39	(a)	46	(a)	53	(b)	60	(b)	67	(a)
5	(b)	12	(c)	19	(c)	26	(a)	33	(a)	40	(a)	47	(c)	54	(a)	61	(b)	68	(a)
6	(d)	13	(a)	20	(d)	27	(c)	34	(a)	41	(a)	48	(b)	55	(c)	62	(c)	69	(a)
7	(b)	14	(c)	21	(d)	28	(b)	35	(c)	42	(a)	49	(a)	56	(c)	63	(b)	70	(b)

Hints & Solutions

1. (c) Tendency of an element to form complexes does not depend on whether the element shows variable O.S. or not.

2. (d) NO_3^- ions are reduced by nascent hydrogen
$Metal + HNO_3 \rightarrow Metal\ nitrate + [H]$
$HNO_3 + 8[H] \rightarrow N_2O + 5H_2O$

3. (b) In MnO_4^{2-} manganese is in +6 oxidation state which is having highest stability.
MnO_4^{2-} disproportionates in neutral or acidic solution.
$$3MnO_4^{2-} + 4H^+ \longrightarrow 2MnO_4^{2-} + MnO_2 + 2H_2O$$

4. (d) Transition metals exhibit variable valency.

5. (b) Zn and Hg have *d*-shells completely filled hence they do not exhibit variable valency.

6. (d) Sc^{3+} is diamagnetic as it does not contain any unpaired electron while others contain.

7. (b) $Cu^{2+}[Ar]3d^9$, $Ti^{4+}[Ar]3d^0$, $Co^{2+}[Ar]3d^7$, $Fe^{2+}[Ar]3d^6$ 1,3,4 are coloured ions hence the answer is b.

8. (d) Magnetic moment $\mu = \sqrt{n(n+2)}$ where n = number of unpaired electrons
$\sqrt{15} = \sqrt{n(n+2)}$, n = 3

9. (d) In $Cu^+[Ar]3d^{10}$ there is no unpaired electron,
$Cu^{2+}[Ar]3d^9$ contains one unpaired electron hence coloured.

10. (b) CuF_2 is both paramagnetic and coloured.

11. (a) The more the oxidation state, the more is the acid character, MnO, Mn_2O_3 are basic, MnO_2 is amphoteric and Mn_2O_7 is acidic.

12. (c) In laboratory, manganese (II) ion salt is oxidised to permagnate ion in aqueous solution by peroxodi-sulphate.
$$2Mn^{2+} + \underset{\text{peroxodisulphate ion}}{S_2O_8^{2-}} + 8H_2O \longrightarrow 2MnO_4^- + 10SO_4^{2-} + 16H^+$$

13. (a) $Cu^0 + 2AgNO_3 \rightarrow \overset{2+}{Cu}(NO_3)_2 + 2Ag$.
It is oxidation of copper $(Cu^0 \rightarrow Cu^{2+})$

14. (c) $2CuSO_4 + 10KCN \longrightarrow 2K_3[Cu(CN)_4] + 2K_2SO_4 + (CN)_2$

15. (b) $2AgNO_3 + K_2CrO_4 \rightarrow \underset{\text{Red precpitate}}{Ag_2CrO_4} + 2KNO_3$

16. (b) $AgCl + 2NaCN \rightarrow Na[Ag(CN)_2] + NaCl$

17. (c) This is cyanide process for gold (M = Au).

18. (a) Metallic bonding in mercury is weak.

19. (c) In acid medium
$MnO_4^- + 8H^+ + 5e^- \rightarrow Mn^{2+} + 4H_2O$
(O.S. of Mn changes form +7 to +2).

20. (d) In MnO_2 (O.S. of Mn is +4) in K_2MnO_4 (O.S.of Mn is +6). Hence O.S. changes by 2.

21. (d) Change in O.S. is by 1. Hence eq. wt. is $\frac{158}{1} = 158$.

22. (b) In KIO_4, O.S of I is +7 and in $KMnO_4$, O.S. of Mn is +7.

23. (d) Green colour of $CuCr_2O_7$ is due to blue colour of Cu^{2+} ions and yellow colour of $Cr_2O_7^{2-}$ ions.

24. (c) $[Fe(CN)_6]^{4-}$
→ [⇅|⇅|⇅|⁚|⁚] [⁚] [⁚|⁚|⁚]
no of unpaired electron = 0
$[MnCl_4]^{2-}$ → [↑|↑|↑|↑|↑] [⁚] [⁚|⁚|⁚]
no of unpaired electron = 5
$[CoCl_4]^{2-}$ → [⇅|⇅|↑|↑|↑] [⁚] [⁚|⁚|⁚]
no of unpaired electron = 3
The greater the number of unpaired electrons, greater the magnitude of magnetic moment. Hence the correct order will be
$[MnCl_4]^{2-} > [CoCl_4]^{2-} > [Fe(CN)_6]^{4-}$

25. (b) The colour exhibited by transition metal ions is due to the presence of unpaired electrons in *d*-orbitals which permits the *d* - *d* excitation of electrons.

In $[TiF_6]^{2-}$ – Ti is in +4 O.S.; $3d^0$ – colourless

In $[CoF_6]^{3-}$ – Co is in +3 O.S; $3d^5$ – coloured

In Cu_2Cl_2 – Cu is in +1 O.S.; $3d^{10}$ – colourless

In $[NiCl_4]^{2-}$ – Ni is in +2 O.S; $3d^8$ – coloured

26. (a) The order of basic character of the transition metal monoxide is TiO > VO > CrO > FeO because basic character of oxides decreases with increase in atomic number.

27. (c) (a), (b) and (d) are correct statements. (c) There is drop in enthalpy of atomization at manganese because of half filled stable electronic configuration leading to less mobilisation of electrons. This results in weaker metallic bonds.

28. (b) Mn^{2+} (d^5) is more stable than Mn^{3+} (d^4), thus $E^-_{Mn^{3+}/Mn^{2+}} = +ve$

29. (c) The solution is acidic due to formation of HCl and brown due to $Fe(OH)_3$.

$$FeCl_3 + 3H_2O \rightarrow \underset{\text{Brown}}{Fe(OH)_3} + 3HCl$$

30. (c) $$\underset{(M)}{Zn} \xrightarrow{\text{dil. } HNO_3} \underset{\substack{\text{Colourless}\\\text{solution}}}{Zn(NO_3)_2} \xrightarrow[\text{aq.}]{NaOH} \underset{\text{White ppt.}}{Zn(OH)_2}$$

$$\xrightarrow[\text{NaOH}]{\text{Excess}} \underset{\text{Soluble}}{Na_2[Zn(OH)_4]} \xrightarrow{H_2S} \underset{\text{White ppt.}}{ZnS}$$

31. (c) $$\underset{(X)}{2MnO_2} + 2K_2CO_3 + \underset{(air)}{O_2} \xrightarrow{\Delta} \underset{\substack{\text{green}\\(Y)}}{2K_2MnO_4} + 2CO_2 \uparrow$$

$$\underset{(Y)}{2K_2MnO_4} + Cl_2 \longrightarrow \underset{(Z)\text{ Pink}}{2KMnO_4} + 2KCl$$

32. (a) $$CuSO_4(aq.) \xrightarrow{H_2S} \underset{\substack{\text{Black ppt.}\\(M)}}{2CuS} \xrightarrow[\text{Excess}]{KCN} \underset{(N)}{[Cu(CN)_4]^{3-}} + \underset{(O)}{(CN)_2} \uparrow$$

33. (a) CrO_4^{2-} (O=Cr(=O)(–O⁻)–O⁻ structure) is tetrahedral

$Cr_2O_7^{2-}$ structure: two CrO_3 units joined by a bridging O, $Cr-O-Cr$ bond in $Cr_2O_7^{2-}$

$Na_2Cr_2O_7$ is used as primary standard in volumetry.

Hydration energy of Na^+ is greater than K^+ because of smaller size of Na^+, $Na_2Cr_2O_7$ is more soluble than $K_2Cr_2O_7$.

34. (a) When KI is added to mercuric iodide it disssolve in it and form complex.

$$\underset{\substack{\text{red, solid}\\\text{(insoluble)}}}{HgI_2} + KI \rightarrow \underset{\text{(soluble)}}{K_2[HgI_4]}$$

On heating HgI_2 decomposes as

$$HgI_2 \rightleftharpoons Hg + \underset{\text{(violet vapours)}}{I_2}$$

35. (c) $CrO_4^{2-}(aq) + 4H_2O + 3e^- \longrightarrow Cr(OH)_3 + 5OH^-$, $E° = -0.13V$

$Cr_2O_7^{2-}(aq.) + 14H^+(aq.) + 6e^- \longrightarrow 2Cr^{3+}(aq.) + 7H_2O$, $E° = 1.33V$

36. (b) $$2CuSO_4 + K_4[Fe(CN)_6] \longrightarrow \underset{\text{Chocolate ppt.}}{Cu_2[Fe(CN)_6]} + 2K_2SO_4$$

37. (b) $Na_2Cr_2O_7$ is hygroscopic.

38. (c) $$Cu^{2+} + 4NH_4OH \rightarrow \underset{\text{deep blue comp}}{[Cu(NH_3)_4]^{2+}} + 4H_2O$$

39. (a) $2CuSO_4 + 4KI \rightarrow Cu_2I_2 + 2K_2SO_4 + I_2$ (not given by KCl).
Fehling solution gives Cu_2O with reducing substances

$$CuSO_4 \xrightarrow[\Delta]{\text{strong}} CuO + SO_3$$

40. (a) $2AgNO_3 + Na_2S_2O_3 \rightarrow Ag_2S_2O_3 + 2NaNO_3$

41. (a) Little bit acid is required for reaction $\therefore Mn^{2+}$ acts as autocatalyst.

42. (a)

	$KMnO_4$	MnO_2	Mn^{2+}	$Mn(OH)_3$	MnO_4^{2-}
O.S of Mn	+7	+4	+2	+3	+6
electrons needed	0	3	5	4	1

43. (b) $\mu = \sqrt{n(n+2)}$

$1.73 = \sqrt{n(n+2)}$

On calculating the value of n we find $n = 1$
No. of unpaired electrons = 1
hence its configuration will be
V(23) = [Ar] $3d^3 4s^2$

$V^{4+} = [Ar]3d^1$

∴ Its chloride has the formula VCl_4

44. (a) Zr (40), Hf (72) have similar properties as they belong to same group and have same valence shell electronic configuration.

45. (a) Aqueous of solution of $CoCl_2$ contains $[Co(H_2O)_6]^{2+}$ which is pinkish in colour so option (d) is incorrect.

Reduction potential of $Co^{3+} \longrightarrow Co^{2+}$ is high so option(b) is incorrect. Co^{2+} does not oxidises easily to Co^{3+}.

It is general case that symmetrical substituted octahedral complexes are less deeper in colour than tetrahedral complexes. So $[CoCl_4]^{2-}$ is deep blue in colour.

46. (a) $K_2Cr_2O_7 + 4H_2SO_4 + 3H_2S \longrightarrow K_2SO_4 + Cr_2(SO_4)_3 + 7H_2O + 3S$

47. (c) Solid potassium dichromate when heated with concentrated sulphuric acid and a soluble chloride gives orange red vapours of a volatile oily liquid CrO_2Cl_2

$K_2Cr_2O_7 + 4NaCl + 6H_2SO_4 \longrightarrow 2KHSO_4 + 4NaHSO_4 + \underset{\text{chromyl chloride}}{2CrO_2Cl_2}$

48. (b) $KMnO_4$ reacts with H_2SO_4 to form Mn_2O_7 which is highly explosive substance.

$2KMnO_4 + H_2SO_4 \longrightarrow K_2SO_4 + Mn_2O_7 + H_2O$

49. (a) Cu_2O is red in colour.

50. (a) Electronic configuration of Mn is $[Ar]3d^5 4s^2$. Being transition metal it has 7 valence electrons and all are involved in bond formation in MnO_4^-. Hence it has no unpaired electron.

51. (d) Manganese exhibits the large number of oxidation states. The most common oxidation states of Mn are +2, +3, + 4, + 6 and +7.

52. (a) The element with outer electron configuration $3d^5 4s^2$ is Mn which exhibits oxidation states from + 2 to + 7.

53. (b) Due to lanthanoid contraction atomic radii of Zr and Hf is almost similar.

54. (a) Option (a) is incorrect as Yb^{3+} is colorless.

55. (c) Gd (64) = [Xe] $4f^7\, 5d^1\, 6s^2$

Gd^{3+} = [Xe] $4f^7\, 5d^0\, 6s^0$

i.e. no. of unpaired electrons = 7

$\mu = \sqrt{n(n+2)} = \sqrt{7(7+2)}$

$= \sqrt{63} = 7.93$ BM

56. (c) Only Ce(Z = 58) shows stable (+4) oxidation state.

57. (b) Gd(64) = [Xe]$4f^7\, 5d^1\, 6s^2$

∴ No. of unpaired electrons = 8

58. (d) We know that lanthanides La, Gd shows +3, oxidation state, while Eu shows oxidation state of +2 and + 3. Am shows +3, +4, +5 and +6 oxidation states. Therefore Americium (Am) has maximum number of oxidation states.

59. (c) Ionic radii $\propto \frac{1}{z}$

Thus, $\frac{z_2}{z_1} \Rightarrow \frac{1.06}{(\text{Ionic radii of } Lu^{3+})} = \frac{71}{57}$

$\Rightarrow$ Ionic radii of $Lu^{3+} = 0.85$ Å

60. (b) Electronic configuration of gadolinium is $[Xe]4f^7 5d^1 6s^2$

61. (b) Eu^{2+} has electronic configuration $[Xe]4f^7$ hence stable due to half filled atomic orbitals.

62. (c) $Sm^{2+}(Z = 62)$

$[Xe]4f^6\, 6s^2$ – 6 unpaired e^- $Eu^{2+}(Z = 63)$

$[Xe]4f^7\, 6s^2$ – 7 unpaired e^- $Yb^{2+}(Z = 70)$

$[Xe]4f^{14}\, 6s^2$ – 0 unpaired e^- $Ce^{2+}(Z = 58)$

$[Xe]4f^1\, 5d^1\, 6s^2$ – 2 unpaired e^-

Only Yb^{2+} is diamagnetic.

63. (b) Amongst the given elements, only Gd is a lanthanide.

64. (b) The oxidation state in both (lanthanide and actinide) is +3. The property of actinide are very similar to those of lanthanide when both are in +3 state.

65. (b) $_{58}Ce$ is lanthanoid. Lanthanoid are from 57 to 71 all present in IIIrd group. Hence $_{58}Ce$ forms $_{56}X$ element on emission of one α particle with belongs to IInd group.

66. (b) Reducing agent

67. (a) Ionic radii decrease across lanthanide series due to lanthanide contraction. As all ions are in +3 O.S., ionic radii will follow the trend of atomic radii.

$\therefore \underset{(Z=57)}{La^{3+}} > \underset{(Z=58)}{Ce^{3+}} > \underset{(Z=61)}{Pm^{3+}} > \underset{(Z=70)}{Yb^{3+}}$

68. (a) $La(OH)_3$ is more basic than $Lu(OH)_3$. This is because ionic size of La^{3+} ion is more than Lu^{3+} ion (Fajan's rule).

69. (a) Gd (Z = 64) $\Rightarrow 4f^7\, 5d^1\, 6s^2$

$Gd^{3+} = 4f^7$

Gd^{3+} is stable due to the presence of exactly half-filled $4f$-subshell.

70. (b) Most of the Ln^{3+} compounds except La^{3+} and Lu^{3+} are coloured due to the presence of f-electrons.

23 Co-ordination Compounds

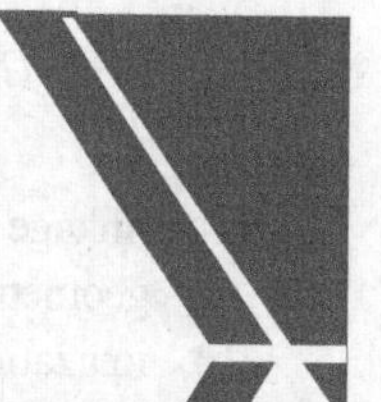

1. AgCl is soluble in NH_4OH solution. The solubility is due to formation of
   (a) AgOH (b) Ag_2O
   (c) $[Ag(NH_3)_2]^+$ (d) NH_4Cl
2. In which of the following coordinate compounds the central metal atom obeys the EAN rule.
   (a) $K_3[Fe(CN)_6]$ (b) $K_4[Fe(CN)_6]$
   (c) $[Cu(NH_3)_4]SO_4$ (d) All of these
3. Which of the following complex will show geometrical as well as optical isomerism (en=ethylenediammine)
   (a) $[Pt(NH_3)_2Cl_2]$ (b) $[Pt(en)Cl_4]$
   (c) $[Pt(en)_3]^{4+}$ (d) $[Pt(en)_2Cl_2]$
4. The total number of possible isomers of the complex compound $[Cu^{II}(NH_3)_4][Pt^{II}Cl_4]$ is
   (a) 3 (b) 6 (c) 5 (d) 4
5. Which of the following will give maximum number of isomers?
   (a) $[Co(NH_3)_4Cl_2]$ (b) $[Ni(en)(NH_3)_4]^{2+}$
   (c) $[Ni(C_2O_4)(en)_2]$ (d) $[Cr(SCN)_2(NH_3)_4]^{2+}$
6. Which of the following will exhibit maximum ionic conductivity ?
   (a) $K_4[Fe(CN)_6]$ (b) $[Co(NH_3)_6]Cl_3$
   (c) $[Cu(NH_3)_4Cl_2]$ (d) $[Ni(CO)_4]$
7. $\left[Fe(H_2O)_6\right]^{2+}$ and $\left[Fe(CN)_6\right]^{4-}$ differ in :
   (a) geometry, magnetic moment
   (b) geometry, hybridization
   (c) magnetic moment, colour
   (d) hybridization, number of d-electrons
8. Which of the following complexes have a maximum number of unpaired electrons?
   (a) $[Ni(CO)_4]$
   (b) $[Co(NH_3)_4(NO_2)_2]^+$
   (c) $[Ag(CN)_2]^-$
   (d) $[CuBr_4]^{2-}$
9. The degeneracy of *d*-orbitals is lost under :
   (I) Strong field ligand
   (II) Weak field ligand
   (III) Mixed field ligand
   (IV) Chelated ligand field
   (a) I, II and IV (b) I and II
   (c) I, II, III and IV (d) I, II and III
10. What is the magnetic moment (spin only) and hybridisation of the brown ring complex $\left[Fe(H_2O)_5NO\right]SO_4$?
   (a) $\sqrt{3}$BM, sp^3d^2 (b) $\sqrt{3}$BM, d^2sp^3
   (c) $\sqrt{15}$BM, sp^3d^2 (d) $\sqrt{15}$BM, d^2sp^3
11. Mn^{2+} forms a complex with Br^- ion. The magnetic moment of the complex is 5.92 B.M. What would be the probable formula and geometry of the complex?
   (a) $[MnBr_6]^{4-}$, octahedral
   (b) $[MnBr_4]^{2-}$, square planar
   (c) $[MnBr_4]^{2-}$, tetrahedral
   (d) $[MnBr_5]^{3-}$, trigonal bipyramidal
12. The complex given is **Toughnut**

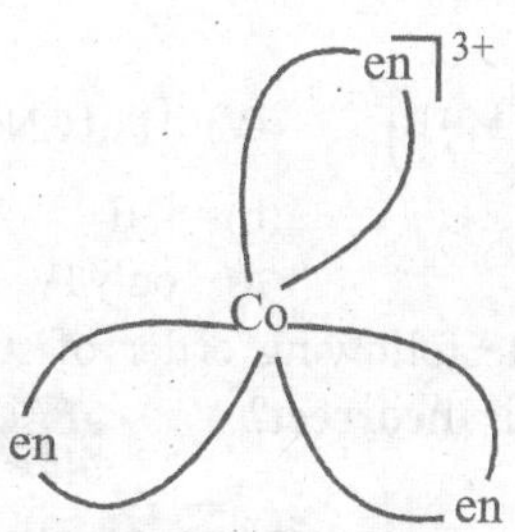

   (i) non-superimposable on its mirror images
   (ii) optically active
   (iii) rotate plane polarised light
   (iv) planar
   (a) (i) and (ii) (b) (i) and (iv)
   (c) (i), (ii) and (iii) (d) (ii) only

13. $[Co(NH_3)_4(NO_2)_2]Cl$ exhibits

(a) linkage isomerism, ionization isomerism and geometrical isomerism
(b) ionization isomerism, geometrical isomerism and optical isomerism
(c) linkage isomerism, geometrical isomerism and optical isomerism
(d) linkage isomerism, ionization isomerism and optical isomerism

14. Coordination number of Cr is six. A complex with $C_2O_4^{2-}$, en and superoxide O_2^- will be in the ratio to make complex $[Cr(C_2O_4)_x, (en)_y(O_2)_z]^-$

	x	y	z
(a)	1	1	1
(b)	1	1	2
(c)	1	2	2
(d)	2	1	1

15. For complex ion/compound formation reactions

(I) $Co^{3+}(aq) + EDTA^{4-} \longrightarrow P$

(II) $Ni^{2+}(aq) + dmg(excess) \xrightarrow{NH_4OH} Q$

(III) $Zn^{2+}(aq) + gly(excess) \longrightarrow R$

(IV) $Pt^{4+}(aq) + en(excess) \longrightarrow S$

which of the following complex ion/compound does not exhibit optical activity?

(a) P (b) Q (c) R (d) S

16. The diamagnetic species is :

(I) $[Cu(CN)_4]^{3-}$ (II) $[Co(NH_3)_6]^{3+}$

(III) $[Ni(NH_3)_6]^{2+}$ (IV) $[Fe(CN)_6]^{3-}$

(a) I, III (b) I, II
(c) III, IV (d) only IV

17. Which of the following order of stability of complex ion is **Incorrect**?

Tricky

(a) $[Fe(C_2O_4)_3]^{3-} < [Fe(H_2O)_6]^{3+}$

(b) $[Fe(edta)]^- > [Fe(en)_3]^{3+}$

(c) $[Ni(en)_2]^{2+} > [Ni(dmg)_2]$

(d) $[Fe(CN)_6]^{3-} > [Fe(CN)_6]^{4-}$

18. The pair in which both species have same magnetic moment (spin only value) is

(a) $[Cr(H_2O)_6]^{2+}$, $[CoCl_4]^{2-}$
(b) $[Cr(H_2O)_6]^{2+}$, $[Fe(H_2O)_6]^{2+}$
(c) $[Mn(H_2O)_6]^{2+}$, $[Cr(H_2O)_6]^{2+}$
(d) $[CoCl_4]^{2-}$, $[Fe(H_2O)_6]^{2+}$

19. In which complex, d_{z^2} orbital of inner shell is not used in the hybridization of central metal cation

(a) $Fe(CO)_5$ (b) $[Cu(NH_3)_5]^{2+}$
(c) $[Co(NH_3)_6]^{2+}$ (d) $[IrF_6]^{3-}$

20. Which of the characteristic is not common between $[Cu(en)_2)^{2+}$ and $[Ni(dmg)_2]$?

(a) Geometry of complexes
(b) Hybridization of central metal cation
(c) Magnetic behaviour
(d) Number of stereoisomers

21. In which of the following complex hybridization of central metal is not same as that of donor atom of ligand

Toughnut

(a) $[Ni(PF_3)_4]$ (b) $[Fe(dmg)_2]$
(c) $[Zn(en)_2]^{2+}$ (d) $[Ni(PMe_3)_4]^{2+}$

22. Which of the following complex compound(s) is/are paramagnetic and low spin?

(I) $K_3[Fe(CN_6)]$ (II) $[Ni(CO)_4]^0$

(III) $[Cr(NH_3)_6]^{3+}$ (IV) $[Mn(CN)_6]^{4-}$

Choose the correct code:

(a) I only (b) II and III
(c) I and IV (d) IV only

23. An aqueous solution of titanium chloride, when subjected to magnetic measurement, measured zero magnetic moment. Assuming the octahedral complex in aqueous solution, the formulae of the complex is:

Critical Thinking

(a) $[Ti(H_2O)_6]Cl_2$ (b) $[Ti(H_2O)_6]Cl_4$

(c) $[TiCl_3(H_2O)_3]$ (d) $[TiCl_2(H_2O)_4]$

24. For octahedral complex, which of the following d^n configurations of metal cation cannot exist in high spin and low spin forms?

(a) d^3 and d^8 (b) d^3 and d^5
(c) d^3 and d^6 (d) d^4 and d^8

25. Which of the following statement is not true for the reaction given below? **Tricky**

$$\left[Cu(H_2O)_4\right]^{2+} + 4NH_3 \rightleftharpoons \left[Cu(NH_3)_4\right]^{2+} + 4H_2O$$

(a) It is a ligand substitution reaction
(b) NH_3 is relatively strong field ligand while H_2O is a weak field ligand
(c) During the reaction, there is a change in colour form light blue to dark blue
(d) $[Cu(NH_3)_4]^{2+}$ has tetrahedral structure and is paramagnetic

26. The *d*-electron configurations of Cr^{2+}, Mn^{2+}, Fe^{2+} and Co^{2+} are d^4, d^5, d^6 and d^7 respectively. Which one of the following will exhibit the lowest paramagnetic behaviour?

(Atomic no. Cr = 24, Mn = 25, Fe = 26, Co = 27).

(a) $[Co(H_2O)_6]^{2+}$ (b) $[Cr(H_2O)_6]^{2+}$
(c) $[Mn(H_2O)_6]^{2+}$ (d) $[Fe(H_2O)_6]^{2+}$

27. An octahedral complex with molecular composition $M.5\,NH_3.Cl.SO_4$ has two isomers, A and B. The solution of A gives a white precipitate with $AgNO_3$ solution and the solution of B gives white precipitate with $BaCl_2$ solution. The type of isomerism exhibited by the complex is: **Toughnut**

(a) Linkage isomerism
(b) Ionisation isomerism
(c) Coordinate isomerism
(d) Geometrical isomerism

28. Type of isomerism which exists between $[Pd(C_6H_5)_2(SCN)_2]$ and $[Pd(C_6H_5)_2(NCS)_2]$ is :

(a) Linkage isomerism
(b) Coordination isomerism
(c) Ionisation isomerism
(d) Solvate isomerism

29. Which of the following complex ions will exhibit optical isomerism? (en = 1, 2-diamine ethane).

(a) $[Cr(NH_3)_2Cl_2]^+$ (b) $[Co(en)_2Cl_2]^+$
(c) $[Co(NH_3)_4Cl_2]^+$ (d) $[Zn(en)_2]^{2+}$

30. The complex ion $[Pt(NO_2)(Py)(NH_3)(NH_2OH)]^+$ will give

(a) 2 isomers (Geometrical)
(b) 3 isomers (Geometrical)
(c) 6 isomers (Geometrical)
(d) 4 isomers (Geometrical)

31. Which of the following complex ions has electrons that are symmetrically filled in both t_{2g} and e_g orbitals ?

(a) $[FeF_6]^{3-}$ (b) $[Mn(CN)_6]^{4-}$
(c) $[CoF_6]^{3-}$ (d) $[Co(NH_3)_6]^{2+}$

32. In $[Co(NH_3)_6]Cl_3$, the number of covalent bonds is

(a) 3 (b) 6 (c) 9 (d) 18

33. The unpaired electrons in $Ni(CO)_4$ are

(a) zero (b) one (c) three (d) four

34. The shape of $[Cu(NH_3)_4]^{2+}$ is

(a) tetrahedral (b) square planar
(c) pyramidal (d) octahedral

35. Consider the following complex $[Co(NH_3)_5CO_3]ClO_4$. The coordination number, oxidation number, number of *d*-electrons and number of unpaired *d*-electrons on the metal are respectively

(a) 6, 3, 6, 0 (b) 7, 2, 7, 1
(c) 7, 1, 6, 4 (d) 6, 2, 7, 3

36. Which of the following statements is incorrect?

(a) $[MnCl_6]^{3-}$ is more paramagnetic than $[Mn(CN)_6]^{3-}$
(b) Both $[Co(C_2O_4)_3]^{3-}$ and $[CoF_6]^{3-}$ are paramagnetic.
(c) $[Fe(CN)_6]^{3-}$ forms inner orbital complex whereas $[FeF_6]^{3-}$ forms outer orbital complex.
(d) Both (a) and (b).

37. Which of the following statements related to crystal field splitting in octahedral coordination entities is incorrect?
(a) The $d_{x^2-y^2}$ and d_{z^2} orbitals has more energy as compared to d_{xy}, d_{yz} and d_{xz} orbitals.
(b) Crystal field splitting energy (Δ_o) depends directly on the charge of the metal ion and on the field produced by the ligand.
(c) In the presence of Br^- as a ligand the distribution of electrons for d^4 configuration will be t_{2g}^3, e_g^1,
(d) In the presence of CN^- as a ligand $\Delta_o < P$.

38. For the reaction of the type $M + 4L \rightleftharpoons ML_4$
(a) larger the stability constant, lower the proportion of ML_4 that exists in solution
(b) larger the stability constant, higher the proportion of ML_4 that exists in solution
(c) smaller the stability constant, higher the proportion of ML_4 that exists in solution
(d) None of the above

39. An octahedral complex of Co^{3+} is diamagnetic. The hybridisation involved in the formation of the complex is:
(a) sp^3d^2 (b) dsp^2 (c) d^2sp^3 (d) sp^3d

40. The correct statement about the magnetic properties of $[Fe(CN)_6]^{3-}$ and $[FeF_6]^{3-}$ is (Z = 26):
(a) both are paramagnetic.
(b) both are diamagnetic.
(c) $[Fe(CN)_6]^{3-}$ is diamagnetic, $[FeF_6]^{3-}$ is paramagnetic.
(d) $[Fe(CN)_6]^{3-}$ is paramagnetic, $[FeF_6]^{3-}$ is diamagnetic.

41. Consider the coordination compound, $[Co(NH_3)_6]Cl_3$. In the formation of this complex, the species which acts as the Lewis acid is:
(a) $[Co(NH_3)_6]^{3+}$ (b) Cl^-
(c) Co^{3+} (d) NH_3

42. Among the following species the one which causes the highest CFSE, Δ_o as a ligand is:
(a) CN^- (b) NH_3 (c) F^- (d) CO

43. Which one of the following complexes will most likely absorb visible light? Toughnut
(At nos. Sc = 21, Ti = 22, V = 23, Zn = 30)
(a) $[Sc(H_2O)_6]^{3+}$ (b) $[Ti(NH_3)_6]^{4+}$
(c) $[V(NH_3)_6]^{3+}$ (d) $[Zn(NH_3)_6]^{2+}$

44. Incorrect match is :

(a) $[Rh(PPh_3)_3Cl]$: Wilkinson's catalyst
(b) $[Co(CO)_4]^-$: Bond order of Co—CO bond is greater than one
(c) $[Zn(NH_3)_4]$ $[Be(OH)_4]$: colorless complex
(d) $[Cr(CN)_6]^{3-}$: Inner orbital low spin complex

45. Which one of the following complexes is an outer orbital complex ?
(a) $[Co(NH_3)_6]^{3+}$ (b) $[Mn(CN)_6]^{4-}$
(c) $[Fe(CN)_6]^{4-}$ (d) $[Ni(NH_3)_6]^{2+}$
(Atomic nos. : Mn = 25; Fe = 26; Co = 27, Ni = 28)

46. Which is not π - bonded complex ?
(a) Zeise's salt
(b) Ferrocene
(c) Dibenzene chromiun
(d) Tetraethyl lead

47. Which of the following carbonyls will have the strongest C – O bond ?
(a) $[Mn(CO)_6]^+$ (b) $[Cr(CO)_6]$
(c) $[V(CO)_6]^-$ (d) $[Fe(CO)_5]$

48. In $Fe(CO)_5$, the Fe – C bond possesses
(a) ionic character
(b) σ-character only
(c) π-character
(d) both σ and π characters

49. Which of the following does not have a metal-carbon bond?
(a) $Al(OC_2H_5)_3$ (b) C_2H_5MgBr
(c) $K[Pt(C_2H_4)Cl_3]$ (d) $Ni(CO)_4$

50. In the silver plating of copper, $K[Ag(CN)_2]$ is used instead of $AgNO_3$. The reason is
(a) a thin layer of Ag is formed on Cu
(b) more voltage is required
(c) Ag^+ ions are completely removed from solution
(d) less availability of Ag^+ ions, as Cu cannot displace Ag from $[Ag(CN)_2]^-$ ion.

51. Pick a poor electrolytic conductor complex in solution
(a) $K_2[PtCl_6]$
(b) $[Co(NH_3)_3](NO_2)_3$
(c) $K_4[Fe(CN)_6]$
(d) $[Co(NH_3)_4]SO_4$

52. Nickel (Z = 28) combines with a uninegative monodentate ligand to form a diamagnetic complex $[NiL_4]^{2-}$. The hybridisation involved and the number of unpaired electrons present in the complex are respectively: Toughnut

(a) sp^3, two (b) dsp^2, zero
(c) dsp^2, one (d) sp^3, zero

53. Which of the following is diamagnetic ?

(a) $[Fe(CN)_6]^{3-}$ (b) $[Co(ox)_3]^{3-}$
(c) $[FeF_6]^{3-}$ (d) $[CoF_6]^{3-}$

54. The magnetic moment of the complex anion $[Cr(NO)(NH_3)(CN)_4]^{2-}$ is :

(a) 5.91 BM (b) 3.87 BM
(c) 1.73 BM (d) 2.82 BM

55. In which of the following octahedral complex species the magnitude of Δ_o will be maximum ?

(a) $[Co(H_2O)_6]^{2+}$ (b) $[Co(CN)_6]^{3-}$
(c) $[Co(C_2O_4)_3]^{3-}$ (4) $[Co(NH_3)_6]^{3+}$

56. The structure of which of the following chloro species can be explained on the basis of dsp^2 hybridization ?

(a) $[PdCl_4]^{2-}$ (b) $[FeCl_4]^{2-}$
(c) $[CoCl_4]^{2-}$ (d) $[NiCl_4]^{2-}$

57. $[Ni(NH_3)_6]^{2+} + en \longrightarrow X + 2NH_3$

$[Ni(NH_3)_6] + 2en \longrightarrow Y + 4NH_3$

$[Ni(NH_3)_6 + 3en \longrightarrow Z + 6NH_3$

Which of them show optical as well as geometrical isomerism? Critical Thinking

(a) X (b) Y
(c) Z (d) All of these

58. 50 mL of 0.2 M solution of a compound with empirical formula $CoCl_3.4NH_3$ on treatment with excess of $AgNO_3(aq)$ yields 1.435g of AgCl. Ammonia is not removed by treatment with concentrated H_2SO_4. The formula of the compound is:

(a) $[Co(NH_3)_4]Cl_3$
(b) $[Co(NH_3)_4Cl_2]Cl$
(c) $[Co(NH_3)_4]Cl_3$
(d) $[CoCl_3(NH_3)](NH_3)_3$

59. Which of the following compounds is not coloured?

(a) $Na_2[CuCl_6]$ (b) $Na_2[CdCl_4]$
(c) $K_4[Fe(CN)_6]$ (d) $K_3[Fe(CN)_6]$

60. The geometry of $Ni(CO)_4$ and $Ni(PPh_3)_2Cl_2$ are

(a) both square planar
(b) tetrahedral and square planar
(c) both tetrahedral
(d) None of these

61. Which one is the most likely structure of $CrCl_3.6H_2O$ if 1/3 of total chlorine of the compound is precipitated by adding $AgNO_3$

(a) $CrCl_3.6H_2O$
(b) $[Cr(H_2O)_3Cl_3].(H_2O)_3$
(c) $[CrCl_2(H_2O)_4]Cl.2H_2O$
(d) $[CrCl(H_2O)_5]Cl_2.H_2O$

62. In the complexes $[Fe(H_2O)_6]^{3+}$, $[Fe(CN)_6]^{3-}$, $[Fe(C_2O_4)_3]^{3-}$ and $[FeCl_6]^{3-}$, more stability is shown by Toughnut

(a) $[Fe(H_2O)_6]^{3+}$ (b) $[Fe(CN)_6]^{3-}$
(c) $[Fe(C_2O_4)_3]^{3-}$ (d) $[FeCl_6]^{3-}$

63. The *d*-electron configurations of Cr^{2+}, Mn^{2+}, Fe^{2+} and Co^{2+} are d^4, d^5, d^6 and d^7 respectively. Which one of the following will exhibit the lowest paramagnetic behaviour?
(Atomic no. Cr = 24, Mn = 25, Fe = 26, Co = 27).

(a) $[Co(H_2O)_6]^{2+}$ (b) $[Cr(H_2O)_6]^{2+}$
(c) $[Mn(H_2O)_6]^{2+}$ (d) $[Fe(H_2O)_6]^{2+}$

64. The correct order of ligands in the spectrochemical series is

(a) $Cl^- > en > CN^- > NCS^-$
(b) $CN^- > en > NCS^- > Cl^-$
(c) $NCS^- > CN^- > Cl^- > en$
(d) $en > CN^- > Cl^- > NCS^-$

65. Which of the following paramagnetic ions would exhibit a magnetic moment (spin only) of the order of 5 BM?
(At. Nos. Mn = 25, Cr = 24, V = 23, Ti = 22)

(a) Mn^{2+} (b) Ti^{2+} (c) V^{2+} (d) Cr^{2+}

66. Square-planar geometry is shown by

(a) $[PtCl_2(NH_3)_2]$ (b) $[NiCl_4]^{2-}$
(c) MnO_4^- (d) CrO_4^{2-}

67. Which molecule/ion among the following cannot act as a ligand in complex compounds?

(a) CH_4 (b) CO (c) CN^- (d) Br^-

68. Which one of the following complexes will consume more equivalents of aqueous solution of $AgNO_3$?

(a) $Na_2[CrCl_5(H_2O)]$ (b) $Na_3[CrCl_6]$
(c) $[Cr(H_2O)_5Cl]Cl_2$ (d) $[Cr(H_2O)_6]Cl_3$

69. Possible isomerism in complexes $[Co(NH_3)_3(NO_2)_3]$ and $[Co(NH_3)_5(NO_2)]Cl_2$, respecitvely are:

(a) Linkage and optical
(b) Geometrical and linkage
(c) Optical and ionization
(d) Linkage and geometrical

70. Which of the following statements is correct ? (Atomic number of Ni = 28)

(a) $Ni(CO)_4$ is diamagnetic and $[NiCl_4]^{2-}$ and $[Ni(CN)_4]^{2-}$ are paramagnetic
(b) $Ni(CO)_4$ and $[Ni(CN)_4]^{2-}$ are diamagnetic and $[NiCl_4]^{2-}$ is paramagnetic
(c) $Ni(CO)_4$ and $[NiCl_4]^{2-}$ are diamagnetic and $[Ni(CN)_4]^{2-}$ is paramagnetic
(d) $[NiCl_4]^{2-}$ and $[Ni(CN)_4]^{2-}$ are diamagnetic and $Ni(CO)_4$ is paramagnetic

71. For $[Co_2(CO)_8]$, what is the total number of metal – carbon bonds and number of metal–metal bonds. **Critical Thinking**

(a) 10,1 (b) 8,2 (c) 8,1 (d) 10,0

72. $[Cr(H_2O)_6]Cl_3$ (atomic number of Cr = 24) has a magnetic moment of 3.83 BM. The correct distribution of 3d-elections in the chromium present in the complex is

(a) $3d_{xy}^1, 3d_{yz}^1, 3d_{zx}^1$
(b) $3d_{xy}^1, 3d_{yz}^1, 3d_{z^2}^1$
(c) $3d^1_{(x^2-y^2)}, 3d^1_{z^2}, 3d^1_{zx}$
(d) $3d_{xy}^1, 3d^1_{(x^2-y^2)}, 3d^1_{xz}$

73. $[NiCl_2\{P(C_2H_5)_2(C_6H_5)\}_2]$ exhibits temperature dependent magnetic behaviour (paramagnetic/diamagnetic). The coordination geometries of Ni^{2+} in the paramagnetic and diamagnetic states are respectively **Toughnut**

(a) tetrahedral and tetrahedral
(b) square planar and square planar
(c) tetrahedral and square planar
(d) square planar and tetrahedral

74. Complexes $[Co(NH_3)_5SO_4]Br$ and $[Co(NH_3)_5Br]SO_4$ can be distinguished by

(a) conductance measurement
(b) using $BaCl_2$
(c) using $AgNO_3$
(d) Both (b) and (c)

75. The correct name for the complex ion $[CoCl(ONO)(en)_2]^+$ is:

(a) chlorobis (ethylenediamine) nitrito-O-cobaltate (III) ion
(b) chlorodiethyldiaminenitrito-O-cobalt (III) ion
(c) chloronitrito-O-diethyldiamine cobaltate (III) ion
(d) chlorobis (ethylenediamine) nitrito-O-cobalt (III) ion

76. Which of the following ions are optically active?

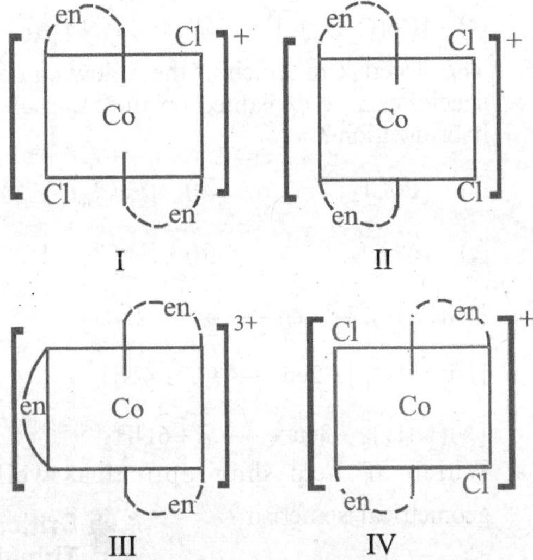

(a) I only (b) II only
(c) II and III (d) IV only

77. Which is **not** a π-acceptor ligand ?

(a) I_3^- (b) NO^+
(c) $(CH_3)_3P$ (d) CN^-

78. Select the complex in which secondary valency is satisfied before the primary valency.

(a) RMgX (b) $[Cr(NH_3)_6]Cl_3$
(c) $K_4[Fe(CN)_6]$ (d) $[Ni(CO)_4]$

79. Select the complex that can be reduced most easily.

(a) $Ni(CO)_4$ (b) $Cr(CO)_6$
(c) $V(CO)_6$ (d) $Fe(CO)_5$

80. The correct order of ligands in the trans-directing series is

(a) $CN^- > CH_3^- > NO_2^- > Br^-$

(b) $CN^- > Br^- > NO_2^- > CH_3^-$

(c) $Br^- > NO_2^- > CN^- > CH_3^-$

(d) $CH_3^- > CN^- > NO_2^- > Br^-$

81. Among the following complexes (K–P) $K_3[Fe(CN)_6]$ (K), $[Co(NH_3)_6]Cl_3$ (L), $Na_3[Co(oxalate)_3]$ (M), $[Ni(H_2O)_6]Cl_2$ (N), $K_2[Pt(CN)_4]$ (O) and $[Zn(H_2O)_6](NO_3)_2$ (P) the diamagnetic complexes are

(a) K, L, M, N (b) K, M, O, P

(c) L, M, O, P (d) L, M, N, O

82. $[Co_2(CO)_8]$ displays :

(a) one Co – Co bond, six terminal CO and two bridging CO

(b) one Co – Co bond, four terminal CO and four bridging CO

(c) no Co – Co bond, six terminal CO and two bridging CO

(d) no Co – Co bond, four terminal CO and four bridging CO

83. In isolated condition C–C bond length of C_2H_4 is x, than the bond length of a C–C bond of C_2H_4 in Zeise's salt is

(a) Greater than x (b) Less than x

(c) Equal to x (d) None of these

84. The number of σ and π-bonds in $Fe_2(CO)_9$, respectively, are

(a) 22σ and 15π (b) 22σ and 16π

(c) 23σ and 15π (d) 15σ and 8π

85. Which of the following type of isomerism is shown by given complex compound?

NH_3, H_3N, NO_2, Co, O_2N, NO_2, NH_3

(a) Facial (b) Meridional

(c) Cis (d) Both (b) and (c)

86. In solid $CuSO_4.5H_2O$ copper is coordinated to

(a) 4 water molecules

(b) 5 water molecules

(c) one sulphate molecule

(d) one water molecule

87. A compound contains 1.08 mole of Na, 0.539 mole of Cu and 2.16 mole of F. Its aqueous solution shows osmotic pressure which is three times that of urea having same molar concentration. The formula of the compound is :

(a) $Na_4[CuF_6]$ (b) $Na[CuF_4]$

(c) $Na_2[CuF_4]$ (d) $Na_2[CuF_3]$

88. Relative to the average energy in the spherical crystal field, the t_{2g} orbitals is tetrahedral field is

(a) Raised by (2/5) Δ_t

(b) Lowered by (2/5) Δ_t

(c) Raised by (3/5) Δ_t

(d) Lowered by (1/5) Δ_t

89. For which value of the x, and y, the following square planar compound shows geometrical isomers $[Pt(Cl)_x(Br)_y]^{2-}$

(a) 1, 3 (b) 3, 1 (c) 2, 2 (d) 1, 1

90. The terahedral complex [M(A)(B)(X)(Y)], where A,B,X and Y are different ligands and M is a metal ion is

(a) optically inactive

(b) rotate plane polarized light

(c) incomplete information

(d) can't be said

91. Incorrect match is :

(a) $K^+[Pt(C_2H_4)Cl_3]^-$: Zeise's salt

(b) $[Co(CO)_4]^-$: Bond order of CO—CO bond is greater than one

(c) $[V(NH_3)_6]^{3+}$ absorb visible light

(d) $[Cr(CN)_6]^{3-}$: Inner orbital low spin complex

92. Which complex of Co^{2+} will have the weakest crystal field splitting –

(a) $[CoCl_6]^{4-}$ (b) $[Co(CN)_6]^{4-}$

(c) $[Co(NH_3)_6]^{2+}$ (d) $[Co(en)_3]^{2+}$

93. The complex showing a spin-only magnetic moment of 2.82 B.M. is :

(a) $Ni(CO)_4$ (b) $[NiCl_4]^{2-}$

(c) $Ni(PPh_3)_4$ (d) $[Ni(CN)_4]^{2-}$

94. Red precipitate is obtained when ethanol solution of dimethylglyoxime is added to ammoniacal Ni(II). Which of the following statements is not true ? **Toughnut**

(a) Red complex has a square planar geometry.
(b) Complex has symmetrical H-bonding
(c) Red complex has a tetrahedral geometry.
(d) Dimethylglyoxime functions as bidentate ligand.

$$\text{dimethylglyoxime} = \left[\begin{array}{l} H_3C-C{=}N{-}OH \\ \quad\;\;\;| \\ H_3C-C{=}N{-}OH \end{array}\right]$$

95. Which of the following statements is correct for the complex $Ca_2[Fe(CN)_5O_2]$ having $t_{2g}{}^6, e_g{}^0$ electronic configuration?

(a) d^2sp^3 hybridised and diamagnetic
(b) sp^3d^2 hybridised and paramagnetic
(c) sp^3d^2 hybridised and diamagnetic
(d) d^2sp^3 hybridised and paramagnetic

96. $[Fe(en)_2(H_2O)_2]^{2+}$ + en → complex (X). The correct statement about the complex (X) is –

(a) it is a low spin complex
(b) it is diamagnetic
(c) it shows geometrical isomerism
(d) (a) and (b) both

97. Which of the following is not chelating agent?

(a) thiosulphato (b) oxalato
(c) glycinato (d) ethylene diamine

98. When HgI_2 is added to excess of aqueous KI, mercury largely exists as.

(a) Hg_2I_2 (b) $[HgI_3]^-$
(c) $[HgI_4]^{2-}$ (d) none of these

99. In nitroprusside ion, the iron and NO exist as Fe (II) and $\overset{+}{N}O$ rather than Fe(III) and NO. This can be established by **Toughnut**

(a) estimating the concentration of iron
(b) estimating the concentration of CN^-
(c) thermally decomposing the compound
(d) measuring the solid state magnetic moment

100. The number of geometrical isomers for octahedral $[Co(NH_3)_2Cl_4]^-$, square planar $AuCl_2Br_2^-$ and $[Co(NO_2)(NH_3)_5]^{2+}$ are

(a) 2, 2, 2 (b) 2, 2, no isomerism
(c) 3, 2, 2 (d) 2, 3, no isomerism

Answer KEYs

1	(c)	11	(c)	21	(b)	31	(a)	41	(a)	51	(d)	61	(c)	71	(a)	81	(c)	91	(d)
2	(b)	12	(c)	22	(c)	32	(d)	42	(d)	52	(a)	62	(c)	72	(a)	82	(a)	92	(a)
3	(d)	13	(a)	23	(b)	33	(a)	43	(c)	53	(b)	63	(a)	73	(c)	83	(a)	93	(b)
4	(d)	14	(d)	24	(a)	34	(b)	44	(d)	54	(d)	64	(b)	74	(d)	84	(a)	94	(c)
5	(d)	15	(b)	25	(d)	35	(a)	45	(d)	55	(b)	65	(d)	75	(d)	85	(b)	95	(d)
6	(a)	16	(b)	26	(a)	36	(b)	46	(d)	56	(a)	66	(a)	76	(c)	86	(a)	96	(d)
7	(c)	17	(c)	27	(b)	37	(d)	47	(a)	57	(b)	67	(a)	77	(b)	87	(c)	97	(a)
8	(d)	18	(b)	28	(a)	38	(b)	48	(d)	58	(b)	68	(d)	78	(d)	88	(a)	98	(c)
9	(c)	19	(b)	29	(b)	39	(c)	49	(a)	59	(b)	69	(b)	79	(c)	89	(b)	99	(d)
10	(c)	20	(c)	30	(b)	40	(a)	50	(d)	60	(c)	70	(b)	80	(a)	90	(b)	100	(b)

Hints & Solutions

1. **(c)** AgCl form soluble complex with NH_4OH which is $[Ag(NH_3)_2]^+$.
2. **(b)** $K_4[Fe(CN)_6]$ obeys the EAN rule.
3. **(d)** $[Pt(en)_2Cl_2]$ is a complex of the type $[M(AA)_2 B_2]$ which is octahedral. Such compounds exhibit optical and geometrical isomerism both.
4. **(d)** For $[Cu^{II}(NH_3)_4][Pt^{II}Cl_4]$ four isomers are possible which are $[Cu(NH_3)_4]$ $[PtCl_4]$, $[CuCl_4][Pt(NH_3)_4]$, $[PtCl_3(NH_3)][Cu(NH_3)_3Cl]$ and $[Pt(NH_3)_3Cl]$ $[Cu(NH_3)Cl_3]$
5. **(d)** The complex ion $[Cr(SCN)_2(NH_3)_4]^{2+}$ can exhibit geometrical and linkage isomerism
6. **(a)** K_4 $[Fe(CN)_6]$ in solution furnishes $4K^+$+ $[Fe(CN)_6]^{4-}$ five ions and exhibits maximum ionic conductivity.

7. **(c)** Fe^{2+}in$\left[Fe(H_2O)_6\right]^{2+}$

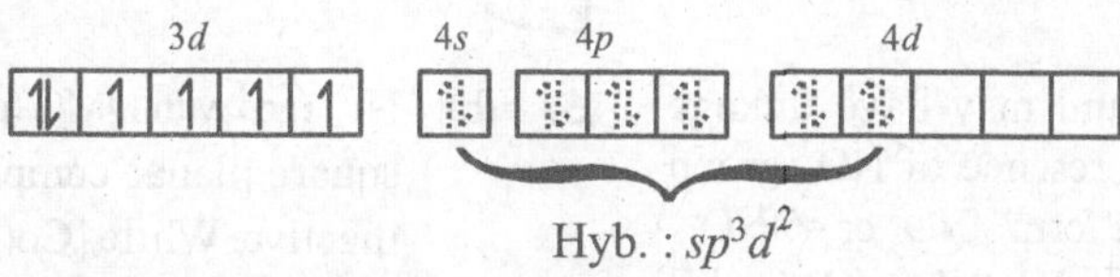

Colour : Pale green $\mu = 4.9$ B.M.; octahedral

Fe^{2+}in$\left[Fe(CN_6)\right]^{4-}$

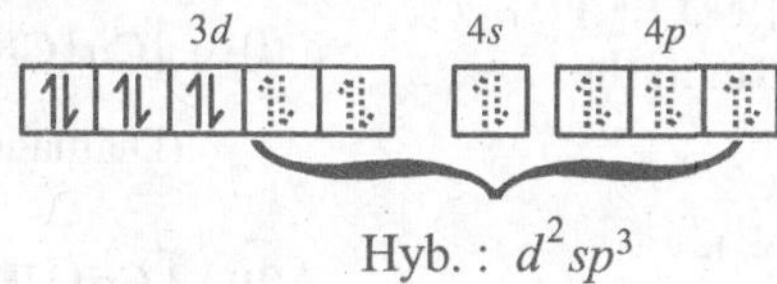

Colour : Yellow; $\mu = 0$; octahedral

8. **(d)** $Ni(CO)_4$; $\mu_{eff} = 0$

$\left[Co(NH_3)_4(NO_2)_2\right]^+$; $\mu_{eff} = 0$

$[Ag(CN)_2]$; $\mu_{eff} = 0$

$[CuBr_4]^{2-}$; $\mu_{eff} = 1.732$ BM

9. **(c)** Degenerate d-orbitals undergo splitting under ligand field created by strong, weak or mixed ligands.
10. **(c)** Number of unpaired electrons = 3

μ_{eff} = 3.9 BM

type of hybridisation = sp^3d^2

11. **(c)** Magnetic moment, $\sqrt{n(n+2)} = 5.92$ B.M. $\Rightarrow n$ (unpaired electrons) = 5

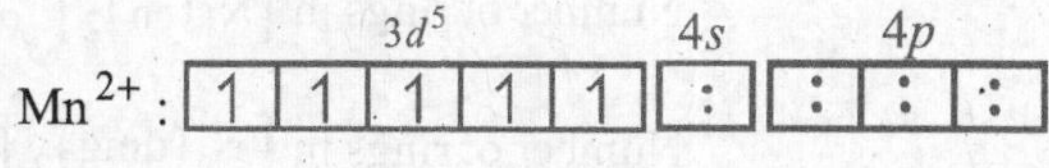

sp^3 hybridization (tetrahedral)

12. (c) Complex is not superimposable on its mirror image hence optically active i.e., rotate plane polarized light.

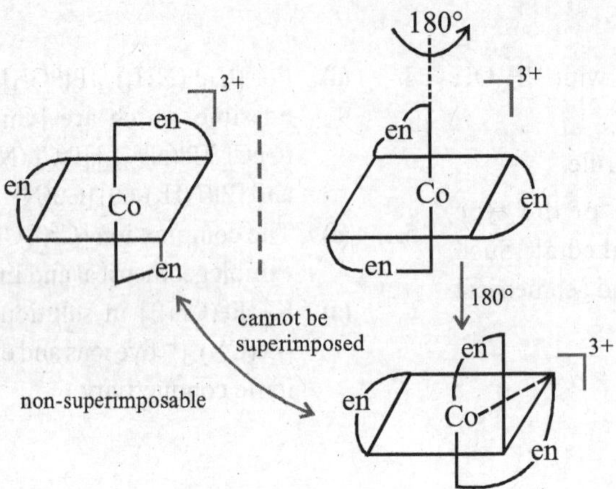

13. (a) The given compound may have linkage isomerism due to presence of NO_2 group which may be in the form $-NO_2$ or $-ONO$.
$[Co(NH_3)_4(NO_2)_2]Cl$ & $[Co(NH_3)_4(ONO)_2]Cl$
It may have ionisation isomerism due to presence of two ionisable groups $-NO_2$ & $-Cl$.
$[Co(NH_3)_4Cl(NO_2)]NO_2$ & $[Co(NH_3)(NO_2)_2]Cl$
It may have geometrical isomerism in the form of *cis trans* form as follows :

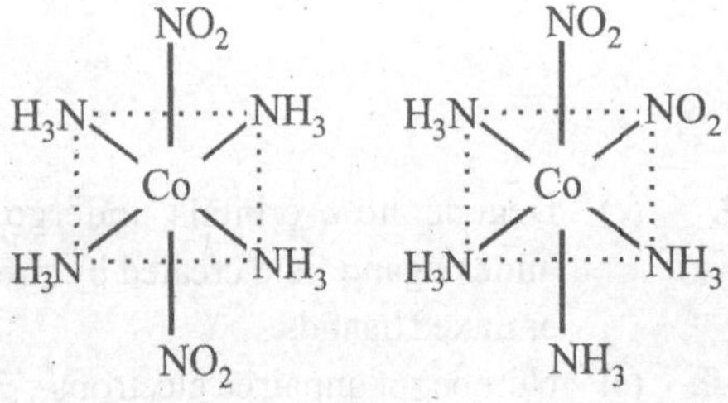

14. (d) $Cr^{3+}, C_2O_4^{2-}, (en)^0, O_2$

$C_2O_4^{2-}$ and $(en)^0$ are bidenate ligand.
CN of $Cr^{3+} = 6$,

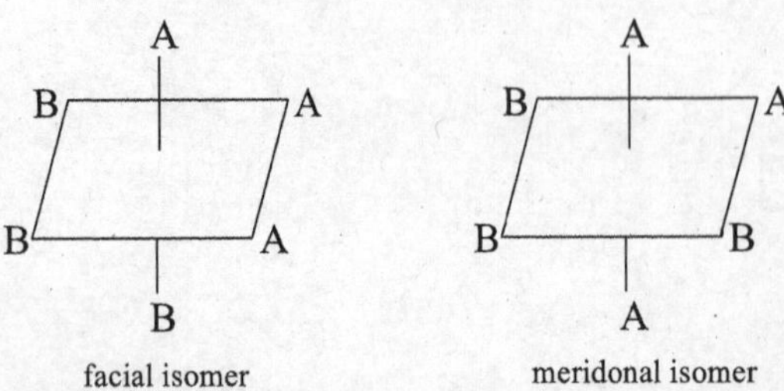

15. (b) Ni^{2+} (aq) with DMG (dimethylglyoxime) form square planar complex which is obtically inactive. While $[Co(EDTA)]^-$, $[Zn(gly)_2]^{2+}$ and $[Pt(en)_3]^{4+}$ are optically active as they do not have P.O.S.

16. (b)

(I) $[Cu(CN)_4]^{3-}$ $\rightarrow \mu = 0$
(Diamagnetic)

(II) $\left[Co(NH_3)_6\right]^{3+}$ $\rightarrow \mu = 0$
(Diamagnetic)

(III) $\left[Ni(NH_3)_6\right]^{2+}$ $\rightarrow \mu = \sqrt{8}$ BM
(Paramagnetic)

(IV) $[Fe(CN)_6]^{3-}$ $\rightarrow \mu = \sqrt{3}$ BM
(Paramagnetic)

17. (c) As number of stable rings increases stability of complex compound also increases due to chelation effect.

Number of rings in $\left[Ni(en)_2\right]^{2+} = 2$

Number of rings in $\left[Ni(dmg)_2\right] = 4$

18. **(b)** Same magnetic moment = Same number of unpaired electrons.

$$\mu = \sqrt{n(n+2)}$$

where n = number of unpaired electrons

$Co^{2+} = 3d^7$, 3 unpaired electrons;

$Cr^{2+} = 3d^4$, 4 unpaired electrons;

$Mn^{2+} = 3d^5$, 5 unpaired electrons;

$Fe^{2+} = 3d^6$, 4 unpaired electrons.

19. **(b)** $[Cu(NH_3)_5]^{2+} \rightarrow$ geometry with C. N. = 5 is sq. pyramidal, dsp^3 hybridisation and $d_{x^2-y^2}$ orbital is used in hybridisation.

(a) $Fe(CO)_5 \rightarrow d_{z^2}sp^3$

(b) $\left[Co(NH_3)_6\right]^{2+} \rightarrow d^2sp^3$

(d) $[IrF_6]^{3-} \rightarrow d^2sp^3$

20. **(c)**

(a) Geometry of complexes $\rightarrow$ Both square planar

(b) Hybridisation of central metal cation $\rightarrow dsp^2$

(c) Magnetic behaviour $[Cu(en)_2)]^2 \rightarrow$ Paramagnetic; $[Ni(dmg)_2] \rightarrow$ Diamagnetic

(d) Number of stereoisomers = 0

21. **(b)**

(a)	$[Ni(PF_3)_4]$	Ni : sp^3 Hybridization P : sp^3 Hybridization
(b)	$[Fe(dmg)_2]$	Fe : dsp^2 Hybridization N-atom of dmg : sp^3 Hybridization
(c)	$[Zn(en)_2]^{2+}$	Zn : sp^3 Hybridization N-atom of en : sp^3 Hybridization
(d)	$[Ni(PMe_3)_4]^{2+}$	Ni : sp^3 Hybridization P : sp^3 Hybridization

22. **(c)** Fe^{3+} in $\left[Fe(CN_6)\right]^{3-}$

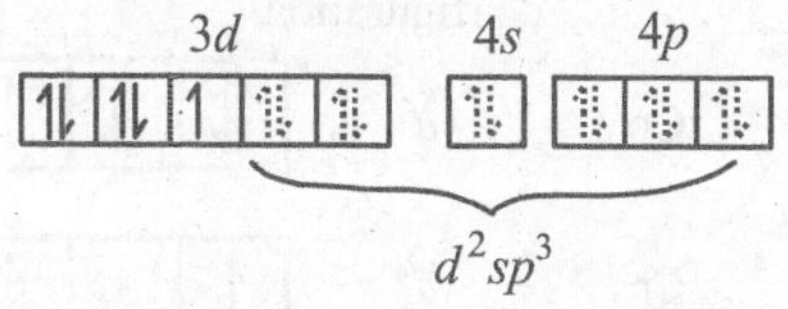

$\Rightarrow$ low spin complex $\mu = 1.732$ BM

Ni in $[Ni(CO)_4]$

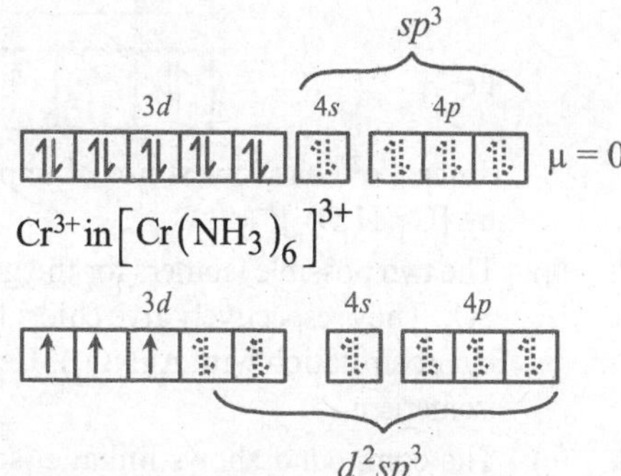

Cr^{3+} in $\left[Cr(NH_3)_6\right]^{3+}$

$\Rightarrow$ low and high spin complex is applicable for d^4 to d^7 configuration

Mn^{2+} in $[Mn(CN)_6]^{4-}$

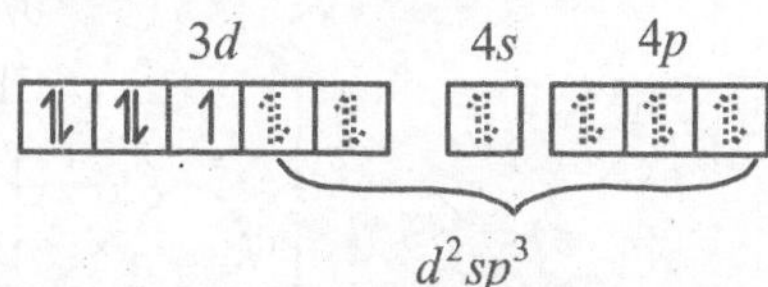

$\Rightarrow$ low spin complex $\mu = 1.732$ BM

23. **(b)** $\left[Ti(H_2O)_6\right]Cl_4$

Coordination number 6 $\Rightarrow$ octahedral complex

Ti is in +4 oxidations state $\Rightarrow$ no unpaired electrons

$\Rightarrow$ magnetic moment = 0 B.M.

24. **(a)** In d^3 and d^8 octahedral complexes number of unpaired e^- s at central metal atom/ion never changes, therefore for such octahedral complexes terms high spin and low spin not used.

25. **(d)** $[Cu(NH_3)_4]^{2+}$ has square planar structure and is paramagnetic.

26. **(a)**

	Electronic configuration		No. of unpaired electrons
Co^{2+}	d^7	⇅ ⇅ ↑ ↑ ↑	3
Cr^{2+}	d^4	↑ ↑ ↑ ↑	4
Mn^{2+}	d^5	↑ ↑ ↑ ↑ ↑	5
Fe^{2+}	d^6	⇅ ↑ ↑ ↑ ↑	4

∴ Since Co^{3+} has lowest no. of unpaired electrons hence lowest paramagnetic behaviour is shown by $[Co(H_2O)_6]^{2+}$

27. **(b)** The two possible isomers for the given octahedral complex are $[M(NH_3)_5 SO_4]$ Cl and $[M(NH_3)_5 Cl]$ SO_4. They respectively give chloride ion (indicated by precipitation with $BaCl_2$) and SO_4 ion (indicated by precipitation with $AgNO_3$). Hence the type of isomerism exhibited by the complex is ionization isomerism.

28. **(a)** The compound shows linkage isomerism because the ligand in the compound is an ambidenate ligand that can bond at more than one atomic site.
i.e., $:NCS^-$ and $:SCN^-$

29. **(b)** Non superimposable mirror images are called optical isomers and may be described as "chiral'. They are also called enantiomers and rotate plane polarised light in opposite directions.

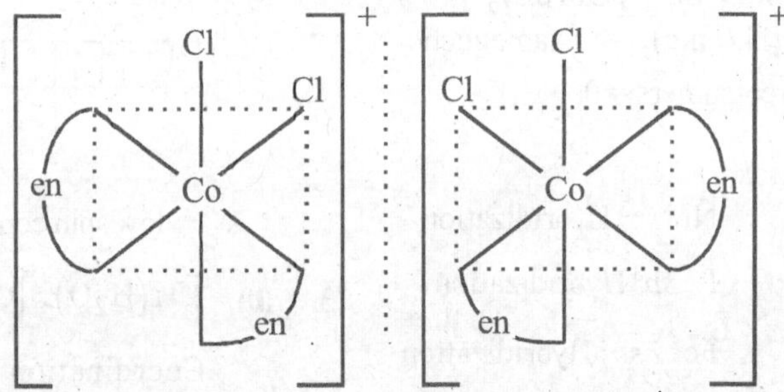

30. **(b)** Complexes of the type M_{ABCD} may exist in three isomeric form. Similarily $[Pt (NO_2)(Py)(NH_3)(NH_2OH)^+]$ may exist in three isomeric form.

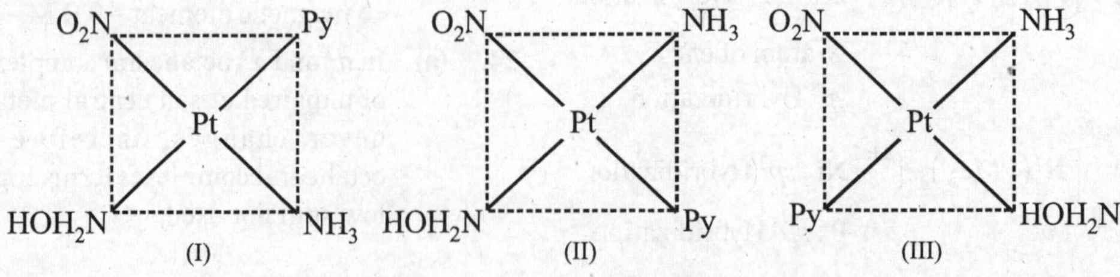

31. (a) Symmetrically filled t_{2g} and e_g are those, which contain equal distribution of electrons.

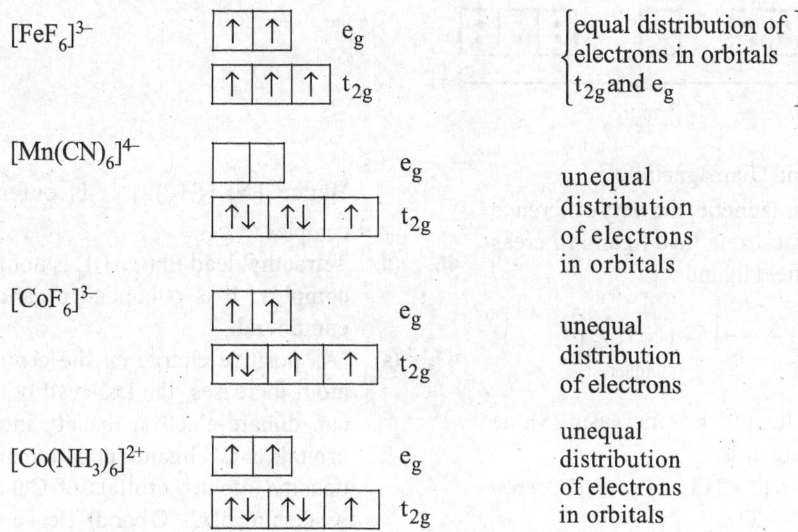

32. (d) The number of ammonia molecules is 6. Each ammonia molecule contains 3 covalent bonds between N and H. Therefore the number of covalent bonds is 18.

33. (a) $Ni(CO)_4$. The O. S. of Ni is zero. Electronic configuration is [Ar] $3d^8\,4s^2\,4p^0$. In presence of strong ligand CO the paring of electrons take place and electronic configuration will be [Ar] $3d^{10}\,4s^0\,4p^0$. Hence unpaired electrons is zero .

34. (b) Shape of $[Cu(NH_3)_4]^{2+}$ is square planar.

35. (a) $[Co(NH_3)_5CO_3]ClO_4$. Six monodentate ligands are attached to Co hence C. N. of Co = 6;
O. N. $= x + 5 \times (0) + 1 \times (-2) + 1 \times (-1) = 0$ ∴ $x = +3$; electronic configuration of Co^{3+}[Ar] $3d^6 4s^0$ hence number of d electrons is 6. All d electrons are paired due to strong ligand hence unpaired electron is zero.

36. (b) $[Co(C_2O_4)_3]^{3-}$ is diamagnetic as oxalate is a strong ligand causing pairing of 3*d* electrons in Co^{3+} thereby leading to d^2sp^3 hybridisation.

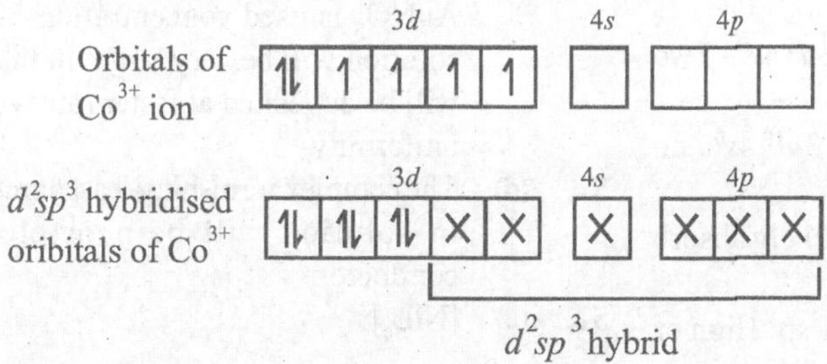

37. (d) CN^- is a strong field ligand and form low spin complexes thus $\Delta_o > P$.

38. (b) For the reaction of the type $M + 4L \rightleftharpoons ML_4$, larger the stability constant, the higher the proportion of ML_4 that exists in solution.

39. (c) $[Co(NH_3)_6]^{3+}$

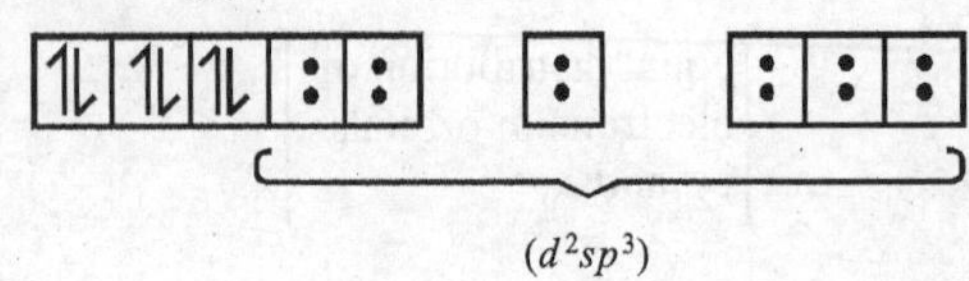

Octahedral and Diamagnetic

40. (a) Both are paramagnetic, the only difference is that CN^- is a strong field ligand whereas F^- is a weak field ligand.

41. (a) $\underset{\text{Lewis acid}}{Co^{3+}} + \underset{\text{Lewis base}}{6NH_3} \longrightarrow \underset{\text{adduct}}{\left[Co(NH_3)_6\right]^{3+}}$

42. (d) Following is the order of increasing value of Δ_o for the ligands.
$I^- < Br^- < Cl^- < F^- < H_2O < Py = NH_3 < en < NO_2^- << CN^- < CO$

43. (c) The absorption of visible light and hence coloured nature of the transition metal cation is due to the promotion of one or more unpaired - *d* - electron from a lower to higher level withing same *d*-subshell. Hence higher will be the number of unpaired electron higher will be the absorpion in visible light.
The electronic configuration of the given elements is
Sc^{3+} (18) = $1s^2\ 2s^2\ 2p^6\ 3s^2\ 3p^6\ 3d^0\ 4s^0$ - no unpaired e^-.
Ti^{4+} (18) = $1s^2\ 2s^2\ 2p^6\ 3s^2\ 3p^6\ 3d^0\ 4s^0$ - no unpaired e^-.
V^{3+} (20) = $1s^2\ 2s^2\ 2p^6\ 3s^2\ 3p^6\ 3d^2\ 4s^0$ - Two unpaired e^-.
Zn^{2+} (28) = $1s^2\ 2s^2\ 2p^6\ 3s^2\ 3p^6\ 3d^{10}\ 4s^0$ - no unpaired e^-.
hence $[V(NH_3)_6]^{3+}$ will most likely absorb visible light.

44. (d) $Cr(CN)_6]^{3-}$: Hybridization : d^2sp^3 High or low spin is not defined because Cr^{3+} is an example of d^3 system of configuration.

45. (d) Hybridisation

$\underset{d^2sp^3}{[Fe(CN)_6]^{4-}}, \underset{d^2sp^3}{[Mn(CN)_6]^{4-}},$

$\underset{d^2sp^3}{[Co(NH_3]^{3+}}, \underset{sp^3d^2}{[Ni(NH_3)_6]^{2+}}$

Hence $[Ni(NH_3)_6]^{2+}$ is outer orbital complex.

46. (d) Tetraethyl lead $Pb(C_2H_5)_4$ is not π bonded complex. It is σ bonded organometallic compound.

47. (a) As positive charge on the central metal atom increases, the less readily the metal can donate electron density into the π^* orbitals of CO ligand (donation of electron density into π^* orbitals of CO result in weakening of C – O bond). Hence, the C – O bond would be strongest in $[Mn(CO)_6]^+$.

48. (d) Due to some backbonding by sidewise overlapping between d-orbitals of metal and p-orbital of carbon, the F–C bond has σ and π character.

49. (a) Triethoxyaluminium has no Al – C linkage

$$Al \begin{cases} O-CH_2CH_3 \\ O-CH_2CH_3 \\ O-CH_2CH_3 \end{cases}$$

50. (d) In silver plating $K[Ag(CN)_2]$ is used which provides constant and required supply of Ag^+ ions as $Ag(CN)_2^-$ is very stable. But if $AgNO_3$ is used concentration of $[Ag^+]$ in solution will be very large. In that case Ag will be deposited at faster rate without any uniformity.

51. (d) The complex furnishing least number of ions in solution will be poor electrolytic conductor.

52. (a) $[NiL_4]^{2-}$

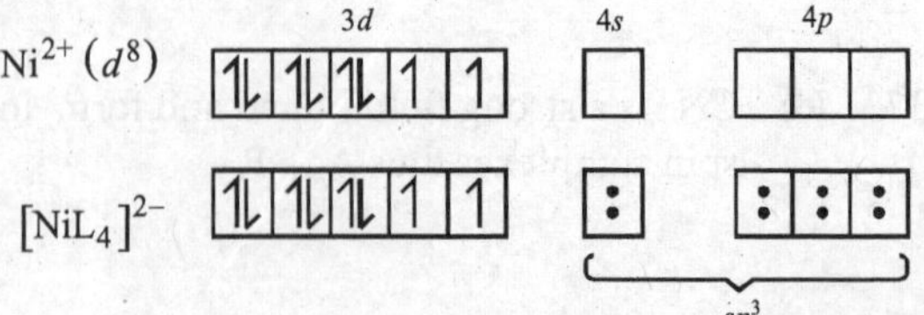

i.e, no. of unpaired electron = 2
hybridization – sp^3.

53. (b) $[Fe(CN)_6]^{3-}$ has magnetic moment of a single unpaired electron whereas $[FeF_6]^{3-}$ has a magnetic moment of five unpaired electrons. $[CoF_6]^{3-}$ is paramagnetic with four unpaired electrons while $[Co(C_2O_4)_3]^3$ is diamagnetic. This anomaly is explained by valence bond theory in terms of formation of inner and outer orbital coordination entities. $[Co(C_2O_4)_3]^{3-}$ is an inner orbital complexes having d^2sp^3 hybridization.

54. (d) In $[Cr(NO)(NH_3)(CN)_4]^{2-}$, $Cr^{2+}(d^4)$ is given as :

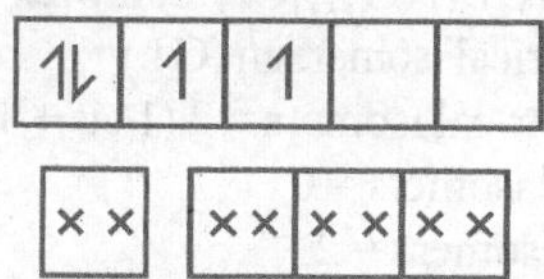

i.e., 2 unpaired electrons

$\mu = \sqrt{2(2+2)} = \sqrt{8} = 2.82$ BM

55. (b) Crystal field splitting depends upon the nature of ligand. The nature of ligand Δ decreases as shown below

$C_2O_4^{2-} < H_2O < NH_3 < CN^-$

hence the crystal field splitting will be maximum for $[Co(CN)_6]^{3-}$

56. (a) $[PdCl_4]^{2-}$ is dsp^2 hybridized and square planar in shape.

57. (b) $X = \left[Ni(en)(NH_3)_4\right]^{2+}$ neither G.I. nor OI

$Y = \left[Ni(en)_2(NH_3)_2\right]^{2+}$ G.I as well as OI

$Z = \left[Ni(en)_3\right]^{2+}$ does not show G.I. while show OI

58. (b) Mole of AgCl $= \dfrac{1.435}{143.5} = 0.01$; 50 mL of 0.2 M complex ≡ 0.01 mole

$$[Co(NH_3)_4Cl_2]Cl \rightarrow [CoNH_3)_4Cl_2]^+ + Cl^-$$

59. (b) $Na_2[CdCl_4]$ does not contain any unpaired electron hence colourless

60. (c)

	$Ni(CO)_4$	$Ni(PPh_3)_2Cl_2$
O.S.	Ni^0	Ni^{2+}
E.C.	$[Ar]3d^84s^2$	$[Ar]3d^84s^0$
Pairing	Pairing of e^-	No pairing of e^-
Hybridization	sp^3 (tetrahedral)	sp^3 (tetrahedral)

61. (c) The ions present in the ionisation sphere are precipited Hence $[CrCl_2(H_2O)_4]Cl.2H_2O$ contains 1/3 Cl in ionisation sphere to be precipited by $AgNO_3$ as AgCl

62. (c) $[Fe(C_2O_4)_3]^{3-}$. The iron is present in the highest oxidation state Fe^{3+} and $C_2O_4^{2-}$ is a chelating ligand. Chelates are always form more stable complexes

63. (a)

		Electronic configuration	No. of unpaired electrons
Co^{2+}	d^7	⇅ ⇅ ↑ ↑ ↑	3
Cr^{2+}	d^4	↑ ↑ ↑ ↑	4
Mn^{2+}	d^5	↑ ↑ ↑ ↑ ↑	5
Fe^{2+}	d^6	⇅ ↑ ↑ ↑ ↑	4

∴ Since Co^{2+} has lowest no. of unpaired electrons hence lowest paramagnetic behaviour is shown by $[Co(H_2O)_6]^{2+}$

64. (b) Correct order is : $CN^- > en > NCS^- > Cl^-$

65. (d)

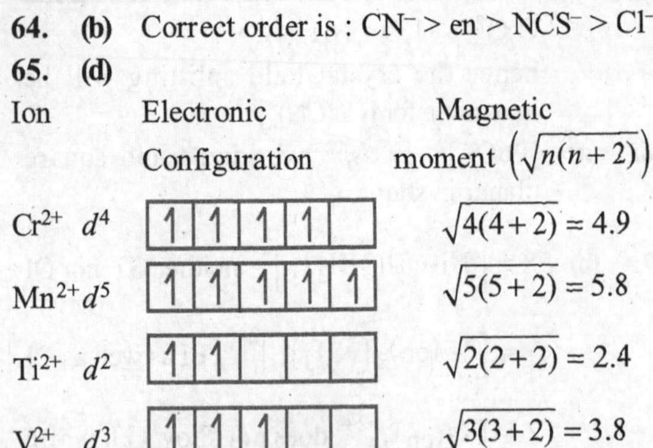

Ion	Electronic Configuration	Magnetic moment $\left(\sqrt{n(n+2)}\right)$
Cr^{2+} d^4	1 1 1 1	$\sqrt{4(4+2)} = 4.9$
$Mn^{2+} d^5$	1 1 1 1 1	$\sqrt{5(5+2)} = 5.8$
Ti^{2+} d^2	1 1	$\sqrt{2(2+2)} = 2.4$
V^{2+} d^3	1 1 1	$\sqrt{3(3+2)} = 3.8$

Hence Cr^{2+} has magnetic moment of the order of 5 B.M.

66. (a) $[PtCl_2(NH_3)_2]$ has square planar geometry.

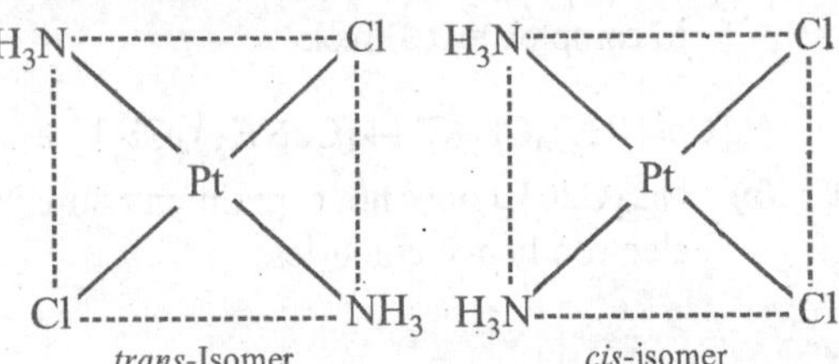

trans-Isomer *cis*-isomer

67. (a) The donor atoms, molecules or anions which donate a pair of electrons to the metal atom or ion and form a coordinate bond with it are called ligands. In methane there is no electrons for donation to central metal atom/ion it is stable with complete octet configuration.

68. (d) More equivalents of $AgNO_3$ aqueous solution will be consumed if complex will furnish more Cl^- ions in solution. Hence complex $[Cr(H_2O)_6]Cl_3$ will consume more equivalents of aqueous solution of $AgNO_3$.

69. (b) $[Co(NH_3)_3(NO_2)_3]Cl_2$ complex exhibits geometrical isomerism (G.I.)
Geometrical isomers = 2 (1 cis + 1 trans)
Optical isomers = 0
Space isomers = 2
$[Co(NH_3)_5(NO_2)]Cl_2$ complex shows linkage and ionization isomerism.

70. (b)

Atom/Ion Complex	Configuration (3*d* | 4*s* | 4*p*)	No. of unpaired electrons	Magnetic nature
Ni^{2+} (d^8)	↑↓ ↑↓ ↑↓ ↑ ↑ | □ | □ □ □	2	Paramagnetic
$[NiCl_4]^{2-}$	↑↓ ↑↓ ↑↓ ↑ ↑ | : | : : : (sp^3)	2	Paramagnetic
$[Ni(CN)_4]^{2-}$	↑↓ ↑↓ ↑↓ ↑↓ : | : | : : □ (Rearrangement, dsp^2)	0	Diamagnetic
Ni (d^8s^2)	↑↓ ↑↓ ↑↓ ↑ ↑ | ↑↓ | □ □ □	2	Paramagnetic
$[Ni(CO)_4]$	↑↓ ↑↓ ↑↓ ↑↓ ↑↓ | : | : : : (Rearrangement, sp^3)	0	Diamagnetic

71. (a) Structure of $[Co_2(CO)_8]$

Total M – C bonds = 10,
Total M – M bonds = 1

72. (a) Magnetic moment indicates that there are three unpaired electrons present in chromium. These must be present in lower energy orbitals which are $3d_{xy}$, $3d_{yz}$ and $3d_{xz}$.

73. (c) In both states (paramagnetic and diamagnetic) of the given complex, Ni exists as Ni^{2+} whose electronic configuration is [Ar] $3d^8 4s^0$.

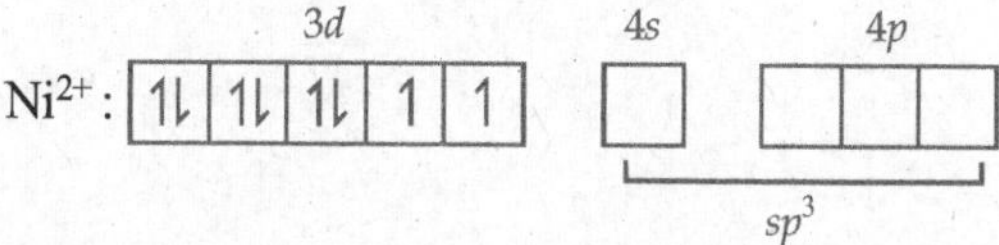

In the above paramagnetic state, geometry of the complex is sp^3 giving tetrahedral geometry. The diamagnetic state is achieved by pairing of electrons in $3d$ orbital.

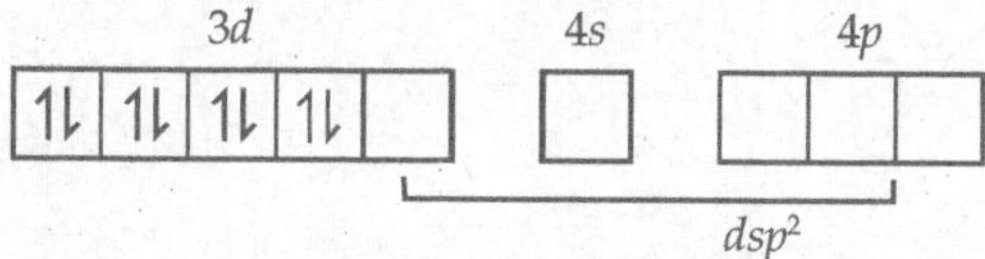

Thus the geometry of the complex will be dsp^2 giving square planar geometry.

74. (d) $[Co(NH_3)_5Br]SO_4$ gives white precipitate of $BaSO_4$ with $BaCl_2(aq)$ whereas $[Co(NH_3)_5SO_4]Br$ gives yellow precipitate (AgBr) with $AgNO_3$(aq).

75. (d) The correct name is chlorobis (ethylenediamine) nitritocobalt (III).

76. (c) Ions I and IV are the same (*trans*), with mirror plane through en groups.

77. (b) NO^+ is not a π-acceptor ligand, because in it nitrogen atom has no vacant orbital to accomodate electron pair from the central metal atom.

78. (d) In metal carbonyls the oxidation state of metal is zero. In this case oxidation state of Ni is zero.

79. (c) In it the E.A.N. of V is 35(23 + 12 = 35) so it can be reduced easily.
In all other cases the E.A.N. is 36.

80. (a) The **trans-effect** may be defined as the labelization of ligands trans- to the other trans-directing ligand.
[Note : The order of ligands in the trans-directing series is as follows :
$CN^- \sim CO \sim NO \sim H^- > CH_3^- \sim SC(NH_2)_2 \sim PR_3 > SO_3H > NO_2^- \sim I^- \sim SCN^- > Br^- > Cl^- > Py > RNH_2 \sim NH_3 > OH^- > H_2O$]

81. (c)

Complex	No. of electrons in outer *d* orbital	No. of unpaired electron (s)
$[Fe(CN)_6]^{3-}$	$3d^5$	1 (CN^- causes pairing of electrons)
$[Co(NH_3)_6]^{3+}$	$3d^6$	0
$[Co(OX)_3]^{3-}$	$3d^6$	0
$[Ni(H_2O)_6]^{2+}$	$3d^8$	2
$[Pt(CN)_4]^{2-}$	$5d^8$	0 (CN^- causes pairing of electrons)
$[Zn(H_2O)_6]^{2+}$	$3d^{10}$	0

Thus L, M, O and P are diamagnetic.

82. (a)

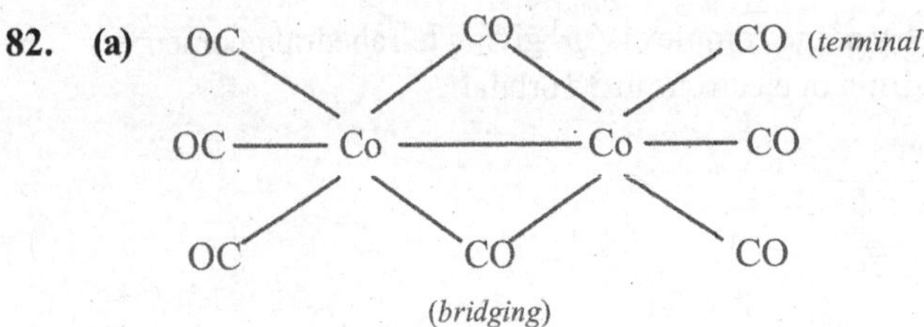

83. (a)

$\overset{+}{K}[PtCl_3(C_2H_4)]^-$

With increase in synergic bonding, C–C bond length increases in Zeise's salt.

84. (a)

$Fe_2(CO)_9$

σ = 22 and π = 15 bonds

85. **(b)** Given compound shows meridional isomerism.

86. **(a)** Four water molecules are coordinated to copper.

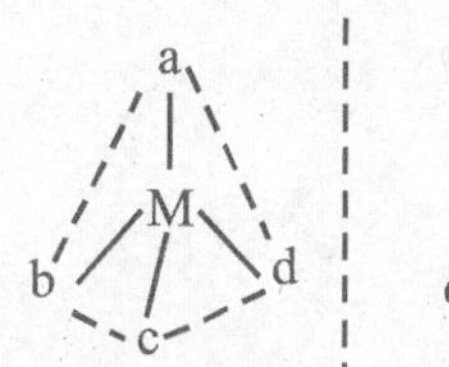

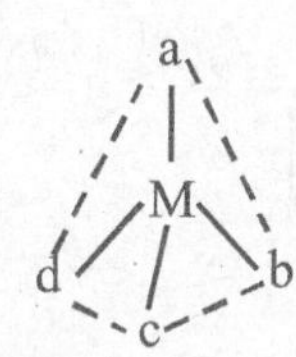

87. **(c)** Mole ratio of Na, Cu and

$$F = \frac{1.08}{0.539} : \frac{0.539}{0.539} : \frac{2.16}{0.539} = 2:1:4$$

Empirical formula $= Na_2CuF_4$: Van't Hoff factor $(i) = 3$ (given)

Hence, formula of the compound :

$$Na_2[CuF_4] \rightarrow 2Na^+ + [CuF_4]^{2-}$$

88. **(a)** t_{2g} orbitals have higher energy by $(2/5)\Delta_t$.

89. **(b)** Geometrical isomers of following type of square planar complexes is possible. Ma_2b_2 type, Ma_2bc type and Mabcd type.

90. **(b)** Non –superimposable mirror images are optically active, hence rotate plane polarized light.

91. **(d)** $Cr(CN)_6]^{3-}$: Hybridization : d^2sp^3 High or low spin is not defined because Cr^{3+} is an example of d^3 system of configuration.

92. **(a)** Cl^- is a weak field ligand.

93. **(b)** $[NiCl_4]^{2-}$, O.S. of Ni = +2

$Ni(28) = 3d^8\,4s^2$

Cl^- being weak ligand it cannot pair up the two electrons present in 3*d* orbital

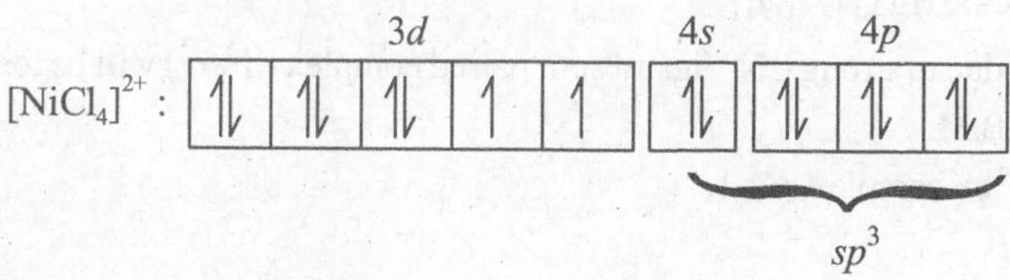

No. of unpaired electrons = 2

Magnetic moment, $\mu = 2.82$ BM.

94. (c) Nickel ions are frequently detected by the formation of red precipitate of the complex of nickel dimethylglyoxime, when heated with dimethylglyoxime.

$$\begin{array}{l} CH_3-C=NOH \\ \quad\;\; | \\ CH_3-C=NOH \end{array} + Ni^{2+} \longrightarrow \begin{array}{c} OH \qquad\qquad O \\ | \qquad\qquad\quad \uparrow \\ CH_3-C=N\searrow \quad\;\; \swarrow N=C-CH_3 \\ |\qquad\qquad Ni \qquad\qquad | \\ CH_3-C=N\nearrow \quad\;\; \nwarrow N=C-CH_3 \\ \downarrow \qquad\qquad\quad | \\ O \qquad\qquad OH \end{array}$$

Dimethylglyoxime — Nickel dimethylglyoxime

95. (d)

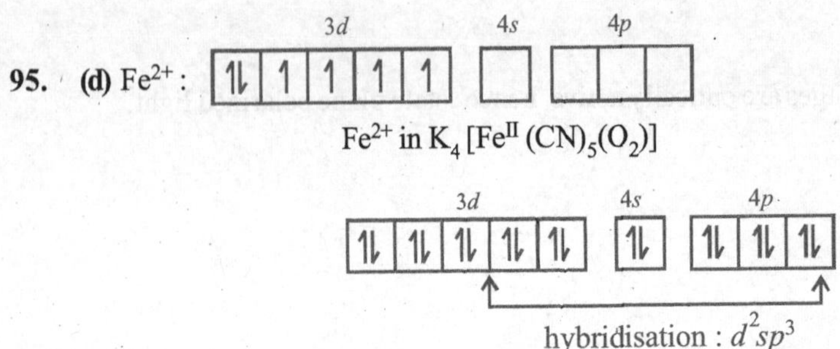

Complex is paramagnetic due to presence of unpaired electron at O_2^- i.e., superoxide acting as ligand.

96. (d) Complex X is $[Fe(en)_3]^{2+}$; as 'en' is a strong field ligand pairing of electrons will take place.

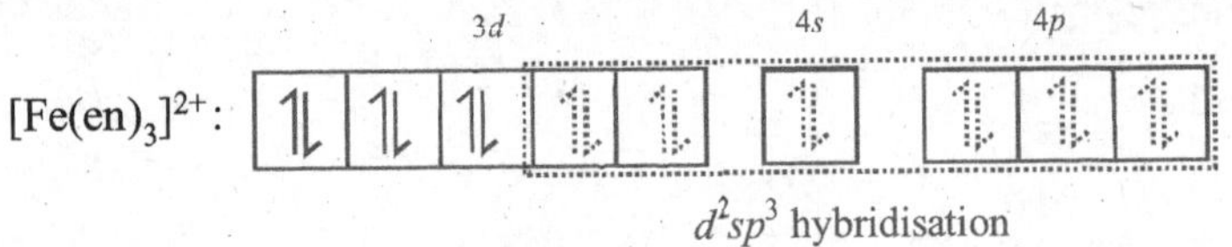

Hence, hybridisation is d^2sp^3 and complex is diamagnetic. As it has 3 bidentate symmetrical 'en' ligands so it will not show geometrical isomerism.

97. (a) $S_2O_3^{2-}$ is monodentate ligand where as other ligands are bidentate.

98. (c) $HgI_2(s) + 2I^-(aq) \rightarrow [HgI_4]^{2-}(aq)$

99. (d) Fe(II) state – $3d^6$; due to strong CN^- ligand spin paired complex (d^2sp^3) will be formed.

Hence $n = 0, \mu = 0$ B.M

Fe(III) state $3d^5$; d^2sp^3 complex, $n = 1$,

$\mu = 1.73$ B. M.

100. (b) The number of geometrical isomers

$[Co(NH_3)_2(Cl)_4]^- - 2$

$AuCl_2Br_2 - 2$

$[Co(NO_2)(NH_3)_5]^{2+}$ – No isomerism

24 Haloalkanes and Haloarenes

1. Arrange the following compounds in order of increasing dipole moment : Toughnut

(I) Toluene

(II) *m*-dichlorobenzene

(III) *o*-dichlorobenzene

(IV) *p*-dichlorobenzene

(a) I < IV < II < III (b) IV < I < II < III

(c) IV < I < III < II (d) IV < II < I < III

2. Pure chloroform is prepared by

(a) distilling chloral hydrate with aqueous sodium hydroxide.

(b) heating ethanol with bleaching powder.

(c) heating acetone with bleaching powder.

(d) reducing carbon tetrachloride.

3. The reaction conditions leading to the best yields of C_2H_5Cl are :

(a) $C_2H_6\text{ (excess)} + Cl_2 \xrightarrow{\text{UV light}}$

(b) $C_2H_6 + Cl_2 \xrightarrow[\text{room temperature}]{\text{dark}}$

(c) $C_2H_6 + Cl_2\text{ (excess)} \xrightarrow{\text{UV light}}$

(d) $C_2H_6 + Cl_2 \xrightarrow{\text{UV light}}$

4. A compound A with molecular formula $C_{10}H_{13}Cl$ gives a white precipitate on adding silver nitrate solution. A on reacting with alcoholic KOH gives compound B as the main product. B on ozonolysis gives C and D. C gives Cannizaro reaction but not aldol condensation. D gives aldol condensation but not Cannizaro reaction. A is :

(a) $C_6H_5-CH_2-CH_2-CH_2-CH_2-Cl$

(b) $C_6H_5-CH_2-CH_2-\underset{\substack{|\\ Cl}}{CH}-CH_3$

(c) $C_6H_5-CH_2-\underset{\substack{|\\ Cl}}{C}(CH_3)_2$

(d) Benzene ring bearing $CH_2-CH_2-CH_3$ and CH_2-Cl on adjacent positions

5. The Wurtz-Fittig reaction involves condensation of

(a) two molecules of aryl halides.

(b) one molecule of each of aryl-halide and alkyl-halide.

(c) one molecule of each of aryl-halide and phenol.

(d) two molecules of aralkyl-halides.

6. For the compounds

CH_3Cl, CH_3Br, CH_3I and CH_3F,

the correct order of increasing C–X bond length is:

(a) $CH_3F < CH_3Cl < CH_3Br < CH_3I$

(b) $CH_3F < CH_3Br < CH_3Cl < CH_3I$

(c) $CH_3F < CH_3I < CH_3Br < CH_3Cl$

(d) $CH_3Cl < CH_3Br < CH_3F < CH_3I$

7. The order of reactivity of the given haloalkanes towards nucleophile is :

(a) RI > RBr > RCl (b) RCl > RBr > RI

(c) RBr > RCl > RI (d) RBr > RI > RCl

8. $AgNO_3$ does not give precipitate with chloroform because:

(a) $CHCl_3$ does not ionise in water.

(b) $CHCl_3$ is insoluble in water.
(c) $AgNO_3$ is insoluble in $CHCl_3$.
(d) $CHCl_3$ is an organic compound.

9. Consider the following anions.

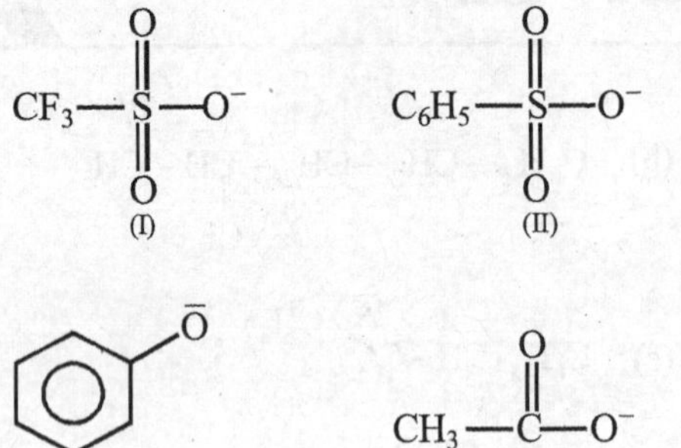

When attached to sp^3-hydridized carbon, their leaving group ability in nucleophilic substitution reaction decreases in the order:

(a) I > II > III > IV (b) I > II > IV > III
(c) IV > I > II > III (d) IV > III > II > I

10. $C_2H_5Br \xrightarrow{AgCN} X \xrightarrow[Zn-Hg/HCl]{Reduction} Y$, Here Y is

(a) Ethyl methyl amine (b) n-propylamine
(c) Isopropylamine (d) Ethylamine

11. The chief reaction product of reaction between *n*-butane and bromine at 573K is :

(a) $CH_3CH_2CH_2CH_2Br$
(b) $CH_3CH_2CHBr(CH_3)$
(c) $CH_3-CH_2CHBr(CH_2Br)$
(d) $CH_3CH_2CBr_2(CH_3)$

12. Aryl fluoride may be prepared from arene diazonium chloride using :

(a) HBF_4/Δ
(b) $HBF_4/NaNO_2, Cu, \Delta$
(c) CuF/HF
(d) Cu/HF

13. When $CH_3CH_2CHCl_2$ is treated with $NaNH_2$, the product formed is Toughnut

(a) $CH_3-CH=CH_2$
(b) $CH_3-C\equiv CH$
(c) $CH_3CH_2CH(NH_2)_2$
(d) $CH_3CH_2CH(Cl)(NH_2)$

14. $CH_3Br + Nu^- \longrightarrow CH_3-Nu + Br^-$

The decreasing order of the rate of the above reaction with nucleophiles (Nu^-) A to D is

[Nu^- = (A) PhO^-, (B) AcO^-, (C) HO^-, (D) CH_3O^-]

(a) A > B > C > D (b) B > D > C > A
(c) D > C > A > B (d) D > C > B > A

15. In a nucleophilic substitution reaction:

$R-Br + Cl^- \xrightarrow{DMF} R-Cl + Br^-$,

Which one of the following undergoes complete inversion of configuration?

(a) $C_6H_5CHC_6H_5Br$ (b) $C_6H_5CH_2Br$
(c) $C_6H_5CHCH_3Br$ (d) $C_6H_5CCH_3C_6H_5Br$

16. Which of the following statements is wrong?

(a) Ethyl chloride on reduction with Zn-Cu couple and alcohol gives ethane.
(b) The reaction of methyl magnesium bromide with acetone gives butanol-2.
(c) Alkyl halides follow the following reactivity sequence on reaction with alkenes.
$R-I > R-Br > R-Cl > R-I$
(d) $C_2H_4Cl_2$ may exist in two isomeric forms.

17. Rate of S_N2 will be negligible in :

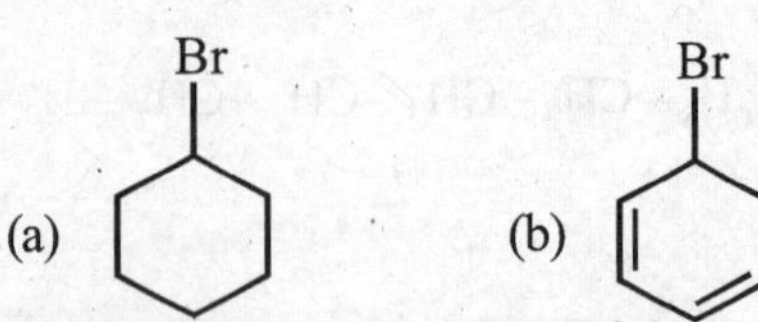

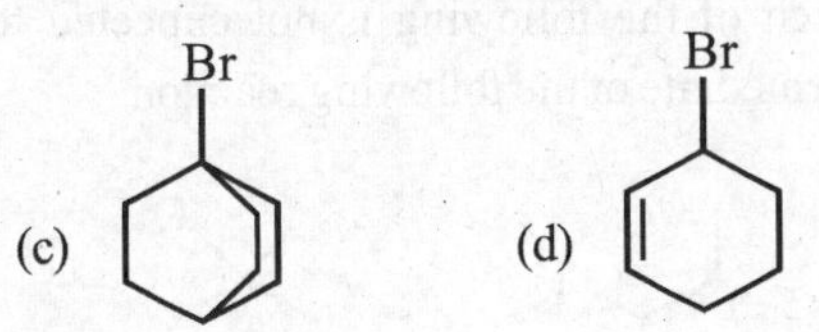

18. A major component of Borsch reagent is obtained by reacting hydrazine hydrate with which of the following ? **Toughnut**

(a) Cl, NO_2 (b) Cl, O_2N, NO_2

(c) Cl, NO_2, NO_2 (d) Cl, O_2N, NO_2, NO_2

19. Chlorobenzne reacts with trichloro acetaldehyde in the presence of H_2SO_4.

$$2\ C_6H_5\text{-}Cl + H-\overset{O}{\overset{\|}{C}}-CCl_3 \xrightarrow{H_2SO_4}$$

The major product formed is:

(a) $Cl-C_6H_4-\underset{Cl}{\overset{Cl}{C}}-C_6H_4-Cl$

(b) $Cl-C_6H_4-\underset{CH_2Cl}{\overset{Cl}{C}}-C_6H_4-Cl$

(c) $Cl-C_6H_4-\underset{CCl_3}{CH}-C_6H_4-Cl$

(d) $Cl-C_6H_4-\underset{Cl}{CH}-C_6H_4-Cl$

20. The total number of acyclic isomers including the stereoisomers with the molecular formula C_4H_7Cl **Toughnut**

(a) 11 (b) 12 (c) 9 (d) 10

21. The major product formed when 1, 1, 1-trichloro-propane is treated with aqueous potassium hydroxide is:

(a) Propyne (b) 1-Propanol

(c) 2-Propanol (d) Propionic acid

22. Which one of the following reagents is not suitable for the elimination reaction ?

(1-bromopropane) Br ⟶ (propene)

(a) NaI/acetone

(b) NaOEt/ EtOH

(c) $NaOH/H_2O$

(d) $NaOH/H_2O-EtOH$

23. Identify Z in the following serie's

$$C_2H_5I \xrightarrow{\text{Alc. KOH}} X \xrightarrow{Br_2} Y \xrightarrow{KCN} Z$$

(a) CH_3CH_2CN (b) $NCCH_2-CH_2CN$

(c) $BrCH_2-CH_2CN$ (d) $BrCH=CHCN$

24. Among the following, the molecule with the lowest dipole moment is

(a) $CHCl_3$ (b) CH_3Cl

(c) CH_2Cl_2 (d) CCl_4

25. Which chloride is least reactive with the hydrolysis point of view?

(a) CH_3Cl (b) CH_3CH_2Cl

(c) $(CH_3)_3CCl$ (d) $CH_2=CH-Cl$

26. The reaction is described as

$$CH_3(CH_2)_5(H)(CH_3)C-Br \xrightarrow{OH^-} HO-C(CH_2)_5CH_3(H)(CH_3)$$

(a) S_E2 (b) S_N1 (c) S_N2 (d) S_N0

27. Replacement of Cl of chlorobenzene to give phenol requires drastic conditions but chlorine of 2, 4-dinitrochlorobenzene is readily replaced. This is because **Tricky**

(a) NO_2 makes the ring electron rich at ortho and para positions.
(b) NO_2 withdraws e^- from meta-position.
(c) NO_2 donates e^- at meta-position.
(d) NO_2 withdraws e^- from ortho/para-positions.

28. $CH_3-CH_2-\underset{\displaystyle Cl}{\underset{|}{CH}}-CH_3$ obtained by chlorination of *n*-butane, will be
(a) meso-form (b) racemic mixture
(c) *d*-form (d) *l*-form

29. Which compound in each of the following pairs is most reactive to the conditions indicated ?

(A) (benzene ring with CH_2Br) and (B) (benzene ring with CH_3 and Br) (KOH in CH_3OH)

(C) (benzene ring with CH_3 and Cl) and (D) (benzene ring with CH_3, Cl and CH_3) ($NaNH_2$ in NH_3)

(a) A and C (b) B and C
(c) A and D (d) B and D

30. 2-Bromopentane is heated with potassium ethoxide in ethanol. The major product obtained is
(a) 2-Ethoxypentane (b) Pentene-1
(c) *cis*-Pentene-2 (d) *trans*-Pentene-2

31. In the following reaction, compound (B) is

(1,2-dibromocyclopentane) $\text{Br} + \underset{\text{(1 eq.)}}{\text{Mg}} \xrightarrow[\text{ether}]{\text{Dry}} A \xrightarrow{\Delta} B$

(a) (cyclopentane with Br and MgBr) (b) (bicyclopentyl with two Br)
(c) (cyclopentene) (d) (bi(cyclopentenyl))

32. Which of the following is not expected to be intermediate of the following reaction?

(1-iodo-2-methylcyclohexane) $\xrightarrow{H_2O}$ (1-methylcyclohexanol, OH)

(a) (carbocation) (b) (carbocation)
(c) (cyclohexane with methyl and $\overset{+}{O}H_2$) (d) (carbocation)

33. Among the given halides, which one will give same product in both S_N1 and S_N2 reactions?

(I) $CH_3-\underset{}{\overset{\displaystyle CH_3}{\overset{|}{CH}}}-CH_2-\underset{\displaystyle Br}{\underset{|}{CH}}-CH_3$

(II) (cyclohexane with CH_3 and Cl) (III) (cyclohexane with Cl)

(IV) $CH_3-\underset{\displaystyle Et}{\underset{|}{CH}}-Br$

(a) (III) only (b) (I) and (II)
(c) (III) and (IV) (d) (I), (III) and (IV)

34. $CH_3CH_2Cl \xrightarrow{NaCN} X \xrightarrow{Ni/H_2} Y \xrightarrow[\text{anhydride}]{\text{Acetic}} Z$

Z in the above reaction sequence is
(a) $CH_3CH_2CH_2NHCOCH_3$
(b) $CH_3CH_2CH_2NH_2$
(c) $CH_3CH_2CH_2CONHCH_3$
(d) $CH_3CH_2CH_2CONHCOCH_3$

35. Which of the following is most reactive towards S_N2 reaction?

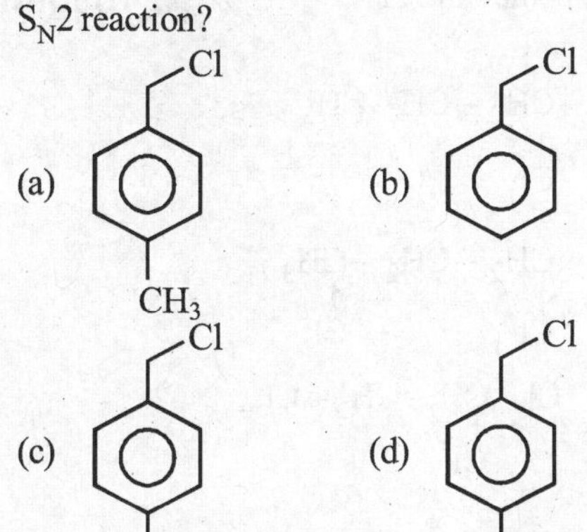

36. Identify correct reactivity order for S_N1 reaction

(i) Cl (ii) Cl (iii) Cl

(a) i > ii > iii (b) ii > iii > i
(c) i > iii > ii (d) iii > ii > i

37. Which compound undergoes nucleophilic substitution with NaCN at the fastest rate?

(a) Br (b) Br

(c) Br (d) Br

38. Identify correct reactivity order for S_N1 reaction

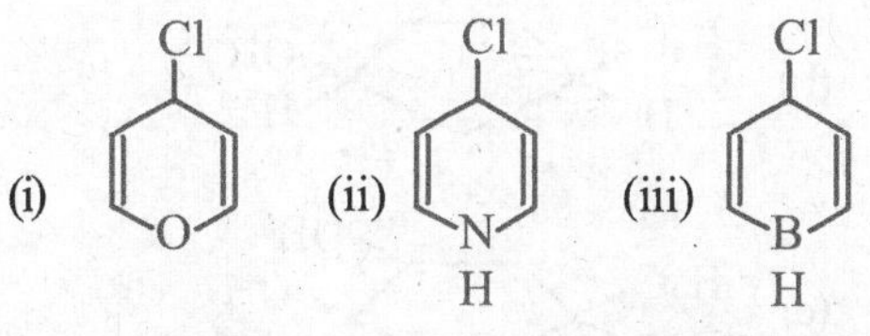

(a) i > ii > iii (b) ii > iii > i
(c) i > iii > ii (d) ii > i > iii

39. Which one of the sequences below is the best synthesis of (E)-3-hexene?

(a) (O, H) $\xrightarrow[\text{(ii) } H_2O]{\text{(i) } \overset{-}{C}H_2\overset{+}{Li}}$ $\xrightarrow{HBr}$ $\xrightarrow[HOC(CH_3)_3]{^-OC(CH_3)_3}$

(b) $^-C{\equiv}CH$ $\xrightarrow{Br}$ $\xrightarrow[\text{(ii) } Br]{\text{(i) NaH}}$ $\xrightarrow{Na/NH_3}$

(c) (O, Me) $\xrightarrow[\text{(ii) } H_2O]{\text{(i) } \overset{-}{C}H_2\overset{+}{Li}}$ $\xrightarrow{HBr}$ $\xrightarrow[HOCH_2CH_3]{^-OCH_2CH_3}$

(d) $^-C{\equiv}CH$ $\xrightarrow{Br}$ $\xrightarrow[\text{(ii) } CH_3Br]{\text{(i) NaH}}$ $\xrightarrow{Na/NH_3}$

40. Which of the following reacts at the fast rate with CH_3OK in CH_3OH ?

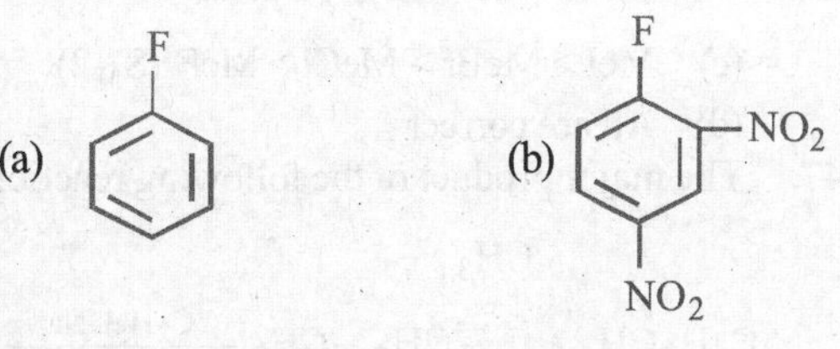

(c) F, O_2N, NO_2, NO_2 (d) F, NO_2

41. In the following sequence of reactions

$$\underset{A}{CH_3CH_2I} \xrightarrow[\text{ether}]{Mg} B \xrightarrow{HCHO} C \xrightarrow{H_2O} D$$

the compound D is

(a) propanal (b) butanal

(c) *n*-butyl alcohol (d) *n*-propyl alcohol

42. The correct product obtained in the reaction

H_3C, H – cyclohexane – Br, H $\xrightarrow{OH^-}$

(a) H_3C, H – cyclohexane – H, OH

(b) H_3C, H – cyclohexane – OH, H

(c) H_3C, H – cyclohexane – OH, OH, H

(d) H_3C, H – cyclohexene

43. 2-phenyl-2-hexanol can be prepared by Grignard synthesis. The pair of compounds giving the desired product is

(a) n-butyl–Br and $Ph-\overset{O}{\overset{\|}{C}}-CH_3$

(b) isobutyl–Br and $Ph-\overset{O}{\overset{\|}{C}}-CH_3$

(c) $Ph-\overset{O}{\overset{\|}{C}}-$pentyl and PhBr

(d) None of these

44. In the following reaction sequence :

$\underset{(C_3H_6Cl_2)}{I} \xrightarrow{KOH(aq)} II \xrightarrow[(ii)\,H_2O/H^+]{(i)\,CH_3MgBr} III$

$\xrightarrow{Anhy.\,ZnCl_2 + Conc.\,HCl}$ given turbidity immediately

The compound I is :

(a) $\underset{Cl}{CH_2}-\underset{Cl}{CH}-CH_3$

(b) $\underset{Cl}{CH_2}-CH_2-\underset{Cl}{CH_3}$

(c) $Cl-\underset{Cl}{CH}-CH_2-CH_3$

(d) $CH_3-\underset{Cl}{\overset{Cl}{C}}-CH_3$

45. Compound that on hydrogenation produces optically inactive compound is Toughnut

(a) $H_3C-CH=CH-C(H)(Br)-CH_3$

(b) $H_2C=CH-C(H)(Br)-CH_2-CH_3$

(c) $H_2C=C(CH_3)-C(H)(Br)-CH_3$

(d) $H_2C=CH-C(Br)(H)-CH_2-C_2H_5$

46. Which of the following order is not correct ?

(a) $MeBr > Me_2CHBr > Me_3CBr > Et_3CBr (S_N2)$

(b) $PhCH_2Br > PhCHBrMe > PhCBrMe_2 > PhCBrMePh (S_N1)$

(c) $MeI > MeBr > MeCl > MeF\ (S_N2)$

(d) All are correct

47. The major product of the following reaction is :

$C_6H_5CH_2-\underset{Br}{\overset{CH_3}{C}}-CH_2-CH_3 \xrightarrow[C_2H_5OH]{C_2H_5ONa}$

(a) $C_6H_5CH_2-\underset{OC_2H_5}{\overset{CH_3}{\underset{|}{\overset{|}{C}}}}-CH_2-CH_3$

(b) $C_6H_5CH=\underset{CH_3}{\underset{|}{C}}-CH_2-CH_3$

(c) $C_6H_5CH_2-\underset{CH_3}{\underset{|}{C}}=CHCH_3$

(d) $C_6H_5CH_2-\underset{CH_2CH_3}{\underset{|}{C}}=CH_2$

48. The compound most reactive towards S_N1 reaction

(a) Me_3COCH_2Cl (b) $MeOCH_2Cl$

(c) $C_6H_5CH_2CH_2Cl$ (d) cyclopropyl–Cl

49. Which chloro derivative of benzene among the following would undergo hydrolysis most readily with an aqueous sodium hydroxide to furnish the corresponding hydroxy derivative?

(a) 2,4,6-trinitrochlorobenzene (O_2N, NO_2, NO_2 substituted benzene with –Cl)

(b) $O_2N-C_6H_4-Cl$

(c) $Me_2N-C_6H_4-Cl$

(d) C_6H_5Cl

50. In the following sequence of reactions, which is incorrect

$$CH_2=CH-CH=CH_2 \xrightarrow[80°C]{HBr} A \xrightarrow{HBr} B$$

$$\xrightarrow{Mg} C$$

Toughnut

(a) A is $CH_3-CH=CH-CH_2Br$

(b) B is $CH_3CHBr-CH_2-CH_2-Br$

(c) C is cyclopropyl methane

(d) B on S_N2 with aqueous NaOH gives $CH_3CHOH.CH_2CH_2Br$ and not $CH_3CHBr.CH_2CH_2OH$

51. The final product obtained in the reaction

$$Br-C_6H_4-CH_2Cl \xrightarrow{Mg\ /\ ether} A \xrightarrow{Heavy\ water}$$

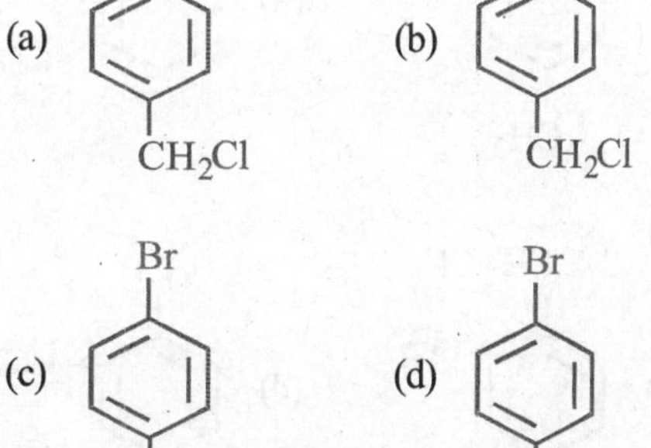

(a) $D-C_6H_4-CH_2Cl$ (b) $DO-C_6H_4-CH_2Cl$

(c) $Br-C_6H_4-CH_2D$ (d) $Br-C_6H_4-CH_2OD$

52. The following reaction proceeds through the intermediate formation of

$$RCOOAg + Br_2 \longrightarrow RBr + CO_2 + AgBr$$

(a) $RCOO^{\bullet}$ (b) $R^{\bullet}$

(c) $Br^{\bullet}$ (d) All of these

53. The compound

$$C_7H_8 \xrightarrow{3Cl_2/\Delta} A \xrightarrow{Br_2/Fe} B \xrightarrow{Zn/HCl} C$$

The compound C is 

(a) *o*-Bromotoluene
(b) *m*-Bromotoluene
(c) *p*-Bromotoluene
(d) 3-Bromo-2, 4, 6-trichlorotoluene

54. The product of 1-bromo-3-chloro cyclobutane with Na in presence of dioxane Tricky

(a) 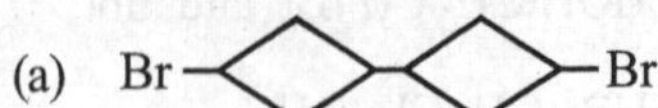

(b) Br–◇◇–Cl

(c) Cl–◇◇–Cl

(d) None of these

55. Which of the following has the weakest carbon - chlorine bond ?

(a) Cl, CH_3 (para)   (b) Cl, CH_3 (meta)

(c) CH_2Cl (d) Cl, CH_3 (ortho)

56. $Cl_3C.CH=CH_2 \xrightarrow{HCl} A$; $Cl_3C.CH=CH_2 \xrightarrow{Br_2} B$ Toughnut

Which of the following is correct ?
(a) A on reaction with aq. KOH gives $HOCH_2CH_2COOK$
(b) B can be resolved into d– and *l* –forms
(c) Both (a) and (b)
(d) Neither (a) nor (b)

57. Although hexafluoroethane (C_2F_6, b.p. –79°C) and ethane (C_2H_6, b.p. – 89°C) differ very much in their molecular weights, their boiling points differ only by 10°C. This is due to

(a) low polarizability of F
(b) nearly similar size of F and H
(c) both (a) and (b)
(d) Neither of the two

58. Which of the following is a key intermediate in the reaction shown below ?

Cl-benzene $\xrightarrow[-33\,°C]{NaNH_2,\ NH_3}$ NH_2-benzene

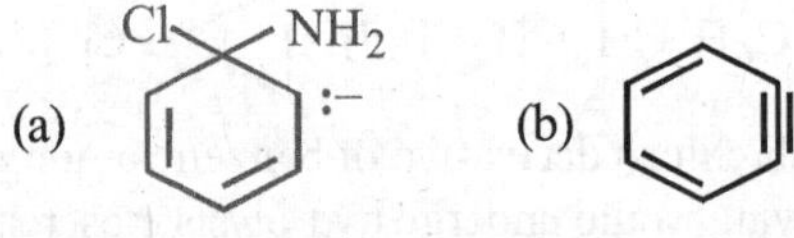

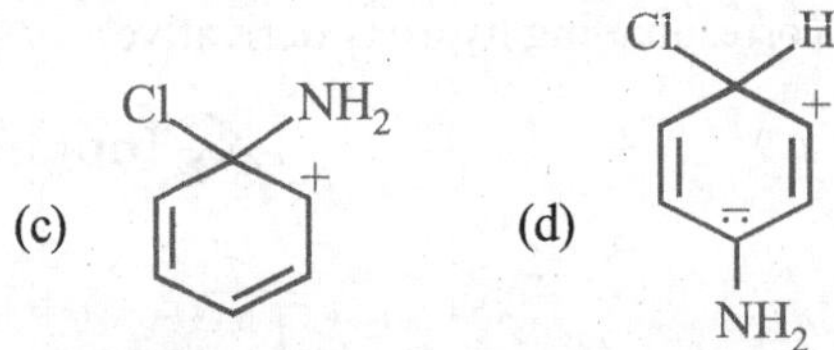

59. Which of the following structure is more stable?

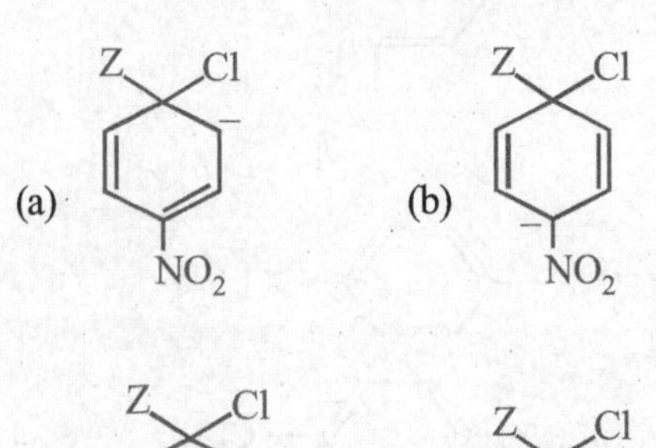

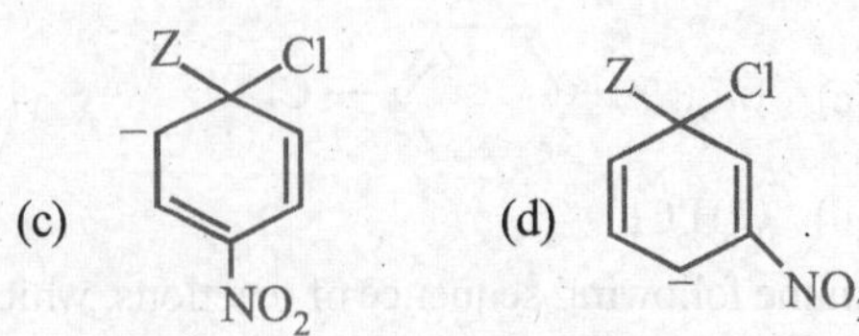

60. Which of the following is (are) true concerning the intermediate in the addition-elimination mechanism of the following reaction ?

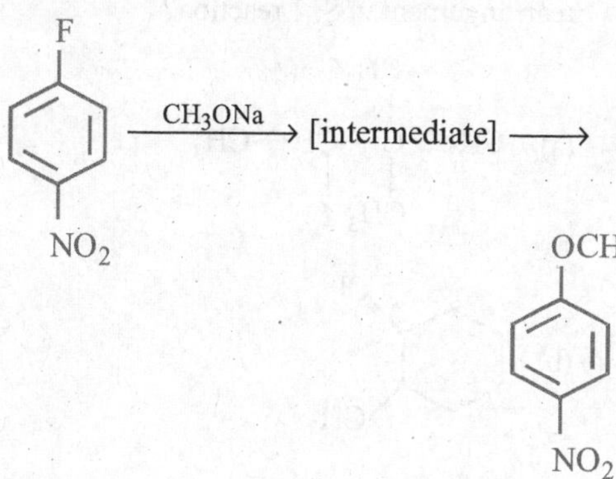

A = The intermediate is aromatic, B = The intermediate is a resonance stabilised anion, C = Electron withdrawing groups on the benzene ring stabilises the intermediate **Toughnut**

(a) only A (b) only B
(c) A and C (d) B and C

61. Identify Z in

$CH_3CH_2CH_2Br \xrightarrow{\text{aq. NaOH}} X \xrightarrow{Al_2O_3} Y \xrightarrow{Cl_2/H_2O} Z$

(a) Mixture of $CH_3CHClCH_2Cl$ and $CH_3CHOHCH_2Cl$
(b) $CH_3CHOHCH_2Cl$
(c) $CH_3CHClCH_2OH$
(d) $CH_3CHClCH_2Cl$

62. CH_2=(cyclohexane ring)=CH_2 + HBr $\xrightarrow{\text{peroxide}}$ A, then A is

(a) CH_3–(ring, C–Br)–CH_2Br
(b) CH_3(Br)–(ring)–(Br)CH_3
(c) CH_3–(ring, Br, Br)–CH_3
(d) $BrCH_2$–(ring)–CH_2Br

63. Silver benzoate reacts with bromine to form

(a) benzene (b) C_6H_5COOBr
(c) m-bromo benzoic acid silver salt ($COOAg$, –Br) (d) C_6H_5Br

64. $PhCOCHBr_2 \xrightarrow{OH^-} A \xrightarrow{OH^-} B \xrightarrow{H^+} C$

The compound C is –

(a) PhCH(OH)CHO
(b) PhCH(OH)COOH
(c) $PhCOCBr_2$ with H on the C
(d) $Ph-\underset{\underset{O}{\|}}{C}-CH_2-OH$

65. $C_6H_5{}^{14}Cl$ + NaOH(aq.) $\xrightarrow{400°C}$ Product is **Toughnut**

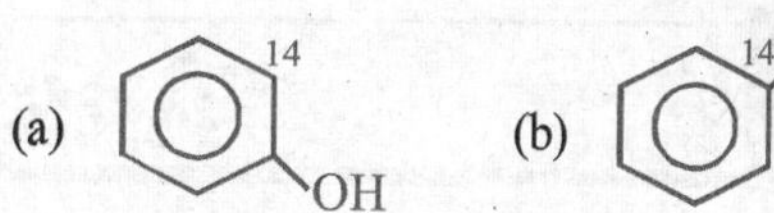

(c) Both (a) and (b) (d) No reaction

66. X in the following reaction is –

$Br_2 + \begin{matrix} CH_3-C-H \\ \| \\ H-C-CH_3 \end{matrix} \xrightarrow{CCl_4} X$

(a) (+) 2, 3-Dibromobutane
(b) (–) 2, 3-Dibromobutane
(c) (±)- 2, 3-Dibromobutane
(d) meso-2, 3-Dibromobutane

67. The missing reagents R_1 and R_2 in the following series of reactions are **Toughnut**

$CH_3CH_2Br \xrightarrow{R_1} [\ ] \xrightarrow{R_2} CH_3\bar{C}HP^+Ph_3$

(a) PhLi and Ph_3P respectively
(b) Ph_3P and PhLi respectively
(c) Ph_3P and C_2H_5ONa respectively
(d) Either (b) or (c)

68. The reaction of toluene with chlorine in presence of ferric chloride gives predominantly :
(a) benzoyl chloride
(b) *m*-chlorotoluene
(c) benzyl chloride
(d) *o*- and *p*-chlorotoluene

69. Arrange the following in order of decreasing rate of solvolysis with aqueous ethanol (fastest → slowest)

(1) $H_2C{=}C(CH_3){-}Br$

(2) 1-bromo-1-methylcyclohexane (ring with CH_3 and Br on the same carbon)

(3) $CH_3{-}CH(Br)CH_2CH(CH_3)_2$

(a) 2>1>3 (b) 1>2>3
(c) 2>3>1 (d) 1>3>2

70. Which of the following alkyl halide undergo rearrangement in S_N1 reaction?

(a) $CH_3{-}C(CH_3)_2{-}CH(I){-}CH_3$

(b) cyclohexane ring with Cl and CH_3 on adjacent carbons

(c) cyclohexene ring with I on the allylic carbon

(d) All of these

Answer KEYs

1	(b)	8	(a)	15	(c)	22	(a)	29	(a)	36	(d)	43	(a)	50	(d)	57	(c)	64	(b)
2	(a)	9	(b)	16	(b)	23	(b)	30	(d)	37	(a)	44	(d)	51	(c)	58	(b)	65	(c)
3	(a)	10	(a)	17	(c)	24	(d)	31	(c)	38	(d)	45	(b)	52	(d)	59	(b)	66	(d)
4	(c)	11	(b)	18	(c)	25	(d)	32	(a)	39	(b)	46	(b)	53	(b)	60	(d)	67	(d)
5	(b)	12	(a)	19	(c)	26	(c)	33	(c)	40	(c)	47	(b)	54	(c)	61	(b)	68	(d)
6	(a)	13	(b)	20	(b)	27	(d)	34	(a)	41	(d)	48	(b)	55	(c)	62	(d)	69	(c)
7	(a)	14	(c)	21	(d)	28	(b)	35	(d)	42	(a)	49	(a)	56	(c)	63	(d)	70	(d)

Hints & Solutions

1. (b) In *p*-dichlorobenzene, the two equal dipoles are in opposite direction, hence the molecule has zero dipole moment. In *o*- and *m*- dichlorobenzenes, the two dipoles are at 60° and 120° apart respectively, and thus according to parallelogram law of forces, the dipole moment of *o*-dichlorobenzene is much higher than that of *m*-isomer. Lastly, toluene with a +I group possesses little dipole moment. Thus the overall order is

   p-dichlorobenzene (IV) < toluene (CH_3) (I) < m-dichlorobenzene (II) < o-dichlorobenzene (III)

   IV < I < II < III

2. (a) $Cl_3CCH(OH)_2 + NaOH \longrightarrow CHCl_3 + HCOONa + H_2O$

3. (a) $C_2H_6\ (excess) + Cl_2 \xrightarrow{\text{UV light}} C_2H_5Cl + HCl$

4. (c) Compound A reacts with alc.KOH to give compound B which on further ozonolysis gives C (does not contains α- H atom) and D (contains α-H atom).This reaction sequence can be achieved by compounds in option (a) and (c). Since compound A gives white ppt. with $AgNO_3$ preferable option will be (c) as tert alkyl reacts with $AgNO_3$ more quickly.

$$C_6H_5-CH_2-C(Cl)(CH_3)_2 \xrightarrow{\text{alc. KOH}} C_6H_5-CH=C(CH_3)_2$$

(A) (it gives white ppt with $AgNO_3$)

$$C_6H_5-CH=C(CH_3)_2 \xrightarrow[\text{(ii) Zn/H}_2\text{O}]{\text{(i) O3}} C_6H_5-\overset{O}{\overset{\|}{C}}-H + CH_3-\overset{O}{\overset{\|}{C}}-CH_3$$

(C) (it gives cannizaro reaction only) (D) (It gives aldol condensation only)

5. (b) Reaction between alkyl halide, aryl halide and sodium in presence of ether is known as Wurtz fitting reaction

$$C_6H_5Cl + 2Na + ClCH_3 \longrightarrow \underset{\text{Toluene}}{C_6H_5CH_3} + 2NaCl$$

6. (a) The correct order of increasing bond length is $CH_3F < CH_3Cl < CH_3Br < CH_3I$

7. (a) For a given alkyl group, the order of reactivity is $\xleftarrow[\text{increasing bond energy}]{R-I > R-Br > R-Cl > R-F}$ decreasing halogen reactivity.
This order depends on the carbon-halogen bond energy; the carbon-fluorine bond energy is maximum and thus fluorides are least reactive while carbon-iodine bond energy is minimum hence iodides are most reactive.

8. (a) Since $CHCl_3$ is covalent compound it does not produce Cl^- ion in H_2O, hence no white ppt is formed during reaction with $AgNO_3$.

9. (b) $CF_3-\overset{O}{\underset{O}{\overset{\|}{\underset{\|}{S}}}}-O^- > Ph-\overset{O}{\underset{O}{\overset{\|}{\underset{\|}{S}}}}-O^- > CH_3-COO^- > PhO^-$

L.G. Ability

10. (a) $\underset{}{C_2H_5Br} \xrightarrow{AgCN} \underset{(X)}{C_2H_5NC} + AgBr$

$(X) \xrightarrow[\text{Zn – Hg / HCl}]{\text{Reduction}} \underset{\text{(Y) ethyl methyl amine}}{C_2H_5NHCH_3}$

11. (b) The reaction proceeds via free radical mechanism.
As 2° free radical is more stable than 1°, so $CH_3CH_2CH(Br)CH_3$ would be formed.

12. (a) $C_6H_5\overset{+}{N}\equiv N\overset{-}{Cl} \xrightarrow{HBF_4} C_6H_5F + N_2 + BF_3 + NaCl$

(Balz-Schiemann's reaction)

13. (b) $CH_3-CH_2-CHCl_2 \xrightarrow[\Delta]{NaNH_2}$

$CH_3-CH=CHCl \xrightarrow[\Delta]{NaNH_2} \underset{\text{Final Product}}{CH_3-C\equiv CH}$

14. (c) The acid character follows the order :
$CH_3COOH > C_6H_5OH > H_2O > CH_3OH$
The basic character will follow the order
$CH_3COO^- < C_6H_5O^- < OH^- < CH_3O^-$
The stronger the acid, the weaker the conjugate base formed.

15. (c) $C_6H_5CHCH_3Br$ being an optically active secondary alkyl bromide undergoes S_N2 nucleophilic substitution reaction. Hence it undergoes complete inversion of configuration.

$$C_6H_5-\overset{H}{\underset{H_3C}{C}}-Br + Cl^- \xrightarrow{DMF} Cl-\overset{H}{\underset{CH_3}{C}}-C_6H_5$$

16. (b) $(CH_3)_2C=O + CH_3CH_2MgBr \longrightarrow (CH_3)_2C(OMgBr)(CH_2CH_3) \xrightarrow{H_2O} (CH_3)_2C(OH)(CH_2CH_3)$

2- methyl butanol-2

(3° alcohol)

17. (c) At bridge head position S_N1 and S_N2 do not takes place.

18. (c) The major component of Borsch reagent is 2,4- dinitrophenyl hydrazine which can be obtained by reaction of 2,4-dinitrochloro benzene and hydrazine

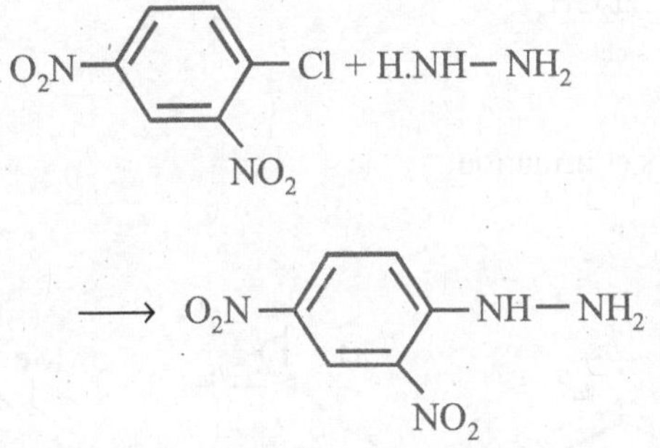

$O_2N-C_6H_3(NO_2)-Cl + H.NH-NH_2 \longrightarrow O_2N-C_6H_3(NO_2)-NH-NH_2$

2,4 -dinitrophenyl hydrazine

19. (c) Chloral on reaction with chlorobenzene in the presence of a catalytic amount of sulphuric acid forms DDT (dichlorodiphenyl Trichloro ethane).

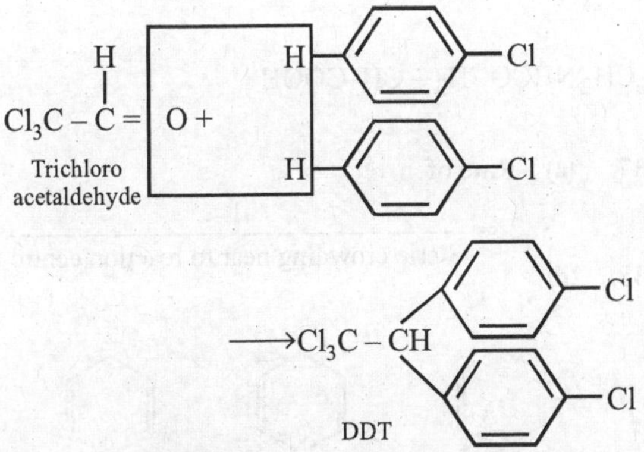

$Cl_3C-C(H)=O + 2\,H-C_6H_4-Cl \longrightarrow Cl_3C-CH(C_6H_4Cl)_2$

Trichloro acetaldehyde

DDT

20. (b) C_4H_7Cl is a monochloro derivative of C_4H_8 which itself exists in three acyclic isomeric forms.

$CH_3CH_2CH=CH_2$

I

(Its four C's are different)

$CH_3CH=CHCH_3$

II

(It has 2 types of carbon)

$CH_3-C(CH_3)=CH_2$

III

(It has 2 types of carbon)

Grand total of acyclic isomers = 6+4+2 = 12

21. (d) $Cl_3C-CH_2CH_3 + KOH \xrightarrow{heat} (OH)_3C-CH_2CH_3 + 3KCl \longrightarrow CH_3CH_2\overset{O}{\overset{\|}{C}}-OH$

22. (a) Alkyl chloride or bromide undergo substitution and get converted to an alkyl iodide on treatment with a solution of sodium iodide in acetone. e.g.

$CH_3CH_2CH_2Br + NaI \xrightarrow{acetone} CH_3CH_2CH_2I + NaBr$

This reaction is also known as Finkelstein Reaction.

23. (b) $C_2H_5I \xrightarrow{alc.\ KOH} CH_2=CH_2 \xrightarrow{Br_2} BrCH_2-CH_2Br \xrightarrow{KCN} CNCH_2.CH_2CN$

24. (d) CCl_4 has lowest (zero) dipole moment. This is due to its symmetrical tetrahedral structure. Due to which dipole moment of one bond is cancelled by opposite dipole moment of the other three bonds.

25. (d) The non-reactivity of chlorine atom in vinyl chloride is due to resonance stabilisation.

$CH_2=CH-\ddot{C}l: \longrightarrow \overset{-}{C}H_2-CH=\overset{+}{C}l:$

26. (c) Inversion in configuration occurs in S_N2 reactions.

27. (d) $-NO_2$ group withdraws electrons from *o*- and *p*-positions and hence activates the $-Cl$ present on that position towards nucleophilic substitution.

28. (b) Chlorination of alkanes is a free-radical reaction. Since the intermediate free radical is planar (sp^2 hybridised) it can be attacked on either side of the face forming racemic mixture.

29. (a) In the first case the reaction gives side chain substitution product which is easier in A. In the second case the reaction will proceed by benzyne mechanism for which ortho position w. r. to Cl must have H-atoms.

30. (d) Potassium ethoxide is a strong base, hence causes elimination reaction. The alkene formed is governed by Saytzeff rule "more substituted an alkene, higher will be its stability".

$$CH_3\underset{\substack{|\\Br}}{CH}CH_2CH_2CH_3 \xrightarrow{^-OC_2H_5} \underset{\text{Pentene–1}}{CH_2=CHCH_2CH_2CH_3} + \underset{\text{Pentene–2 (more stable)}}{CH_3CH=CHCH_2CH_3}$$

Further pentene-2 shows geometrical isomerism, *cis*- and *trans*- in which *trans*-isomer having bulkier groups away from each other is more stable than the *cis*-isomer.

$$\underset{\textit{cis}\text{-Pentene-2}}{(CH_3)(H)C=C(CH_2CH_3)(H)} \qquad \underset{\textit{trans}\text{-Pentene-2 (more stable)}}{(CH_3)(H)C=C(H)(CH_2CH_3)}$$

31. (c) 2-Bromocyclopentylmagnesium bromide $\xrightarrow{\Delta}$ cyclopentene; heat favours elimination.

32. (a) 1-iodo-1-(2-methylcyclohexyl)... $\xrightarrow[\text{solvolysis}\ -I^-]{H_3O^+}$ carbocation $\xrightarrow{H^+\ \text{Shift}}$ 1-methylcyclohexyl cation $\xrightarrow{H_2O}$ oxonium ion $\xrightarrow{-H^+}$ 1-methylcyclohexanol (Me, OH)

33. (c) S_N2 and S_N1 same, if C^+ not rearrange.

34. (a) $CH_3CH_2Cl \xrightarrow{NaCN} \underset{(X)}{CH_3CH_2CN} \xrightarrow{Ni/H_2}$

$$\underset{(Y)}{CH_3CH_2CH_2NH_2} \xrightarrow[(CH_3CO)_2O]{} \underset{(Z)}{CH_3CH_2CH_2NHCOCH_3} + CH_3COOH$$

35. (d) 4-nitrobenzyl chloride is likely to react by the S_N2 mechanism as the strongly electron-withdrawing nitro group would destabilize the carbocation intermediate of the S_N1 mechanism, a benzylic chloride that disfavours the S_N1 mechanism.

36. (d) Stablity

cyclopropyl cation < cyclobutyl cation < cyclopentyl cation

37. (a) Rate of

$$S_N2 \propto \frac{1}{\text{Steric crowding near to reaction centre}}.$$

38. (d)

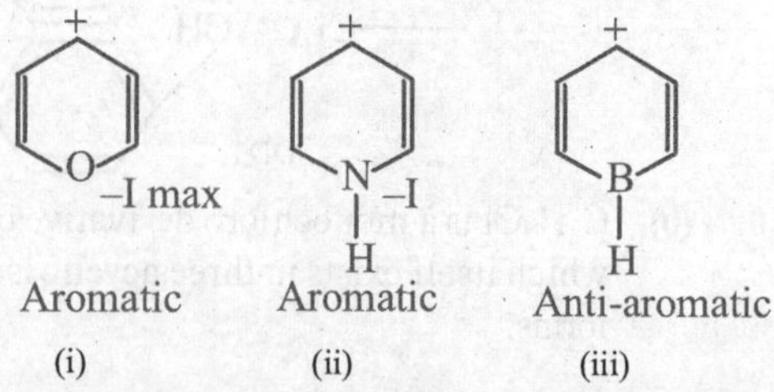

39. (b) $^{-}C\equiv CH \xrightarrow{\text{(ethyl bromide)}} H_5C_2-C\equiv CH \xrightarrow[\text{(ii) ethyl bromide}]{\text{(i) NaH}}$

$H_5C_2-C\equiv C-C_2H_5 \xrightarrow{Na/NH_3}$ (E) – 3 – hexene

40. (c) The electron withdrawing nitro groups weaken the C – F bond by inductive effect and resonance.

41. (d) $\underset{(A)}{CH_3CH_2I} \xrightarrow[\text{Ether}]{Mg} \underset{(B)}{CH_3CH_2MgI} \xrightarrow{HCHO} \underset{(C)}{H-C(CH_2CH_3)(H)-OMgI} \xrightarrow{H_2O} \underset{(D)}{H-C(CH_2CH_3)(H)-OH}$

n–Propyl alcohol

42. (a) In 2° halides of this type the product formed has inverted configuration (S_N2 mechanism).

43. (a) $Ph-CO-CH_3$ + n-butyl–Mg Br $\longrightarrow$ Ph–C(OMgBr)(CH_3)(n-butyl) $\xrightarrow{H_2O}$ Ph–C(OH)(CH_3)(n-butyl)

2 – phenyl -2-hexanol

44. (d) $\underset{\substack{(C_3H_6Cl_2)\\(I)}}{H_3C-CCl_2-CH_3} \xrightarrow[\text{(aq.)}]{KOH} \underset{\text{Unstable}}{CH_3-C(OH)_2-CH_3} \xrightarrow{-H_2O} \underset{(II)}{CH_3-CO-CH_3} \xrightarrow[\text{(ii) } H_2O/H^{\oplus}]{\text{(i) } CH_3MgBr}$

$\underset{\substack{3° \text{ alcohol}\\(III)}}{CH_3-C(OH)(CH_3)-CH_3} \xrightarrow[\text{Lucas Test}]{\text{Anhy. } ZnCl_2 + HCl}$ Gives turbidity immediately

45. (b) (a) $CH_3CH=CH-CH(Br)-CH_3 \xrightarrow{H_2/Ni} CH_3CH_2CH_2-C^{*}H(Br)-CH_3$ (Optically active)

(b) $CH_2=CH-CH(Br)-CH_2CH_3 \xrightarrow{H_2/Ni} CH_3CH_2-CH(Br)-CH_2CH_3$ (Optically inactive)

(c) $H_2C=C(CH_3)-CH(Br)-CH_3 \xrightarrow{H_2/Ni} H_3C-CH(CH_3)-C^{*}H(Br)-CH_3$ (Optically active)

(d) $H_2C=CH-CH(Br)-C_2H_5 \xrightarrow{H_2/Ni} H_3C-CH_2-C^{*}H(Br)-C_2H_5$ (Optically active)

46. (b) The more is the stability of intermediate (carbonium ion), the more is the chance of S_N1 mechanism. The intermediates obtained will be $Ph\overset{+}{C}H_2$ (i), $Ph\overset{+}{C}H-Me$ (ii), $Ph\overset{+}{C}-Me_2$ (iii), $Ph\overset{+}{C}MePh$ (iv). The stabilty is of the order iv > iii > ii > i.

47. (b) More acidic $C_2H_5O^-$

$C_6H_5-CH(H)-C(CH_3)(Br)-CH(H)-CH_3$

(–I effect) (+I effect)

$\xrightarrow{C_2H_5ONa,\ C_2H_5OH}$

$C_6H_5-CH=C(CH_3)-CH_2-CH_3$

48. (b) $MeOCH_2Cl \xrightarrow{Slow} Me\ddot{\overset{+}{O}}CH_2 + Cl^-$ (i) $\updownarrow$ $Me-\overset{+}{\ddot{O}}=CH_2$ (ii)

Though (ii) contains +ve charge on oxygen. Since octet around each atom does complete the structure II is more stable than I.

49. (a) Cl in 2, 4, 6-trinitrochlorobenzene is activated by three NO_2 groups at *o*, and *p*-positions and hence undergoes hydrolysis most readily.

50. (d) A is 1, 4-addition product due to conjugation B is obtained by further addition of HBr. $(CH_3CHBr.CH_2.CH_2Br)$. B gives C by cyclisation $CH_3-CH-CH_2$ (ring with CH_2). If B reacts by S_N2 mechanism,

Br on 1° carbon must be replaced by OH^- to give $CH_3CHBr-CH_2-CH_2OH$.

51. (c) $BrC_6H_4CH_2Br \xrightarrow{Mg/ether} BrC_6H_4CH_2MgBr \xrightarrow{D_2O.} BrC_6H_4CH_2D + DOMgBr$

52. (d) Mechanism of Hunsdiecker's reaction is

$R-COO^-\overset{+}{Ag} \xrightarrow[-AgBr]{Br_2} RCOOBr \longrightarrow RCO\dot{O} + \dot{Br} \longrightarrow R^{\bullet} + CO_2;$

$$R^{\bullet} + RCOO\,Br \longrightarrow R-Br + RCO\dot{O}$$

53. (b) $C_6H_5CH_3 \xrightarrow[\text{heat}]{3Cl_2} C_6H_5CCl_3 \xrightarrow{Br_2/Fe} m\text{-}BrC_6H_4CCl_3 \xrightarrow{Zn/HCl} m\text{-}BrC_6H_4CH_3$

54. (c) Bond strength follows the order
$C-F > C-Cl > C-Br > C-I$

$$2\ \text{(1-bromo-3-chlorocyclobutane)} + 2Na \xrightarrow{\text{dioxane}} Cl-C_4H_6-C_4H_6-Cl + 2NaBr$$

55. (c) **Nuclear** substituted halogen has double bond character due to resonance.

56. (c) The compound A is

$$Cl_3C.CH_2.CH_2Cl \xrightarrow{\text{aq.KOH}} (OH)_3C.CH_2CH_2OH \xrightarrow{-2H_2O} HO\ CH_2.CH_2COOK.$$

The compounds B is $Cl_3C-\overset{*}{C}H(Br)-CH_2Br$ has chiral centre and can be resolved into d- and *l*- form.

57. (c) Small difference in boiling points of C_2F_6 and C_2H_6 is due to the fact that (i) the F atom is slightly larger than H, and (ii) F has low polarizability

58. (b) The benzyne is an intermediate of this type of reaction.

59. (b) Structure (b) is more stable because here negative charge is dispersed to the maximum extent as it is present on the carbon atom bearing an elecron-withdrawing ($-NO_2$) group

60. (d) $p\text{-}FC_6H_4NO_2 + CH_3O^- Na^+ \longrightarrow$ [resonance-stabilised carbanion intermediates bearing CH_3O, F and NO_2] $\longleftrightarrow \ldots \longleftrightarrow \ldots \longrightarrow p\text{-}CH_3OC_6H_4NO_2 + F^-$

61. (b) $CH_3CH_2CH_2Br \xrightarrow{\text{aq. NaOH}} CH_3CH_2CH_2OH \xrightarrow[\text{heat}]{Al_2O_3}$

$$CH_3CH = CH_2 \xrightarrow{Cl_2/H_2O} CH_3.CHOH.CH_2Cl$$

62. (d) Addition product is formed by anti-Markownikoff's rule.

63. (d) $C_6H_5COOAg + Br_2 \xrightarrow{\text{Hunsdiecker reaction}} C_6H_5Br + CO_2 + AgBr$

64. (b) $\underset{(A)}{PhCOCHBr_2} \xrightarrow{OH^-} PhCOCHO \xrightarrow{OH^-} \underset{(B)}{Ph-\overset{O^-}{\underset{H}{C}}-C(=O)OH} \xrightarrow{H^+} \underset{(C)}{PhCH(OH)COOH}$

65. (c) $C_6H_5{}^{14}Cl \xrightarrow[\text{(-HCl)}]{\text{NaOH(aq.), high temp.}}$ Benzyne (14) $\xrightarrow{OH^-}$ $C_6H_5{}^{14}OH$ + phenol (OH ortho to ^{14}C)

66. (d)

67. (d) $CH_3CH_2Br + Ph_3P \longrightarrow CH_3CH_2P^+Ph_3Br^- \xrightarrow[C_2H_5ONa]{C_6H_5Li \text{ or}}$

$$CH_3CH = PPh_3 \longleftrightarrow CH_3\overset{-}{C}H\overset{+}{P}Ph_3$$

68. (d) The given reaction is an example of electrophilic subsitution. Further, CH_3 group in toluene is o,p-directing

69. (c) Rate of solvolysis $\propto$ stability of C^+.

70. (d) Unstable C^+ has general tendency to rearrange to more stable C^+.

25 Alcohols, Phenols and Ethers

1. Which one of the following statements is not correct?
   (a) Alcohols are weaker acids than water.
   (b) Acid strength of alcohols decreases in the following $RCH_2OH > R_2CHOH > R_3COH$.
   (c) Carbon-oxygen bond length in methanol, CH_3OH is shorter than that of C – O bond length in phenol.
   (d) The bond angle C–O–H in methanol is 108.9°.

2. What is the structure of the major product when phenol is treated with bromine water ?
   (a) 2,4,6-tribromophenol (OH; Br, Br, Br)
   (b) cyclohexadienone (O; Br, Br; H, Br)
   (c) 4-bromophenol (OH; Br)
   (d) 2-bromophenol (OH; Br)

3. In the given pair of alcohol, in which pair second alcohol is more reactive than first towards hydrogen bromide?
   (a) 1-methylcyclopentanol (CH_3, OH) and cyclohexanol (H, OH)
   (b) 1-ethylcyclopentanol (OH) and cyclopentyl–$\underset{\displaystyle OH}{CH}$–$CH_3$
   (c) $CH_3-\underset{\displaystyle OH}{CH}-CH_2-CH_3$ and $CH_3-CH_2-\underset{\displaystyle CH_3}{CH}-CH_2-OH$
   (d) $CH_3-\underset{\displaystyle OH}{CH}-CH_2-CH_3$ and $(CH_3)_2-\underset{\displaystyle OH}{C}-CH_2-CH_3$

4. 2-(hydroxymethyl)cyclohexanol (OH, OH) $\xrightarrow{PCC}$ (A) $\xrightarrow[H^+]{\text{OH} \quad \text{OH (ethylene glycol)}}$ (B) $\xrightarrow[(2)\ H_3O^+]{(1)\ MeMgBr}$ (C) $\xrightarrow{NaBH_4,\ EtOH}$ (D)

   Product (D) in above reaction is:

   Toughnut

   (a) (OH, OH; methyl-substituted cyclohexane)
   (b) (OH, OH; methyl-substituted cyclohexane)
   (c) (OH, OH; methyl-substituted cyclohexane)
   (d) (OH, OH; methyl-substituted cyclohexane)

5. Which of the following reagents would carry out the following transformation? (D = ^2H)

   $C_6H_5-\overset{\displaystyle O}{\overset{\|}{C}}CH_3 \xrightarrow{?} C_6H_5-\underset{\displaystyle D}{\overset{\displaystyle OH}{C}}CH_3$

(a) $NaBD_4$ in CH_3OH (b) $LiAlH_4$, then D_2O
(c) $NaBD_4$ in CH_3OD (d) $LiAlD_4$, then D_2O

6. In the Victor-Meyer's test, the colour given by 1°, 2° and 3° alcohols are respectively:
(a) Red, colourless, blue
(b) Red, blue, colourless
(c) Colourless, red, blue
(d) Red, blue, violet

7. Rate of dehydration of alcohols follows the order:
(a) $2° > 1° > CH_3OH > 3°$
(b) $3° > 2° > 1° > CH_3OH$
(c) $2° > 3° > 1° > CH_3OH$
(d) $CH_3OH > 1° > 2° > 3°$

8. In Williamson synthesis of mixed ether having a primary and a tertiary alkyl group if tertiary halide is used, then :
(a) Rate of reaction will be slow due to slow cleavage of carbon-halogen bond.
(b) Alkene will be the main product.
(c) Simple ether will form instead of mixed ether.
(d) Expected mixed ether will be formed.

9. An ether (A), $C_5H_{12}O$, when heated with excess of hot concentrated HI produced two alkyl halides which when treated with NaOH yielded compounds (B) and (C). Oxidation of (B) and (C) gave a propanone and an ethanoic acid respectively. The IUPAC name of the ether (A) is:

Toughnut

(a) 2-ethoxypropane (b) ethoxypropane
(c) methoxybutane (d) 2-methoxybutane

10. Allyl phenyl ether can be prepared by heating:
(a) $C_6H_5Br + CH_2=CH-CH_2-ONa$
(b) $CH_2=CH-CH_2-Br + C_6H_5ONa$
(c) $C_6H_5-CH=CH-Br + CH_3-ONa$
(d) $CH_2=CH-Br + C_6H_5-CH_2-ONa$

11. Amongst the following alcohols which would react fastest with conc. HCl and $ZnCl_2$?
(a) pentanol (b) 2-methyl butanol
(c) 2-pentanol (d) 2-methyl butan-2-ol

12. Reagent used to convert allyl alcohol to acrolein is
(a) MnO_2 (b) H_2O_2
(c) OsO_4 (d) $KMnO_4$

13. Which are not cleaved by HIO_4?
I. glycerol
II. glycol
III. 1, 3 propenediol
IV. methoxy-2-propanol
(a) I, II, III, IV (b) I, II
(c) II, III (d) III, IV

14. The reaction of phenol with benzoyl chloride to give phenyl benzoate is known as :
(a) Claisen reaction
(b) Schotten-Baumann reaction
(c) Reimer-Tiemann reaction
(d) Gatterman-Koch reaction

15. OH (1-tetralol) $\xrightarrow[\text{Pyridine}]{\text{TsCl}}$ (A) $\xrightarrow{LiAlH_4}$ (B)

Product (B) of the above reaction is:

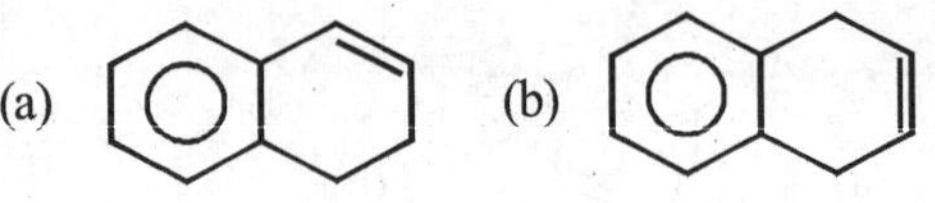

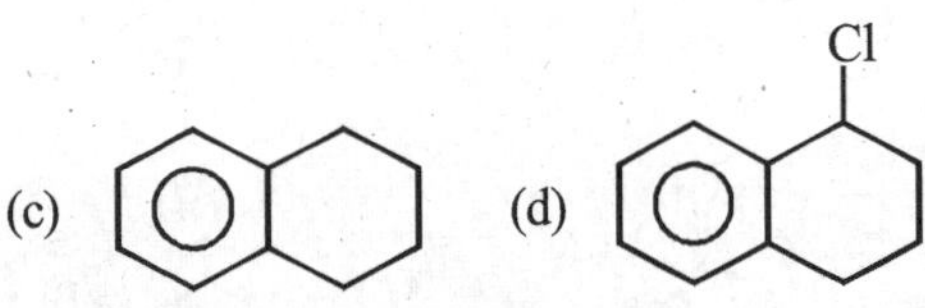

16. Which one of the following compounds will not be soluble in sodium bicarbonate?
(a) 2, 4, 6 - Trinitrophenol
(b) Benzoic acid
(c) *o* - Nitrophenol
(d) Benzene sulphonic acid

17. Which one of the following substituents at *para*-position is most effective in stabilizing the phenoxide ($C_6H_5-O^-$) ion?
(a) $-CH_3$ (b) $-OCH_3$
(c) $-COCH_3$ (d) $-CH_2OH$

18. [4-substituted cyclohexylmethanol: CH_2OH and $CH_3—C(OH)—CH_3$ on the ring] $\xrightarrow{\text{Pyridinium } HCrO_3Cl^-, (PCC), CH_2Cl_2}$

Product of the reaction is:

(a) [ring with CH_2OH and $=C(CH_3)_2$] (b) [ring with CHO and $C(CH_3)_2OH$]

(c) [ring with CO_2H and $C(CH_3)_2OH$] (d) [cyclohexanone with $C(CH_3)_2OH$]

19. Consider the following alcohols,

(I) $C_6H_5CH_2OH$

(II) O_2N–C_6H_4–CH_2OH

(III) CH_3O–C_6H_4–CH_2OH

(IV) Br–C_6H_4–CH_2OH

The order of decreasing reactivities of these alcohols towards nucleophilic substitution with HBr is:

(a) III > I > IV > II (b) III > I > II > IV
(c) I > III > IV > II (d) I > III > II > IV

20. Williamson synthesis of ether is an example of:
(a) Nucleophilic addition
(b) Electrophilic addition
(c) Electrophilic substitution
(d) Nucleophilic substitution

21. Phenol on heating with $CHCl_3$ and NaOH gives salicylaldehyde. The reaction is called :
(a) Reimer - Tiemann reaction
(b) Claisen reaction
(c) Cannizzaro's reaction
(d) Hell - Volhard - Zelinsky reaction

22. In which of the following reactions hydrogen gas will not be evolved?

(a) $CH_3—CH_2—OH \xrightarrow{Na}$
(b) $CH_3—CH_2—OH \xrightarrow{K}$
(c) $CH_3—CH(CH_3)—OH \xrightarrow{Al}$
(d) $CH_3—CH(CH_3)—OH \xrightarrow{CH_3MgBr}$

23. Which is the best reagent to convert isopropyl alcohol to isopropyl bromide?

$$CH_3—CH(CH_3)—OH \xrightarrow{?} CH_3—CH(CH_3)—Br$$

(a) HBr (b) $SOBr_2$
(c) Br_2 (c) CH_3MgBr

24. How many isomers of $C_5H_{11}OH$ will be primary alcohols?

(a) 5 (b) 4 (c) 2 (d) 3

25. On reaction with sodium, 1 mol of a compound X gives 1 mol of H_2. Which one of the following compounds might be X?
(a) $CH_3CH_2CH=CH_2$
(b) $CH_3COOCH_2CH_3$
(c) $CH_3CH_2CH_2OH$
(d) $CH_2(OH)CH_2CH_2CH_2OH$

26. Which of the following reactions would convert 2-butanol into deuterated compound

$CH_3 - CH_2 - \underset{\displaystyle D}{\underset{|}{CH}} - CH_3$? **Toughnut**

(a) 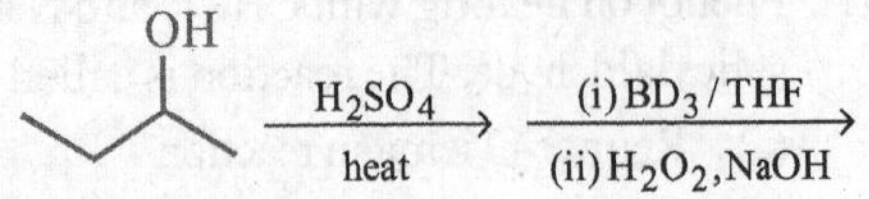 2-butanol (OH) $\xrightarrow[\text{heat}]{H_2SO_4}$ $\xrightarrow[\text{(ii) } H_2O_2, NaOH]{\text{(i) } BD_3/THF}$

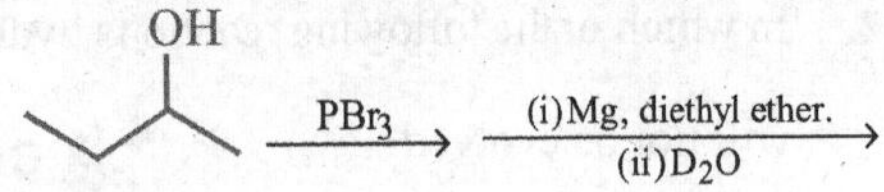

(b) 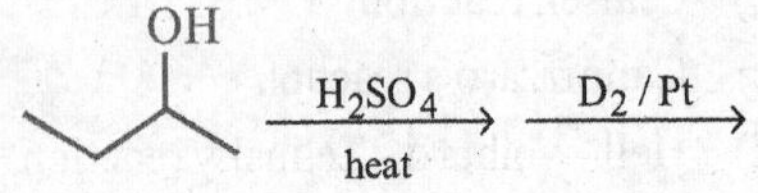 2-butanol (OH) $\xrightarrow[\text{heat}]{H_2SO_4}$ $\xrightarrow{D_2/Pt}$

(c) 2-butanol (OH) $\xrightarrow{PBr_3}$ $\xrightarrow[\text{(ii) } D_2O]{\text{(i) Mg, diethyl ether.}}$

(d) 2-butanol (OH) $\xrightarrow{PBr_3}$ $\xrightarrow{NaOD,\ D_2O}$

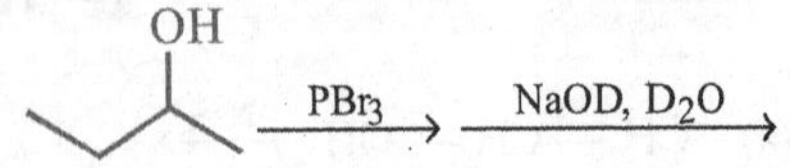

27. To prepare 3-ethylpentan-3-ol, the reagents needed are:

(a) $CH_3CH_2MgBr + CH_3COCH_2CH_3$

(b) $CH_3MgBr + CH_3CH_2CH_2COCH_2CH_3$

(c) $CH_3CH_2MgBr + CH_3CH_2COCH_2CH_3$

(d) $CH_3CH_2CH_2MgBr + CH_3COCH_2CH_3$

28. What is X in the following reaction ?

$$H_2C\overset{O}{\frown}\underset{\displaystyle CH_3}{\underset{|}{C}} - CH_3 \xrightarrow{X} \underset{}{\overset{\displaystyle OH}{\overset{|}{H_2C}}} - \underset{\displaystyle CH_3}{\underset{|}{\overset{\displaystyle OCH_3}{\overset{|}{C}}}} - CH_3$$

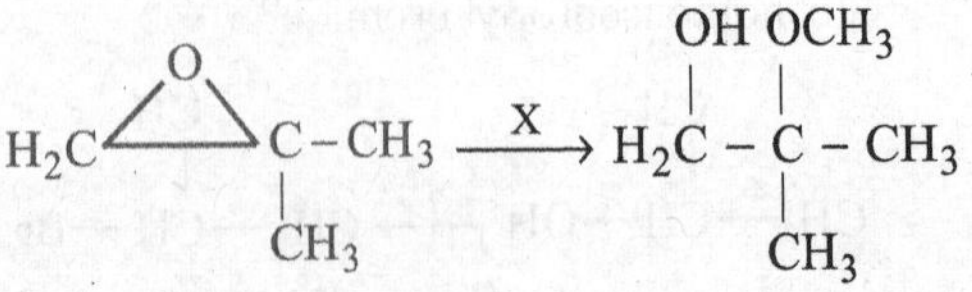

(a) CH_3OH, H_2SO_4 +

(b) $CH_3OH, CH_3O^- Na$

(c) H_2O/H_2SO_4 followed by CH_3OH

(d) CH_3MgBr / ether followed by H_3O^+

29. What is the major product of the following reaction?

$$CH_3-\overset{\displaystyle OH}{\overset{|}{CH}}-CH_2-CH_2-OH \xrightarrow[\text{Pyridine cold}]{CrO_3} \text{Product}$$

(a) $CH_3-\overset{\displaystyle OH}{\overset{|}{CH}}-CH_2-\overset{\displaystyle O}{\overset{\|}{C}}-H$

(b) $CH_3-\overset{\displaystyle O}{\overset{\|}{C}}-CH_2-\overset{\displaystyle O}{\overset{\|}{C}}-H$

(c) $CH_3-\overset{\displaystyle O}{\overset{\|}{C}}-CH_2-\overset{\displaystyle O}{\overset{\|}{C}}-OH$

(d) $CH_3-\overset{\displaystyle OH}{\overset{|}{CH}}-CH_2-\overset{\displaystyle O}{\overset{\|}{C}}-OH$

30. A compound of the formula $C_4H_{10}O$ reacts with sodium and undergoes oxidation to give a carbonyl compound which does not reduce Tollen's reagent, the original compound is

(a) Diethyl ether (b) *n*-Butyl alcohol

(c) Isobutyl alcohol (d) *sec*-Butyl alcohol

31. What amount of bromine will be required to convert 2g of phenol into 2, 4, 6-tribromophenol?

(a) 4.0 (b) 6.0 (c) 10.22 (d) 20.44

32. Identify X in the following sequence:

$$\underset{(X)}{C_3H_8O} \xrightarrow[H_2SO_4]{K_2Cr_2O_7} C_3H_6O \xrightarrow[I_2+NaOH]{\text{warm}} CHI_3$$

(a) $CH_3CH_2CH_2OH$ (b) $CH_3CH(OH)CH_3$

(c) $CH_3OCH_2CH_3$ (d) CH_3CH_2CHO

33. *tert*-Butyl ethyl ether can't be prepared by which reaction? **Toughnut**

(a) *tert* – Butanol + ethanol $\xrightarrow{H^+}$

(b) *tert*-Butyl bromide + sodium ethoxide →

(c) Sodium *tert*-butoxide + ethyl bromide →

(d) Isobutene + ethanol $\xrightarrow{H^+}$

34. Succinic acid

$\xrightarrow{\Delta}$ (A) $\xrightarrow[\Delta]{NH_3}$ (B) $\xrightarrow[KOH]{Br_2}$ (C) ;

Product (C) will be:

(a) $\underset{\displaystyle CH_2-CH_2-NH_2}{\underset{|}{CH_2-CO_2H}}$ (b) $\underset{\displaystyle CH_2-NH_2}{\underset{|}{CH_2-CO_2H}}$

(c) $CH_2—CO_2^- K^+$ / $CH_2—NH_2$ (d) $CH_2—CO_2H$ / $CH_2—CH_2—Br$

35. $(CH_3)_2C(OH)—CH_2—CH_3$ (3° alcohol) $\xleftarrow{(x)}$ $(CH_3)_2C—CH—CH_3$ (epoxide, O bridging) $\xrightarrow{(y)}$ $(CH_3)_2CH—CH(OH)—CH_3$ (2° alcohol)

Find missing reagents.

(a) $x = LiAlH_4, y = NaBH_4$
(b) $x = LiAlH_4/AlCl_3, y = LialH_4$
(c) $x = LiAlH_4, y = LiAlH_4/AlCl_3$
(d) $x = H_2/Ni, y = H_2/Pt$

36. Diethyl ether reacts, inspite of its usual inert nature, with :

(a) Dilute suphuric acid
(b) Dilute sodium hydroxide
(c) Boron trifluoride
(d) Metallic sodium

37. Give the best conditions for this transformation:

$H_3C—CH_2—CH_2—CHO \longrightarrow H_3C—CH_2—CH_2—CH(OCH_3)_2$

(a) CH_3OH, H^+(cat), heat
(b) H_2O, H^+ (cat.), heat
(c) Mg, ether, CH_3OH
(d) $SOCl_2, CH_3OH$

38. The order of reactivity of the following alcohols towards conc. HCl is

(I) F–CH₂–CH₂–CH(OH)–CH₃ skeleton: F, OH, CH_3
(II) F, CH_3, OH
(III) CH_3–CH(OH)–CH_3
(IV) Ph–CH₂–CH₂–OH

(a) I > II > III > IV (b) I > III > II > V
(c) IV > III > II > I (d) IV > III > I > II

39. The end product of the following sequence of reactions

Pentan-1-ol (OH) $\xrightarrow[H_2O]{K_2Cr_2O_7, H_2SO_4}$ $\xrightarrow[H^+ (Cat)]{CH_3OH\ (excess)}$ $\xrightarrow[(ii)\ H_3O^+]{(i)\ 2CH_3MgBr}$

Toughnut

(a) (skeletal structure with OH) (b) (skeletal structure with OCH_3)
(c) (skeletal structure with OH) (d) (skeletal structure)

40. An optically active alcohol of formula $C_9H_{12}O_2$ produced the following compound when refluxed with $KMnO_4$.

Critical Thinking

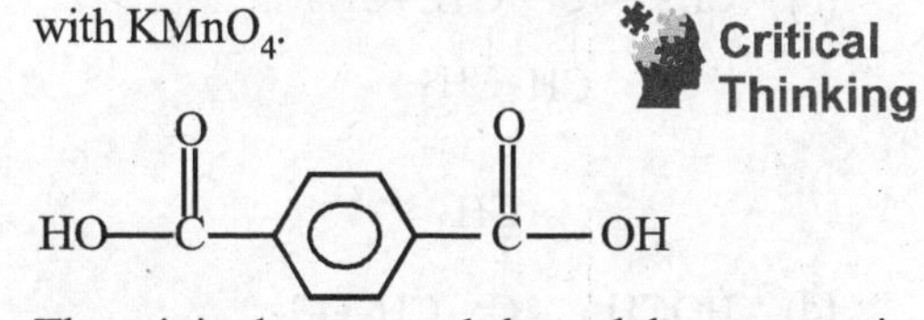

The original compound showed these properties also:

$C_9H_{12}O_2$ (A) $\xrightarrow{Na}$ H_2 liberated

$\xrightarrow{Br_2}$ no rapid reaction

$\xrightarrow[cool]{CrO_3/H^+}$ $C_9H_8O_3$

What is structure of (A)?

(a) CH_3–CH(OH)– on benzene ring, para CH_2OH
(b) HO–CH(OH)– on benzene ring, para CH_3
(c) OH on substituent of benzene ring, para CH_3
(d) both (a) and (b)

41. The product of the following reaction,

$$CH_3-\underset{CH_3}{\overset{CH_3}{\underset{|}{\overset{|}{C}}}}-CH=CH_2 \xrightarrow[\text{(ii) } NaBH_4]{\text{(i) } Hg(OAc)_2,\ H_2O} \text{is}$$

(a) $CH_3-\underset{CH_3}{\overset{CH_3}{\underset{|}{\overset{|}{C}}}}-\underset{OH}{\underset{|}{CH}}-CH_3$

(b) $CH_3-\underset{CH_3}{\overset{CH_3}{\underset{|}{\overset{|}{C}}}}-CH_2-CH_2OH$

(c) $CH_3-\underset{CH_3}{\overset{OH}{\underset{|}{\overset{|}{C}}}}-\underset{CH_3}{\underset{|}{CH}}-CH_3$

(d) $HOCH_2-\underset{CH_3}{\overset{CH_3}{\underset{|}{\overset{|}{C}}}}-CH_2CH_3$

42. An aromatic ether is not cleaved by HI even at 525 K. The compound is

(a) $C_6H_5OCH_3$ (b) $C_6H_5OC_6H_5$

(c) $C_6H_5OC_3H_7$ (d) Tetrahydrofuran

43. Which of the following reactions will not result in the formation of anisole?

(a) Phenol + dimethyl sulphate in presence of a base

(b) Sodium phenoxide is treated with methyl iodide

(c) Reaction of diazomethane with phenol

(d) Reaction of methylmagnesium iodide with phenol

44. Which of the following is not true in case of reaction with heated copper at 300°C?

(a) Phenol ⟶ Benzyl alcohol

(b) Primary alcohol ⟶ Aldehyde

(c) Secondary alcohol ⟶ Ketone

(d) Tertiary alcohol ⟶ Olefin

45. Which of the following synthesis gives 3-methyl-1-hexanol? **Toughnut**

(a) 2 - bromohexane $\xrightarrow[C_2H_5OC_2H_5]{Mg} \xrightarrow[\text{(ii) } H_3O^+]{\text{(i) } H_2C=O}$

(b) 2 - bromopentane $\xrightarrow[C_2H_5OC_2H_5]{Mg} \xrightarrow[\text{(ii) } H_3O^+]{\text{(i) ethylene oxide}}$

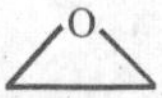

(c) 3 - bromopentane $\xrightarrow[C_2H_5OC_2H_5]{Mg} \xrightarrow[\text{(ii) } H_3O^+]{\text{(i) } CH_3CH=O}$

(d) 1 - bromobutane $\xrightarrow[C_2H_5OC_2H_5]{Mg} \xrightarrow[\text{(ii) } H_3O^+]{\text{(i) } CH_3COCH_3}$

46. What is the product of the following sequence of reactions? **Toughnut**

$$(CH_3)_2C=CH.CH_2CH_3 \xrightarrow[\text{(ii) } H_2O_2,OH^-]{\text{(i) } BH_3/THF} \xrightarrow[CH_2Cl_2]{PCC} \xrightarrow[\text{(ii) } H_3O^+]{\text{(i) } CH_3MgBr}$$

(a) 2, 4 - dimethyl - 3 - pentanol

(b) 2, 3 - dimethyl - 3 - pentanol

(c) 2, 3 - dimethyl - 2 - pentanol

(d) 2, 2 - dimethyl - 3 - pentanol

47. Which of the following are intermediates in the reaction of excess of CH_3MgBr with $C_6H_5COOC_2H_5$ to make 2-phenyl - 2-propanol?

A. $C_6H_5-\underset{CH_3}{\overset{OMgBr}{\underset{|}{\overset{|}{C}}}}-OCH_2CH_3$

B. $C_6H_5-\overset{O}{\overset{||}{C}}-CH_3$

C. $C_6H_5-\underset{CH_3}{\overset{OMgBr}{\overset{|}{\underset{|}{C}}}}-CH_3$

(a) A and B (b) A, B and C
(c) A and C (d) B and C

48. Which of the following compound can react with hydroxylamine? **Tricky**

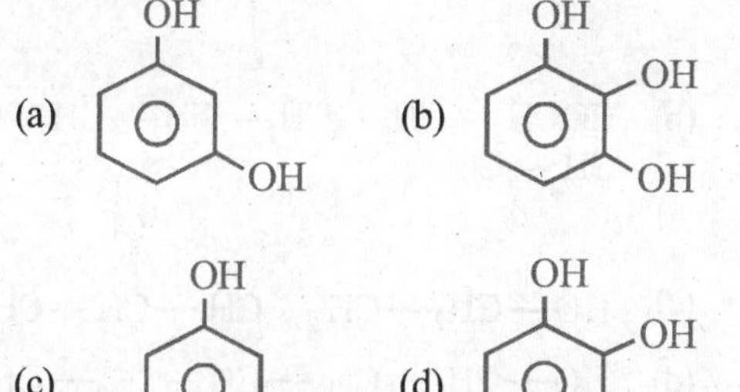

49. Acetic anhydride reacts with diethyl ether in the presence of anhydrous $AlCl_3$ to give
(a) CH_3COOCH_3 (b) $CH_3CH_2COOCH_3$
(c) $CH_3COOCH_2CH_3$ (d) CH_3CH_2OH

50. In the following reaction.

$$C_2H_5OC_2H_5 + 4H \xrightarrow{\text{Red P + HI}} 2X + H_2O;$$

X is
(a) ethane (b) ethylene
(c) butane (d) propane

51. *n*-propyl alcohol and isopropyl alcohol can be chemically distinguished by which reagent
(a) PCl_5
(b) reduction
(c) oxidation with potassium dichromate
(d) ozonolysis

52. An organic compound **A** reacts with methyl magnesium iodide to form an addition product which on hydrolysis forms the compound **B**. Compound **B** gives blue colour salt in Victor Meyer's test. The compounds **A** and **B** are respectively **Toughnut**
(a) acetaldehyde, tertiary butyl alcohol
(b) acetaldehyde, ethyl alcohol
(c) acetaldehyde, isopropyl alcohol
(d) acetone, isopropyl alcohol

53. Identify the major product, **Critical Thinking**

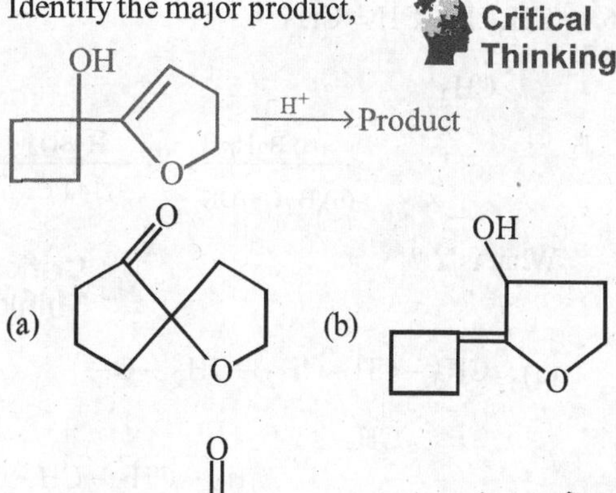

54. (A) $\xrightarrow{PBr_3}$ (C) $\xrightarrow{\text{Mg, ether}}$ Grignard reagent
(A) $\xrightarrow[H_2SO_4]{Na_2Cr_2O_7}$ (B)
(3, 4-dimethyl 3-hexanol) $\xleftarrow{H_3O^+}$ (D)

When Grignard reagent reacts with (B) product (D) will obtained. Reactant (A) of the above reaction is: **Toughnut**

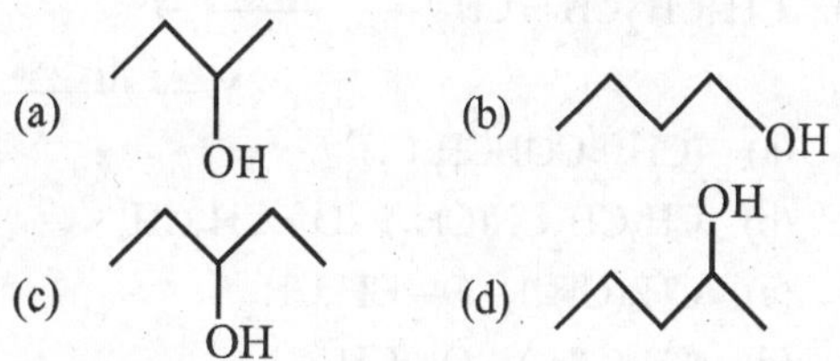

55. Diols (I-IV) which react with CrO_3 in aqueous H_2SO_4 and yield products that readily under go dercarboxylation on heating, are: **Toughnut**

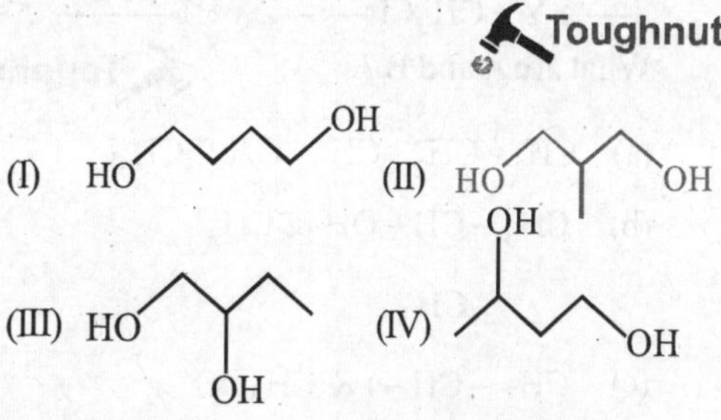

(a) I and II (b) II and III
(c) II and IV (d) I and IV

56. $CH_3CH-CH=CH_2$ (with CH_3 on the second carbon)

$\xrightarrow[\text{(ii) } H_2O_2/OH^-]{\text{(i) } B_2H_6} X \xrightarrow[140°C]{H_2SO_4} Y$.

What is Y ?

Critical Thinking

(a) $CH_3-CH(CH_3)-CH_2-CH_2-O-CH_2-CH_2-CH(CH_3)-CH_3$

(b) $CH_3-CH(CH_3)-CH=CH_2$

(c) $CH_3-CH(CH_3)-CH(CH_3)-O-CH(CH_3)-CH(CH_3)-CH_3$

(d) $CH_3-C(CH_3)(C_2H_5)-O-C(CH_3)(C_2H_5)-CH_3$

57. Identify Z in the sequence:

$CH_3CH_2CH=CH_2 \xrightarrow{HBr/H_2O_2} Y \xrightarrow{C_2H_5ONa} Z$

(a) $(CH_3)_2COHCH_2CH_3$
(b) $CH_3CH_2CH(CH_3)-O-CH_2CH_3$
(c) $CH_3(CH_2)_3-O-CH_2CH_3$
(d) $CH_3(CH_2)_4-O-CH_3$

58. $CH_3-CH=CH_2 \xrightarrow[\text{(ii) } NaBH_4]{\text{(i) } Hg(OAc)_2/H_2O} X + Na \longrightarrow Y + CH_3Cl \longrightarrow Z + HI \xrightarrow{0°C} A + B$

What are A and B ?

Toughnut

(a) $CH_3-CH_2-CH_2-OH$ & CH_3I
(b) $CH_3-CH(CH_3)-OH$ & CH_3I
(c) $CH_3-CH(CH_3)-I$ & CH_3OH
(d) $CH_3-CH_2CH_2I$ & CH_3OH

59. $H_2C=CH-CH_2-CH_2-CH(OH)-CH_3$

$\xrightarrow[\text{Pyridine}]{SOCl_2} (A) \xrightarrow[(H_2O)]{O_3/Zn} \underset{C_5H_9ClO}{(B)} \xrightarrow{NaBH_4} (C)$

Compound (C) is

Critical Thinking

(a) $CH_3-CH(OH)-CH_2-CH(Cl)-CH_3$
(b) $HOCH_2-CH_2-CH_2-CH_2-CH_2-CH_2-Cl$
(c) $HO-CH_2-CH_2-CH_2-CH(Cl)-CH_3$
(d) $HO-CH_2-CH_2-CH(Cl)-CH_2-CH_3$

60. $\underset{(C_5H_{10}O)}{(X)} \xrightarrow{H_3O^+} Y + Z$ (Y and Z both give the Iodoform test). The compound X is –

(a) $CH_3-CH=CH-O-CH_2-CH_3$
(b) $CH_3-CH(CH_3)-O-CH_2-CH_3$ (H on the central carbon)
(c) $CH_3-C(=CH_2)-O-CH_2-CH_3$
(d) Both (a) and (c)

61. What is the major product of the following reaction ?

p-Bromotoluene (CH_3 and Br para on benzene ring) $\xrightarrow[CCl_4]{\text{NBS, benzoyl peroxide}} \xrightarrow[CH_3CH_2OH]{CH_3CH_2ONa}$

(a) Benzene ring with CH_2Br and, para, OCH_2CH_3

(b) $CH_2OCH_2CH_3$ / Br

(c) $CH_2OCH_2CH_3$ / OCH_2CH_3

(d) CH_3 / Br / OCH_2CH_3

62. Dehydration of alcohols by conc. H_2SO_4 takes place according to following steps :

Critical Thinking

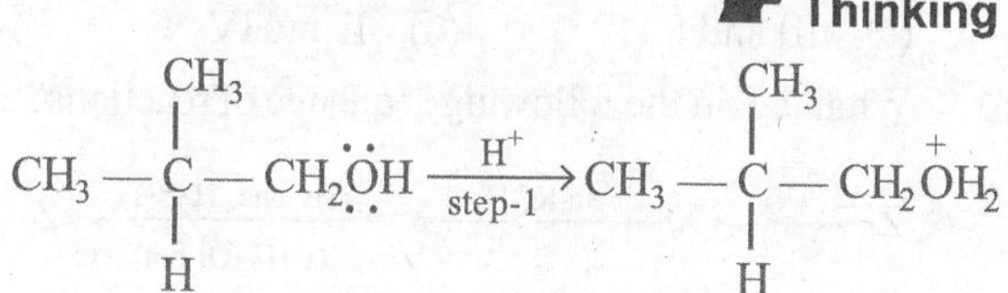

$CH_3-\underset{H}{\overset{CH_3}{C}}-CH_2\ddot{O}H \xrightarrow[\text{step-1}]{H^+} CH_3-\underset{H}{\overset{CH_3}{C}}-CH_2\overset{+}{O}H_2$

$\xrightarrow[\text{step-2}]{-H_2O} CH_3-\underset{H}{\overset{CH_3}{C}}-\overset{+}{C}H_2$

$\xrightarrow[\text{step-3}]{\text{1, 2-H}^- \text{ shift}} CH_3-\underset{+}{\overset{CH_3}{C}}-CH_3 \xrightarrow[\text{step-4}]{-H^+} CH_2=\overset{CH_3}{C}-CH_3$

The slowest and fastest steps in the above reaction are

(a) step 1 is slowest, while 3 is fastest.
(b) step 2 is slowest, while 3 is fastest.
(c) step 2 is slowest, while 4 is fastest.
(d) all steps proceed at equal rate.

63. In order to get 2-hydroxybenzaldehyde from phenol, which of the following reagents is required ?

(a) $(CH_3CO)_2O, H_2SO_4$ (b) $CHCl_3/NaOH$
(c) CO_2, NaOH (d) $CCl_4/NaOH$

64. The structure of the product formed in the reaction given below is:

Critical Thinking

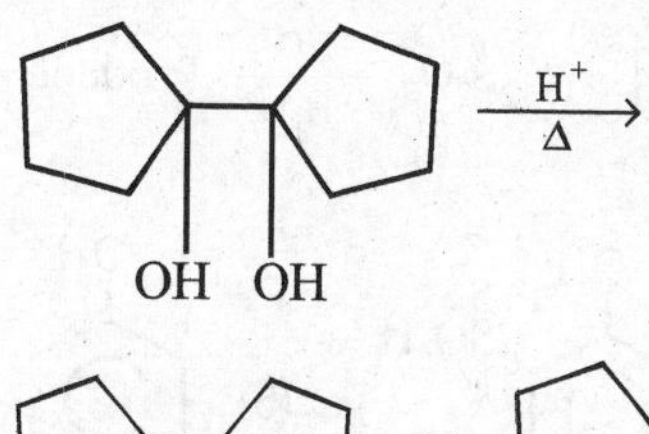

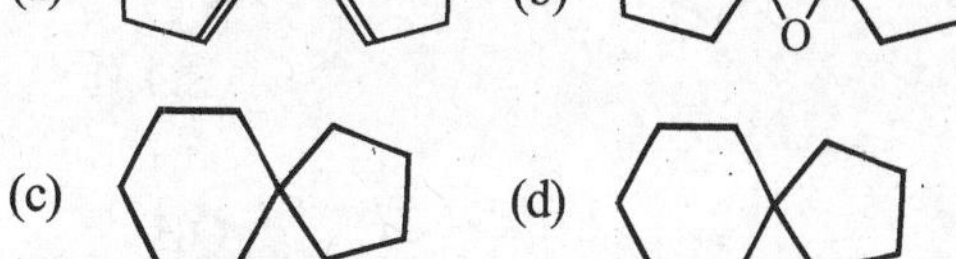

(a) (b) (c) (d)

65. The most probable product in the reacion given below is

$\xrightarrow[\text{heat}]{H_2SO_4}$ OH

(a) (b) (c) (d)

66. Which of the following alcohols is dehydrated most readily with conc. H_2SO_4 ?

Critical Thinking

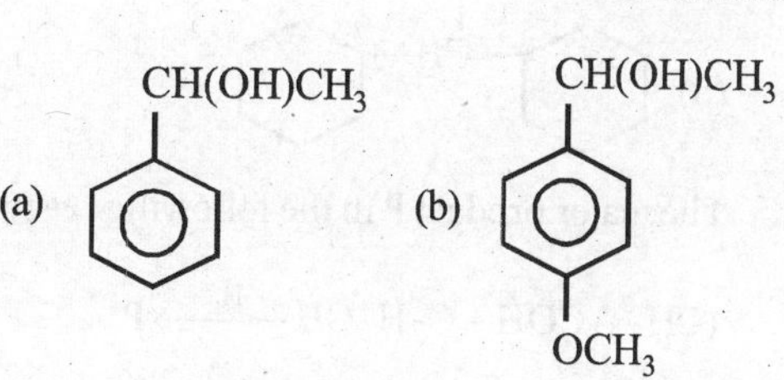

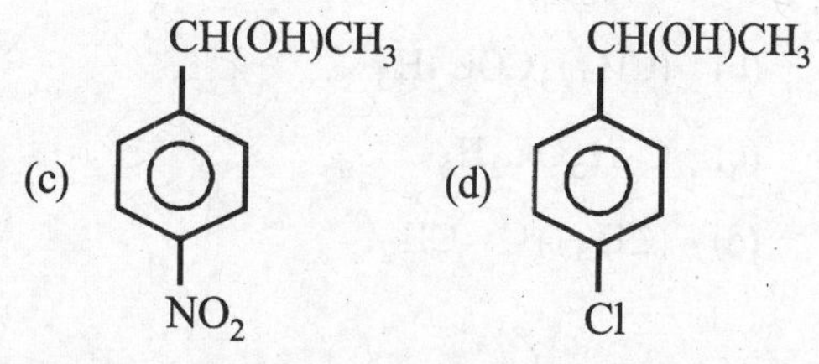

67. Identify the nature of product in the following reaction

OH (phenol) $+ K_2S_2O_8 \xrightarrow[H_2O]{HO^-}$ Product

(a) OH, SO_3H (o-hydroxybenzenesulphonic acid)
(b) OH, SO_3H (p-hydroxybenzenesulphonic acid)
(c) OH, OH (catechol)
(d) OH, OH (hydroquinone)

68. Which one of the following compounds has the most acidic nature?

(a) C_6H_5–CH_2OH
(b) C_6H_5–OH
(c) C_6H_{11}–OH
(d) C_6H_{11}–CH(OH)–C_6H_{11}

69. The major product P in the following reaction is

$(CH_3)_3COH + C_2H_5OH \xrightarrow{H^+} P$

(a) $(CH_3)_3COC(CH_3)_3$
(b) $(CH_3)_3COC_2H_5$
(c) $C_2H_5OC_2H_5$
(d) $(CH_3)_2C=CH_2$

70. Correct increasing order of acidity is as follows

(a) H_2O, C_2H_2, H_2CO_3, phenol
(b) C_2H_2, H_2O, H_2CO_3, phenol
(c) phenol, C_2H_2, H_2CO_3, H_2O
(d) C_2H_2, H_2O, phenol, H_2CO_3

71. Which one/ones of the following reactions will yield 2-propanol?

I. $CH_2=CH-CH_3 + H_2O \xrightarrow{H^+}$

II. $CH_3-CHO \xrightarrow[\text{(ii) } H_2O]{\text{(i) } CH_3MgI}$

III. $CH_2O \xrightarrow[\text{(ii) } H_2O]{\text{(i) } C_2H_5MgI}$

IV. $CH_2=CH-CH_3 \xrightarrow{\text{Neutral } KMnO_4}$

(a) I and II (b) II and III
(c) III and I (d) II and IV

72. What is Z in the following sequence of reactions?

$Z \xrightarrow{PCl_5} X \xrightarrow{\text{Alc. KOH}} Y \xrightarrow[\text{(ii) } H_2O;\text{ boil}]{\text{(i) Conc. } H_2SO_4} Z$

(a) $CH_3CH_2CH_2OH$ (b) $CH_3CH(OH)CH_3$
(c) $(CH_3CH_2)_2CHOH$ (d) $CH_3CH=CH_2$

73. Which of the following has lowest boiling point?

(a) *p*-Nitrophenol (b) *m*-Nitrophenol
(c) *o*-Nitrophenol (d) Phenol

74. Arrange the following in increasing order of their acidity?

(a) *o*–cresol, (b) salicylic acid, (c) phenol

(a) c < a < b (b) b < c < a
(c) a < b < a (d) a < c < b

75. Which of the following diols would cleave into two fragments with HIO_4

(a) 1, 3-hexanediol (b) 2, 4-hexanediol
(c) 1, 6-hexanediol (d) 3, 4-hexanediol

76. In the reaction:

$$CH_3-\underset{\displaystyle |}{\overset{\displaystyle CH_3}{CH}}-CH_2-O-CH_2-CH_3+HI\xrightarrow{\text{Heated}}$$

Which of the following compounds will be formed?

(a) $CH_3-\underset{\displaystyle CH_3}{\underset{|}{CH}}-CH_3+CH_3CH_2OH$

(b) $CH_3-\underset{\displaystyle CH_3}{\underset{|}{CH}}-CH_2OH+CH_3CH_3$

(c) $CH_3-\overset{\displaystyle CH_3}{\overset{|}{CH}}-CH_2OH+CH_3-CH_2-I$

(d) $CH_3-\overset{\displaystyle CH_3}{\overset{|}{CH}}-CH_2-I+CH_3CH_2OH$

77. The following change can be carried out with

O

OH $\xrightarrow{?}$

OH

(a) $NaBH_4$ (b) $LiAlH_4$

(c) H_2/Pt (d) PCC/CH_2Cl_2

78.  $\xrightarrow{O_3}$ [ ] $\xrightarrow{NaBH_4}$ Product is

(a) 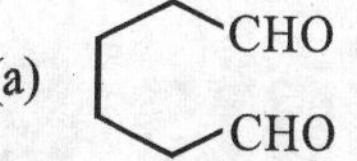

(b) COOH COOH

(c) mixture of (a) and (b)

(d) HO OH

79. Ethanol can be prepared more easily by which reaction ?

(i) $CH_3CH_2Br+H_2O\longrightarrow CH_3CH_2OH$

(ii) $CH_3CH_2Br+Ag_2O$ (in boiling water) $\longrightarrow CH_3CH_2OH$

(a) by (i) reaction

(b) by (ii) reaction

(c) Both reactions proceed at same rate

(d) by none

80. 3 moles of ethanol react with one mole of phosphorus tribromide to form 3 moles of bromoethane and one mole of X. Which of the following is X?

(a) H_3PO_4 (b) H_3PO_2

(c) HPO_3 (d) H_3PO_3

Answer KEYs

1	(c)	9	(a)	17	(c)	25	(d)	33	(b)	41	(a)	49	(c)	57	(c)	65	(c)	73	(c)
2	(a)	10	(a)	18	(b)	26	(c)	34	(c)	42	(b)	50	(a)	58	(d)	66	(b)	74	(d)
3	(d)	11	(d)	19	(a)	27	(c)	35	(c)	43	(d)	51	(c)	59	(c)	67	(d)	75	(d)
4	(b)	12	(a)	20	(d)	28	(a)	36	(c)	44	(a)	52	(c)	60	(c)	68	(b)	76	(c)
5	(a)	13	(d)	21	(a)	29	(b)	37	(a)	45	(b)	53	(a)	61	(b)	69	(b)	77	(b)
6	(b)	14	(b)	22	(d)	30	(d)	38	(c)	46	(b)	54	(a)	62	(c)	70	(d)	78	(d)
7	(b)	15	(c)	23	(b)	31	(c)	39	(a)	47	(b)	55	(c)	63	(b)	71	(a)	79	(b)
8	(b)	16	(c)	24	(b)	32	(b)	40	(d)	48	(c)	56	(a)	64	(c)	72	(b)	80	(d)

Hints & Solutions

1. (c) The C – O bond length in alcohols is 142 pm and in Phenol it is 136 pm. The C – O bond length in phenol is shorter than methanol it is due to the conjugation of unshared pair of electrons on oxygen with the ring, which imparts double bond character to the C – O bond.

Methanol: $H_3C \overset{142\ pm}{—} \ddot{\underset{\cdot\cdot}{O}}—H$ Phenol: $C_6H_5 \overset{136\ pm}{—} \ddot{\underset{\cdot\cdot}{O}}—H$

2. (a) Phenol has activating (electron releasing)–OH group and bromine water supplies Br^+ ion easily, hence under such conditions reaction does not stop at monobromo or dibromo stage but a fully brominated (2,4,6,-tribromophenol) compound is the final product.

$$C_6H_5OH + 3\,Br_2 \longrightarrow \text{2, 4, 6-Tribromophenol}$$

2, 4, 6-Tribromophenol

3. (d) $CH_3—\underset{OH}{\overset{CH_3}{C}}—CH_2CH_3 \xrightarrow{H^+} CH_3—\underset{+}{\overset{CH_3}{C}}—CH_2CH_3$ (3° carbocation)

$CH_3—\underset{OH}{CH}—CH_2CH_3 \xrightarrow{H^+} CH_3—\underset{+}{CH}—CH_2CH_3$ (2° carbocation)

Stability order of carbocation 3° > 2°.

4. (b) (A) 2-oxocyclohexane with CHO (B) cyclohexanone with CH(–O–)₂ dioxolane (C) cyclohexane with OH, CH₃ and CHO (D) cyclohexane with OH, CH₃ and CH₂—OH

5. (a) $C_6H_5\text{—}\overset{O}{\overset{\|}{C}}CH_3 \xrightarrow[CH_3OH]{NaBD_4} C_6H_5\text{—}\underset{D}{\overset{OH}{C}}CH_3$

(Nucleophilic addition takes place)

6. (b) ***Victor Meyer's test:*** The various steps involved are

(i) $RCH_2OH \xrightarrow{HI} RCH_2I \xrightarrow{AgNO_2} RCH_2NO_2 \xrightarrow{HNO_2} R-\underset{\underset{NOH}{\|}}{C}-NO_2 \xrightarrow{KOH}$ Blood red colour

Nitrolic acid

(ii) $RCHOHR \xrightarrow{HI} R_2CHI \xrightarrow{AgNO_2} R_2CHNO_2$

$R_2C(H)NO_2 \xrightarrow[H_2O]{HONO} R_2C(NO)NO_2 \xrightarrow{KOH}$ blue colour

(iii) $R_3COH \xrightarrow{HI} R_3CI \xrightarrow{AgNO_2} R_3CNO_2 \xrightarrow{HNO_2}$ No reaction $\xrightarrow{KOH}$ Colourless

7. (b) The order of dehydration among three type of alcohols is 3° > 2° > 1° > CH_3OH. This behaviour is related to the relative stabilities of carbocations (3° > 2° > 1°).

8. (b) The tertiary alkyl halide undergoes elimination reaction to give alkenes.

$$CH_3-\underset{CH_3}{\overset{CH_3}{C}}-X + NaOC_2H_5 \longrightarrow CH_3-\overset{CH_3}{C}=CH_2$$

2-methyl propene

9. (a) $\underset{(A)}{CH_3CH_2OCH(CH_3)_2} \xrightarrow{\text{Conc. HI}} \underset{(B)}{CH_3CH_2I} + \underset{(C)}{(CH_3)_2CHI}$

$CH_3CH_2I \xrightarrow{NaOH} CH_3CH_2OH \xrightarrow{[O]} CH_3COOH$

$(CH_3)_2CHI \xrightarrow{NaOH} (CH_3)_2CHOH \xrightarrow{[O]} (CH_3)_2CO$

hence the IUPAC name of compound is $CH_3-CH_2-O-\overset{2}{C}H(\overset{3}{C}H_3)-\overset{1}{C}H_3$

2–ethoxy propane

10. (a) $C_6H_5Br + NaO-CH_2-CH=CH_2 \longrightarrow C_6H_5-O-CH_2-CH=CH_2 + NaBr$

Allyl phenyl ether

11. (d) $ZnCl_2$ + conc. HCl is Lucas reagent. Lucas reagent reacts fastest with tertiary alcohol.
3° alcohol + Lucas reagent = Immediate turbidity

12. (a) MnO_2 being a mild oxidising agent stops the oxidation of $-CH_2OH$ group at aldehyde stage.

13. (d) HIO_4 will not oxidise diol from 1, 3 atom and cannot used for cleavage of ether.

14. (b) Phenol + benzoyl chloride ⟶ Phenyl benzoate

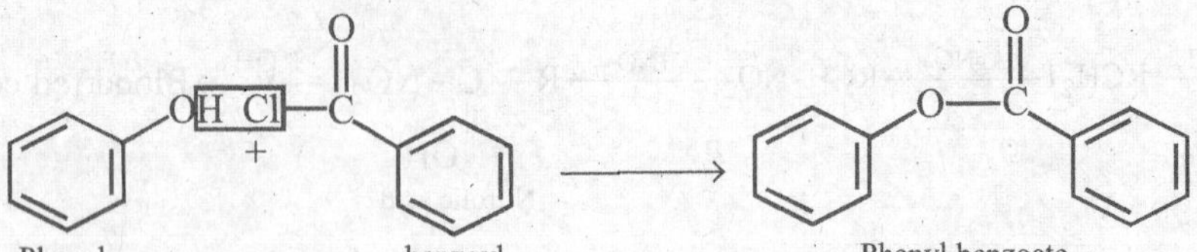

15. (c) $\xrightarrow{TsCl}$ (A) $\xrightarrow{LiAlH_4}$ (B)

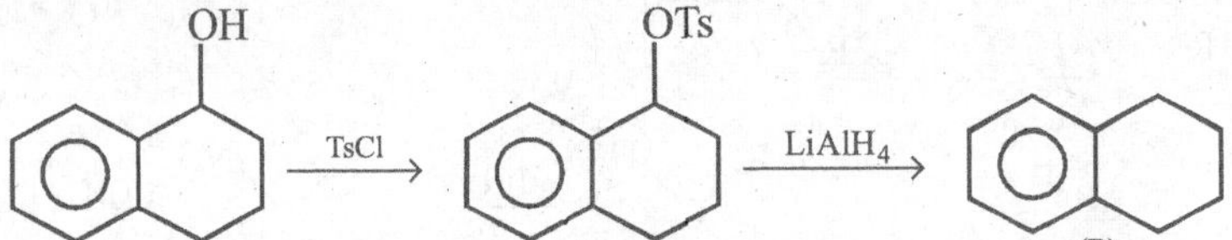

16. (c) Due to intramolecular hydrogen bonding this will not be soluble in sodium bicarbonate.

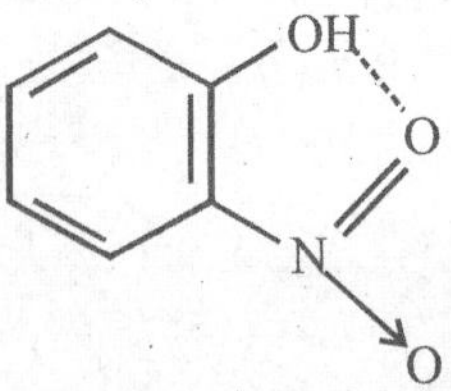

17. (c) Electron withdrawing group stabilises the benzene ring due to delocalisation of charge.
$-CH_3$ and $-CH_2OH$ are electron donating group and hence decrease the stability of benzene ring $-OCH_3$ is weaker electron withdrawing group than $-COCH_3$. Hence $-COCH_3$ group more stabilize the phenoxide ion at *p*-position.

18. (b) $\xrightarrow{PCC}$

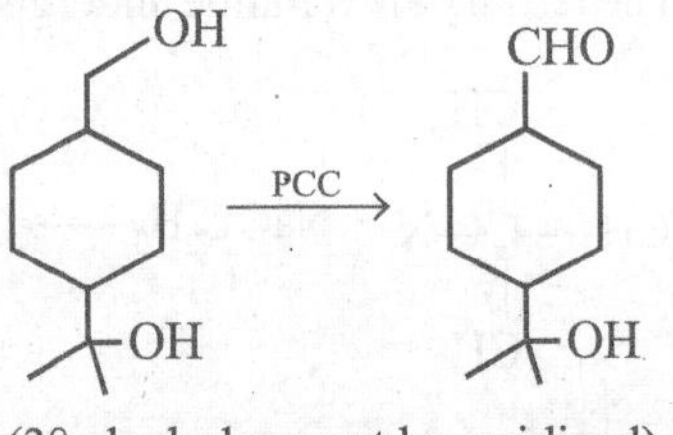

(3° alcohol cannot be oxidized)

19. (a) More stable carbocation, more is the rate toward HBr (acid).

20. (d) This method is suitable for the preparation of a wide variety of unsymmetrical ethers. The nucleophilic substitution of halides with alkoxide leads to desired product.

21. (a) Phenol $\xrightarrow[\text{aq. NaOH, 57°C}]{CHCl_3}$ [intermediate with $CHCl_2$] ⟶ ⟶ Salicyaldehyde (chief product) + *p*-hydroxy benzaldehyde

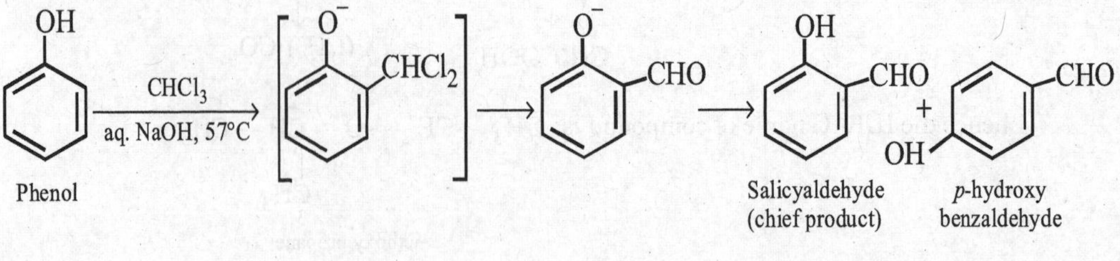

22. (d) $CH_3-\underset{\underset{CH_3}{|}}{CH}-\overset{-}{O}-\overset{+}{H} \xrightarrow{\overset{-}{C}H_3\overset{+}{Mg}Br} CH_4$ (Major product)

Acid-base reaction takes place.

23. (b) $CH_3-\overset{\overset{CH_3}{|}}{CH}-OH \xrightarrow{SOBr_2} CH_3-\overset{\overset{CH_3}{|}}{CH}-Br + SO_2 + HBr$

Side products are gases.

In case of HBr elimination also favours to produce alkene.

24. (b) (i) $CH_3CH_2CH_2CH_2CH_2OH$

(ii) $CH_3CH_2\underset{\underset{CH_3}{|}}{CH}CH_2OH$

(iii) $CH_3\overset{\overset{CH_3}{|}}{CH}CH_2CH_2OH$

(iv) $(CH_3)_3CCH_2OH$

25. (d) Since 1 mole of compound X reacts with Na to evolve 1 mole of H_2 gas, therefore the compound should have 2 active hydrogen atoms per mole which is possible only in option *d*.

$$CH_2OHCH_2CH_2CH_2OH + 2Na \longrightarrow NaOCH_2CH_2CH_2CH_2ONa + 2H$$

26. (c) $C_2H_5.CHOH.CH_3 \xrightarrow{PBr_3} C_2H_5-\overset{\overset{Br}{|}}{CH}-CH_3 \xrightarrow[D_2O]{Mg,\ ether} CH_3-\overset{\overset{D}{|}}{CH}-CH_3$

27. (c) First write down the structure of the required compound, observe the nature of functional group,

$$CH_3CH_2-\underset{\underset{OH}{|}}{\overset{\overset{C_2H_5}{|}}{C}}-CH_2CH_3$$

and review the memory for preparing the required functional group. Here the required compound is a 3° alcohol having three ethyl groups, so only combination given in option *c* will introduce three ethyl groups.

$$CH_3CH_2MgBr + CH_3CH_2COCH_2CH_3 \longrightarrow CH_3CH_2\underset{\underset{OH}{|}}{\overset{\overset{CH_2CH_3}{|}}{C}}CH_2CH_3$$

28. (a) $CH_2\overset{O}{\frown}\underset{\underset{CH_3}{|}}{C}-CH_3 \xrightarrow[H_2SO_4]{CH_3OH} H_2\overset{\overset{OH}{|}}{C}-\underset{\underset{CH_3}{|}}{\overset{\overset{OCH_3}{|}}{C}}-CH_3$

29. (b) CrO_3 / pyridine is mild oxidising agent.

1° Alcohol → Aldehyde

2° Alcohol → Ketone ;

$$CH_3-\underset{\substack{|\\OH}}{CH}-CH_2-CH_2-OH \xrightarrow{CrO_3} CH_3-\overset{\substack{O\\||}}{C}-CH_2-\overset{\substack{O\\||}}{C}-H$$

30. (d) Since the compound ($C_4H_{10}O$) react with sodium, it must be alcohol (option *b, c, or d*). As it is oxidised to carbonyl compound which does not reduce Tollen's reagent, the carbonyl compound should be a ketone and thus $C_4H_{10}O$ should be a secondary alcohol, i.e. *sec*-butyl alcohol; other two given alcohols are 1°.

31. (c)

$$\underset{94\text{ g}}{C_6H_5OH} + \underset{3\times160=480\text{ g}}{3Br_2} \longrightarrow \text{2,4,6-}Br_3C_6H_2OH$$

Thus 94 g of phenol require $Br_2 = 480$ g

2 g of phenol require $Br_2 = \frac{480}{94} \times 2 = 10.22$ g

32. (b) Since the compound X (C_3H_6O) gives iodoform test, it must have $-COCH_3$ groups hence it should be CH_3COCH_3. Therefore X should be $CH_3CH(OH)CH_3$.

33. (b) $(CH_3)_3CBr + NaOC_2H_5$ can't be applied for synthesising the ether because sod. ethoxide, being a strong base, will preferentially cause elimination reaction.

$$(CH_3)_3CBr \xrightarrow{^-OC_2H_5} (CH_3)_2C=CH_2 + HBr$$

In isobutene + ethanol, isobutene will form *tert*-butyl cation which reacts with ethanol, a nucleophile to form ether.

$$(CH_3)_2C=CH_2 \xrightarrow{H^+} (CH_3)_2\overset{+}{C}CH_3 \xrightarrow[\text{(ii) }-H^+]{\text{(i) }CH_3CH_2OH} (CH_3)_3COCH_2CH_3$$

34. (c)

$$\begin{matrix}CH_2COOH\\ |\\ CH_2COOH\end{matrix} \xrightarrow[-H_2O]{\Delta} \text{(A)} \xrightarrow[\Delta]{H_2-NH} \text{(B)} \xrightarrow[KOH]{Br_2} \begin{matrix}CH_2CO\overset{-}{O}\overset{+}{K}\\ |\\ CH_2NH_2\end{matrix}$$

(A) succinic anhydride

(B) Succinamide

35. (c) $(CH_3)_2C(-O-)CH-CH_3 \xrightarrow{H^-} (CH_3)_2C(O^-)-CH_2-CH_3 \xrightarrow{H^+} (CH_3)_2C(OH)-CH_2CH_3$

Less hindered carbon

$(CH_3)_2C(-O-)CH-CH_3 \xrightarrow{AlCl_3} (CH_3)_2C(-O^+(AlCl_3)-)CH-CH_3$

$\longrightarrow (CH_3)_2\overset{+}{C}-CH(OAlCl_3)-CH_3 \xrightarrow{H^-} CH_3-CH(CH_3)-CH(OH)-CH_3$

x = $LiAlH_4$

y = $LiAlH_4/AlCl_3$

36. (c)

37. (a) $CH_3CH_2CH_2CHO \xrightarrow[H^+]{CH_3OH} CH_3CH_2CH_2CH(OCH_3)(OH) \xrightarrow[H^+]{CH_3OH} CH_3CH_2CH_2CH(OCH_3)_2$ (Acetal)

38. (c) The order of reactivity depends upon the stability of the carbocations formed.

From (I) 2° carbocation	From (II) 2°	From (III) 2°	From (IV). 1°, Most stable benzyl carbocation

Remember that presence of electron-withdrawing group intensifies i.e., destabilises the carbocation thus (i) and (ii) are less stable than (iii). Further (i) is less stable than (ii) because –I effect is more pronounced in (i) due to less distance between F and positive charge. Thus the stability order of the four carbocations and reactivity of their parent alcohols will be

IV > III > II > I

39. (a) $CH_3CH_2CH_2CH_2OH \xrightarrow[H_2O]{K_2Cr_2O_7, H_2SO_4} CH_3CH_2CH_2COOH \xrightarrow[H^+]{CH_3OH} CH_3CH_2CH_2COOCH_3$

$\xrightarrow[\text{Excess}/H_3O^+]{CH_3MgBr} CH_3CH_2CH_2C(OH)(CH_3)-CH_3$

40. (d) 4-(1-hydroxyethyl)benzyl alcohol $\xrightarrow{CrO_3}$ 4-acetylbenzoic acid (CO_2H), 1-(4-methylphenyl)ethane-1,2-diol $\xrightarrow{CrO_3}$ 4-methylphenylglyoxylic acid ($CH_3C_6H_4-C(=O)-CO_2H$)

41. (a) Recall that oxymercuration-reduction of alkenes lead to hydration in Markovnikov's way. Thus

$$(CH_3)_3CCH=CH_2 \xrightarrow[\text{(ii) } NaBH_4]{\text{(i) } Hg(OAc)_2, H_2O} (CH_3)_3C\overset{OH}{\overset{|}{C}}HCH_3$$

42. (b) Due to greater electronegativity of sp^2-hybridized carbon atoms of the benzene ring, diaryl ethers are not attacked by nucleophiles like I^-.

43. (d) Phenol has active (acidic) hydrogen so it reacts with CH_3MgI to give CH_4, and not anisole

$$C_6H_5OH + CH_3MgI \longrightarrow CH_4 + C_6H_5OMgI$$

44. (a) Phenol does not react with heated Cu at 300°C.

45. (b) $$CH_3-\underset{Br}{\underset{|}{CH}}-CH_2-CH_2-CH_3 \xrightarrow[(C_2H_5)_2O]{Mg}$$

$$CH_3-\underset{MgBr}{\underset{|}{CH}}-CH_2-CH_2-CH_3 \xrightarrow{\text{ethylene oxide}} CH_3-\underset{CH_2CH_2OMgBr}{\underset{|}{CH}}-CH_2-CH_2-CH_3$$

$$\xrightarrow{H_3O^+} CH_3-\underset{CH_2.CH_2OH}{\underset{|}{CH}}-CH_2-CH_2-CH_3$$

46. (b) $$(CH_3)_2C=CH.CH_2CH_3 \xrightarrow[H_2O_2, OH^-]{BH_3/THF} (CH_3)_2CH-\underset{OH}{\underset{|}{CH}}-CH_2CH_3 \xrightarrow{[O]}$$

$$(CH_3)_2CH-\overset{O}{\overset{||}{C}}-CH_2-CH_3 \xrightarrow[H_3O^-]{CH_3MgBr} (CH_3)_2CH-\underset{CH_3}{\underset{|}{\overset{OH}{\overset{|}{C}}}}-CH_2-CH_3$$

2, 3 - Dimethyl -3 - pentanol

47. (b) $C_6H_5-\overset{O}{\overset{\|}{C}}-OC_2H_5 + CH_3MgBr \longrightarrow C_6H_5-\underset{CH_3}{\underset{|}{\overset{OMgBr}{\overset{|}{C}}}}-OC_2H_5 \longrightarrow C_6H_5-\underset{CH_3}{\underset{|}{\overset{O}{\overset{\|}{C}}}} + C_2H_5OMgBr$

(A) (B)

$\xrightarrow{CH_3MgBr} C_6H_5-\underset{CH_3}{\underset{|}{\overset{OMgBr}{\overset{|}{C}}}}-CH_3 \xrightarrow{H_3O^+} C_6H_5-\underset{CH_3}{\underset{|}{\overset{OH}{\overset{|}{C}}}}-CH_3$

(C)

48. (c)

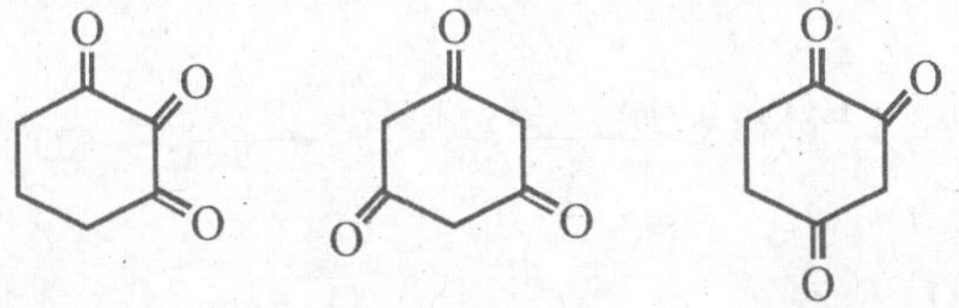

Unstable Stable Unstable

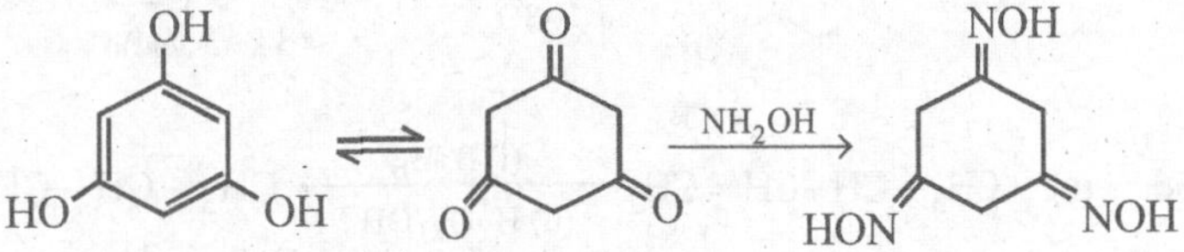

$\xrightarrow{NH_2OH}$ HO OH NOH HON NOH

49. (c) $\underset{\text{Ether}}{(CH_3CH_2)_2O} + \underset{\text{Acid anhydride}}{(CH_3CO)_2O} \xrightarrow{\text{anhy. } AlCl_3} \underset{\text{Ester}}{CH_3COOCH_2CH_3}$

50. (a) Ethers are reduced by red P and HI to alkanes through alkyl iodides

$$C_2H_5OC_2H_5 \xrightarrow{\text{red P+HI}} 2C_2H_5I \xrightarrow{\text{red P+HI}} 2C_2H_6$$

51. (c) In such reactions, first write down the sequence of reactions and then proceed backward/forward from a known compound, reagent, or known fact (test).

$$A \xrightarrow[\text{(ii) } H_2O]{\text{(i) } CH_3MgI} B \xrightarrow{\text{Victor Meyer test}} \text{Blue colour}$$

Appearance of blue colour in Victor Meyer test indicates that B is 2° alcohol and thus A must be an aldehyde other than HCHO. Thus among the given options, only option (*c*) is correct.

52. (c)

53. (a) OH $\xrightarrow{H^+}$:OH ⊕ O $\longrightarrow$ O O

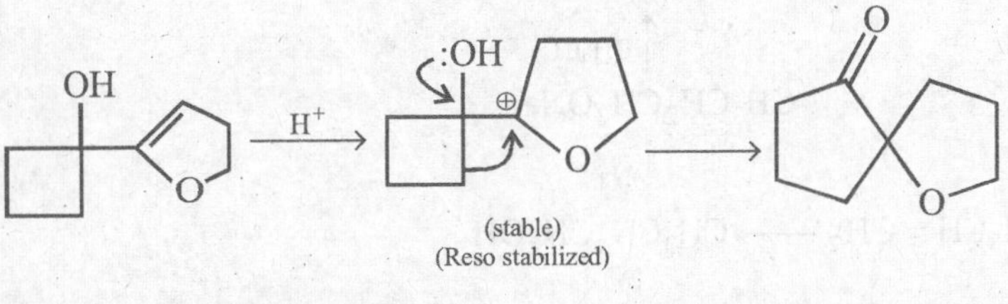

(stable)
(Reso stabilized)

54. (a) (A) (3-methyl-2-butanol, skeletal: OH) $\xrightarrow{PBr_3}$ (Br) $\xrightarrow{Mg}$ (MgBr) Grignard reagent

(A) $\xrightarrow[H_2SO_4]{Na_2Cr_2O_7}$ (B) (ketone, O) ; MgBr + O $\longrightarrow$ (D) OMgBr $\xrightarrow{H_3O^+}$ OH (3,4-dimethyl 3-hexanol)

55. (c) (II) HO–(skeletal)–HO $\xrightarrow[aq.\ H_2SO_4]{CrO_3}$ CO_2H–(skeletal)–CO_2H (Gem) $\xrightarrow[(-CO_2)]{\Delta}$ $CH_3-CH_2-CO_2H$

(IV) (skeletal diol: OH, OH) $\xrightarrow[aq.\ H_2SO_4]{CrO_3}$ $CH_3-\overset{O}{\overset{\|}{C}}-CH_2-\overset{O}{\overset{\|}{C}}-OH$ (β-kero acid) $\xrightarrow{\Delta}$ $CH_3-\overset{O}{\overset{\|}{C}}-CH_3+CO_2$

56. (a) $CH_3-\underset{CH_3}{\underset{|}{CH}}-CH=CH_2 \xrightarrow[(ii)\ H_2O_2/OH^-]{(i)\ B_2H_6} CH_3-\underset{CH_3}{\underset{|}{CH}}-CH_2-\underset{OH}{\underset{|}{CH_2}}$

$\downarrow H_2SO_4/140°C$

$CH_3-\underset{CH_3}{\underset{|}{CH}}-CH_2-CH_2-O-CH_2-CH_2-\underset{CH_3}{\underset{|}{CH}}-CH_3$ (Y)

57. (c) $CH_3CH_2CH=CH_2 \xrightarrow{HBr/H_2O_2} CH_3CH_2CH_2CH_2Br \xrightarrow{C_2H_5ONa} CH_3(CH_2)_3OC_2H_5$

In presence of H_2O_2, HBr adds in *anti*-Markovnikov's way (peroxide effect).

58. (d) Going backward from A and B

$\underset{(A\ and\ B)}{CH_3CH_2CH_2I+HOCH_3} \xleftarrow{HI} \underset{(Z)}{CH_3CH_2CH_2OCH_3}$

$\uparrow CH_3Cl$

$CH_3CH_2CH_2ONa$

$\uparrow Na$

$CH_3CH=CH_2 \longrightarrow \underset{(X)}{CH_3CH_2CH_2OH}$

59. (c) $H_2C=CH-CH_2-CH_2-\underset{\displaystyle OH}{\underset{|}{CH}}-CH_3 \xrightarrow[\text{Pyridine}]{SOCl_2}$

$$H_2C=CH-CH_2-CH_2-\overset{\displaystyle Cl}{\overset{|}{CH}}-CH_3$$

$$\downarrow O_3/Zn(H_2O)$$

$$H-\overset{\displaystyle O}{\overset{||}{C}}-CH_2-CH_2-\overset{\displaystyle Cl}{\overset{|}{CH}}-CH_3$$

$$\downarrow NaBH_4$$

$$HO-CH_2-CH_2-CH_2-\overset{\displaystyle Cl}{\overset{|}{CH}}-CH_3$$

60. (c) $\underset{(X)}{CH_3-\underset{\displaystyle CH_2}{\underset{||}{C}}-O-CH_2-CH_3} \xrightarrow{H_3O^+}$

$$CH_3-\underset{\displaystyle CH_2}{\underset{||}{C}}-OH + \underset{(z)}{HO-CH_2-CH_3}$$

$$\rightleftharpoons$$

$$\underset{(y)}{CH_3-\underset{\displaystyle O}{\underset{||}{C}}-CH_3}$$

61. (b) p-Bromotoluene (CH_3, Br) $\xrightarrow[\text{CCl}_4\text{, heat, allylic bromination}]{\text{NBS, benzoyl peroxide}}$ p-Bromobenzyl bromide (CH_2Br, Br) $\xrightarrow[CH_3CH_2OH]{CH_3CH_2ONa}$ ($CH_2OCH_2CH_3$, Br)

62. (c) Step 2 involves the formation of carbonium ion by the loss of weakly basic H_2O molecule. It is slowest step. Step 4 involves the conversion of an unstable (or intermediate) into a quite stable product, hence it is fastest step.

63. (b) 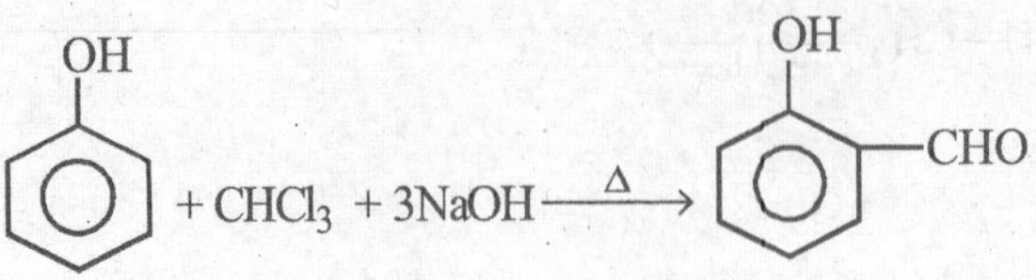 Phenol (OH) $+ CHCl_3 + 3NaOH \xrightarrow{\Delta}$ salicylaldehyde (OH, CHO)

64. (c) 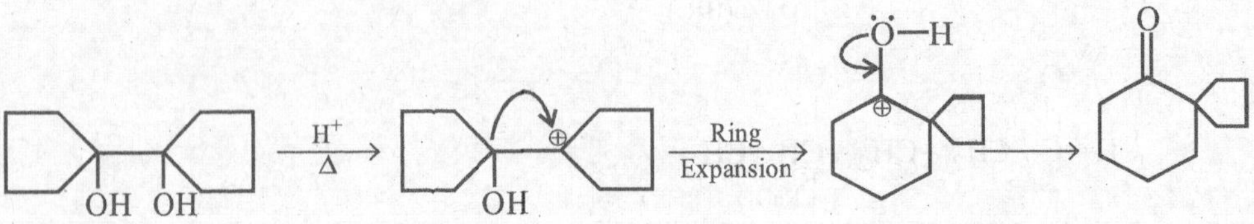 OH OH $\xrightarrow[\Delta]{H^+}$ OH ⊕ $\xrightarrow[\text{Expansion}]{\text{Ring}}$ Ö—H ⊕ $\longrightarrow$ O

65. (c) 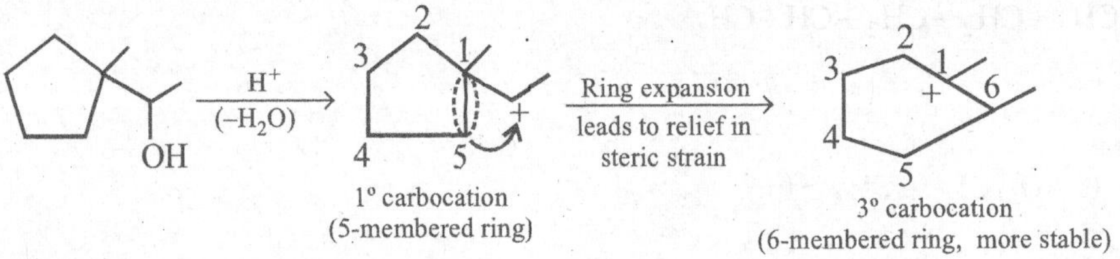 OH $\xrightarrow[(-H_2O)]{H^+}$ 3, 2, 1, 4, 5, + $\xrightarrow[\text{leads to relief in steric strain}]{\text{Ring expansion}}$ 3, 2, 1, +, 6, 4, 5

1° carbocation (5-membered ring)

3° carbocation (6-membered ring, more stable)

or 3, 2, 1, +, 6, 4, 5 $\xrightarrow{-H^+}$

3° carbocation

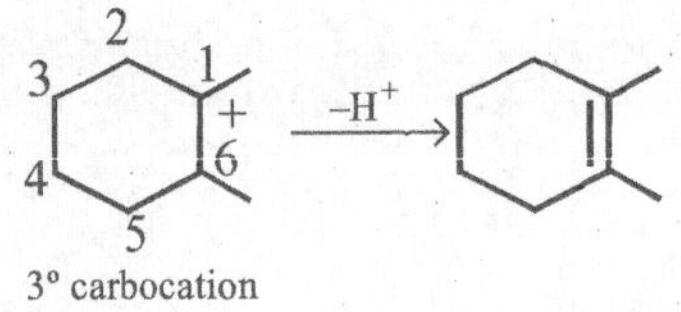

66. (b) The reaction involves the formation of carbocation as intermediate. Hence more the stability of the carbocation, more will be the rate of reaction. Let us draw the structure of the corresponding carbocation and observe the relative stability of the four benzyl carbocations.

The relative stability of the four carbocations is

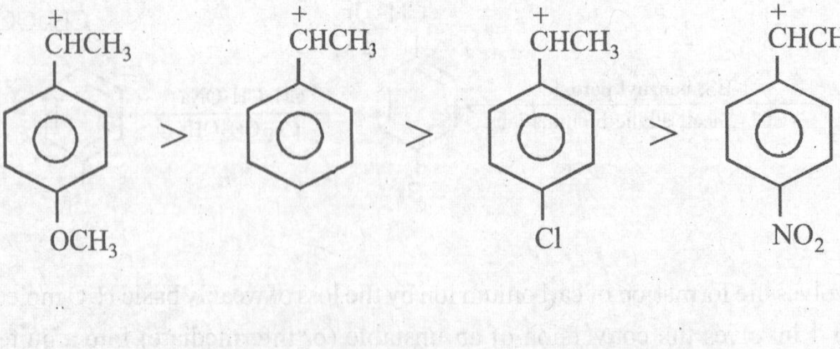

$\overset{+}{C}HCH_3$ (p-OCH_3) > $\overset{+}{C}HCH_3$ (benzene) > $\overset{+}{C}HCH_3$ (p-Cl) > $\overset{+}{C}HCH_3$ (p-NO_2)

67. (d) Oxidation of phenol by $K_2S_2O_8$ is an example of Elbs persulphate oxidation.

68. (b) Phenol is most acidic because its conjugate base is stabilised due to resonance, while the rest three compounds are alcohols, hence, their corrosponding conjugate bases do not exhibit resonance.

69. (b) *tert* - Butyl alcohol is preferentially protonated because its corresponding carbocation is highly stable than that of C_2H_5OH

$$(CH_3)_3COH \xrightarrow{H^+} (CH_3)_3C\overset{+}{O}H_2 \xrightarrow{-H_2O} \underset{\text{Stable}}{(CH_3)_3\overset{+}{C}} \xrightarrow{CH_3CH_2\ddot{O}H}$$

$$(CH_3)_3C\overset{H}{O^+}CH_2CH_3 \xrightarrow{-H^+} (CH_3)_3COCH_2CH_3$$

The *tert*-butyl cation is preferentially attacked by CH_3CH_2OH rather than bulky $(CH_3)_3COH$

70. (d) Such questions can be solved by considering the relative basic character of their conjugated bases which for H_2O, C_2H_2, H_2CO_3 and C_6H_5OH are

$$^-OH,\ HC \equiv C\bar{:},\ HCO_3^-,\ C_6H_5O^-$$

More the possibility for the dispersal of the negative charge, weaker will be the base. Thus the relative basic character of the four bases is

$$HCO_3^- \ <\ C_6H_5O^- \ <\ ^-OH \ <\ HC \equiv C^-$$

Eqivalent Non-equivalent resonating structures — Due to resonance; Oxygen can accommodate negative charge easily

Thus the acidic character of the four corresponding acids will be

$$H_2CO_3 > C_6H_5OH > H_2O > HC \equiv CH$$

71. (a) I. $CH_3CH = CH_2 + H_2O \xrightarrow{H^+} CH_3\underset{}{\overset{OH}{|}}{C}HCH_3$

II. $CH_3CHO \xrightarrow[\text{(ii) } H_2O]{\text{(i) } CH_3MgI} CH_3\overset{OH}{|}{C}HCH_3$

III. $CH_2O \xrightarrow[\text{(ii) } H_2O]{\text{(i) } C_2H_5MgI} C_2H_5CH_2OH$

IV. $CH_2CH = CH_2 \xrightarrow[\text{neutral}]{KMnO_4} CH_3\underset{OH}{C}H\underset{OH}{C}H_2$

72. (b) Reagents used in the various steps indicate that the compound Z has an alcoholic group. This set of reactions is possible only when Z is $CH_3CHOHCH_3$.

$$\underset{(Z)}{CH_3CH(OH)CH_3} \xrightarrow{PCl_5} \underset{(X)}{CH_3CHClCH_3} \xrightarrow{\text{alc.KOH}} \underset{(Y)}{CH_3CH = CH_2}$$

(Y → Z: (i) Conc. H_2SO_4, (ii) H_2O, boil — Markovnikov's hydration)

In options *a* and *c*, Y cannot be converted back into d by the given series of reactions.

73. (c) *o*-Nitrophenol has intramolecular H-bonding.

74. (d) o-Cresol (a) — Salicylic acid (b) — Phenol (c)

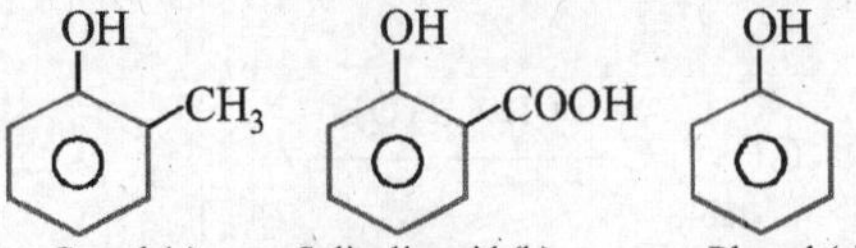

Electron releasing groups ($-CH_3$, $-OCH_3$, $-NCH_3$ etc) intensify the negative charge of phenoxide ion, i.e., destablises it hence decreases ionization of parent phenol, therefore decreases acidity. While electron withdrawing groups ($-NO_2$, $-COOH$, $-CHO$ etc.) increases acidity.

75. (d) $$CH_3-CH_2-\underset{OH}{\underset{|}{CH}}-\underset{OH}{\underset{|}{CH}}-CH_2-CH_3 \xrightarrow{HIO_4} 2\,CH_3-CH_2-CHO$$

76. (c) In the cleavage of mixed ethers having two different alkyl groups, the alcohol and alkyl iodide formed depends on the nature of alkyl groups. When primary or secondary alkyl groups are present, it is the lower alkyl group that forms alkyl iodide therefore

$$CH_3-\underset{CH_3}{\underset{|}{CH}}-CH_2-O-CH_2-CH_3+HI \xrightarrow{\Delta} CH_3-\overset{CH_3}{\overset{|}{CH}}-CH_2OH+CH_3CH_2I$$

77. (b) $LiAlH_4$ will give the desired compound.

78. (d) Ozonide $\xrightarrow[\text{(reduction)}]{NaBH_4}$ (diol, OH / OH)

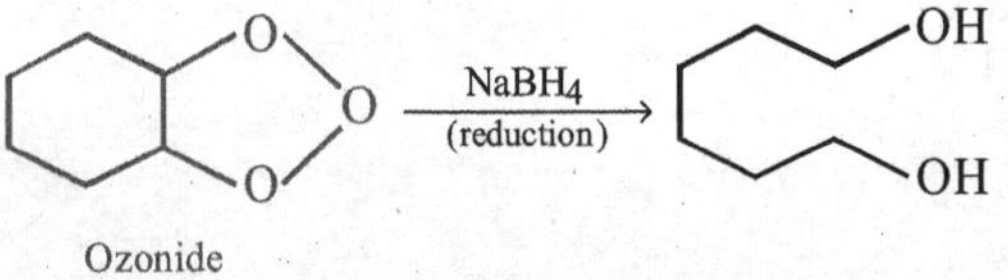

79. (b) Heavy metal ions, particularly Ag^+, catalyse S_N1 reaction because of the presence of empty orbital.

$$CH_3CH_2Br + Ag^+ \rightleftharpoons [CH_3CH_2\cdots X\cdots Ag]^+ \xrightarrow[(-AgX)]{\text{slow}} CH_3\overset{+}{C}H_2 \xrightarrow[\text{from } H_2O]{\overset{-}{O}H} CH_3CH_2OH$$

80. (d) $$3C_2H_5OH+PBr_3 \longrightarrow 3C_2H_5Br+\underset{[X]}{H_3PO_3}$$

26 Aldehydes, Ketones and Carboxylic Acids

1. 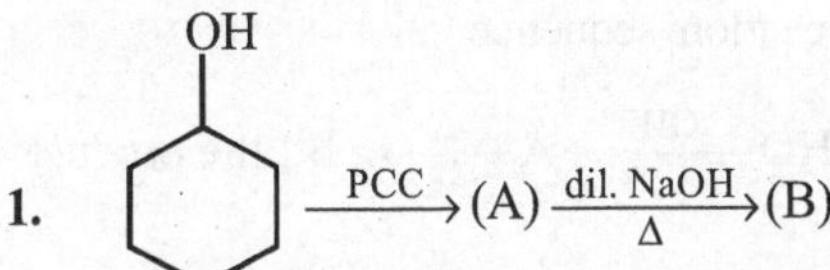

Product (B) is:  Tricky

(a) (b)

(c) (d)

2. (P) $\xrightarrow[\Delta]{\text{EOH}}$

(Q) $\xrightarrow[\Delta]{\text{EOH}}$ $Ph-CH_2-OH + Ph-CO_2$

(R) $\xrightarrow{O_3}$ P + Q, Structure of (R) is:

Toughnut

(a) $Ph-CH=CH-CH_3$

(b) $Ph-CH=C(CH_3)_2$

(c) $Ph-C(CH_3)=CH-CH_3$

(d) $Ph-C(CH_3)=CH_2$

3. A $\xrightarrow{\text{Ozonolysis}}$ X + Y

B $\xrightarrow{\text{Ozonolysis}}$ C + D

Critical Thinking

From the ozonolysis products, the two isomers A and B can be distinguished with the help of

(a) Fehling solution (b) Tollen's reagent

(c) Haloform test (d) only spectroscopy

4. The correct set of reagents to carry out the following conversion

$C_6H_5-CH=CH_2 \xrightarrow{?} C_6H_5-CH_2CHO$

(a) B_2H_6/diglyme, H_2O_2/NaOH, PCC/CH_2Cl_2

(b) Br_2, $NaNH_2$, $HgSO_4/H_2SO_4$

(c) H_2O/H_2SO_4, CrO_3/H_2SO_4

(d) None of the above

5. Which of the following gives ethanal CH_3CHO on acid hydrolysis?

(a) (b) (c) (d)

6. The following reaction

(phenol) $+ HCN + HCl \xrightarrow[ZnCl_2]{Anhyd.}$ (2-hydroxybenzaldehyde: OH, CHO)

is known as
(a) Perkin reaction
(b) Gattermann aldehyde synthesis
(c) Kolbe's reaction
(d) Gattermann-Koch reaction

7. What is X in the following conversion ?

(cyclopentenyl–CH_2OH) $\xrightarrow{X}$ (cyclopentenyl–$\overset{O}{\overset{\|}{C}}-H$)

(a) $K_2Cr_2O_7$, H_2SO_4/H_2O
(b) HIO_4
(c) PCC/CH_2Cl_2
(d) OsO_4, $(CH_3)_3C.COOH$, $(CH_3)_3COH$, OH^-

8. What is the product of the following reaction ?

(2-pentanone) $\xrightarrow[DMSO]{(C_6H_5)_3\,P^+ - \ddot{C}H_2^-}$

(a) 2 - methyl - 1- pentene
(b) 4 - methyl - 1- pentene
(c) 2 - methyl - 2- propyloxirane
(d) 1 - pentene

9. 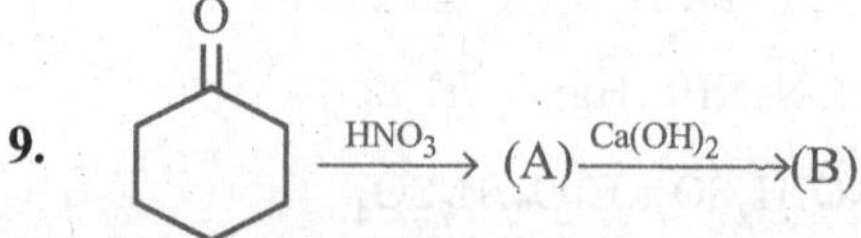

Product (B) in this reaction is: 

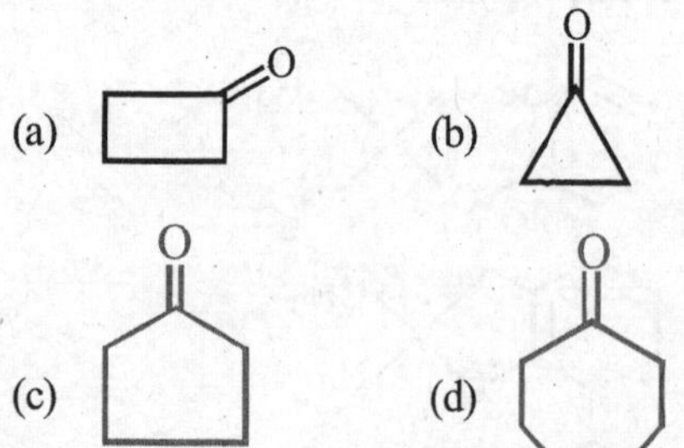

10. Which of the following pairs of reactants is most effective in forming an enamine?

(a) $CH_3CH_2\overset{O}{\overset{\|}{C}}H + [(CH_3)_2CH]_2\,NH$

(b) (cyclohexanone) $+ CH_3—NH— CH_3$

(c) $(CH_3)_3C\overset{O}{\overset{\|}{C}}H + (CH_3)_2\,NH$

(d) None of these form an enamine.

11. In the reaction sequence

$2CH_3CHO \xrightarrow{OH^-} A \xrightarrow{\Delta} B$; the product B is :

(a) $CH_3- CH_2 - CH_2- CH_2- OH$
(b) $CH_3-CH = CH - CHO$
(c) $CH_3-CH_2-CH_2-CH_3$
(d) $CH_3 - \overset{O}{\overset{\|}{C}} - CH_3$

12. Which one of the following reactions will not result in the formation of carbon-carbon bond?
(a) Reimer-Tieman reaction
(b) Friedel Craft's acylation
(c) Wurtz reaction
(d) Cannizzaro reaction

13. Which is major product formed when acetone is heated with iodine and potassium hydroxide?
(a) Iodoacetone (b) Acetic acid
(c) Iodoform (d) Acetophenone

14. (cyclopentanone) $\xrightarrow{NH_2OH} (A) \xrightarrow{H^+} (B) \xrightarrow{LAH} (C)$; product (C) of the reaction is:

Critical Thinking

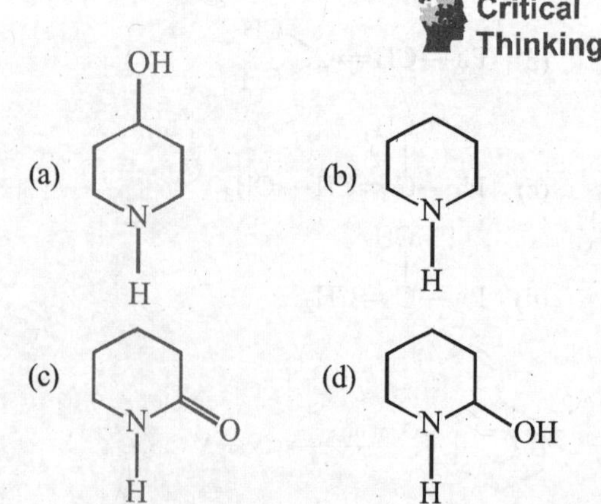

15. End product of the following sequence of reactions are: Toughnut

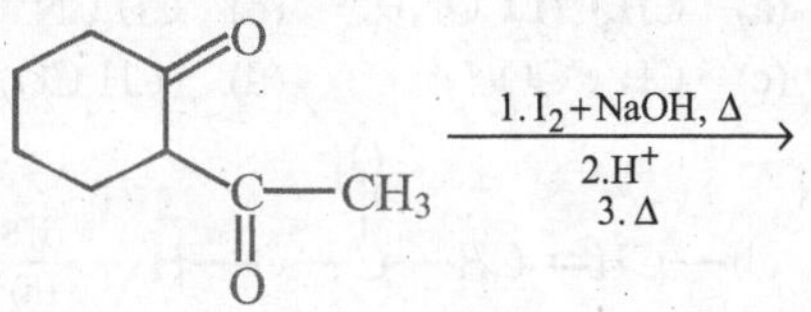

(2-acetylcyclohexanone) $\xrightarrow[\text{2.}H^+ \text{ 3.}\Delta]{\text{1.}I_2+NaOH,\ \Delta}$

(a) yellow ppt. of CHI_3, (2-oxocyclohexane-COOH)

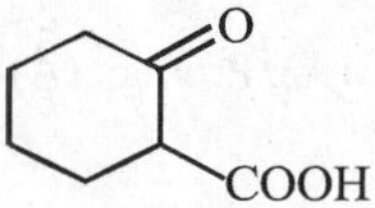

(b) yellow ppt. of CHI_3, (2-oxocyclohexane-CHO)

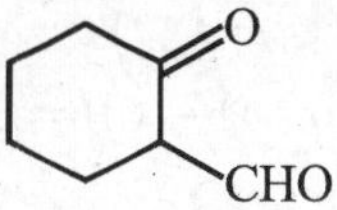

(c) yellow ppt. of CHI_3, (cyclohexanone)

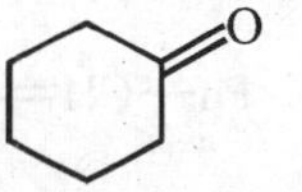

(d) yellow ppt. of CHI_3, (cyclohexane-1,2-dicarboxylic acid: COOH, COOH)

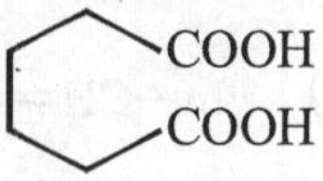

16. (cyclohexane-C(=O)–H) $\xrightarrow{a}$ (1-bromocyclohexane-1-CHO) $\xrightarrow{b}$ (cyclohex-1-ene-1-CHO)

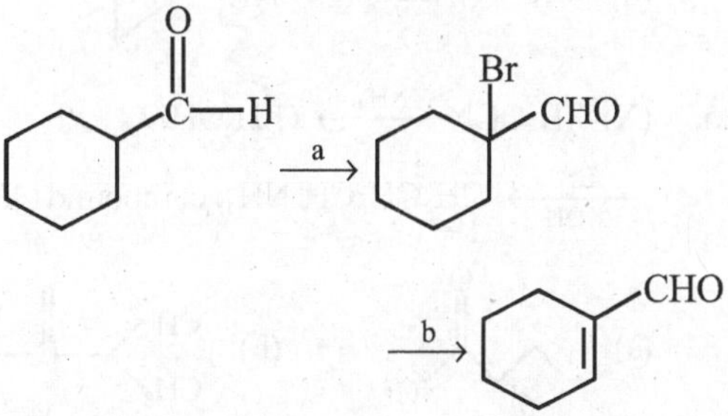

Identify appropriate reagents for the above reaction: Tricky

(a) a = Br_2/CCl_4, b = aq. KOH
(b) a = Br_2/H^+, b = aq. KOH
(c) a = Br_2/H^+, b = alc. KOH
(d) a = Br_2/HO^-, b = aq. KOH

17. The starting compounds needed to make the following compound ?

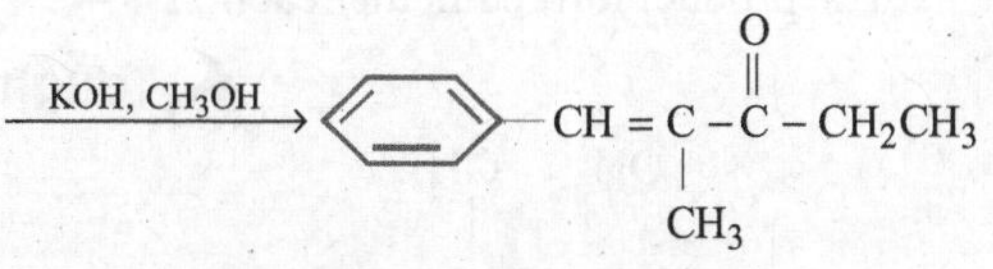

$\xrightarrow{KOH,\ CH_3OH}$ $C_6H_5-CH=C(CH_3)-C(=O)-CH_2CH_3$

(a) benzaldehyde and 3-pentanone
(b) acetophenone and 2-butanone
(c) acetophenone and butanal
(d) benzaldehyde and 2-pentanone

18. 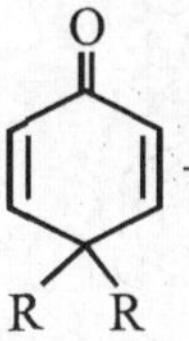 (4,4-dialkylcyclohexa-2,5-dienone, R R) $\xrightarrow{H^+}$ X, most likely the compound X is

(a) 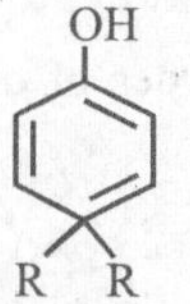 (OH; R, R) (b) 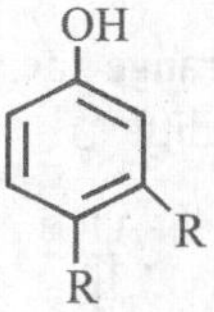 (OH; R, R) (c) 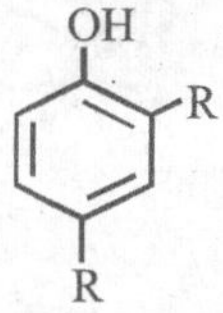 (OH; R, R) (d) 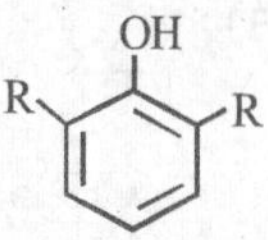 (OH; R, R)

19. The reagents employed to carry the following transformation Toughnut

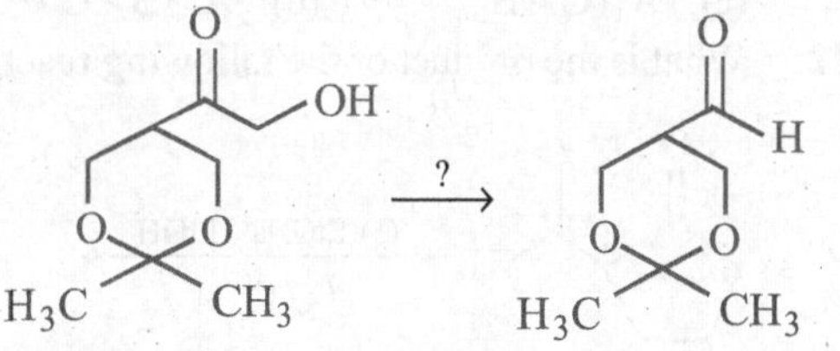

(O, OH; O, O; H_3C, CH_3) $\xrightarrow{?}$ (O, H; O, O; H_3C, CH_3)

(a) $LiAlH_4, H_2SO_4$ / heat
(b) PCC/CH_2Cl_2 followed by HIO_4
(c) $NaBH_4/CH_3OH$ followed by HIO_4
(d) O_3 followed by $(CH_3)_2S$

20. What is the product of the following reaction ?

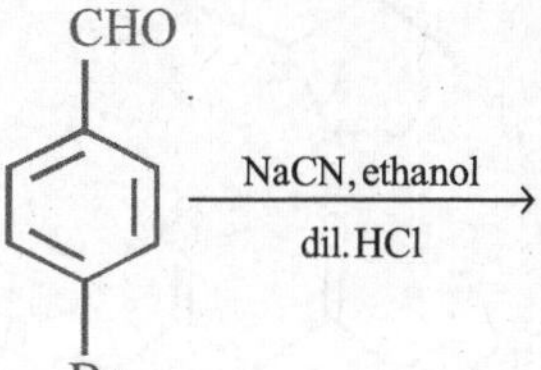

(CHO; Br) $\xrightarrow[\text{dil. HCl}]{\text{NaCN, ethanol}}$

(a) 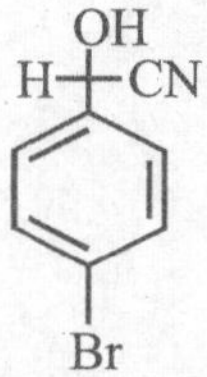 (OH, H–C–CN; Br) (b) 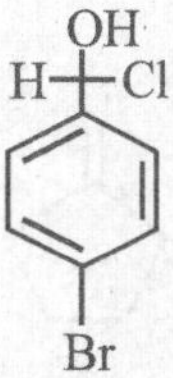 (OH, H–C–Cl; Br)

(c) CN, H—C—H on a benzene ring with Br in the para position

(d) OH, H—C—Cl on a benzene ring with Br in the para position

21. Arrange the following in order of decreasing acidity

(A) Cyclopentanone (C=O)

(B) Cyclopentane-1,3-dione (O=, =O)

(C) Methylenecyclopentane (CH_2=)

(a) C>A>B (b) C>B>A

(c) A>C>B (d) A>B>C

22. What is the product of the following reaction ?

2-Formylcyclopentanone (O, O, H) $\xrightarrow[\text{(2) } CH_3I]{\text{(1) EtONa , EtOH}}$

(a) 2-Methyl-2-formylcyclopentanone (O, O, H, CH_3)

(b) OH, O, H_3C, H

(c) H_3C, O, O, H

(d) O, OH, CH_3, H

23. Identify Z in the sequence

$CH_3COONH_4 \xrightarrow{\Delta} X \xrightarrow[\Delta]{P_2O_5} Y \xrightarrow{H_2O/H^+} Z$

(a) $CH_3CH_2CONH_2$ (b) CH_3CN

(c) CH_3COOH (d) $(CH_3CO)_2O$

24. Ph—CH=CH—C(=O)—O—H $\xrightarrow[\text{(ii) } NH_2\text{-cyclopropyl}]{\text{(i) } SOCl_2}$

(A); Product (A) of the reaction is:

Toughnut

(a) Ph—CH=CH—C(=O)—CH_2—NH—cyclopropyl

(b) Ph—CH=CH—C(=O)—NH—cyclopropyl

(c) Ph—CH=CH—C(=O)—H

(d) Ph—CH=CH—NH—cyclopropyl

25. (X)$C_4H_7OCl \xrightarrow{NH_3} C_4H_9OH$

$\xrightarrow[KOH]{Br_2} CH_3CH_2CH_2NH_2$; compound (X) is:

(a) Butanoyl chloride (O, Cl)

(b) $(CH_3)_2CH$—C(=O)—Cl

(c) Cl—, —OH

(d) Cl, CHO

26. Succinic anhydride-type cyclic anhydride (O, O, O) $\xrightarrow[\text{(1 eq.)}]{PhMgBr} \xrightarrow[\text{(1 eq.)}]{CH_3MgBr} \xrightarrow[\text{(1 eq.)}]{H^+/H_2O}$

Product.

The product formed in the reaction is –

Toughnut

(a) OH, OH, Me, Ph, O

(b) $Me-\overset{O}{\overset{\|}{C}}-\overset{O}{\overset{\|}{C}}-Ph$

(c) $Me-\underset{Ph}{\underset{|}{\overset{OH}{\overset{|}{C}}}}-COOH$

(d) $Ph-\overset{O}{\overset{\|}{C}}-\overset{O}{\overset{\|}{C}}-O-Me$

27. 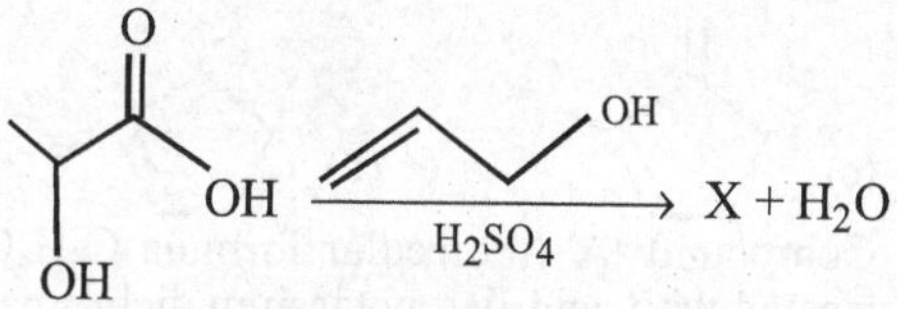 $CH_3CH(OH)COOH \xrightarrow[H_2SO_4]{CH_2=CH-CH_2OH} X + H_2O$

Product (X) of the reaction is–

(a) 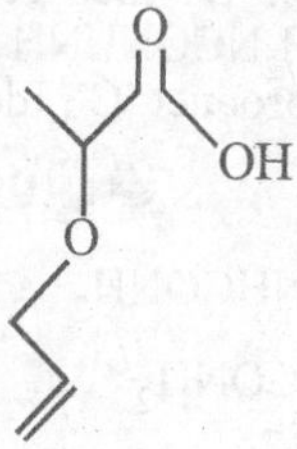 O, OH, O (allyl ether of lactic acid)

(b) 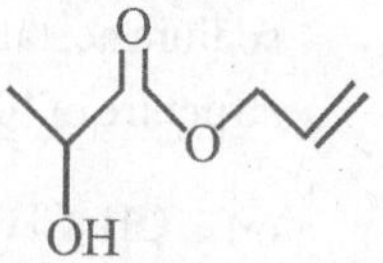 O, O, OH (allyl ester of lactic acid)

(c) 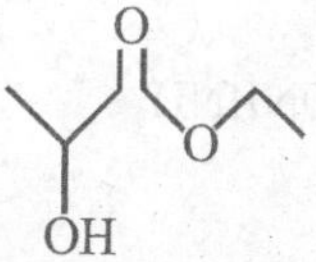 O, O, OH

(d) 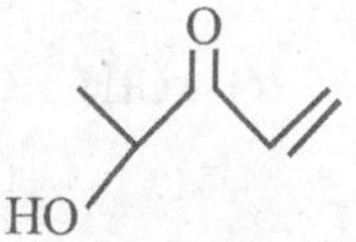 O, HO

28. In a set of reactions, acetic acid yielded a product D.

$$CH_3COOH \xrightarrow{SOCl_2} (A) \xrightarrow[AlCl_3]{Benzene} (B)$$

$$\xrightarrow{HCN} (C) \xrightarrow{HOH} (D)$$

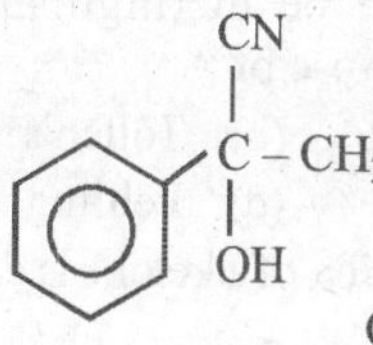

The structure of (D) would be –

Toughnut

(a) 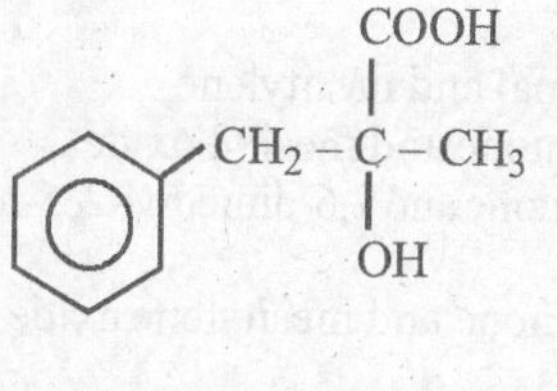 $C_6H_5-CH_2-\underset{OH}{\underset{|}{\overset{COOH}{\overset{|}{C}}}}-CH_3$

(b) $C_6H_5-\underset{OH}{\underset{|}{\overset{CN}{\overset{|}{C}}}}-CH_3$

(c) $C_6H_5-CH_2-\underset{CN}{\underset{|}{\overset{OH}{\overset{|}{C}}}}-CH_3$

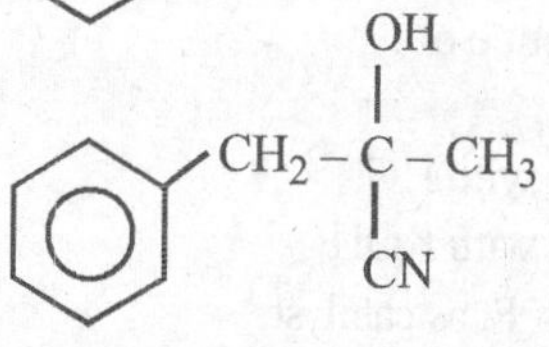

(d) 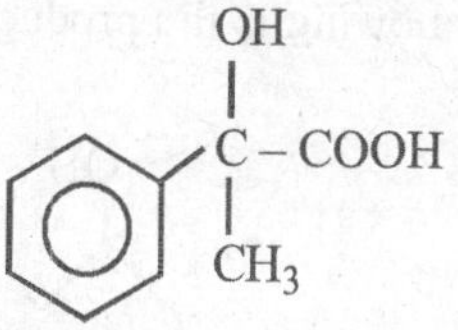 $C_6H_5-\underset{CH_3}{\underset{|}{\overset{OH}{\overset{|}{C}}}}-COOH$

29. Calculate number of molecules of Grignard reagent consumed by 1 molecule of following compound.

Tricky

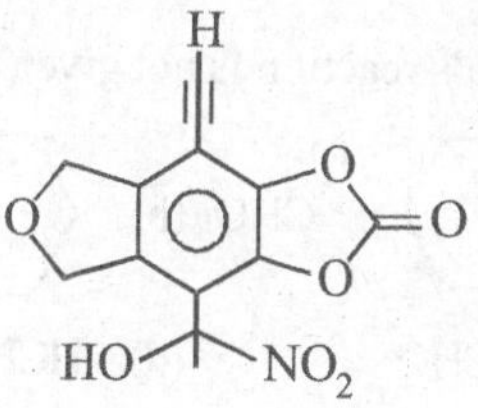

H, O, O, O, =O, HO, NO_2

(a) 5 (b) 2 (c) 3 (d) 1

30. Which of the following will not give butyl acetate when treated with 1- butanol

Critical Thinking

(a) $CH_3-\overset{O}{\overset{\|}{C}}-Cl(\text{pyridene})$

(b) $CH_3-\overset{O}{\overset{\|}{C}}-OH(H_2SO_4)$

(c) $CH_3-\overset{O}{\overset{\|}{C}}-O-\overset{O}{\overset{\|}{C}}-CH_3(\text{pyridene})$

(d) $CH_3-\overset{O}{\overset{\|}{C}}-H(H_2SO_4)$

31. Tischenko reaction is a modification of:

(a) Aldol condensation
(b) Claisen condensation
(c) Cannizzaro reaction
(d) Pinacol-pinacolon reaction

32. Formaldehyde can be distinguished from acetaldehyde by the use of :
(a) Schiff's reagent (b) Tollen's reagent
(c) I_2/Alkali (d) Fehling's solution

33. Clemmensen reduction of a ketone is carried out in the presence of :
(a) $LiAlH_4$
(b) Zn-Hg with HCl
(c) Glycol with KOH
(d) H_2 with Pt as catalyst

34. Which of the following is the product of aldol condensation?
(a) O, OH (b) OH O
(c) HO, O (d) O, OH

35. Cannizaro's reaction is not given by :
(a) (benzene ring)–CHO (b) (cyclohexane ring with CHO and CH_3)
(c) CH_3CHO (d) HCHO

36. (cyclohexane ring with two CHO groups) $\xrightarrow[\text{(ii) } H_2O/H^+]{\text{(i) KOH}/\Delta}$ [X]. What is [X]?
(a) CH_2-OH, COOH
(b) CHO, COOH
(c) CH_2-OH, CHO
(d) CH_2-OH, COO^-K^+

37. What is the product of the following reaction.
H, O, O, H $\xrightarrow[\text{heat}]{KOH,\ CH_3OH}$?

Toughnut

(a) CHO (b) O
(c) O (d) CHO

38. Compound 'A' (molecular formula C_3H_8O) is treated with acidified potassium dichromate to form a product 'B' (molecular formula C_3H_6O). 'B' forms a shining silver mirror on warming with ammonical silver nitrate. 'B' when treated with an aqueous solution of $H_2NCONHNH_2$.HCl and sodium acetate gives a product 'C'. Identify the structure of 'C'

Toughnut

(a) $CH_3CH_2CH=NNHCONH_2$
(b) $CH_3-\underset{\underset{CH_3}{|}}{C}=NNHCONH_2$
(c) $CH_3-\underset{\underset{CH_3}{|}}{C}=NCONHNH_2$
(d) $CH_3CH_2CH=NCONHNH_2$

39. The most reactive compound towards formation of cyanohydrin on treatment with KCN followed by acidification is
(a) benzaldehyde
(b) *p*-nitrobenzaldehyde
(c) phenylacetaldehyde
(d) *p*-hydroxybenzaldehyde

40. An organic compound 'A' has the molecular formula C_3H_6O. It undergoes iodoform test. When staturated with HCl it gives 'B' of molecular formula $C_9H_{14}O$. 'A' and 'B' respectively are

(a) Propanal and mesitylene
(b) Propanone and mesityl oxide
(c) Propanone and 2,6-dimethyl-2, 5-heptadien-4-one
(d) Propanone and mesitylene oxide

41. What is D in the following sequence of reactions?

Cyclopentanone (=O) $\xrightarrow[CH_3OH]{NaBH_4}$ A $\xrightarrow{HBr}$ B $\xrightarrow[\text{(ii) } H_2C=O \text{ (iii) } H_3O^+]{\text{(i) Mg, } Et_2O}$ C $\xrightarrow[CH_2Cl_2]{PCC}$ D

(a) cyclopentyl–CHO (b) cyclopentyl–COOH

(c) 1-(OH)(CHO)cyclopentane (d) 1-(CH_3)(OH)cyclopentane

42. Biphenyl-2,2',6,6'-tetracarbaldehyde (CHO, OHC, CHO, OHC) $\xrightarrow[\text{(ii) } H^+/H_2O]{\text{(i) NaOH/100°C}}$?

The major product in the above reaction is

(a) biphenyl with COOH, HOOC, COOH, HOOC

(b) biphenyl with CH_2OH, HOOC, COOH, HOCH_2

(c) dilactone (O, O, O, O)

(d) biphenyl with CH_2OH, HOH_2C, CH_2OH, HOH_2C

43. Identify the missing reagent in the reaction given below

Cyclopentanone + ? $\xrightarrow[\text{(ii) } H_3O^+]{\text{(i) EtONa}}$ 2-(COOEt)cyclopentanone

(a) Diethyl carbonate $(EtO)_2C=O$

(b) Ethyl acetate CH_3COOEt

(c) Ethyl formate HCOOEt

(d) Diethyl oxalate Et. OOC. COOEt

44. The final product of the following sequence of reactions is

4-CH_3-1-$C(CH_3)_3$-benzene $\xrightarrow[\text{benzoyl peroxide}]{NBS, CCl_4, \text{ heat}}$ $\xrightarrow[CO_2]{\text{Mg/ether}}$ $\xrightarrow{H_3O^+}$?

(a) CH_3–C_6H_4–$C(CH_3)_2$–CH_2COOH ($H_3C-C-CH_3$, CH_2COOH)

(b) CH_2COOH–C_6H_4–$C(CH_3)_3$

(c) HOOC–, CH_3, $C(CH_3)_3$ substituted benzene

(d) COOH–C_6H_4–$C(CH_3)_3$

45. Identify Z in the following reaction sequence.

$$CH_3CHO \xrightarrow[Dil.H_2SO_4]{MnO_4^-} X \xrightarrow{SOCl_2} Y \xrightarrow[\Delta]{CH_3COONa} Z$$

(a) CH_3COCH_2COONa
(b) $(CH_3CO)_2O$
(c) $CH_3CO-O-COCH_2Cl$
(d) $CH_3CO-OCOCHCl_2$

46. In the following sequence of reactions, what is D? Toughnut

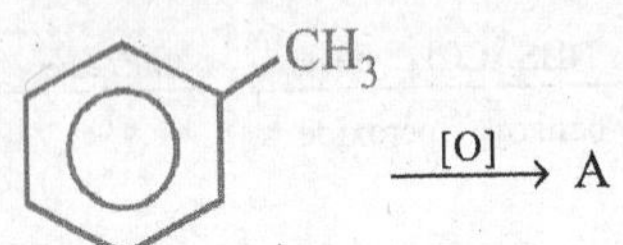

$\xrightarrow{[O]}$ A

$$A \xrightarrow{SOCl_2} B \xrightarrow{NaN_3} C \xrightarrow{heat} D$$

(a) Primary amine
(b) An amide
(c) Phenyl isocyanate
(d) A higher hydrocarbon

47. Which one of the following compounds undergoes decarboxylation upon heating ?

(a) [structure: $CH_3CH_2COCH_2CH_2COOH$]
(b) [structure: $CH_3CH_2CH_2CH_2COCOOH$]
(c) [structure: $CH_3COCH_2CH_2COOH$]
(d) [structure: $CH_3CH_2CH_2COCH_2COOH$]

48. Which lactone is formed by heating the following hydroxy acid ?

[structure: $CH_3CH(OH)CH_2CH(CH_3)CH_2COOH$]

(a) [lactone structure] (b) [lactone structure]
(c) [lactone structure] (d) [lactone structure]

49. The correct product of the following sequence of reactions — Critical Thinking

$$2\ CH_3CH_2CH_2COOEt \xrightarrow[H_3O^+]{EtONa} \xrightarrow[H_3O^+]{NaOH, H_2O, \Delta} \xrightarrow{\Delta} ?$$

(a) 4 - heptanone
(b) 4 - methyl - 3 - hexanone
(c) 2 - ethyl pentanoic acid
(d) 2 - propyl butanoic acid

50. The correct product of the following reactions

$$EtOOC-CH_2-COOEt \xrightarrow[2\ CH_3I]{2\ EtONa} \xrightarrow[(ii)\ H_2O]{(i)\ LiAlH_4}$$

(a) 2, 2-dimethyl propane diol
(b) 2-methyl-1-propanol
(c) 2, 2-dimethyl propanedioic acid
(d) 2-methyl propanoic acid

51. Arrange the following carbonyl compounds in decreasing order of their reactivity in nucleophilic addition reaction.

(i) C_6H_5CHO (ii) p-$NO_2C_6H_4CHO$ (iii) $C_6H_5COCH_3$ (iv) p-$CH_3C_6H_4CHO$

(a) (ii) > (iii) > (i) > (iv)
(b) (ii) > (i) > (iv) > (iii)
(c) (iii) > (ii) > (i) > (iv)
(d) (iii) > (i) > (iv) > (ii)

52. [cyclohexyl keto acid: O, H–O, O] $\xrightarrow{\Delta}$ (A) $\xrightarrow{Zn(Hg)/HCl}$ (B)

In the above reaction, product (B) is:

(a) (b)

(c) (d) OH

53. Correct order of reactivity of following compounds towards Grignard reagent?

$CH_3-\overset{O}{\overset{\|}{C}}-H$ (I) $H-\overset{O}{\overset{\|}{C}}-H$ (II)

$CH_3-\overset{O}{\overset{\|}{C}}-C_6H_5$ (III)

(a) I > II > III (b) II > I > III
(c) II > III > I (d) I > III > II

54. What is the final product (B) of this sequence?

CH_3 (toluene) $\xrightarrow[\text{hight}]{Br_2}$ A $\xrightarrow[2.\ H_3O^+,\ \text{heat}]{1.\ KCN}$ B

(a) CH_3, CO_2H (b) CO_2H

(c) CH_3, CO_2H (d) CH_2CO_2H

55. 1-Phenylethanol can be prepared by reaction of benzaldehyde with
(a) methyl bromide
(b) ethyl iodide and magnesium
(c) methyl bromide and aluminium bromide
(d) methyl iodide and magnesium

56. Oxidation of acetaldehyde with selenium dioxide produces
(a) ethanoic acid (b) methanoic acid
(c) glyoxal (d) oxalic acid

57. Which of the following will respond to Cannizzaro's reaction ?
(a) 2,2-Dimethylpropanal
(b) Acetaldehyde
(c) Propionaldehyde
(d) Cinnamaldehyde

58. If 3-hexanone is reacted with $NaBH_4$ followed by hydrolysis with D_2O, the product will be

(a) $CH_3CH_2CH(OH)CH_2CH_2CH_3$
(b) $CH_3CH_2CD(OH)CH_2CH_2CH_3$
(c) $CH_3CH_2CH(OD)CH_2CH_2CH_3$
(d) $CH_3CH_2CD(OD)CH_2CH_2CH_3$

59. Propanal on reaction with dilute NaOH forms
(a) $CH_3CH_2CH_2CH_2CH_2CHO$
(b) $CH_3CH_2CH(OH)CH_2CH_2CHO$
(c) $CH_3CH_2CH_2CH(OH)CH_2CHO$
(d) $CH_3CH_2CH(OH)CH(CH_3)CHO$

60. Which is not true about acetophenone?
(a) It reacts with 2,4-dinitrophenylhydrazine to form 2, 4-dinitrophenylhydrazone
(b) It reacts with Tollen's reagent to form silver mirror
(c) It reacts with I_2/NaOH to form iodoform
(d) On oxidation with alkaline $KMnO_4$ followed by hydrolysis it gives benzoic acid

61. The order of decreasing ease of reaction with ammonia is

Critical Thinking

(a) anhydrides, esters, ethers
(b) anhydrides, ethers, esters
(c) ethers, anhydrides, esters
(d) esters, ethers, anhydrides

62. $CH_3CH_2COOH \xrightarrow[\text{Red P}]{Br_2} X \xrightarrow{NH_3\ (alc.)} Y$

Y in the above reactions is
(a) lactic acid (b) ethylamine
(c) propylamine (d) alanine

63. In the mechanism of Hoffmann reaction, which intermediate rearranges to alkyl isocyanate ?
(a) Bromamide (b) Nitrene
(c) Nitroso (d) Amide

64. A colourless liquid, at room temperature, reacts with soda-lime to form sodium salt of a carboxylic acid and ammonia gas. The liquid is
(a) propanoic acid (b) formamide
(c) propanamide (d) methyl enthanoate

65. Which of the following on oxidation followed by hydrolysis gives pyruvic acid ?
(a) Acetaldehyde cyanohydrin
(b) Acetone cyanohydrin
(c) Formaldehyde cyanohydrin
(d) None of these

66. Identify the correct order of boiling points of the following compounds :
$CH_3CH_2CH_2CH_2OH$ (1) $CH_3CH_2CH_2CHO$ (2)
$CH_3CH_2CH_2COOH$ (3)
(a) $1>2>3$ (b) $3>1>2$
(c) $1>3>2$ (d) $3>2>1$

67. Carboxylic acids are more acidic than phenol and alcohol because of
(a) intermolecular hydrogen bonding
(b) formation of dimers
(c) highly acidic hydrogen
(d) resonance stabilization of their conjugate base

68. Which of the following compounds will react with $NaHCO_3$ solution to give sodium salt and carbon dioxide ?
(a) Acetic acid (b) *n*-Hexanol
(c) Phenol (d) Both (b) and (c)

69. Which of the following acids has the smallest dissociation constant ?
(a) $CH_3CH(F)COOH$ (b) FCH_2CH_2COOH
(c) $BrCH_2CH_2COOH$ (d) $CH_3CH(Br)COOH$

70. Which of the following orders of relative strengths of acids is correct ?
(a) $FCH_2COOH > ClCH_2COOH > BrCH_2COOH$
(b) $ClCH_2COOH > BrCH_2COOH > FCH_2COOH$
(c) $BrCH_2COOH > ClCH_2COOH > FCH_2COOH$
(d) $ClCH_2CO_2H > FCH_2COOH > BrCH_2COOH$

71. Acetic acid on heating in presence of P_2O_5 gives
(a) acetic anhydride (b) acetylene
(c) peracid (d) no reaction

72. Among the given compounds, the most susceptible to nucleophilic attack at the carbonyl group is
(a) MeCOCl (b) MeCHO
(c) MeCOOMe (d) MeCOOCOMe

73. Ethyl acetate is obtained when methylmagnesium iodide reacts with
(a) ethyl formate (b) ethyl chloroformate
(c) acetyl chloride (d) carbon dioxide

74. An ester is boiled with KOH. The product is cooled and acidified with concentrated HCl. A white crystalline acid separates. The ester is
(a) methyl acetate (b) ethyl acetate
(c) ethyl formate (d) ethyl benzoate

75. Hydrolsis of an ester gives a carboxylic acid which on Kolbe's electrolysis yields ethane. The ester is
(a) ethyl methanoate (b) methyl ethanoate
(c) methyl methanoate (d) methyl propanoate

76. What is the missing reagent in the synthesis shown below **Toughnut**

$EtOOC-CH_2-COOEt \xrightarrow[\text{(ii) Reagant}]{\text{(i) EtONa}} \xrightarrow[\Delta]{NaOH,\ H_2O} \xrightarrow[\text{(ii) }\Delta]{\text{(i) }H_2O,\ H^+}$ cyclopentanecarboxylic acid (COOH)

(a) bromocyclopentane
(b) 1, 5 - dibromo pentane
(c) 1, 4 - dibromo butane
(d) 1, 1 - dibromo cyclopentane

77. The correct product of the following sequence of reactions **Critical Thinking**

$(CH_3)_2CHCOOH \xrightarrow[\text{(ii) }H_2O]{\text{(i) }LiAlH_4} \xrightarrow{PBr_3} \xrightarrow[DMSO]{KCN} \xrightarrow[\Delta]{H_2O^+,\ H^+}$

(a) $(CH_3)_2CHCHBr.COOH$
(b) $(CH_3)_2CHCH_2COOH$
(c) $(CH_3)_2CHCH_2CH_2NH_2$
(d) $(CH_3)_2C=CHCOOH$

78. Carboxylic acid group does not give the usual addition and elimination reactions of aldehydes and ketones because

(a) O–H bond is more polar than $>C=O$ group
(b) carboxylate ion gets ionised
(c) carboxylate ion gets stabilised by resonance
(d) it exists as – COOH and there is no carbonyl group

79. Which one of the following orders of acid strength is correct?

(a) $RCOOH > HOH > ROH > HC\equiv CH$
(b) $RCOOH > HOH > HC\equiv CH > ROH$
(c) $RCOOH > HC\equiv CH > HOH > ROH$
(d) $RCOOH > ROH > HOH > HC\equiv CH$

80. Which of the following is incorrect ?

(a) $FeCl_3$ is used in detection of phenols.
(b) Fehling solution is used in detection of glucose.
(c) Tollen's reagent is used in detection of unsaturation.
(d) $NaHSO_3$ is used in detection of carbonyl compounds.

81. $(CH_3)_2CO \xrightarrow[(HCl)]{NaCN} A \xrightarrow[\Delta]{H_3O^+} B$

In the above sequence of reactions A and B are

(a) $(CH_3)_2C(OH)CN, (CH_3)_2C(OH)COOH$
(b) $(CH_3)_2C(OH)CN, (CH_3)_2C(OH)_2$
(c) $(CH_3)_2C(OH)CN, (CH_3)_2CHCOOH$
(d) $(CH_3)_2C(OH)CN, (CH_3)_2C=O$

82. Acetophenone when reacted with a base, C_2H_5ONa, yields a stable compound which has the structure.

Critical Thinking

(a) $C_6H_5-C(CH_3)=CH-C(=O)-C_6H_5$

(b) $C_6H_5-CH(CH_3)-CH_2C(=O)-C_6H_5$

(c) $C_6H_5-C(CH_3)(OH)-C(CH_3)(OH)-C_6H_5$

(d) $C_6H_5-CH(OH)-CH(OH)-C_6H_5$

83. In Cannizzaro reaction given below

$2PhCHO \xrightarrow{:OH^-} PhCH_2OH + PhC\ddot{O}_2^-$

the slowest step is :

(a) the transfer of proton to the carbonyl group
(b) the abstraction of proton from the carboxylic group
(c) the deprotonation of Ph CH_2OH
(d) the attack of : $\bar{O}H$ at the carboxyl group

84. $C_6H_5-CH_3 \underset{Y}{\overset{X}{\rightleftharpoons}} C_6H_5-COOH$

In the above sequence of reaction X and Y are respectively

(a) H_2/Pt ; Br_2
(b) $KMnO_4$; H_2/Pt
(c) $KMnO_4$ (aq); HI/P
(d) NH_2-NH_2/KOH, HI/P

85. Consider the following transformations :

$CH_3COOH \xrightarrow{CaCO_3} A \xrightarrow{heat} B \xrightarrow[NaOH]{I_2} C$

The molecular formula of C is

(a) $CH_3-C(OH)(I)-CH_3$ (b) ICH_2-COCH_3

(c) CHI_3 (d) CH_3I

86. Silver benzoate will react with bromine in CCl_4 to form :

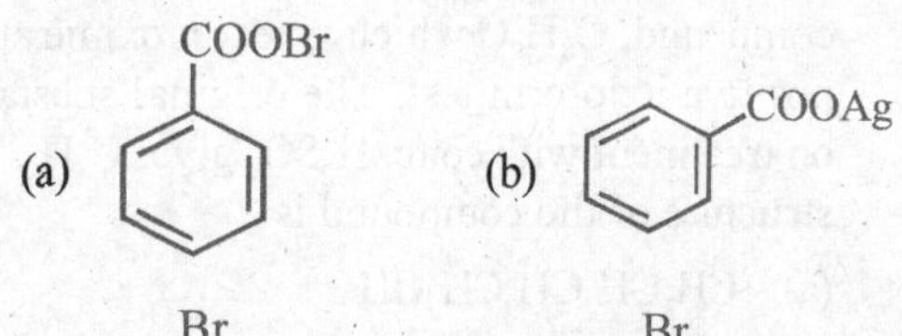

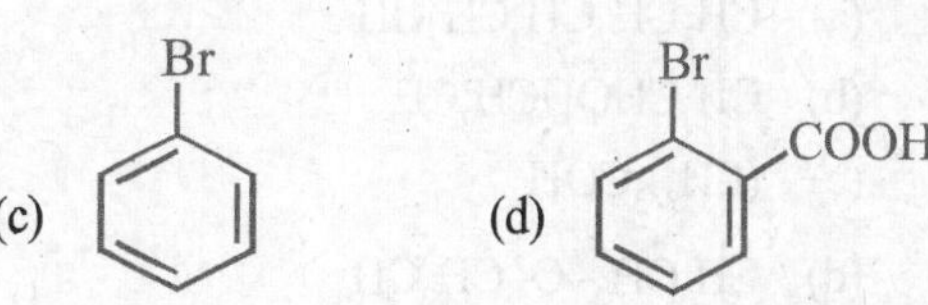

87. The compound that undergoes decarboxylation most readily under mild condition is

(a) 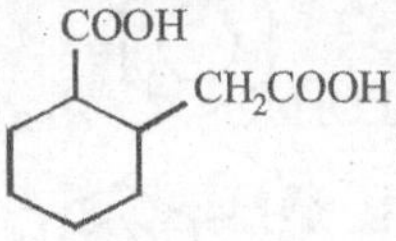 Cyclohexane with COOH and CH_2COOH on adjacent carbons
(b) 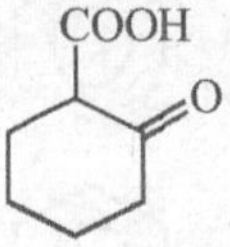 Cyclohexanone with COOH at the α-carbon
(c) 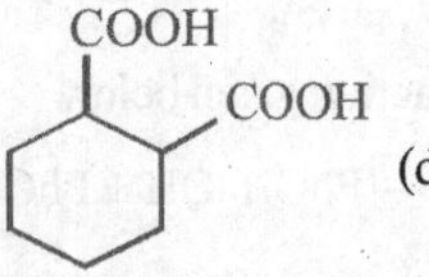 Cyclohexane with two COOH groups on adjacent carbons
(d) 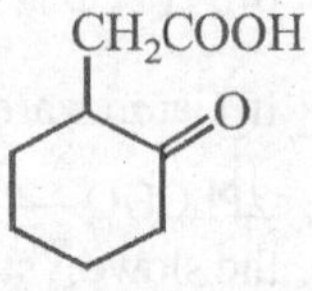 Cyclohexanone with CH_2COOH at the α-carbon

88. Which of the following reaction can produce R – CO – Ar ?

(a) $ArCOCl + H - Ar \xrightarrow{AlCl_3}$
(b) $RCOCl + ArMgX \longrightarrow$
(c) $ArCOCl + RMgX \longrightarrow$
(d) $RCOCl + H - Ar \xrightarrow{AlCl_3}$

89. The end product (C) in the following sequence of reactions is **Toughnut**

$$HC \equiv CH \xrightarrow[20\%\ H_2SO_4]{1\%\ HgSO_4} A \xrightarrow[H_2O]{CH_3MgX} B \xrightarrow{[O]} (C)$$

(a) acetic acid (b) isopropyl alcohol
(c) acetone (d) ethanol

90. Predict the product 'B' in the sequence of reactions

$$CH \equiv CH \xrightarrow[\%\ HgSO_4]{30\%\ H_2SO_4} A \xrightarrow{NaOH} B$$

(a) CH_3COONa
(b) CH_3COOH
(c) CH_3CHO
(d) $CH_3 - CH(OH) - CH_2 - CHO$

91. A substance $C_4H_{10}O$ yields on oxidation a compound, C_4H_8O which gives an oxime and a positive iodoform test. The original substance on treatment with conc. H_2SO_4 gives C_4H_8. The structure of the compound is

(a) $CH_3CH_2CH_2CH_2OH$
(b) $CH_3CHOHCH_2CH_3$
(c) $(CH_3)_3COH$
(d) $CH_3CH_2–O–CH_2CH_3$

92. Which of following compound is hemiacetal?

(a) 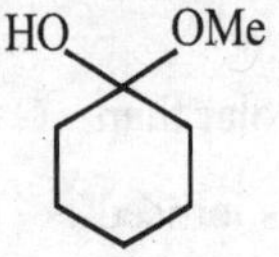 Cyclohexane with HO and OMe on the same carbon
(b) 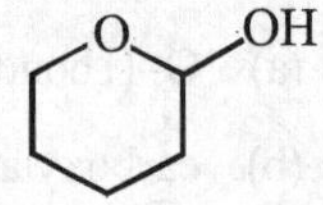 Tetrahydropyran with OH on the carbon adjacent to ring O
(c) 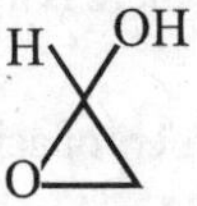 Epoxide carbon bearing H and OH
(d) All of these

93. $C_{16}H_{16} \xrightarrow{O_3} (B) \xrightarrow[HO^-/\Delta]{NH_2-NH_2}$ ethylbenzene;
(A) C_8H_8O

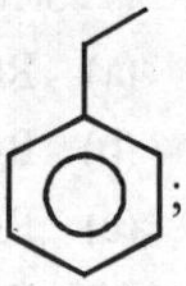

reactant (A) in this reaction is: **Critical Thinking**

(a) 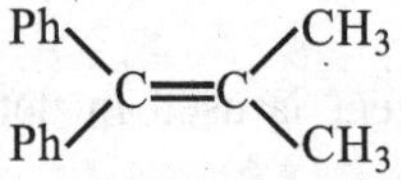 $(Ph)_2C=C(CH_3)_2$
(b) Ph(CH$_3$)C=C(CH$_3$)Ph (Ph and CH$_3$ trans)
(c) Ph(CH$_3$)C=C(CH$_3$)Ph (Ph groups cis)
(d) both (b) and (c)

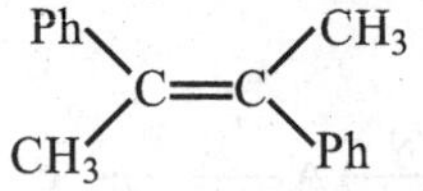

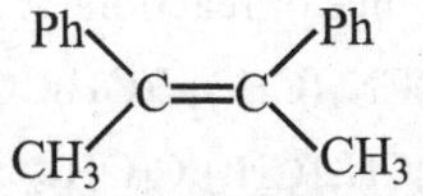

94. $$R-\overset{O}{\overset{\|}{C}}-R \underset{(catalyst)}{\overset{HCN}{\rightleftharpoons}} R-\overset{OH}{\overset{|}{\underset{CN}{\underset{|}{C}}}}-R$$

Which of following can be used as a catalyst in the above reaction?

(a) Cl^- (b) $CH_3-\overset{O}{\overset{\|}{C}}-O^-$
(c) EtO^- (d) HSO_4^-

95. Which of the following reagent(s) used for the conversion?

Critical Thinking

O ‖ (butanone chain)–CO_2Et ⟶ O ‖ (butanone chain)–OH

(a) glycol/$LiAlH_4$/H_3O^+
(b) glycol/NaH/H_3O^+
(c) $LiAlH_4$
(d) $NaBH_4$

96. Among the following the order of reactivity towards nucleophilic addition is

(a) $CH_3CHO > CH_3COCH_3 > HCHO$
(b) $HCHO > CH_3CHO > CH_3COCH_3$
(c) $CH_3CHO > HCHO > CH_3COCH_3$
(d) $CH_3COCH_3 > CH_3CHO > HCHO$

97. The reaction,

$$CH_3CHO \xrightarrow{Zn(Hg)/Conc.\ HCl\ [H]} CH_3CH_3$$

is

(a) Cannizaro's reaction
(b) Rosenmund reduction
(c) Wolf-Kishner reduction
(d) Clemmenson reduction

98. Tollen's reagent and Fehling solutions are used to distinguish between

(a) acids and alcohols
(b) alkanes and alcohols
(c) ketones and aldehydes
(d) *n*-alkanes and branched alkanes

99. In the presence of a small amount of phosphorous, aliphatic carboxylic acids react with chlorine or bromine to yield a compound in which α-hydrogen has been replaced by halogen. This reaction is known as :

(a) Wolff - Kishner reaction
(b) Rosenmund reaction
(c) Etard reaction
(d) Hell - Volhard - Zelinsky reaction

100. Phthalic acid reacts with resorcinol in the presence of concentrated H_2SO_4 to give:

(a) Phenolphthalein (b) Alizarin
(c) Coumarin (d) Fluorescein

Answer KEYs

1	(b)	11	(b)	21	(a)	31	(c)	41	(a)	51	(b)	61	(a)	71	(a)	81	(a)	91	(b)
2	(b)	12	(d)	22	(a)	32	(c)	42	(b)	52	(b)	62	(d)	72	(a)	82	(a)	92	(d)
3	(c)	13	(c)	23	(c)	33	(b)	43	(a)	53	(b)	63	(b)	73	(b)	83	(a)	93	(d)
4	(a)	14	(b)	24	(b)	34	(b)	44	(b)	54	(d)	64	(b)	74	(d)	84	(c)	94	(c)
5	(a)	15	(c)	25	(a)	35	(c)	45	(b)	55	(d)	65	(a)	75	(b)	85	(c)	95	(a)
6	(b)	16	(c)	26	(c)	36	(a)	46	(c)	56	(c)	66	(b)	76	(c)	86	(c)	96	(b)
7	(c)	17	(a)	27	(b)	37	(a)	47	(d)	57	(a)	67	(d)	77	(b)	87	(b)	97	(d)
8	(a)	18	(b)	28	(d)	38	(a)	48	(c)	58	(c)	68	(a)	78	(c)	88	(a)	98	(c)
9	(c)	19	(c)	29	(a)	39	(b)	49	(a)	59	(d)	69	(c)	79	(a)	89	(c)	99	(d)
10	(c)	20	(a)	30	(d)	40	(d)	50	(a)	60	(b)	70	(a)	80	(c)	90	(d)	100	(d)

Hints & Solutions

1. (b) Cyclohexanol $\xrightarrow{PCC}$ Cyclohexanone (A) $\xrightarrow[\Delta\ (aldol)]{HO^-}$ 2-Cyclohexylidenecyclohexanone (B)

2. (b) Acetone (P) $\xrightarrow[\Delta\ (aldol)]{KOH}$ $CH_3COCH=C(CH_3)_2$

$$\underset{(Q)}{Ph-CHO} \xrightarrow[\Delta\ (Cannizzaro)]{KOH} PhCH_2OH + PhCOO^-$$

$$\underset{(R)}{Ph-CH=C(CH_3)_2} \xrightarrow{O_3} \underset{(Q)}{PhCHO} + \underset{(P)}{(CH_3)_2C=O}$$

3. (c) $\underset{A}{C_6H_5CH=CHCH_3} \xrightarrow{O_3} C_6H_5CHO + CH_3CHO$ (Responds haloform test)

$\underset{B}{C_6H_5CH_2CH=CH_2} \xrightarrow{O_3} C_6H_5CH_2CHO + CH_2O$

4. (a) $C_6H_5-CH=CH_2 \xrightarrow[H_2O_2,\ NaOH]{B_2H_6/\text{diglyme.}} C_6H_5-CH_2-CH_2OH \xrightarrow[CH_2Cl_2]{PCC} C_6H_5-CH_2CHO$

5. (a) $C_6H_4O_2C(CH_3)(H) \xrightarrow{H^+/HOH} (HO)_2C(CH_3)(H) \xrightarrow{-H_2O} CH_3CHO$

6. (b) $C_6H_6 + HCN + HCl \xrightarrow{\text{anhy. } ZnCl_2} [C_6H_5CH=NH] \longrightarrow C_6H_5CHO$

7. (c) Pyridinium chloro chromate selectively oxidises a primary alcohol to an aldehyde

e.g. $RCH_2OH \xrightarrow[CH_2Cl_2]{PCC} RCHO$

8. (a) $(H_3C)(n\text{-}H_7C_3)C=O + CH_2=P(C_6H_5)_3 \longrightarrow (H_3C)(n\text{-}H_7C_3)C=CH_2 + O=P(C_6H_5)_3$

9. (c) Cyclohexanone $\xrightarrow[\text{(Oxidising agent)}]{HNO_3}$ (enol, OH) $\longrightarrow$ hexanedioic acid (COOH, COOH) $\xrightarrow{Ca(OH)_2}$ cyclopentanone $+ CaCO_3$

10. (c) 2° amine must be less sterically hindered. In option (b) and (c) 2° amine is being used but due to more reactivity of aldehyde compared to cyclohexanone. Option (c) is correct.

11. (b) $CH_3-\overset{O}{\overset{||}{C}}-H + OH^{\ominus} \rightleftharpoons \left[\overset{\ominus}{C}H_2-\overset{O}{\overset{||}{C}}-H \longleftrightarrow H_2C=\overset{O^{\ominus}}{\overset{|}{C}}-H \right]$

$CH_3-\overset{O}{\overset{||}{C}}-H + \overset{\ominus}{C}H_2-\overset{O}{\overset{||}{C}}-H \longrightarrow CH_3-\underset{H}{\underset{|}{\overset{OH}{\overset{|}{C}}}}-CH_2-\overset{O}{\overset{||}{C}}-H \xrightarrow{\Delta} CH_3-CH=CH-\overset{O}{\overset{||}{C}}-H$

(Aldol condensation)

12. (d) Canizzaro's reaction is a disproportionation reaction of aldehyde in which one molecule of aldehyde is reduced to alcohol whereas other is oxidised to salt of carboxylic acid

$C_6H_5CHO + KOH \longrightarrow C_6H_5CH_2OH + C_6H_5COOK$

13. (c) $CH_3COCH_3 + 3I_2 + 4KOH \longrightarrow CHI_3 + CH_3COOK + 3HI + 3H_2O$

14. (b) Cyclopentanone $\xrightarrow{NH_2OH}$ oxime (N–OH) $\xrightarrow[\text{(Beckmann rearrangement)}]{H^+}$ cyclic amide (NH, C=O) $\xrightarrow{LAH}$ piperidine (N–H)

amide amine

15. (c) 2-acetylcyclohexanone ($\overset{O}{\overset{||}{C}}-CH_3$) $\xrightarrow[\text{(ii) } H^+]{\text{(i) } I_2 + NaOH}$ 2-oxocyclohexane carboxylic acid (CO_2H) $\xrightarrow{\text{(iii) } \Delta}$ cyclohexanone $+ CO_2$

16. (c) Cyclohexane-CHO $\xrightarrow{H^+}$ cyclohexylidene =CH–OH $\xrightarrow{Br_2}$ 1-bromocyclohexane-1-CHO (CHO, Br) $\xrightarrow[\text{alc.KOH}]{E_2}$ cyclohex-1-ene-1-CHO

17. (a) $C_6H_5-\underset{H}{\underset{|}{\overset{O}{\overset{||}{C}}}} + \underset{CH_3}{\underset{|}{C}H_2}-\overset{O}{\overset{||}{C}}-CH_2-CH_3 \xrightarrow[\Delta]{KOH,\ CH_3OH}$

$C_6H_5-\overset{OH}{\overset{|}{C}H}-\underset{CH_3}{\underset{|}{C}H}-\overset{O}{\overset{||}{C}}-CH_2CH_3 \xrightarrow{-H_2O} C_6H_5-CH=\underset{CH_3}{\underset{|}{C}}-\overset{O}{\overset{||}{C}}-CH_2CH_3$

18. (b) Reaction of 4,4-dialkylcyclohexadienone with an acid to form phenol with the migration of one of the alkyl groups to the adjacent carbon is known as **dienone-phenol rearrangement**.

19. (c)

$\xrightarrow[CH_3OH]{NaBH_4}$ $\xrightarrow{HIO_4}$

20. (a) Only –CHO group reacts with CN^- ion and the reaction is

$$>C=O \text{ (H)} \xrightarrow[\text{dil. HCl}]{\text{NaCN, ethanol}} >C(H)(OH)(CN)$$

21. (a) (C) contains the most reactive methylene group followed by (A) then (B).

22. (a)

$\xrightarrow{C_2H_5ONa,\ EtOH}$ $\xrightarrow[-NaI]{CH_3I}$

23. (c) $$CH_3COONH_4 \xrightarrow[(-H_2O)]{heat} \underset{X}{CH_3CONH_2} \xrightarrow[(-H_2O)]{P_2O_5} \underset{Y}{CH_3CN} \xrightarrow[(-NH_3)]{H_3O^+} \underset{Z}{CH_3COOH}$$

24. (b) $$PH-CH=CH-\overset{O}{\overset{||}{C}}-OH \xrightarrow{SOCl_2} Ph-CH=CH-\overset{O}{\overset{||}{C}}-Cl \quad \triangleright-\ddot{N}H_2$$

$$\downarrow$$

$$Ph-CH=CH-\overset{O}{\overset{||}{C}}-NH-\triangleleft$$

25. (a) $$X = CH_3-CH_2-CH_2-\overset{O}{\overset{||}{C}}-Cl + \ddot{N}H_3 \longrightarrow CH_3-CH_2-CH_2-\overset{O}{\overset{||}{C}}-NH_2 \xrightarrow{Br_2 + KOH} CH_3-CH_2-CH_2-NH_2$$

26. (c)

$$\xrightarrow{PhMgBr} Ph-\overset{O}{\overset{||}{C}}-\overset{O}{\overset{||}{C}}-OMgBr \xrightarrow{MeMgBr} Ph-\underset{Me}{\overset{OMgBr}{C}}-\underset{O}{\underset{||}{C}}-OMgBr \xrightarrow{H^+/H_2O} Ph-\underset{Me}{\overset{OH}{C}}-COOH$$

27. (b) $CH_3-\underset{OH}{\overset{H}{C}}-CO\boxed{OH+H}O-CH_2-CH=CH_2$

$$\downarrow$$

$CH_3-\underset{OH}{\overset{H}{C}}-CO-O-CH_2-CH=CH_2$

X

28. (d) $CH_3COOH \xrightarrow{SOCl_2} \underset{A}{CH_3-\overset{O}{\overset{\|}{C}}-Cl} \xrightarrow[AlCl_3]{Benzene}$

$\underset{B}{C_6H_5-\overset{O}{\overset{\|}{C}}-CH_3} \xrightarrow{HCN} \underset{C}{C_6H_5-\underset{CN}{\overset{OH}{C}}-CH_3} \xrightarrow{HOH} \underset{D}{C_6H_5-\underset{COOH}{\overset{OH}{C}}-CH_3}$

29. (a) [1], [1]→ HO, NO_2, [3], O, H, C, C

31. (c) Tishchenko reaction is a modification of Canizzaro's reaction. This reaction involves disproportionation of an aldehyde lacking a hydrogen atom in the alpha position in the presence of an alkoxide. The reaction product is an ester. Catalysts are aluminium alkoxide or sodium alkoxide.

In Cannizzaro's reaction the base is sodium hydroxide and the oxidation product is a carboxylic acid and the reduction product is an alcohol.

$2\,C_6H_5CHO \xrightarrow[50-60°,\ 4\ hr]{C_6H_5CH_2ONa} C_6H_5COOCH_2C_6H_5$ (90%)

30. (d) Acid chlorides, anhydrides, acids give ester with alcohol

$ROH + R'COCl \longrightarrow ROCOR' + HCl$

$ROH + (R'CO)_2O \longrightarrow ROCOR' + RCOOH$

$ROH + R'COOH \longrightarrow ROCOR' + H_2O$

32. (c) Only acetaldehyde and methyl ketones give iodoform test.

33. (b) $>C=O \xrightarrow{\text{Zn-Hg/HCl}} >CH_2$

Ketone

34. (b) $CH_3-\underset{}{\overset{CH_3}{\overset{|}{C}}}=O + CH_3COCH_3 \longrightarrow CH_3-\overset{CH_3}{\overset{|}{\underset{OH}{\underset{|}{C}}}}-CH_2-\overset{O}{\overset{||}{C}}-CH_3$

4–Hydroxy–4–methyl–2–pentanone

35. (c) Only those aldehydes which do not have α-H atom undergo Cannizaro's reaction. Hence CH_3CHO will not undergo Cannizaro's reaction as it has 3 αH atoms.

36. (a) It is an example of intramolecular cannizzaro reaction.

37. (a) It is an example of intramolecular aldol condensation

$OHC-CH_2CH_2CH_2CH_2-CHO \xrightarrow[\text{heat}]{\text{KOH, } CH_3OH}$ 2-hydroxycyclopentanecarbaldehyde (CHO, OH) $\xrightarrow{-H_2O}$ cyclopent-1-enecarbaldehyde (CHO)

38. (a) $\underset{A}{C_3H_8O} \xrightarrow{K_2Cr_2O_7/H^+} \underset{B\ (-CHO)}{C_3H_6O} \xrightarrow{H_2NCONHNH_2} C$

Since B reduces Tollen's reagent, it indicates that it has an –CHO group, so it must be CH_3CH_2CHO. Hence

$$\underset{[A]}{CH_3CH_2CH_2OH} \rightarrow \underset{[B]}{CH_3CH_2CHO} \xrightarrow{H_2NNHCONH_2} \underset{[C]}{CH_3CH_2CH=NNHCONH_2}$$

39. (b) $-NO_2$ group is electron withdrawing which enchances the addition of CN^- ion (a nucleophilic additon reaction).

40. (d) $CH_3COOH \xrightarrow{CaCO_3} \underset{[A]}{(CH_3COO)_2Ca} \xrightarrow{\text{heat}} \underset{[B]}{CH_3\overset{O}{\overset{||}{C}}CH_3} \xrightarrow{NH_2OH} \underset{[C]\ \text{Acetoxime}}{CH_3\overset{NOH}{\overset{||}{C}}CH_3}$

41. (a) Cyclopentanone (C=O) $\xrightarrow[CH_3OH]{NaBH_4}$ cyclopentyl–OH $\xrightarrow{HBr}$ cyclopentyl–Br

$\xrightarrow[\text{(ii) } H_2C=O]{\text{(i) Mg. } Et_2O}$ cyclopentyl–CH_2OH $\xrightarrow[CH_2Cl_2]{PCC}$ cyclopentyl–CHO

42. **(b)** This is intramolecular Cannizzaro's reaction.

43. **(a)** Cyclopentanone–H + EtO–C(=O)–OEt $\xrightarrow{Et\bar{O}}$ 2-(ethoxycarbonyl)cyclopentanone, –C(=O)–OEt

44. **(b)** 4-tert-butyltoluene (CH_3, $C(CH_3)_3$) $\xrightarrow[\text{benzoyl peroxide}]{\text{NBS, } CCl_4\text{, heat}}$ CH_2Br, $C(CH_3)_3$ $\xrightarrow[\text{(ii) } CO_2,\ H_3O^+]{\text{(i) Mg/ether}}$ CH_2COOH, $C(CH_3)_3$

Allylic subsitution

45. **(b)** $CH_3CHO \xrightarrow{MnO_4^-/H^+} \underset{X}{CH_3COOH} \xrightarrow{SOCl_2} \underset{Y}{CH_3COCl} \xrightarrow[\text{heat}]{CH_3COONa} \underset{Z}{CH_3COOCOCH_3}$

46. **(c)** $C_6H_5CH_3 \xrightarrow{[O]} \underset{[A]}{C_6H_5COOH} \xrightarrow{SOCl_2} \underset{[B]}{C_6H_5COCl} \xrightarrow{NaN_3} \underset{[C]}{C_6H_5CON_3}$

$\xrightarrow{heat} \left[C_6H_5-\overset{O}{\overset{\|}{C}}-\ddot{N}: \right] \longrightarrow C_6H_5NCO$

Unstable Benzoyl nitrene — Phenyl isocyanate, [D]

47. **(d)** β - Keto acids are decarboxylated by following mechanism

H_7C_3–CO–CH_2–C(=O)–O–H $\xrightarrow{\Delta}$ H_7C_3–C(OH)=CH_2 $\longleftrightarrow$ H_7C_3–C(=O)–CH_3

48. **(c)** 5-hydroxy-3-methylhexanoic acid (C1–C5 numbered, OH, O, OH) $\xrightarrow{-H_2O}$ lactone (numbered 1–5)

49. **(a)** 2 $CH_3CH_2CH_2$–C(=O)–OEt $\xrightarrow[H_3O^+]{EtONa}$ $CH_3CH_2CH_2$–C(=O)–CH(C_2H_5)–C(=O)–OEt $\xrightarrow[H^+]{NaOH}$

$CH_3CH_2CH_2$–C(=O)–CH(C_2H_5)–C(=O)–OH $\xrightarrow[\text{decarboxylation}]{\Delta}$ 4 – Heptanone (C7–C1 numbered)

50. **(a)** 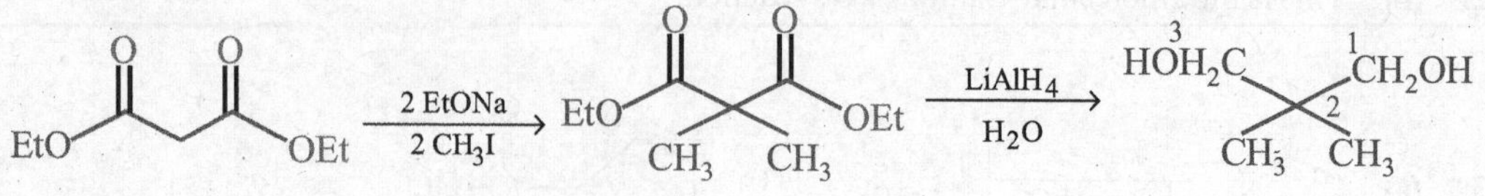 $EtO-C(=O)-CH_2-C(=O)-OEt \xrightarrow[2\,CH_3I]{2\,EtONa} EtO-C(=O)-C(CH_3)_2-C(=O)-OEt \xrightarrow[H_2O]{LiAlH_4} HOH_2C-C(CH_3)_2-CH_2OH$ (carbons numbered 3, 2, 1)

51. **(b)** Aldehyde will be more reactive than ketone.

52. **(b)** β-keto acid $\xrightarrow{\Delta}$ cyclohexyl-$COCH_3$ $\xrightarrow[\text{Reduction}]{\text{Clemmensen}}$ ethylcyclohexane

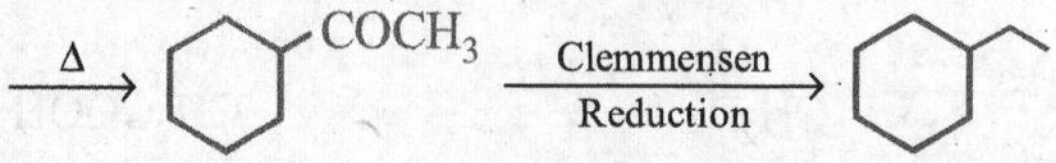

53. **(b)** $-CH_3$ group give $+M$ and $+I$ but $+M$ work only at ortho and para position.
Reactivity ∝ positive charge on carbonyl carbon so that's why reactivity II > I > III.

54. **(d)** (A) $Ph-CH_2-Br \xrightarrow{KCN} Ph-CH_2-C\equiv N \xrightarrow{H_3O^+} Ph-CH_2-CO_2H$ (B)

55. **(d)** 1-Phenylethanol ($CH_3CH(C_6H_5)OH$) is a 2° alcohol and can be prepared by the reaction of benzaldehyde with Grignard reagent CH_3MgI.

56. **(c)** Selenium dioxide oxidises aldehydes and ketones at α-position to carbonyl group.

$$CH_3CHO \xrightarrow{SeO_2} OHC.CHO$$

57. **(a)** (a) $H_3C-C^{\alpha}(CH_3)_2-CHO$ 2,2-dimethylpropanal (no α-H)
(b) $H_3C^{\alpha}-CHO$ acetaldehyde (α-H)
(c) $CH_3\overset{\alpha}{C}H_2CHO$ propionaldehyde (α-H)
(d) $C_6H_5CH=\overset{\alpha}{C}HCHO$ cinnamaldehyde (α-H)

58. **(c)** Remember that during reduction with $NaBH_4$, a hydride ion is transferred by $NaBH_4$ to carbonyl carbon and a proton (or deutron) is transferred from the solvent (in the second step) to carbonyl oxygen.

$$\underset{\text{3-Hexanone}}{C_2H_5-\overset{O}{\overset{\|}{C}}-C_3H_7} \xrightarrow{NaBH_4} C_2H_5-\overset{O^-}{\overset{|}{C}H}-C_3H_7 \xrightarrow[(-OD^-)]{D_2O} C_2H_5-\overset{OD}{\overset{|}{C}H}-C_3H_7$$

59. **(d)** $CH_3CH_2CHO + CH_2(CH_3)CH(=O) \xrightarrow{\text{dil. } OH^-} \overset{\beta}{C}H_3-\underset{\alpha}{C}H_2-CH(OH)-CH(CH_3)CHO$
Propanal (2 molecules)

Note that it is the α-carbon (and not β–) that is adding on the carbonyl oxygen of the other propanal molecule.

60. **(b)** Acetophenone ($C_6H_5COCH_3$) being a ketone, does not reduce Tollen's reagent.

61. **(a)** Weaker the base, better the leaving group and hence more is the reactivity of the compound. The relative basic character of the four groups is

$$NH_2^- > OR^- > OCOCH_3^- > Cl^-$$

Thus the relative leaving ability of the four groups is

$NH_2^- < OR^- < OCOCH_3^- < Cl^-$

Hence the relative reactivity of the four acyl derivatives toward nucleophile is

$$\underset{\text{Amide}}{R-\overset{O}{\overset{\|}{C}}-NH_2} < \underset{\text{Ester}}{R-\overset{O}{\overset{\|}{C}}-OR} < \underset{\text{Acid anhydride}}{R-\overset{O}{\overset{\|}{C}}-OCOCH_3} < \underset{\text{Acid chloride}}{R-\overset{O}{\overset{\|}{C}}-Cl}$$

62. (d) $CH_3CH_2COOH \xrightarrow{\text{red P, } Br_2} CH_3CHBrCOOH \xrightarrow{NH_3} \underset{\alpha\text{-Aminopropionic acid (alanine)}}{CH_3CHNH_2COOH}$

63. (b) $R-\overset{O}{\overset{\|}{C}}-NH_2 \xrightarrow{Br_2/KOH} \left[\underset{\text{Acylnitrene}}{R-\overset{O}{\overset{\|}{C}}-\ddot{N}:}\right] \xrightarrow{\text{Rearrangement}} \underset{\text{Alkyl isocyanate}}{R-N=C=O}$

64. (b) Only amides (but not acids and esters) undergo hydrolysis in presence of soda-lime to form sodium salt of a carboxylic acid and ammonia gas. Further, since the given compound is a liquid, it must be formamide, because propanamide is a solid.

65. (a) $\underset{\text{Acetaldehyde cyanohydrin}}{CH_3\overset{OH}{\overset{|}{C}}HCN} \xrightarrow{[O]} CH_3-\overset{O}{\overset{\|}{C}}-CN \xrightarrow{H^+,\, H_2O} CH_3\overset{O}{\overset{\|}{C}}-COOH$

66. (b) Due to –I effect of the –COOH group, H-bonds in acids are much stronger than in alcohols; while aldehydes do not exhibit H-bonding.

67. (d) In carboxylates (conjugate base of carboxylic acids), resonance is more significant because the two resonating structures are similar, while in phenoxide, the resonating structures are not equivalent, alkoxide ions do not show resonance.

68. (a) Only acetic acid is more acidic than carbonic acid (conjugate acid of $NaHCO_3$) hence it dissolves in $NaHCO_3$,

$$\underset{\text{stronger acid}}{CH_3COOH} + NaHCO_3 \longrightarrow CH_3COO^-Na^+ + \underset{\text{weaker acid}}{H_2CO_3}\ (H_2O + CO_2)$$

while phenol and *n*-hexanol are less acidic than carbonic acid and hence do not dissolve in $NaHCO_3$.

69. (c) Bromine is less electronegative than F, further in $BrCH_2CH_2COOH$, Br is more away from the –COOH group than in $CH_3CHBrCOOH$.

70. (a) $F-CH_2-\overset{O}{\overset{\|}{C}}-OH > Cl-CH_2-\overset{O}{\overset{\|}{C}}-OH > Br-CH_2-\overset{O}{\overset{\|}{C}}-OH$

(Electron-withdrawing character decreasing from L to R)

71. (a) P_2O_5 is a dehydrating agent and removes a molecule of water from two molecules of acetic acid

$$CH_3COOH + HOOCCH_3 \xrightarrow{P_2O_5} CH_3COOCOCH_3$$

72. (a) More the magnitude of positive charge on the carbonyl carbon, higher will be its reactivity toward nucleophilic attack.

$$CH_3-\overset{O}{\overset{||}{C}}-Cl \quad CH_3-\overset{O}{\overset{||}{C}}-H \quad CH_3-\overset{O}{\overset{||}{C}}-OCOCH_3 \quad CH_3-\overset{O}{\overset{||}{C}}-OCH_3$$

Chlorine being more electron-withdrawing than O develops more positive charge on carbonyl carbon. The order of reactivity of the various carboxyl derivatives toward nucleophile is

$$\underset{\text{Most reactive}}{RCOCl} > RCHO > (RCO)_2O > \underset{\text{Least reactive}}{RCOOR}$$

73. (b) $$\underset{\text{Ethylchloroformate}}{Cl-\overset{O}{\overset{||}{C}}-OC_2H_5} \xrightarrow{CH_3MgBr} \left[Cl-\underset{CH_3}{\underset{|}{\overset{OMgBr}{\overset{|}{C}}}}-OC_2H_5\right] \longrightarrow \underset{\text{Ethyl acetate}}{\underset{CH_3}{\underset{|}{\overset{O}{\overset{||}{C}}}}-OC_2H_5}$$

74. (d) Ethyl benzoate on hydrolysis gives benzoic acid which is a solid, other esters give CH_3COOH and HCOOH, both of which are liquids.

75. (b) Ethane is obtained by the electrolysis of sodium or potassium salt of ethanoic acid (CH_3COOH), hence the ester must be ethanoate CH_3COOR.

76. (c) $$EtO-\overset{O}{\overset{||}{C}}-CH_2-\overset{O}{\overset{||}{C}}-OEt \xrightarrow{EtONa} EtO-\overset{O}{\overset{||}{C}}-C(Na)_2-\overset{O}{\overset{||}{C}}-OEt \xrightarrow{Br-CH_2CH_2CH_2CH_2-Br}$$

$$\text{(1,1-cyclopentane)}(COOEt)_2 \xrightarrow[\text{(iii) }\Delta]{\text{(i) NaOH}/\Delta \;\; \text{(ii) }H_3O^+/H^+} \text{cyclopentane-}COOH$$

77. (b) $$-COOH \xrightarrow[H_2O]{LiAlH_4} -CH_2OH \xrightarrow{PBr_3} -CH_2Br \xrightarrow[DMSO]{KCN} -CH_2CN \xrightarrow[H_2O]{H^+} -CH_2COOH$$

78. (c) Due to resonance in carboxylate ion, the double bond character of C = O bond in carboxylic acids is greatly reduced as compared to that in aldehydes and ketones.

$$-C\begin{matrix}\nearrow O\\ \searrow O^-\end{matrix} \longleftrightarrow -C\begin{matrix}\nearrow O^-\\ \searrow O\end{matrix} \equiv -C\begin{matrix}\nearrow O\\ \searrow O\end{matrix}\Bigg\}^-$$

79. (a) RCOOH > ROH > HC ≡ CH. Water is more acidic then alcohol.

80. (c) Unsaturation is not detected by Tollen's reagent, a reagent for differentiating between aldehydes and ketones.

81. (a) $$\underset{[A]}{(CH_3)_2CO} \xrightarrow{NaCN/HCl} \underset{[B]}{(CH_3)_2C\begin{matrix}OH\\ CN\end{matrix}} \xrightarrow{H_3O^+} (CH_3)_2C\begin{matrix}OH\\ COOH\end{matrix}$$

82. (a) $C_6H_5-\overset{O}{\overset{\|}{C}}-CH_3 \xrightarrow[-C_2H_5OH]{C_2H_5O^-} C_6H_5-\overset{O}{\overset{\|}{C}}-\bar{C}H_2$

$C_6H_5-\overset{O}{\overset{\|}{C}}-\bar{C}H_2 + \underset{CH_3}{\underset{|}{\overset{O}{\overset{\|}{C}}}}-C_6H_5 \longrightarrow C_6H_5-\overset{O}{\overset{\|}{C}}-CH_2-\underset{CH_3}{\underset{|}{\overset{O^-}{\overset{|}{C}}}}-C_6H_5 \xrightarrow[-OH^-]{H_2O} C_6H_5-\overset{O}{\overset{\|}{C}}-CH_2-\underset{CH_3}{\underset{|}{\overset{OH}{\overset{|}{C}}}}-C_6H_5$

$\xrightarrow{\Delta,\ -H_2O} C_6H_5-\overset{O}{\overset{\|}{C}}-CH=\underset{CH_3}{\underset{|}{C}}-C_6H_5$

83. (a) $Ph-\overset{O}{\overset{\|}{C}}-H + OH^- \underset{}{\overset{fast}{\rightleftharpoons}} Ph-\underset{OH}{\underset{|}{\overset{O^-}{\overset{|}{C}}}}-H \xrightarrow[slow]{Ph-\overset{O}{\overset{\|}{C}}-H} Ph-\underset{OH}{\underset{|}{\overset{O}{\overset{\|}{C}}}} + Ph-\underset{H}{\underset{|}{\overset{O^-}{\overset{|}{C}}}}-H$

$\xrightarrow[fast]{H^+ \text{ exchange}} Ph-\underset{O^-}{\underset{|}{\overset{O}{\overset{\|}{C}}}} + Ph-\underset{H}{\underset{|}{\overset{OH}{\overset{|}{C}}}}-H$

84. (c) $KMnO_4$ converts $-CH_3$ group of toluene into $-COOH$ while HI reduces $-COOH$ group into $-CH_3$ group.

85. (c) $CH_3COOH + CaCO_3 \rightarrow \underset{(A)}{(CH_3COO)_2Ca} \xrightarrow{Heat} \underset{(B)}{(H_3C)_2C{=}O} \xrightarrow{I_2+NaOH} \underset{(C)}{CHI_3}$

86. (c) This is Hunsdiecker's reaction.

$\underset{\text{Silver benzoate}}{C_6H_5COOAg} + Br_2 \xrightarrow{CCl_4} \underset{\text{Bromobenzene}}{C_6H_5Br} + AgBr + CO_2$

87. (b) β-Ketoacids undergo decarboxylation easily.

88. (a) ArCOR can be prepared by the combination of ArH + RCOCl and not by ArCOCl + RMgX because here the ArCOR formed will further react with RMgX to form 3° alcohol, $ArC(OH)R_2$ as the final products.

89. (c) $HC\equiv CH \xrightarrow[20\%H_2SO_4]{1\%HgSO_4} \underset{[A]}{CH_3CHO} \xrightarrow[H_2O]{CH_3MgX} \underset{[B]}{CH_3CHOHCH_3} \xrightarrow{[O]} \underset{\text{Acetone }[C]}{CH_3COCH_3}$

90. (d) $CH\equiv CH \xrightarrow[HgSO_4]{30\%H_2SO_4} \underset{[A]}{CH_3CHO} \xrightarrow{NaOH} \underset{[B]}{CH_3CHOHCH_2CHO}$

91. (b) $C_4H_8 \xleftarrow[(-H_2O)]{Conc.H_2SO_4} C_4H_{10}O \xrightarrow{Oxidation} C_4H_8O\ (R-COCH_3)$

Thus C_4H_8O should be $CH_3CH_2COCH_3$, hence $C_4H_{10}O$ should be $CH_3CH_2CHOHCH_3$

92. (d) Hemiacetal → presence of alcohol and ether on same carbon.

93. (d) (A) $Ph(CH_3)C=C(Ph)CH_3$ (cis & trans) $\xrightarrow{O_3}$ (B) $Ph-\overset{O}{\overset{||}{C}}-CH_3 \xrightarrow{NH_2-NH_2,\ HO^-,\ \Delta} Ph-CH_2-CH_3$

94. (c) Base stronger than cyanide is used as a catalyst.

95. (a)

$CH_3COCH_2CH_2COOC_2H_5 \xrightarrow{H_3O^+} CH_3COCH_2CH_2COOH$

$\xrightarrow{CH_2OH-CH_2OH}$ (ketal of ketone with ethylene glycol)$-CH_2CH_2COOH \xrightarrow[\text{(ii) } H_3O^+]{\text{(i) } LiAlH_4} CH_3COCH_2CH_2CH_2OH$

96. (b) Presence of alkyl group in carbonyl compounds decreases their reactivity towards nucleophilic addition. Further greater the number of such groups lesser will be the reactivity towards nucleophilic addition ∴ correct order is
$HCHO > CH_3CHO > CH_3COCH_3$

97. (d) The reaction given is a Clemmenson reduction.

98. (c) All aldehydes and ketones show reaction with Tollen's reagent and Fehling solutions, but ketones do not show this reaction.
Note: Benzaldehyde do not give reaction with Fehling solution.

99. (d) **Hell-Volhard-Zelinsky Reaction :**
It is α-substitution when a carboxylic acid having α-hydrogens is treated with chlorine or bromine in presence of small amount of red phosphorous.

$$R-CH_2COOH + X_2 \xrightarrow{P} R\overset{X}{\overset{|}{C}}H-COOH + HX (X = Cl, Br)$$

100. (d) Phthalic acid ($C_6H_4(COOH)_2$) + 2 Resorcinol (HO–C_6H_4–OH) $\xrightarrow{\text{Conc. } H_2SO_4}$ Fluorescein + H_2O

Fluorescein

27 Amines

1. The conversion of benzene diazonium chloride to bromobenzene can be accomplished by
   (a) Reimer-Tiemann reaction
   (b) Friedel-Crafts reaction
   (c) Gattermann reaction
   (d) Azo-Coupling reaction
2. C_7H_9N has how many isomeric forms that contain a benzene ring?

Critical Thinking

   (a) 4 (b) 5 (c) 6 (d) 7
3. The amine that does not react with acetyl chloride is
   (a) CH_3NH_2 (b) $(CH_3)_2NH$
   (c) $(CH_3)_3N$ (d) None of these
4. In the diazotisation of arylamines with sodium nitrite and hydrochloric acid, an excess of hydrochloric acid is used primarily to
   (a) supress the concentration of free aniline available for coupling
   (b) supress hydrolysis of phenol
   (c) ensure a stoichiometric amount of nitrous acid
   (d) neutralise the base liberated
5. Nitrosoamines ($R_2N - N = O$) are insoluble in water. On heating with conc. H_2SO_4, they give secondary amines. The reaction is called
   (a) Liebermann nitroso reaction
   (b) Etard reaction
   (c) Fries reaction
   (d) Perkin reaction
6. Aniline reacts with phosgene and KOH to form
   (a) C_6H_5-OH
   (b) $C_6H_5-\overset{O}{\overset{||}{C}}-Cl$
   (c) $C_6H_5-NHCOCl$
   (d) C_6H_5-NCO
7. In a reaction of aniline a coloured product C was obtained.

$$\underset{A}{C_6H_5NH_2} \xrightarrow[HCl]{NaNO_2} B \xrightarrow[Cold]{C_6H_5-N(CH_3)_2} C$$

The structure of C would be :
   (a) $C_6H_5-N=N-CH_2-\underset{CH_3}{\underset{|}{N}}-C_6H_5$
   (b) $(CH_3)C_6H_4-N=N-C_6H_4(CH_3)$ (with CH_3 on each ring, ortho)
   (c) $C_6H_5-NH-NH-C_6H_4-N(CH_3)_2$
   (d) $C_6H_5-N=N-C_6H_4-N(CH_3)_2$
8. Reaction of cyclohexanone with dimethylamine in the presence of catalytic amount of an acid forms a compound if water during the reaction is continuously removed. The compound formed is generally known as

Critical Thinking

   (a) an amine (b) an imine
   (c) an enamine (d) a Schiff's base

9. [Structure: 1-(N-n-butyl-N-ethyl-N-methylammonio)-6-methylcyclohexene with $\overset{\ominus}{O}H$] $\xrightarrow{\Delta}$

The alkene formed as a major product in the above elimination reaction is

(a) 1-methylcyclohexene (b) 3-methylcyclohexene

(c) Me–CH₂–CH=CH₂ (d) $CH_2 = CH_2$

10. Aniline (NH_2) $\xrightarrow[H_2O]{Br_2}$ (A) $\xrightarrow[(ii)\ H_3PO_2]{(i)\ NaNO_2/HCl}$ (B),

Product (B) in this reaction is:

(a) 1,2,4-tribromobenzene (b) 1,3,5-tribromobenzene

(c) 1,2,3-tribromobenzene (d) bromobenzene

11. *p*-Chloroaniline and anilinium hydrogen chloride can be distinguished by:
(a) Sandmeyer reaction
(b) Carbylamine reaction
(c) Hinsberg's reaction
(d) $AgNO_3$

12. Consider the following sequence of reactions :

$$\text{Compound [A]} \xrightarrow{\text{Reduction}} [B] \xrightarrow{HNO_2} CH_3CH_2OH$$

The compound [A] is
(a) CH_3CH_2CN (b) CH_3NO_2
(c) CH_3NC (d) CH_3CN

13. What will be the final product in the following reaction sequence?

$$CH_3CH_2CN \xrightarrow{H^+/H_2O} A \xrightarrow[\Delta]{NH_3} B \xrightarrow{NaOBr} C$$

(a) $CH_3CH_2CONH_2$ (b) CH_3CH_2COBr
(c) $CH_3CH_2NH_2$ (d) $CH_3CH_2CH_2NH_2$

14. $CH_3-CH_2C \equiv N \xrightarrow{X} CH_3CH_2CHO$.
The compound X is
(a) $SnCl_2 / HCl / H_2O$, boil
(b) $H_2 / Pd - BaSO_4$
(c) $LiAlH_4$ / ether
(d) $NaBH_4$ / ether / H_3O^+

15. Tautomerism will be exhibited by
(a) $(CH_3)_3CNO$ (b) $(CH_3)_2NH$
(c) R_3CNO_2 (d) RCH_2NO_2

16. Hydrolysis of $CH_3CH_2NO_2$ with 85% H_2SO_4 gives
(a) CH_3CH_2OH (b) C_2H_6
(c) $CH_3CH = NOH$ (d) CH_3COOH

17. Towards electrophilic substitution, the most reactive will be **Critical Thinking**
(a) Nitrobenzene
(b) Aniline
(c) Aniline hydrochloride
(d) N-Acetylaniline

18. 3, 5-dibromotoluene can be best synthesised by

(a) p-toluidine (CH_3, NH_2) $\xrightarrow{Br_2/H_2O}$ $\xrightarrow[H_3PO_2]{HNO_2}$

(b) toluene (CH_3) $\xrightarrow[H_2SO_4]{\text{Fuming } HNO_3}$ $\xrightarrow{NH_3}$ $\xrightarrow[CuBr]{HONO}$

(c) toluene (CH_3) $\xrightarrow[\Delta]{Br_2/Fe}$

(d) p-bromotoluene (CH_3, Br) $\xrightarrow[\Delta]{CH_3Cl/AlCl_3}$

19. Which of the following compounds is an enamine?

(a) 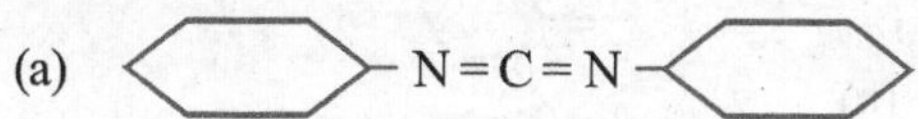

(b) $-NH_2$

(c) $-N$

(d) $=N-$

20. Which of the following reactions will not give N, N- dimethyl benzamide ?

(a) $C_6H_5-COOC_2H_5 + (CH_3)_2NH \longrightarrow$

(b) $C_6H_5-CONH_2 + CH_3MgI \longrightarrow$

(c) $C_6H_5-COCl + (CH_3)_2NH \longrightarrow$

(d) $C_6H_5-COOCO-C_6H_5 + (CH_3)_2NH \longrightarrow$

21. The product – (C) obtained in the following sequence of reactions is **Toughnut**

NH_2, Br, Br, NO_2 (substituted benzene) $\xrightarrow[(2)\ CuCl]{(1)\ HONO} A \xrightarrow{Sn/HCl} B \xrightarrow[(2)\ H_3PO_2]{(1)\ HONO} C$

(a) Br, Br, NO_2 (substituted benzene) (b) Cl, Br, Br (substituted benzene)

(c) Cl, Br, Br (substituted benzene) (d) None of these

22. Which of the following reagents will be useful to distinguish between

$H_3C-C_6H_4-NH_2$ and $C_6H_5-CH_2NH_2$?

(a) Dilute HCl

(b) $C_6H_5SO_2$ and OH^-/H_2O

(c) HONO then β-naphthol

(d) $AgNO_3$ in H_2O

23. Which of the following is the strongest base ?

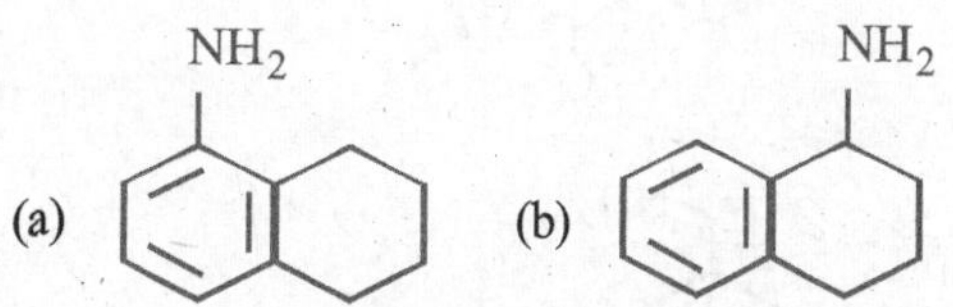

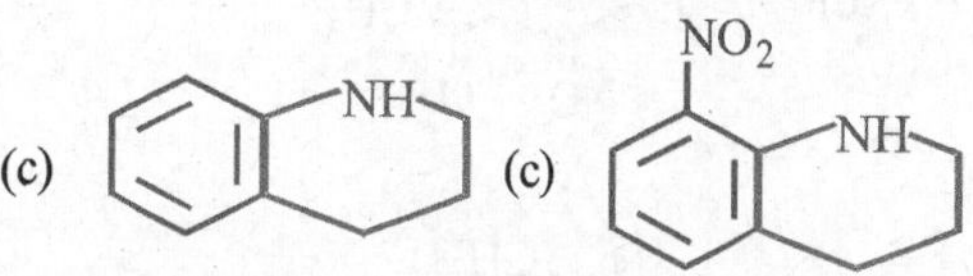

24. The cyclobutyl methylamine with nitrous acid gives **Toughnut**

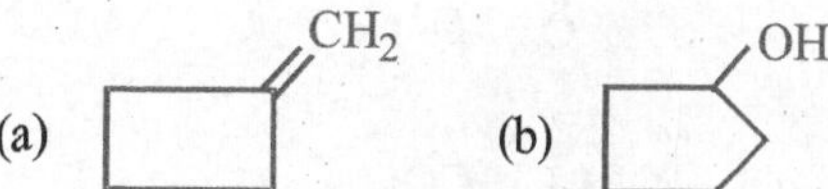

(c) (cyclopentene) (d) All of these

25. Which of the following amines will react with cyclohexanone to give enamine ?

(a) $(CH_3)_2NH$ (b)

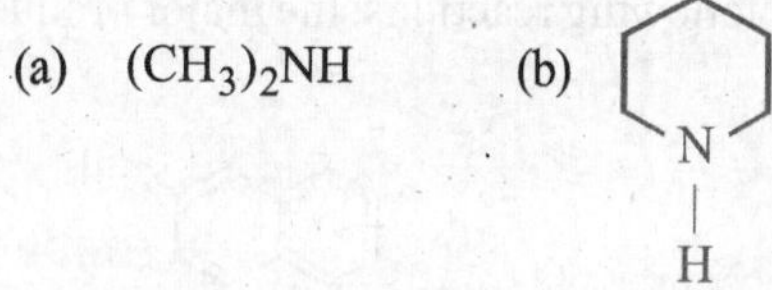

(c)

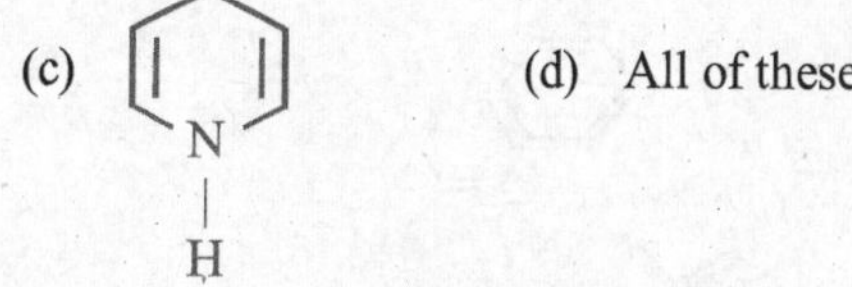

(d) All of these

26. In the reaction shown below, the major product(s) formed is/are

Toughnut

NH_2, Z, NH_2, O $\xrightarrow[CH_2Cl_2]{\text{acetic anhydride}}$ Product(s)

(a) H, N, CH_3, O, NH_2, O + CH_3COOH

(b) NH_2, H, N, CH_3, O, O + CH_3COOH

(c) H, N, CH_3, O, H, N, CH_3, O, O + H_2O

(d) $\overset{+}{N}H_3CH_3CO\bar{O}$, H, N, CH_3, O, O

27. In the following reactions, the major product W is

NH_2 $\xrightarrow[0°C]{NaNO_2\text{ , HCl}}$ V $\xrightarrow{\text{OH}\text{ , NaOH}}$ W

(a) N = N, OH

(b) OH, N = N

(c) OH, N = N

(d) HO, N = N

28. In the following reactions, the product S is

H_3C $\xrightarrow[\text{ii. Zn, } H_2O]{\text{i. } O_3}$ R $\xrightarrow{NH_3}$ S

(a) H_3C, N

(b) H_3C, N

(c) N, H_3C

(d) N, H_3C

29. In the reaction,

$CH_3CN + 2H \longrightarrow X \xrightarrow{\text{Boiling } H_2O} Y$, the term Y is

(a) acetone (b) ethanamine

(c) acetaldehyde (d) dimethylamine

30. In the reaction $C_6H_5NH_2 + CS_2 \xrightarrow[\Delta]{HgCl_2}$ the product obtained is
(a) phenyl isocyanide
(b) phenyl cyanide
(c) *p*-amino benzene sulphonic acid
(d) phenyl isothiocyanate

31. Arrange the following amines in the order of increasing basicity.
(a) $CH_3NH_2 < C_6H_5NH_2 < p\text{-}CH_3OC_6H_4NH_2 < p\text{-}O_2NC_6H_4NH_2$
(b) $p\text{-}CH_3OC_6H_4NH_2 < p\text{-}O_2NC_6H_4NH_2 < C_6H_5NH_2 < CH_3NH_2$
(c) $p\text{-}O_2NC_6H_4NH_2 < C_6H_5NH_2 < p\text{-}CH_3OC_6H_4NH_2 < CH_3NH_2$
(d) $C_6H_5NH_2 < p\text{-}CH_3OC_6H_4NH_2 < p\text{-}O_2NC_6H_4NH_2 < CH_3NH_2$

32. The final product formed when Methyl amine is treated with $NaNO_2$ and HCl is:
(a) Diazomethane (b) Methylalcohol
(c) Methylcyanide (d) Nitromethane

33. The order of basicity of amines in gaseous state is :
(a) $1° > 2° > 3° > NH_3$ (b) $3° > 2° > NH_3 > 1°$
(c) $3° > 2° > 1° > NH_3$ (d) $NH_3 > 1° > 2° > 3°$

34. Carbylamine forms from aliphatic or aromatic primary amine via which of the following intermediates ?
(a) Carbanion (b) Carbene
(c) Carbocation (d) Carbon radical

35. The most basic compound among the following is
(a) Acetanilide (b) Benzylamine
(c) *p*-Nitro aniline (d) Aniline

36. Which of the following compounds gives dye test?
(a) Aniline (b) Methylamine
(c) Diphenylamine (d) Ethylamine

37. Acetanilide on nitration followed by alkaline hydrolysis mainly gives
(a) *o*-Nitroacetanilide
(b) *p*-Nitroaniline
(c) *m*-Nitroaniline
(d) 2, 4, 6-Trinitroaniline

38. $p\text{-}CH_3C_6H_4NH_2 \xrightarrow{(CH_3CO)_2O} A \xrightarrow{Br_2/CH_3COOH} B \xrightarrow{H^+/H_2O} X.$

What is X? **Toughnut**

(a)
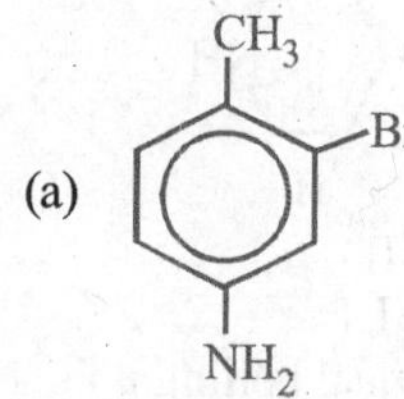

(b) CH_3, NH$_2$, Br (Br ortho to NH_2)

(c) CH_3, $COCH_3$ (ortho to CH_3), NH_2

(d) CH_3, $COCH_3$ (ortho to NH_2), NH_2

39. Which reaction sequence would be best to prepare 3-chloroanilne from benzene ?
(a) Chlorination, nitration, reduction
(b) Nitration, chlorination, reduction
(c) Nitration, reduction, chlorination
(d) Nitration, reduction, acetylation, chlorination, hydrolysis

40. A compound 'A' when treated with HNO_3 (in presence of H_2SO_4) gives compound 'B', which is then reduced with Sn and HCl to aniline? The compound 'A' is
(a) toluene (b) benzene
(c) ethane (d) acetamide

41. An organic compound (A) on reduction gives compound (B). (B) on treatment with $CHCl_3$ and alcoholic KOH gives (C). (C) on catalytic reduction gives N-methylaniline. The compound A is
(a) Methylamine (b) Nitromethane
(c) Aniline (d) Nitrobenzene

42. Predict about the relative boiling point of the following two amines.

[Structures: I — cyclic NCH$_3$; II — H_3C–ring–NH]

(a) Boiling point of I > II
(b) Boiling point of II > I
(c) Both should have equal boiling points
(d) It can't be predicted

43. Which statement is true regarding the following structure?

[Structure: N bonded to CH_3, C_3H_7, C_2H_5]

(a) It is a chiral molecule
(b) It exists in two resolvable optically active forms
(c) Both (a) and (b)
(d) Neither (a) nor (b)

44. Which one of the following is not an oxidation product of a primary amine?
(a) A hydroxylamine
(b) A nitroso compound
(c) A nitro compound
(d) All of these

45. Identify Z in the following sequence of reactions–

$$CH_3COONH_4 \xrightarrow{\Delta} X \xrightarrow{P_2O_5} Y \xrightarrow{H_2O/H} Z$$

(a) $CH_3-CH_2-CO-NH_2$
(b) CH_3-CN
(c) $(CH_3CO)_2O$
(d) CH_3-COOH

46. The order of basicity of the compounds

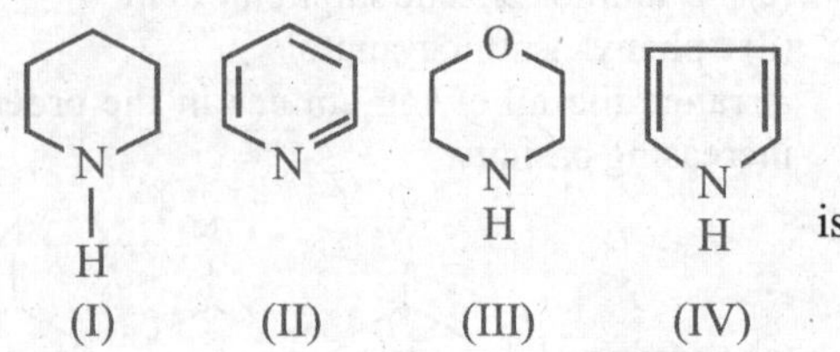

is

Toughnut

(a) IV > I > III > II (b) I > III > II > IV
(c) III > I > IV > II (d) II > I > III > IV

47. Complete reduction of benzene-diazonium chloride with Zn/HCl gives:
(a) Aniline (b) Phenylhydrazine
(c) Azobenzene (d) Hydrazobenzene

48. Conversion of benzene diazonium chloride to chlorobenzene is an example of which of the following reactions?
(a) Claisen (b) Friedel-craft
(c) Sandmeyer (d) Wurtz

49. In a set of reactions *p*-nitrotoluene yielded a product E.

Toughnut

[Structure: p-nitrotoluene (CH_3, NO_2)] $\xrightarrow[FeBr_3]{Br_2}$ B $\xrightarrow{Sn/HCl}$ C $\xrightarrow[HCl]{NaNO_2}$ D $\xrightarrow[HBr]{CuBr}$ E

The product E would be:

(a) [Structure: CH_3 with Br, Br at 2,6 and Br at 4]
(b) [Structure: CH_3 with Br at 3 and Br at 4]
(c) [Structure: CH_3 with Br at 2 and Br at 3]
(d) [Structure: CH_2Br with Br at 4]

50. Which of the following will produce isopropyl amine– **Tricky**

(I) $(CH_3)_2CO \xrightarrow{NH_2OH} X \xrightarrow{LiAlH_4}$

(II) $CH_3-CH_2-CHO \xrightarrow[heat]{NH_3} X \xrightarrow{LiAlH_4}$

(III) $(CH_3)_2CH-OH + PCl_5 \longrightarrow X \xrightarrow{NH_3}$

(IV) $CH_3-CH_2-CH_2-NH_2 \xrightarrow{heat}$

(a) I, II (b) II, III
(c) I, III (d) IV only

51. Acetaldoxime reacts with P_2O_5 to give
(a) Methyl cyanide (b) Methyl cyanate
(c) Ethyl cyanide (d) None of these

52. Identify the prdouct C in the series

$CH_3CN \xrightarrow{Na/C_2H_5OH} A \xrightarrow{HNO_2} B \xrightarrow{Cu/573K} C$

(a) CH_3COOH (b) CH_3CH_2NHOH
(c) CH_3CONH_2 (d) CH_3CHO

53. An organic compound 'A' having molecular formula C_2H_3N on reduction gave another compound 'B'. Upon treatment with nitrous acid, 'B' gave ethyl alcohol. On warming with chloroform and alcoholic KOH, B formed an offensive smelling compound 'C'. The compound 'C' is

(a) $CH_3CH_2NH_2$ (b) $CH_3CH_2N \overset{\rightarrow}{=} C$
(c) $CH_3C \equiv N$ (d) CH_2CH_2OH

54. Ethyl isocyanide on hydrolysis in acidic medium generates
(a) propanoic acid and ammonium salt
(b) ethanoic acid and ammonium salt
(c) methylamine salt and ethanoic acid
(d) ethylamine salt and methanoic acid

55. Primary nitro compounds react with nitrous acid to form nitrolic acids which dissolve in NaOH giving
(a) yellow solution
(b) blue solution
(c) colourless solution
(d) red solution

56. What is the product of the following series of reactions?

NH_2, Cl $\xrightarrow[H_2O,\ 0\,°C]{NaNO_2,\ HCl} \xrightarrow{CuCN} \xrightarrow[(2)\ H_2O,\ H^+]{(1)\ CH_3MgBr}$

(a) $O=C-CH_3$, Cl
(b) $COOCH_3$, Cl
(c) $HO-CH-CH_3$, Cl
(d) CH_2CN, Cl

57. The starting reagents needed to make the azo compound shown below

$CH_3-CH_2-C_6H_4-N=N-C_6H_4-OH$

(a) NH_2, OH + ethylamine
(b) C_2H_5, NH_2 + OH
(c) NH_2, OH + NH_2, C_2H_5
(d) NH_2 + OH, C_2H_5

58. Which of the following does not give N–ethyl cyclopentylamine as major product ?

(a) Cyclopentyl–$NH_2 + CH_3CHO \xrightarrow{H_2/Pt}$

(b) Cyclopentanone $= O + CH_3CH_2NH_2 \xrightarrow{H_2/Pt}$

(c) Cyclopentyl–$\overset{O}{\overset{\|}{C}} - Cl \xrightarrow[\text{Pyridine}]{CH_3CH_2NH_2} \xrightarrow[\text{(ii) } H_2O]{\text{(i) } LiAlH_4,\ Et_2O}$

(d) Cyclopentyl–$NH_2 \xrightarrow[\text{Pyridine}]{CH_3COCl} \xrightarrow[\text{(2) } H_2O]{\text{(1) } LiAlH_4}$

59. Match the compounds in List I with their nature from List II, as seen in aqueous medium

List I		List II	
I.	Acetamide	A.	Acidic
II.	Benzonitrile	B.	Basic
III.	Triethylamine	C.	Neutral
IV.	Phenol		

(a) I–C; II–C; III–B; IV–A
(b) I–B; II–C; III–C; IV–A
(c) I–C; II–B; III–B; IV–C
(d) I–A; II–A; III–C; IV–B

60. The correct order of basicities of the following compounds is

1: $CH_3-C(=NH)-NH_2$

2: $CH_3-CH_2-NH_2$

3: $(CH_3)_2NH$

4: $CH_3-\overset{O}{\overset{\|}{C}}-NH_2$

(a) $2>1>3>4$ (b) $1>3>2>4$
(c) $3>1>2>4$ (d) $1>2>3>4$

61. In the following reaction, X is

$X \xrightarrow{\text{Bromination}} Y \xrightarrow{NaNO_2/HCl} Z \xrightarrow[C_2H_5OH]{\text{Boiling}} \text{Tribromobenzene}$

(a) benzoic acid (b) salicylic acid
(c) phenol (d) aniline

62. Aniline is reacted with bromine water and the resulting product is treated with an aqueous solution of sodium nitrite in presence of dilute HCl. The compound so formed is converted into tetrafluoroborate which is subsequently heated dry. The final product is
(a) *p*-Bromofluorobenzene
(b) *p*-Bromoaniline
(c) 2, 4, 6-Tribromofluorobenzene
(d) 1, 3, 5-Tribromobenzene

63. Arrange the following three compounds in order of decreasing acidity

(A) $C_6H_5\overset{+}{N}H_3$ (B) $p\text{-}NO_2C_6H_4\overset{+}{N}H_3$ (C) $p\text{-}CH_3C_6H_4\overset{+}{N}H_3$

(a) B>A> C (b) B>C>A
(c) C>B> A (d) C>A>B

64. In the reaction

Cyclopropyl–$CH_2CN \xrightarrow[NaNH_2, NH_3, -80°C]{CH_3Br}$

the products obtained are

(a) Cyclopropyl–$\overset{CH_3}{\overset{|}{C}H} - C \equiv N$

(b) Cyclopropyl–$CH_2.NH_2$

(c) Cyclopropyl–$CH_2.OH$

(d) None of these

65. In the acidic reduction of nitrobenzene, which of the following is the intermediate?
(a) $C_6H_5-N=O$
(b) $C_6H_5NH-NHC_6H_5$
(c) $C_6H_5-N=N-C_6H_5$
(d) $C_6H_5-\overset{O}{\overset{\uparrow}{N}}=N-C_6H_5$

66. 'Z' in the following sequence of reactions is

$C_6H_6 \xrightarrow[\Delta]{HNO_3/H_2SO_4} W \xrightarrow{Zn/HCl} X \xrightarrow[HCl]{NaNO_2} Y \xrightarrow{H_2O/H_3PO_2} Z$

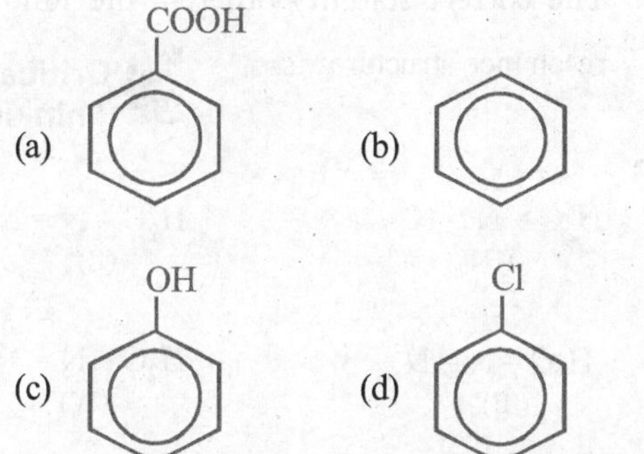

67. An organic compound (A) on reduction gives compound (B). (B) on treatment with $CHCl_3$ and alcoholic KOH gives (C). (C) on catalytic reduction gives N-methylaniline. The compound A is
(a) Methylamine (b) Nitromethane
(c) Aniline (d) Nitrobenzene

68. $Cl^-H_3N^+$ ~~~ $N^+H_3Cl^-$ $\xrightarrow{heat}$ Z; Z is

(a) 

(b) 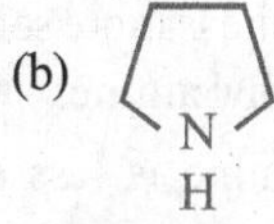

(c) 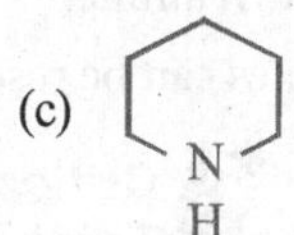

(d) 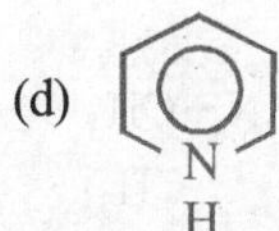

69. Pyrrole and pyridine both are basic and form salts with acids? **Tricky**

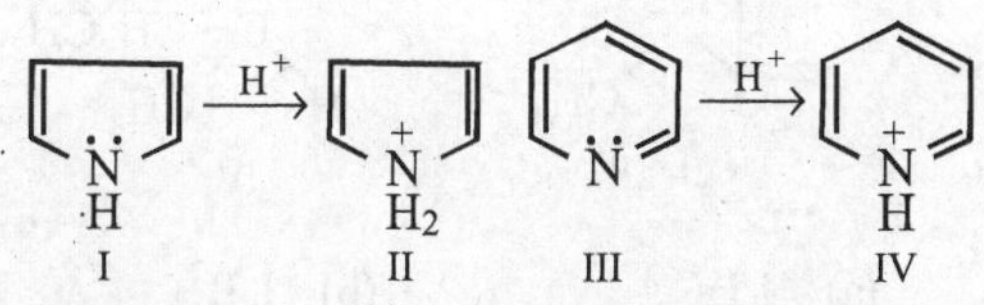

Which of the following statement is true regarding the aromatic character of the four species?
(a) All the four are aromatic
(b) I, III and IV are aromatic
(c) I, II and III are aromatic
(d) I and III are aromatic

70. Ethylene can be prepared in good yield by
(a) $CH_3CH_2N^+(CH_3)_3I^- \xrightarrow{heat} CH_2=CH_2+(CH_3)_3N+HI$
(b) $CH_3CH_2N^+(CH_3)_3OH^- \xrightarrow{heat} CH_2=CH_2+(CH_3)_3N+H_2O$
(c) Both (a) and (b)
(d) $CH_3CH_2NH_2 \xrightarrow{heat} CH_2=CH_2+NH_3$

71. Which of the following leads to carbon-carbon double bond?
(a) 1° Amine + RCHO →
(b) 2° Amine + R_2CO →
(c) 2° Amine + RCHO →
(d) Both (b) and (c)

72. In Hofmann bromamide degradation, one of the important steps is the migration of
(a) an alkyl group without its electron pair to electron deficient N atom.
(b) an alkyl group with its electron pair to electron deficient O atom.
(c) an alkyl group with its electron pair to electron rich N atom.
(d) an alkyl group with its electron pair to electron deficient N atom.

73. $(CH_3)_3C-\overset{O}{\overset{\|}{C}}-NH_2 \xrightarrow[(ii) D_2O]{(i) OD^-/Br_2}$ Product P is
(a) $(CH_3)_3CNH_2$ (b) $(CH_3)_3CNHD$
(c) $(CH_3)_3CND_2$ (d) no reaction

74. Identify product D in the following reaction sequence: Tricky

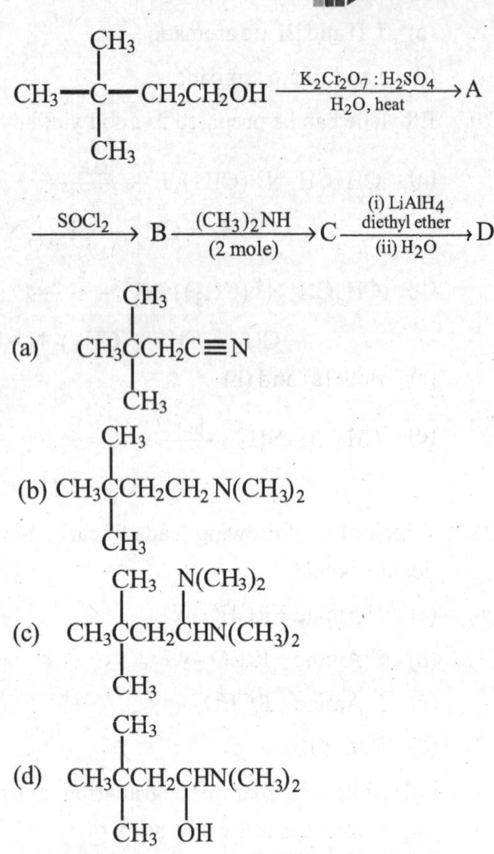

75. The correct order of decreasing basic character of the three aliphatic primary amines is

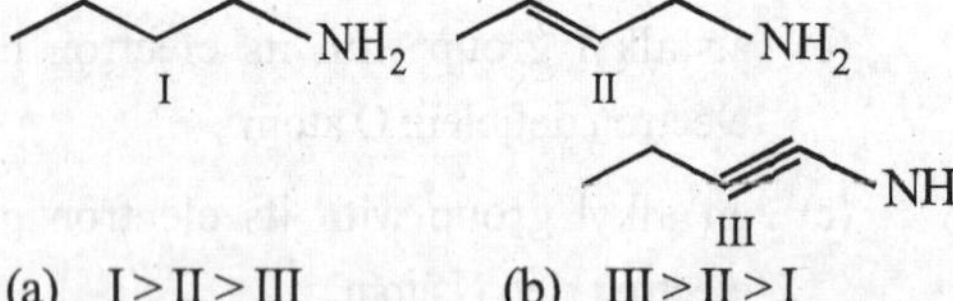

(a) I > II > III (b) III > II > I
(c) I > II ≈ III (d) I = II ≡ III

76. Which of the following can undergo electrophilic substitution when treated with nitrous acid at 0º C?
(a) $C_6H_5NH_2$ (b) $C_6H_5NHCH_3$
(c) $C_6H_5N(CH_3)_2$ (d) None

77. When aniline reacts with oil of bitter almonds (C_6H_5CHO) condensation takes place and benzal derivative is formed. This is known as
(a) Million's base (b) Schiff's reagent
(c) Schiff's base (d) Benedict's reagent

78. The correct stability order of the following resonance structures is Critical Thinking

$H_2C=\overset{+}{N}=\overset{-}{N}$ (I) $\quad H_2\overset{+}{C}-N=\overset{-}{N}$ (II)

$H_2\overset{-}{C}-\overset{+}{N}\equiv N$ (III) $\quad H_2\overset{-}{C}-N=\overset{+}{N}$ (IV)

(a) (I) > (II) > (IV) > (III)
(b) (I) > (III) > (II) > (IV)
(c) (II) > (I) > (III) > (IV)
(d) (III) > (I) > (IV) > (II)

79. Introduction of a methyl group in ammonia markedly increases the basic strength of ammonia in aq. solution, introduction of the second methyl group increases only marginally the basic strength of methyl amine in water. This is due to
(a) different type of hybridisation in the two amines.
(b) protonated dimethyl amine is more solvated than methyl amine.
(c) protonated dimethyl amine is more solvated than the protonated methyl amine.
(d) protonated dimethyl amine is less stable than the protonated methyl amine.

80. Which of the following amines can be resolved into two enantiomers?

I: $H-\ddot{N}(CH_2CH_3)(CH_3)$ $\quad$ II: $H-\ddot{N}(CH_3)(CH_3)$

III: aziridine with H_3C on ring carbon and CH_3 on $\ddot{N}$ $\quad$ IV: $H-C(\ddot{N}H_2)(CH_2CH_3)(CH_3)$

(a) I, IV (b) I, II
(c) I, III, IV (d) III, IV

Answer KEYs

1	(c)	9	(d)	17	(b)	25	(d)	33	(c)	41	(d)	49	(b)	57	(b)	65	(a)	73	(c)
2	(b)	10	(b)	18	(a)	26	(a)	34	(b)	42	(b)	50	(c)	58	(c)	66	(b)	74	(b)
3	(c)	11	(d)	19	(c)	27	(a)	35	(b)	43	(a)	51	(a)	59	(a)	67	(d)	75	(a)
4	(a)	12	(d)	20	(b)	28	(a)	36	(a)	44	(d)	52	(d)	60	(b)	68	(c)	76	(c)
5	(a)	13	(c)	21	(c)	29	(c)	37	(b)	45	(d)	53	(b)	61	(d)	69	(b)	77	(c)
6	(d)	14	(a)	22	(c)	30	(d)	38	(b)	46	(b)	54	(d)	62	(c)	70	(b)	78	(b)
7	(d)	15	(d)	23	(b)	31	(c)	39	(b)	47	(a)	55	(d)	63	(a)	71	(d)	79	(d)
8	(c)	16	(d)	24	(d)	32	(b)	40	(b)	48	(c)	56	(a)	64	(a)	72	(d)	80	(d)

Hints & Solutions

1. **(c)** Diazonium salts react with copper powder and hydrogen bromide to form bromobenzene. This reaction is known as **Gattermann reaction.**

$$C_6H_5\overset{+}{N}\equiv N\overset{-}{Cl} \xrightarrow[HBr]{Cu} C_6H_5Br$$

2. **(b)** C_7H_9N having one benzene ring may be represented as $C_6H_5CH_4N$, may be in the form of 1° and 2° amines in the following five isomeric forms.

(I) o-$CH_3C_6H_4NH_2$ (II) m-$CH_3C_6H_4NH_2$ (III) p-$CH_3C_6H_4NH_2$ (IV) $C_6H_5CH_2NH_2$ (V) $C_6H_5NHCH_3$

3. **(c)** The compounds containing active H-atoms (H atoms attached to N, O or S) react with CH_3COCl to form acetyl derivatives.

4. **(a)** Excess of HCl is used to convert free aniline to aniline hydrochloride otherwise free aniline would undergo coupling reaction with benzenediazonium chloride.

5. **(a)** $R_2NNO + H_2O \rightarrow R_2NH + HNO_2$

6. (d) $C_6H_5NH_2 + COCl_2 + KOH \rightarrow C_6H_5NH.COCl + HCl \xrightarrow{\Delta} C_6H_5NCO + HCl$

7. (d) The reaction can be completed as follows:

$$\underset{\text{(Aniline)}}{\underset{\text{'A'}}{C_6H_5NH_2}} \xrightarrow[\text{(diazotisation)}]{NaNO_2/HCl} \underset{\text{Benzene diazonium chloride}}{\underset{\text{'B'}}{C_6H_5N_2Cl}} + \underset{\text{N, N- Dimethylaniline}}{C_6H_5N(CH_3)_2} \xrightarrow{\text{cold}} \underset{p\text{-Dimethylaminazobenzene}}{\underset{\text{'C'}}{C_6H_5-N=N-C_6H_4-N(CH_3)_2}}$$

8. (c) Cyclohexanone $= O + HN(CH_3)_2 \longrightarrow$ 1-hydroxy-1-$N(CH_3)_2$ cyclohexane $\xrightarrow{-H_2O}$ 1-$N(CH_3)_2$ cyclohexene

Enamine

9. (d) **Hofmann's rule :** When theoretically more than one type of alkenes are possible, the alkene containing least alkylated double bond is formed. Hence

$$\overset{-}{OH}\ \text{(2-Me-cyclohexenyl)}\overset{+}{N}(Me)(n\text{-Bu})(Et) \xrightarrow{\Delta} \text{(2-Me-cyclohexenyl)}\ddot{N}(Me)(n\text{-Bu}) + CH_2=CH_2 + H_2O$$

10. (b) (A) 2,4,6-tribromoaniline (NH_2, Br, Br, Br) (B) 1,3,5-tribromobenzene (Br, Br, Br)

11. (d) *p*-Chloroaniline and anilinium hydrogen chloride can be distinguished by $AgNO_3$. Anilinium hydrogen chloride will give white ppt of AgCl on reaction with $AgNO_3$, but *p*-chloronoaniline will not react with it because Cl is directly attached to benzene nucleus.

12. (d) $\underset{(A)}{CH_3CN} \xrightarrow{\text{Reduction}} \underset{(B)}{CH_3-CH_2-NH_2} \xrightarrow{HONO} CH_3CH_2OH$

13. (c) $CH_3CH_2CN \xrightarrow{H^+/H_2O} \underset{(A)}{CH_3CH_2COOH} \xrightarrow[\Delta]{NH_3}$

$$\underset{(B)}{CH_3CH_2CONH_2} \xrightarrow[NaOBr]{\text{Hoffmann bromamide reaction}} \underset{(C)}{CH_3CH_2NH_2}$$

14. (a) It is Stephen's reaction.

$$CH_3CH_2C\equiv N \xrightarrow{SnCl_2/HCl} CH_3CH_2CH=NH.HCl \xrightarrow{H_2O} CH_3CH_2CHO + NH_4Cl$$

15. (d) Nitro compounds having α-hydrogen show tautomerism

$$RCH_2-\overset{+}{N}(=O)O^- \rightleftharpoons RCH=\overset{+}{N}(OH)O^-$$

16. (d) 1° Nitroalkanes on hydrolysis with boiling 85% H_2SO_4 give acids.

$$CH_3CH_2NO_2 + H_2O + H_2SO_4 \xrightarrow{heat} CH_3COOH + [NH_3OH]^+ HSO_4^-$$

17. (b) $C_6H_5NO_2$ (Nitrobenzene), $C_6H_5\ddot{N}H_2$ (Aniline), $C_6H_5\overset{+}{N}H_3\overset{-}{Cl}$ (Aniline hydrochloride), $C_6H_5-\ddot{N}H-C(=O)-CH_3$ (N- Acetylaniline)

Nitrobenzene and aniline hydrochloride have electron- withdrawing ($-NO_2$ and $-\overset{+}{N}H_3$) groups, hence these will undergo electrophilic substitution with difficulty. Aniline and N– acetylaniline (acetanilide) have electron– releasing groups, however $-NHCOCH_3$ is less electron- releasing than $-NH_2$ due to delocalisation of lone pair of electron on N toward carbonyl group. Hence aniline (having $-NH_2$) will undergo electrophilic substitution most easily.

18. (a) p-$CH_3C_6H_4NH_2$ $\xrightarrow{Br_2/H_2O}$ 4-CH_3-2,6-$Br_2C_6H_2NH_2$ $\xrightarrow[H_3PO_2]{HONO}$ 3,5-$Br_2C_6H_3CH_3$

19. (c) Enamine (ene for carbon-carbon double bond and amine for amine group). $>C=C-N<$

20. (b) (a) $C_6H_5COOC_2H_5 + (CH_3)_2NH \rightarrow C_6H_5CON(CH_3)_2 + C_2H_5OH$

(b) $C_6H_5CONH_2 + CH_3MgI \rightarrow C_6H_5CONHMgI + CH_4$

(c) $C_6H_5COCl + (CH_3)_2NH \rightarrow C_6H_5CON(CH_3)_2 + HCl$

(d) $C_6H_5CO.O.COC_6H_5 + (CH_3)_2NH \rightarrow C_6H_5CON(CH_3)_2 + C_6H_5COOH$

21. (c) 2,6-Dibromo-4-nitroaniline $\xrightarrow[\text{Sandmeyer}]{\text{HONO, CuCl}}$ 1-chloro-2,6-dibromo-4-nitrobenzene $\xrightarrow{Sn/HCl}$ 4-chloro-3,5-dibromoaniline $\xrightarrow[\text{elimination of } NH_2]{\text{(i) HONO (ii) } H_3PO_2}$ 1-chloro-2,6-dibromobenzene

22. (c) The first compound will give azo dye

$H_3C-C_6H_4-N=N-$(naphthol, OH)

23. (b) Aliphatic amines are stronger base than aromatic amines. Further the order 2° > 1° > 3° > NH_3

24. (d) Cyclobutyl$-CH_2\ddot{N}H_2$ + $:N=O$ (with OH) $\longrightarrow$ cyclobutyl$-CH_2-\overset{\oplus}{N}H_2-N(OH)-\bar{O}$ $\xrightarrow{H^{\oplus}}$

cyclobutyl$-CH_2-\overset{\oplus}{N}H_2-N(OH)_2$ $\xrightarrow{-2H_2O}$ cyclobutyl$-CH_2-\overset{\oplus}{N}\equiv N:$ $\longrightarrow$ cyclobutyl$-\overset{+}{C}H_2 + :N\equiv N:$

$\xrightarrow{-H^+}$ methylenecyclobutane ($=CH_2$)

(ring expansion) $\longrightarrow$ cyclopentyl cation ($\oplus$) $\xrightarrow{-H^+}$ cyclopentene

cyclopentyl cation $\xrightarrow{OH^-}$ cyclopentanol (OH)

25. (d) Carbonyl compounds containing at least one α-hydrogen atom react with a secondary amine to give enamine

$$>CH-\overset{|}{C}=O + HN<^{R}_{R} \xrightarrow[-H_2O]{} >C=\overset{|}{C}-N<^{R}_{R}$$

Primary amines also form enamines

$$>CH-\overset{|}{C}=O + H_2N-R \longrightarrow \underset{\text{enamine}}{>C=\overset{|}{C}-N<^{H}_{R}} \rightleftharpoons \underset{\text{imine}}{>CH-\overset{|}{C}=N-R}$$

In second case enamine – amine tautomerism is possible and equilibrium lies completely on the imine side.

26. (a) — $\ddot{N}H_2$ group is acetylated by acetic anhydride in methylene chloride (solvent).

$$-CH_2-\ddot{N}H_2 + (CH_3CO)_2O \xrightarrow{CH_2Cl_2} -CH_2-\underset{|}{N}H(COCH_3) + CH_3COOH \quad -\overset{O}{\overset{\|}{C}}-\ddot{N}H_2 \longleftrightarrow -\overset{O^-}{\overset{|}{C}}=\overset{+}{N}H_2$$

Note that —$CONH_2$ group does not undergo acetylation because here lone pair of electrons is delocalised.

27. (a) Aniline ($C_6H_5NH_2$) $\xrightarrow[0\ °C]{NaNO_2,\ HCl}$ $C_6H_5\overset{+}{N}\equiv N\overset{-}{Cl}$ (V) $\xrightarrow{\text{2-naphthol, NaOH}}$ 1-phenylazo-2-naphthol (W)

28. (a) 5-methylindene $\xrightarrow[(ii)\ Zn,\ H_2O]{(i)\ O_3}$ dialdehyde (CHO and CH_2CHO) $\xrightarrow{NH_3}$ cyclic diol amine (OH, NH, OH) $\xrightarrow{-2H_2O}$ methylisoquinoline

29. (c) $CH_3CN + 2H \rightarrow \underset{X}{CH_3CH = NH} \xrightarrow{H_2O,\ boil} \underset{Y}{CH_3CHO}$

Note that cyanides are reduced only by 2H atoms, so reduction stops at aldimine stage which on hydrolysis with water gives aldehyde; it is an example of Stephen's reaction.

30. (d) $C_6H_5NH_2 + CS_2 \rightarrow C_6H_5NH.CS.SH \xrightarrow{HgCl_2} C_6H_5N = C = S + HgS + 2HCl$

The reaction is called mustard oil reaction.

31. (c) Aliphatic amines are more basic than aromatic amines thus methylamine is most basic. Electron donating groups increase the basicity whereas electron withdrawing groups decrease the basicity of the aromatic amines. Thus *p*-methoxyaniline is more basic then aniline which is further more basic then *p*-nitroaniline.

p-nitroaniline (NO_2) < aniline < p-methoxyaniline (OCH_3) < $CH_3-\ddot{N}H_2$

32. (b) $CH_3NH_2 + HNO_2 \longrightarrow CH_3OH + N_2 + H_2O$

33. (c) The correct order of relative basicity of amines in the gas phase is $3° > 2° > 1° > NH_3$
The alkyl group releases electron and thus, tends to disperse the positive charge of the alkyl ammonium ion and therefore stabilises it. Since, NH_4^+ (from NH_3) has no such alkyl group, it is not stabilised to such an extent as alkylammonium ion.

34. (b) $RNH_2 + CHCl_3 + 3KOH \longrightarrow RNC + 3KCl + 3H_2O$
Mechanism

$$CHCl_3 \xrightarrow[\alpha-\text{elimination}]{OH^-/\Delta} \ddot{C}Cl_2 \text{ (Dichlorocarbene)}$$

$$R-\ddot{N}H_2 + \ddot{C}Cl_2 \longrightarrow R-\overset{\oplus}{N}H_2-\overset{\ominus}{C}Cl_2 \longrightarrow R-\underset{H}{\overset{..}{N}}-CHCl_2 \longrightarrow R-\overset{\oplus}{N}H=C(H)Cl + Cl^- \longrightarrow R-\ddot{N}=C(H)Cl \longrightarrow R-\overset{\oplus}{N}\equiv C-H \xrightarrow{\overset{\ominus}{O}H} R-\overset{\oplus}{N}\equiv\overset{\ominus}{C} + HOH$$

35. (b) $C_6H_5-CH_2-NH_2$ compound is most basic due to presence of localized lone pair of electron on nitrogen atom while other compounds have delocalized lone pair of electron.

36. (a) Only 1° aromatic amines undergo coupling reactions to form a dye

37. (b) $C_6H_5NHCOCH_3 \xrightarrow{\text{Nitration}} p\text{-}O_2NC_6H_4NHCOCH_3 \xrightarrow{H_2O/OH^-} p\text{-}O_2NC_6H_4NH_2$ (*p*-Nitroaniline)

38. (b) $p\text{-}CH_3C_6H_4NH_2 \xrightarrow{(CH_3CO)_2O} p\text{-}CH_3C_6H_4NHCOCH_3 \xrightarrow{Br_2/CH_3COOH}$ 3-Br-4-($NHCOCH_3$)-$C_6H_3CH_3$ $\xrightarrow{H^+}$ 3-Br-4-(NH_2)-$C_6H_3CH_3$

(–$NHCOCH_3$ is more electron-releasing than –CH_3 group)

39. (b) $C_6H_6 \xrightarrow[333\text{ K}]{HNO_3,\ H_2SO_4} C_6H_5NO_2 \xrightarrow{Cl_2/Fe} m\text{-}ClC_6H_4NO_2 \xrightarrow{\text{reduction}} m\text{-}ClC_6H_4NH_2$

(a) When chlorination is done earlier than nitration, chlorobenzene formed at first step would introduce – NO_2 group in ortho-position, not in *m*-position.

(b) Again if –NO_2 group is reduced earlier than the chlorination step, –NH_2 group formed on reduction will again introduce –Cl in *o*-position.

40. **(b)** $A \xrightarrow{HNO_3/H_2SO_4} B \xrightarrow{Sn/HCl} C_6H_5NH_2$

This indicates that B is $C_6H_5NO_2$ and hence A is C_6H_6

41. **(d)** $\underset{(A)}{C_6H_5NO_2} \xrightarrow{\text{reduction}} \underset{(B)}{C_6H_5NH_2} \xrightarrow[\text{alc. KOH}]{CHCl_3} \underset{(C)}{C_6H_5NC} \xrightarrow{H_2/\text{Catalyst}} \underset{\text{N-Methylaniline}}{C_6H_5NHCH_3}$

42. **(b)** In II, intermolecular H-bonding between similar molecules is possible, while this is not so in I.

43. **(a)** In amines, N is sp^3 hybridised and thus has pyramidal shape. In the given structure, since the three alkyl groups are different, and the fourth corner of the pyramid is occupied by lone pair of electrons, the molecule is chiral. However, the two enantiomers of the amine are not resolvable because of their rapid interconversion through a transition state having planar structure (sp^2 hybridised nitrogen)

44. **(d)** All of the three are oxidation products of a 1° amine.

$$\underset{1^\circ\text{ Amine}}{R-\ddot{N}H-H} \xrightarrow{[O]} \underset{\text{Hydroxylamine}}{R-\ddot{N}(OH)-H} \xrightarrow{[O]} \underset{\text{Nitroso}}{R-\ddot{N}=O} \xrightarrow{[O]} \underset{\text{Nitro}}{R-\overset{+}{N}(=O)-O^-}$$

45. **(d)** $CH_3COONH_4 \xrightarrow{\Delta} \underset{(X)}{CH_3CONH_2} \xrightarrow{\Delta/P_2O_5} \underset{(Y)}{CH_3CN} \xrightarrow{H_3O^{\oplus}} \underset{(Z)}{CH_3COOH}$

46. **(b)** The order of basicity is I > III > II > IV.

The lone pair of electrons on N is more readily available for protonation in I and III then in II. III contains an oxygen atom which has – I effect due to which it is less basic than I. In compound IV lone pair of e^-s on N–atom is contributed towards the aromatic sextet formation and hence is not at all available for protonation. Hence option (b) is correct.

47. **(a)** On reduction with Zn and HCl, $C_6H_5N_2Cl$ forms aniline as the main product.

48. **(c)** $\underset{\text{Benzene diazoniumchloride}}{C_6H_5N_2Cl} \xrightarrow[\text{or Cu+HCl}]{CuCl/HCl} \underset{\text{Chloro benzene}}{C_6H_5Cl} + N_2$

The above reaction is known as Sandmayer's reaction.

49. (b) 4-nitrotoluene $\xrightarrow[FeBr_3]{Br_2}$ 2-bromo-4-nitrotoluene (CH_3, Br, NO_2) $\xrightarrow{Sn/HCl}$ 2-bromo-4-aminotoluene (CH_3, Br, NH_2) $\xrightarrow[HCl]{NaNO_2}$ diazonium salt (CH_3, Br, $N^+\equiv N\bar{C}l$) $\xrightarrow[CuBr]{HBr}$ 2,4-dibromotoluene (CH_3, Br, Br)

50. (c) $(CH_3)_2CO \xrightarrow{NH_2OH} (CH_3)_2CH = N - OH \xrightarrow{LiAlH_4} (CH_3)_2CHNH_2$

$(CH_3)_2CHOH \xrightarrow{PCl_5} (CH_3)_2CHCl \xrightarrow{NH_3} (CH_3)_2CHNH_2$

51. (a) $CH_3CH = NOH \xrightarrow{P_2O_5} CH_3CN + H_2O$

52. (d) $CH_3CN \xrightarrow{Na/C_2H_5OH} \underset{(A)}{CH_3CH_2NH_2} \xrightarrow{HNO_2} \underset{(B)}{CH_3CH_2OH} \xrightarrow[573\ K]{Cu} \underset{(C)}{CH_3CHO}$

53. (b) $\underset{(C_2H_3N)}{A} \xrightarrow{reduction} B \xrightarrow{HNO_2} C_2H_5OH$

$B \xrightarrow{\text{Carbylamine reaction}}$ Offensive smell (C)

Given reactions indicate that B has 1° NH_2 group, and thus A, C_2H_3N, should be $CH_3C \equiv N$. Hence C should be CH_3CH_2NC

$$\underset{A}{CH_3C \equiv N} \xrightarrow{reduction} \underset{B}{CH_3CH_2NH_2} \xrightarrow[KOH]{CHCl_3} \underset{C}{CH_3CH_2N \stackrel{\rightarrow}{=} C}$$

54. (d) $CH_3CH_2N \stackrel{\rightarrow}{=} C + H_2O \xrightarrow{H^+} CH_3CH_2NH_2 + HCOOH$

55. (d) Red solution

56. (a) 2-chloroaniline (NH_2, Cl) $\xrightarrow[H_2O,\ 0°C]{NaNO_2/HCl}$ (N_2Cl, Cl) $\xrightarrow{CuCN}$ (CN, Cl)

$\xrightarrow{(i)\ CH_3MgBr}$ ($CH_3 - C = N - MgBr$, Cl) $\xrightarrow{(ii)\ H_2O/H^+}$ ($CH_3 - C = O$, Cl) $+ NH_3 + HOMgBr$

57. (b) $C_2H_5-C_6H_4-NH_2 \xrightarrow[0-5°C]{NaNO_2-HCl} C_2H_5-C_6H_4-\overset{+}{N}\equiv N-\overset{-}{Cl}$

$C_2H_5-C_6H_4-\overset{+}{N}\equiv N-\overset{-}{Cl} + C_6H_5-OH \rightarrow$

$C_2H_5-C_6H_4-N=N-C_6H_4-OH.$

58. (c) $C_5H_9-COCl + CH_3CH_2NH_2 \xrightarrow{Pyridine}$

$C_5H_9-CONHCH_2CH_3 \xrightarrow[(ii)H_2O]{(i)LiAlH_4\ ,Et_2O} C_5H_9-CH_2NHCH_2CH_3$

N-ethyl cyclopentyl methyl amine

59. (a) Acetamide is basic due to the presence of lone pair of electrons of N; it is also acidic because its conjugate base shows resonance.

$$CH_3-\overset{O}{\overset{\|}{C}}-NH_2 \xrightarrow{base} CH_3-\overset{O}{\overset{\|}{C}}-NH^- \longleftrightarrow CH_3-\overset{O^-}{\overset{|}{C}}=NH$$

(more stable due to –ve charge on O)

Benzonitrile ($C_6H_5C\equiv N$) acts as an electrophile (Lewis acid) due to electron deficiency of C of CN as well as nucleophile (Lewis base) in nature due to presence of lone pair electrons on N; hence it is neutral Triethylamine and phenol are basic and acidic in nature respectively.

60. (b) Guanidine is most basic because its conjugate acid is stabilized by two equivalent resonance structures,.

$$CH_3-\overset{NH}{\overset{\|}{C}}-NH_2 \xrightarrow{H^+} CH_3-\overset{NH_2}{\overset{|}{\overset{+}{C}}}-NH_2 \longleftrightarrow CH_3-\overset{^+NH_2}{\overset{\|}{C}}-NH_2 \longleftrightarrow CH_3-\overset{NH_2}{\overset{|}{C}}=\overset{+}{N}H_2$$

Further 2° amines are more basic than 1°, while amides are least basic due to delocalisation of the lone pair of electrons on N

$$CH_3-\overset{:O:}{\overset{\|}{C}}-\ddot{N}H_2 \longleftrightarrow CH_3-\overset{:\ddot{O}:^-}{\overset{|}{C}}=\overset{+}{N}H_2$$

Thus the correct order of basicities is

$$CH_3-\overset{NH}{\overset{\|}{C}}-NH_2 > CH_3-\overset{CH_3}{\overset{|}{N}}H > CH_3CH_2NH_2 > CH_3-\overset{O}{\overset{\|}{C}}-NH_2$$

61. (d) Proceed backward ; tribromobenzene is produced by boiling compound Z with C_2H_5OH ; Z in turn is obtained by diazotisation of Y, so Y and Z should have $-NH_2$ and $-N_2Cl$ groups respectively, in addition to three Br atoms. Hence X should be $C_6H_5NH_2$

$$\underset{X}{C_6H_5NH_2} \xrightarrow{\text{bromination}} \underset{Y}{2,4,6\text{-}Br_3C_6H_2NH_2} \xrightarrow[HCl]{NaNO_2} \underset{Z}{2,4,6\text{-}Br_3C_6H_2N_2Cl} \xrightarrow[\text{boil}]{C_2H_5OH} 1,3,5\text{-}C_6H_3Br_3$$

62. (c)

$$C_6H_5NH_2 \xrightarrow{Br_2\text{ water}} 2,4,6\text{-}Br_3C_6H_2NH_2 \xrightarrow[HCl]{NaNO_2} 2,4,6\text{-}Br_3C_6H_2N_2Cl \xrightarrow[\text{(ii) heat}]{\text{(i) }BHF_4} 2,4,6\text{-}Br_3C_6H_2F$$

63. (a) The electrons withdrawing groups decrease the basic character and increase the acid character by inductive effect and resonance.

64. (a) Nitriles having α -hydrogen atom form alkyl derivatives with RBr in presence of $NaNH_2 / NH_3$.

65. (a) Nitrobenzene is reduced in the following manner through various intermediates to form aniline as the final product.

$$\underset{\text{nitrobenzene}}{C_6H_5NO_2} \xrightarrow{2[H]} \underset{\text{nitrosobenzene}}{C_6H_5NO} \xrightarrow{2[H]} \underset{\text{phenylhydroxylamine}}{C_6H_5NHOH} \xrightarrow{2[H]} \underset{\text{aniline}}{C_6H_5NH_2}$$

66. (b)

$$C_6H_6 \xrightarrow[H_2SO_4]{HNO_3} \underset{W}{C_6H_5NO_2} \xrightarrow[\text{(acidic reduction)}]{Zn/HCl} \underset{X}{C_6H_5NH_2} \xrightarrow[HCl]{NaNO_2} \underset{Y}{C_6H_5N_2Cl} \xrightarrow[\text{(reduction)}]{H_3PO_2} \underset{Z}{C_6H_6}$$

67. (d)

$$\underset{(A)}{C_6H_5NO_2} \xrightarrow{\text{reduction}} \underset{(B)}{C_6H_5NH_2} \xrightarrow[\text{alc. KOH}]{CHCl_3} \underset{(C)}{C_6H_5NC} \xrightarrow{H_2/\text{Catalyst}} \underset{\text{N-Methylaniline}}{C_6H_5NHCH_3}$$

68. (c)

$$\text{(ring bis-ammonium salt, } \overset{+}{N}H_3Cl^- \text{ / } Cl^-H_3\overset{+}{N}\text{)} \xrightarrow{\text{heat}} \text{piperidine (ring N–H)} + NH_4Cl$$

69. **(b)** 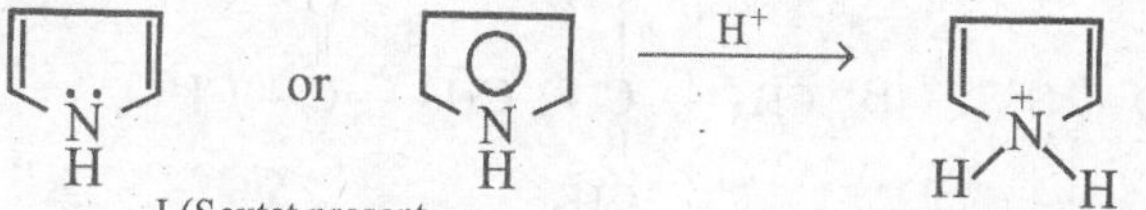 Pyrrole (I) or $\xrightarrow{H^+}$ II

I (Sextet present, hence aromatic) (*e* pair on N forms a part of aromatic sextet)

II (not aromatic) (Aromatic sextet is not present because *e* pair is involved in protonation)

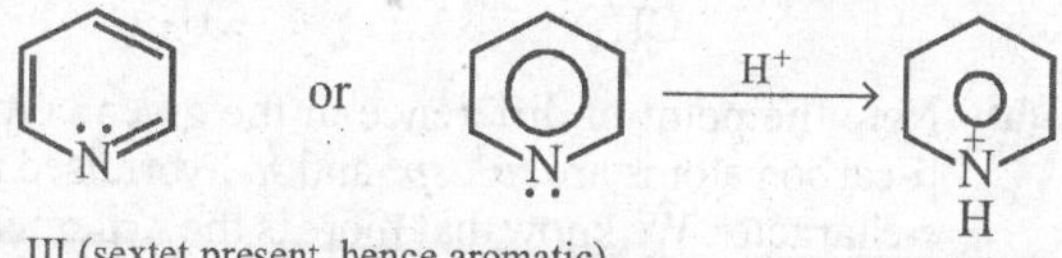

Pyridine (III) or $\xrightarrow{H^+}$ IV

III (sextet present, hence aromatic) (*e* pair on N, not a part of sextet)

IV (sextet present, hence aromatic)

70. **(b)** This is an example of Hofmann elimination which generally takes place by E2 mechanism and the latter requires a strong base (recall that OH^- is a strong base than I^-). The NH_2^-, being a strong base, can't be eliminiated easily.

71. **(d)**

$$\underset{1^\circ \text{ Amine}}{RNH_2} + \underset{\text{Aldehyde/ketone}}{>C=O} \longrightarrow \underset{\text{Carbinolamine}}{>C(OH)(NHR)} \xrightarrow{-H_2O} \underset{\substack{\text{Imine}\\ \text{(Schiff's base)}}}{>C=NR}$$

$$\underset{2^\circ \text{ Amine}}{R_2NH} + \underset{\substack{\text{Aldehyde / ketone}\\ (R = H)\ (R=\text{Alkyl})}}{CH_3(R)C=O} \longrightarrow \underset{\text{Carbinolamine}}{CH_3(R)C(NR_2)(OH)} \longrightarrow \underset{\text{Enamine}}{CH_2=C(R)-NR_2}$$

72. **(d)** For this one should remember that Hofmann's degradation involves migration to electron deficient nitrogen atom, hence the alkyl group will migrate with its bonding pair of electrons.

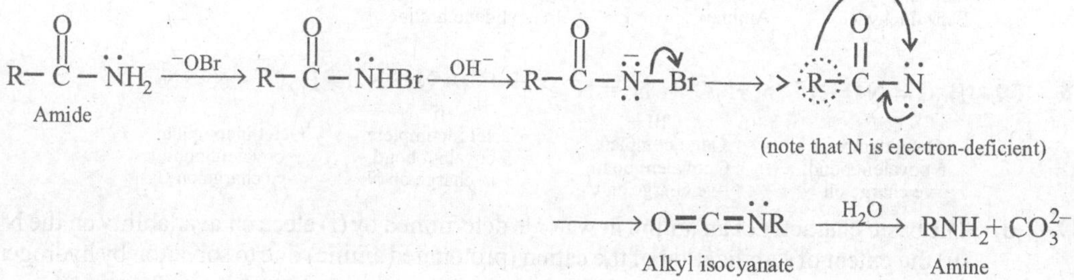

$$\underset{\text{Amide}}{R-\overset{O}{\overset{\|}{C}}-\ddot{N}H_2} \xrightarrow{^-OBr} R-\overset{O}{\overset{\|}{C}}-\ddot{N}HBr \xrightarrow{OH^-} R-\overset{O}{\overset{\|}{C}}-\ddot{\underset{..}{N}}-Br \longrightarrow R-\overset{O}{\overset{\|}{C}}-\ddot{N}$$

(note that N is electron-deficient)

$$\longrightarrow \underset{\text{Alkyl isocyanate}}{O=C=\ddot{N}R} \xrightarrow{H_2O} \underset{\text{Amine}}{RNH_2} + CO_3^{2-}$$

73. **(c)** Remember that in Hofmann rearrangement, the two original H atoms of the $-CONH_2$ group are removed by base (OH^-) and new H's are derived from H_2O.

$$(CH_3)_3C-\overset{O}{\overset{\|}{C}}-NH_2 \xrightarrow[\text{(ii) } D_2O]{\text{(i) } OD^-/Br_2} (CH_3)_3C-ND_2$$

74. **(b)** (A) $CH_3-\underset{CH_3}{\overset{CH_3}{|}}{C}-CH_2-\overset{O}{\overset{\|}{C}}-O-H$ (B) $CH_3-\underset{CH_3}{\overset{CH_3}{|}}{C}-CH_2-\overset{O}{\overset{\|}{C}}-Cl$

(C) $CH_3-\underset{CH_3}{\overset{CH_3}{|}}{C}-CH_2-\overset{O}{\overset{\|}{C}}-\underset{CH_3}{N}-CH_3$ (D) $CH_3-\underset{CH_3}{\overset{CH_3}{|}}{C}-CH_2-CH_2-N\begin{smallmatrix}CH_3\\CH_3\end{smallmatrix}$

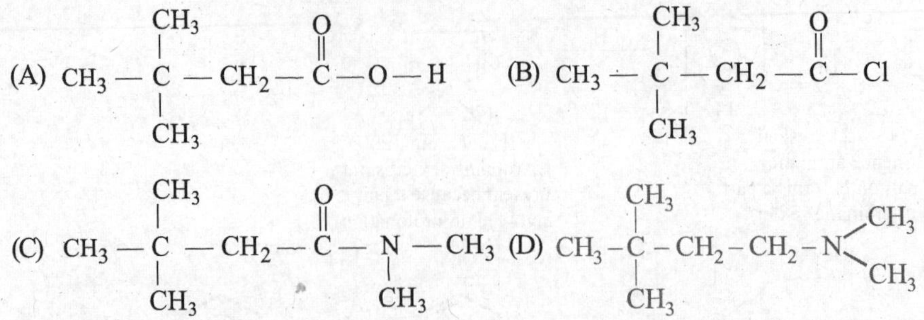

75. **(a)** Note the point of difference in the given compounds which here lies at β-carbon. In I, II, III, the β-carbon atoms are sp^3, sp^2 and *sp* hybridised respectively which in turn cause the difference in their *s*-character. We know that more is the *s* character of an atom, greater will be its electron-withdrawing nature. Thus *sp* (50% *s* character) hybridised carbon is most electron-withdrawing, while sp^3 (25% *s*-character) is least electron-withdrawing. Further, we know that presence of an electron-withdrawing group decreases basicity of an amine. Thus

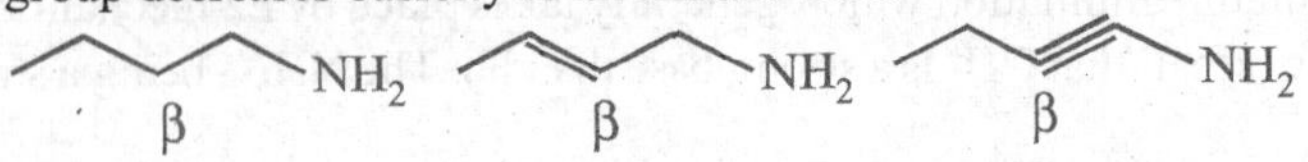

NH_2 β NH_2 β NH_2 β

I. (β – sp^3 hybridised)   II. (β – sp^2)   III. (β – C sp^2)

76. **(c)** Nitrosonium ion, $\overset{+}{N}O$ from HONO is a weak electrophile, hence it can attack only on highly activated benzene nucleus, provided proper position, *p*- or *o*- is free.

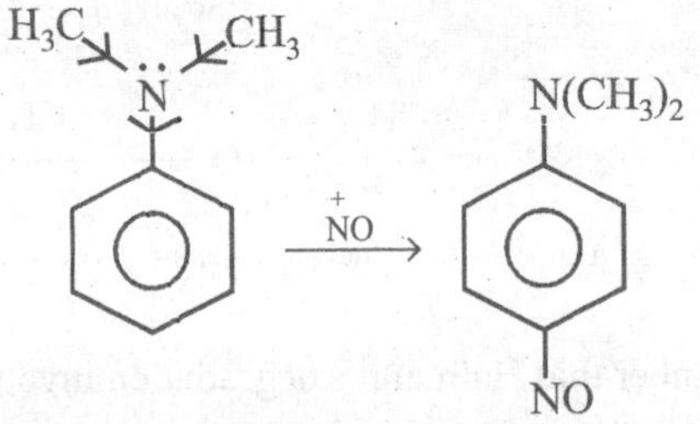

$$C_6H_5N(CH_3)_2 \xrightarrow{\overset{+}{N}O} p\text{-}(NO)C_6H_4N(CH_3)_2$$

77. **(c)** Benzaldehyde reacts with primary aromatic amine to form schiff's base

$$\underset{\text{Benzaldehyde}}{C_6H_5CH=O}+\underset{\text{Aniline}}{C_6H_5NH_2}\longrightarrow\underset{\text{Benzylidene aniline}}{C_6H_5CH=NC_6H_5}$$

78. **(b)** $H_2C=N^+=N^-$ > $H_2C^--N^+\equiv N$ > $H_2C^+-N=N^-$ > $H_2C^--N=N^+$

I: Octet complete, 6 covalent bond, –ve charge on N
III: Octet complete, 6 covalent bond, –ve charge on C
II: Octet incomplete, 5 covalent bond, –ve charge on N
IV: Octet incomplete, 5 covalent bond, –ve charge on C

79. **(d)** The basic character of an amine in water is determined by (i) electron availability on the N atom and (ii) the extent of stabilization of the cation (protonated amine) due to solvation by hydrogen bonding

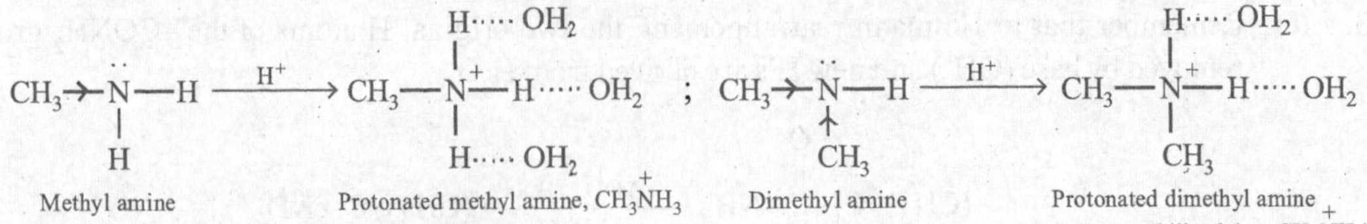

$$\underset{\text{Methyl amine}}{CH_3\rightarrow\ddot{N}H_2}\xrightarrow{H^+}\underset{\text{Protonated methyl amine, }CH_3\overset{+}{N}H_3\text{ (Highly stabilized)}}{CH_3-\overset{+}{N}(-H\cdots OH_2)_3}\;;\;\underset{\text{Dimethyl amine}}{CH_3\rightarrow\ddot{N}H\leftarrow CH_3}\xrightarrow{H^+}\underset{\text{Protonated dimethyl amine (Lesser stabilized than }CH_3\overset{+}{N}H_3)}{(CH_3)_2\overset{+}{N}(-H\cdots OH_2)_2}$$

80. **(d)** Most of the simple amines those having smaller alkyl (groups) can't be resolved, although they may contain three different alkyl groups.

28 Biomolecules

1. In fructose, the possible optical isomers are
(a) 12 (b) 8 (c) 16 (d) 4

2. Glucose molecule reacts with X number of molecules of phenylhydrazine to yield osazone. The value of X is
(a) three (b) two (c) one (d) four

3. Which L-sugar on oxidation gives an optically active dibasic acid (2 COOH groups)?

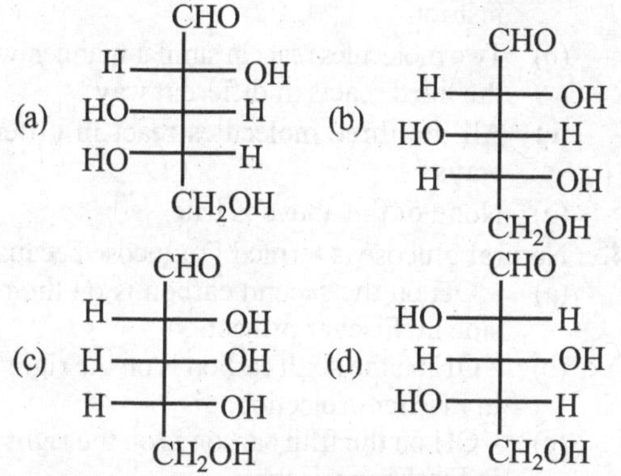

4. In cells the net production of ATP molecules generated from one glucose molecule is
(a) 46 (b) 32 (c) 36 (d) 40

5. Which is correct statement?

(a) Starch is a polymer of α-glucose
(b) In cyclic structure of fructose, there are four carbons and one oxygen atom
(c) Amylose is a component of cellulose
(d) Proteins are composed of only one type of amino acids

6. Vitamin D is also known as
(a) sunshine vitamin
(b) ascorbic acid
(c) growth vitamin
(d) reproductive vitamin

7. Night blindness is caused by deficiency of
(a) vitamin B_{12} (b) vitamin A
(c) vitamin C (d) vitamin E

8. The hormone which controls the processes of burning of fats, proteins and carbohydrates to liberate energy in the body is
(a) cortisone (b) adrenaline
(c) thyroxine (d) insulin

9. Two vitamins absorbed from intestine along with fats are
(a) A, D (b) A, B (c) A, C (d) D, B

10. Which one of the following is an amine hormone?
(a) Thyroxine (b) Oxypurin
(c) Insulin (d) Progesterone

11. Complete hydrolysis of starch gives :
(a) glucose only
(b) galactose and fructose in equimolar amounts
(c) glucose and galactose in equimolar amounts
(d) glucose and fructose in equimolar amounts

12. Accumulation of which of the following molecules in the muscles occurs as a result of vigorous exercise ?
(a) Glycogen (b) Glucose
(c) Pyruvic acid (d) L-lactic acid

13. Which of the following will not show mutarotation?
(a) Maltose (b) Lactose
(c) Glucose (d) Sucrose

14. An organic compound with the formula $C_6H_{12}O_6$ forms a yellow crystalline solid with phenylhydrazine and gives a mixture of sorbitol and mannitol when reduced with sodium. Which among the following could be the compound?

Toughnut

(a) fructose (b) glucose
(c) mannose (d) sucrose

15. When α-D-glucose and β-D-glucose are dissolved in water in two separate beakers I and II respectively and allowed to stand, then –

(a) specific rotation in beaker I will decrease while in II will increase upto a constant value

(b) the specific rotation of equilibrium mixture in two beakers will be different
(c) the equilibrium mixture in both beakers will be leavorotatory
(d) the equilibrium mixture in both beakers will contain only cyclic form of glucose

16. The pK_a values for the three ionizable groups X, Y and Z of glutamic acid are 4.3, 9.7 and 2.2 respectively **Tricky**

$$\underset{X}{HO_2C}—CH_2—\underset{\underset{Y}{\overset{|}{+NH_3}}}{CH}—\underset{Z}{CO_2H}$$

The isoelectric point for the amino acid is:
(a) 7.00 (b) 3.25 (c) 4.95 (d) 5.95

17. A vitamin that contains both N and P is
(a) Vitamin C (b) Vitamin K
(c) Vitamin B_{12} (d) Vitamin D

18. Match List I (name of vitamin) with List II (deficiency result/disease) and select the correct answer using the codes given below the lists :

	List I		List II
I.	Ascorbic acid	A.	Beri-beri
II.	Retinol	B.	Cracked lips
III.	Riboflavin	C.	Scurvy
IV.	Thiamine	D.	Night blindness

(a) I - B, II - A, III - C, IV - D
(b) I - A, II - B, III - C, IV - D
(c) I - D, II - C, III - B, IV - A
(d) I - C, II - D, III - B, IV - A

19. Nucleotides in DNA are linked by–
(a) hydrogen bond
(b) 3', 5'-phosphodiester bond
(c) glycosidic bond
(d) peptide bond

20. If one strand of DNA has the sequence ATGCTTGA, the sequence in the complimentary strand would be
(a) TACGAACT (b) TCCGAACT
(c) TACGTACT (d) TACGTAGT

21. Which of the following pairs can be distinguished by Fehling's solution ? **Toughnut**
(a) Glucose and fructose
(b) Glucose and sucrose
(c) Methanal and ethanal
(d) Hydroxypropanone and benzaldehyde.

22. Which of the following statement(s) is (are) true? **Toughnut**
(i) All amino acids contain one chiral center.
(ii) Some amino acids contain one, while some contain more chiral center or even no chiral center.
(iii) All amino acids found in proteins have L configuration.
(iv) All amino acids found in proteins have 1° amino group.
(a) (ii), (iii) & (iv) (b) (ii) & (iii)
(c) (i), (iii) & (iv) (d) (i) & (iv).

23. In osazone formation, glucose reacts with three molecules of phenylhydrazine. Which statement is true regarding this?

(a) All the three molecules react in similar fashion
(b) Two molecules react in similar manner, while the third reacts in different way
(c) All the three molecules react in different ways
(d) None of the above is true

24. Natural glucose is termed D-glucose because :
(a) – OH on the second carbon is on the right side in Fischer projection
(b) – OH on the sixth carbon is on the right side in Fischer projection.
(c) – OH on the fifth carbon is on the right side in Fischer projection.
(d) It is dextrorotatory.

25. Which of the following statement is not correct?
(a) Amylopectin is a branched polymer of α - glucose.
(b) Cellulose is a linear polymer of β-glucose
(c) Glycogen is the food reserve of plants
(d) All proteins are polymers of α - amino acids.

26. Amylopectin is a polymer of
(a) $\alpha - D -$ glucose (b) amino acid
(c) $\beta - D -$ glucose (b) amylase.

27. Which of the following is a non-reducing sugar?
(a) Lactose (b) Fructose
(c) Sucrose (d) Maltose

28. Glycosidic linkage is actually an :
(a) Carbonyl bond (b) Ether bond
(c) Ester bond (d) Amide bond

29. Which of the following does not reduce Benedict's solution?
(a) Glucose (b) Fructose
(c) Sucrose (d) Aldehyde

30. Hydrolysis of sucrose is called
(a) hydration (b) saponification
(c) esterification (d) inversion

31. Assume that a particular amino acid has an isoelectric point of 6.0. In a solution at pH 1.0, which of the following species will predominate?

(a) 
(b) $H_2N\overset{\displaystyle R}{\overset{|}{C}}HCO_2H$

(c) 
(d) $H_2N\overset{\displaystyle R}{\overset{|}{C}}HCO_2^-$

32. A mixture of two amino acids having pI 9.60 and 5.40 can be separated **Toughnut**
(a) by adjusting the pH of the solution at 9.60
(b) by adjusting the pH of the solution at 4.20
(c) by adjusting the pH of the solution at 7.0
(d) by adjusting the pH of the solution at 7.5.

33. The correct statement in respect of protein haemoglobin is that it
(a) acts as an oxygen carrier in the blood
(b) forms antibodies and offers resistance to diseases
(c) enzymes are specific biological catalysts that can normally function at very high temperature (T ~ 1000K)
(d) enzymes are specific biological catalysts that possess well-defined active sites

34. Which of the following carbohydrates is not related to (+)–glucose?
(a) Amylopectin (b) Amylose
(c) Inulin (d) Glycogen

35. What will happen when D-(+)-glucose is treated with methanolic —HCl followed by Tollens' reagent ? **Toughnut**
(a) A black ppt. will be formed
(b) A red ppt. will be formed
(c) A green colour will appear
(d) No characteristic colour or ppt. will be formed.

36. Glucose-D has a great tendency to be converted into cyclic isomer. Which two carbon atoms get joined through 'O' to form this hemiacetal ?
(a) C_1 and C_4 (b) C_1 and C_5
(c) C_1 and C_6 (d) C_2 and C_6

37. Among the three compounds shown below, two yield the same product on reaction with warm HNO_3. The exception is:

(a)
CHO
HO—H
H—OH
H—OH
CH_2OH

(b)
CHO
HO—H
HO—H
H—OH
CH_3

(c)
CHO
HO—H
HO—H
H—OH
CH_2OH

(d) None of these

38. The enantiomer of α-D-(+)-glucose is
(a) β-D-(+)-glucose (b) α-D-(–)-glucose
(c) α-L-(–)-glucose (d) β-L-(–)-glucose.

39.

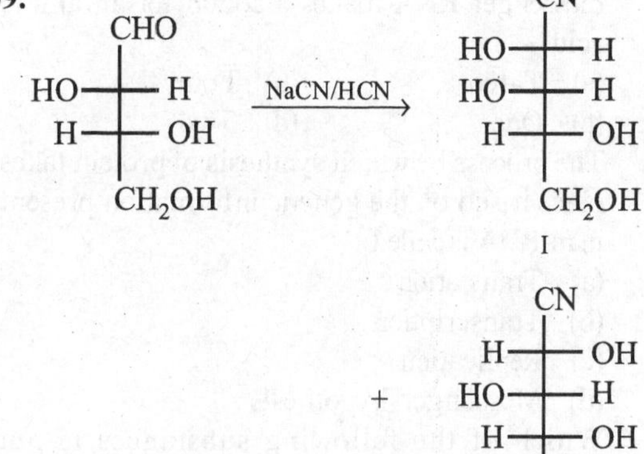

CH_2OH
II

Compounds I and II may be grouped as
(a) diastereomers (b) epimers
(c) C–2 epimers (d) all of the three.

40. Which of the following evolves carbon dioxide, on oxidation with periodate ?

(a)
CHO
|
CHOH
|
CHOH
|
CH_2OH

(b)
CH_2OH
|
CO
|
CHOH
|
CH_2OH

(c) Both (d) None

41. Imino acid among these compounds is
(a) serine (b) proline
(c) tyrosine (d) lysine

42. Which amino acid is achiral?
(a) alanine (b) valine
(c) proline (d) None of these

43. Which is not a true statement? Critical Thinking
(a) α-carbon of α-amino acid is asymmetric
(b) All proteins are found in L-form
(c) Human body can synthesize all proteins they need
(d) At pH = 7 both amino acids and carboxylic groups exist in the ionised form

44. Which of the following is a steroid hormone?
(a) Cholesterol (b) Adrenaline
(c) Thyroxine (d) Progesterone

45. Vitamin B_6 is known as
(a) pyridoxine (b) thiamine
(c) tocopherol (d) riboflavin

46. A sequence of how many nucleotides in messenger RNA makes a codon for an amino acid?
(a) Three (b) Four
(c) One (d) Two

47. The process by which synthesis of protein takes place based on the genetic information present in m-RNA is called
(a) Translation
(b) Transcription
(c) Replication
(d) Messenger hypothesis

48. Which of the following substances is not present in nucleic acids?
(a) Cytosine (b) Adenine
(c) Thymine (d) Guanidine

49. Which of the following is not present in a nucleotide?
(a) Guanine (b) Cytosine
(c) Adenine (d) Tyrosine

50. Iso-electric point of alanine is (pH = 6). At which pH, maximum concentration of zwitter ion of alanine will be present?
(a) pH > 6 (b) pH < 6
(c) pH = 6 (d) pH = 7

51. $H-C\equiv C-H \xrightarrow[H_2SO_4]{HgSO_4} (A) \xrightarrow[(2)\ H_3O^+]{(1)\ NH_3 + HCN} (B);$

Product (B) of given reaction is:  Toughnut
(a) Glycine (b) Alanine
(c) valine (d) Leucine

52. The dipeptide, Gly. Ala has structure –
(a) $H_3\overset{+}{N}CH_2\overset{\overset{O}{||}}{C}NH\underset{\underset{CH_3}{|}}{C}HC\bar{O}_2$
(b) $H_3\overset{+}{N}\underset{\underset{CH_3}{|}}{C}H\overset{\overset{O}{||}}{C}NHCH_2C\bar{O}_2$
(c) $\bar{O}_2C\underset{\underset{CH_3}{|}}{C}H-NH-\overset{\overset{O}{||}}{C}-CH_2\overset{+}{N}H_3$
(d) $\bar{O}_2CCH_2NH-\overset{\overset{O}{||}}{C}\underset{\underset{CH_3}{|}}{C}H\overset{+}{N}H_3$

53. All of the following statements apply to proteins except
(a) Proteins generally have no definite melting point
(b) Proteins contain the grouping —CONH—
(c) Proteins have high molecular weight
(d) Proteins can only contain the elements C, H, O and N.

54. The reason for double helical structure of DNA is the operation of:
(a) Electrostatic attractions
(b) van der Waals forces
(c) Dipole - Dipole interactions
(d) Hydrogen bonding

55. Among the following organic acids, the acid present in rancid butter is:
(a) Pyruvic acid (b) Lactic acid
(c) Butyric acid (d) Acetic acid

56. Which of the following enzyme converts starch into maltose?
(a) Diastase (b) Maltase
(c) Zymase (d) Invertase

57. Which statement is incorrect about peptide bond? Critical Thinking
(a) C–N bond length in proteins is longer than usual C–N bond length
(b) Spectroscopic analysis shows planar structure of $-\underset{\underset{O}{||}}{C}-NH-$ bond

(c) C–N bond length in proteins is smaller than usual C–N bond length
(d) None of these

58. D-(+)-Glucose $\xrightarrow{5(CH_3CO)_2O}$ D-(+)-Glucose pentaacetate
Which statement is true about glucose pentaacetate ? **Toughnut**

(a) It will react with phenylhydrazine but not with Tollens' reagent.
(b) It will react with Tollens' reagent but not with phenylhydrazine.
(c) It will react with both of the above mentioned reagents.
(d) It will react neither with phenylhydrazine nor with Tollens' reagent.

59. Which of the following structures represents thymine ?

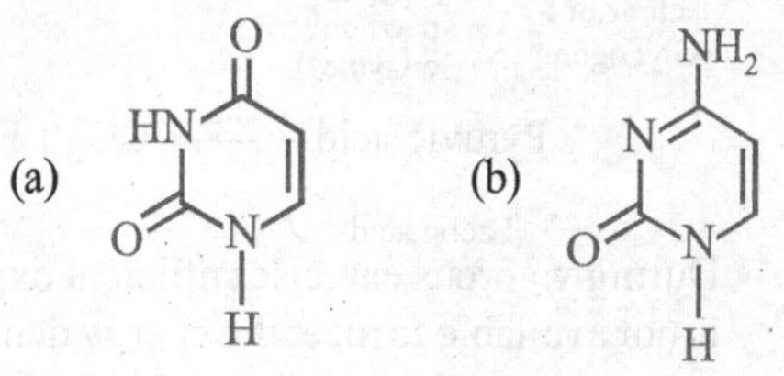

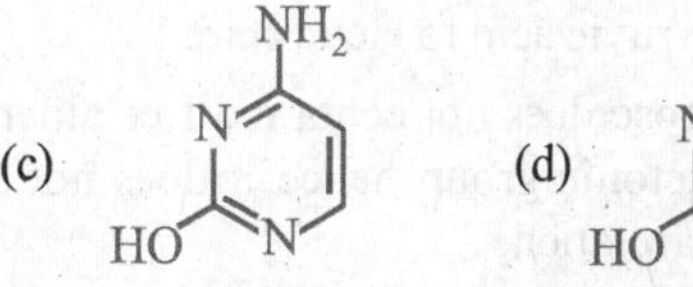

60. Among the following vitamins the one whose deficiency causes rickets (bone deficiency) is
(a) Vitamin A (b) Vitamin B
(c) Vitamin D (d) Vitamin C

Answer KEYs

1	(b)	7	(b)	13	(d)	19	(b)	25	(c)	31	(a)	37	(b)	43	(b)	49	(d)	55	(c)
2	(a)	8	(d)	14	(a)	20	(a)	26	(a)	32	(a)	38	(c)	44	(d)	50	(c)	56	(a)
3	(a)	9	(a)	15	(a)	21	(b)	27	(c)	33	(a)	39	(d)	45	(a)	51	(b)	57	(c)
4	(c)	10	(a)	16	(b)	22	(b)	28	(b)	34	(c)	40	(b)	46	(a)	52	(a)	58	(d)
5	(a)	11	(a)	17	(c)	23	(b)	29	(c)	35	(d)	41	(b)	47	(a)	53	(d)	59	(d)
6	(a)	12	(d)	18	(d)	24	(c)	30	(d)	36	(b)	42	(d)	48	(d)	54	(d)	60	(c)

Hints & Solutions

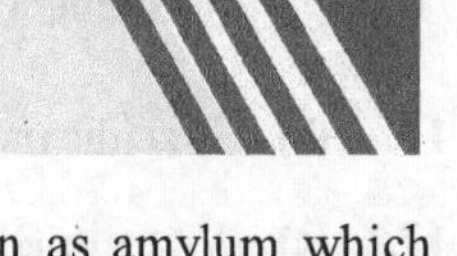

1. **(b)** Fructose has 3 chiral centres and hence number of optical isomers are $2^3 = 8$

2. **(a)**

$$\begin{array}{l} CHO \\ | \\ CHOH \\ | \\ (CHOH)_3 \\ | \\ CH_2OH \end{array} \xrightarrow{3PhNHNH_2} \begin{array}{l} CH{=}N.NHPh \\ | \\ CH{=}N.NHPh + PhNH_2 + NH_3 \\ | \\ (CHOH)_3 \\ | \\ CH_2OH \end{array}$$

3. **(a)** (a) and (d) are L^- sugar but (a) gives an optically active dibasic acid.

4. **(c)** $C_6H_{12}O_6 + 6O_2 + 2ATP \longrightarrow 6CO_2 + 6H_2O + 38ATP$

$\therefore$ Net ATP molecules evolved = 36

5. **(a)** Starch is also known as amylum which occurs in all green plants. A molecule of starch $(C_6H_{10}O_5)_n$ is built of a large number of α-glucose rings joined through oxygen atom.

6. **(a)** Vitamin D is also known as sunshine vitamin.

7. **(b)** Vitamin A.

8. **(d)** Insulin.

9. **(a)** Vitamin A, D are absorbed from the intestine by fats.

10. **(a)** Thyroxine is an amine hormone.

11. **(a)** Starch is a mixture of amylose & amylopectin polysaccharides and monomer is glucose. Thus on complete hydrolysis it gives only glucose.

12. (d) Glucose $\xrightarrow[\text{(does not need oxygen; need only enzymes)}]{}$ Pyruvic acid $\xrightarrow{O_2}$ $CO_2 + H_2O$

stored in the form of ↓ Glycogen

Pyruvic acid ↓ Lactic acid

During vigorous exercise sufficient oxygen is not available to meet the energy demand so, energy is derived through conversion of pyruvic acid to lactic acid.

13. (d) Sucrose does not contain a free aldehydic or ketonic group hence it does not show mutarotation.

14. (a) Since the compound forms a yellow crystalline solid, i.e. osazone with phenylhydrazine, it may be an aldohexose or a ketohexose. Further, since on reduction, compound forms a mixture of sorbitol and mannitol, it must be a ketohexose, i.e. fructose. Recall that glucose on reduction gives only one alcohol glucitol (Sorbitol)

15. (a) α-D-glucose or β-D-glucose when dissolved in water and allowed to stand, following equilibrium is established.

$$\underset{(+111^\circ)}{\alpha\text{-D-glucose}} \rightleftharpoons \text{Open chain form} \rightleftharpoons \underset{(+19^\circ)}{\beta\text{-D-glucose}}$$

Specific rotation of α-form falls, while that of β-form increases until a constant value of +52.5° is reached at equilibrium. This phenomenon is known as **mutarotation**.

16. (b) In acidic medium $\frac{4.3+2.2}{2} = 3.25$

17. (c) Vitamin B_{12} contains both N and P

18. (d) Ascorbic acid - Scurvy
Retinol - Night blindness
Riboflavin - Cracked lips
Thiamine - Beri-Beri
Thus option (d) is correct.

19. (b) Phosphate is linked to 3rd carbon of 1 and 5th carbon of the other sugar molecules.

20. (a) On the basis of structure of guanine and complementary bases present in them, we can say that if the sequence of bases in one strand of DNA is I, then the sequence in the second strand should be II

A:T:G:C:T:T:G:A I
T:A:C:G:A:A:C:T II

21. (b) Glucose is a hemiacetal, so in presence of a base (alkaline medium is provided by Fehling's solution) it can develop —CHO group in the form of open chain structure which responds Fehling's solution. Sucrose is a glycoside (acetal), *i.e.* its hemiacetal OH groups (one due to glucose and another due to fructose) are not free, so it can't attain —CHO group. Hence it will not respond Fehling's solution.

22. (b) Although D-alanine is a constituent of a bacterial cell walls, it is not found in proteins.

23. (b) First molecule of phenylhydrazine unedrgoes nucleophilic addition on carbonyl (–CHO in glucose and >CO in fructose) group. Second molecule of the reagent oxidizes –CHOH– at position 2(in aldoses) or $-CH_2OH$ at position 1 (in ketoses) to form –CHO or $>C=O$ respectively. The third molecule again undergoes nucleophilic addition on the newly developed carbonyl group to form osazone.

24. (c) Fischer gave the prefix "D" to compounds whose bottom chiral has its OH to the right. So natural glucose is called D-glucose or dextrose.

Structure of D-Glucose :

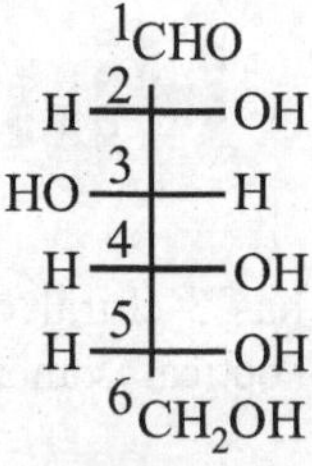

D-Glucose

25. (c) Glycogen is called animal starch and is found in all animal cells. It constitutes the reserve food material.

26. (a) Amylopectin is a polymer of α-*D*-glucose.

27. (c) Sucrose is a non-reducing sugar as it does not reduce Fehling reagent and Tollen's reagent.

28. (b) Glycosidic linkage is actually an ether bond as the linkage forming the rings in an oligosaccharide or polysaccharide is not

just one bond, but the two bonds sharing an oxygen atom e.g. sucrose

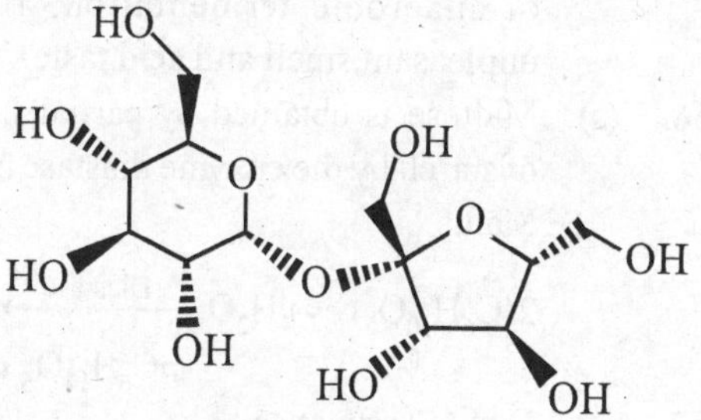

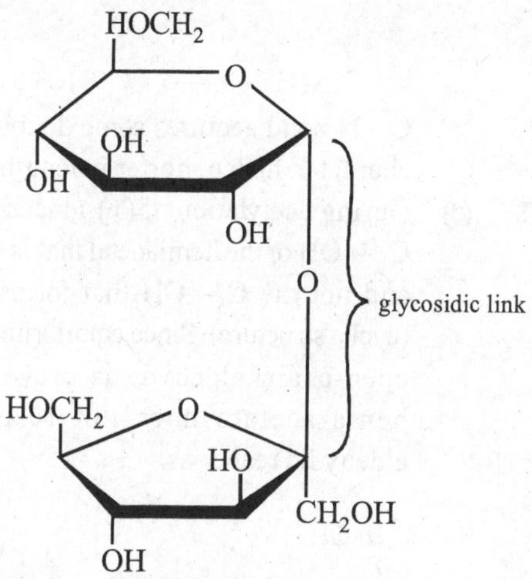

glycosidic link

29. (c) Sucrose, being a non-reducing sugar, does not reduce Benedict's solution. Remember that fructose has an α-hydroxy ketonic group, which is also reducing group (different from ordinary ketonic group)

30. (d) Since sucrose is dextrorotatory while hydrolysis product of sucrose, having equimolar mixture of glucose and fructose, is laevorotatory. Hence the hydrolysed product of sucrose is known as invert sugar and the hydrolysis of sucrose is known as inversion.

31. (a) In acidic medium Zwitter ion convert into

$$\overset{+}{N}H_3-\underset{\underset{R}{|}}{CH}-COO^- + H^+ \longrightarrow \overset{+}{N}H_3-\underset{\underset{R}{|}}{CH}-COOH$$

32. (a) Every amino acid exists exclusively as dipolar ion when the pH of the solution is equal to its isoelectric point (pI), hence at this pH it does not migrate to either electrode, while at other pH, an amino acid migrates either to cathode or to anode depending upon its pI. Thus at pH 9.60, amino acid with pI 5.40 will exist as an anion and migrate to anode; while that with pI 9.60 will not migrate to any electrode.

33. (a) Haemoglobin acts as an oxygen carrier in the blood since it reacts with oxygen to form unstable oxyhaemoglobin which easily breaks to give back haemoglobin and oxygen.

34. (c) Inulin is a polysaccharide of fructose, while all others are polysaccharides of glucose ; although all the four has same molecular formula, $(C_6H_{10}O_5)_n$.

35. (d) Reaction of D-(+)-glucose with methanolic —HCl leads to formation of methyl glucoside (C_1—OH group is methylated) which, being acetal, is not hydrolysable by base, so it will not respond Tollen's reagent.

36. (b) Glucose exists as glucopyranose (a six membered ring structure) which is formed through C_1 and C_5.

37. (b) a and c give the same product.

38. (c) In the D family the more dextrorotatory anomer is named α-D-. In the L family the more laevorotatory anomer is named α L.

39. (d) When structures I and II are C–2 epimers, it implies that these are epimers and diastereomers too.

40. (b) A ketonic group having —OH group on both sides is removed as CO_2 during oxidation with periodate.

41. (b) Proline contains imino (secondary amino), >NH group

42. (d) All are chiral

43. (b) All proteins are not found in L-form but they may be present in form of D or L.

44. (d) Progesterone (Gestogens) is a steroid hormone, which controls the development and maintainance of pregnancy. Thryoxine and adrenaline are amine hormones.

45. (a) Vitamin B_6 is called pyridoxine. It is found in fruits, green-vegetables, milk, etc.

46. (a) The sequence of bases in mRNA are read in a serial order in groups of three at a time. Each triplet of nucleotides (having a specific sequence of bases) is known as codon. Each codon specifies one amino acid. Further since, there are four bases, therefore, $4^3 = 64$ triplets or codons are possible.

47. (a) Synthesis of polypeptide is known as translation. For this process three type of RNA are essential.

48. (d) Guanidine, $H_2N-\overset{\overset{NH}{||}}{C}-NH_2$, is neither a purine nor a pyrimidine base

49. (d) Tyrosine is an α-amino acid, and neither a purine nor a pyrimidine base

50. (c) At isoelectric point, conc. of Zwitter ion will be maximum.

51. (b) $HC\equiv CH \xrightarrow[H_2SO_4]{HgSO_4} CH_3-\overset{\overset{O}{||}}{C}H$

$$\xrightarrow[H^+]{NH_3+HCN} CH_3-\overset{\overset{NH_2}{|}}{CH}-COOH$$

52. (a) By convention, the amino acid with the free amino group (N-terminal) is written at the left end and the one with the unreacted carboxyl group (C-terminal) at the right end. Thus, the structure of Gly. Ala is

$$H_3\overset{+}{N}CH_2\overset{\overset{O}{||}}{C}NH\underset{\underset{CH_3}{|}}{CH}C\overset{-}{O_2}$$

53. (d) Statement (d) is not correct. Some proteins also contain S, with C, H, O and N.

54. (d) The two polynucleotide chains of DNA molecules are twisted around a common axis but run in opposite directions to form a right handed helix. The two chains are joined together by specific hydrogen bonds.

55. (c) Butyric acid also known as butanoic acid is found in milk and butter and is a product of anaerobic fermentation. It has an unpleasant smell and acid taste.

56. (a) Maltose is obtained by partial hydrolysis of starch by the enzyme diastase present in Malt.

$$2(C_6H_{10}O_5)_n + nH_2O \xrightarrow{Diastase} nC_6H_{12}O_6 \text{ (Maltose)}$$

57. (c) Due to resonance,

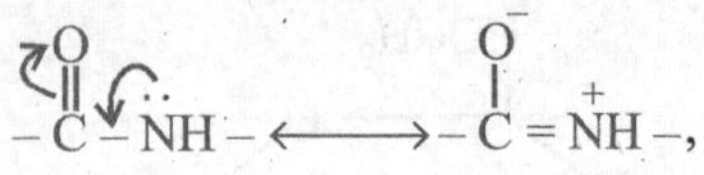

$$-\overset{\overset{O}{||}}{C}-\ddot{N}H- \longleftrightarrow -\overset{\overset{O^-}{|}}{C}=\overset{+}{N}H-,$$

C – N bond acquires some double bond character, hence shorter in length

58. (d) During acetylation of (+)-glucose, it is the C_1—OH of the hemiacetal that is acetylated and not the C_5—OH that forms the ring (cyclic structure). Since equilibrium with the open-chain aldehyde is prevented, the penta-acetate does not respond the aldehydic reactions.

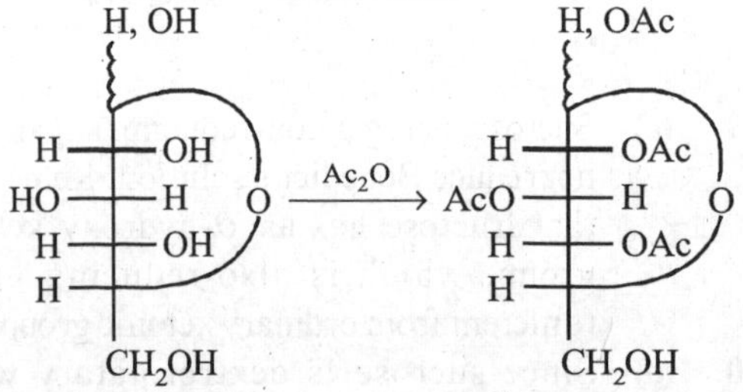

H, OH; H—OH; HO—H; H—OH; H; CH_2OH (ring O) $\xrightarrow{Ac_2O}$ H, OAc; H—OAc; AcO—H; H—OAc; H; CH_2OH (ring O)

D-(+)-Glucose   α or β-Glucose penta acetate

59. (d) The correct structure of thymine is

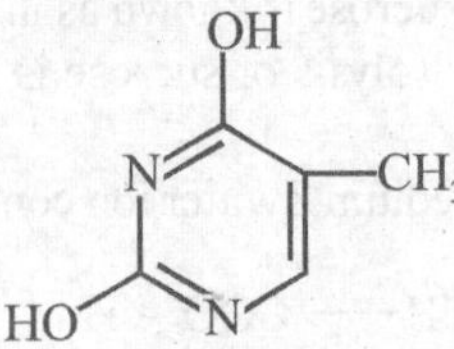

OH, N, CH_3, HO, N

Thymine (T)

60. (c) Deficiency of vitamin D causes rickets.

29 Polymers

1. Which of the following is not an example of addition polymer?
   (a) Polystyrene (b) Nylon
   (c) PVC (d) Polypropylene
2. Which of the following is not a biopolymer?
   (a) Proteins (b) Rubber
   (c) Cellulose (d) RNA
3. Thermosets are:
   (a) cross-linked polymers
   (b) don't melt or soften on heating
   (c) cross-linking is usually developed at the time of moulding where they harden reversibly
   (d) all of the above
4. Which of the following statements is not correct for fibres?

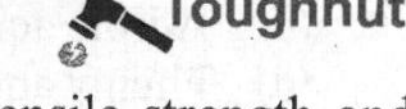
Toughnut

   (a) Fibres possess high tensile strength and high modulus.
   (b) Fibres impart crystalline nature.
   (c) Characteristic features of fibres are due to strong intermolecular forces like hydrogen bonding.
   (d) All are correct.
5. Perlon is
   (a) Rubber (b) Nylon-6
   (c) Terylene (d) Orlon
6. Which one of the following is a chain growth polymer?
   (a) Nucleic acid (b) Polystyrene
   (c) Protein (d) Starch
7. Structure of some important polymers are given. Which one represents Buna-S?
   (a) $-\!\!\left(CH_2 - \underset{}{\overset{CH_3}{\overset{|}{C}}} = CH - CH_2\right)\!\!-_n$
   (b) $-\!\!\left(CH_2 - CH = CH - CH_2 - \underset{C_6H_5}{\underset{|}{CH}} - CH_2\right)\!\!-_n$
   (c) $-\!\!\left(CH_2 - CH = CH - CH_2 - \underset{CN}{\underset{|}{CH}} - CH_2\right)\!\!-_n$
   (d) $-\!\!\left(CH_2 - \overset{Cl}{\overset{|}{C}} = CH - CH_2\right)\!\!-_n$
8. Low density polythene is prepared by
   (a) Free radical polymerisation
   (b) cationic polymerisation
   (c) anionic polymerisation
   (d) Ziegler-Natta polymerisation
9. The polymer used for optical lenses is :
   (a) polypropylene
   (b) polyvinyl chloride
   (c) polythene
   (d) polymethyl methacrylate
10. Polymer used in bullet proof glass is
    (a) lexan (b) PMMA
    (c) nomex (d) kevlar
11. Glyptals are chiefly employed in
    (a) toy making
    (b) surface coating
    (c) photofilm making
    (d) electrical insulators
12. Which of the following is currently used as a tyre cord?
    (a) Terylene (b) Polyethylene
    (c) Polypropylene (d) Nylon - 6
13. The plastic household crockery is prepared by using

Critical Thinking

    (a) melamine and tetrafluoroethane
    (b) malonic acid and hexamethyleneamine
    (c) melamine and vinyl acetate
    (d) melamine and formaldehyde
14. Mark out the most unlike form of polymerization of $CH_2 = CH - CH = CH_2$
    (a) $\left(\begin{array}{c} H \\ -CH_2 \end{array} \!\!\!>\! C = C \!<\!\!\! \begin{array}{c} CH_2- \\ H \end{array} \right)_n$

(b) $\left[\begin{matrix} H \\ CH_2 \end{matrix} \!\!>\! C = C \!<\!\! \begin{matrix} H \\ CH_2 \end{matrix} \right]_n$

(c) $\left(\begin{matrix} & CH = CH_2 & & CH = CH_2 \\ CH_2 - & CH & - CH_2 - & CH \end{matrix} \right)_n$

(d) $\left(\begin{matrix} CH_2 & CH_2 \\ \| & \| \\ C & — C \end{matrix} \right)_n$

15. The condensation of hexamethylenediamine with sebacoyl chloride at 525 K gives
(a) nylon-6,20 (b) nylon-6,01
(c) nylon-6,10 (d) None of these

16. Example of condensation polymer is
(a) Formaldehyde $\rightarrow$ meta-formaldehyde
(b) Acetaldehyde $\rightarrow$ para-aldehyde
(c) Acetone $\rightarrow$ mesityl oxide
(d) Ethene $\rightarrow$ polyethene

17. The mass average molecular mass & number average molecular mass of a polymer are 40,000 and 30,000 respectively. The polydispersity index of polymer will be

Critical Thinking

(a) <1 (b) >1 (c) 1 (d) 0

18. Given the polymers (i) Nylon-6, 6; (ii) Buna-S; (iii) Polythene. Arrange these in increasing order of their inter-molecular forces (lower to higher)

Toughnut

(a) (i) > (ii) > (iii) (b) (ii) > (iii) > (i)
(c) (ii) < (iii) < (i) (d) (iii) < (i) < (ii)

19. What is the percentage of sulphur used in vulcanization of rubber?
(a) 05% to 30% (b) 03% to 25%
(c) 10% to 20% (d) 05% to 25%

20. Which of the following is used in vulcanization of rubber ?
(a) S_8 (b) CF_4 (c) Cl_2F_2 (d) C_2F_2

21. Which one of the following statement is *not true*?
(a) In vulcanization the formation of sulphur bridges between different chains make rubber harder and stronger.
(b) Natural rubber has the *trans* -configuration at every double bond
(c) Buna-S is a copolymer of butadiene and styrene
(d) Natural rubber is a 1,4 - polymer of isoprene

22. Which of the following organic compounds polymerizes to form the polyester Dacron?
(a) Propylene and para HO—(C_6H_4)— OH
(b) Benzoic acid and ethanol
(c) Terephthalic acid and ethylene glycol
(d) Benzoic acid and para HO-(C_6H_4)—OH

23. Biodegradable polymer which can be produced from glycine and aminocaproic acid is :
(a) PHBV (b) Buna - N
(c) Nylon 6, 6 (d) Nylon 2- nylon 6

24. Caprolactum is used for the manufacture of:

(a) Nylon - 6 (b) Teflon
(c) Terylene (d) Nylon - 6,6

25. If a polythene sample contains two monodisperse fractions in the ratio 2 : 3 with degree of polymerization 100 and 200, respectively, then its weight average molecular weight will be :
(a) 4900 (b) 4600 (c) 4300 (d) 5200

26. Synthetic polymer bakelite can be prepared from following compounds
(a) Styrene and vinyl chloride
(b) Acrylonitrile and vinyl chloride
(c) Adipic acid and ethylene glycol
(d) Phenol and formaldehyde

27. Among cellulose, polyvinyl chloride, nylon and natural rubber, the polymer in which the intermolecular force of attraction is weakest is
(a) Nylon (b) Polyvinyl chloride
(c) Cellulose (d) Natural Rubber

28. Which one of the following is an example of thermosetting polymers?
(a) Neoprene (b) Buna-N
(c) Nylon 6, 6 (d) Bakelite

29. Which of the following polymer is a polyamide?
(a) Terylene (b) Nylon
(b) Rubber (d) Vulcanised rubber

30. Ebonite is a
(a) natural rubber
(b) synthetic rubber
(c) highly vulcanized rubber
(d) polypropene

31. Melamine plastic crockery is a codensation polymer of
(a) HCHO and melamine
(b) HCHO and ethylene
(c) melamine and ethylene
(d) None of these

32. Nylon 6,6 is a polyamide obtained by the reaction of
(a) $COOH(CH_2)_4COOH + NH_2C_6H_4NH_2-(p)$
(b) $COOH(CH_2)_4COOH + NH_2(CH_2)_6NH_2$
(c) $COOH(CH_2)_6COOH + NH_2(CH_2)_4NH_2$
(d) $COOHC_6H_4COOH-(p) + NH_2(CH_2)_6NH_2$

33. Which of the following is not correctly represented?
(a) Terylene
$$\left[-OCH_2-CH_2-\overset{O}{\overset{||}{C}}-C_6H_4-\overset{O}{\overset{||}{C}}- \right]_n$$
(b) Neoprene
$$\left[-CH_2-\underset{Cl}{\underset{|}{C}}=CH-CH_2- \right]_n$$
(c) Nylon-6, 6
$$\left[-NH-(CH_2)_6-NH-\overset{O}{\overset{||}{C}}-(CH_2)_4-\overset{O}{\overset{||}{C}}-O- \right]_n$$
(d) PMMA $\left[-CH_2-\underset{COOCH_3}{\underset{|}{\overset{CH_3}{\overset{|}{C}}}}- \right]_n$

34. The monomer of the polymer;

Toughnut

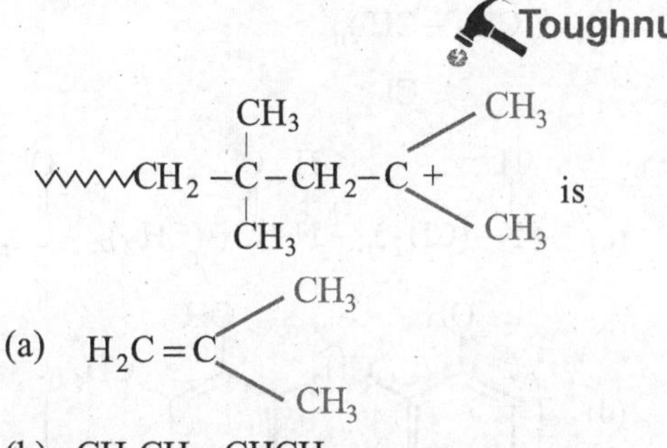

$$\sim\sim CH_2-\underset{CH_3}{\underset{|}{\overset{CH_3}{\overset{|}{C}}}}-CH_2-\overset{+}{C}(CH_3)_2 \text{ is}$$

(a) $H_2C=C(CH_3)_2$
(b) $CH_3CH=CHCH_3$
(c) $CH_3CH=CH_2$
(d) $(CH_3)_2C=C(CH_3)_2$

35. Which of the following polymers do not involve cross linkages?
(a) Melmac (b) Bakelite
(c) Polythene (d) Vulcanised rubber

36. Number average molecular mass, $\overline{M}_n$ and weight average molecular mass $(\overline{M}_w)$ of synthetic polymers are related as

Critical Thinking

(a) $\overline{M}_n = (\overline{M}_w)^{1/2}$ (b) $\overline{M}_n = \overline{M}_w$
(c) $\overline{M}_w > \overline{M}_n$ (d) $\overline{M}_w < \overline{M}_n$

37. Polymer formation from monomers starts by
(a) condensation or addition reaction between monomers
(b) coordinate reaction between monomers
(c) conversion of monomer to monomer ions
(d) hydrolysis of monomers.

38. $n(CF_2 = CF_2) \xrightarrow[(NH_4)_2S_2O_8]{\text{Benzoyl peroxide or}} X$

Here, X is :
(a) RMMA (b) PVC
(c) PAN (d) None of these

39. Which pair of polymers have similar properties ?

(a) Nylon, PVC
(b) PAN, PTFE
(c) PCTFE, PTFE
(d) Bakelite, alkyl resin

40. Among the following, the wrong statement is
(a) PMMA is plexiglass
(b) SBR is natural rubber
(c) PTFE is teflon
(d) LDPE is low density polythene

41. When condensation product of hexamethylenediamine and adipic acid is heated to 525 K in an atmosphere of nitrogen for about 4-5 hours, the product obtained is
(a) solid polymer of nylon 6, 6
(b) liquid polymer of nylon 6, 6
(c) gaseous polymer of nylon 6, 6
(d) liquid polymer of nylon 6

42. In which of the following polymers, empirical formula resembles with monomer?
(a) Bakelite (b) Teflon
(c) Nylon-6, 6 (d) Dacron

43. Which is a polymer of three different monomers?
(a) ABS (b) SBR
(c) NBR (d) Nylon-2, 6

44. The polymer which has conducting power is
(a) polyethylene (b) polybutadiene
(c) polystyrene (d) polyacetylene

45. Which of the following compound is used for preparation of melamine formaldehyde polymer?

(a) 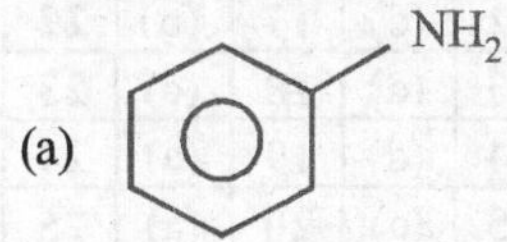

(b) N N N, H_2N, NH_2, NH_2

(c) H_2N, NH_2, N, N, N, NH_2

(d) NH_2, N, H_2N

46. Match the polymers in column - A with their main uses in column - B and choose the correct answer:

Column - A	Column - B
(A) Polystrene	(i) Paints and lacquers
(B) Glyptal	(ii) Rain coats
(C) Polyvinyl chloride	(iii) Manufacture of toys
(D) Bakelite	(iv) Computer discs

(a) (A) - (ii), (B) - (i), (C) - (iii), (D) - (iv)
(b) (A) - (ii), (B) - (iv), (C) - (iii), (D) - (i)
(c) (A) - (iii), (B) - (iv), (C) - (ii), (D) - (i)
(d) (A) - (iii), (B) - (i), (C) - (ii), (D) - (iv)

47. The compound which cannot be used as a plasticizer, is
(a) di-n-butylphthalate
(b) tricresyl phosphate
(c) di-n-octyphthalate
(d) diethyl phthalate

48. Which of the following statements is false?
(a) Artificial silk is derived from cellulose.
(b) Nylon-6, 6 is an example of elastomer.
(c) The repeat unit in natural rubber is isoprene.
(d) Both starch and cellulose are polymers of glucose.

49. Which one of the following sets forms the biodegradable polymer? **Critical Thinking**
(a) $CH_2=CH-CN$ and $CH_2=CH-CH=CH_2$
(b) H_2N-CH_2-COOH and $H_2N-(CH_2)_5-COOH$
(c) $HO-CH_2-CH_2-OH$ and HOOC–C_6H_4–COOH
(d) C_6H_5–$CH=CH_2$ and $CH_2=CH-CH=CH_2$

50. Which one of the following is an example of a thermosetting polymer?
(a) $\left(CH_2-\underset{\displaystyle Cl}{\underset{|}{C}}=CH-CH_2\right)_n$
(b) $\left(CH_2-\underset{\displaystyle Cl}{\underset{|}{CH}}\right)_n$
(c) $\left(\overset{\displaystyle H}{\overset{|}{N}}-(CH_2)_6-\overset{\displaystyle H}{\overset{|}{N}}-\overset{\displaystyle O}{\overset{\|}{C}}-(CH_2)_4-\overset{\displaystyle O}{\overset{\|}{C}}\right)_n$
(d) OH, CH_2, OH, CH_2, n

Answer KEYs

1	(b)	6	(b)	11	(b)	16	(c)	21	(b)	26	(d)	31	(a)	36	(c)	41	(b)	46	(d)
2	(b)	7	(b)	12	(d)	17	(b)	22	(c)	27	(d)	32	(b)	37	(a)	42	(b)	47	(d)
3	(d)	8	(a)	13	(d)	18	(c)	23	(d)	28	(d)	33	(a)	38	(d)	43	(a)	48	(b)
4	(d)	9	(d)	14	(d)	19	(a)	24	(a)	29	(b)	34	(a)	39	(c)	44	(d)	49	(b)
5	(b)	10	(b)	15	(c)	20	(a)	25	(a)	30	(c)	35	(c)	40	(b)	45	(c)	50	(d)

Hints & Solutions

1. (b) Nylon is a condensation polymer
2. (b) Since proteins, cellulose and RNA control various activities of plants and animals, they are called biopolymers. They are found in living organisms.
3. (d) These are characteristics of thermosets.
4. (d) All the given statements about fibres are correct.
5. (b) Perlon is nylon-6. It is prepared from a single monomer (caprolactam) having a potential amino group at one end and a potential carbonyl group at other end.
6. (b) Polystyrene is a chain growth polymer.
7. (b) $nCH_2 = CH - CH = CH_2 + n\,C_6H_5CH = CH_2 \xrightarrow[\text{Polymerisation}]{\text{Na, Heat}} \left(CH_2 - CH = CH - CH_2 - CH(C_6H_5) - CH_2\right)_n$

   1, 3-Butadiene; Styrene; Butadiene - Styrene copolymer (SBR or BUNA - S)

8. (a) Ethene on free radical polymerisation gives low density polythene.
9. (d) Polymethyl methacrylate is hard, fairly rigid. It is used for optical lenses.
10. (b) PMMA and polycarbonate are used in bullet proof glass
11. (b) Glyptal is an alkyd resin obtained from polyhydric alcohols and polybasic organic acids or their anhydrides. They are used for surface coatings.
12. (d) Nylon tyre cord is made from high tenacity continuous filament yarn by twisting and plying.
13. (d) The unbreakable plastic household crockery is made from copolymer of formaldehyde (HCHO) and melamine.

   Melamine (2,4,6-triamino-1,3,5-triazine, $C_3N_3(NH_2)_3$) $+ \underset{\text{Formaldehyde}}{HCHO} \xrightarrow{\text{Polymerisation}}$ Melmac

14. (d) In polymerisation of 1,3-butadiene either 1,4-polymerisation or 1,2-polymerisation occurs. In case of 1,4-polymerisation, the double bond shifts at C_2 and C_4 carbon, while the chain propagates from C_1 and C_4 end. In this either trans or cis polymeric chain is formed.
   Option (a) and (b) represent 'trans' and 'cis' 1,4 polymerisation respectively.
   Option (c) resembles 1,2 polymerisation, where as option (d) most unlikely to happen.

15. (c) $H_2N-(CH_2-)_6NH_2 + Cl-\overset{O}{\overset{||}{C}}-(CH_2)_8-\overset{O}{\overset{||}{C}}-Cl \xrightarrow{\Delta} \left(HN-(CH_2)_6-(NH-\overset{O}{\overset{||}{C}}-(CH_2)_8-\overset{O}{\overset{||}{C}}\right)_n$

Hexamethylene diamine — Sebacoyl chloride — Nylon 6, 10

16. (c) Polymers formed by condensation process with elimination of small molecule like H_2O, CO_2 etc. are known as condensation polymers.

Eg. $(CH_3)_2C=O + H_2CH-\underset{\underset{O}{||}}{C}-CH_3 \xrightarrow[HCl]{dil.} (CH_3)_2C=CH-\overset{O}{\overset{||}{C}}-CH_3 + H_2O$

Acetone — Mesityl oxide

17. (b) Average number molecular weight $\overline{M_n} = 30,000$

Average mass molecular weight $\overline{M_w} = 40,000$

$$\text{Polydispersity index (PDI)} = \frac{\overline{M}_w}{\overline{M}_n} = \frac{40,000}{30,000} = 1.33$$

18. (c) Nylon – 6, 6 is a fibre with hydrogen bonds thus have strongest intermolecular forces. Buna-S is a elastomer whereas polythene is thermoplastic. the intermolecular forces of attraction in thermoplastics is intermediate between elastomer and fibres. Therefore forces in elastomers are weakest.

19. (a) Vulcanization is a process in which natural rubber is treated with sulphur and certain organic compounds which accelerate the reaction between the rubber and sulphur. Thus about 5% S is used for making tyre rubber, 20-25% S for making ebonite and 30% S for making battery case rubber.

20. (a) SF_6 is used in the vulcanisation of rubber. Sulphur is heated with polymer to introduce cross-linking and thus, form tough polymer.

21. (b) $-CH_2(CH_3)C=C(H)CH_2-CH_2(CH_3)C=C(H)CH_2-CH_2(CH_3)C=C(H)CH_2-$

Natural rubber
(All cis configuration)

All statements except (b) are correct

22. (c) $n\,HOCH_2CH_2OH + n\,HOOC-C_6H_4-COOH \xrightarrow{\Delta} \left(OCH_2CH_2-O-\overset{O}{\overset{||}{C}}-C_6H_4-\overset{O}{\overset{||}{C}}\right)_n$

Ethylene glycol — Terephthalic acid — Terylene

It is resistant to mineral and organic acids. It is used for blending with wool to provide better crease, in safety helmets and aircraft battery boxes.

23. (d) $H_2N—CH_2—COOH + H_2N—(CH_2)_5—COOH \rightarrow \left(HN—CH_2—CO—NH—(CH_2)_5—CO\right)$

Glycine — Aminocapric acid — Nylon-2-nylon-6

24. (a) Caprolactum (cyclic NH, C=O) $\xrightarrow{\text{(Caprolactum)}}$ $[-NH-(CH_2)_5-CO-]_n$ Nylon–6

25. (a)

26. (d) Phenol and formaldehyde undergo condensation polymerisation under two different conditions to give a cross linked polymer called bakelite.

27. (d) Nylon and cellulose, both have intermolecular hydrogen bonding, polyvinyl chloride has dipole-dipole interaction, while natural rubber has van der Waal forces which are weakest.

28. (d) Polymers which change irreversibly into hard and rigid material on heating are known as thermosetting polymers e.g bakelite.

29. (b) Nylon is a polyamide fibre. It is prepared by the condensation polymerisation of adipic acid ($HOOC(CH_2)_4COOH$) and hexamethylene diamine ($H_2N(CH_2)_6NH_2$).

30. (c) Ebonite is a hard highly vulcanized rubber, containing 20-25%, sulphur.

31. (a) Melamine plastic crockery is a copolymer of HCHO and Melamine.

32. (b) $$\underset{\text{Adipic acid}}{n\,HOOC(CH_2)_4COOH} + \underset{\text{Hexamethylene diamine}}{n\,H_2N(CH_2)_6NH_2}$$

$$\xrightarrow[\text{Polymerisation}]{525K} \underset{\text{Nylon 6, 6}}{[\overset{O}{\overset{||}{C}}-(CH_2)_4-\overset{O}{\overset{||}{C}}-NH-(CH_2)_6-NH]_n}$$

33. (a) Terylene is prepared by condensing terephthalic acid and ethylene glycol

$$\underset{\text{Terephthalic acid}}{n\,HOOC-C_6H_4-COOH} + \underset{\text{Ethylene glycol}}{n\,HOCH_2CH_2OH} \longrightarrow \underset{\text{Terylene}}{\left[\overset{O}{\overset{||}{C}}-C_6H_4-\overset{O}{\overset{||}{C}}-OCH_2-CH_2-O\right]_n}$$

34. (a) Addition of monomers follows isoprene rule

$$\underset{H\qquad\qquad T}{(CH_3)_2C=CH_2} + \underset{H\qquad\qquad T}{(CH_3)_2C=CH_2} \longrightarrow (H_3C)_2\overset{+}{C}-CH_2-\underset{CH_3}{\overset{CH_3}{\underset{|}{\overset{|}{C}}}}-CH_2 \sim\sim$$

35. (c) Polythene is a linear polymer

36. (c) $PDI = \dfrac{\overline{M}_w}{\overline{M}_n}$

For synthetic polymer, PDI > 1

$\therefore \overline{M}_w > \overline{M}_n$

37. (a) Polymerisation starts either by condensation or by addition reactions between monomers

38. (d) $$n\,(CF_2=CF_2)\xrightarrow[(NH_4)_2S_2O_8]{\text{Benzoyl peroxide or}} \underset{\underset{(X)}{\text{PTFE}}}{-(CF_2-CF_2)_n-}$$

Here X is polytetrafluoroethylene. So, none of these i.e., option (d) is correct choice.

39. **(c)** PCTFE and PTFE both have same carbon backbone.

$$\text{+ClFC}-\text{CF}_2\text{+}_n \qquad \text{+F}_2\text{C}-\text{CF}_2\text{+}_n$$

PCTFE (polymonochloro trifluoroethylene) PTFE (poly tetrafluoro ethylene)

40. **(b)** SBR is styrene-butadiene rubber, is a synthetic rubber.

41. **(b)** The condensation polymerisation of hexamethylene diamine and adipic acid is done in solution form by interface technique. In this, liquid nylon polymer is obtained.

42. **(b)** In addition homopolymers such as teflon, empirical formula resembles with monomer.

43. **(a)** ABS is acrylonitrile-butadiene-styrene rubber which is obtained by copolymerisation of acrylonitrile, 1, 3-butadiene and styrene.

$$\underset{\text{Acrylonitrile}}{\underset{|\atop \text{CN}}{CH_2 = CH}} + \underset{\text{1,3-Butadiene}}{\underset{|\atop CH=CH_2}{CH_2 = CH}} + \underset{\text{Styrene}}{CH_2 = CHC_6H_5} \longrightarrow$$

$$\left[CH_2 - \underset{|\atop CN}{CH} - CH_2CH = CHCH_2 - CH_2 - \underset{|\atop C_6H_5}{CH} \right]_n$$

ABS rubber

44. **(d)** Polyacetylene, due to presence of double bonds, is a conducting polymer.

45. **(c)**

46. **(d)**

Column-A	**Column-B**
Polymer	**Use**
Polystyrene	Manufacture of toys
Glyptal	Paints and lacquers.
P.V.C.	Rain Coats
Bakelite	Computer discs

47. **(d)** Generally high boiling esters or haloalkanes act as plasticizer.

48. **(b)** Nylons (polyamides) are fibres.

49. **(b)** Biodegradable polymer is nylon-2, 6 which is copolymer of glycine (H_2N-CH_2-COOH) and amino caproic acid ($H_2N-(CH_2)_5-COOH$).

$$\underset{\text{Glycine}}{n\,H_2N-CH_2-COOH} + \underset{\text{Amino caproic acid}}{nH_2N-(CH_2)_5-COOH} \longrightarrow \underset{\text{Nylon}-2,\,6}{\left(HN-CH_2-\overset{O}{\overset{\|}{C}}-NH-(CH_2)_5-\overset{O}{\overset{\|}{C}} \right)_n}$$

50. **(d)** Glyptal is used in the manufacture of paints and lacquers.

30 Chemistry in Everyday Life

1. Terfenadine is commonly used as a/an
   (a) tranquilizer (b) antihistamine
   (c) antimicrobial (d) antibiotic
2. Which of the following term means pain killer?
   (a) Antibiotic (b) Analgesic
   (c) Antipyretic (d) Penicillin
3. Which of the following is a bactericidal antibiotic?
   (a) Ofloxacin (b) Tetracycline
   (c) Chloramphenicol (d) Erythromycin
4. Which one of the following is used as Antihistamine?
   (a) Omeprazole (b) Chloranphenicol
   (c) Diphenhydramine (d) Norethindrone
5. Which of the following statements about aspirin is **not** true?
   (a) It is effective in relieving pain.
   (b) It is a neurologically active drug.
   (c) It has antiblood clotting action.
   (d) It belongs to narcotic analgesics.
6. Sulphonamides act as
   (a) Antiseptic (b) Analgesic
   (c) Antimicrobials (d) Antipyretic
7. Various phenol derivatives, tincture of iodine (2 – 3%) I_2 in (water / alcohol) and some dyes like methylene blue are

   (a) antiseptics (b) disinfectants
   (c) analgesics (d) antipyretics
8. Which of the following is a hypnotic drug?
   (a) luminal (b) salol
   (c) catechol (d) chemisol
9. An antibiotic with a broad spectrum
   (a) kills the antibodies
   (b) acts on a specific antigen
   (c) acts on different antigens
   (d) acts on both the antigens and antibodies
10. An ester used as medicine is
   (a) ethyl acetate (b) methyl acetate
   (c) methyl salicylate (d) ethyl benzoate
11. Select the incorrect statement.
   (a) Equanil is used to control depression and hypertension.
   (b) Mifepristone is a synthetic steroid used as "morning after pill".
   (c) 0.2 per cent solution of phenol is an antiseptic while its 1.0 per cent solution is a disinfectant.
   (d) A drug which kills the organism in the body is called bacteriostatic.
12. Which of the following acts as an antioxidant in edible oils ?
   (a) Vitamin B (b) Vitamin C
   (c) Vitamin D (d) Vitamin E
13. H_1 – Receptor antagonists is a term associated with :
   (a) Antiseptics (b) Antihistamins
   (c) Antacids (d) Analgesics
14. The structure given below is known as

Critical Thinking

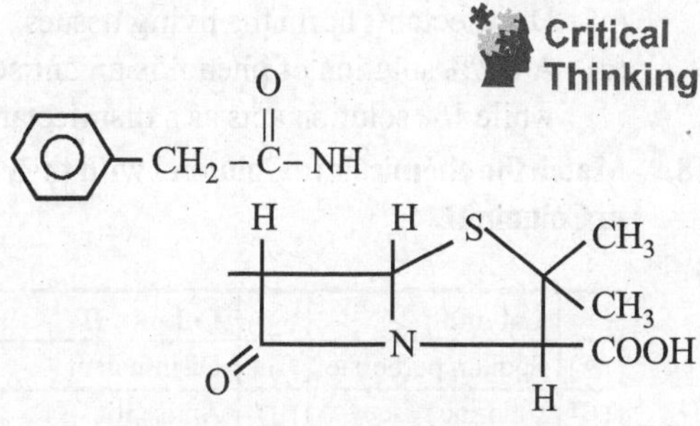

   (a) Penicillin F (b) Penicillin G
   (c) Penicillin K (d) Ampicillin

15. The correct structure of the drug paracetamol is

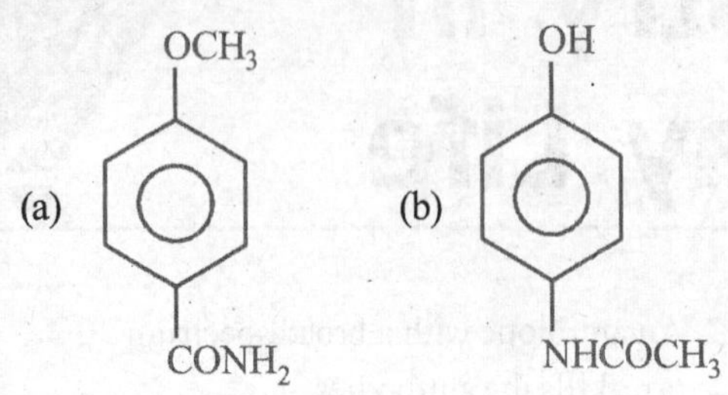

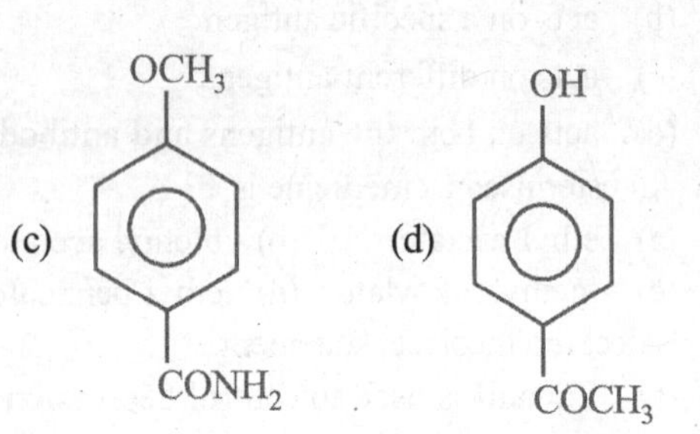

16. Which one of the following can possibly be used as analgesic without causing addiction and mood modification?
(a) Diazepam
(b) Morphine
(c) N-acetyl-para-aminophenol
(d) Tetrahydrocannabinol

17. Antiseptics and disinfectants either kill or prevent growth of microorganisms. Identify which of the following statements **is not true:**

(a) Chlorine and iodine are used as strong disinfectants.
(b) Dilute solutions of boric acid and hydrogen Peroxide are strong antiseptics.
(c) Disinfectants harm the living tissues.
(d) A 0.2% solution of phenol is an antiseptic while 1% solution acts as a disinfectant.

18. Match the chemicals in Column I with their uses in Column II.

	Column I		Column II
(A)	Sodium perborate	(I)	Disinfectant
(B)	Chlorine	(II)	Antiseptic
(c)	Bithional	(III)	Milk bleaching agent
(D)	Potassium stearate	(IV)	Soap

(a) A - I, B - II, C - III, D - IV
(b) A - II, B - III, C - IV, D - I
(c) A - III, B - I, C - II, D - IV
(d) A - IV, B - I, C - II, D - III

19. An antipyretic is
(a) quinine (b) paracetamol
(c) luminal (d) piperazine

20. Salol can be used as
(a) antiseptic (b) antipyretic
(c) analgesic (d) None of these

21. Which one of the following is employed as a tranquilizer?
(a) Naproxen (b) Tetracycline
(c) Chlorpheninamine (d) Equanil

22. The insecticide containing 99% γ – isomer of benzene hexachloride is known as
(a) lindane (b) TNT
(c) malathion (d) methoxychlor

23. The following compound is used as

O
‖
O – C – CH_3
COOH

(a) an anti-inflammatory compound
(b) analgesic
(c) hypnotic
(d) antiseptic

24. Amoxycillin is semi-synthetic modification of
(a) penicillin (b) streptomycin
(c) tetracyclin (d) chloroampheniol

25. Which of these is a hypnotic ?
(a) metaldehyde (b) acetaldehyde
(c) paraldehyde (d) None of these

26. Which of the following is used as an antibiotic?
(a) ciprofloxacin (b) paracetamol
(c) ibuprofen (d) tocopherol

27. The drug used for prevention of heart attacks is
(a) aspirin (b) valium
(c) chloramphenicol (d) cephalosporin

28. Omeoprazole and lansoprazole are used as –

(a) antifertility (b) antiallergic
(c) antibiotic (d) antacid

29. Sulpha drugs are used for
(a) precipitating bacteria
(b) removing bacteria
(c) decreasing the size of bacteria
(d) stopping the growth of bacteria

30. The drug which is effective in curing malaria is
(a) quinine (b) aspirin
(c) analgin (d) equanil

31. Heroin is a derivative of
(a) cocaine (b) morphine
(c) caffeine (d) nicotine

32. The formulation of dettol contains
(a) chloroxylenol (b) terpineol
(c) alcohol (d) All of these

33. Which of the following is an antidiabetic drug?
(a) Insulin (b) Penicillin
(c) Chloroquine (d) Aspirin

34. When salicylic acid is treated with acetic anhydride we get
(a) aspirin (b) paracetamol
(c) salol (d) none of these

35. Which artificial sweetener contains chlorine ?
(a) Sucralose (b) Alitame
(c) Aspartame (d) Saccharin

36. Structuraly biodegradable detergents, should contain

(a) normal alkyl chain
(b) branched alkyl chain
(c) phenyl side chain
(d) cyclohexyl side chain

37. Substance used for the preservation of coloured fruit juices is
(a) benzene
(b) benzoic acid
(c) phenol
(d) sodium meta bisulphite

38. End of detergents have
(a) ester group (b) sodium sulphate
(c) aldehyde (d) amine group

39. Which of the following represents a synthetic detergent?
(a) $C_{15}H_{31}COOK$
(b) $CH_3[CH_2]_{16}COONa$
(c) $C_{12}H_{25}-C_6H_4-SO_3Na$
(d) None of these

40. Which of the following represents soap ?
(a) $C_{17}H_{35}COOK$ (b) $C_{17}H_{35}COOH$
(c) $C_{15}H_{31}COOH$ (d) $(C_{17}H_{35}COO)_2Ca$

41. An antibiotic contains nitro group attached to aromatic nucleus. It is
(a) penicillin (b) streptomycin
(c) tetracycline (d) chloramphenicol

42. Aspirin can be prepared by the reaction of
(a) Salicyldehyde with acetic anhydride in presence of H_2SO_4
(b) Salicylic acid with methanol in presence of H_2SO_4
(c) Salicylic acid with acetic anhydride in presence of H_2SO_4
(d) Cinnamic acid with acetic anhydride in presence of H_2SO_4

43. Which is correct about vanillin?
(a) A flavouring agent having vanila flavour
(b) *p*-hydroxy-*m*-methoxy benzaldehyde
(c) A food additive
(d) All of these

44. Which is correct about saccharin?
(a) It is 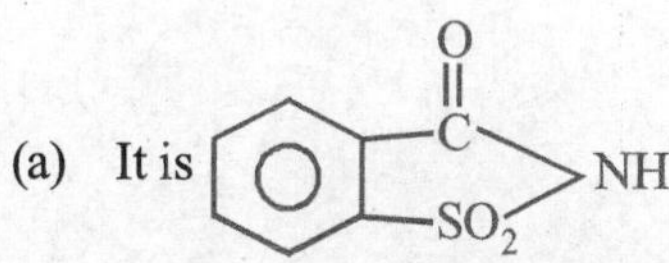

(b) It is 600 times sweeter than sugar
(c) It is used as sweetening agent
(d) All of these

45. Benzalkonium chloride is a 
(a) cationic surfactant and antiseptic
(b) anionic surfactant and soluble in most of organic solvents
(c) cationic surfactant and insoluble in most of organic solvents
(d) cationic surfactant and antimalarial

46. Which one of the following is employed as antihistamine?
(a) Chloramphenicol
(b) Diphenylhydramine
(c) Norothindrone
(d) Omeprazole

47. Chloroamphenicol is an :
(a) antifertility drug
(b) antihistaminic
(c) antiseptic and disinfectant
(d) antibiotic-broad spectrum

48. Artificial sweetner which is stable under cold conditions only is :
(a) Saccharine (b) Sucralose
(c) Aspartame (d) Alitame

49. Which one of the following compounds is an anti-fertility drug? 
(a) Aspirin (b) Chloromycetin
(c) Saheli (d) Penicillin

50. Chemically heroin is
(a) morphine monoacetate
(b) morphine dibenzoate
(c) morphine diacetate
(d) morphine monobenzoate

Answer KEYs

1	(b)	6	(c)	11	(d)	16	(c)	21	(d)	26	(a)	31	(b)	36	(b)	41	(d)	46	(b)
2	(b)	7	(a)	12	(d)	17	(b)	22	(a)	27	(a)	32	(d)	37	(b)	42	(c)	47	(d)
3	(a)	8	(a)	13	(b)	18	(c)	23	(b)	28	(d)	33	(a)	38	(b)	43	(d)	48	(c)
4	(c)	9	(c)	14	(b)	19	(b)	24	(a)	29	(d)	34	(a)	39	(c)	44	(d)	49	(c)
5	(d)	10	(c)	15	(b)	20	(a)	25	(c)	30	(a)	35	(a)	40	(a)	45	(a)	50	(c)

Hints & Solutions

1. (b) Terfenadine is commonly used as antihistamine.
2. (b) Analgesic means pain killer.
3. (a) Bactericidal are the drugs that kills bacteria. Ofloxacin works by stopping the growth of bacteria. This antibiotic treats only bacterial infections.
4. (c) Diphenhydramine is used as antihistamine.
5. (d) Aspirin is an non-narcotics analgesic.
6. (c) Sulphonamides act as antimicrobials.
7. (a) Antiseptic drugs cause destruction of micro-organism that produce septic diseases e.g. Dettol, savlon, boric acid, phenol, iodoform, $KMnO_4$ and some dyes such as methylene blue, genatian violet.
8. (a) These drugs induce sleep and are habit forming, common example of hypnotic drugs are luminal and saconal.
9. (c) Broad spectrum antibiotics act on different antigens.
10. (c) Oil of winter green or methyl salicylate is used as medicine.
11. (d) Bacteriostatic drugs inhibit the growth of organisms while bactericidal drugs kill the microorganisms.
12. (d) Vitamin E is an antioxidant present in edible oils.
13. (b) The term "antihistamine" refers only to H_1 antagonists, which is also known as H_1-receptor antagonists and H_1-antihistamine.
14. (b) It is the known structure of penicillin G or benzyl penicillin
15. (b)
16. (c) We know that N-acetyl-para-aminophenol (or paracetamol) is an antipyretic which can also be used as an analgesic to relieve pains.
17. (b) Dilute solutions of boric acid and hydrogen peroxide are weak antiseptics.
18. (c) The correct matching is as follows :

	Column I	Column II
(A)	Sodium perborate	Milk bleaching agent
(B)	Chlorine	Disinfectant
(C)	Bithional	Antiseptic
(D)	Potassium stearate	Soap

19. (b) Paracetamol is an antipyretic
20. (a) Salol is phenyl salicylate used as antiseptic.
21. (d) Equanil is an important medicine used in depression and hypertension.
22. (a) Lindane or gammexane is γ isomer of BHC.
23. (b) It is acetyl salicylic acid i.e., aspirin which is used as analgesic and antipyretic.
24. (a) Amoxycillin is semisynthetic modification of penicillin
25. (c) Paraldehyde is a hypnotic.
26. (a) Ciprofloxacin is used as an antibiotic, while paracetamol, ibuprofen and tocopherol are respectively antipyretic, pain killer and vitamin E.
27. (a) Due to anti-blood clotting action of aspirin, it is used to prevent heart attack.
28. (d) Antacids decrease acidity in stomach.
29. (d) Sulpha drugs (antibacterials and antibiotics) are group of drugs which are derivatives of sulphanilamide.
30. (a) Substances used for the treatment of malaria are antimalarials, e.g. quinine, chloroquine.
31. (b) Heroin is acyl derivative of morphine.
32. (d) Dettol (antiseptic) is a mixture of 4.8% chloroxylenol + 9.9% terpineol and absolute alcohol.
33. (a) Insulin is antidiabetic drug.
34. (a)

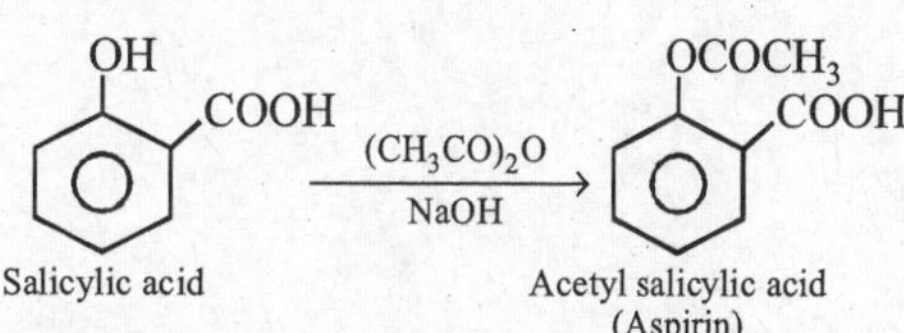

35. (a) [sucralose structure: HO, Cl, O, OH, HO, Cl, HO, O, OH, Cl, O] is sucralose.

36. (b) Structurally, biodegradable detergents should contain branched alkyl chain.

37. (b) Benzoic acid is used as preservative as sodium benzoate.

38. (b) A detergent molecule consists of a large hydrocarbon group that is non-ionic and a sulphonate ($SO_3^-Na^+$) or a sulphate ($SO_4^-Na^+$) group that is ionic. Examples of detergents are sodium-*n*-dodecyl benzene sulphonate, sodium dodecyl sulphate etc.

39. (c) The most widely used domestic detergent is the sodium dodecyl benzene sulphonate (SDS).

$CH_3-(CH_2)_{11}-C_6H_4-SO_3Na$

40. (a) Soaps are the sodium or potassium salts of higher fatty acids e.g., $C_{17}H_{37}COOK$ (potassium stearate). These are obtained by alkaline hydrolysis of oils and fats. The reaction is called saponification.

41. (d) Chloramphenicol is

$O_2N-C_6H_4-CH(OH)-CH(CH_2OH)-NH-C(=O)-CHCl_2$

42. (c) Asprin is prepared by reaction of salicylic acid with acetic anhydride in presence of H_2SO_4.

$$C_6H_4(OH)(COOH) + CH_3-CO-O-CO-CH_3 \xrightarrow{H_2SO_4} C_6H_4(COOH)(OCOCH_3)$$

Acetyl salicylic acid (Aspirin)

43. (d) All are characteristics of vanillin.

44. (d) All are characteristics of saccharin.

45. (a) Benzalkonium chloride, also known as alkyldimethylbenzylammonium chloride is nitrogenous cationic surface active agent belonging to the quaternary ammonium group. It is used as antiseptic.

$C_6H_5CH_2-N^+(CH_3)_2-C_nH_{2n+1}\ Cl^-$

$n = 8, 10, 12, 14, 16, 18$

46. (b) Diphenylhydramine also known as benadryl is an antihistamine.

47. (d) Chloroamphenicol is a broad spectrum antibiotic.

48. (c) Aspartame is stable under cold conditions.

49. (c) The antifertility drugs are those chemical compounds which prevent pregnancy in woman. These, drugs controls the menstrual cycle and ovulation and so used as birth control drug or contraceptives.

50. (c) Morphine diacetate.

31

Analytical Chemistry

1. An aqueous solution of a salt X turns blood red on treatment with CNS^- and blue on treatment with $K_4[Fe(CN)_6]$. X also gives a positive chromyl chloride test. The salt X is :
   (a) $CuCl_2$ (b) $FeCl_3$
   (c) $Cu(NO_3)_2$ (d) $Fe(NO_3)_3$
2. $KMnO_4$ reacts with oxalic acid as :

Critical Thinking

$$MnO_4^- + C_2O_4^{2-} + H^+ \longrightarrow Mn^{2+} + CO_2 + H_2O$$

Hence, 50 ml of 0.04 M $KMnO_4$ is acidic medium is chemically equivalent to
   (a) 100 ml of 0.1 M $H_2C_2O_4$
   (b) 50 ml of 0.2 M $H_2C_2O_4$
   (c) 50 ml of 0.1 M $H_2C_2O_4$
   (d) 25 ml of 0.1 M $H_2C_2O_4$
3. For detection of sulphur in an organic compound sodium nitroprusside is added to the sodium extract. A violet colour is obtained due to formation of
   (a) $Fe(CN)_2$
   (b) $K_3Fe(CN)_5NS$
   (c) $Na_4[Fe(CN)_5NOS]$
   (d) $Na_4Fe(CN)_6$
4. 0.45 g of acid (molecular weight 90) is neutralised by 20 ml of 0.5N caustic potash. The basicity of acid is
   (a) 1 (b) 2 (c) 3 (d) 4
5. The cation that will not be precipitated by H_2S in the presence of dil. HCl is :
   (a) Pb^{2+} (b) Cu^{2+} (c) Co^{2+} (d) As^{3+}
6. A pink coloured salt turns blue on heating. The presence of which cation is most likely ?
   (a) Co^{2+} (b) Cu^{2+} (c) Zn^{2+} (d) Fe^{2+}
7. 0.5 g mixture of $K_2Cr_2O_7$ and $KMnO_4$ was treated with excess of KI in acidic medium. I_2 liberated required 100 cm^3 of 0.15 N $Na_2S_2O_3$ solution for titration. The percentage amount of $K_2Cr_2O_7$ in the mixture is Toughnut
   (a) 85.36 % (b) 14.64 %
   (c) 58.63 % (d) 26.14 %
8. In Kjeldahl's method, $CuSO_4$ acts as
   (a) oxidising agent (b) reducing agent
   (c) hydrolysing agent (d) catalytic agent
9. In Lassaigne's test, the organic compound is fused with a piece of sodium metal in order to
   (a) increase the ionisation of the compound.
   (b) decrease the melting point of the compound.
   (c) increase the reactivity of the compound.
   (d) convert the covalent compound into a mixture of ionic compounds.
10. A white sodium salt dissolves readily in water to give a solution which is neutral to litmus. When silver nitrate solution is added to the solution, a white precipitate is obtained which does not dissolve in dil. HNO_3. The anion could be
    (a) CO_3^{2-} (b) Cl^- (c) SO_4^{2-} (d) S^{2-}
11. An aqueous solution of a substance gives a white precipitate on treatment with dil. HCl which dissolves on heating. When H_2S is passed through the hot acidic solution, a black precipitate is obtained. The substance is a Tricky
    (a) Hg_2^{2+} salt (b) Cu^{2+} salt
    (c) Ag^+ salt (d) Pb^{2+} salt
12. A is a lighter phenol and B is an aromatic carboxylic acid. Separation of mixture of A and B can be carried out easily by using a solution of
    (a) sodium hydroxide
    (b) sodium sulphate
    (c) calcium chloride
    (d) sodium bicarbonate
13. A metal chloride dissolves appreciably in cold water. When placed on a platinum wire in Bunsen flame no distinctive colour is noticed, the cation would be
    (a) Mg^{2+} (b) Ba^{2+} (c) Pb^{2+} (d) Ca^{2+}

14. 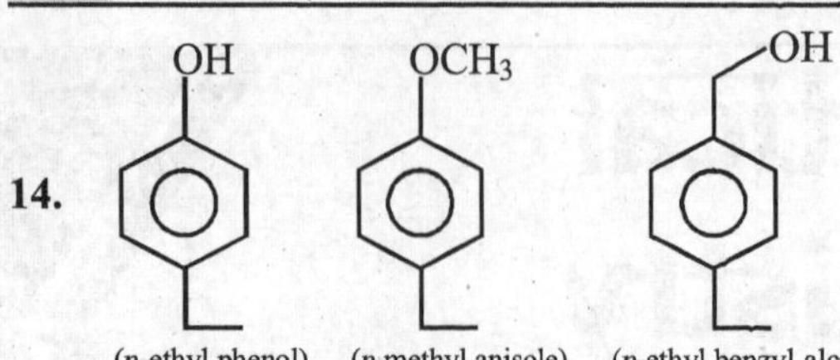

Above compounds can be differentiated by using the reagent:
(a) NaOH, Tollen's reagent, $FeCl_3$
(b) CrO_3, Tollen's reagent, $FeCl_3$
(c) Tollen's reagent, CrO_3, $FeCl_3$
(d) Na, Tollen's reagent, $FeCl_3$

15. Copper wire test for halogens is known as
(a) Duma's Test (b) Beilstein's Test
(c) Liebig's Test (d) Lassigne's Test

16. In the separation of Cu^{2+} and Cd^{2+} in 2nd group qualitative analysis of cations, tetrammine copper (II) sulphate and tetrammine cadmium (II) sulphate react with KCN to form the corresponding cyano complexes. Which one of the following pairs of the complexes and their relative stability enable the separation of Cu^{2+} and Cd^{2+}?
(a) $K_3[Cu(CN)_4]$ more stable and $K_2[Cd(CN)_4]$ less stable
(b) $K_2[Cu(CN)_4]$ less stable and $K_2[Cd(CN)_4]$ more stable
(c) $K_2[Cu(CN)_4]$ more stable and $K_2[Cd(CN)_4]$ less stable
(d) $K_3[Cu(CN)_4]$ less stable and $K_2[Cd(CN)_4]$ more stable

17. Sodium carbonate cannot be used in place of $(NH_4)_2CO_3$ for the identification of Ca^{2+}, Ba^{2+} and Sr^{2+} ions (in group V) during mixture analysis because :
(a) Mg^{2+} ions will also be precipitated.
(b) Concentration of CO_3^{2-} ions is very low.
(c) Sodium ions will react with acid radicals.
(d) Na^+ ions will interfere with the detection of Ca^{2+}, Ba^{2+}, Sr^{2+} ions.

18. A solution containing As^{3+}, Cd^{2+}, Ni^{2+} and Zn^{2+} is made alkaline with dilute NH_4OH and treated with H_2S. The precipitate obtained will consist of

Toughnut

(a) As_2S_3 and CdS (b) CdS, NiS and ZnS
(c) NiS and ZnS (d) Sulphide of all ions

19. 3.92 g of ferrous ammonium sulphate react completely with 50 ml $\frac{N}{10}$ $KMnO_4$ solution. The percentage purity of the sample is
(a) 50 (b) 78.4 (c) 80 (d) 39.2

20. 1.25 g of a sample of Na_2CO_3 and Na_2SO_4 is dissolved in 250 ml solution. 25 ml of this solution neutralises 20 ml of 0.1N H_2SO_4. The % of Na_2CO_3 in this sample is
(a) 84.8% (b) 8.48% (c) 15.2% (d) 42.4%

21. An aqueous solution of $FeSO_4$, $Al_2(SO_4)_3$ and chrome alum is heated with excess of Na_2O_2 and filtered. The materials obtained are:-
(a) A colourless filtrate and a green residue.
(b) A yellow filtrate and a green residue.
(c) A yellow filtrate and a brown residue.
(d) A green filtrate and a brown residue.

22. The formula of gas is $[CO]_x$. If its vapour density is 140, the value of x will be:
(a) 2.5 (b) 3.0 (c) 5.0 (d) 6.0

23. If 0.2 gram of an organic compound containing carbon, hydrogen and oxygen on combustion, yielded 0.147 gram carbon dioxide and 0.12 gram water. What will be the content of oxygen in the substance ?
(a) 73.29% (b) 78.45%
(c) 83.23% (d) 89.50%

24. Which of the following statements is incorrect ?
(a) Fe^{2+} ion also gives blood red colour with SCN^- ion.
(b) Fe^{3+} ion also gives blood red colour with SCN^- ion.
(c) On passing H_2S into Na_2ZnO_2 solution a white ppt of ZnS is formed.
(d) Cupric ion reacts with excess of ammonia solution to give deep blue colour of $[Cu(NH_3)_4]^{2+}$ ion.

25. An aqueous solution of colourless metal sulphate M gives a white precipitate with NH_4OH. This was soluble in excess of NH_4OH. On passing H_2S through this solution a white ppt. is formed. The metal M in the salt is

(a) Ca (b) Ba (c) Al (d) Zn

26. Volume of 3% solution of sodium carbonate necessary to neutralise a litre of 0.1 N sulphuric acid
(a) 176.66 ml (b) 156.6 ml
(c) 116.0 ml (d) 196.1 ml

27. The Lassaigne's extract is boiled with dil. HNO_3 before testing for halogens because

Toughnut

(a) silver halides are soluble in HNO_3
(b) Na_2S and NaCN are decomposed by HNO_3
(c) Ag_2S is soluble in HNO_3
(d) AgCN is soluble in HNO_3

28. For preparing 250 mL of N/20 solution of Mohr's salt, the amount of Mohr's salt needed is
(a) 9.8 g (b) 4.9 g (c) 19.6 g (d) 3.2 g

29. Potassium chromate solution is added to an aqueous solution of a metal chloride. The precipitate thus obtained is insoluble in acetic acid. When precipitate is subjected to flame test the colour of the flame is
(a) lilac (b) apple green
(c) crimson red (d) golden yellow

30. Three separate samples of a solution of a single salt gave these results. One formed a white precipitate with excess ammonia solution, one formed a white precipitate with dil. NaCl solution and one formed a black precipitate with H_2S. The salt could be
(a) $AgNO_3$ (b) $Pb(NO_3)_2$
(c) $Hg(NO_3)_2$ (d) $MnSO_4$

31. Compound $CH_3-CH(OEt)_2$ (P) (Acetal) and $CH_3-CH_2-O-CH_2-CH_2$ (Q) can be differentiated by:
(a) $H_3O^{\oplus}$, Na (b) $H_3O^{\oplus}$, Tollens' test
(c) $H_3O^{\oplus}$, Fehling test (d) All of these

32. The best method for the separation of naphthalene and benzoic acid from their mixture is:
(a) Distillation (b) Sublimation
(c) Chromatography (d) Crystallisation

33. A salt on treatment with dil. HCl gives a pungent smelling gas and a yellow precipitate. The salt gives green flame when tested. The solution gives a yellow precipitate with potassium chromate. The salt is
(a) $NiSO_4$ (b) BaS_2O_3
(c) PbS_2O_3 (d) $CuSO_4$

34. 0.24 g of a volatile liquid on vaporization gives 45 ml of vapours at NTP. What will be the vapour density of the substance ?
(Density of $H_2 = 0.089\ g\ L^{-1}$)
(a) 95.39 (b) 39.95 (c) 99.53 (d) 59.93

35. The salt used for performing 'bead' test in qualitative inorganic analysis is

Critical Thinking

(a) $K_2SO_4.Al_2(SO_4)_3.24H_2O$
(b) $FeSO_4.(NH_4)_2SO_4 4H_2O$
(c) $Na(NH_4)HPO_4.4H_2O$
(d) $CaSO_4.2H_2O$

36. A gas "X" is passed through water to form a saturated solution. The aqueous solutions on treatment with the $AgNO_3$ gives a white preciptate. The saturated aqueous solution also dissolves magnesium ribbon with evolution of a colourless gas "Y". Identify 'X' and 'Y'.
(a) $X = CO_2, Y = Cl_2$ (b) $X = Cl_2, Y = CO_2$
(c) $X = Cl_2, Y = H_2$ (d) $X = H_2, Y = Cl_2$

37. In Kjeldahl's method for the estimation of N_2, potassium sulphate and copper sulphate are used. On the basis of their functions which of the following statement(s) is/are correct?
I. Potassium sulphate raises the bpt. and ensures complete reaction.
II. Copper sulphate acts as catalyst.
III. Potassium sulphate acts as catalyst and copper sulphate raises the bpt.
(a) Only III is correct (b) I and II are correct
(c) Only II is correct (d) None is correct

38. Aniline is usually purified by
(a) chromatographic technique
(b) steam distillation
(c) by addition of oxalic acid
(d) fractional crystallization

39. Prussian blue is formed when
(a) ferrous sulphate reacts with $FeCl_3$
(b) ferric sulphate reacts with $K_4Fe(CN)_6$
(c) ferrous ammonium sulphate reacts with $FeCl_3$
(d) ammonium sulphate reacts with NO_2

40. 0.59 g of an organic substance when treated with caustic soda evolved ammonia which required 20 cc of N/2 sulphuric acid for neutralisation. The percentage of nitrogen is
(a) 26.32% (b) 40%
(c) 53.6% (d) 63.6%

41. An organic compound has C and H percentage in the ratio 6 : 1 and C and O percentage in the ratio 3:4 the compound is **Critical Thinking**
(a) HCHO (b) CH_3OH
(c) CH_3CH_2OH (d) $(COOH)_2$

42. The presence of magnesium is confirmed in the qualitative analysis by the formation of a white crystalline precipitate of
(a) $Mg(HCO_3)_2$ (b) $MgNH_4PO_4$
(c) $MgNH_4(HCO_3)_3$ (d) $MgCO_3$

43. A salt which gives CO_2 with hot conc.H_2SO_4 and also decolourizes acidified $KMnO_4$ on warming is:
(a) HCO_3^- (b) CO_3^{2-}
(c) Oxalate (d) Acetate

44. 2.79 g of an organic compound when heated in Carius tube with conc. HNO_3 and H_3PO_4 formed is converted into $MgNH_4.PO_4$ ppt. The ppt. on heating gave 1.332 g of $Mg_2P_2O_7$. The percentage of P in the compound is **Toughnut**
(a) 23.33% (b) 13.33%
(c) 33.33% (d) 26.66%

45. $[X] + H_2SO_4 \longrightarrow [Y]$ a colourless gas with irritating smell,
$[Y] + K_2Cr_2O_7 + H_2SO_4 \longrightarrow$ green solution.
[X] and [Y] is:
(a) SO_3^{2-}, SO_2 (b) Cl^-, HCl
(c) S^{2-}, H_2S (d) CO_3^{2-}, CO_2

46. 0.75 g platinic chloride of a mono- acidic base on ignition gives 0.245 g platinum. The molecular weight of the base is
(a) 75.0 (b) 93.5 (c) 100 (d) 80.0

47. The sodium extract prepared from sulphanilic acid, contains SCN^-. It gives blood red colouration with
(a) a mixture of Na_2S and CS_2
(b) $FeCl_3$
(c) $FeSO_4$
(d) Na_2SO_3

48. An orange precipitate of II group is dissolved in conc. HCl; the solution when treated with excess of water turns milky due to formation of
(a) Sn(OH)Cl (b) $Sb(OH)Cl_2$
(c) SbOCl (d) $Sb(OH)_2Cl$

49. A substance on treatment with dil.H_2SO_4 liberates a colourless gas which produces
(I) turbidity with baryta water and
(II) turns acidified dichromate solution green.
The reaction indicates the presence of
(a) CO_3^{2-} (b) S^{2-} (c) SO_3^{2-} (d) NO_3^-

50. Which of the following gives a precipitate with $Pb(NO_3)_2$ but not with $Ba(NO_3)_2$?
(a) Sodium chloride
(b) Sodium acetate
(c) Sodium nitrate
(d) Sodium hyrogen phosphate

Answer KEYs

1	(b)	6	(a)	11	(d)	16	(a)	21	(b)	26	(a)	31	(d)	36	(c)	41	(a)	46	(b)
2	(c)	7	(b)	12	(d)	17	(a)	22	(c)	27	(b)	32	(b)	37	(b)	42	(b)	47	(b)
3	(c)	8	(d)	13	(a)	18	(d)	23	(a)	28	(b)	33	(b)	38	(b)	43	(c)	48	(c)
4	(b)	9	(d)	14	(b)	19	(a)	24	(a)	29	(b)	34	(d)	39	(b)	44	(b)	49	(c)
5	(c)	10	(b)	15	(b)	20	(a)	25	(d)	30	(b)	35	(c)	40	(a)	45	(a)	50	(a)

Hints & Solutions

1. (b) Fe^{3+}, radical gives blood red with CNS^-

$$Fe^{3+} + 3\,CNS^- \longrightarrow \underset{\text{(Blood red colouration)}}{\underset{\text{Ferric sulphocyanide}}{Fe(CNS)_3}}$$

Fe^{3+} gives blue colour on treatment with $K_4[Fe(CN)_6]$

$$4Fe^{3+} + 3K_4[Fe(CN)_6] \longrightarrow 12K^+ + \underset{\text{(Prussian blue)}}{\underset{\text{Ferricferrocyanide}}{Fe_4[Fe(CN)_6]_3}}$$

Cl^- radical gives chromyl chloride test.

2. (c) Equiv. mass of $KMnO_4$

$$MnO_4^- = \frac{\text{molar mass}}{7-2} = \frac{\text{molar mass}}{5}.$$

Equiv. mass of oxalic acid

$$C_2O_4^{2-} = \frac{\text{molar mass}}{2(4-3)} = \frac{\text{molar mass}}{2}$$

Meq. of $KMnO_4 = 50 \times 5 \times 0.04 = 10 =$ meq of $H_2C_2O_4 = 50 \times 2 \times 0.1 = 10$. Hence (c).

3. (c) $$\underset{}{Na_2S} + \underset{\text{Sodium nitroprusside}}{Na_2[Fe(CN)_5NO]} \longrightarrow \underset{\text{Sodium thionitroprusside}}{Na_4[Fe(CN)_5NOS]}$$

4. (b) Eq. of acid = Eq of base,

$$\therefore \frac{0.45}{E.wt} = \frac{20 \times 0.5}{1000}, = E.wt = 45$$

$$\text{Basicity} = \frac{M.wt}{E.wt} = \frac{90}{45} = 2$$

5. (c) Co^{+2} ion is precipitated by H_2S in presence of NH_4OH which is a group reagent of group IV in cationic analysis.

6. (a) Zn^{2+} salts are white. Usually Fe^{2+} salts are rarely pink. Cu^{2+} salts are usually blue in hydrated form. Co^{2+} is pink in aqueous solution.

7. (b) Let the amount of the $K_2Cr_2O_7$ in the mixture be x g, then amount of $KMnO_4$ will be $(0.5 - x)$ g

$$\therefore \left(\frac{x}{49} + \frac{0.5-x}{31.6}\right) = \frac{100 \times 0.15}{1000}$$

where 49 is Eq. wt. of $K_2Cr_2O_7$ and 31.6 is Eq. wt. of $KMnO_4$.

On solving, we get $x = 0.073$ g

Percentage of $K_2Cr_2O_7$

$$= \frac{0.0732 \times 100}{0.5} = 14.64\%$$

8. (d) Kjeldahl's method depends upon the fact that most of the organic compounds containing nitrogen are quantitatively decomposed to give $(NH_4)_2SO_4$ when heated strongly with conc. H_2SO_4. In this method $CuSO_4$ acts as catalytic agent.

9. (d) To convert covalent compounds into ionic compounds such as NaCN, Na_2S, NaX, etc.

10. (b) NaCl is a salt of strong acid and strong base hence on dissolution will give neutral solution.

$$NaCl + AgNO_3 \longrightarrow AgCl\downarrow + HNO_3$$

11. (d) $PbCl_2$ is insoluble in cold water, soluble in hot water and PbS is black ppt in acidic medium.

12. (d) Carboxylic acids dissolve in sodium bicarbonate, while phenol does not.

13. (a) $MgCl_2$ is soluble in water and does not give flame test. Other give flame test.

14. (b)

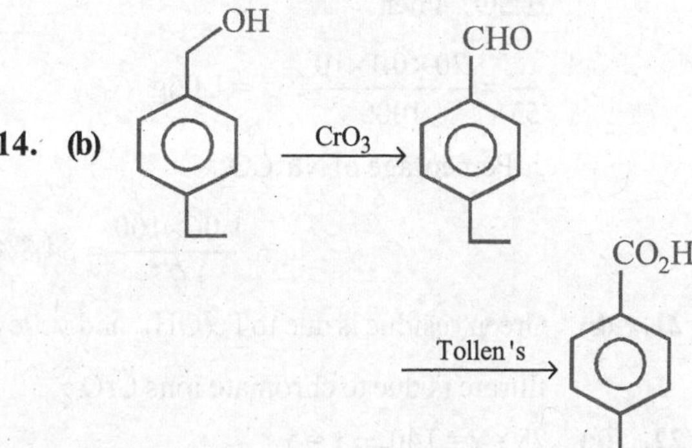

15. (b) Beilstein's test: Organic compounds containing halogens when heated over Cu wire loop give blue or green colour flame due to formation of volatile copper halides.

16. (a) $K_3[Cu(CN)_4]$ is more stable whereas $K_2[Cd(CN)_4]$ is less stable.

17. (a) If Na_2CO_3 is used in place of $(NH_4)_2CO_3$. It will precipitate group V radicals as well as magnesium radicals. The reason for this is the high ionization of Na_2CO_3 in water into Na^+ and CO_3^{2-}. Now the higher concentration of CO_3^{2-} is available which exceeds the solubility product of group V radicals as well as that of magnesium radicals.

18. (d) As^{3+} and Cd^{2+} are the radicals of group II, whereas Ni^{2+} & Zn^{2+} are the radicals of group IV. The solubility product of group IV radicals is higher as compared to group II. NH_4OH increases the ionisation of H_2S by removing H^+ of H_2S as unionisable water.

$H_2S \rightleftharpoons 2H^+ + S^{2-}$;

$$H^+ + OH^- \longrightarrow H_2O$$

Thus excess of sulphide ions are present which leads to the precipitation of all the four ions.

Note : HCl decreases ionisation of H_2S whereas NH_4OH increases the ionisation of H_2S.

19. (a) Eq of $KMnO_4$ used $= \dfrac{50\times1}{1000\times10} = 0.005$

$\therefore$ Eq of FAS reacted = 0.005

$\therefore$ weight of FAS needed

$= 0.005\times392 = 1.96$ g

20. (a) Let the amount of Na_2CO_3 present in the mixture be x g. Na_2SO_4 will not react with H_2SO_4. Then

$$\frac{x}{53} = \frac{20\times0.1\times10}{1000} \quad \therefore x = 1.06\text{g}$$

$\therefore$ Percentage of Na_2CO_3

$$= \frac{1.06\times100}{1.25} = 84.8\%$$

21. (b) Green residue is due to $Fe(OH)_2$ and yellow filtrate is due to chromate ions CrO_4^{2-}

22. (c) $28\times x = 140 \Rightarrow x = 5$

23. (a) $\%\text{ of C} = \dfrac{12}{44}\times\dfrac{0.147}{0.2}\times100 = 20.045\,\%$

$$\%\text{ of H} = \frac{2}{18}\times\frac{0.12}{0.2}\times100 = 6.666\,\%$$

$$\%\text{ of O} = 100-(20.045+6.666) = 73.29\%$$

24. (a) Only Fe^{3+} ions give blood red colouration with SCN^- ions.

$$Fe^{3+} + SCN^- \longrightarrow \underset{\text{(dark red)}}{[Fe(SCN)]^{2+}}$$

25. (d) $Zn^{2+} + 2NH_4OH \longrightarrow \underset{\text{White ppt.}}{Zn(OH)_2} + 2NH_4^+$

$$Zn(OH)_2 + 2NH_4OH \longrightarrow \underset{\text{Soluble}}{(NH_4)_2ZnO_2} + 2H_2O$$

$$(NH_4)_2ZnO_2 + H_2S \longrightarrow \underset{\text{White ppt.}}{ZnS} + 2NH_4OH$$

26. (a) Normality of 3% Na_2CO_3.

$$N = \frac{3\times100}{53\times100} = 0.566N$$

For H_2SO_4 sol. $N_1 = 0.1$, $V_1 = 100$ mL

For Na_2CO_3 sol. $N_2 = 0.566$. N_2?

Now apply $N_1V_1 = N_2V_2$

$$V_2 = \frac{N_1V_1}{N_2} = \frac{0.1\times100\,\text{mL}}{0.566} = 176.66\,\text{mL}$$

27. (b) Na_2S and NaCN, formed during fusion with metallic sodium, must be removed before adding $AgNO_3$, otherwise black ppt. due to Na_2S or white precipitate due to AgCN will be formed and thus white precipitate of AgCl will not be identified easily.

$$Na_2S + 2AgNO_3 \longrightarrow 2NaNO_3 + \underset{\text{Black}}{Ag_2S\downarrow}$$

$$NaCN + AgNO_3 \longrightarrow NaNO_3 + \underset{\text{White}}{AgCN\downarrow}$$

$$NaCl + AgNO_3 \longrightarrow NaNO_3 + \underset{\text{white}}{AgCl\downarrow}$$

$$Na_2S + 2HNO_3 \xrightarrow{\text{boil}} 2NaNO_3 + H_2S\uparrow$$

$$NaCN + HNO_3 \xrightarrow{\text{boil}} NaNO_3 + HCN\uparrow$$

28. (b) The ionic equation for oxidation of Mohr's salt is

$$Fe^{2+} \longrightarrow Fe^{3+} + e^-$$

Now Eq. of Mohr's salt $= \frac{392}{1} = 392$

Strength = Normality × Eq. mass

$$= \frac{1}{20} \times 392 = 19.6 g/lit$$

Thus for preparing 250 ml of N/20 Mohr's salt solution, Mohr's salt needed

$$= \frac{19.6}{1000} \times 250 = 4.9 g$$

29. (b) $BaCl_2 + K_2CrO_4 \rightarrow BaCrO_4 + 2KCl$. $BaCrO_4$ is insoluble in acetic acid and Ba gives apple green colour in flame test.

30. (b) $Pb(NO_3)_2 + 2NH_4OH \rightarrow$

$$Pb(OH)_2 \downarrow + 2NH_4NO_3 \text{ (white ppt)}$$

$$Pb(NO_3)_2 + 2HCl \rightarrow PbCl_2 \downarrow + 2HNO_3 \text{ (white ppt)}$$

$$Pb(NO_3)_2 + H_2S \rightarrow PbS \downarrow + 2HNO_3 \text{ (black)}$$

31. (d) $CH_3 - CH(OEt)_2 \xrightarrow{H_3O^{\oplus}}$

(P) (Acetal)

CH_3CHO; differentiated by Na, Fehling Tollen's test

$CH_3 - CH_2 - O - CH_2 - CH_3$ (Q) $\xrightarrow{H_3O^{\oplus}}$ EtOH

32. (b) Among the given compounds naphthelene is volatile but benzoic acid is non-volatile (it forms a dimer). So, the best method for their separation is sublimation,

33. (b) Gas evolved is SO_2

34. (d) $V.D. = \frac{\text{Wt. of 45 ml. of vapours at NTP}}{\text{Wt. of 45 ml. of } H_2 \text{ at NTP}}$

$$= \frac{0.24 g}{45 ml. \times 0.000089 g\ ml^{-1}} = 59.93$$

35. (c) Microcosmic salt. $Na(NH_4)HPO_4.4H_2O$ is used for bead test

$$Na.(NH_4)HPO_4 \rightarrow NaPO_3 + NH_3 + H_2O$$

$$NaPO_3 + CoO \rightarrow NaCo.PO_4 \text{ (blue)}$$

36. (c) $Cl_2 + H_2O \rightarrow HCl + HClO$

$$HCl + AgNO_3 \rightarrow AgCl \downarrow + HNO_3$$

White ppt

$$Mg + 2HCl \rightarrow MgCl_2 + H_2 \uparrow$$

Hence X is Cl_2 and Y is H_2

37. (b) K_2SO_4 raises the bpt. and ensure complete reaction and $CuSO_4$ acts as catalyst.

38. (b) Aniline is miscible in steam and has b.p. lower than water, hence it is collected as distillate along with steam.

39. (b) $2Fe_2(SO_4)_3 + 3K_4[(Fe(CN)_6] \rightarrow$

$$Fe_4[Fe(CN)_6]_3 + 6K_2SO_4$$

prussian blue

40. (a) Percentage of nitrogen by Kjeldahl's method

$$= \frac{1.4 \times N \times V}{\text{Wt. of organic compound}}$$

$$= \frac{1.4 \times 20}{0.532} \times \frac{1}{2} = 26.32\%$$

41. (a) % ratio of C : H is 6 : 1 and C : O is 3 : 4 or 6 : 8

∴ % ratio of C : H : O is 6 : 1 : 0 (Total = 15)

% of C $= \frac{6}{15} \times 100 = 40$

$40/12 = 3.33 = 1$

% of H $= \frac{1}{15} \times 100 = 6.6$

$6.6/1 = 6.6 = 2$

% of O $= \frac{8}{15} \times 100 = 53.3$

$53.3/16 = 3.3 = 1$

simple ratio: CH_2O

∴ The compound is HCHO

42. (b) $MgCl_2 + Na_2HPO_4 + NH_4OH \rightarrow$

$$Mg(NH_4)PO_4 + 2NaCl + H_2O$$

white ppt.

43. (c) $Na_2C_2O_4 + H_2SO_4 \rightarrow Na_2SO_4 + H_2O + CO + CO_2$

CO burns with blue flame $2CO + O_2 \rightarrow 2CO_2$

$Na_2C_2O_4 + H_2SO_4(dil) \rightarrow Na_2SO_4 + H_2C_2O_4$

$2KMnO_4 + 3H_2SO_4 + 5H_2C_2O_4 \rightarrow K_2SO_4 + 2MnSO_4 + 8H_2O + 10CO_2$

44. (b) Percentage of P

$$= \frac{62}{222} \times \frac{\text{wt.of } Mg_2P_2O_7}{\text{wt.of compound}} \times 100$$

$$= \frac{62}{222} \times \frac{1.332}{2.79} \times 100 = 13.33\%$$

45. (a) SO_2 and H_2S both being reducing agents, can turn acidified dichromate solution green. SO_2 can be obtained by the action of acid upon sulphite while H_2S is evolved by the action of acid upon sulphide. However, SO_2 has a burning sulphur smell which is irritating. H_2S has rotten egg like smell.

46. (b) Apply the formula M.wt

$$= \frac{\text{Acidity}}{2}\left(\frac{\text{wt of Pt salt} \times 195}{\text{wt of Pt}} - 410\right)$$

$$\therefore \text{M.wt} = \frac{1}{2}\left(\frac{0.75 \times 195}{0.245} - 410\right) = 93.5$$

47. (b) $SCN^- + Fe^{3+} \longrightarrow \underset{\text{blood red ppt.}}{Fe(SCN)_3}$

48. (c) $Sb_2S_3 + 6HCl \rightarrow 2SbCl_3 + 3H_2S$

$SbCl_3 + 6H_2O \rightarrow \underset{\text{ppt (White)}}{SbOCl} + 2HCl$

49. (c) $Na_2SO_3 + 2HCl(dil) \rightarrow 2NaCl + H_2O + SO_2$

$K_2Cr_2O_7 + H_2SO_4 + 3SO_2 \rightarrow K_2SO_4 + \underset{\text{Green}}{Cr_2(SO_4)_3} + H_2O$

50. (a) $Pb(NO_3)_2 + 2NaCl \rightarrow PbCl_2 \downarrow + 2NaNO_3$; $BaCl_2$ is soluble in water